Bioarchaeology

Bioarchaeology

The Contextual Analysis of Human Remains

Edited by

Jane E. Buikstra

Arizona State University

Lane A. Beck

Arizona State Museum
University of Arizona

Routledge
Taylor & Francis Group
LONDON AND NEW YORK

Cover Design: Eric DeCicco

First published 2006 by Left Coast Press, Inc.

Published 2016 by Routledge
2 Park Square, Milton Park, Abingdon, Oxon OX14 4RN
711 Third Avenue, New York, NY 10017, USA

Routledge is an imprint of the Taylor & Francis Group, an informa business

Library of Congress Cataloging-in-Publication Data
Bioarchaeoogy : the contextual analysis of human remains / Jane E. Buikstra and Lane A. Beck, editors.
p. cm.
Includes bibliographical references and index.
ISBN 0-12-369541-4 (alk. paper)
1. Human remains (Archaeology) 2. Human skeleton–Analysis. 3. Paleopathology. 4. Paleoanthropology. I. Buikstra, Jane E. II. Beck, Lane A.

CC79.H85B56 2006
930.1–dc22

2006045979

British Library Cataloguing-in-Publication Data
A catalogue record for this book is available from the British Library.

ISBN 13: 978-1-59874-450-7 (hbk)

Dedication

As illustrated in this book, bioarchaeology is a product of concepts and standards that have been developed over several generations of scholarship. Many of the individuals who contributed to this concept are discussed at length in the individual chapters that follow.

We dedicate this book to those who came before us and laid the foundations on which we have built. To Earnest Hooton who begat the first generation of anthropologically trained biological anthropologists in the United States. To Aleš Hrdlička who created a center for research in human osteology at the Smithsonian Institution. To Larry Angel who integrated aspects of the heritage of both Hooton and Hrdlička. To Chuck Merbs who supervised Buikstra's graduate training. To Bob Blakely who organized the symposium where this usage of the term bioarchaeology was introduced. To all who came before us, we thank you for opening the door.

Jane E. Buikstra
Lane A. Beck

Contents

Contributors

Lane Anderson Beck, PhD, Associate Curator, Arizona State Museum, The University of Arizona, Tucson, AZ, USA

Georgieann Bogdan, MA, Teacher, Guilford Day School, Greensboro, NC, USA

Jane E. Buikstra, PhD, Professor, School of Human Evolution & Social Change, Arizona State University, Tempe, AZ, USA

Mario M. Castro, PhD, Assistant Professor, Department of Anthropology, University of Chile, Santiago, Chile; Anatomy Unit, Department of Morphology, Faculty of Medicine, University of Chile, Santiago, Chile

Della Collins Cook, PhD, Professor, Department of Anthropology, Indiana University, Bloomington, IN, USA

Dolores L. Burke, PhD, Adjunct Assistant Professor, Department of Anthropology, University of Arkansas, Fayetteville, AR, USA

Susan R. Frankenberg, PhD, Research Associate Professor and Curator, Department of Anthropology, The University of Tennessee, Knoxville, TN, USA

Lynne Goldstein, PhD, Professor, Department of Anthropology, Michigan State University, East Lansing, MI, USA

Patrick D. Horne, PhD, Pathologist, Department of Pathology, York County Hospital, Newmarket, Ontario, Canada

David R. Hunt, PhD, Assistant Collections Manager, Department of Anthropology, Smithsonian Institution, NMNH, Washington, DC, USA

Keith P. Jacobi, PhD, Associate Professor, Blount Fellow of the College of Arts and Sciences, Department of Anthropology, Alabama Museum of Natural History, The University of Alabama, Tuscaloosa, AL, USA

Lyle W. Konigsberg, PhD, Professor, Department of Anthropology, The University of Tennessee, Knoxville, TN, USA

Richard T. Koritzer, DDS, MLA, PhD, Adjunct Research Associate Volunteer, Biomedical Sciences, Dental School, University of Maryland, Glen Burnie, MD, USA

Clark Spencer Larsen, PhD, Distinguished Chair and Professor, Department of Anthropology, The Ohio State University, Columbus, OH, USA

Glaucia Aparecida Malerba Sene, MA, Archaeologist, Instituto de Arqueologia Brasilieira, Rio de Janeiro, Brazil; PhD Student, Museum of Archaeology and Ethnology, University of San Paulo, Brazil

Sheila Ferraz Mendonça de Souza, PhD, Senior Researcher, Department of Endemic Diseases Samuel Pessoa, National School of Public Health Sérgio Arouca, Oswaldo Cruz Foundation, Rio de Janeiro, Brazil

George R. Milner, PhD, Interim Head and Professor, Department of Anthropology, Pennsylvania State University, University Park, PA, USA

Osbjorn M. Pearson, PhD, Assistant Professor, Department of Anthropology, University of New Mexico, Albuquerque, NM, USA

Mary Lucas Powell, PhD, Newsletter Editor, Paleopathology Association, Lexington, KY, USA

Gordon F. M. Rakita, PhD, Assistant Professor, Department of Sociology and Anthropology, University of North Florida, Jacksonville, FL, USA

Charlotte A. Roberts, PhD, Professor, Department of Archaeology, University of Durham, UK

Jerome C. Rose, PhD, Chair and Professor, Department of Anthropology, University of Arkansas, Fayetteville, AR, USA

Mary Kay Sandford, PhD, Professor, Anthropology Department, The University of North Carolina, Greensboro, NC, USA

Lorraine P. Saunders, PhD, Research Associate, Rochester Museum & Science Center, Rochester, NY, USA

Lynne Sullivan, PhD, Associate Research Professor and Curator, Department of Anthropology, Frank H. McClung Museum, The University of Tennessee, Knoxville, TN, USA

John J. Swetnam, PhD, Professor, Department of Anthropology, University of Nevada, Las Vegas, NV, USA

Douglas H. Ubelaker, PhD, Curator, Department of Anthropology, Division of Physical Anthropology, Smithsonian Institution, NMNH, Washington, DC, USA

Foreword

Why are humans so intensely interested in our past? We invest very substantial resources in a quest to reconstruct what our ancestors looked like and how they lived. We spend equally substantial resources to leave behind something of ourselves for those who follow us. This preoccupation with the past is pervasive throughout human societies today. It is manifest in specialists ranging from people who memorize and can repeat the folk traditions and history of small ethnic groups to historians with endowed chairs at major universities who publish heavy tomes on the subject. This focus on the past presumably has been part of human culture for at least the past 10,000 years. A partial reason for this interest is that who we are today rests on the accumulated knowledge and innovations of our ancestors. This means that an understanding of our past informs and empowers our present and gives us a sense of the future for ourselves and our descendants.

Historical documents of various kinds provide us with much of what we know about written human history. As important as this dimension of human history is, much about our knowledge of past human societies depends on other types of information. Written history tends to highlight the social and political elite. If we wish to know about other aspects of past human societies, other sources of information must be utilized. Archaeology provides a different view of these societies that generally is less specific than written history. However, archaeological data tend to be more representative of the total population. It also gives us access to knowledge of past human societies for which there are no written records or where these records may be inadequate. One focus of archaeological excavation is architectural structures, which usually have associated cultural artifacts. Another emphasis is on cemeteries where funerary artifacts and human skeletons provide a rich lode of data on past societies. Discard deposits, such as the often large shell middens of eastern North America, are another important source of information, particularly on food resources as revealed by animal remains. The physical remains of humans, nonhuman animals, and plants associated with past human societies provide a major source of data on those human groups.

The emergence and application of new methods, including remarkable technological innovations, in the study of past human groups particularly in the last quarter century has provided insights that were certainly beyond my imagination when I began my career as a biological (physical) anthropologist more than 40 years ago. In the current study of osteology, powerful statistical methods have become critical tools in advancing our understanding of biological changes that have occurred. Analytical procedures for identifying isotopes and biomolecules in biological materials excavated from archaeological sites offer remarkable strategies for exploring new dimensions of past human relationships and activities. All of these diverse sources of data are best interpreted in a theoretical context that integrates biological and cultural data.

Perhaps the most important development in the study of past human societies is the emergence of bioarchaelogy as an interpretative framework for the diverse data obtained today. In the Preface to this book, Buikstra discusses the somewhat different emphases of bioarchaeological research in European versus American endeavors, but for scholars in both areas, integration of cultural and biological data is central. Because culture is such an important component of human society, human biology must be understood in the context of the associated culture. This linkage brings a far richer understanding of biological data than the latter alone.

The emergence of bioarchaeology as an important interpretative framework has its roots in earlier research. One immediately recalls the remarkable publication of the Pecos human remains from the American Southwest in 1930 by Earnest Hooton of Harvard University. More recently, one thinks of the publications of J. Lawrence Angel, a student of Hooton, on past Greek societies in which biological change was discussed in the context of historical and cultural change. These and other scholarly works provide a clear direction leading to today's emphasis regarding the impact of changing culture on human biology. There are few topics more salient than the complex relationship between biology and culture. Clearly a book that integrates the perspectives of history, cultural dynamics, and archaeology in the development of bioarchaeology is an important milestone. The inclusion of chapters, by many of today's leading practitioners of bioarchaeology, that highlight the knowledge and insight regarding the application of bioarchaeology to contemporary research in both archaeology and biological anthropology provides an important source in the development of this framework. Such a book is, if anything, overdue. All scientists and scholars who use or plan to use the interpretative framework of bioarchaeology to inform their understanding of the data they collect are indebted to Drs. Buikstra and Beck for investing the time to assemble a collection of chapters that provides an accessible, authoritative, and convenient source of information that will permit all of us to make further progress in understanding the synchronic and diachronic dynamics of culture and biology.

Donald J. Ortner

Preface

What is "bioarchaeology?" Increasingly visible during the waning years of the 20th century, the term is now firmly embedded in anthropological and archaeological literatures. It is even legitimized within the pages of Webster's *New World Medical Dictionary*.[1] Positions for bioarchaeologists are regularly advertised and centers dedicated to bioarchaeological study are found in both museums and universities. The term has also spread widely into non-English speaking venues, as *bioarcheología, bioarchéologie,* and *bioarqueología* characterize curricula and research programs across the globe.

The term "bioarchaeology" arose independently, with distinctive definitions, in the United Kingdom and the United States during the 1970s. The UK version boldly appeared in the title of Grahame Clark's *Starr Carr: A Case Study in Bioarchaeology* (1972). By "bioarchaeology," Clark meant inferences derived from the study of archaeologically recovered faunal remains. Today, in the United Kingdom and beyond, the word has taken on additional meanings. It has, for example, been broadened to describe the study of all biological materials from archaeological contexts, especially flora (paleobotany) and fauna (paleozoology), as at Cambridge University (UK) (Department of Archaeology Web site, Bioarchaeology). In contrast, at Bradford University (UK), the term explicitly references the "reconstruction of human activity, health and disease," with coursework in "human osteoarchaeology"[2] being part of the undergraduate

[1] The dictionary definition is the "use of a range of biological techniques on archaeological material in order to learn more about past populations" (MedicineNet.com).

[2] In 1973, Vilhelm Møller-Christensen (1973, 1978) adamantly took ownership of the term osteoarchaeology, which he linked to an excavation method he had been developing since 1935. Complaining of the manner in which archaeologists excavated cemetery sites, Møller-Christensen (1973:412) emphasized that during osteoarchaeological excavations, "the main idea . . . is to treat any part of a tiny and fragile bone just as carefully as the archaeologists treat jewels, gold and pearls . . . ." He (Møller-Christensen, 1973:413) went on to declare that the "osteo-archaeologic examination of a burial place is therefore something quite different from the taking-up of skeletons by archaeologists."

curriculum (University of Bradford, Module Catalogue). Bioarchaeology and human osteoarchaeology are also linked in the charter of the British Association for Biological Anthropology and Osteoarchaeology (BABAO), which has held annual conferences since 1999 (BABAO conference page).[3] A recently developed graduate prospectus in skeletal and dental bioarchaeology at the University College London (UK) also emphasizes the study of human remains, with options to study animal bones (UCL Web site, Graduate Prospectus 2006). The analysis of human bones and animal bones is also linked in the definition of osteoarchaeology used by the *Journal of Osteoarchaeology*, published in the United Kingdom, beginning in 1991. Thus, from an original emphasis on faunal remains, the term "bioarchaeology" is now applied variously in the United Kingdom, sometimes linked to "osteoarchaeology." All archaeologically recovered biological materials or a subset, including human remains, may be specified. Similar differences exist in many other parts of the world.[4]

In 1976, a "bioarchaeology" distinctly different from that of Clark was proposed in the United States.[5] It was coined in 1976 at the 11th annual meeting of the Southern Anthropological Society and published the following year (Blakely, 1977a; Buikstra, 1977). In a paper entitled "Biocultural Dimensions of Archeological Study: A Regional Perspective," Buikstra (1977:67) defined a multidisciplinary, bioarchaeological research program that integrated human osteologists with other scholars in addressing a series of topics, including (1) burial programs and social organization; (2) daily activities and division of labor; (3) paleodemography, including estimates of population size and density; (4) population movement and genetic relationships; and (5) diet and disease. Explicitly influenced by the "New Archaeology" (Binford, 1962) and population-based, ecological studies, this approach emphasized that all participating scholars should be engaged in both research design and execution. As in the "New Physical Anthropology" (Washburn, 1951, 1953), bioarchaeology emphasized anthropological problem solving rather than descriptive data collection. Buikstra (1977:70) also argued that the sequence in which bioarchaeological topics should be addressed

---

[3]BABAO's stated objective "is to promote the study of human bioarchaeology and for the purpose of understanding the human condition from the past to the present" (BABAO Web site).

[4]Compare, for example, the nonhuman biological usage for the Bioarchéologie at the University of Geneva (Chevalier, 1997, ILLAPA Web site) and Bioarcheologia at San Vincenzo al Volturno (San Vincenzo al Volturno, 2001, Bioarcheologia Web page) with the bioarqueología taught at the Autonomous University of the Yucatán (UADY) (www.uady.mx/sitios/antropol/arqueologia/index.html).

[5]In the United States, animal bones are typically studied by archaeozoologists or zooarchaeologists, archaeologists with special training in identification and analysis of animal bone. Biological anthropologists specialize in the study of human bone. Occasionally, "bioarchaeology" is glossed with the United Kingdom definition, e.g., University of Texas at El Paso's Laboratory for Environmental Biology (UTEP Laboratory for Environmental Biology Bioarchaeological References Web page).

was not arbitrary, in that selective burial programs can bias archaeological samples and that health-related issues should be addressed only after the researcher understands whether groups being compared are or are not closely related, due to possible differences in genetic predisposition for disease. Referencing earlier work (Cook and Buikstra, 1973; see also Cook and Buikstra, 1979), Buikstra (1977:79) also underscored differences between living communities and cemetery samples, thus reflecting concerns that would later become part of "the osteological paradox" (Wood *et al.*, 1992; discussed in Chapter 11). The history of this American "bioarchaeology" departs from that in the United Kingdom in that it has almost exclusively been focused upon the reconstruction of human histories, with emphasis on anthropological problem solving and the integration of archaeological data. As discussed by Roberts in Chapter 16, it has influenced the nature of human osteoarchaeology (bioarchaeology) in the United Kingdom.[6]

Why this singular focus on the study of human remains and reconstructing human life histories and population structure? Certainly, American anthropology's four-field approach is one significant influence, encouraging practitioners to integrate information about languages, biology, human cultural diversity, and archaeological interpretations. The multiethnic nature of America should also be cited, as well as its rich and relatively untouched archaeological record where systematic research on changing landscapes and their human inhabitants encouraged multidisciplinary collaborations. In this context there emerged a bioarchaeological emphasis upon peopling the past that remains influential today.

The overall goal of this volume therefore is to explore the history and future of "bioarchaeology," as it is understood and practiced in North America at the turn of the 21st century. Nuanced differences between bioarchaeology and related approaches, such as social biology (Angel, 1946a), osteobiography (Saul, 1972), and "l'anthropologie de terrain" (Leroi-Gourhan *et al.*, 1962; Masset, 1972; Duday, 1978), are also considered. As bioarchaeology's history is closely linked to that of North American archaeology, most contributions focus on the study of North American native peoples.

While the definition of American bioarchaeology is a late 20th-century phenomenon, the integration of skeletal biological data with archaeological problem sets has, however, much deeper intellectual roots. Therefore, Section I of this volume, "People and Projects," focuses on 19th- and early 20th-century studies of human skeletal remains from archaeological contexts. These investigations consider subjects ranging from craniology to the early population-based approaches such as that of Earnest Hooton, whose 1930 report on the Pecos Pueblo collection

---

[6]Certainly, important influences from the United Kingdom and elsewhere have affected the course of American studies of archaeologically derived human remains, e.g., Brothwell's handbook (1963c) and broadly based scholarship. British archaeologists, however, actively imported American "bioarchaeology," beginning in the late 1970s, e.g., Chapman *et al.* (1981).

is widely recognized as a primary influence on bioarchaeology today. We also consider the seminal contributions of the extensive Works Progress Administration excavations to museum collections, some of which are only now being studied. Chapter 7 reports the all-too-frequently underappreciated contributions of late 19th- and 20th-century women to the study of human remains from archaeological contexts.

Section II, "Emerging Specialties," develops the themes initially defined by Buikstra (1977) and of continuing importance in contemporary bioarchaeological research. This section is prefaced by introductory comments that review mortuary theories and ritual studies that have informed bioarchaeology. As contextually sensitive, problem-oriented research must be both biological and archaeological, theoretical developments in social science, especially theories of mortuary behavior, are crucial foundations for skeletal biological study. We also briefly consider the history of sexing and aging methods that serve as the fundamentals for paleodemography. The specialty themes are developed in Chapters 8–12, which treat topics ranging from behavioral interpretations to dental anthropology.

The volume closes with Section III, "On to the 21st Century." In this section, authors consider late 20th-century bioarchaeological achievements and challenges in the 21st century. We end with Charlotte Roberts' thoughtful reflection upon Americanist bioarchaeology from a European perspective.

Jane E. Buikstra

Acknowledgments

The editors first thank the chapter authors for their patience during the long germination of this volume. The shift in editorial responsibilities forced by Lane Beck's illness and the subsequent delay following closure of the Smithsonian Institution Press have all taken their temporal toll. Charlotte Roberts deserves special appreciation for taking on her chapter during the final stages of manuscript preparation, when she was exceedingly busy with other matters. We are indebted to Mary Powell for compiling Chapter 7, and several folks joined the search for the Joseph Jones collection: Susan Anton, Eric Baker, Della Cook, Rachel Griffin, Mary Manhein, Christine McGee, Kevin Smith, Jamie Suskewicz, David Hurst Thomas, and John Verano. Others read sections and offered cogent comments: Jerry Cybulski, Lynne Goldstein, and Tiffany Tung. Our enduring gratitude to you all!

We extend very special and heartfelt appreciation to Melissa LaLiberte, who diligently and effectively read and checked references prior to manuscript submission. Early stages of preparation and writing were supported by an American Fellowship to Beck from the American Association of University Women and a summer Fellowship at the School of American Research.

Jane E. Buikstra
Lane A. Beck

Section I

People and Projects: Early Landmarks in American Bioarchaeology

Introduction by Jane E. Buikstra

Recognizing that 20th- and 21st-century bioarchaeology rests on earlier foundations, Chapter 1 begins by considering selected 18th- and 19th-century examples, chosen because these studies of skeletal remains are keenly grounded in problem-oriented research. As restudy to validate the results of earlier research is essential in scientific inquiry, the curation histories of these collections and the Harvard Peabody holdings are traced to illustrate the highly variable strategies of that period. The most satisfying approaches emerge as those that described remains individually, keeping burial-specific information on grave lots and conserving records, remains, and objects within the same institution. Such strategies facilitate reexamination, as well as the exploration of new problem sets.

Chapter 1 then turns to two figures, Aleš Hrdlička and Earnest A. Hooton, whose significant contributions to early 20th-century physical anthropology included both the recovery of human remains from archaeological contexts and the creation of significant research collections. Even so, there were key differences between them in problem orientation and their consideration of archaeological contexts. These disparities are illustrated here, important because they keenly influenced later 20th-century scholarship.

Subsequent chapters in Section I are ordered chronologically, beginning with Cook's penetrating discussion of typological, craniological approaches to interpreting the origins and diversity of American Indians in Chapter 2, "The Old Physical Anthropology and the New World: A Look at the Accomplishments of an Antiquated Paradigm." Importantly, at the outset, she urges an evaluation of our intellectual ancestors within their social and historical contexts rather than within our own.

Cook's careful, contextually sensitive evaluations of primary texts in relationship to recent critiques deftly free Johann Blumenbach from the weight of racist and sexist attributions. Blumenbach's view of human variation emphasized continuity, and his widely cited ranking of races in his 1775 *De generis humani varietate nativa* appears overinterpreted by most scholars, including Gould (1994). Cook argues that Blumenbach was innovative in the way he conceptualized human variation as well as in the manner through which he studied it, also contributing the modern use of the term "anthropology" to the literature.

Influenced by Blumenbach, Samuel Morton appears to Cook much less the phrenological, polygenetic, racist than many critics would have us believe. To her mind, Morton's refutation of the Moundbuilder myth was his major contribution. She also notes that his argument for two races among American Indians was influenced by John Collins Warren's 1822 monograph.

Cook also reports that the late 19th-century craniological contributions of scholars such as Daniel Wilson, J. Aitken Meigs, Harrison Allen, Frederick Ward Putnam, and Putnam's students tend to be neglected. While Putnam himself did not publish extensive craniological treatises, he clearly encouraged the craniology of students and colleagues.

In her careful treatment of Hrdlička's craniology, Cook emphasizes his continued dedication to a fundamental unity in the American "race." This never wavered throughout his long and distinguished career. Hooton opposed this stance and just as steadfastly maintained that the diversity he observed could only be explained through multiple migrations.

Cook also cites the little known and comprehensive work of Paul Rivet on facial prognathism. Rivet, working in France during the first part of the 20th century, examined facial angle variation in a wide range of primates, primarily but not exclusively humans. He effectively demolished any scientific basis for ranking races by facial angle.

Cook then turns to Bruno Oetteking, who trained in the German tradition with Rudolf Martin and collaborated with Franz Boas at both the American Museum of Natural History and at Columbia University. As Cook emphasizes, however, Oetteking's craniology lacks the innovative elements of Boas' anthropological approaches.

Finally, Cook considers the typologist Georg Neumann, her faculty predecessor at Indiana University. Neumann was, indeed, the last firmly committed typologist whose monumental dissertation was widely published and critiqued. Cook's penetrating observations close with a brief discussion of the shift toward a statistically sophisticated perspective on human variation. She ends by explicitly addressing the manner in which scholars who study inheritance and human variation, including the typologists, have been glossed as "racist."

Chapter 3, "The Changing Role of Skeletal Biology at the Smithsonian," by Douglas Ubelaker, begins with a discussion of Hrdlička's background, the breadth of his research interests, and his contributions in building the Division of Physical Anthropology at the Smithsonian. Hrdlička was succeeded by T. Dale Stewart, who both expanded collections and published extensively on subjects ranging from paleoanthropology to paleopathology. J. Lawrence Angel, a student of Hooton's and, like Hooton, classically trained, then joined the department. Both Stewart and Angel's work moved Smithsonian scholarship away from its prior focus upon race and craniology. Others who made key contributions to skeletal biological research at the Smithsonian include Lucile St. Hoyme, Marshall Newman, and William Bass. Ubelaker closes his discussion with a description of the early 21st-century status of skeletal biology at the Smithsonian, including the impact of repatriation legislation.

As Beck underscores in Chapter 4, "Kidder, Hooton, Pecos, and the Birth of Bioarchaeology," Hooton's approach to skeletal biological study at Pecos and his association with A. V. Kidder have been enormously influential. Beck's chapter considers the Pecos project in detail, beginning with Hooton's participation in archaeological fieldwork. She argues that Hooton was one of the first to consider how the human community, reflected in the Pecos burials, changed over time. In her detailed discussion of Hooton's Pecos report, she cites multiple methodological and inferential advances, such as explicit concerns for taphonomy, innovative statistical approaches to paleodemographic reconstructions, and population-based discussions of health status.

Rakita's Chapter 5, "Hemenway, Hrdlička, and Hawikku: A Historical Perspective on Bioarchaeological Research in the American Southwest," reports that despite promising early contributions by Matthews and Hooton, an integrated bioarchaeology, as understood today, has been late to arrive in this region. Presenting in detail the collaborative efforts of Cushing's Hemenway Expedition, Rakita underscores the significance of the Matthews, Wortman, and Billing's (1893) report and physical anthropologist ten Kate's (1892) craniological research. Hrdlička's studies of both living groups and archaeological samples are summarized, as are Hodge's excavations at both historic and prehistoric sites, which were sponsored by the Heye Foundation and the Museum of the American Indian in New York. Hodge's failure to include physical anthropological expertise within his field crew, the division of the collection between museums located in the United States and abroad, and selective skeletal recovery are cited as key factors limiting the quantity and quality of bioarchaeological research possible using these collections.

Rakita notes that the 1930s and 1940s saw important biological distance studies designed to test conventional archaeological models that specified population replacement at the Basketmaker–Pueblo transition. These craniological investigations supported an alternative interpretation, one of genetic continuity. Few bioarchaeological studies were conducted in the Southwest during the 1940s through the 1960s. With renewed interest during the late 1960s through the 1980s, biodistance research again prevailed, along with growing interest in population health and disease. During this period, the development of large multidisciplinary projects encouraged integrated bioarchaeological research. Additional studies in the closing decade of the 20th century have included fine-grained investigations of inheritance, health, diet, warfare, and cannibalism. Rakita closes with a call for increased collaboration among bioarchaeologists, archaeologists, and Native American communities.

As stressed by Milner and Jacobi in Chapter 6, "A New Deal for Human Osteology," vast archaeological field projects, including burial excavations, were completed during the period from 1933 to 1942. When World War II abruptly closed processing and analytical laboratories, many materials had not yet been cleaned, let alone studied. Most of the thousands of skeletons recovered by WPA teams originated from the southeastern United States, including well-known collections from Indian Knoll and Moundville. Milner and Jacobi describe the quality of field data and skeletal recovery as fairly good, given the standards of the time. Osteological reports tended to be descriptive, emphasizing tables of measurements and observations of pathological conditions. Hooton's students were conspicuous among the physical anthropologists involved in WPA work, where standardization of observation protocols became a stated goal. Unfortunately, typological perspectives drove research designs. Cultural and physical types were thought to coincide and differences were assumed to reflect migrations. Milner and Jacobi close their discussion by emphasizing the excellent potential WPA remains hold for today's bioarchaeological problem solving.

Section I closes with Chapter 7, by Powell and colleagues, which examines the contributions of women to late 19th- and 20th-century bioarchaeological research. Included here are brief biographies of individuals whose contributions to bioarchaeological research are commonly under reported and/or underappreciated. The first woman to be so recognized is Cordelia Studley, mentored by

Putnam, whose 5-year association with the Peabody involved both archaeological field recovery and skeletal analysis (Studley, 1884). Other firsts are Susanna Boyle-Hamilton, the first Canadian female physical anthropologist, and Juliane Dillenius, the first woman to earn a doctorate in physical anthropology in the Americas. Ruth Wallis studied with both Hooton and Boas; influenced by the latter, she contributed bioarchaeological research that questioned the simplistic craniological race constructs of the 19th and early 20th centuries. Mentored by Harris Hawthorne Wilder at Smith, Marian Knight Steckel contributed both to the craniological literature and to the facial reconstructions published by Wilder. Mildred Trotter and Alice Brues, who served as the first and second woman presidents of the American Association of Physical Anthropologists, respectively, are best know for developing bioanthropological methods (Trotter) and theories (Brues) that subsequently figured heavily in bioarchaeological research. Brues also wrote insightfully concerning infectious lesions observed in ancient skeletons that could be attributed to syphilis, although it is the work of Adelaide Bullen that is most visible in reports on American treponemal disease. Madeline Kneberg, as noted in Chapter 6, figured prominently in WPA archaeological initiatives.

More recent bioarchaeological scholarship has been notable for its interpretations based upon multiple lines of evidence, including that of Lucile St. Hoyme and Bullen, whose sensitive use of ethnohistoric materials enhanced their respective interpretations of early peoples of Virginia and Florida. Mary Frances Eriksen's histological research pioneered comparisons between modern anatomical materials and ancient remains. While Sheilagh T. Brooks' bioarchaeological efforts focused primarily on paleopathological analyses of archaeological series from the Great Basin, she is most widely known for questioning the Todd standards for age estimation and working with her colleague, Judy Suchey, to develop a more accurate system for evaluating the pubic symphysis. Louise Robbins' career exemplifies a scholar who adapted fully to changing intellectual climates as she moved from thesis research in the craniological mode of her mentor, Neumann, to a broader biocultural perspective (e.g., Robbins, 1977). Audrey Sublett's brief career was innovative in its emphasis upon excavating and recording historic period Indian remains and developing collaborative initiatives with Native Americans. Her article (with Rebecca Lane, Lane and Sublett, 1972) on kinship and residence among the historic period Seneca continues to be cited as a creative, pioneering effort.

The chapter authors also provide briefer sketches of other women scholars, perhaps less well known but with active careers in bioarchaeology. Katharine Bartlett's original interest in physical anthropology was redirected primarily toward education and museum collections curation, although she also analyzed numerous skeletal series from archaeological excavations during her long and productive career at the Museum of Northern Arizona. Grete Mosny, an Austrian by birth, devoted her career to education and museum conservation at the University and National Museum of Chile, her adopted country. Marília Carvalho de Mello e Alvim, a scholar associated with the Museu Nacional in Rio de Janeiro, taught large numbers of students at the University of Rio de Janeiro. Her scholarly contributions include research on the early South Americans from the Lagoa Santa site. Lilia Maria Cheuiche Macado, another Brazilian scholar, spent her long productive career championing an integrated biocultural approach to the study of past peoples.

Chapter 1

A Historical Introduction

Jane E. Buikstra

I. BEFORE 1900: THE ARCHAEOLOGICAL RECORD AND MUSEUMS AS EARLY RESEARCH CONTEXTS

In North America, the systematic observation of ancient excavated skeletal materials to investigate alternative interpretations of past lifeways can be traced to Thomas Jefferson (1853), one of the Founding Fathers and the third President of the United States.[1] Jefferson opened a burial mound located on his property in order to explore different contemporary explanations for their purpose. Commenting upon the "Barrows" or mounds "found all over this country," Jefferson outlined alternative interpretations:

> That they were repositories of the dead, has been obvious to all; but on what particular occasion constructed, was matter of doubt. Some have thought they covered the bones of those who have fallen in battles fought on the spot of interment. Some ascribed them to the custom, said to prevail among the Indians, of collecting, at certain periods, the bones of all their dead, wheresoever deposited at the time of death. Others again supposed them the general sepulchres for towns. (Jefferson, 1853:104)

Did the mounds contain the dead accrued from ancient battles or were they ossuaries or perhaps community cemeteries? Indian traditions, Jefferson stated,

[1] The significance of Jefferson's excavations has been noted previously (Lehmann-Hartleben, 1943; Silverberg, 1968; Willey and Sabloff, 1980; Wheeler, 1954). In each case, Jefferson's careful, modern, scientific approach was lauded. Willey and Sabloff (1980:32) marveled that "he excavated at all" and underscored his problem orientation. Lehmann-Hartleben (1943:163) further emphasized that "most amazing of all" was his ability to realize the significance of excavating to "the virgin soil" and the recognition of superimposed strata, "which reveal the inner structure of the mound." To this list should be added the integration of human biological and archaeological information in his careful, problem-oriented approach.

Bioarchaeology: The Contextual Analysis of Human Remains, Buikstra and Beck (eds.)

supported the last alternative, whereby mounds were said to contain sequential burials placed erect and then covered with earth. Jefferson's excavations, however, led him to define four superimposed ossuary episodes, based on: "1. The number of bones. 2. Their confused position. 3. Their being in different strata. 4. The strata in one part having no correspondence with those in another. 5. The different states of decay in these strata, which seem to indicate a difference in the time of inhumation. 6. The existence of infant bones among them." In reaching his conclusion, Jefferson thus utilized information that combined the observation of human remains within an archaeological context to select between alternative interpretative models. Today this approach would be considered bioarchaeological.

Following Jefferson there were many important 19th-century contributions that integrated the study of human remains within broader investigations of American Indians. Three examples from this period are cited here, beginning with a cautionary tale. These three examples serve to emphasize the fundamental importance of contextually based interpretations.

In 1839, Samuel George Morton, M.D. published *Crania Americana*, a work designed to address an important 19th-century issue — the physical diversity of the American Indian (see also Chapter 2). Assuming that similarities in skeletal morphology reflected heritage relationships, Morton considered, for example, the identity of the "Moundbuilders" of North America (Buikstra, 1979). Were the "authors" of the prominent tumuli that lined the major river systems of eastern North America ancestors of the Indians encountered by early explorers and settlers or were they associated with Old World creators of monumental architecture, perhaps the Egyptians (Silverberg, 1968; Stanton, 1960)?

Based on his observations, which have been the subject of intense 20th-century criticism [Gould (1978a,b, 1981); but see Michael (1988) and Chapter 2], Morton emphasized the fundamental unity of the "American race." His five Moundbuilder skulls were grouped with other Toltecan builders of monuments in Mexico and Peru. According to Morton (1839), these North American Toltecans were driven south by migrants from the north, the true ancestors of living Indians.

Morton's work was acclaimed by many contemporary medical and natural scientists (Grant, 1852; Meigs, 1851; Patterson, 1854; Wood, 1853). In his eulogy read before Philadelphia's Academy of Nature Sciences, for example, Charles D. Meigs (1851:20) described Morton as "America's Humbolt." Morton's conclusions were, however, criticized by the archaeologists Squier and Davis (1848), who argued that the true Moundbuilders were much more distinct from other American Indians than Morton had claimed. Morton, lacking definitive spatial and temporal data for the materials he studied, was led to reverse his opinion in a posthumous publication, which concluded that due to the great age of the mounds one would rarely find preserved remains. He had probably never seen the skull of a mound-building Indian, he opined (Morton, 1852). Thus, because his skulls had been procured without detailed archaeological information, he was forced to equivocate. This underscores the importance of contextual knowledge in bioarchaeological study.

The pursuit of ancient American skeletal materials led another medical doctor, Joseph Jones (see also Chapter 11), to engage directly in archaeological recovery of remains from "Mounds, Earthworks, and Stone-graves" from Tennessee during 1868 and 1869 (Jones, 1876:v). His observation that representations of the so-called "pigmy race" were, quite simply, children's remains was astute (Jones, 1876:9–11), but it was his careful diagnosis of syphilis in pre-Columbian remains that drew 20th-century attention to his work (Baker and Armelagos, 1988; Cook, 1976; Jarcho, 1966a; Powell, 1988, 2000). As Jarcho remarked (1966a:9), Jones' "diagnoses rest upon gross criteria that are nearly identical with those in use today, and the conclusions stand as well as any other landmark in this battleground of controversy." In reaching his conclusions, Jones (1876) not only relied on gross observations of skeletal material, but also observed thin sections and conducted experiments with hydrochloric acid designed to address the relative age and overall antiquity of the interments. Jones continued his archaeological interests until his death in 1896, expanding his observations to shell mounds of the Louisiana and Mississippi (1878). As a result of his ongoing observation of skeletal disease in excavated materials, coupled with scholarly evaluations of published sources, he formulated a well-informed theory specifying New World origins for a syphilis that "was a pestilential fever, which was communicable through the genitals, and otherwise" (Jones, 1878:932). He thus endorsed the presence of a nonvenereal syphilis in the pre-Columbian New World, a view substantiated by more recent studies (Cook, 1976; Powell and Cook, 2005; Chapter 11). Clearly, Jones' careful attention to archaeological contexts, especially their antiquity, markedly enhanced his conclusions.

Another 19th-century medical doctor, Washington Matthews, included not only archaeological data but also information derived from ethnology, ethnohistory, and oral traditions in his analysis of ancient human remains (see also Chapter 5). Research developed through his collaboration with the ethnologist–archaeologist Frank Hamilton Cushing exemplifies Matthews' integrative approach.

In 1897, Cushing embarked upon an archaeological expedition grounded in his more than 5 years' experience with the living Zuñi[2] (Cushing, 1890; Hinsley and Wilcox, 1996, 2002). Cushing's remarkably creative intellect had been immersed in Zuñi culture and he now wanted to learn of their unwritten past through archaeological investigations. Having gained support from Mrs. Mary Hemenway, he began fieldwork in February 1887 near present-day Tempe, Arizona. This site was ultimately named "Los Muertos" due to the presence of large numbers of

---

[2]Cushing's preferred annotation.

human remains (Cushing, 1890; Hinsley and Wilcox, 1996, 2002; Matthews *et al.*, 1893; Merbs, 2002b).

Cushing interpreted his archaeological discoveries in relationship both to the living Zuñi and to broader theoretical and historical issues (Cushing, 1890; Hinsley and Wilcox, 1996, 2002; Matthews *et al.*, 1893). He was convinced that oral traditions and hence his ethnographic knowledge were valid bases for interpreting ancient pasts (Hinsley and Wilcox, 2002:89). He also sought interpretations from members of the Zuñi community. According to Cushing: "[The Zuñi] also taught me the importance of testing whether the myth of the Lost Others was founded on fact, as there seemed reason to suppose, and in this the Zuñi could be of great help, confirming, for instance, my identification of ruins and the symbolic meanings of such pictographic and other art remains as might be found" (Hinsley and Wilcox, 2002:89).

Cushing also defined ancient and modern American Indians in terms of a primordial *Idea*, which was fundamental to all human groups and emanated from the "living soul of a dead culture." All American Indians had developed the *Idea* from the ancient Zuñi (Cushing, 1890:151). Peruvians, according to Cushing, were quite closely associated to the Zuñi, based on observations of guanaco pictographs and excavated terra cotta figurines, bolas, *quipus*, and Inca bones (Cushing, 1890; Matthews *et al.*, 1893; Merbs, 2002b).

Due to Cushing's poor health, the U.S. surgeon general was petitioned by Mary Hemenway to send Dr. Washington Matthews to visit the excavations in Arizona in August of 1887. The month that Matthews spent at Camp Hemenway led him to ask that the anatomist of the Army Medical Museum (AMM), Dr. J. C. Wortman, be sent out to conserve the fragile human remains. Following recovery, the remains were sent to the AMM in Washington, DC for further study (Matthews *et al.*, 1893; Merbs, 2002b).

The final osteological report (Matthews *et al.*, 1893) was notable for several reasons. While subtly distancing himself from Cushing's more controversial interpretations (Cushing, 1890; Hinsley and Wilcox, 1996, 2002), Matthews discussed both the rationale for the excavations and the preliminary archaeological results as background for his skeletal biological analysis. Thus, they reported Cushing's identification of vertical status distinctions between those interred without cremation (sacerdotal elite) and those from the "pyral" or cremation cemeteries (commoners), even though their skeletal analysis was not similarly partitioned.[3] Cushing's interpretation of "killed" vs unblemished grave goods in terms of soul release was also discussed, as was his assertion of alleged Peruvian–southwestern U.S. (Saladoan) links. The latter was formally tested by Matthews, Wortman,

[3]Harrison Allen's 1898 study of Hawaiian skulls from caves and coastal sites appears to be among the first to explicitly compare physical features of different status groups that may be contemporary.

and Billings (1893) through observations of the *os inca*[4] in a worldwide sample. They concluded, based on similar high frequencies of the Inca bone, that the Saladoans "out-inca['d] the incas" (Matthews *et al.*, 1893:190; see also Matthews, 1889; Merbs, 2002b). Matthews and colleagues' (1893) analysis also carefully evaluated measurement methods for estimating sex in skeletal remains and considered the impact of cradle boarding on cranial shape and development. They invoked both environmental and cultural observations in associating platycnemia (tibial medio-lateral flattening) with carrying heavy loads, explicitly rejecting other biomechanically plausible interpretations, such as long distance running or moving over vertically differentiated landscapes. Septal apertures[5] were also interpreted behaviorally as possibly being due to repetitive grinding of maize upon a metate. Thus, Matthews' emphasis on contextually sensitive analyses and interpreting skeletal analyses in a rigorous, problem-orientated manner serves as a valuable precursor to 20th-century bioarchaeology.

Recovery and systematic laboratory analysis of prehistoric skeletal material postdates Jefferson's explorations in Virginia. Increased numbers of archaeological initiatives, such as the Hemenway Expedition, encouraged institutional support for the collection, preservation, and study of excavated materials, including human remains. While the retrieval, analysis, and curation of skeletons and associated artifacts would become controversial during the 20th century (see Chapter 15), 19th-century museums actively sought and retained human remains as a valuable source of information about the unwritten past. Although "cabinets of curiosities" and local historical societies had previously held and displayed such materials, it was the systematic collections begun in cities such as Boston, Philadelphia, and Washington, DC during the 1800s that set the stage for 20th-century bioarchaeology in North America. However, the manner in which archaeological collections were accessioned, maintained, traded, or dispersed varied considerably.

The Morton collection, which numbered 867 (Meigs, 1851:23), 951 (Patterson, 1854:xxx), or 968 (Hrdlička, 1914a:513) human skulls at the time of Morton's death in 1851, still survives today,[6] unlike many other contemporary collections. According to Patterson, by 1854 it included an additional 51 human crania, along with the skulls of other mammals (278), birds (271), reptiles, and fishes (88). By that time, the citizens of Philadelphia purchased the collection for $4000 and donated it to the Philadelphia Academy of Sciences (Patterson, 1854:xxx;

---

[4] *Os inca*, or "Inca bones," are separate bones appearing in the posterior aspect of the skull due to nonpathological failure in suture closure. High frequencies of Inca bones have been reported in Andean skeletal series.

[5] These are ossification failures adjacent to the articular surfaces of the distal humeri.

[6] The Morton collection has been transferred to the University of Pennsylvania (1966), where it is curated today.

Hrdlička, 1914a:523). While the Philadelphia Academy attempted to engage two scientists, Joseph Leidy and J. Aitkin Meigs, in furthering Morton's work, their ability to achieve this goal was limited (Hrdlička, 1914a). By 1914, Hrdlička (p. 523) described the collection as a "sad relic" still held by the Philadelphia Academy of Sciences. Hrdlička (1914a:523) attributed the failed research legacy to a lack of scholarly attention or "other reasons," among them the failure to gain further systematic accessions. Such "other reasons" certainly must have also included the lack of detailed provenience data, incomplete sampling of archaeological sites,[7] and the bias introduced by restricting the sample to adult skulls. These factors limited the Morton collection's value for 20th-century bioarchaeological research.

Joseph Jones' collection appears to have been dispersed and lost. Following Jones' death in 1896, his widow, Susan Polk Jones, corresponded (1901–1909) with a number of institutions attempting to sell his archaeological collection (Jones and Jones, 1901–1909). The "Abstract of the Catalogue of the Archaeological Collection of Joseph Jones" indicates that the "greater part" of the collections was composed of materials from Jones' 1868–1869 excavations (Jones, 1901). While artifacts apparently formed the core of the catalog, the caption for Figure 100, the final illustration in the catalog, indicates that this "collection contains a large number of Moundbuilders' skulls, carefully numbered and measured . . . also a large number of those of various nationalities of modern times" (Jones, 1901:35). Sectioned postcranial bones are not mentioned, although they appear to have been present at Jones' presentation to the New Orleans Medical and Surgical Association on Saturday evening, April 27th, 1878 (Jones, 1878). As reported by Susan and Joseph Jones' grandson, Stanhope Bayne-Jones:

> The bones were kept in the house in New Orleans for a long time. When my grandfather died in 1896 the family needed a little cash and the specimens were sold for what they would fetch. They were bought for the Heye Museum in New York. Most of them now are in a warehouse in Brooklyn; the rest are distributed through the display cases in accordance with some artistic scheme, I think, rather than with a palaeopathological scheme. It seems to me that this whole large collection ought to be re-examined by the new methods. But in this case there is a little difficulty. My grandfather pasted paper labels on the foreheads of the skulls and on the bones. These labels don't withstand climatic changes too well and most of them have come off. It is going to be difficult to identify them again, but I don't know why they couldn't be examined by these new methods to see what sorts of lesions they present. (Bayne-Jones, 1996:39)

While restudy would indeed be a desirable goal, facilitating the resolution of Jones' diagnosis of widespread syphilis and placating his skeptics, it appears that

[7]In fact, incomplete sensitivity to historical and ethnological contexts, as well as selective sampling, were critiques leveled against the Morton collection by Daniel Wilson (1876) during the 19th century.

the collection was dispersed following acquisition in 1906 by the Heye Foundation/Museum of the American Indian (MAI). According to records currently at the National Museum of the American Indian (NMAI), Mrs. Jones received $4500 for the collection (Rachel Griffin, personal communication, 2004). The history of dispersal is unclear, although the "physical collection" of the Heye was of sufficient significance for Franz Boas (1972), at the recommendation of Bruno Oetteking, to approach George G. Heye with a request for a long-term loan to Columbia University, a transaction that never occurred. However, the Jones collection was said by Herbert U. Williams to be composed solely of skulls and only one with syphilitic changes by the time of his observations (Williams, 1932). However, given the relatively rare occurrence of diagnostic cranial lesions in North American skeletal materials presenting postcranial changes attributable to treponemal infection, Williams' observations cannot be considered definitive proof that the Jones collection of crania had dispersed by that time. The records of the MAI that were transferred to the NMAI after 1989 indicate that nine skulls from the Jones collection had been deaccessioned to the New York University (NYU) School of Dentistry in 1956, while two were still present in the NMAI collections, as of 2004 (Rachel Griffin, personal communication, 2004). While records at NYU indicate that materials were received from the MAI in 1956, the Jones collection is not specified (Eric Baker, personal communication, 2004). Thus, from the "large number" reported in 1901, the only skeletal materials remaining from Jones' collection appear to be the two skulls housed with the NMAI in Suitland, Maryland.

Although excavated in the Southwest, the skeletons from the Hemenway Expedition were also curated in East Coast museums. Wortman returned to the AMM with human remains in June of 1888 (Lamb, 1915:630), thus significantly increasing the museum's holdings. The AMM had been founded on May 21, 1862, during the Civil War, beginning as "a set of three dried and varnished bones resting above an inkstand on the desk of Brigade Surgeon John Hill Brinton," the first curator of the Museum (Henry, 1964:1). Medical officers were subsequently directed to collect and forward to the Office of the Surgeon General "all specimens of morbid anatomy, surgical or medical, which may be regarded as valuable; together with projectile and foreign bodies removed, and such other matters as may prove of interest in the study of military medicine or surgery" (Henry, 1964:1; Sledzik and Barbian, 2001:227). The mission of the museum, however, was soon to expand.

In 1868, George A. Otis, in charge of the AMM anatomical collection, sent letters to medical officers, noting that the museum already had 143 (Indian) skulls and requested concerted collection of more (Lamb, 1915). The stated purpose of such efforts was "to aid the progress of anthropological science" through cranial measurement of American Indians (Henry, 1964:58). The next year, Otis reached an agreement with secretary Joseph Henry of the Smithsonian Institution (SI) such that the Smithsonian's skeletal materials would be transferred to the AMM in exchange for ethnographic and artifactual items.[8] Transfers began the same year and, by April 12, 1870, Otis reported to the National Academy of Sciences that the AMM possessed over 900 crania, 376 of these having been transferred from the SI. Cranial measurement and evaluation of cranial deformation were emphasized in Otis' papers, although septal aperture morphology was mentioned in 1878 (Lamb, 1915). Clearly, at this time, the AMM-collecting strategy, conducted as it was by remote medical officers who lacked archaeological expertise, focused on broad regional coverage rather than detailed provenience data. Curation of remains and artifacts in administratively and spatially removed institutions also would have encouraged interpretations of skeletal remains only generally linked to their cultural or environmental contexts. Thus, these early SI and AMM collections suffered many of the same limitations as those of Samuel Morton, after whom Otis modeled his collecting strategies (Lamb, 1915).

Following Otis' death (in 1881), J. S. Billings became curator of the museum (in 1884) and was soon joined by Washington Matthews. Both Billings and Matthews actively engaged in descriptive and methodological study of the collections. For example, prior to AMM involvement with the Hemenway Expedition in 1887, Matthews experimented with methods for measuring cranial capacity and cranial form, while Billings investigated composite photography and its craniological applications (Lamb, 1915; Billings and Matthews, 1885; Matthews, 1898). By 1886, Matthews had also accumulated 21 years of medical experience among Indians within a dozen states and territories. His systematic observations of consumption frequencies among Indians with distinctive lifeways (Matthews, 1886, 1887, 1888) foreshadowed Hrdlička's (1909a) more recent study of tuberculosis among five Indian tribes.

Why was the report on the Hemenway Expedition materials much more culturally and behaviorally nuanced than earlier AMM studies? As also emphasized in Chapter 5, both Matthews and Wortman had actually visited Los Muertos and Wortman is described by the home secretary to the expedition,

---

[8]Under Hrdlička's influence, most human remains were returned to the SI from the AMM during 1898, with only a few crania retained along with items of pathological significance (Sledzik and Barbian, 2001:228). Lamb (1915:631) reported that the transfer involved 2206 Indian crania, followed by a few additional skulls from the archaeologist C. B. Moore. He goes on to state that "in May, 1899, 115 boxes of bones from the Hemenway Expedition that had remained at the Army Medical Museum, were transferred to the National Museum, and in January, 1904, nearly 600 skulls, pelves, and two Indian brains were likewise transferred" (Lamb, 1915:631). Following this transfer, the materials were lodged within the Division of Physical Anthropology, rather than in the Division of Mammals, where they had been accessioned prior to 1869. During these early years, only those human bones showing cultural modifications such that they could be classified as artifacts, e.g., bone flutes, were accessioned and stored with the archaeological collections (D. Hunt, personal communication, 2004) in what was called "its ethnological department" (reference for quotes: Smithsonian Institution, NMNH Web site Department of Anthropology: A History of the Department, 1897–1997).

Sylvester Baxter (1889:33), as one of "two Doctors[9] . . . found grubbing in the pits, industriously at work over the skeletons, over whose anatomical characteristics their enthusiasm is aroused to a high pitch. They are intent on securing and saving every bone, and are regardless of personal discomfort, not only their clothes being covered with the dust, but their faces begrimed and their hair and beards thoroughly powdered, making them look like some strange burrowing animals. The result of their painstaking is one of the finest and most complete collections of ancient skeletons ever brought together . . . ."

Matthews, Billings, and Wortman's 1893 report undoubtedly also benefited from Cushing's ethnographic knowledge of the Zuñi and his stimulating speculations, as well as Bandalier's ethnohistoric information. Cushing's emphasis on understanding the lives of the people who created Los Muertos, rather than simply reporting architectural plans and finely crafted material culture, is aptly summarized by Baxter's statement: "It will be seen that the results of the Hemenway Expedition are of importance, not so much through what has been found, as by what has been *found out* in the progress of the work" [Baxter (1888), reported in Hinsley and Wilcox (1996:134)]. This individualized approach to the past led Cushing to keep grave lots together and separately identified.

> By a study of these accompaniments to each burial (which I at once determined to keep the identity and interrelation of distinct), the sex, often the condition in life, and in fact many other personal items relating to the individual buried may be definitely made known when these collections, if ever, are minutely studied by me, and cannot fail to give vivid, as it were, even historic knowledge of the people and phase of culture represented by these wasted and buried cities. [Cushing, reported in Hinsley and Wilcox (2002:200)]

Thus, even though Cushing would never develop his detailed report, the manner in which he collected grave lots facilitated later scholarship, including dissertations by Haury (1934) and Brunson (Brunson, 1989; Hinsley and Wilcox, 2002).

Unfortunately, however, by the mid-20th century, no records existed to link the Hemenway's inhumed remains with burial contexts (Brunson, 1989). Cushing's personal secretary, Frederick W. Hodge, had indicated that he was creating such a list, but it has never been discovered. When the skeletons were transferred from the AMM to the Smithsonian Institution, accession numbers were assigned

---

[9]The other was Dr. Herman F. C. ten Kate (Baxter, 1889). Dr. ten Kate [Ph.D., M.D. from the University of Leyden, The Netherlands (Hrdlička, 1919)] was the physical anthropologist hired by Cushing for the expedition. As it turned out, he arrived at camp on November 18, 1887, only 1 week before Wortman. Brunson (1989) reported that after initial tensions, the two men actively engaged in field work designed to maximize recovery of the fragile skeletal remains. Dr. ten Kate published reports on the hyoid bone (with Wortman *et al.*, 1888) and also investigated the somatology of living Southwestern Indians in relationship to the ancient remains from Los Muertos. He concluded that the ancient remains from the Salado valley and sites near Zuñi most closely resembled the living Zuñis (ten Kate, 1892).

without reference to field contexts. In 1957, T. Dale Stewart, curator of physical anthropology, reported that skeletons could not be associated with either field locations or grave lots (Brunson, 1989:146).

Billings was lauded by Lamb (1915:631) for holding "tenaciously to the belief . . . that it was best to hold the osteological collections together until a complete study could be made of them, and not to scatter the specimens by donations and exchanges." This goal was not, however, fully met for the Hemenway Expedition skeletal collections. There were two exceptions. First, only 35 of 200 remains recovered from the circum-Zuñi sites traveled to Washington (Matthews *et al.*, 1893). Presumably, some 25 (Peabody Museum, 2001) of these are referred to by Putnam in his curator's report to the trustees of the Peabody Museum for 1890 (1891:90): "It is with pleasure that I also mention a collection of human crania from the ruins near Zuñi, collected by the Hemenway Southwestern Archaeological Expedition, and kindly presented by Mrs. Mary Hemenway." A second group of bones arrived at the Peabody Museum with less fanfare. Numerous "incendiary urns" recovered from the bases of Cushing's (1890) "pyral mounds" were part of the ~5000 specimens from the Hemenway Expedition loaned to the Peabody Museum of Salem upon completion of the expedition in 1888. They were transferred to the George Peabody Museum at Harvard in 1894 upon the death of Mary Hemenway (Haury, 1945). The fact that the cremated remains from Los Muertos ($n = 129$) and Las Acequias[10] ($n = 1$) were not accessioned into the Harvard Peabody collections until 1946 (accession #46-73: Peabody Museum), the year after Haury's study[11] of Los Muertos and neighboring ruins, including the cremations, was published, suggests that the bones may have entered the museum with or within the burial urns. Haury (1945:45) reported that "the contents of many of the funerary jars were saved," but not the point in time when they were removed from their containers. Nor is there a collection date for the cremations given in the Peabody accession records (Peabody Museum, 2001).[12]

While published documents do not directly state why the burned and intact skeletal samples were divided between the AMM and the Peabody museums,

[10]Las Acequias is spelled "las acquias" in the George Peabody accession list (Peabody BIOCAT database), but as it is reported for the Salt River Valley, it appears that the two are identical.

[11]The 1945 publication was a revision of Haury's 1934 dissertation. Haury reported examining 134 lots of bones, while there are only 130 accessions within #46-73 (Peabody Museum, 2001). There were three examples of double cremations (Haury, 1945:45), which could render the lot numbers and the accession records nearly identical (131/130).

[12]Brunson (1989:146) reported that distinctly different numbering systems were used for grave lots with inhumations and with cremations. Accompaniments with inhumations were numbered sequentially, whereas all items, including human cremains, associated with cremation urns were assigned a single number, followed by sequential alphabetic designations, e.g., 1a, 1b, 1c. This would have facilitated attempts to reassociate skeletal remains with their interment contexts.

the following quotation from the Matthews, Wortman, and Billings report leads to the inference that 19th-century researchers considered only intact, unburned remains suitable for study.

> It is believed that those of the priestly race were not cremated because they had the power to release their own souls from their bodies while the laity, having no such power, had to have their bodies burned to effect the desired release. Whatever may have been the creed that thus preserved some bodies for simple interment, anthropology owes it gratitude, *for without it the unique skeletons of this archaic race would not have been preserved for modern study and comparison* (italics added). (Matthews *et al.*, 1893:150)

This argument is reinforced by Haury's observation that Earnest Hooton encouraged him to "examine the cremated human remains, evidence which heretofore has been generally neglected because of its supposed uselessness" (Haury, 1945:xii). Thus, at this time in the history of bioarchaeological study, practitioners felt that there was little to be gained through the study of cremated remains. Fortunately, the cremations were retained and were thus available at Harvard's Peabody museum when Haury developed his dissertation project.

While other sets of skeletal remains associated with Harvard University clearly have a longer history in the anatomical collections of the Boston Society for Medical Improvement[13] and the Warren Anatomical Museum (Jackson, 1847, 1870; Beecher and Altschule, 1977), systematic curation of human skeletal materials from archaeological contexts dates to the founding of the Peabody Museum with Jeffries Wyman as the curator. In the first annual report to the trustees, Wyman characterizes a rather unpretentious inception:

> On the 9th of November 1866, a collection of various objects pertaining to the purposes of this Museum was begun, and temporarily deposited in one of the cases of the Museum of Comparative Anatomy, in Boylston Hall. The collection consisted of crania and bones of North-American Indians, a few casts of crania of other races, several kinds of stone implements, and a few articles of pottery, — in all, about fifty specimens. Of these, about one-half belonged to Harvard College, and with the consent of the President, were transferred to this Museum, the others were from the collections of the Curator. (Wyman, 1868a:5)

Wyman was the curator for the Peabody from 1868 until his death in 1874. His approach to collections acquisition, contextual detail, and the significance of human remains, as represented in his curator's reports and his publications on Florida shell mounds, appeared to be well ahead of its time. In the first

---

[13]The collection of the Anatomical Museum of the Boston Society for Medical Improvement was accompanied by a detailed catalogue of the cabinet, which was summarized by Jackson (1847). The collection was developed from specimens presented to the society at bimonthly meetings. Materials useful for teaching, those with authentic case histories, and unusual examples were preferred. Thus, the collection included a "very beautiful French preparation" of a disarticulated skull (p. 2, #14), alongside an albatross humerus (p. 3, #25) and a diseased mink skull (p. 11, #79).

annual report (Wyman, 1868a), for example, he expressed enthusiasm over two relatively large collections of skulls, one from Peru and another from Hawaii. In the former, provenience data are considered important and, in the second case, concern over the age of the specimens is expressed. His discussion of explorations of the East Coast of Florida, beginning with the first report, are remarkable for their integrated observations of stratigraphy, artifacts, faunal and floral remains, and human bone. While he may not necessarily have collected all bones from all skeletons excavated at a site, there is a clear emphasis beyond the cranium. For example, Wyman (1869:18), in discussing the collections made and transferred to the Peabody by S. S. Lyon from "ancient mounds in Kentucky," noted that the materials included "large numbers of crania and extensive collections of the more important bones of the skeletons. . . ."

In an analysis of crania and "other parts of the skeleton" presented in the fourth annual report, Wyman (1871:10) concluded ". . . brain measurement cannot be assumed as an indication of the intellectual position of races any more than of individuals.[14] From such results the question is very naturally forced upon us whether comparisons, based upon cranial measurements of capacity as generally made, are entitled to the value usually assigned them." This prescient statement came at a time when other American and European scholars were obsessed with craniometry as a measure of innate intelligence (Gould, 1981). In addition, he reported several postcranial attributes, including the first observations of "flattening of the tibia," in remains from the Western Hemisphere.

Wyman's most extensive discussion of human skeletal materials appeared in his consideration of cannibalism from the St. John's River "shell heaps" (Wyman, 1874, 1875). Wyman approached the issue with characteristic rigor, having noticed a number of human bones found under "peculiar circumstances" and not interred in articulation (1874:60, 1875:26). He details broken remains that were treated just as were animal bones from the same deposits and fragmented in a predictable pattern (1874:60, 1875:27). In another example, he reported a femur that was separated into proximal and distal segments by first "cutting a groove around the circumference of the bone and thus weakening it and then breaking the remainder. This is a common method of dividing [animal] bones used by Indians" (Wyman, 1874:63). Wyman (1873) supplemented his arguments with ethnographic and ethnohistoric observations of cannibalism, primarily in the Americas. While his analyses are not as detailed as those of more recent scholars (Turner and Turner, 1999; White, 1992), evidence of cannibalism among ancient American Indians remained a highly visible and controversial topic in the late 20th century (Billman *et al.*, 2000; Bullock, 1991; Darling, 1999; Dongoske *et al.*, 2000; Walker, 1998).

---

[14]Other 19th-century scientists who questioned race-based craniology included Allen (1895a, 1898) and Wilson (1876; see also Trigger, 1966).

The remarkably "modern" nature of Wyman's contextualized and problem-oriented approach to the study of ancient human remains is underscored by comparison with that of his successor, Frederic Ward Putnam. For example, during Putnam's tenure, accessions 13116–13565 were described as "[a] number of human skeletons and a large and valuable collection of implements and ornaments of stone, bone and shell of native manufacture . . ." (Putnam 1878a:216). Such statements stand in marked contrast to Wyman's (1873:6) discussion of "... the large and very valuable collection of ancient Peruvian skulls, obtained by the Hassler Expedition ..." and his conclusion that "[t]he most important addition to the Museum during the year is the archaeological and craniological collection of Dr. Giustimano Nicolucci, of the Island of Sora, Naples." For Wyman, human remains were of at least equal significance to artifacts. In parallel, although Putnam solicited, influenced, and published reports on skeletal remains, e.g., "Measurements of Crania Received during the Year" (Putnam, 1878a:221), "Observations on the Crania from the Stone Graves in Tennessee" (Carr, 1878), "Notes on the Anomalies, Injuries, and Diseases of the Bones of the Native Races of North America" (Whitney, 1886), and "The Madisonville Prehistoric Cemetery: Anthropological Notes" (Langdon, 1881), these studies lack the innovative contextual sensitivity and problem orientation of Wyman's earlier work. Jarcho (1966a:12), for example, characterized Whitney's study as a "systematic compendium" and Langdon's as "good of its time."

Thus, by the turn of the 20th century, medical doctors, anatomists, and other scientists were addressing several issues that would be reflected in later bioarchaeological studies. Biological distance investigations, particularly those on a continental scale and focused primarily on the origin of American Indians, continued to dominate the field. The numerous studies of artificial cranial deformation, a form of cultural modification, reflect an emphasis on the cranium. Although most biological distance inferences were based on cranial measurements, Matthews *et al.* (1893) also considered a non-metric variant, the Inca bone, in their comparisons of Los Muertos skeletons with those from Peru. Paleodemographic inferences of population numbers and age distributions had been important for Jefferson's remarkable 18th-century investigation. During the 19th century, Broca's six-stage age grades (first period of childhood: birth to 6th year; second period of childhood: 7–14 years; youth: 14–25 years; adult age: 25–40 years; ripe age: 40–60 years; and senility: 60+ years) had been imported and detailed by scholars such as Matthews *et al.* (1893). The Matthews investigation also addressed metric evaluations of sex differences in the bony pelvis (Hoyme, 1957b; Matthews *et al.*, 1893; Stewart, 1979a[15]). Melding contemporary medical knowledge with the study of the past was well illustrated in

[15]Both Hoyme (1957) and Stewart (1979) cited the Matthews, Wortman, and Billings (1893) report as Matthews and Billings (1891).

Jones' (1876) investigations of syphilis among the stone box grave interments from Tennessee. While Wyman (1871) was the first to note "flattening of the tibia" among American Indians, the relationship of long bone shape to daily activity stresses, including sex-based differences, was initially emphasized in the study of Los Muertos (Matthews *et al.*, 1893). Dietary issues were also addressed in the American Southwest, with the authors concluding that "[w]ith this evidence before us it can not [*sic*] said that a meat diet is injurious to the teeth or a vegetable diet especially beneficial" (Matthews *et al.*, 1893:201). The way was thus paved for the development of a topically diverse bioarchaeology in the 20th century.

In closing this overview of 18th- and 19th-century scientists, their problem orientation, and their collections, it is important to underscore a point also made by Cook in Chapter 2. While these contributions were made by North American scholars, working with North American materials, their perspective was not provincial. Jefferson traveled extensively and corresponded globally. Morton's second medical degree was awarded in Edinburgh (1823) and he also studied in France. He was well respected in Europe and building his collection generated an extensive international correspondence. While Washington Matthews may have never ventured from the boundaries of the region that would become the United States, he was well informed concerning methods and conclusions being generated in Europe. Cushing and Matthews' colleague, Baxter [1888, reprinted in Hinsley and Wilcox (1996:133)], wrote about parallels between the Los Muertos discoveries and those made in Almería, Spain, both excavated during 1887.[16] The Harvard Peabody scholars were similarly well informed and widely respected. While the political and religious climate developing in the new nation doubtless affected the way in which some research problems were framed, this American scholarship should not be viewed as insular.[17]

---

[16]"An indication of the possible age of these remains may be found in a consideration of the remarkable archaeological discoveries reported from the Spanish province of Almeria, made last summer, so shortly after these of Los Muertos as to be almost simultaneous. The account of those reads like a repetition of the story of these, for there, too, it was a stone-age culture whose remains have been brought to light; that people also practiced both cremation and house-burial, and there, as here, the house-burials often included both husband and wife, or at least man and woman, side by side. As the conditions of soil and climate in southern Spain and our Southwest are remarkably alike, both regions being dry, hot and desert-like, and conducive to the long preservation of burial remains, it is quite possible for relics of the past to last as long here as there. And for European archaeology there is set an interesting task in estimating the possible period of a stone-age civilization on the borders of the Mediterranean, in a land subject to the influences of the iron-age Latin cultures and the bronze-age pre-Latin people. It is a striking fact, that at nearly the same time there should be discovered the remains of two cultures so closely resembling each other in their institutions, both in new Spain and in old." [Baxter (1888), cited in Hinsley and Wilcox (1996:133)]

[17]The same may not have pertained to the Canadian Daniel Wilson. Trigger (1966) believed that one reason Wilson, born in Scotland, did not pursue his anthropological interests more vigorously was his Toronto location, remote physically from like-minded scholars in Europe and isolated

II. HRDLIČKA AND HOOTON: CONTEXTS AND CONTRASTS

Two major figures, Aleš Hrdlička and Earnest A. Hooton, dominated early 20th-century physical anthropology in the United States. These two scholars, however, had very different orientations to archaeological contexts and anthropological problem solving. Given their enormous visibility and prominence, their contrasting attitudes toward archaeological contexts and bioarchaeological problem orientations are addressed here in detail.

T. Dale Stewart (1940a:20), mentored by Hrdlička from his undergraduate days and chosen to succeed Hrdlička as curator of physical anthropology at the Smithsonian Institution, notes nine "underlying aims" of Hrdlička's work. Among these were determining the range of normal variation for the human skeleton, preserving older American skeletal materials, and resolving the subject of geologically ancient man in the New World. This last-mentioned goal was undoubtedly influenced by Hrdlička's Smithsonian Institution mentor, W. H. Holmes (Spencer, 1979:289).

Hrdlička actively conducted fieldwork to measure remote skeletons and skeletal series, to gather collections for the Smithsonian, and to evaluate archaeological sites reputed to contain evidence of early man. It was during his critical review of early man sites in the New World that Hrdlička was most contextually sensitive. He actively collaborated with geologists in evaluating stratigraphy (Hrdlička, 1912a, 1916, 1917a, 1937a) and based arguments critical of proposed early finds on soil formation processes.

> The occurrence of isolated fossil animal bones or fragments in contact with, or even above, the human skeleton would have no significance. In digging a grave the earth thrown out might well contain fossils even of considerable size, which, after the body was introduced, would be thrown in about or above it. The apparently undisturbed condition of the partial and irregular sandy layers which occur in the muck where the skeleton No. II was discovered would hardly be regarded as sufficient proof that the bones were not introduced from above. The muck and sand thrown in over a body would tend in the course of time so completely to assume the appearance and characteristics of the original deposits that distinction between the two would be quite impossible. (Hrdlička, 1917:48)

While Hrdlička also evaluated allegedly ancient New World skeletal remains for evidence of morphological features distinctive from those of more recent Indians, his critiques frequently began with contextual information. For example, in considering skeletal remains attributed to early man in North America, he argued that such materials should "be photographed in situ, and should be

---

intellectually from the racist climate of the United States. Wilson's contributions are considered by Cook in Chapter 2.

examined by more than one man of science, including especially a geologist familiar with the particular formations involved; and the chemical and somatological characters of the bones should receive the closest attention with the view of determining their bearing on questions involving the antiquity of the remains" (Hrdlička, 1907:11). In his summary statement for the same article, he argued that the identification of human remains as those of early man "demands indisputable stratigraphical evidence, some degree of fossilization of the bones, and marked serial somatological distinctions in the more important osseous parts" (Hrdlička, 1907:13).[18]

Such concerns for fine contextual control are not conspicuous, however, in Hrdlička's work with more recent skeletal materials. When he was just beginning to amass the collection that would grow from approximately 3000 skulls and other bones in 1904 (Lamb, 1915) to more than 15,000 human skulls or skeletons by the time of his death in 1943 (Schultz, 1945), Hrdlička (1904:22) sought to enlist the aid of "medical men," especially those "who travel, or have charge of hospitals, colleges, dissecting rooms, and remedial institutions." Reminiscent of assistant surgeon general George A. Otis' 1868 memorandum to army medical officers "to aid in the progress of anthropological science (Lamb, 1915:625)," Hrdlička (1904:22) created a set of instructions for the collection of skeletal remains by foreign missionaries and teachers, explorers, miners, prospectors, surveyors, and engineers of railroads, "men engaged in trades that take them into virgin regions; and travelers of means and leisure." Designed to "help the science of physical anthropology" (Hrdlička, 1904:5), this pamphlet discussed the manner in which collections should be made. "The further back in time we recede from the actual period, the more essential become the preservation of the specimens and of all objects associated with them, and the correct localization of all with reference to geological formations ... [a]ll explorations for the skeletal remains of early man should be intrusted [*sic*] to thoroughly trained men only" (Hrdlička, 1904:11). Hrdlička's reports of his own collecting expeditions to such

[18]Hrdlička also emphasized stratigraphy and other contextual features in evaluating evidence for early man in South America (1912a:385). "The main defects of the testimony thought to establish the presence of various representatives of early man and his precursors in South America are (1) imperfect geologic determinations, especially with regard to the immediate conditions under which the finds were made; (2) imperfect consideration of the circumstances relating to the human remains, particularly as to possibilities of their artificial or accidental introduction into the older terrains, and as to the value of their association from the standpoint of zoopaleontology; (3) the attributing of undue weight to the organic and inorganic alternations exhibited by the human bones; and (4) morphological consideration of the human bones by those who were not expert anthropologists, who at times were misled in the important matter of placing and orienting the specimens and who accepted mere individual variations or features due to artificial deformation as normal and specifically distinctive characters."

venues as Peru and Alaska do not provide clear evidence of concern for careful recording of contextual information (Hrdlička, 1915, 1927b,c, 1941a,b).[19] In creating his remarkable collection of recent materials, the acquisition of large quantities of skeletons and mummies rather than archaeological control clearly remained Hrdlička's priority throughout his professional life. "Everywhere and at all times he indulged in his absorbing passion for collecting knowledge and potential new data in [*sic*] form of specimens. To the very last of his field-trips he derived the keenest happiness from every new skull which he could carry back to his boat to be added to the thousands of others he had already amassed at home" (Schultz, 1945:314).

Hrdlička did, however, eschew the selective collecting patterns of the 19th century.

> All parts of the body, from all stages of life, are fit subjects for physical anthropology, because racial, tribal, or other group differences are found in all of them. Thus far, however, but little attention has been paid in the United States to anything besides the racial, particularly Indian, skeletal constituents, and especially the skull, objects which are of more general interest, more abundant, and comparatively easy of collection and transportation. But even with the skulls and skeletons no systematic collection on a large scale, or a collection comprehending all the important elements of the population, has ever been attempted. All of this explains the condition of our collections and should indicate the way to their improvement. (Hrdlička, 1904:7)

In sum, Hrdlička must be credited with amassing a remarkable collection of human skeletal remains and beginning a tradition of skeletal study at the

---

[19]William Laughlin, following his first year at Willamette University (1938), spent 3 months excavating with Hrdlička in Alaska and Siberia (Commander Islands). Based on that experience, Laughlin (interview reported in Krupnik, 2003:211) expressed reservations about Hrdlička's archaeological procedures: "Hrdlička was not good at excavating. He did not keep any records, and he just dug to get any skeletons he needed. I remember he always kept two separate boxes on this trip (of 1938): for 'good' artifacts and for 'other' (bad) ones. At the seminars, he used to show the skeleton of a 17-year-old girl and every time he added: It's only a girl. He always degraded women, even in the skeletons. At Amchitka, we left everything we excavated at the beach. As the tide came in, it washed the objects over every morning and we picked them up again — whatever was left behind. But of course we missed lots of things or recovered them much later. That was his style . . . ." Other comments critical of Hrdlička's excavation methods can be found in de Laguna (1956), Bray and Killion (1994), and Scott (1992), who termed Hrdlička an "incautious" archaeologist. In his review of Hrdlička's early 1930s excavations at the Uyak site (Larsen Bay, Kodiak Island, AK), Speaker (1994:56) reported: "Hrdlička, although known as a meticulous physical anthropologist, approached the archaeological excavations of the Uyak site with the primary intent of recovering as many skeletal remains as possible (Hrdlička, 1941:1; 1944a:3, 141). He made no systematic attempt to record archaeological information at the site, and his field records are limited to anecdotal notes, sketch maps, and rudimentary profile drawings. Hrdlička divided the midden deposits into three strata. He referred to these as the black, red, and blue levels. Artifacts and skeletons were marked with colored pencils to indicate the stratum from which the item had been removed. Only rarely was any more precise provenience information recorded."

Smithsonian Institution. Given his background as a "medical man" and as an anatomist, his emphasis on description and variation is understandable. As Schultz (1945:311) remarked, Hrdlička "concerned himself properly and exclusively with the primary question: *What* are the variations of man? He left the secondary, though more fascinating, questions, beginning with *how* and *why*, to his successors." And indeed, his successors, such as T. Dale Stewart, Lucile St. Hoyme, J. Lawrence Angel, Donald Ortner, Douglas Ubelaker, and Douglas Owsley, have all developed a more nuanced approach to the archaeological record.

Despite his dedication to the development of the institution's holdings, Hrdlička's relationship with archaeologists who contributed to the Smithsonian's collections was not without tension. For example, when archaeologist Gerard Fowke sent his materials from Missouri mounds to the Smithsonian for analysis, Hrdlička grumbled about the condition of the remains.

> On the whole the material is very defective; there is not an entire skull, and there are only a few entire long bones. The specimens were damaged for the most part during excavation, as shown by fresh breaks, and in most cases important parts thus broken off were lost. More than nine-tenths of the bones of the skeletons are missing altogether. Moreover, the surfaces of the skulls were treated with a glue-like substance which has since begun to crack and scale off, doing further damage. (Hrdlička, 1910:103)

Fowke, however, had expressed an opinion critical of physical anthropologists and their approach to the study of skeletal remains. In describing the contributions of physical anthropologists to archaeological issues, Fowke opined:

> It is a beautiful scheme; the only trouble with it is that no one has ever been able to reduce it to a system from which it is possible to obtain any certain or definite results. When this difficulty is overcome—no special progress appears yet to have been made in that direction—we may look for the announcement of some interesting discoveries. (Fowke, 1902:132)

A closer working relationship between archaeologists and physical anthropologists is exemplified in the early 20th-century research of Earnest Hooton. In contrast to Hrdlička's general concern for human variability, Hooton's work with skeletal collections was closely linked to focused, regional questions currently being asked by archaeologists. He himself had conducted archaeological recovery of both artifacts and human remains in the Canary Islands (1915) as part of the Harvard African Series (Hooton, 1925). In discussing this project, Hooton emphasized its problem orientation and his engagement in field research.

> The first field effort incidental to the production of this series was an expedition to the Canary Islands, designed to clear up the much argued question of the affinities of the Guanches, an extinct race of cave dwellers alleged to be remnants of the famous Cro-Magnon artists of Upper Palaeolithic Europe. Upon this expedition the writer was forced to gather his own data, both physical and archaeological. . . . On the whole, this

> study seemed to confirm the necessity of an intimate cooperation of the archaeologist with the physical anthropologist. (Hooton, 1935:503)

Hooton's emphasis on integration of archaeological and physical anthropological knowledge had also been encouraged by his experience with North American remains housed in the Peabody. His frustrating experience with the Madisonville Cemetery collection is recounted in Chapter 4 (Hooton, 1935:501).

As the Pecos project developed, Hooton spent 2 months during 1920 working with the archaeologist A. V. Kidder at the site. Kidder took advantage of Hooton's presence to open additional trenches in burial-rich areas (Kidder, 1924; Givens, 1992). As emphasized in Chapter 2, Kidder thus became an advocate for physical anthropology, underscoring the importance of temporal control in studying changing demographic and heritage patterns over time. He also expressed interest in issues such as infant mortality, length of life, and the impact of disease (Kidder, 1924:33).

Hooton's emphasis on anthropological integration was also reflected in the training program at Harvard.

> The development of physical anthropological studies at Harvard and the advance of research in other anthropological fields have resulted in progressive encroachments of each specialty upon the preserves of the others, in order to secure significant explanatory data. Thus the archaeologist is driven further and further into interpretation of his cultural data in connection with skeletal material; the physical anthropologist finds himself perforce delving deeper and deeper into the collection and correlation of sociological data or of archaeological facts; the ethnologist advances steadily into the fields of his colleagues for the same reason. It is a significant fact that almost none of the anthropologists recently trained at Harvard can be forced to relinquish to their specialist colleagues the data in allied fields which they themselves have collected in expeditions. Thus the archaeologist insists upon working up the skeletal material which he has exhumed; the ethnologist prefers to correlate his own anthropometric data with his cultural findings, and the physical anthropologist raids in all directions and utilizes miscellaneous booty. Such a development is most healthy. . . . (Hooton, 1935:511)

Thus, with Hooton as with Jeffries Wyman before him, the Harvard/Peabody scholars presented a tight integration between physical anthropology and archaeological problem solving. Physical anthropologists participated in fieldwork and learned contemporary field excavation methods.

Chapter 2

The Old Physical Anthropology and the New World: A Look at the Accomplishments of an Antiquated Paradigm

Della Collins Cook

It is indeed difficult to imagine an all-wise Providence, after having by the Deluge destroyed all mankind excepting the family of Noah, should leave these to combat, and with seemingly uncertain and inadequate means, the various external causes that tended to oppose the great object of their dispersion: we are left to the reasonable conclusion that each Race was adapted from the beginning to its peculiar local destination. In other words, it is assumed, that the physical characteristics which distinguish the different Races are independent of external causes. (Morton 1839:3)

Craniology in the work of the 18th- and 19th-century anthropologist–physicians Blumenbach, Morton, and Warren serves largely as a descriptive tool, and analysis for these early typologists was confined to evaluating individual specimens. Variability was unimportant, and the approach is primarily one of classification. The typological study of Indian and Eskimo crania became the dominant enterprise as American physical anthropology emerged as a profession around 1900. The contributions of Hooton, Hrdlička, Rivet, Oetteking, and Neumann are reviewed. Among these, Hrdlička and Rivet built on the

Bioarchaeology: The Contextual Analysis of Human Remains, Buikstra and Beck (eds.)

19th-century French school that begins with the work of Paul Broca. Oetteking and Neumann built on the Boasian school, and through it as well as independently on the German school. American craniology is distinct from similar work in Europe in the degree to which these researchers interacted with archaeologists, in part because the Boasian race–language–culture model encouraged such interaction. The cultural and historical questions that motivated the typologists remain with us today.

I. INTRODUCTION

The twin problems of the origins and diversity of American Indians emerged in the earliest European accounts of the New World (Arensberg, 1995). Were Indians fundamentally similar or were they diverse? Were they closely related to one or to several peoples of the Old World? The most balanced and detailed account of this history remains that of Juan Comas (1960, 1974); he presented Hrdlička's model for a single northwest Asian origin for American Indians as a novel formulation that contrasts the various hypotheses for multiple origins that had and continue to have considerable currency in Latin America. Hrdlička's model has been and remains the dominant or only model among North American anthropologists (Stewart, 1960a, 1981; Stewart and Newman, 1951; Crawford, 1998). Thus, key issues that were formulated in the earliest literature in our field have persisted to the present day, despite theoretical and methodological transformations that might have been expected to influence them. This chapter focuses on some issues of method in the typological research.

The typological paradigm in physical anthropology gave way in the middle of the 20th century to a concept of human variation grounded, on the one hand, in the emerging field of population genetics and, on the other, in the powerful new statistical tools of biological distance. From its origin our field was wedded to typological thinking. In rejecting this outdated paradigm, we have turned away from much of what our discipline accomplished before the latter half of the 20th century. This chapter reviews and reevaluates this past.

Much of what has been written about the typological era in physical anthropology has been couched in a disciplinary critique of racism in the latter half of the 20th century. The focus—often implicit rather than stated—has been on the cultural freight of White/Black or, more accurately, White/other racism that the typologists brought to their science. These are important issues, but the result is a sort of presentism. In holding our intellectual ancestors to the standards of the present, the rhetoric of late 20th-century social context distracts us from an appreciation of the questions that motivated the craniologists, questions that are peculiar to Americanist anthropology. Where did the Indians come from? How diverse are they and how is that diversity related to their origins? How is their biological diversity related to linguistic, cultural, and ecological diversity in the New World?

II. JOHANN FRIEDRICH BLUMENBACH 1752–1840

Typological characterization of the newly discovered American peoples appears in the earliest work that we can label physical anthropology: the craniological research of the German anatomist Johann Friedrich Blumenbach. Blumenbach has been the focus of considerable attention in recent work on the history of science regarding the origins of the biological concept of race and critiques of scientific racism.

Blumenbach's work is couched in the degenerationist paradigm that dominated biology in his day. His understanding of race combined elements from the work of Kant and Buffon (Larson, 1994). Emmanuel Kant had attributed human variability to the effects of climate on an ideal, created or ancestral type. Variability was thus the result of the degeneration — here an accommodation to local conditions that foreshadows the modern concept of adaptation — of a single original type that was of intermediate skin color. Kant recognized four Old World races and considered the Americas too recently settled to have given rise to a constant type. Larson summarized the resultant concept of variation thus: "A race was a class or series of individuals issued from one another and distinguished by a variation that had become constant. Naturalists considered these subtypes distinct branches, in spite of their common origin, and recognized that yet further subtypes might arise from them" (Larson, 1994:63). Georges-Louis Leclerc, Comte de Buffon, introduced the concept of reproductive isolation as the defining feature of species, and he expected to find infertility in crosses between human races. His view of the Americas was degenerationist in a different — and quite negative — sense. He argued that the American fauna, the Indians included, was smaller, weaker, and less vigorous than its Old World counterparts (Larson, 1994; De Waal Malefijt, 1974). His discussion of human diversity also contributed to a third sense in which the degeneration came to be used in the 19th century, which Stepan has labeled "the race out of place" (1985). Buffon expected rapid change in migrant populations toward the characteristics of native groups. Marks (1995) argued that Buffon's concept of human variation was adaptationist and thus modern, in contrast to the misguided "anti-anthropological, anti-biological and anti-historical" (1995:52) typological concept of Linneaus and Blumenbach. Any Americanist will find this rosy view of Buffon difficult to reconcile with Buffon's highly negative view of American Indians.

Blumenbach generated a remarkably modern account of the continuous and trivial nature of human variation. A succinct statement of his race concept is this quotation from the English translation of his 1775 work *De generis humani varietate nativa*: "The variations of skin color, stature, body proportions, etc., which we have been able to observe, considerable though they may appear at first sight, have no absolute value; they all merge gradually one into another and, accordingly, classification into human races is arbitrary" [Bendyshe (1865), quoted in Comas (1960:16)]. A race concept of underlying unity did not prevent him from defining five races: the Caucasian, the Mongolian, the Ethiopian, the American, and the Malayan, corresponding to five skulls he illustrated as exemplars of these races. Gould (1994) and Marks (1995) point out that the Malay race is a late addition; the first edition of *De generis* presents only four categories, as do the classifications of Blumenbach's contemporaries, Cuvier and Linneaus. Gould inferred that the addition was made for reasons of symmetry in fitting the scheme to degenerationist theory, serving as "the transitional form between Europeans and Africans" (1994:69). It seems equally likely that Blumenbach did not have a skull from the Pacific in 1775. In his third edition (1795) of *De generis*, Blumenbach thanks Joseph Banks for providing him with a skull from Botany Bay, and he dedicates this edition to Banks. Banks, the British botanist and patron of scientific explorations, had accompanied Captain James Cook to Australia in 1770, and the Cook expedition represented the first opportunity for Europeans to accommodate Australian Aborigines in their accounts of natural history.

The ordering of these categories has been a subject of much recent discussion. While Blumenbach discusses Camper's facial angle at some length, he rejects it as a criterion for assigning skulls to races. Nevertheless, he illustrates his five races in order of facial projection: that is to say Mongolian, American, Caucasian, Malayan, and Ethiopian. He discusses his five races in a different order, beginning in the middle with the Caucasian, proceeding to the extremes of flat Mongolian and projecting Ethiopian faces, and ending with the intermediate American and Malayan faces (Fig. 1). Gould argued (1994) that this is the first ranking of races in science and that the ranking itself is perniciously hierarchical, even if Blumenbach himself was not racist. This seems to be an inappropriately quantitative reading of Blumenbach's work, and it is perhaps not trivial that Gould's critique appears in a popular magazine in a collection of essays responding to *The Bell Curve*. Gould singles out the facial angle in a way that Blumenbach explicitly rejects in this passage from the 1795 edition of *De generis*: "It very often happens that the skulls of the most different nations, who are separated as they say by the whole heaven from one another, have still one and the same direction of the facial line: and on the other hand many skulls of one and the same race, agreeing entirely with a common disposition, have a facial line as different as possible. We can form but a poor opinion of skulls when seen in profile alone, unless at the same time account be taken of their breadth" (Bendyshe 1865:235). Blumenbach's most important contribution to theory in biology is the concept of *habitus* in systematics (Farber, 1982). The whole organism, not a

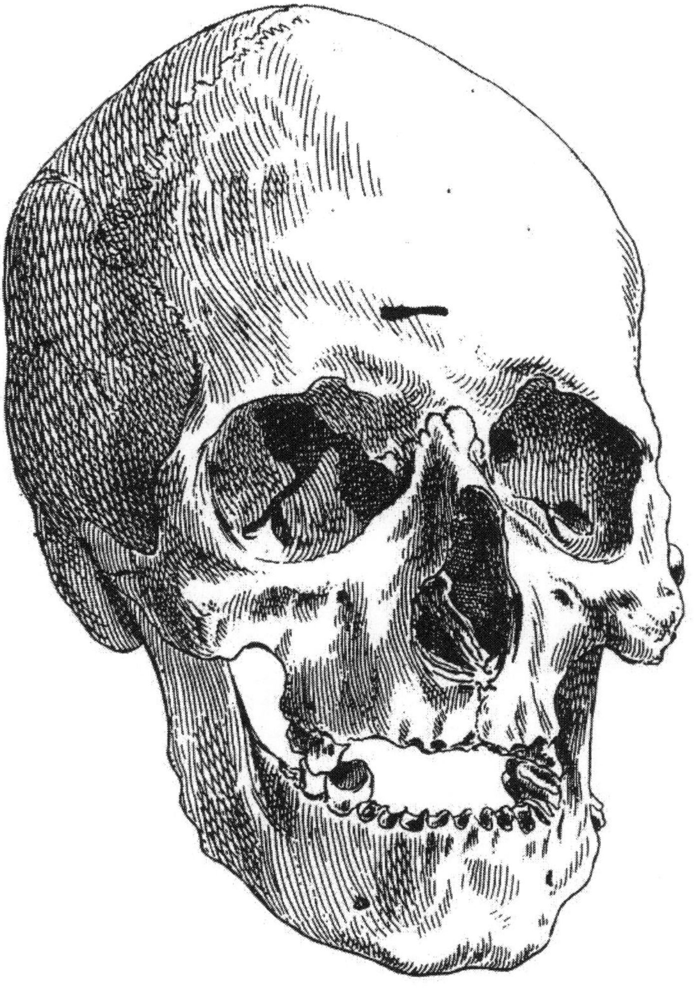

XXXVIII . *Americani Illinoici.*

Figure 1 Blumenbach's cranium from Illinois (Blumenbach, 1800).

single character, should be used in assessing affinities. It is thus particularly inappropriate to represent him as having ranked races on a single scale.

The language that Blumenbach uses in *De generis* has been a lively subject in recent literature on scientific racism. Schiebinger (1993) reads his choice of skulls for description as a complex text that conflates religious meanings attached to the mountains of central Asia with lubricious accounts of the Turkish slave trade, expressing a species of sexism that she finds pervading Enlightenment science. Gould takes Blumenbach to task for the language he uses in describing his Caucasian exemplar: "Blumenbach's descriptions are pervaded by his subjective sense of relative beauty, presented as though he were discussing an objective and quantifiable property not subject to doubt or disagreement" (Gould, 1994:69) and he quotes the description of the Georgian female skull as if it were the description of the whole Caucasian race. Here both Gould and Schiebinger have missed Blumenbach's allusion to the historical context in which *De generis* was written. Blumenbach's colleagues at Göttingen University included S. T. Soemmerring, an anatomist who had dissected the cadavers of Africans, arguing that they were intermediate between apes and Europeans, and the philosopher C. Meiners, who ranked the races on relative beauty in building a justification for slavery (Jahoda, 1999). Female skeletons were similarly aestheticized and stereotyped in anatomical literature until the early 20th century (Fee, 1979), and Buffon and other contemporaries of Blumenbach used aesthetic language in describing human variation. The author prefers to read the gushing language Blumenbach applies to the skull of his Georgian woman as irony aimed at these colleagues, a reading that Jahoda supports from Blumenbach's correspondence (Jahoda, 1999).

The concept of variation expressed in the 1795 text quoted earlier does not accord with the prevailing 17th-century definition of races as constant varieties, and in this regard it approaches variability in a novel way. Blumenbach's method was also novel, a novelty for which he used the term *anthropology* for the first time in its modern sense because he tested his models of human variation using observations on skulls. Most recent discussions to the contrary, Blumenbach did not measure skulls (Bowles, 1976; Ubelaker, 1982; Burke, 1998; Joyce, 2001). Rigorously defined measurements and tools for making them are a product of 19th-century anthropology. He proposed the *norma verticalis* as the best perspective from which to view the skull, but the calipers and the craniophor were still in the future. While many features of the skulls are described, only the relative projection of the face is treated analytically, and Blumenbach's observations were visual, not metric.

All recent scholarship on Blumenbach of which the author is aware has focused on his *De generis.* This book is a natural history in the sense that it belongs to a genre of science writing in which the writer presents a comprehensive, literary account of humans of the natural world. Natural histories were popular in a way that is difficult for modern readers to comprehend in our age of scientific specialization. The 19th-century translation of this work into the major scholarly languages reflects this popular audience. The exclusive focus on *De generis* misrepresents Blumenbach's methods. His science lies in his craniology. He collected skulls by corresponding with travelers to various parts of the world and stimulated scientific collection on the part of travelers. This method, if we wish to use the word, was novel. In his fascinating study of travel and natural history, Liebersohn says of Blumenbach that "As a scientific entrepreneur, he linked the burgeoning interest in travel to university learning and powerful patrons" (1998:135). By the end of his career, Blumenbach had amassed a collection of 245 skulls at Göttingen, 43 of them from the Americas (Bendyshe 1865:348). Between 1790 and 1828 he published a series of detailed descriptions of 65 crania, including provenience information and an engraved illustration of each. The title of the series varies somewhat: *Decas prima collectionis sua craniorum diversarum gentium illustrata* appeared in 1790 and *Nova pentas collectionis suae craniorum diversarum gentium* in 1828. The author refers to this work collectively as *Decas*.

The *Decas* includes nine crania of American Indians. Specimen 9, which Blumenbach describes among the 10 presented in the first installment of the *Decas* 1790, is a Cherokee sent to him by a Dr. Michaelis of Philadelphia. Blumenbach comments on cranial deformation and on the size of the nasal aperture, relating the volume of the nasal cavity and the complexity of the turbinals to reports of the acuity of the sense of smell among Indians. Numbers 10 and 20 are Caribs from the island of St. Vincent contributed by Joseph Banks. Cranial deformation is again noted. Skull 38 is from Illinois near Cahokia, contributed by a Dr. Barton. This is perhaps Benjamin Smith Barton (1766–1815) of Philadelphia, who had studied at Göttingen. Blumenbach remarks on Caucasian features in his Illinois specimen, thus prefiguring the Kennewick Man controversy in the 1990s. Blumenbach's illustration of this skull appears in Fig. 1. Specimen 46 is a skull from the upper Orinoco donated by Alexander von Humboldt, 47 is a decorated trophy head, 48 a native woman, origin unspecified, and 57 is a Coroa woman, all from Brazil. Specimen 58 is a Botocudo from Brazil donated by Maximillian, Prince of Wied, the ethnographer and explorer (Liebersohn, 1998). Specimen 65 is a deformed Inca skull excavated by Alexander Caldcleugh, a British diplomat and travel writer. There are in addition four Eskimo, two from the North American Arctic and two from Greenland, an Aleut, and several representatives of Siberian peoples. Blumenbach has been credited as the first scholar to recognize the Asian affinities of the Eskimo and Aleut [Harper and Laughlin (1982:282); Szathmary and Ossenberg (1978) pointed out that Dav̆id Cranz made the same inference a decade earlier]. They had been previously understood by the natural historians as most closely related to Europeans and appear in the earliest anthropological literature among the Hyperboreans along with the Lapps, Picts, and Scots.

The only portion of the *Decas* that appears to have been translated into English is an excerpt from the description of the Botocudo skull that appears in Samuel Morton's *Crania Americana*:

> The age of this man was about five and twenty. During the war between the Botocudos and the Portuguese, he was accustomed to join his countrymen in their hostile incursions; but after the hostilities ceased, he frequently visited the garrison on the Rio Doce, where he not long after fell sick and died. The cranium, which is large, is also very ponderous from the thickness of the bones, and their dense and hard texture: and as a whole, if you disregard for a moment the under jaw, the figure and interval of the orbits, the elevated nasal spine, and other particulars peculiar to man, the general aspect approaches nearer to that of the Orang Outang than any other skull from a barbarous nation to be seen in my collection. I have indeed one or two specimens of the Negro, in which the upper jaw is more projecting; but this skull differs from them in other respects, besides having the cheek bones more prominent, and a greater swell of the parietal bones. But what deserves particular notice is an indentation, shaped like the point of the finger on wax, which remains after the loss of the front teeth, the sockets of which are compressed, or rather completely absorbed. So universally, the Prince of Wied assures me, does this happen to the youth of this nation from wearing the wooden lip-ornament, already mentioned, that you will scarcely find one of them arrived at the age of thirty who retains these teeth. (Morton, 1839:140)

This passage illustrates the character of Blumenbach's descriptions. The attention to provenience is typical of his work. This description is unusual in remarking on resemblances to a nonhuman primate. In sharp contrast to the work of his contemporaries, the likeness he draws is limited, qualified, and without any suggestion of affinity. The passage is also interesting for its notice of pathological conditions. Blumenbach has been credited as the founder of craniology and calumnied as the inventor of racial classification. Perhaps we ought also to claim him as an early contributor to paleopathology, as this passage in the *Decas sexta* of 1820 is the first published description of alveolar pathology resulting from the wearing of a wooden labret or botoque, for which the Botocudo were named (see Fig. 2). Blumenbach's collections eventually contained many American Indian crania not included in *Decas*, e.g., two Arikara skulls collected for him by Karl Bodmer in 1834 (Bass *et al.*, 1971).

III. SAMUEL GEORGE MORTON 1799–1851

Blumenbach's work served as the model for the efforts of the Philadelphia physician and anthropologist Samuel G. Morton. His 1839 *Crania Americana* tested then-prevalent accounts of New World peoples that attributed the ancient monuments of high civilization to an extinct race of immigrants from Europe or elsewhere in the Old World (Silverberg, 1968; Buikstra, 1979). Morton's research

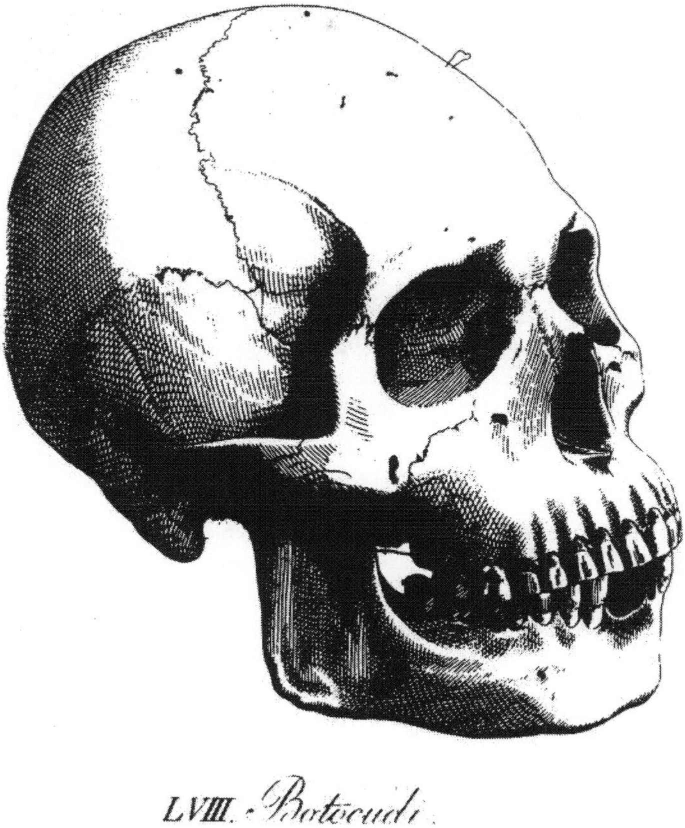

Figure 2 Blumenbach's Botocudo skull (Blumenbach, 1820).

soundly discredited this "Moundbuilder" myth. He grouped specimens from Peru, Mexico, and the ancient earthworks of the Ohio Valley into his Toltecan Race and found this group to be essentially similar to the other American crania in his collection, which he assigned to the category Barbarous Nations. He concluded that "the American nations, excepting the Polar tribes, are of one Race and one species, but of two great Families, which resemble each other in physical, but differ in intellectual character" (1839:260). Much recent scholarship on Morton has largely focused on constructing his methods as racist and has reduced his work to a ranking of races (Browne, 2000; Bruce, 1988; Joyce, 2001; Gould, 1996; Worden, 2002). The author argues that in the context of his day, his research was grounded in ethnology, and his view of the unity of ancient and recent American

Indians was antithetical to that of many of his contemporaries, who denigrated Indian intellectual and cultural capacity in the long tradition extending from the *Book of Mormon* to the alien fantasies of van Daniken.

The literature on scientific racism has largely ignored Morton's scientific contributions, but physical anthropologists claim him as an intellectual ancestor. Brace (1982) traced a genealogy from Blumenbach to Morton and from Morton to the French founder of physical anthropology, Paul Broca (1824–1880). Building on Morton's concept of anthropology, Broca professionalized the discipline. The profession was then returned to the Americas through the efforts of Aleš Hrdlička in Brace's view (Brace, 1982). Hrdlička himself (1918) stressed the detailed continuity between Morton's techniques and the standards for craniometry that emerged at the end of the 19th century. Ten linear measurements, one angle, and an internal capacity with four component measurements are defined in *Crania Americana,* and four instruments, a "facial goniometer" (Morton, 1839:252), a graduated cylinder, a device for finding partial cranial volumes (Morton, 1839:254), and a "craniograph" (Morton, 1839:294) for drawing skulls are described.

The illustrations in *Crania Americana* are remarkable for their beauty and precision. Unlike the illustrations in Blumenbach's *Decas*, they have great anatomical detail, perhaps because Morton used his craniograph to do the rough drawings. Lithographs prepared by John Collins are among the earliest examples of the use of lithography for scientific illustration in the United States and reflect the rapid improvement of visual presentation of natural science during the 19th century (Blum, 1993). Folio publication was funded by William Maclure, the last of several lavish publications he supported (Porter, 1986). Figure 3 is Collins' lithograph of the same Botocudo skull that appears in Blumenbach's engraving in Fig. 2. The illustrations in *Crania Americana* are detailed and anatomically precise. Most are lateral views. Several three-quarter views show exaggerated shallow perspective that reflects the use of the craniograph, an equivalent of the camera obscura (see Hockney, 2001). The many plates drawn from "nature" are printed at 1:1 scale, and measurements in the text that can be checked on the plates are remarkably accurate, an innovation comparable to the scaled-down engravings in Cuvier's *Le Règne Animal: Races Humaines* (1836). Pathological changes, such as trephination (Plate 11D), taphonomic alterations, e.g., rodent gnawing (Plate 68), and anatomical variants, e.g., epipteric bones (Plates 34 and 37), are illustrated, although not necessarily noted in the text.

Morton's race concept is founded on Blumenbach's in the sense that he uses Blumenbach's five races as the framework for his analysis. Morton elides these five distinct races with the three sons of Noah in the introductory pages of *Crania Americana* (1839:1). He divides these races into 22 families, with the Toltecan and the Barbarous Nations representing the two subdivisions of the American race at the level of the "family." The American race is contrasted with the

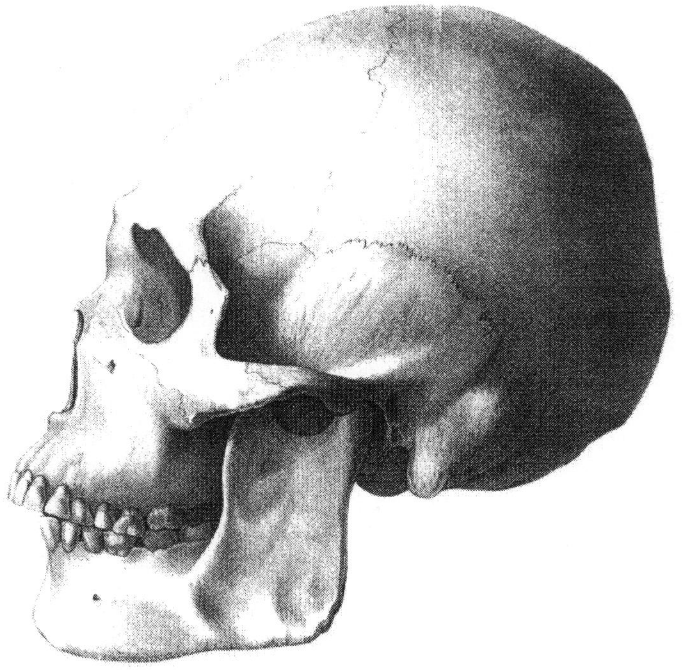

BOTOCUDO of BRAZIL

Drawn from Nature by W.Eberlein On Stone by John Collins

Figure 3 Morton's Botocudo skull (Morton, 1839).

Mongol-Americans or "Esquimaux." The Barbarous Nations are further subdivided into Appalachian, Brazilian, Patagonian, and Fuegian branches, the first of these accounting for all of North American north of Mexico.

A question that requires further investigation is the source of the model for the presence of several races in Native North America that Morton confronts in *Crania Americana*. Gruber points out that Morton was a Quaker and that the prominent Quaker intellectuals senior to and contemporary with Morton had argued that the Indians were remnants of the lost tribes of Israel (Gruber, 1967). Silverberg credits Caleb Atwater's 1820 report on Ohio Hopewell remains as the first scientific claim that more than one race was present in ancient North America. Atwater contrasted crania from the mounds with contemporary Indians: "Their foreheads were low, cheekbones rather high; their faces were short and broad; their eyes were very large; and, they had broad chins. . . . The limbs of our fossils are short and very thick and resemble the Germans, more than any Europeans with whom I am acquainted" (Silverberg, 1968:107). Morton corresponded with Atwater, and cited his work, and it is certainly possible that Atwater's formulation of the problem of who the Moundbuilders were provided the stimulus for Morton's work. Atwater was not alone in his opinions, as Silverberg has shown, and it is argued later that Morton was stimulated by his contemporary and scientific rival, Dr. J. C. Warren, who followed Atwater closely in claiming that the Indians represented more than one race. Hrdlička's assessment of Morton's contribution deserves emphasis: in finding that the Indians constituted a single race, Morton "subverted the numerous loosely formed but commonly held theories respecting the racial complexity of the American natives, as well as those of a racial separateness of the "Moundbuilders" from the rest of the American Indians" (1918:141).

Was Morton a polygenist? Morton's ideas concerning race origins were unremarkable in 19th-century America and are a very minor part of his work. As Arensburg (1995) has shown, notions of separate creation of races can be traced back as far as Columbus and were especially pervasive in the Iberoamerican world. The radical notions of separate creation of the races expounded in Nott and Gliddon's account of Morton's work are primarily Nott's work, not Morton's (Brace, 1974; Porter, 1986). Morton has surprisingly little to say on the subject of polygenesis, given the extent to which recent accounts have stressed his adherence to this model for human diversity. His strongest statements in this regard are found in his correspondence with Nott (Horsman, 1987). His findings, both in *Crania Americana* and in his smaller parallel study of Egyptian antiquities (1844a), are cautiously phrased and limited to the observation that ancient crania are as distinct racially as are recent ones, so much so that it is difficult to find a passage in his published work that clearly expresses a commitment to the concept of polygenesis. Stanton quotes a statement from his correspondence: a skull obtained from Squier's excavations in the Ohio mounds was "a perfect type" of the race "indigenous to the American continent, having been planted there by the hand of Omnipotence" [Stanton (1960:84), quoting Morton to Squier 1947]. In *Crania Americana* there is only one allusion to the concept of separate creation of human races, and it follows a discussion of the conflict of the five-race and four-race models of Blumenbach and Cuvier with the Biblical three-race model. This passage appears as the epigram of the present chapter (Morton, 1839:3). Divine providence is otherwise notably absent from the remainder of the text of *Crania Americana*.

Was Morton a phrenologist? Spencer finds evidence for a long-term commitment to phrenology as well as polygeny in Morton's doctoral thesis written at Edinburgh in 1822 (Spencer, 1983). In the author's view, Spencer's case is circumstantial: the portions of Morton's thesis that he chooses to translate make no claims concerning race. For example, stoicism in American Indians is placed in a context of human nature in general: "all over the world examples have been found of people suffering . . . without uttering a single moan . . . among the aborigines of America, a prisoner, condemned by the enemy to torture and slow death, sings his funeral song unmoved . . ." (Spencer, 1983:335). A similar passage appears in *Crania Americana* (1839:77), where Morton likens the courage of Indian captives to that of European martyrs and denies that Indians are less sensitive to pain than others.

Similarly, De Waal Malefijt (1974) suggested that Morton became interested in craniology and the relationship between skull size and shape and mental ability because he was a correspondent and colleague of the phrenologist George Combe. This scenario also seems unlikely. Morton's acquaintance with Combe began rather late, shortly before *Crania Americana* was published, and Morton had begun to collect skulls in 1820 (Stanton, 1960:27). Hrdlička characterizes Morton as an "investigator" of phrenology rather than as a "promoter" (1918:138). Combe's assessment of the phrenology of Morton's collection is appended to Morton's study rather than integrated with it. While Morton occasionally remarks on the development of one of the phrenological landmarks in some of his specimens (1839:169, 202) he does not cite Combe in the text, except to acknowledge him as the donor of several Eskimo and Plains Indian specimens. The broader question of cerebral localization did not begin with Combe, and it was very much normal science during Morton's career (Young, 1990). It is an anachronism to view localization or, for that matter, phrenology as the bizarre pseudoscience it seems today.

In contrast to his circumspect treatment of phrenology, Morton discusses Blumenbach's craniology extensively, and he cites both *De generis* and several of the descriptions from the *Decas*. If we trust Morton's own account of the beginnings of his interest in the subject, he wanted specimens to illustrate his anatomy lectures on the varieties of mankind (Stanton, 1960:27), an enterprise he shared with many less ambitious anatomists of his day.

Was Morton a racist? Stephen J. Gould has misled many to a conception of *Crania Americana* that centers on cranial capacity and on its use of the relative ranking of races (Gould, 1978a, 1981, 1996). For example, a recent history of anthropology claims "[i]n 1839 Morton published *Crania Americana*, in which the inherent capabilities of a race of people was scientifically determined by skull size and capacity" (Joyce, 2001:8). Another author opines with more generosity "the Quaker physician inadvertently opened the door for others to associate cranial shape with brain size and brain size with mental capacity and social station" (Porter, 1986:70). Others dismiss Morton as a racist skull collector or an apologist for slavery, citing only Gould (Bruce, 1988; Blakey, 1987). These summations, like many in recent literature, seriously misrepresent Morton's work. The overwhelming majority of Morton's text is concerned with natural history. An "introductory essay" of 95 pages is devoted to a lengthy ethnological and historical discussion. Description of the skull collection occupies 253 pages, including detailed accounts of provenience and funeral customs. What we would now call metric methods and results occupy only 12 pages, and Combe's phrenology results another 7. Cranial capacity is 1 of 12 measurements Morton tabulated in his collection. While it is true that only cranial capacity is analyzed in detail, Morton's measurements are supportive of his typological analysis rather than central to it. It remains a puzzle that Morton devoted his energies to the other 11 measurements, but failed to discuss the results. His complaints about the accuracy of calculations conducted by his assistants (1849) hint that the sheer magnitude of the task was a factor!

Was Morton a cheat? Gould accuses Morton of conscious or subconscious falsification of his data through the use of grouped means, both in *Crania Americana* and in his later essay on cranial capacity. Gould also suggests that the measurements may have been manipulated in favor of the hypothesis of Caucasian superiority. Gould's supposition that the measurements may have been manipulated consciously or unconsciously was tested directly by Michael (1988), who replicated Morton's measurements for a portion of the collection. It is noteworthy that in the second edition of *Mismeasure of Man*, Gould failed to respond to Michael's demonstration that Morton's measurements were accurate. Gould's most interesting argument concerns his allegation that Morton manipulated his data through the use of different proportions of males and females and of large-statured and small-statured peoples in the groups he compared. Gould recognizes that Morton's discovery that Peruvian mummies had smaller crania capacities than other Indians, particularly those of his so-called Barbarous Races, contradicted the hypothesis that cranial capacity constrains cultural capacity, and that this discovery argues for Morton's scientific objectivity, but he remains convinced that "Morton's summaries are a patchwork of fudging and finangling in the clear interest of controlling a priori convictions" (1996:86). The author finds Gould's argument unpersuasive because it views Morton's work through the lens of 20th-century quantitative sophistication. Morton worked before the invention of statistical methods appropriate to his research. While it may be difficult for anyone educated in the sciences today to understand that Morton may have been blind to the effects of sexual dimorphism and body size differences on his means, the author's experience in teaching Gould's paper to undergraduates has been that Gould unfairly brands Morton as racist. The concept of grouped means is exceptionally difficult for students who lack a quantitative bent. Sorting out the relative contributions of sex, body size, latitude, and subsistence on cranial capacity has required multivariate statistics, as well as samples far beyond Morton's considerable efforts in collecting crania. Indeed, anthropologists did not complete this task until the late 20th century (Beals *et al.*, 1984; Smith and Beals, 1990), an accomplishment that Gould also fails to note in his second edition. The necessary statistical tools were unavailable to Morton.

Morton's principal anthropological accomplishment was the demonstration that the Moundbuilders were Indians and that American Indians constituted a single race. As Stanton showed more than 40 years ago, this was an antiracist point of view. It credited the Indians with the capacity for high culture. Stanton's case that Morton was motivated by a desire to refute the popular culture claims for various migration legends is less convincing. The scale of his research suggests a more scholarly, scientific target. It seems more plausible that Morton's concept of two races among American Indians was stimulated instead by the work of his Boston contemporary, John Collins Warren.

IV. JOHN COLLINS WARREN 1778–1856

Dr. John Collins Warren was a Boston physician, surgeon, and anatomist whose family had a long association with Harvard University. Hrdlička acknowledges Warren as a pioneer: "Inspired evidently by Blumenbach's works, Professor Warren began to collect and examine skulls of different races, and in 1822 he published an *Account of the Crania of some of the Aborigines of the United States*, the first publication in this field on the continent . . . while of no permanent value scientifically . . . is nevertheless remarkable for the systematic, technical description of the specimens" (1918:136). With such faint praise he has consigned Warren to relative obscurity. Hrdlička gives pride of place to Morton as the first American physical anthropologist, but fails to explore any connections between Morton and Warren, apart from pointing out that Morton had read Warren. Perhaps because Hrdlička is dismissive of Warren's physical anthropology, the historical literature on Morton and on race in the Americas has ignored Warren's earlier work.

Both Morton and Warren were natural historians. Warren wrote a natural history of an anatomical region, the nervous system, whereas Morton wrote two natural histories of human races. Anthropology is a secondary concern in Warren's work, whereas it is the primary focus of Morton's. Both were institution builders, but Warren was the more prominent in this regard.

The publication that Hrdlička cites is an appendix to a monograph, *A Comparative View of the Sensorial and Nervous Systems in Man and Animals* (1822). Warren's theory of the multiple origins of North American Indian populations is presented in this brief appendix. The author argues that Warren's theory provided a motive for Morton's work and focused Morton's attention on skulls. Silverberg (1968) has shown that the multiple origins of North American Indians, specifically the attribution of all high culture in the New World to an Old World immigrant group (Atlantean, Egyptian, Phoenician, Israelite, or whatever), were pervasive in the United States in the early 19th century, but curiously omits Warren from his account. We will see that Morton had read Warren carefully.

Warren's *A Comparative View of the Sensorial and Nervous Systems in Man and Animals* (1822) is a natural history of the neurological system. Most of the work consists of a literature review with strong preference for the work of Lamarck and follows Lamarck in viewing the brain as the prime mover in anatomy. The section of interest to anthropologists is Warren's original contribution, a neurology of several New World forms. It includes dissections of a lobster, a centipede, and an oyster, together with descriptions of four human skulls: a "Caucasian," two Indians from the Columbia River, and a "South Sea Islander" from the Marquesas. American Indian skulls from the vicinity of Boston and from Marietta, Ohio, are discussed, but are not illustrated.

Warren's theory of the peopling of the New World appears in his footnotes: "All who have turned their attention to the subject have, I believe, satisfied themselves that the ancient inhabitants of the Ohio and the Mississippi, of the middle and southern part of the United States, were a different people from the aborigines found here by our ancestors" (Warren, 1822:138). He gives a lengthy account of the Heckwelder's version of the Delaware or Lenni Lenape migration legend, in which three linguistically distinct migrations account for the diversity among the Indians.

The collections of the Warren Anatomical Museum at Harvard were assembled in part by the Boston Phrenological Society 1832–1842 (Bowles, 1976) and in part by Warren himself. Warren and Morton acquired skulls from many of the same sources. For example, Schoolcraft collected for Warren (Hrdlička, 1918) as well as Morton. Robert Bieder (1986) has criticized Morton's collecting practices and those of 19th-century anthropology as a whole as racist, but fails to explore the extent to which the various collectors and institutions were competitors or collaborators. Stanton discusses Warren as a member of the scientific community that appreciated Morton's research and states without citation or elaboration that Morton and Warren exchanged specimens (1960). Examination of their published accounts and illustrations shows that this was not the case. Morton presents figures of eight Northwest Coast and Columbia River specimens. Morton's plate 42 is similar to Warren's plate 6 (Fig. 4), but lacks postcoronal depression that Warren notes and has a canine that is missing in Warren's specimen. Morton's plate 43 is similar to Warren's plate 6 in both these regards, but details, e.g., the form of pterion, do not match. Morton's plate 48 Clickitat shares a fissure and missing anterior teeth with Warren's plate 7, but pterion is dissimilar. Morton notes the fissure as a healed fracture (1839:214). Warren credits T. H. Perkins for Columbia River specimens, whereas Morton credits J. K. Townshend. If one compares Morton's Naumkeag (1839:plate 33) and Warren's Nahant, the descriptions of the skulls do not match, although the descriptions of mortuary practices are very similar. Morton's robust male (plate 63) from a cave near Marietta, Ohio, is clearly not the female skull described by Warren from Marietta, even though both credit Dr. Hildreth for specimens. While both Morton and Warren appear to be describing the same cave site, Morton credits the skull he figures to Andrews. There is thus no direct evidence that Morton and Warren shared specimens. Warren's illustrations lack the detail and accuracy of Morton's.

Warren had studied in Paris in 1799–1801 and he encouraged his son to seek out Cuvier during his son's tenure in Paris in 1832–1835, a period that coincides with Morton's most intense collecting activities. Warren and his son exchanged a lively series of letters that include many references to acquisitions of crania and other anatomical specimens (Jones, 1978). There are no references to Morton in the correspondence. This is surprising if Warren and Morton were advancing each other's collections, because Morton's many articles in the *Proceedings of the Academy of Natural Sciences of Philadelphia* in this era make the nature of his research quite clear. Similarly, Morton makes few references to Warren in the *Proceedings*. Strikingly, the only quotation from Warren's work in *Crania Americana* is a description of the intellectual abilities of American Indian students at Harvard (Morton, 1839:82). Warren is not cited in Morton's discussion of crania from the Columbia River region. Warren cites Heckwelder's work on Delaware migration legends from the *Transactions of the American Philosophical*

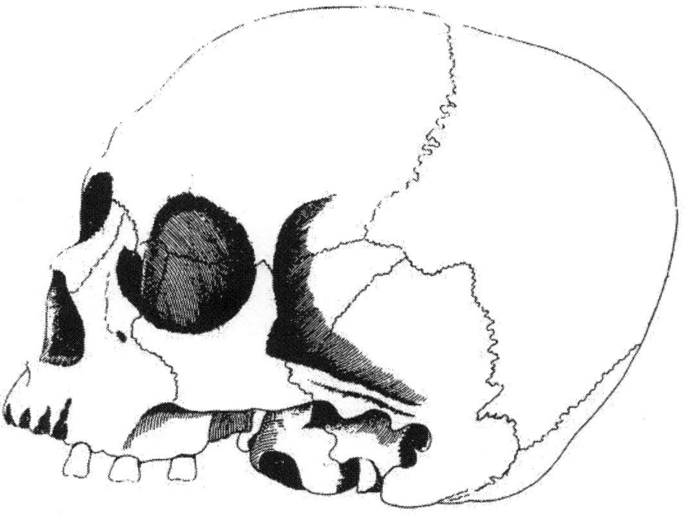

(a)

Figure 4 (a) Warren's Columbia River skull, plate 6 (Warren, 1822), and (b) Morton's Chinook skull (Morton, 1839).

(*continues*)

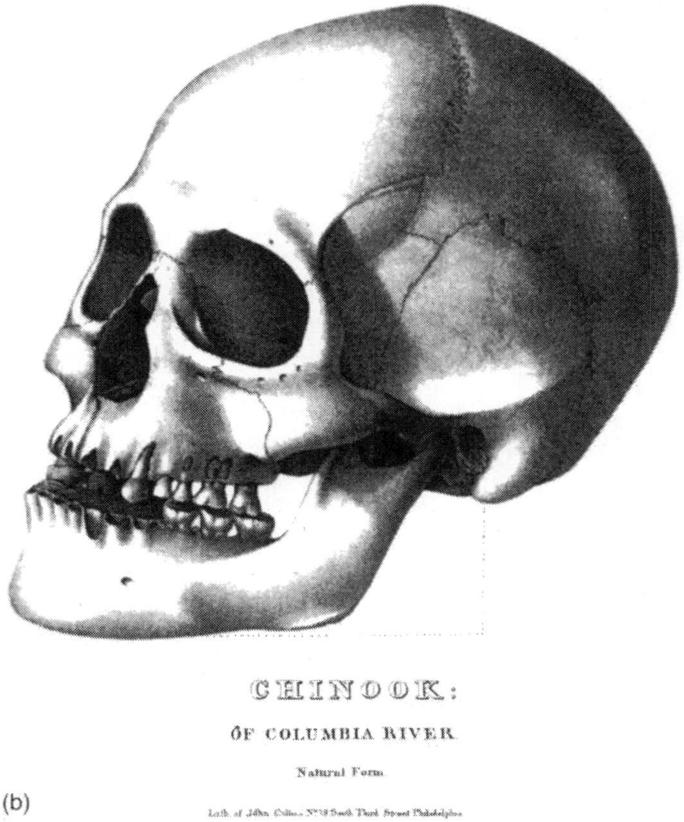

(b)

Figure 4 (*Continued*)

Society and must have seen Morton's many contributions to that journal. Warren and Morton were aware of one another's projects, but may have been competitors more than colleagues, much as were the larger competing scientific communities of Boston and Philadelphia.

V. LATE 19TH-CENTURY CRANIOLOGY IN NORTH AMERICA

In tracing a scientific genealogy linking Blumenbach to Morton, Morton to Broca, and Broca to Hrdlička, Brace has argued that Darwinism and the Civil

War resulted in "the effective eclipse of interest in Morton's work in America" (Brace, 1982:18). In so doing he has deflected attention from a group of scholars who contributed to craniology in the latter half of the 19th century and the first decade of the 20th. This group of scholars is interesting in their diverse perspectives and in their responses to new ideas reaching them from Europe. This chapter is limited to North America, but a similar history could be traced in South America. That history would begin with the Brazilian anthropologist João Batista de Lacerda (1846–1915), who cited Morton and—through Morton—Blumenbach in discussing crania in the collections of the Museu Nacional in Rio de Janeiro (Lacerda and Rodrigues, 1876).

Daniel Wilson (1816–1892), a professor of literature at University of Toronto and founder of the discipline of anthropology in Canada (Wilson, 1863; Trigger, 1966; Kehoe, 2002), contested Morton's claims for the racial unity of American Indians, stressing cranial index as the most important variable. Wilson collected eight linear measurements following Morton and Warren, making comparative use of their published data. Like theirs, his measurements are reported in inches. He collected crania throughout eastern Canada. Crania from a given region are grouped into "dolichocephalic" and "brachycephalic" types and means are reported for these groups. These two terms, with no intermediate category, come from the work of the Anders Retzius (1796–1860), who had earlier proposed three American races: the dolichocephalic Eskimo allied to the peoples of Northeast Asia, a round-headed race allied to peoples of the Pacific, and a long-headed one allied to the Guanches of the Canary Islands and perhaps the Lost Tribes of Israel via Atlantis (Retzius, 1859). Wilson cites Retzius as claiming that "it is scarcely possible to find a more distinct separation into dolichocephalic and brachycephalic races than in America" (Wilson, 1863:244), but he neglects to note Retzius' extreme diffusionism. It is noteworthy regarding Stephen J. Gould's criticisms of Morton that Wilson seems as oblivious to the effects of grouped data on comparisons of means as Morton was.

Language groups are a prominent feature in Wilson's analysis. He compared Huron, Iroquois, and Algonquian crania with Morton's moundbuilders and found the differences comparable to those seen among disparate European groups. He was particularly interested in cranial deformation as a contributor to differences between groups. He summarizes his reservations about the existence of a unitary American Indian race:

> But the legitimate deduction from such a recognition, alike of extreme diversities of cranial form and of many intermediate gradations, characterizing the nations of the New World as well as the Old, is not that cranial formation has no ethnic value, but that the truths embodied in such physiological data are as little to be eliminated by ignoring or slighting all diversities from the predominant form, and assigning it as the sole normal type, as by neglecting the many intermediate gradations, and dwelling exclusively on the examples of extreme divergence from any prevailing type. (Wilson, 1863:264)

Wilson argued that there had been three migrations to the New World: from Asia via the Bering Strait, from Polynesia across the Pacific, and from Europe across the Atlantic. Kehoe (1999) has related this model to his earlier research on the possibility that there had been an ancient group in Britain that preceded the migration of Celtic peoples.

A second proponent of diversity among New World populations was James Aitken Meigs (1829–1879). Meigs was Samuel Morton's student and successor as curator of the collection Morton had amassed. Like Morton, Meigs was a Philadelphia physician and professor of medicine and an active participant in the Philadelphia scientific community. His earliest publication on Morton's collection is a lengthy review of international craniological literature and is largely hagiographic in its account of Morton's work (Meigs, 1857). It appears in Nott and Gliddon's *Indigenous Races of the Earth*, but pointedly fails to engage in the polygenist and racist agenda of the remainder of their volume (Horsman, 1987). A noteworthy point is Meig's objection to the label "Mongol-American" for Eskimo as misleading. He argues that there is no close resemblance with crania one might associate with the historical Mongol peoples.

By 1866 Meigs had assimilated Retzius' and Wilson's critiques of Morton's work (Meigs, 1866). He reanalyzed Morton's collection, classifying the American Indian crania as dolichocephalic, mesocephalic, or brachycephalic, and according to eight skull shape categories, six of these applying to the most common long-headed, or dolichocephalic skulls. Individual crania or small series of crania identified by tribe are his unit of analysis. Measurements are not presented directly. He concludes "that these ethnical or typical groups are founded upon osteological differences as great as those which, in Europe, suffice to separate the Germanic and Celtic stocks on the one hand, from the Ugrian, Turkish and Sclavonian, on the other" (Meigs, 1866:235). In his 1866 paper he again situates his study with respect to international literature in anthropology. For example, he points out that d'Orbigny's *L'Homme Américain* appeared in the same year as *Crania Americana*, but reached the opposite conclusions about the diversity of Indians. For Meigs the radical aspect of Morton's work was his argument for the unity of American Indians. Meigs even suggests an interesting link between Morton and Benjamin Smith Barton, a Philadelphia physician and academic who wrote a philological treatise arguing that all American Indian languages sprang from a single ancestor, an insight that deserves further investigation.

Meigs' most surprising paper is a description of a low, heavy-browed skull from Illinois that suggests that he was familiar with the Neanderthal find just a decade after its discovery: "If the position in which it was discovered be any evidence of its age, it belongs, in all probability, to an earlier inhabitant of the American continent than the present race of Indians" (Meigs, 1867:415). Meigs is thus an early contributor to the claims for great antiquity in the Americas that Hrdlička would spend his career combating.

Neither Wilson nor Meigs was in any way isolated from developments in anthropology in the Old World, and this is equally true for late 19th-century scholars. The legacy of Morton and Meigs in Philadelphia passed to Harrison Allen (1841–1897), a physician and professor of medicine. Allen published several anatomical and pathological studies of the skull (Hrdlička, 1918; Spencer, 1997c). His principal contribution to our subject is his monograph on five crania from Moore's shell mound excavations in Florida (Allen, 1895a). It rivals Morton's work in its beautiful 1:1 engravings, with four views of each skull, and in its meticulous descriptions following the conventions of Broca's French school. Crania representing 17 tribes from the Philadelphia collections and from the Columbia University medical department are used in comparison. He notes a moderate frequency of metopism and cites—without making behavioral inferences—the work of Lombroso and others on the very high frequencies of this condition among European criminals.

Meanwhile in Boston, Frederic Ward Putnam (1839–1915) gathered around him at the Peabody Museum several physical anthropologists who worked in Broca's paradigm (Mark, 1980; Brew, 1968). Cordelia A. Studley (1855–1887) published a single paper (1884), a description of skeletons from four caves in Coahuila, Mexico, from the museum collections. Her craniology consists of 62 measurements, including angles, indices, and cranial capacity. Skulls are grouped as "dolichocephali," "mesaticephali," and "brachycephali" following Retzius, and means and ranges are reported separately for these three groups. She is perhaps the first person to point out that crania from cave site mummy bundles in the greater Southwest are markedly more dolichocephalic than recent peoples of the region. Lucien Carr (1829–1915) may be better remembered for his archaeological explorations on behalf of Putnam's museum, but he also produced three descriptive papers on craniology that are similar in method to Studley's (Carr, 1878, 1879, 1880).

The first American doctorate in our field was awarded in 1896 to Putnam's student Frank Russell (1868–1903). Russell was curator of physical anthropology at the Peabody Museum until his early death (Hrdlička, 1914a; Brew, 1968). He contributed two craniological papers: a comparison of New England Indian and Labrador Eskimo crania and long bones (Russell and Huxley, 1899) and an application of what we now call discrete trait analysis to the problem of American Indian diversity. The latter paper used "nearly two thousand skulls in the Peabody Museum at Harvard University" (Russell, 1900:737) and has sample sizes of 1200 to 1500 for most comparisons. He presents frequencies for nine characters across nine regional series, but concludes: "I hope that the facts presented may prove suggestive and interesting, but do not expect them to establish firmly any hypotheses regarding the origin or affinities of the Amerinds" (1900:743). His series include all those that Studley and Carr had measured.

Further afield, the Southwestern studies of Washington Matthews (1843–1905) posited a relationship among the Zuni, the Hohokam, and ancient Peru, a pet theory of Matthews' colleague Frank Cushing (Merbs, 2002). The Inca bone and brachycephalization are the keystones in Matthews' edifice (Matthews *et al.*, 1893). Matthews' work is indirectly connected to Harvard via Putnam's encouragement, support, and curation of some of the materials, and Putnam encouraged Cushing in his racial theories. George Langford's (1876–1964) demonstration that dolichocrany was older than brachycrany on stratigraphic grounds in Illinois mounds (Langford, 1927) was perhaps inspired in part by Putnam (Browman, 2002:261; Kullen, 2000). A descriptive craniology of Ontario Indians produced by Susanna Boyle (1869–1947) is similarly connected to and influenced by Putnam (Boyle, 1892; Killan, 1983). Even Harrison Allen (1895a), who was Meigs' successor in Philadelphia, produced his monograph on Florida crania with encouragement from Putnam via Putnam's support for the archaeologist C. B. Moore.

Putnam articulated his vision of American Indians in an 1899 address before the AAAS: "The facts show diversity—of race" (Putnam, 1899:12). He recognized nine types: Eskimo type, northern and central so-called Indian type, Northwest brachycephalic type, Southwestern brachycephalic type, Antillean type, Toltecan brachycephalic type, Ancient Brazilian, Fuegan, and pre-Inca (1899:8). He thus retains Morton's three races and adds six more. Putnam's commitment was clearly to a model of multiple origins of American Indians, and his influence is visible among the late 19th-century craniologists with whom he interacted. Putnam encouraged both Hooton and Hrdlička in their early 20th-century efforts in craniology. Hrdlička even refers to himself as one of Putnam's "boys" (1918:155). Hrdlička went on to rebut the concept of multiple origins of American Indians (Spencer, 1979), while Hooton endeavored to reinforce it.

VI. ALEŠ HRDLIČKA 1869–1943

The typological study of Indian and Eskimo crania became the dominant enterprise as American physical anthropology emerged as a profession after 1900. Aleš Hrdlička, founder of the American Association of Physical Anthropologists and first editor of its journal in 1918, as well as first curator of physical anthropology at the Smithsonian Institution, developed the new science using the race concepts of the 19th-century French school of anthropology, as Brace has shown (1982).

What did Hrdlička import from France? Brace argued that Hrdlička's resistance to Darwinian explanations and his static race concept are attributable to his admiration for Broca's anthropology. Paul Broca wrote little about the

New World, and his views were largely based on Morton's work. For example, Broca's disciple Paul Topinard devotes only a few sentences to the subject:

> If one trusts the cranial capacity method followed by Morton, the American cranium is one of the least capacious among humans. It is more often dolichocephalic than brachycephalic, with respect to the collection in Philadelphia. Judging by the collection of the Museum, it would be on the contrary mesaticephalic, what could be had from a mixture in equal proportions of brachycephals and dolichocephals. . . . Dolichocephaly is more extensive, following Morton, among the tribes that originally lived east of the Alleghenies, and brachycephaly among those west of the Mississippi. The same condition is reproduced on the coasts of South America. (Topinard, 1876:507, author's translation)

At the time the collections of the Musée de l'Homme were largely South American. Topinard follows Morton in excluding the Eskimo from the American race as defined earlier. In the early 20th century, American craniology becomes distinct from similar work in Europe in the degree to which these researchers interacted with archaeologists and others, in part because the Boasian race–language–culture model encouraged such interaction and in part because Hrdlička's work was grounded in the multidisciplinary perspective encouraged by the institution with which he was affiliated throughout his career.

Hrdlička's concept of race in the Americas is difficult to characterize. He stressed the relative unity and recent origins from North Asia throughout his career, but the details vary. In an early paper heavily influenced by Putnam, he admits the possibility that there was a low-vaulted race that preceded the historic peoples of the Delaware valley, assigning two crania from one of several sites studied by Russell to this group (Hrdlička, 1902c). As late as 1912 he opined that "it is also probable that the western coast of America, within the last two thousand years, was on more than one occasion reached by small parties of Polynesians, and that the eastern coast was similarly reached by small groups of whites, but these accretions have not modified greatly, if at all, the mass of the native population (Hrdlička, 1912b:12). By 1917 he recognized four subtypes scattered among the native populations of the Americas: dolichocephals, eastern brachycephals, western brachycephals, and the Eskimo (Hrdlička, 1917b; Rivet, 1943:57). That this model is little advanced beyond Broca's is readily apparent, but it is less complex than Putnam's. He consistently minimized New World variability: "There are, it is true, subraces of the American Indians, a number of them; but the differences between them are less than the differences between, for instance, the Italian and the Scandinavian in Europe" (Hrdlička, 1928:815). Near the end of his life he summarized his views: "The Chinese present at least two types, the American Indians five or six, the Eskimo two, but these do not deserve the name 'races,' unless the use of the term be much stretched" (Hrdlička, 1941:184).

In later life his extensive fieldwork in Alaska led him to complexities. Harper and Laughlin (1982) point out that Hrdlička's view of the relationships among Eskimo, Aleut, and Indian peoples was novel, a concept that they label the Eskimo wedge hypothesis. Hrdlička saw the eastern Eskimo as quite distinct from Indians, but found less evidence for distinctiveness in the western Arctic, arguing for a common ancestry separate from most Asian peoples, and further differentiation in the Americas.

Letters exchanged during Hrdlička's lifetime by Georg Neumann and Charles E. Snow criticize Hrdlička for ignoring archaeological provenience and lumping crania by state (Jacobi, 2002). Hrdlička was remarkably insensitive to subtleties of archaeological provenience and there is a frank recent literature critical of his field technique (Krupnik, 2003; Loring and Prokopec, 1994). Hooton refers tongue in cheek to the dogma of isolation of the New World from the Old as "a sort of *ex post facto* Monroe Doctrine" (1973:133):

> In fact, it seems glaringly improbable that the Bering Straits and the Aleutian Islands should have strained out all prospective incomers except Mongoloids . . . there was no Dr. Hrdlička standing on the Aleutian equivalent of Ellis Island, acting as Prehistoric Commissioner of Immigration to enforce an alien exclusion act applicable to all save Mongoloids. (Hooton, 1946:650)

Hrdlička's single mindedness regarding the racial prehistory of the New World is difficult to overstress. In the lengthy essay on the history of physical anthropology in the United States that appears in the first volume of *American Journal of Physical Anthropology*, he says of Herman ten Kate (1858–1931), a Dutch Americanist trained by Broca, "He has the distinction of being perhaps the last living anthropologist of note who defends the theory of a multiplicity of races on the American continent, though this is largely if not entirely due to his interpretation of the term 'race'" (Hrdlička, 1918:379). Hrdlička's assessment was premature.

VII. EARNEST ALBERT HOOTON 1887–1954

E. A. Hooton is the most quixotic of the prominent contributors to physical anthropology in North America. Unlike his predecessors, his training was in classics, not medicine. His doctoral thesis betrays interest neither in physical anthropology nor in human variability (Hooton, 1911). Hooton's sojourn in England as a Rhodes scholar was a watershed experience. He studied first with Robert R. Marett and then with Arthur Keith. He returned to the United States to teach physical anthropology at Harvard for the remainder of his life, mentoring most of the prominent contributors to mid-20th-century physical anthropology (Giles, 1997; Garn and Giles, 1995; Shapiro, 1981). Oddly, those students have written relatively little about his contributions to our understanding of the diversity of Native American peoples.

Hooton's first contribution to craniology relevant to the peopling of the Americas appeared as the third article in the first volume of Hrdlička's new periodical, the *American Journal of Physical Anthropology*. It is remarkably modern in its method and tone. He compared Viking remains from Iceland with Eskimo, California, Chukchi, Italian, and Libyan crania with respect to frequencies of mandibular torus, palatine torus, thickened tympanic plate, and sagittal keel, attributing similarities between Icelanders and Eskimo to functional consequences of "habitual chewing of very tough food" (Hooton, 1918:76). Here we see an adaptationist perspective on variation: features shared by disparate groups in similar environments are adaptive and not useful in assessing affinities.

Hooton's descriptions of skeletal remains from two sites in southern Ohio, Madisonville (1920), a Fort Ancient Late Prehistoric site, and the Turner Group (1922), a Hopewell mound complex, constitute his contribution to the physical anthropology of the ancient Midwest. Both are quite conventional in tone and content, proceeding from age and sex composition to cranial measurements, discrete variation, and postcranial metrics. There is little discussion of pathology. Hooton's study of Turner incorporates Cordelia Studley's unpublished observations, although he did not find her measurements useful. The intrusive component at Turner is differentiated from primary series in lacking brachycephaly but exhibiting cranial deformation. The salient point for this paper comes at the end of the Turner Group paper:

> The primary and secondary series resemble each other much more closely than either resembles the Madisonville series. It may be said positively that the people of the Turner Group show practically no physical affinities with the people who live on the Madisonville site, beyond those which are common to all Indians. (Hooton, 1922:132)

Here, Hooton demonstrates the morphological distinctiveness of Middle Woodland and Mississippian populations in the Midwest in a context in which cultural distinctiveness and chronological sequence, if not its magnitude, are clear from the accompanying archaeological analysis.

Hooton's most substantial contribution to Americanist anthropology is his *Indians of Pecos Pueblo* (1930). Several aspects of this project are discussed at length in Lane Beck's contribution to this volume and are not repeated here. In it, he takes the theme of variability among American Indians to an extreme that has stimulated a continuing critique of his work as racist, and his version of the typological paradigm is certainly remarkable in its complexity and eccentricity. The 129 suitable male crania from Pecos were sorted into seven morphological types: Basket Makers[1], Pseudo-Negroids, Pseudo-Australoids, Plains Indians,

---

[1] This is the archaic spelling. The current accepted spelling is Basketmaker.

Long-faced Europeans, Pseudo-Alpines, and Large hybrids, as well as residuals not accommodated in the seven types, without regard for cranial deformation. The "validity" of these types was established by comparing means for cranial measurements and indices with means for the group as a whole: if the type mean deviates by more than one standard error from the group mean in several variables, Hooton accepted the type as valid (1930:203). The types were then compared to one another and to crania from elsewhere in the Americas and the world, again by examination of means. While all types persist throughout the site sequence, proportions of the types shift through time, with dolichocephalic types predominating in the earlier horizons.

What all this meant to Hooton is a puzzle. Clearly, he demonstrated that the Pecos crania are quite variable and that the variability is not explained by cranial deformation or by change in stature through time. In *Indians of Pecos Pueblo*, several interpretations are presented, for example:

> Of course, if one wishes, he may argue with considerable plausibility that the earliest strata of American Indians may have carried among other strains some of the Australoid blood and that these Pecos "Pseudo-Australoids" represent a segregation of such strains. Candidly however, I do not think that our Pecos Australoids sufficiently resemble real Australians to justify even this moderate opinion. Large brow-ridges and platyrrhine noses together with short, broad faces may not always mean Australians, although they suggest such a type. The total absence of prognathism in our "Pseudo-Australoids" is a strong argument against the identification. I am much more impressed with the resemblance of our "Pseudo-Australoids" to the Ainu, since here the indicial similarities are very marked. (Hooton, 1930:262)

As in this quotation, the comparison groups are selected consistent with the racial identification of each type. No attempt is made to compare each type to the whole range of comparison groups. "Basket Makers" are compared with crania from the Coahuila Caves, California and Egypt, "Pseudo-Negroids" with groups ranging from the Andaman Islands to Zulu, "Pseudo-Australoids" to Tasmanians and Peruvians, "Plains Indians" to Arikara and Illinois Algonkians, "Long-faced Europeans" to Eskimo and Chinese, "Pseudo-Alpines" to Burmese and Tibetans, and "Large hybrids" to Tennessee Stone Grave and Madisonville crania, among many others. Hrdlička's *Catalog of Crania* (1924, 1927, 1928) is conspicuous among the citations.

Hooton's Harvard colleague Roland B. Dixon (1875–1934), whose work is clearly a source for a portion of the Pecos typology, recognized three North American races more or less consistent with Hrdlička's: Northeastern Dolichocephals (including Hrdlička's Eskimo), Southwestern Dolichocephals, and Central Brachycephals, and recognized within these varying proportions of his eight Old World types (Dixon, 1923), despite hewing to the Bering Straits orthodoxy. Dixon (1923:419) first noted on an Alpine and a Proto-Negroid type at Pecos and in the Coahuila crania. In what is perhaps a response to criticism,

Hooton comments on their relationship 3 years after the publication of *Indians of Pecos Pueblo*:

> "The method which I have employed in segregating cranial types differs quite radically from that of my colleague, Professor Dixon. He utilized only combinations of the conventional subdivisions of the length-breadth, length-height, and nasal indices . . . I, on the contrary, used morphological judgments in selecting the types, and, after establishing their statistical integrity, sought their affinities with other crania by utilizing the means of all available cranial measurements and indices and appraising the sum total of significant differences" (1973:161) and he insists that "the American race is a composite race . . . composed of heterogenous strains welded together by mixture, not of wonderfully adapted types made out of common clay by a creative environment." (Hooton, 1973:162)

Hooton insisted on the multiple origins on American Indians throughout his life (Hooton, 1946). Pecos was a lens through which he saw a grand and unorthodox scheme for the peopling of the Americas. Modern morphometric studies support his views at several levels. For example, Brace still finds evidence linking the most ancient American specimens with the Ainu on the one hand and Oceania on the other (Brace *et al.*, 2001), and a recent reanalysis of Hooton's craniometric and discrete data finds considerable variability and little evidence for change through time among the chronological components at Pecos [Weisensee (2001); see Beck's demographic reanalysis in this volume].

Hooton's method at Pecos and in his more general schemes for race classification is at root an application bertillonage, a primitive form of multivariate classification. While he may have borrowed the technique from Dixon, whose three indices, each trichotomized, yield 27 possible types, or from Francis Galton (1822–1911), who had adapted the methods of criminologist Alphonse Bertillon (1853–1914) in his study of dermatoglyphics (Gillham, 2001), the similarity is clear. Bertillon appears to have invented this antecedent of contingency table analysis. Hooton also appears to have borrowed the method of composite photographs of the Pecos types from Galton, although Galton is not cited in the bibliography, which is quite limited. Connections via Hooton's extensive work on criminal typology are likely. Howells argued that Dixon's work was largely independently developed and that Dixon's influence on Hooton flowed largely through their earlier collaboration on racial assessment of crania from the Canary Islands. Dixon himself cited sources with regard to data rather than ideas, suggesting that his analysis is largely original. However, Dixon's race labels correspond closely to those in general use in Europe. His method of casting crania from a single site or region into a series of types is unusual, and Hooton adopted this practice. Both represented a group of people as consisting of various percentages of types. As Dixon's colleagues noted at his death "he was the first anthropologist to show by scientific data the composite character of the American Indians as being primarily Mongolian but with admixtures which can be affiliated with early white and negroid strains. Recent archaeological investigations have borne out this thesis" (Tozzer and Kroeber, 1945:105). Those investigations were surely Hooton's.

The last of Hooton's contributions relevant to the present topic is his paper on skeletal material from the Cenote of Sacrifice at Chichen Itza (1940). There is a lengthy discussion of various mechanisms for cranial deformation and some interesting paleopathology antithetical to the interpretation of the remains as sacrifices that has been largely overlooked by Mesoamerican archaeologists. Echoing Matthews and Putnam on brachycephaly and high civilization in the Americas, Hooton remarks on the similarity of Peruvian, Maya, and Southwestern remains. He reiterates his distain for the Bering Land Bridge model and suggests Armenian and even Toda contributions to the remote ancestry of the Maya!

It is noteworthy that Hooton's work extends the projects of Putnam's protégés at the turn of the century. Cordelia Studley had begun a study of the Turner material, although her AAAS address on this series was never published. Hooton's publications on the Turner and Madisonville series build directly on her work, although she is not acknowledged through citation. Similarly, Hooton's Iceland study mirrors Russell's Labrador paper in its logical structure (1899). *Indians of Pecos Pueblo* cites none of Putnam's protégés, but Studley's Coahuila series and Russell's California series are used in comparisons. Dixon might be added to this list.

The larger connections of Hooton's craniology are similarly difficult to trace through his citations. Hooton's mentor Sir Arthur Keith is central to recent critiques of scientific racism, and some have seen close correspondences in their work (e.g., Brace, 1982; Barkan, 1992). Whatever one's opinion of his racial politics, Keith was certainly a taxonomic splitter whose ideas are often congruent with Hooton's. Keith accepted the Punin skull from Ecuador as evidence for "a pleistocene invasion of America by an Australoid people" (Keith, 1931:312), and Hooton refers to it as "a skull any competent craniologist would identify as Australian in type" (Hooton, 1946:650).

Much of the work of Hooton and Dixon seems fanciful to modern readers. It is a useful corrective to presentism to note that T. D. Stewart took their case for multiple late migrations accounting for brachycephalization in the New World to develop his own argument that these late migrants brought with them, not only round, high heads, but the practice of cranial deformation and the pathogen responsible for syphilis (Stewart, 1940f).

While there are several modern summaries of Hooton's work, some written with great affection (Garn and Giles, 1995; Shapiro, 1981), there is as yet no full biography of this remarkably interesting figure. Critical assessments of Hooton's work are astonishingly varied in their focus. Wolpoff and Caspari (1997) call him racist and Lamarckian, and his work polygenism, blaming his association with both Dixon and Keith for these faults. Brace (1981) calls Hooton's scheme "'polyphyletic,' not 'polygenetic,'" and suggests that "Hooton and his students were less than fully conscious of the strains of romantic racism that constituted a major part of their background" (Brace, 1982:15). One of the few positive recent assessments is Stewart's (1981) demonstration that Hooton demolished Hrdlička's claims for morphological dating. Stewart also pointed out that despite their apparent intellectual differences, Hrdlička counted Hooton as his closest friend. It is perhaps less surprising than some have found it (Wolpoff and Caspari, 1997) that Boas enlisted Hooton and Hrdlička in trying to move American physical anthropologists to speak out against Hitler's race policies.

VIII. PAUL RIVET 1876–1958

Hooton's work may look a bit more mainstream when viewed from the perspective of contemporaneous work in France. Paul Rivet was a polymath anthropologist of the Boasian style, publishing in all four subfields. He was particularly influential in South America and contributed to the organization of physical anthropology as a discipline in Mexico, Ecuador, Bolivia, and Brazil (Leon, 1977).

Christine Lauriere (2000) has written an insightful analysis of Rivet's early career. He became interested in anthropology while serving as an army physician in Ecuador. Between 1906 and 1912 he established himself as a professional anthropologist. His research on prognathism was part of a campaign aimed at securing a position, first at Societe d'Americanistes, then at the Museum d'Histoire Naturelle, and eventually as founder of the new Musee de l'Homme. Lauriere showed that Rivet's demonstration that the facial angle produced no systematic hierarchy of races was strategic as much as scientific and was a key element in his rejection of Paul Broca's 19th-century physical anthropology. Rivet left Broca's Société d'Anthropologie in 1911 and founded, with colleagues who constitute a roster of the memorable figures in French social thought, a new Institute Français d'Anthropologie that integrated all the human sciences. Lauriere summarizes the importance of the prognathism studies: "He had to construct for himself a most convincing curriculum vitae in looking toward the next candidacy at the museum that he knew from experience was very attached to the pre-eminence of the biological over the cultural."

However, once nominated, Paul Rivet took advantage of the global conception of anthropology defended by Paul Broca and Armand de Quatrefages to take his work in a completely different direction: "He devoted himself henceforth to studies of American Indian linguistics, ethnography and archaeology" (Lauriere, 2000:20, author's translation). The parallels to Franz Boas' career path in the United States are remarkable: the legitimizing role of early research in physical anthropology, a revolutionary concept of an integrated field, and an emphasis on institution building are shared features of Boas and Rivet.

Rivet's four papers on prognathism are a remarkable tour de force, both with regard to sample size and with regard to exhaustiveness (1909b,c, 1910a,b).

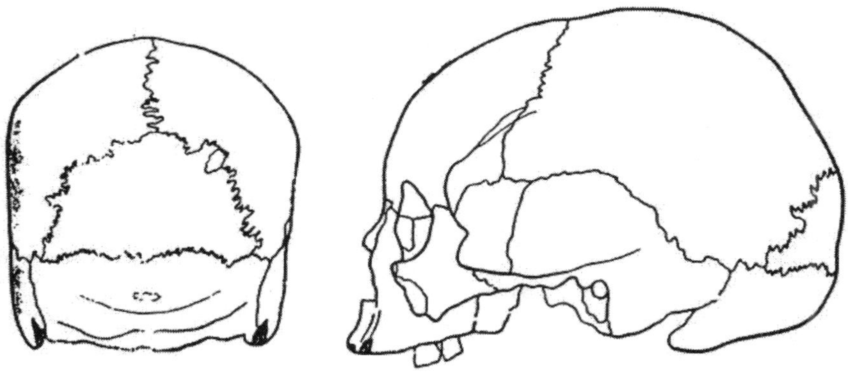

Figure 5 Rivet's Pericue skull XVI (Rivet, 1909a).

He compared the several measures of facial angle, beginning with 5615 humans, 151 apes, and 334 monkeys (1909c) and adding series as the study progressed. A table in the final study includes 665 crania from the Americas: 11 Amazonians, 18 Zuni, 30 Ancient Peruvians, 17 Ancient Mexicans, 73 Ancient Ecuadorians, 29 "Peaux-Rouges" (presumably Plains Indians), 44 Eskimo, 25 Tierra del Fuegans, 31 Moundbuilders (including Hrdlička's series from Arkansas and Louisiana), 21 Andeans from Argentina, 18 Pampians, 36 Northwest Coast, 17 continental California, 17 Pericue (Baja California; Fig. 5), 240 Channel Islands, California, 21 Aleuts, 9 Carib-Arawaks, and 7 Yucatecs, listed in order of facial angle from mesognathic to prognathic (Rivet, 1910b:642)! He summarizes:

> In America, a great center of prognathism occupies the Northwest Coast, represented by the Aleuts, the Californians and the Indians of the Northwest. In the Eskimos and above all in the Peaux-Rouges prognathism diminishes clearly. It is the same with the Zuni and the ancient Mexicans. On the other hand, the Moundbuilders and above all the Yucatecs are distinguished by the small size of their naso-alveolo-basilar angle. (Rivet, 1910b:648, author's translation)

The prognathism papers are of interest here because this large sample constitutes the experiential basis for Rivet's concept of race as expressed in the skull. One notes a bit of bias toward California. Statistical analysis is limited to comparison of means and ranges by inspection, and extensive use is made of tripartite categorization of continuous measures, e.g., orthognathic–mesognathic–prognathic for the facial angle. Nevertheless, the scope, energy, and complexity of Rivet's study is impressive: he demonstrated that facial angle varies with age and sex, that it has no consistent relationship to cranial index and facial index, that geographical races include populations that differ enormously in facial angle, and that the various measures of facial projection are far from equivalent one to another, thus laying to rest the enterprise begun by Blumenbach: arranging races in order of facial projection.

Rivet's other contribution to the craniology of North American groups is his description of 18 skulls from five localities in Baja California (1909a). It is the third of a series of studies, with the earlier two concerning ancient crania from Paltacalo, Ecuador, and Lagoa Santa, Brazil (Rivet, 1908). This paper makes extensive use of bivariate plots of the principal cranial indices to separate Baja California from other North American Indian groups, making use of published and unpublished data from Carr, Allen, and Hrdlička, among others. He then links the Baja California series, first, with the ancient population from Lagoa Santa in Brazil, and thence with Melanesia and Australia in a *type hypsistenocephale*, characterized by a high, narrow skull. This type is contrasted with American Indians. He then proposes a trans-Pacific migration accounting for his findings:

> I have searched without success for an explanation of how Melanesian migrations could have reached the coast of California, whether voluntarily, or by way of sea currents. It suffices to recall that numerous and indisputable observations have shown the possibility of great voyages, even for uncivilized populations. Besides, we are more or less completely ignorant of the exact configuration of the north Pacific in the geological period that followed the appearance of humans. (Rivet, 1909a:247)

Spencer (1997c) implies that Rivet used Mendes Correa's map of Antarctic migration routes, but this is an anachronism, perhaps misunderstood from Stewart's (1973) popular account of these ideas. Recent work revisiting the question of the affinities of Baja California populations using modern morphometric techniques have resurrected Rivet's thesis (Gonzáles-José *et al.*, 2003).

Rivet's ideas about the peopling of the New World are laid out in their fullest in a monograph he produced late in his career that integrates his craniology with his ethnographic and linguistic research. *Les Origines de l'Homme Americain* was published simultaneously in translation in Mexico as well as in Canada (Rivet, 1943). Neither historical linguistics nor ancient DNA supports his views today, but this does not lessen their historical interest, and he has been cited quite frequently in recent literature on the peopling of South America.

Like Hooton, Rivet confronted Hrdlička's dogma of a single migration of American Indians across the Bering land bridge. In his view the Indians were too diverse to fit Hrdlička's model. He recognized, marshaling craniometric and linguistic evidence, late Asian affinities in the Eskimo, Polynesian, and Indonesian affinities in the Hokans, and Australian affinities in the peoples of Tierra del Fuego and Lagoa Santa in Brasil (Rivet, 1943). Rivet's ideas have been revived recently in the controversial new morphometric studies surrounding the "Luiza" specimen from Brazil (Powell and Neves, 1999), and his views have continued to be accepted as mainstream in Latin America (Comas, 1960, 1974).

Rivet's craniology is more quantitative than Hrdlička's and more facile in its use of geometric techniques than Hooton's (e.g., Rivet, 1909a,b,c). Rivet's use of bivariate plots of the principal cranial indices is an intriguing precursor to multivariate statistics in that they visually summarize three or four linear variables at once. Indices were of course central to late 19th- and early 20th-century craniology and to a quest for measures of shape—of morphology—independent of size. They have largely disappeared from our science, partly because of their refractoriness to statistical analysis and partly because multivariate methods have supplanted them.

Historians of anthropology are discomfited by the failure of early 20th-century anthropologists whom they regard as liberal and antiracist to reject the concept of race, and Rivet is no exception in this regard. Despite his liberal role in the history of French anthropology and his heroism in the Resistance, he was paternalistic toward his ethnographic subjects and opposed the decolonialization of Algeria (Lauriere, 2000; Reynaud-Paligot, 2001).

IX. BRUNO OETTEKING 1871–1960

If the French and British traditions in early 20th-century physical anthropology reached the New World in such diverse forms, we may expect similar variety from the German tradition. Bruno Oetteking did his doctoral work under Rudolf Martin (1864–1925) at the University of Zurich, completing his dissertation on the craniology of ancient Egyptians in 1908. He held positions at several German institutions. In 1913, Franz Boas (1858–1942) recruited Oetteking for his research group at the American Museum of Natural History, which focused on the Arctic and Northwest Coast collections. Oetteking moved with Boas to Columbia University in 1920, where he held an appointment as lecturer until his abrupt dismissal 1938. He also served as curator of physical anthropology at the Museum of the American Indian. At Columbia, Oetteking taught the physical anthropology courses that made Boas' program a four-field department (Weiant, 1960).

Oetteking's work is meticulously, perhaps obsessively, descriptive. His most ambitious project was the study of skulls collected by the Jesup North Pacific expedition, *Craniology of the North Pacific Coast* (1930a), published in a volume shaped by Boas' interests and published under Boas' editorship. The series of 560 skulls is divided among four groups: those evidencing Cowichan, Chinook, or Koskimo styles of cranial deformation and undeformed crania. Ethnic groups are distributed unevenly across these four categories, and the last category includes Siberian Eskimo and Chuckchee crania, a strategy that makes comparison among groups problematic. The analysis is grounded in Boas' article on cultural patterns of cranial deformation (Boas, 1890). Oetteking's principal question is metric and non-metric distinctions among the three varieties of artificial cranial shaping. There are 107 figures illustrating discrete variations in exhaustive detail. Most recent citations of Oetteking's work draw on his descriptions of variants and on his somewhat questionable demonstration that cranial deformation affects the frequency of many discrete traits. The question of race occupies a small part of this monograph. The undeformed group is compared with data from Oetteking's own previous studies of Egyptian and Californian crania at the Museum of the American Indian (1925) and Eskimo crania in the collections at Dresden, as well as Hrdlička's data on Mongol crania. A form of pattern profile analysis is an interesting innovation in these comparisons (Oetteking, 1930a). His conclusion is remarkably brief and qualified:

> Of a number of crossproducts the narrowing of the face and nose have been recognized as progressive and would have to be attributed in our case to the blending with another morphologically different and, as it were, superior racial group, such as early caucasoid elements .... It was not intended by the author to draw into his study of a rather limited but at the same time all the more important anthropologic domain, the problem of Polynesian or other origin. From his present investigations, however, he derives the conviction of North Asiatic migration, the Mongolian affinity, the premigratory cross-breeding with distant (precaucasid?) elements, and finally the phaenotypical differentiation of the American Indian on American soil. (Oetteking, 1930a:376)

Oetteking's reluctance to reach conclusions afflicts his other typological publications to an even greater degree than in this tortured prose (1925, 1930b, 1931, 1934, 1945). Certainly, the labeling of features as "primitive" or "superior" strikes a discordant note in the work of a protégé of Boas, and the failure to develop an ethnic or linguistic dimension to the analysis is surprising. His many publications in *Indian Notes and Monographs* seldom venture beyond description, and he was notably slow in producing them.

The extensive literature on the career of Franz Boas is essentially silent on his relationship with Oetteking. One is curious about their long professional association and the issues that led to Oetteking's dismissal. The focus on cranial deformation in *Craniology of the North Pacific Coast* is certainly consistent with Boas' agenda of demonstrating environmental plasticity in skull shape (Holloway, 2002). One looks in vain for the statistical sophistication that characterized Boas' publications in physical anthropology (Tanner, 1959; Howells, 1959). There are many eccentricities in statistical language. For example, Oetteking uses the term "correlation" for tables reporting means of facial measurements grouped by tripartite categorizations of cranial length, cranial breadth, and cranial index (Oetteking, 1930a:102), but he uses no correlation statistics.

In a late publication on Arctic crania he assigns three skulls from a single site to three different types: "the crudest and most robust ones as the morphologically most inferior and belonging to an old, perhaps primarily pre-Columbian ethnic stratum." He attributes the lack of "homogenous racial integration" (1945:307) to Russian admixture and "extraneous derivation" (1945:308), citing Georg Neumann (1942) on low vaults. He omits the cranium with the most gracile face and rounded vault as pathological, diagnosing hydrocephalus to explain the anomaly. The Dayak are used as a comparison group in this study. This hardly strikes one as the work of a committed Boasian!

While Oetteking's overview papers, many in German language journals, advocate the orthodox view—Boas and Hrdlička's Bering Strait scenario (1928, 1932)—one finds some remarkably old-fashioned claims, e.g., orthogenesis: "Nature herself always progresses from the crude to the more refined, and from the simple to the more complex" (1928:817). He flirts with ideas from Hooton, Rivet, and ten Kate in his choice of comparative samples and in his hair-splitting, tentative approach to typology. If we remember him for nothing more, he documented a great many skulls now threatened with deaccession, and he did so in a very transparent way.

Boas' extensive correspondence contains a few hints concerning his rapport with Oetteking (Boas, 1972). Oetteking wrote to Boas on January 18, 1936, protesting his dismissal. There is no letter in response, and other correspondence shows Boas negotiating his own retirement and arranging lectureships for others, notably the ethnomusicologist George Herzog and the ethnographer Frans Olbrechts, both refugees from Nazi Europe. A letter to Dean H. L. McBain dated March 12, 1936, rates these two candidates among others, including Rivet: "Rivet does not speak English. He is an agreeable dabbler in many different subjects and has a good knowledge of the archaeology of the most northern part of South America." In an intriguing letter to Boas dated Feb. 27, 1936, Alfred Tozzer of Harvard wrote "I am delighted to learn that you are not going to have Columbia humiliated by the presence of our ex-tutor and instructor." This person is not identified, but one suspects that Tozzer refers to Oetteking.

X. GEORG KARL NEUMANN 1907–1971

Georg K. Neumann was among the last physical anthropologists committed to the typological concept. His dissertation *Racial Differentiation in the American Indian* was the last grand effort at defining races in native North America. The University of Chicago dissertation was accepted in 1950. Preliminary versions were circulated and cited earlier (Neumann, 1941; Martin *et al.*, 1947). The dissertation was immediately and widely published (1952, 1954a,b) and was critically reviewed (Angel, 1954; Stewart and Newman, 1951, 1954; Comas, 1960). It remained the paradigm for the remainder of Neumann's scholarly output. This work is particularly situated in culture history and language, reflecting Neumann's mentors.

Neumann's training in physical anthropology began under Fay-Cooper Cole (1881–1961), even though Krogman rather than Cole supervised his dissertation. Neumann was a student in Cole's archaeological field program at the University of Chicago, and he excavated cemetery sites for Cole from 1928 through 1934.

Cole's field projects put North American archaeology on its modern footing, establishing standards for data collection and excavation, although Browman has recently questioned whether Putnam deserves credit for many of Cole's (2002) innovations. Cole has received less attention than one might expect in the history of anthropology. Cole was one of the founding members of the American Association of Physical Anthropologists and was a father figure for North American archaeologists. He was the lone archaeologist among Boas' successful students, and he transplanted the Boasian program to the University of Chicago. Griffin discusses his central role in establishing archaeology as a scientific discipline in the United States (Griffin, 1996). Krogman (1981:470) writes of Cole as a teacher: "Dr. Cole was almost 100% a disciple of Rudolf Martin's osteometry and somatometry. We who majored in physical anthropology became first 'measurers of man' in purely osteological and morphometric terms and only later in functional terms: physiological, biochemical, and genetic. But these latter were not taught to us in depth, for their relevance to physical anthropology was yet to be clarified and developed." Krogman adds that Cole sent him to study with T. Wingate Todd at Western Reserve University. Similarly, Cole sent Neumann to work with Todd in 1932–1933. Neumann's relationship with Todd was apparently problematic; e.g., he was the uncredited anonymous illustrator of the Todd pubic phases (see Stewart, 1979a:159). W. M. Krogman eventually served as Neumann's dissertation supervisor. We now think of Krogman as a pioneer of growth studies, but early in his career he made several contributions to physical anthropology and archaeology in the Midwest. At the time he was a partisan of the view that "the American Indian — the First American — has also emerged from a racial 'melting-pot'" (Krogman, 1941:812).

Neumann's 1937–1942 sojourn at the University of Michigan and the early years of his employment at Indiana University were supported by Eli Lilly (1885–1977), a philanthropist who was deeply interested in American archaeology. He was a founder and major supporter of the Indiana Historical Society. On the advice of the archaeologist James B. Griffin, Lilly supplied funding to Indiana University to hire Neumann, the linguist Carl Voegelin, and the archaeologist Glenn Black as faculty. He later helped establish a department of anthropology for them. Erminie Wheeler Voegelin, an ethnohistorian and specialist in Indian land claims, was hired in the history department; finally the ethnologist Harold Driver and ethnomusicologist George Herzog were recruited in anthropology (Griffin, 1972; Jones, 1976). Mr. Lilly had a project in mind for his department: the authentication of the *Walam Olum*, a purported Delaware migration legend, and in 1954 his document was published along with essays written by each of his anthropologists (Voegelin and Rafinesque, 1954). Most modern scholars regard the *Walam Olum* as a forgery, and the scholarly essays and their authors have been ridiculed (Oestreicher, 1996, 2002); however, the Delaware remain convinced of its authenticity and accept it as a true account of their ancestry (McCutchen, 1993). In defense of the author's own institutional ancestors, the scholarly essays are best read as exercises in stating one's contradictory conclusions in a manner designed to give as little offense as possible to one's sponsor.

To the end of his life Lilly remained convinced that the *Walam Olum* would eventually prove to be authentic. At a 1974 lecture celebrating the Glenn A. Black Laboratory of Archaeology at Indiana University, Black's successor, James A. Kellar, suggested that the team had shown it to be inauthentic. Mr. Lilly rose and said that he considered "the jury to be still out" (author's notes on the lecture).

Neumann's contribution to the *Walam Olum* project is confined to a comparison of 10 male putative Munsee skulls with 20 male Seneca skulls. He finds that the Munsee (a Delaware Algonquian-speaking group) differ from the Seneca (an Iroquoian-speaking group), but that the former also differ from other presumed Algonquian speakers in ways consistent with details of the *Walam Olum* text and, one notes, equally consistent with the multiple-group model for late prehistoric peoples that Griffin favored.

In *Archaeology and Race in the American Indian*, Neumann fused race, language, and culture in a manner that reflects the culture-history interests of the archaeologists who were his mentors and colleagues. The type, not the population, was Neumann's unit of analysis, and he defined eight such types using the term variety. This effort was a refinement of the taxonomy produced by Egon von Eickstedt in the German tradition reaching back to Blumenbach. Neumann's variety is the penultimate taxon in what must be the ultimate splitter's taxonomy. The species is divided at five levels: subspecies, series, pars, varietas, and subvarietas (von Eickstedt, 1940:65). These would replace the varieties von Eickstedt proposed for North American Indians: Pacifid, Centralid, Silvid, and Margid, as well as Eskimid from the Arctic series. We can thus understand Neumann, via the Cole genealogy, as the flowering of the Boasian four-field concept in physical anthropology and connect him to Virchow and thence to liberal, monogenist German physical anthropology of the 19th century (Massin, 1996), but his taxonomic choices ally him with the polygenists via Gmelin and Haeckel to Linneaus.

The eight varieties were Otamid, widespread and ancient but surviving in Coastal Texas and the eponymous Tohono O'Odham; Iswanid, also widespread and ancient, typified by Archaic Indian Knoll and named linking it to Catawba; Ashiwid in the Southwest; Walcolid in the Southeast extending to the Midwest in Adena and Mississippian groups and to the Pacific Coast; Lenapid in the Northeast; Inuid for the eponymous Inuit and their precursors; Deneid for Aleut and wide-ranging Athabaskans; and Lakotid for peoples of the northern Plains. Note that the Eskimo are not set apart from other New World groups and that distinct Asian connections are discussed for Otamid, Iswanid, Inuid, and Deneid. In summarizing, the author has touched on the range of each type, but failed to convey the fluidity and complexity of Neumann's concept. This intricate picture of population movements is all the more remarkable in that at that time archaeology as a whole was in the process of purging itself of migrationism (Adams *et al.*, 1978) and that James B. Griffin played a major role in this process.

Neumann discusses the work of Hooton and Hrdlička extensively. Madisonville is Walcolid, whereas Hopewell is Lenapid. Hrdlička's unitary views of the Plains and the Northeast are dissected. Rivet is not cited, but his Pericue are assigned to Otamid. *Archaeology and Race* was published just as radiocarbon dating was becoming available, and some sense of the ferment this engendered is reflected in the frequent discussion of chronological relationships. Neumann may have felt some ambivalence about the lack of securely dated early series, and there are several interesting conjectures in this vein. One, a putative Paleoindian skeleton from Clark's Fork, Idaho, assigned to the Otamid variety, has been confirmed as ancient, if not quite as old as Neumann believed (Pennefather-O'Brien and Strezewski, 2002).

Oddly, given the importance of archaeological context and time depth in Neumann's scheme, site, population, and specimen identifiers were not salient. Indeed one of the frustrations of dealing with his output is that after 1928 he did not publish the detail that would allow one to know which crania were measured in any study. While he measured crania thoroughly—one might think obsessively—his analysis was limited to tabular presentation of means and standard deviations. He did not use the multivariate techniques that became the standard for biological distance studies during his lifetime. While he measured female skulls, his typological analyses used males exclusively. As for Hooton, the type was the unit of analysis, but unlike Hooton, an archaeological site or component was expected to yield a single type. Measurements are used only to support findings of the typologist's eye. *Archaeology and Race in the American Indian* was criticized for the subjectiveness of the types, for the arbitrariness of his choice of just 471 crania from the 10,000 he claims to have studied, and above all for his delay in publishing (Angel, 1954; Stewart and Newman, 1954). Stewart and Newman are remarkably sanguine in their account of Neumann's work, given their own investment in adaptationist models for change in skull shape (Newman, 1953, 1962). They accept much of the typology and point out improvements over Hrdlička's scheme, but argue that Iswanid and Ashiwid are not sufficiently different: "such evidence leads us to conclude again that these particular varieties have more archaeological rather than craniological validity" (Stewart and Neumann, 1954:141).

Neumann tinkered with his types over time. Varieties were renamed and subdivided chronologically into an ancestral Paleoamerind series and a descendent Mesoamerind series. Lenapid was renamed Ilinid, perhaps in response to doubts about the authenticity of the *Walam Olum*, Otamid branched off

Lenid in the east, Lakotid became Dakotid, and Walcolid became Muskogid (Neumann, 1960, 1966; H. Neumann, 1960a,b; Robbins and Neumann, 1972). A Uinicid variety for the Maya and Nootchid for the Great Basin were added, and the Deneid and Inuid varieties were put in a separate Cenoamerind series for the most recent immigrants (Neumann, 1960). Interestingly, the illustrated specimen for Neumann's Lenapid in 1952 becomes Lenid for Robbins and Neumann in 1972. In Fig. 6, the author arranges the illustrations from *Archaeology and Race in the American Indian*, plus one Ilinid illustration from Robbins and Neumann, to illustrate this scheme.

Neumann channeled most of his graduate students into craniometric dissertation projects aimed at testing details of his typology. He asked that his students work with measurements Neumann himself had taken as part of his dissertation project and insisted that he measure any new material side by side with the student. His students Constance Omoto (1960), Holm Neumann (1960a,b), Kenneth Smail (1964), David Skomp (1965), James F. Metress (1971), Ralph Alexander (1971), Robert Blakely (1971, 1973), Louise Robbins (1964, 1968; Robbins and Neumann, 1972), Elizabeth Glenn (1965, 1974), and Judith Droessler (1975) published local or regional studies that evaluated boundaries between Neumann's types using modern statistical techniques. Robbins participated in adding a variety, Illinid, to the later prehistory of the Midwest (Robbins and Neumann, 1972; Neumann, 1966). Three of his students addressed Neumann's typology as a whole. Joseph Long (1966) tested the eastern North American types using multivariate analysis, a project that began as a University of Kentucky M.S. thesis directed by Neumann's close colleague, Charles Snow, and found limited support for the typology, if not for Neumann's interpretations of his types as evidence for migrations. Kenneth Smail (1964) asked whether female crania supported the model Neumann proposed for male crania and found mixed results, with females showing clearer Plains or Oneota affinity than males. He interpreted these findings as reflecting gender differences in the population structure. Matthew Brennan and W. W. Howells, in an unpublished paper meant for the ill-fated physical anthropology volume of the *Handbook of North American Indians*, used principal components analysis to discern groups among 68 series of Siberian and North American crania measured by Hrdlička. Brennan had been an undergraduate student of Neumann's, and this project was part of his graduate work under Howells at Harvard. They conclude (Brennan and Howells n.d.:33):

> These results do not coincide particularly with older attempts to classify North Americans. . . . Our groups do, however, correspond quite well with varieties discerned by the Experienced [loc. cit.] G. K. Neumann. . . . On the basis of mean figures and general morphology, he examined many samples large and small (as here), and selected particular ones which seemed both representative and clearly characterized, and suggested the distribution, origin and final development of each. Here we approach

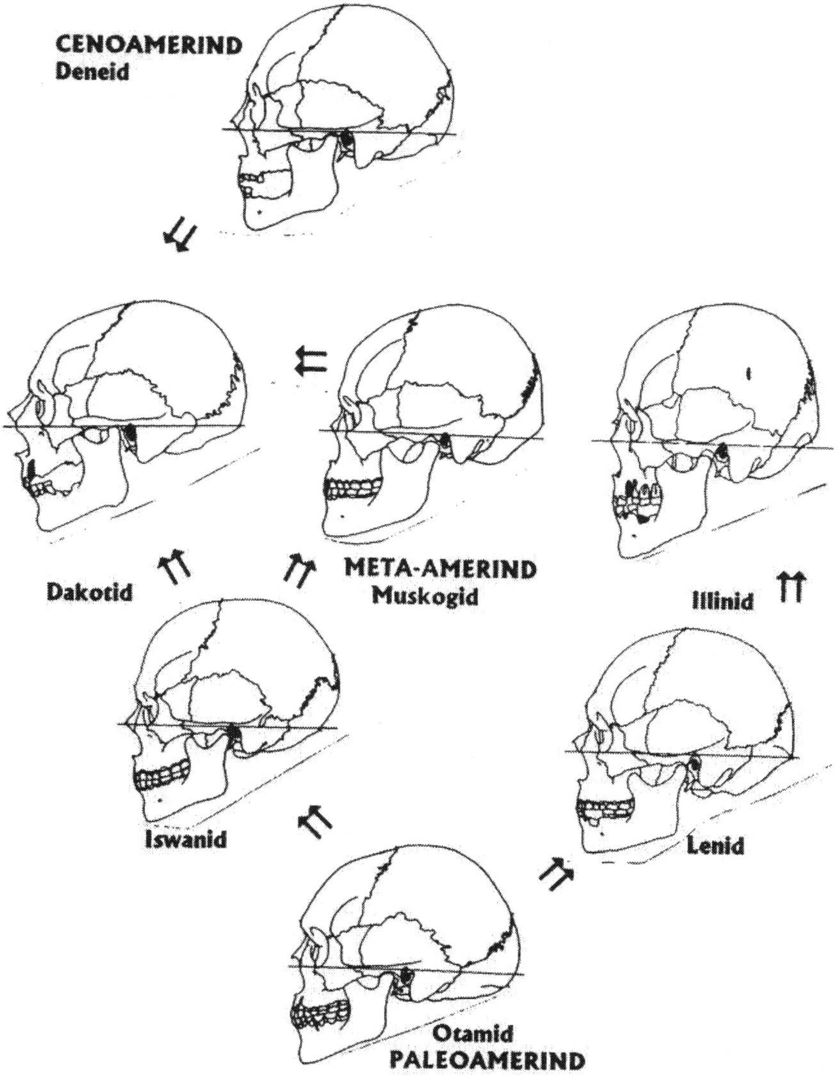

Figure 6 Neumann's varieties arranged to correspond to the evolutionary scenario proposed by Neumann and Robbins (Neumann, 1952; Robbins and Neumann, 1972).

similar series from the other direction, letting groups form (the essence of the study) and then examining their characters and relationships.

Howells and Brennan found five clusters and analogized them to Neumann's Inuid, Lakotid, Deneid, and Walcolid varieties. The fifth group "General United States . . . seems to merge Neumann's Iswanid, Ashiwid, and Lenapid varieties, though not closely fitting his descriptions, especially the last" (Brennan and Howells n.d.:35). They attribute this failure to limitations of the series they analyzed and to Neumann's use of temporal distinctions. A more recent discussion of this study suggests that unrecognized cranial deformation contributed to conflating the latter varieties (Howells n.d.).

Perhaps the most widely cited of Neumann's (1942) works is his paper on types of cranial deformation. It bears an interesting relationship to his racial typology; in order to assign a group to a variety using Neumann's scheme, one had to omit deformed skulls, a major factor in the reduction of his study series from 10,000 to just 471. The deformation study is itself typological in that it assumes discontinuities among the eight types, an assumption that does not stand up to rigorous testing (Droessler, 1981). The type is communicated primarily through craniophor drawings of typical exemplars, much like the varieties in Neumann's larger study. Oddly, he chose a skull with a bipartite parietal that Putnam (1884) had published as abnormal as one of his exemplars. The problem — accounting for intentional cranial shaping as well as positional plagiocephaly — remains a vexing one in metric studies of the cranium and is still generally handled typologically.

The last of the racial typologists, Carlton S. Coon, cited Neumann's (1965) work as the authoritative bibliography on North American Indians. This is a surprising choice because Coon, a radical splitter in other regions, adhered to Hrdlička's dogma of a single migration across the Bering land bridge. Neumann's work is still cited as normal science, often in some surprising places (e.g., Wolpoff and Caspari, 1997:393, n.123; Stewart, 1981; Haskell, 1987; Ousley, 1995). There has been relatively little recent assessment of his contributions to our literature (but see Buikstra, 1979; Crawford, 1998; Griffin, 1996; Howells n.d.). Perhaps his most important role was in salvaging the human skeletal collections when Sherwood Washburn dismantled Cole's laboratory at the University of Chicago and in providing a home for skeletal collections from Gregory I. Perino's excavations for the Gilcrease Institute of American Indian History and Art. As his successor, the author is grateful.

Robert Meier (personal communication, 2004) recalls a conversation he had with Neumann in 1968: "He did ask me as we were driving to the AAPA meetings held in Michigan if I thought that the typological approach would be supplanted by the population/variation approach, and when I said that I was sure that it would, he simply shrugged and seemed not very keen to contest the statement on what he probably considered the inevitable."

XI. GÖTTERDÄMMERUNG

The end of the typological paradigm was very much in sight during the careers of Hrdlička, Hooton, Rivet, Oetteking, and Neumann. The first application of multivariate statistics to the question of American Indian races is a 1938 paper that Gerhardt von Bonin (1899–1964) and Geoffrey M. Morant (1890–1979) published in Karl Pearson's journal *Biometrika.* Neither author rated an entry in Spencer's *History of Physical Anthropology: An Encyclopedia* (1997c), an oversight that speaks to the unfortunate provincial biases of American physical anthropology. When the paper was written, von Bonin was a neuroanatomist at University of Illinois, Chicago, and participated in Fay Cooper Cole's circle. Morant spent a long career at the Galton laboratory and was a prolific contributor to the literature on anthropometry and craniometry.

Their paper applies Pearson's coefficient of racial likeness to data from Hrdlička's *Catalogue of Crania* and Hooton's *Indians of Pecos Pueblo* to comparisons among American Indian series and to comparisons with Asian and Eskimo series. While the language is still typological, the analysis is a biological distance answer to the question of New World affinities. Some highlights include the discovery that Hrdlička's Kentucky Algonkin differed markedly from other Algonkin and Iroquois series. One would now point out that the Indian Knoll series is archaic, several thousand years older than the others (see Neumann, 1952), and that its linguistic affiliations are a surmise at best. Von Bonin and Morant found that it resembled a Japanese series among those included in the larger analysis. California crania were found to differ from other U.S. series, and "the Pecos Pueblo series was not included in the second group because its standard deviations are obviously peculiar . . . its peculiarity may be due either to the fact that the measurements selected because they were believed to be unaffected by artificial deformation were not uninfluenced by this disturbing factor, or to the fact that the population represented was racially more heterogenous than all the others" (von Bonin and Morant, 1938:124). Some California crania were linked to Ainu and other Japanese series. "A surprising diversity is found among the Indian populations of the country. . . . On this account it will be necessary to have considerably more material than that available at present to reveal their interrelationships in a completely satisfactory way" (von Bonin and Morant, 1938:127).

An appendix to the paper analyzes Neumann's data from Cole's excavations in Fulton County, Illinois (Neumann, 1937), and concludes that "the total series must hence be supposed racially heterogenous" (von Bonin and Morant, 1938:128) and fairly distinct from all other groups included in their study except Algonkin East-Central. Because the series includes Archaic, Early, Middle, and Late Woodland and Mississippian components, the heterogeneity is hardly surprising. Neumann had not yet developed his typology in 1937, and his analysis linked the earlier components to Hooton's Pseudo-Australoids.

Later Neumann (1952) would assign the chronological components variously to his Otamid, Lenapid, and Walcolid varieties.

Hrdlička actively resisted statistical innovations as editor of his journal. Hooton wrote of his statistical objectivity with obvious pride, but ignored the first studies in the new biometric paradigm. His later work does not cite von Bonin and Morant. Neumann (1952) cited them, but he discussed only their Indian Knoll and Eskimo results. He ignored the appendix reanalyzing his own work and made no mention of the paper's statistical advances. In contrast, von Eickstedt (1940) devoted several pages to von Bonin and Morant and reproduced their graphics. Both *Biometrika* and *Die Forschung am Menschen* are available in Indiana University's library, but there is neither evidence that Neumann used them nor that he encouraged his students to do so. He cited and taught from von Eickstedt's (1937) earlier *Rassenkunde und Rassengeschichte der Menschheit*, a work not available here. Perhaps Neumann did not know that von Bonin and Morant had reanalyzed his Fulton County data. Perhaps he was unready to face the paradigm change. It was left to Hooton's student and successor at Harvard, W.W. Howells, to champion the biometric paradigm in the United States. Howells begins an early foray (using data provided by Morant) thus:

> It is surprising that the natural variation in recent human head form—and let us consider particularly the cephalic index—remains a generally uncomprehended phenomenon. Many of the functional explanations offered can only be called fantastic today; and in general these, and phylogenetic explanations as well, fail to give an answer to the really notable differences in this prominent characteristic, especially as between populations of the same racial stock such as the European. Even extreme forms, such as that of the most long-headed Eskimo groups, have not been given any satisfactory explanation, in spite of some celebrated discussion. (Howells, 1957:19)

The eclipse of the typological concept had begun.

XII. WHERE HAVE WE BEEN?

In a forum very different from this one, Adam Gopnik (2000) contrasted "sizzlist" histories with "steakist" histories. "Sizzlist" histories are written from the perspective of social constructionism and address various contemporary social agendas as means of illuminating the past. In contrast, "steakist" histories are written from a technical perspective and emphasize, to use a concept from the vocabulary of anthropology, processual explanation. Of the former, Gopnik (2000) writes: "The trouble with this kind of reading . . . is that it vastly underestimates the difficulty of doing things as opposed to thinking about them." The latter are what historians often label — perjoratively — as "insider histories," and they are prone to positivistic bias. This essay is an insider history and it has focused on the craft of doing typology.

Measurement in the work of both Blumenbach and Morton serves largely as a descriptive tool, and detailed analysis in each researcher's work is confined to a single variable. Variability is unimportant, and the approach is primarily one of classification. Variability becomes the important focus among the late 19th-century practitioners of the typological paradigm. However, the uses of measurement from these early efforts through the mid-20th century are curiously limited and secondary to the definition of types or varieties. As Andrew Lang may have quipped about politicians, they used "statistics as a drunken man uses lampposts—for support rather than for illumination" (Ratcliffe, 2000).

The typological era was anything but monolithic in its paradigm. There was lively controversy over the origins of North American Indians that is certainly not settled today, as the contributions to this volume on morphometrics and mitochondrial DNA witness. There was remarkable disagreement about many issues. Was the unit of analysis the individual, the population, the site, or the type? Should both male and female crania be evaluated? Should one exclude deformed skulls? If so, what was the appropriate threshold?

The typological paradigm did, however, set the rules of the game. There was a shared sense of what needed to be measured and of shared methods, thanks to the craniometric conferences at the turn of the century. The typologists shared collections, and the 20th-century figures discussed here even shared forms for collecting craniometric data. For example, Neumann used Harvard University/Peabody Museum craniometric data forms, and Snow's and Angel's forms are only slightly modified versions of the Harvard model.

The typological paradigm had certain advantages we may have difficulty appreciating: one could type a fragmentary or immature skull, or a small series that cannot be evaluated using biological distance techniques. It is to that extent inappropriate to expect a morphometric study to validate a typological one given the same data base because the statistical requirements for sample size and preservation are such that efforts such as Long's are compromised at the outset.

Early 20th-century physical anthropology was a very small field. Its practitioners knew one another better and corresponded more extensively than we do today. Teaching methods and research methods were widely shared. For example, Neumann taught a version of Fay Cooper Cole's excavation manual throughout his career, and among his legacies to his department was a file drawer full of 19-page course handouts on Hooton's racial taxonomy from *Up from the Ape*. Paul Gebhard's notes from Hooton's 1948 course in physical anthropology at Harvard show that Hooton returned the compliment. His students read a preliminary summary of Neumann's dissertation project that included a version of Neumann's eight varieties.

The grand, old-fashioned typological studies of the pre-Columbian peoples of North America failed to discover ethnic or tribal boundaries because their statistical tools were inadequate and because they had no real concept of populations and little chronological control. On the one hand, most modern biological distance studies have been either too local (e.g., Szathmary and Ossenberg, 1978; Steadman, 2001) or too global (e.g., Howells, 1989; Brace *et al.*, 2001) to model ancient populations as cultural systems in the way that Neumann attempted. In part this is a technical limitation of the population paradigm: morphometric statistical techniques require samples orders of magnitude larger than the typologist's eye. On the other hand, the obsession with remote origins and with a concept of race as stable through time deflected the attention of the typologists from such anthropologically meaningful concepts as ethnic or tribal boundaries that have become the focus of much recent biological distance research.

Were Indians fundamentally similar or diverse? Were they closely related to one or to several peoples of the Old World? These questions are racially charged—perhaps all questions in American social life have some racial valence—but to reduce the work of the physical anthropologists who practiced the typological paradigm to mere racism is to lose its meaning. The Moundbuilder myth was a species of racism, and we should celebrate Morton for undermining it.

Is the study of race necessarily racist? There is controversy within and beyond physical anthropology. Most of us have given up the word "race" for less loaded formulations such as 'population history' or 'ancestry,' although the meaning of this trend is itself controversial [Cartmill and Brown, 2003; see Bocquet-Appel (1989) for an earlier parallel in France]. The typological paradigm rested on a concept of races as having discernible boundaries and persisting through time as bounded entities; this concept has been abandoned, but the questions that motivated typological anthropologists are still very much with us.

Massin (1996) has written of the "crisis of classical physical anthropology" in the context of German science at the beginning of the 20th century. Cranial measurements, whether taken singly or as indices, failed to differentiate races. We have seen a similar developmental sequence in North America. The extreme diversity of assumptions and race concepts in the work of Hrdlička, Hooton, Rivet, and Neumann is a symptom of this crisis. Massin and others write as if craniology had disappeared after the middle of the 20th century. Indeed, several authors with insiders' knowledge of physical anthropology adopted similar language, as if wishing it so would make what continues to constitute a major focus of research in our discipline disappear (Adams *et al.*, 1978; Armelagos and van Gerven, 2003). The crisis was resolved through the shift from the race concept to the population concept and through the introduction of multivariate statistical techniques that continue to generate detailed and rigorous accounts of the natural history of our species.

Despite Foucault's argument that biology replaced natural history in the mid-19th century (Larson, 1994), natural history persists to the present day as an organizing concept in anthropology (cf. Cavalli-Sforza, 1997). All of the work reviewed here is natural history. If that paradigm is an increasingly contested one in modern anthropology — witness the schisms at Connecticut and Berkeley, among others — it continues to be a richly productive one, and it lies at the heart of Boasian, or four-field, anthropology. Foucault has emphasized institutions in the rise of natural history and its replacement by specialized disciplines, and most anthropologists writing about our history have likewise emphasized the role of institutions — departments, associations, journals — in professionalization. The building of collections that were publicly held, properly curated, and accessible to researchers was an equally important condition for professionalization (see Farber, 1982). Everyone whose work the author has reviewed contributed to building collections and relied on the collections and data of his predecessors. As we witness the wholesale destruction of these resources through repatriation, we must insist on the importance of study and restudy in our science (Buikstra and Gordon, 1981).

Where did the Indians come from? How diverse are they, and how is that diversity related to their origins? How is their biological variability related to linguistic, cultural, and ecological systems in the New World? Twenty-first-century answers to these questions await us. Let us hope that adequate collections will remain to permit these studies.

ACKNOWLEDGMENTS

I write this essay as a nonspecialist in the question of American Indian origins. I have never published a biological distance study, and I spent much of my early career trying to distance myself from Georg Neumann's brand of anthropology at my home institution. The task of documenting our collections for NAGPRA compliance required me to read Neumann's work carefully and to understand it in its historical context. The task has deepened my respect for him and his contemporaries. My visits to Brazil and to Museo Nacional and Museo do Homen Americano helped me approach Rivet and his critique of the Bering Strait dogma that was central to my training with an open mind, as well as a forum to present a discussion of Morton that is the basis for a portion of this chapter. I thank the Fulbright Foundation and CNPq for their support. My colleagues Robert Meier and Paul Gebhard provided helpful comments.

Chapter 3

The Changing Role of Skeletal Biology at the Smithsonian

Douglas H. Ubelaker

I. INTRODUCTION

The history of physical anthropology at the Smithsonian Institution is closely linked with the development of American physical anthropology. The Smithsonian chapter in this story effectively began in 1903 when officials decided that physical anthropology should be represented in the already established anthropology effort. An ambitious, young physician turned physical anthropologist named Aleš Hrdlička (1869–1943) was hired to inaugurate this effort at the Smithsonian.

Physical anthropology had long been established in Europe as the comparative science of humankind through the work of Johann Blumenbach (1752–1840), Paul Broca (1824–1880), and others. This effort included new methodology (e.g., Blumenbach's standard positioning of crania for comparative viewing and Broca's craniometric techniques and designs of new measuring equipment), training (e.g., Broca's Institute), and attempts to build comparative skeletal collections (e.g., Blumenbach's collection of human crania, Spencer, 1997a,b).

By the time Hrdlička (Fig. 1) became a professional in the late 19th century, physical anthropology and collection building had already begun in the United States. Hrdlička himself credits Samuel G. Morton (1799–1851) of Philadelphia for initiating this effort (Hrdlička, 1918, 1943a).

Aleš Hrdlička was born in Humpolec, Bohemia (now located in the southern Czech Republic). After immigrating with his family to the United States in 1881, he received his M.D. degree from New York Eclectic Medical College in 1892. Hrdlička also received training at the New York Homeopathic Medical College

Bioarchaeology: The Contextual Analysis of Human Remains, Buikstra and Beck (eds.)

Figure 1 Aleš Hrdlička.

and exposure to techniques of physical anthropology and legal medicine in Paris. After working in private medical practice and with the New York Middleton State Homeopathic Hospital for the Insane, the Pathological Institute, and the American Museum of Natural History in New York, he joined the Smithsonian in 1903, where he spent the remainder of his career (Spencer, 1979; Stewart, 1940a; Ubelaker, 1999).

Like Morton, Hrdlička recognized the scientific need for comparative collections of human remains. Much of his pre-Smithsonian research had focused on the biological basis of abnormal human behavior. He had amassed extensive data on abnormal individuals but realized that to make sense of them he needed comparative information from normal individuals (Stewart, 1940a). Following Morton's lead, Hrdlička worked to build the collections that would make this comparative research possible. Initially, this involved collaboration with George S. Huntington, anatomist with the College of Physicians and Surgeons in New York, in assembling and conducting research on skeletons derived from medical school dissection (Stewart, 1940a).

As physical anthropology achieved growing visibility, the Smithsonian Institution recognized the need to add this speciality to its anthropology staff. Prior to that time, as discussed in Chapter 1, human remains acquired by the Smithsonian were transferred to the Army Medical Museum in Washington where they had received relatively little curatorial attention (Stewart, 1940a). In 1902, Smithsonian anthropologist William Henry Holmes requested that a Division of Physical Anthropology be established within the Department of Anthropology of the National Museum. According to Holmes, the purpose of this effort was "the comprehensive biological study of the many and diverse racial elements of the American nation, and the application of the results to promoting the welfare of the NATION" (Spencer, 1979:248). Hrdlička was hired in 1903 as the first physical anthropologist of this division.

Although Hrdlička likely viewed the new Smithsonian position as offering valuable potential for his collections and research interests, the necessary resources were not immediately available. According to Stewart (1940a:12) "[o]n taking up his work in the National Museum Dr. Hrdlička found himself assigned to a small section of one of the galleries in the Old Museum building. His whole equipment consisted of an old kitchen-table, chair, a pen rack, inkwell, a pen and a pencil. Nevertheless, he was again in the position where he could plan the future course of an institutional branch of physical anthropology. He proceeded to build up his Division until it has come to rival in size and importance of collections the oldest and best in the Old World."

It is important to note that at this early period in the development of American physical anthropology, Hrdlička conducted research and published in all major areas of the discipline (e.g., Hrdlička, 1894, 1895, 1896, 1897, 1899b, 1900, 1901; Lumholtz and Hrdlička, 1897, 1898). Gradually, his medical interests in the biological basis of abnormal behavior shifted toward a more comparative, anthropological focus (Hrdlička, 1902a,b,c). As he collaborated with archaeologists or conducted excavations himself, his intellectual engagement evolved with these new experiences. He published not only on the bones, but also on archaeological and ethnological topics (e.g., Hrdlička, 1903a,b, 1904a,b,c,d, 1905a,b,c,d, 1906a,b), as well as even more general ones (Hrdlička, 1909b, 1912c, 1919, 1920b, 1921a).

Hrdlička noted that the cornerstone of the developing field of physical anthropology consisted of the assemblage of large, well-documented collections of human remains from diverse sources. Through his own work and others, by 1918, he was able to report substantial progress. In the lead-off article of the first issue of the *American Journal of Physical Anthropology* (founded by Hrdlička), he remarked on collections available 50 years before in the United States and Europe:

> ... all this material was limited to crania, and was useful in arousing curiosity and false expectations rather than in leading to definite progress in our science. It required years of assiduous excavation and collecting before scientific work of any extent could anywhere be attempted. Such collecting, fortunately, has been carried on in a diligent and continued way to this day, until there are in this country alone several great and many lesser gatherings of identified skeletal and other anthropological material, led by that of the U.S. National Museum. Yet even now we are far from the goal in this direction; that is, from collections comprising adequate series of bones of the entire skeleton, besides those of other normal important parts of the body; collections that would enable us to determine the complete range of variation in these parts in at least the most significant groups of mankind. The requirements in this direction will appear more clearly when it is appreciated that, to determine the total range of variation in a single long-bone, such as the humerus, in any group to be studied, there are needed the remains of hundreds of adult individuals of each sex from that group. As it is, even the greatest collections we possess still fall short of the requirements, consequently our investigations can be seldom perfect or final. (Hrdlička, 1918:10)

The collection goals of Hrdlička, like his contemporaries in physical anthropology, were primarily to acquire comparative collections of normal individuals. Research on these collections was mostly aimed at providing "normal" perspective for other data on abnormal individuals and documenting the range of variation for skeletal attributes. Although Hrdlička made important contributions to paleopathology, despite his medical training, he did not concentrate his research in this area. Still, he recognized the need to curate the entire skeleton, not just the skull, the importance of documentation and dating of remains, and the need for large samples. Collections assembled by Hrdlička with these points in mind paved the way for future research in bioarchaeology.

II. T. DALE STEWART

Hrdlička retired in 1942 and died the following year. He was succeeded at the Smithsonian by his long-time assistant T. Dale Stewart (1901–1997). Like his predecessor, Stewart held a medical degree (Johns Hopkins, 1931), but from the beginning maintained a distinct skeletal focus (Stewart, 1930, 1931b). Stewart also encouraged the assembling of well-documented collections and published research in paleoanthropology (Stewart, 1959a,b, 1960a, 1961a,b, 1962a,b,c, 1963c, 1964) and other areas of anthropology (Stewart, 1953, 1954a). In contrast to Hrdlička, Stewart published regularly on paleopathology

(Stewart, 1950a, 1966, 1969, 1974, 1979b, 1984a; Stewart and Quade, 1969; Stewart and Spoehr, 1952; Tobin and Stewart, 1952) and forensic anthropology (McKern and Stewart, 1957; Stewart, 1948a, 1954b, 1959c, 1968, 1970a, 1972, 1973a, 1978, 1979a,c,d, 1982, 1983, 1984b; Stewart and Trotter, 1955), emphasizing the importance of collections in this research. By improving storage and accessibility to collections, he increasingly made them available to outside researchers, enabling them to include Smithsonian collections in their own research designs. In the area of bioarchaeology, Stewart routinely analyzed human remains at the request of archaeologists and collaborated in studying direct cultural effects on the skeleton, such as cranial deformation (Stewart, 1939a, 1941a, 1948b) and intentional dental alterations (Stewart, 1941b, 1942; Stewart and Titterington, 1944, 1946). Like many of his colleagues of that time, Stewart tended to publish the results of his studies of remains from archaeological excavations as appendices of the archaeological reports (Stewart, 1940b,c, 1941c,d, 1943a,b, 1950b, 1951a, 1959d,e,f). However, his work included bioarchaeological investigation of ossuaries in the vicinity of the Smithsonian (Stewart, 1939b; 1940d,e, 1941e, 1992; Stewart and Wedel, 1937) and utilizing results of skeletal studies to address larger issues of population history (Stewart, 1973b). Stewart's chapter in the saga of Smithsonian physical anthropology also demonstrates intellectual movement away from an emphasis on racial typology and classification of head shape toward problem-oriented detailed research.

When Stewart became a museum director in 1962, the Smithsonian hired J. Lawrence Angel (1915–1986) from the Daniel Baugh Institute of Anatomy of the Jefferson Medical College in Philadelphia (St. Hoyme, 1988; Ubelaker, 1989) (Fig. 2). Angel received his Ph.D. in 1942 from Harvard where he had worked extensively with Earnest Hooton (1887–1954). Like Hrdlička, Hooton's interests were broad and included research into racial topology and the biological basis of criminal behavior. However, Hooton also published in skeletal biology. As noted in Chapters 2 and 4, his classic work, *The Indians of Pecos Pueblo* (1930), demonstrates an unprecedented intellectual interplay between skeletal analysis and archaeological observations that sets the stage for more recent studies of bioarchaeology.

Unlike Hrdlička, Hooton's long tenure at Harvard generated many students who in turn greatly influenced the development of American physical anthropology. Hooton's 19th Ph.D. student was J. Lawrence Angel. Working mostly in the eastern Mediterranean area, Angel expanded on Hooton's ideas and methodology in bioarchaeology. Although Angel published a number of site reports and appendices, he also demonstrated how data amassed from such works could be used to address key anthropological issues of paleodemography and correlations of disease and culture, which he termed "social biology." From his work emerged a sense that physical anthropologists involved in the excavation and analysis of

Figure 2 T. D. Stewart and J. L. Angel with a portrait of Angel painted by Stewart.

human skeletal remains cannot only provide useful data to the archaeologists and use the samples in studies of human variation and paleopathology, but can also directly address broader anthropological issues.

Note that throughout his career at the Smithsonian, Angel worked just down the hall from Stewart. When Stewart returned to the Department of Anthropology in 1966 from his duties as director of the Museum of Natural History, he began a long period of research and writing, largely free of administration. This period also overlapped the career of skeletal biologist Lucile St. Hoyme, whose research included issues of bioarchaeology (Hunt, 2004).

Marshall T. Newman (1911–1996) worked at the Smithsonian in physical anthropology from 1941 to 1942 and then again between 1946 and 1962. Newman received his Ph.D. in 1941 from Hooton at Harvard, but left the Smithsonian to expand his teaching experience.

The summers of 1956 through 1959 also found William M. Bass working at the Smithsonian for the River Basin Surveys. During this time Bass conducted laboratory research in Washington and supervised mortuary site excavations in South Dakota. Bass pioneered bioarchaeology in the Plains and taught many students who have become leaders in this area of research.

III. CURRENT ACTIVITY IN BIOARCHAEOLOGY

The current Division of Physical Anthropology at the Smithsonian's Department of Anthropology of the National Museum of Natural History maintains a strong focus in areas of physical anthropology relating to bioarchaeology, although other areas of physical anthropology, such as population genetics and growth and development, are not well represented. This area of emphasis reflects hiring practices that have recognized the value and continued needs of the collections as well as areas of traditional strength.

Present staff of physical anthropology in the Smithsonian's Department of Anthropology pursue research that combines archaeological technique and interpretation with the specialized anatomical knowledge of skeletal biology. This translates into more precision in measurement and disease diagnosis than was possible just a few decades ago, coupled with sophisticated integration with archaeological information aimed at anthropological interpretation. This work is possible because of the collections assembled by past workers with different problem orientations and because of the changing methodology of the field at large. The diversity of activity during Hrdlička's time has been sacrificed in favor of more intense, detailed effort within the areas represented. Smithsonian skeletal biologists have managed mortuary site excavation and analysis with the aim of maximizing the amount of information retrieved in field recovery. Laboratory analysis enables information about disease, demography, stature, and other biological attributes to be correlated with site information. This research is consistent with that of colleagues throughout skeletal biology who also integrate mortuary site excavation information with that derived from laboratory analysis of human remains.

IV. REPATRIATION

Back in 1918, Hrdlička called attention to the special nature of human remains and how public sentiments about them can dramatically affect collections and related research. He noted:

> The difficulties in gathering the requisite material, and even the crude data alone, have been and are still very great; in fact they are sometimes insurmountable. Religious beliefs, sentimentality and superstition, as well as love, nearly everywhere invest the bodies of the dead with sacredness or awe which no stranger is willingly permitted to disturb. It is seldom appreciated that the remains would be dealt with and guarded with the utmost care, and be used only for the most worthy ends, including the benefit of the living. The mind of the friends sees only annoyance and sacrilege, or fears to offend the spirits of the departed. This may not apply to older remains, but these in turn are frequently defective; yet even old remains are sometimes difficult to acquire. . . . (Hrdlička, 1918:11)

Hrdlička likely would be shocked to learn just how far those sentiments recently have gone to shape bioarchaeological research. As also discussed in Chapter 15, legislation and policy formation have not only limited the acquisition and study of human remains, but have forced the transfer and loss to science of large collections of North American human remains of archaeological origin that already had been curated. United States Public Law 101-601, Native American Graves Protection and Repatriation Act, Hawaiian Natives, Historic Preservation, H.R. 5237, 25 USC 3001, Nov. 16, 1990, addresses human remains, associated funerary objects, unassociated funerary objects, sacred objects, and objects of cultural patrimony that can be culturally affiliated with a present-day Indian tribe or Native Hawaiian organization. Upon request, such materials must be transferred to the appropriate group.

Although the Smithsonian Institution was exempted from the NAGPRA law summarized earlier, it was targeted by another similar law, the National Museum of the American Indian Act, Public Law 101-185, Nov. 28, 1989, 103 Stat. 1336, 20 USC 80q. This legislation requires the Smithsonian to identify the tribal origins (cultural affiliation) of human remains and funerary objects in its collections and, if requested, transfer them to the appropriate group.

Responding to federal legislation, the Smithsonian's National Museum of Natural History formed an Office of Repatriation. A large staff is employed to assess the collections to determine which represent those factors targeted by legislation. Many of the human skeletal remains originating from archaeological sites within the United States are potentially affected, including material that Hrdlička collected and studied. A physical anthropology component of the laboratory collects standard information from the human remains in order to help determine the cultural affiliation and to salvage scientific information. Although many of these collections likely will be unavailable for future analyses, in the short term, the issue has forced attention to those remains, producing data collected in a standard format that may enable enhanced synthetic biocultural interpretation.

V. SUMMARY

The history of American activity in bioarchaeology research has recorded major changes and shifts of interest. In the 19th century, ancient mortuary sites were generally regarded by physical anthropologists as resources to be mined for comparative collections. These collections were desperately needed to document human variation and to test medically oriented theories. The Smithsonian's Hrdlička was initially attracted to archaeological mortuary sites, not to understand ancient ways of life but to obtain the "normal" sample for his comparative studies of the biological basis of human behavior. Gradually, as he became involved in the necessary fieldwork, he became intellectually involved in the problems presented by the sites and the collections themselves.

Largely through the work of Hooton and his students, bioarchaeology evolved with the understanding that skeletal analysis could be coupled with mortuary site excavation to reach a greater understanding of past human populations. At the Smithsonian, this effort was championed by one of Hooton's students, J. Lawrence Angel, especially through his work in the eastern Mediterranean. Also at the Smithsonian, T. D. Stewart demonstrated how careful research design, an attention to detail, and a problem orientation could enhance diagnosis of disease from bone and bioarchaeology research in general.

The 20th century also witnessed remarkable developments in the recovery and curation of human remains of archaeological origin. Through the early efforts of the Works Progress Administration (WPA)-sponsored archaeological projects, the Smithsonian-affiliated River Basin Surveys, and other archaeological investigation, well-documented human remains from archaeological contexts were assembled and available for research. Much of this material was deposited in the collections of the Division of Physical Anthropology of the Smithsonian because of the federal status of the Smithsonian and its traditional interest in such materials. Physical anthropologists such as William M. Bass not only increased cooperation with archaeologists in the excavation of human remains, but were available to excavate them directly.

By the 1970s, collections of well-documented human remains were available for research and of such size and documentation that remarkable research was possible on ancient biocultural patterns. It appeared that Hrdlička's dream of adequate comparative collections would finally be realized.

However, the 1970s also witnessed an increase in concern on the part of contemporary American Indians and others about the appropriateness of maintaining those collections (Ubelaker and Grant, 1989). This concern led to law and policies that have forced a transfer of aspects of those collections to contemporary groups.

Despite these developments, research in bioarchaeology remains strong and increasingly synthetic and interdisciplinary. The Smithsonian Institution continues involvement in bioarchaeological issues not only by meeting the challenges of the repatriation legislation, but through vigorous research aimed at a greater understanding of past populations.

ACKNOWLEDGMENTS

I thank Erica B. Jones for her assistance in manuscript preparation. Both illustrations are provided courtesy of the Smithsonian Institution.

Chapter 4

Kidder, Hooton, Pecos, and the Birth of Bioarchaeology

Lane Anderson Beck

I. INTRODUCTION

Bioarchaeology, put simply, is the contextual analysis of human populations from archaeological sites (Buikstra, 1977). It uses skeletal biology and archaeology in combination to ask questions not about how people died, but about how they lived. It does this through focusing on the osteobiography of individuals and the biocultural adaptations of populations as viewed through the lens of archaeological context. Although use of the term bioarchaeology is relatively recent, the precepts of the field have deep roots in American archaeology.

In 1930 a report by E. A. Hooton on the people of Pecos Pueblo was published. Hooton's emphasis on the analysis of human remains in reference to their archaeological context emerged, not as a tentative step, but as mature, integrative form of analysis. This project, when examined in detail, reveals a partnership between A. V. Kidder and E. A. Hooton as pioneers in developing an integrated, interdisciplinary perspective on the past. As Schwartz (2000:19) emphasizes, "Hooton's work on the human remains was significant, for Kidder was realizing that the only way he was going to obtain the essence of the settlement's cultural development was by using insights from a wide range of other social, natural, and environmental disciplines. This multidisciplinary approach to his archaeology became a centerpiece of Kidder's research design."

II. THE PECOS EXCAVATIONS

In 1915 the Department of Archaeology of Phillips Academy, Andover, began excavations at Pecos Pueblo, carried out under the direction of A. V. Kidder.

Bioarchaeology: The Contextual Analysis of Human Remains, Buikstra and Beck (eds.)

As the site had been occupied continuously for a period of several hundred years, one of Kidder's primary objectives was to identify temporally sequential cultural units through the analysis of ceramic and stratigraphic data (Kidder and Kidder, 1917; Kidder, 1924; Hooton, 1930). His success in this endeavor remains a landmark in the history of American archaeology. Using his chronological sequence, Kidder assigned temporal associations to over 2000 Pecos burials, enabling Hooton to investigate changing patterns in demography and disease over time. This is perhaps the largest series ever recovered from a scientifically excavated, stratified site in the New World (Kidder, 1924). Early in his excavations of the middens at Pecos, Kidder actively sought burials:

> Some human bones had been found on the surface, and a few had come from the digging. We were most anxious to discover burials; so a reward of twenty-five cents was offered to the workmen for every skeleton uncovered. The next day one appeared, the following day six; the reward was reduced to ten cents; this brought fifteen more, and in the course of a week or so we were forced to discontinue the bonus or go into bankruptcy. The higher we got uphill the deeper grew the rubbish and the more crowded became the skeletons. (Kidder, 1924:94)

As the second season at Pecos began, Kidder discovered that burials were not limited to the midden areas on the sides of the mesa, but were also located throughout the mesa top, amid all the structures. As the number of interments expanded, Kidder recognized that he needed a physical anthropologist to step in and assume responsibility for burial analyses. Kidder believed that an osteologist must begin analysis in the field and not merely wait in the laboratory for burials to arrive (Kidder, 1924). As a result, he arranged to have Earnest Hooton, the physical anthropologist for this project, assist in excavation as well as in laboratory analysis. Hooton joined the field crew for 2 months during the 1920 season (Hooton, 1930).

Skeletons were subsequently shipped from Pecos to Boston. One of the first shipments was mistakenly delivered to the Peabody Museum rather than to the warehouse. During delivery, the crates were tossed from the truck, down the steps, and into the museum's basement. Many of the crates broke open. Perhaps this circumstance led to Hooton's complaints to Kidder about recent damage to the bones. Kidder reports that Hooton felt:

> . . . I had been kicking about my skeletons. He said they had an awful lot of fresh breaks on the bones. In the Southwest, a bone will often crack and not come apart. When you take it out, it comes in two pieces and, it looks like a fresh break but, it isn't a fresh break. I tried to explain that to Earnest, but that didn't do any good, so I said, "You come out and dig some skeletons yourself." So he did. Then I discovered that Earnest had done practically no excavation at all. He had worked a little in a long barrow and then he had been to the Canaries and worked in a cave but as far as digging skeletons, in bad conditions, he knew very little about it. He would clean out a long bone and put his knife under it and pry and the damn thing would break. It was very interesting having him there, because he gave us a lot of information about

> the age of children, the dentition, and he made out a whole lot of tables for us, of one sort or another. (Givens, 1992:141)

The tables referred to here were slates of standards for the determination of age and sex. Both Hooton and Kidder state that the in-field assessments for the later years at Pecos became very close to matching Hooton's analysis in the laboratory (Hooton, 1930).

Just as Kidder's excavations at Pecos are of major significance in the development of American archaeology, so is Hooton's analysis a landmark in American physical anthropology. Hooton was among the first to explicitly use archaeological context as a guide to the questions he asked. This enabled him to raise intrasite research inquiries rather than being limited to total sample as the unit of analysis. Speaking in 1935, Hooton described the Pecos project as a turning point in physical anthropology that allowed research to go beyond a "mere description of bones" and facilitated studies of change within a population. He went on to speak of "the necessity of an intimate cooperation of the archaeologist with the physical anthropologist" (Hooton, 1935:503):

> In the pre-war period the first research efforts of a physical anthropologist attached to a museum were likely to be studies of skeletal remains deposited by archaeologists as a result of their excavations. The job of the physical anthropologist was to describe these remains and to make some sort of a racial diagnosis. Usually the archaeologist prepared and published his report without any reference to the skeletal finds. Most were so conscious of their virtue in preserving the bones that they considered their scientific responsibilities fully discharged when the skeletons had been dumped in a museum. The present writer undertook several such tasks, mostly relating to the bones of American Indians. From them he learned the folly of dissociating excavation reports from the study of the skeletal material which they produce. One example will suffice. The Peabody Museum excavated a large Indian cemetery at Madisonville, Ohio, in spasmodic efforts beginning in 1882 and ending in 1911. It devolved upon this unfortunate to study the bones. In order to make such a study intelligible, he was forced to spend an entire summer struggling with the field notes and records of three generations of archaeologists who had worked the site. He had to patch together by collation and speculation some sort of consecutive account of the excavations. All evidence as to the relative ages of the different portions of the cemeteries had been lost, and stratigraphy was absent or unrecorded.
>
> The physical anthropologist had to content himself with a consideration of the remains as of one period. Apart from the mere description of the bones, the only advance in anthropological method resulting from this effort was a fairly successful attempt to deduce the size of the population and its probable annual death rate from an examination of the proportions of each age and sex represented in the skeletal material. (Hooton, 1935:501)

Pecos was the first major archaeological sample to be so fully studied. The quality of the field notes combined with the good preservation of the human remains enabled Hooton to apply contemporary, standard approaches to skeletal analysis and to pioneer new methods. He combined demography, pathology, morphological, and metric data to examine changes in a community over time — time as defined by Kidder's work on the archaeological context.

In his analysis of the people from Pecos, Hooton departed from the mainstream of skeletal biology. Although he did measure skulls and generate typologies, he did not stop there. Working closely with data generated by Kidder and his own observations made in the field, Hooton was among the first to seriously examine questions from the perspective of the archaeological context. Instead of focusing on the site as the generalized, single unit of analysis, Hooton subdivided the sample, utilizing Kidder's chronology, and asked questions about how a human community had changed over time. In the 1920s this was not a routine procedure but instead a highly innovative approach.

III. LABORATORY ANALYSIS OF THE PECOS COLLECTION

Rather than simply saying that Hooton's publication on Pecos Pueblo is a landmark study, one should take a detailed look at just what this report includes and how that relates to the scientific foundations of its time. All laboratory observations of the Pecos skeletons were made by Hooton, with the exception of cranial capacity estimates, which he assigned to two of his assistants.

Hooton begins the Pecos volume with a report of the excavations that summarizes their extent at the time of publication and provides an overview of the significance of the site and the work being done there (Chapter 1, pp. 3–13). He explicitly states that the work is still ongoing and that this analysis includes only the burials excavated by the end of the 1924 field season.

The first analytical portion of the Pecos report deals with post-depositional changes, "state of preservation" (pp. 14–15), which we would today term taphonomy. Hooton clearly recognized that the demographic pattern of the recovered burials was somewhat skewed and discussed the possibilities that such factors as age and sex of the deceased, as well as microenvironmental factors, could have led to differential bone preservation. He emphasized that the bones of infants and young children were less fully ossified than those of adults and that the bones of the elderly may be relatively thin and more porous than those of other adults. Presaging more recent studies of taphonomy and demographic bias (e.g., Walker *et al.*, 1988), he also noted that women's bones are generally smaller and lighter than those of men. He further reported that certain features of specific graves may alter the patterns of preservation and that the drainage of different types of soils and microenvironments created by burial associations can create situations of better or worse preservation. This is a very early report of taphonomy and its implications for burial analysis (see also Wilder, 1923).

Following these cautionary notes on potential biases, Hooton proceeded to reconstruct demographic patterning (pp. 16–32). He began with the estimation of a mortality profile. At the date of this research, the first detailed standards for assessment of age-at-death from skeletal remains were just beginning to emerge. For example, T. Wingate Todd at Washington University in St. Louis was assembling an anatomical collection that included skeletal remains with documented sex and age-at-death. From this work, Todd proposed a series of pubic symphyseal phases that could be used to estimate age-at-death (Todd, 1920). Hooton contacted Todd to assist with the Pecos Pueblo analysis. Todd provided Hooton with a series of photographs that illustrated the age changes in the pubic symphysis (Hooton, 1930:21).

In addition to his own assessment of age-at-death and sex, Hooton also arranged for Todd to personally assess age and sex for the remains from Pecos (Hooton, 1930:18; Todd, 1927:494). Field estimates had also been recorded on each burial feature form. In analyzing age and sex data for Pecos, Hooton compared the three profiles, thus explicitly addressing the issue of interobserver error, a pioneering effort (Hooton, 1930:18).

Following his overview of paleodemography, Hooton compared the Pecos mortality profile to national death rates from various countries (pp. 24–25). Although it may seem odd that Hooton chose to compare Pecos to European data, it must be remembered that this was the first study of a large sample from a single archaeological site. There were no other well-provenienced North American series with which Hooton could compare Pecos. As a result of these and other analyses, Hooton emphasized juvenile underenumeration, which he interpreted as an artifact of the archaeological context rather than as a measure of community health (p. 24).

Next Hooton assessed cranial deformation patterns (pp. 33–39). He first categorized cranial deformation by form and degree and then examined temporal sequences for systematic changes over time. Patterning was interpreted in terms of ethnographic reports on cradleboarding as well as studies of infant behavior. He concluded that variability in form and degree of cranial deformation resulted from an interaction of infant behavior and skull shape. In his scenario, doliocephalic infants tended to rotate their head slightly toward one side, whereas brachyocephalic infants are more likely to lie flat. Hooton also proposed that the greater tendency toward flattening of the right side of the back of the skull in doliocephalic infants was related to handedness and a tendency of the infant to face toward its dominant side (p. 38).

Having evaluated the degree and form of cranial deformation, Hooton was able to begin his analysis of craniometric data (pp. 38–78). In order to include as many skulls as possible, Hooton had one of his graduate students, Harry Shapiro, statistically evaluate measurements of the deformed and undeformed crania from Pecos. Through correlation and regression analysis, Shapiro developed a correction formula that facilitated analysis of all measurable crania. Hooton generally reported measurements for both deformed and undeformed separately due to what he viewed as the probable imprecision in the correction formula (pp. 38–39).

Following 19th-century traditions, Hooton reported copious amounts of craniometric data. While a century earlier Morton (1839) had developed a suite of 10 cranial measurements for his research, Hooton used an expanded list of 29 measurements and 9 calculated indices. For each sex, means and standard deviations are reported for the total series and for the four temporal groups defined by Kidder's field records. Within each temporal division, the measurements were subdivided into deformed and undeformed categories, resulting in a total of 61 tables and 35 plates that illustrated typical patterns. Hooton reported only summary statistics because he planned to publish a supplement containing all raw data. He concluded that the earliest population samples at Pecos were more variable than the later occupants of the pueblo.

The volume of craniometric calculations is remarkable, with the total number of summary statistics involving nearly 1000 data sets. In 1935, 2 years before the first computers became available, Hooton reported that he had just purchased several electronic calculators to facilitate his biostatisitics work at Harvard (Hooton, 1935). Thus, the Pecos statistics, generated during the 1920s, must have been calculated by hand. Furthermore, the remaining chapters in the book often contained large numbers of calculations. The table of contents describes 362 statistical tables, in addition to 26 figures and 97 photographic plates. While there was no attempt to apply multivariate approaches, such as the newly available coefficient of racial likeness (Pearson, 1926), Hooton's emphasis upon making his data available to others set a high standard for future studies.

Following his section on skull measurement, Hooton presented information on cranial morphology (pp. 80–132). He observed 31 morphological characteristics, scored by form and degree of expression. These were then sorted by sex and time. Dental variants were also included: molar cusp variation for both arcades, degree of incisor shoveling, and a variety of other dental anomalies. Also within this section were data on dental wear, dental eruption, caries, abscesses, and antemortem tooth loss.

Some morphological attributes included features we now use for other purposes, such as the degree of dental wear. Others are morphological variables standardized today for estimation of sex, such as size of the mastoid process, while some are among those included in listings of discrete traits generally collected today for biological distance analyses. As with the measurements, Hooton provides detailed statistical examination of each feature as reported in 1 or more of 66 tables. He also noted that there is a "certain spurious correlation" in certain traits due to the traits sometimes being those used for identification of sex (p. 131), an issue we take seriously in statistical analyses today (e.g., Konigsberg and Buikstra, 1995).

Continuing his detailed analyses, Hooton next turned from the skull to the postcranial skeleton (Chapter 5, pp. 133–184). He presented data on a bone-by-bone basis, with metric data and morphological observations both reported. For example, Hooton reported seven femur measurements and two indices, sides reported separately and partitioned by sex and temporal assignments. He reported changes over time, with the earliest groups being larger than the later groups. He also commented on patterns of bilateral asymmetry overall and in comparisons of male and females.

Morphological features of the postcranial skeleton are also reported. For example, six sets of morphological data characterizing the femur are presented and partitioned. Data range from description of the size of the linea aspera to observations on squatting facets. He noted that squatting facets are more common among males than among females and that the frequency in females diminishes over time while that for males is stable. After completing his report on the femur, including 15 summary tables, Hooton proceeded to offer similar analysis for the tibia, fibula, humerus, radius, ulna, clavicle, scapula, pelvis, and lumbar portion of the spinal column.

Calculations of stature followed Hooton's discussion of morphology (p. 178). Employing a weighted combination of femur and tibia lengths, Hooton used Pearson's formulae as the basis for his estimates, with males and females considered separately. Recognizing the limitations of these formulae when applied to Native American contexts, he argued that, even so, they were the best method available at that time. Results are reported for the pooled sample, as well as for the four temporal divisions partitioned by sex. While noting a slight decrease in stature for the more recent samples, Hooton remarked that the change was slight and may have been an artifact of sampling bias. He then compared the Pecos stature estimates to those for a variety of living populations and observed that the people of Jemez Pueblo were very close to estimates for Pecos (pp. 178–180).

Following this analysis of osteometric and morphological data, Hooton was still dissatisfied with the high degree of heterogeneity he found at Pecos and his inability to identify consistent temporal trends. At this point he shifted his reference point from the archaeological record to the morphological (Chapters 6–8, pp. 185–288).

> Nevertheless it is apparent to the craniologist that the skeletal population of Pecos was at no time markedly homogeneous in type. On the contrary the handling and measurement of the Glaze subgroups leaves the impression of a number of markedly diverse cranial types, found in varying proportions at all periods. Therefore upon the conclusions of the study of period groups it was decided to reanalyze the material, relying upon morphological rather than upon archaeological criteria for the differentiation of groups. (Hooton, 1930:185)

In developing this approach, Hooton attempted to move beyond the comparisons typically made at that time and to generate a more complex method, one approaching multivariate analysis.

To establish his morphological types, Hooton chose 129 skulls of adult males. He placed them on a laboratory table and then grouped them visually into homogeneous subsets. His final sorting resulted in eight morphological types, which he named according to what he saw as gross patterns of distinctive features (pp. 185–186). If he had simply referred to these as morphological groups rather than types and had numbered them rather than naming them, this aspect of his work would be less vulnerable to recent attributions of racism (Armelagos *et al.*, 1982; Armelagos and Van Gerven, 2003).

While Hooton's language concerning race and morphology is very much a product of its time, he emphasized that the names he assigned his groups were somewhat arbitrary. In naming his first, "Basket Maker" type, he notes "a general resemblance, perhaps fancied, to the veritable Basket Makers of the Arizona caves" (Hooton, 1930:185). He then emphasized that the "second type was styled 'Pseudo-Negroid,' not because of any theory of the observer as to the presence of a Negroid strain in the American Indian" but rather because of certain features reminiscent of those typically attributed in those days to individuals of African descent (Hooton, 1930:185). Following statistical validation of his types and in the context of alternative models for peopling of the New World, Hooton does, however, link type resemblances to heredity. He argues, for example, in a discussion of the "sequence of immigrant types, especially at Pecos" that the presence of the "Pseudo-Negroid Type" was due to an ancient admixture prior to the migrations from Asia (p. 356). He further states that the craniometric validation of morphological types reflects "the segregation of features in occasional individuals," not the migration of distinctive groups from homeland regions (p. 357). His was clearly a typological approach uninformed by population genetics (see Chapter 2, by Cook, this volume).

After creating these types through visual sorting, Hooton then attempted to test their validity through observations of cranial deformation and craniometric comparisons. He thus concluded that six of his eight groups are internally more homogeneous than any of the temporal segments described in earlier chapters. One of the remaining two groups included some burials from the Pecos church, which are probably Spanish rather than Indian. He also compared stature across these types and again found homogeneity greater than that for the temporal divisions. As always, Hooton attempted to make his data fully available to the reader and included 31 photographic plates to illustrate his types. From these photographs, it is clear that several of the crania he chose to illustrate appear to be female rather than male, e.g., plates VI-4 and VI-6. Given that these fall primarily within a single type, coupled with the fact that stature sorted well across types, it would appear that a key variable in Hooton's typology was relative size — a visual parallel to the first component of principal components analysis.

Having confirmed these types through craniometric analysis, Hooton then considered variation in cranial morphology and chronology. Three of his types dominated the earliest time periods and were absent or relatively underrepresented in the later phases. Two types reverse this trend and are dominant in the later periods. One group is most common in the middle periods. The final two groups are consistent in frequency over time. In other words, Hooton's typological analysis did identify temporal patterning that was not otherwise evident.

After a comparison of Pecos craniometric data with that from other sites, which proved inconclusive, Hooton turned to intrasite correlations of metric and morphological data (Chapter 9, pp. 289–305). He reported, for example, that dental wear and abscesses were highly correlated. Such close examination of correlation among variables did not become prominent in osteological studies until after the rise of computer-aided multivariate analysis during the second half of the 20th century. Hooton was again decades ahead of the profession.

The next chapter, which focuses on paleopathology (Chapter 10, pp. 306–330), is most commonly cited as Hooton's landmark contribution to bioarchaeology (Jarcho, 1966a; Ubelaker, 1982). Prior to 1930, paleopathology tended to focus on diagnoses of obviously deformed remains, often divorced from archaeological contexts. Hooton's study was both sensitive to population dynamics and the archaeological context. Both Jarcho and Ubelaker argue, for example, that Hooton's Pecos Pueblo report is the precursor to later paleoepidemiological approaches (Jarcho, 1966a:22; Ubelaker, 1982:342). For each category of disease that he or his medical collaborators identified, Hooton discussed the frequency with which it was observed and also discussed the pattern of its distribution in the Pecos sample in terms of age, sex, and temporal horizon. He also consulted on individual cases with at least eight medical doctors, including both radiologists and pathologists. Table X-11, which summarizes the pathological observations, was in fact compiled by Dr. G. D. Williams, with supplemental notes from other physicians (Hooton, 1930:305).

On page 305, Hooton indicates that he took notes on diseased bones as he conducted his morphological and metric surveys. He indicates that these notes will appear in Appendix III, but unfortunately there is no such appendix in the volume.

Hooton divided his discussion of pathology into four gross categories: arthritis, inflammatory lesions, trauma, and miscellaneous. Within the category "arthritis," he separated spinal osteophytosis from degenerative joint disease. He noted that the frequency for both forms was somewhat lower in the earliest time periods and that the majority of the cases involve older adults. He also pointed out that many of the cases of arthritis in the long bones are associated with fractures.

Under the heading "inflammatory lesions," Hooton discussed periostitis and osteomyelitis in both the cranial and the postcranial skeleton. Three cases of possible syphilis were submitted to two physicians, James Ewing of Cornell Medical College and H. U. Williams of the University of Buffalo. Ewing said none of the cases were syphilis, while Williams said all three cases were probably syphilis. Hooton concluded that the evidence was inconclusive. One of the cases is from an apparent historic context.

Hooton reported that postcranial fractures were most common in the femur and humerus. The frequency was greater in older people and occurred on the right side more often than on the left. He found the frequency of cranial trauma highest in mature to older adult males. Fractures of any sort occurred more frequently in the most recent time period. Hooton concluded that the pattern suggested injuries to males in conflict and that the recent intervals might reflect an increase in violence (p. 315).

Hooton's classic discussion of the condition he called "osteoporosis symmetrica" appeared under the heading of miscellaneous pathology. He also included cribra orbitalia as an early and milder stage of the disorder and claimed that it was clear that whatever produced these lesions was limited to childhood and adolescence. His description of this condition is worth repeating.

> The typically honeycombed condition which involves a hyperostosis of the diploe and a destruction of the external table of compact bone, frequently extends to the parietals, where it may be seen in symmetrical patches sometimes extending over the greater portion of both bones and causing a thickening of ten to fifteen millimeters in the middle portion of the area affected. On the base of the skull it may be observed in the form of numerous small pits on the palatine roof and on the wings of the sphenoid. Traces of the condition may also be observed in some crania on the temporal bones just above the auditory meatus. The most pronounced osteoporotic conditions, in my experience, are found in the crania of immature subjects from ancient Peru and from the Sacred Cenote of Chichen Itza in Yucatan. (Hooton, 1930:316)

He also provided further details on active versus inactive conditions, their general appearance, and their radiographic attributes. One point that is often overlooked is that most of Hooton's discussion of osteoporosis symmetrica does not focus upon remains from Pecos Pueblo. The crania he submitted to the doctors for radiographic analysis and morphological examination were instead from Chichen Itza, a Maya site in the Yucatan Peninsula. At Pecos Hooton felt that he saw traces of this condition, which appears in a more extreme expression in the Cenote collection (p. 316).

Within his discussioin of osteoporosis symmetrica appears information seldom acknowledged by subsequent scholars. He reported that Dr. Percy Howe of Forsyth Dental Infirmary showed him crania from monkeys fed on a scorbutic diet and that the crania of these monkeys exhibited a pattern highly similar to that in the Cenote collection (p. 317). He noted that the condition is restricted to bones that form intramembranously rather than cartilaginous bones. He also pointed out that the lesions were restricted to the diploe and outer table of the skull in regions outside areas of muscle attachment. His medical consultants both suggested that, if dietary, the condition looks most like scurvy or rickets but that it most closely resembles congenital anemias that we today would call thalassemia. On returning the discussion from Chichen Itza to Pecos, Hooton noted that the frequency there was approximately 3%.

Other lesions noted in the Pecos collection included Pott's disease (tuberculosis) in one historic and one pre-contact set of remains. A case of cancer from the earlier part of the archaeological sequence is also noted in a young adult female.

The final analytical chapter in the main body of the Pecos report reconstructs changes in population dynamics over time (Chapter 11, pp. 331–343). In so doing, he explicitly divided his discussion into known, deduced, and unknown factors. Known factors included population estimates recorded in the historic literature and the number of burials excavated for each ceramic period. The two deduced variables involved chronology: the founding date for Pecos and the chronology for the ceramic sequence. The final group of variables Hooton described as "unknown factors." These included the annual death rate and the percentage of total burial area that was excavated. He then presented various calculations for estimating population size at different times in the pueblo history that most closely match his known factors. His critical review of this work was once again expressed with total candor.

> In this chapter I have built up a house of cards. The assumptions made and the methods employed are all questionable, perhaps erroneous. The reader need not attach much importance to this effort, nor rely at all upon its conclusions. I have merely attempted to reach a plausible solution of an impossible problem. (Hooton, 1930:340)

At the end of the complex chapter on population estimates, Hooton and Kidder each inserted an addendum (pp. 342–343). Tree ring dates by A. E. Douglass had just arrived, which reduced the duration of occupation at Pecos by 300 years. Hooton found this reduction astounding. His original estimates had focused on historic reports of a community of over 2000 at Pecos and he felt that the reduced time frame elevated this number to an unreasonable figure. He acknowledged that part of the problem was finding a way to deal more precisely with issues of growth.

The summary chapter for the Pecos book provided an encompassing overview (Chapter 12, pp. 344–363). Hooton cautioned the reader about potential imprecision and then turned toward the larger issues of what Pecos revealed about peopling of the New World. Here he drew on contemporary data about early man in the New World, as well as recent studies of blood group distributions. He concluded that humans arrived in the Americans not by one migration but by several, with the Eskimo being the most recent arrival before Euro-Americans.

IV. HOOTON AND PECOS TODAY

Far beyond the case studies or summary reports of that era, Hooton asked questions about populations and how they change over time. He used every method available to him and presented innovative approaches to the study of population health and demography. The labels Hooton used for his morphological types have received considerable recent criticism, even though Hooton himself regarded these labels as somewhat arbitrary.

Hooton's analysis of Pecos remains today one of the most comprehensive presentations of bioarchaeology ever generated. The breadth of data reported, the statistical comparisons both internally and externally, and the variety of experts consulted all form a model for analysis that has seldom been achieved. That his interpretations would be modified by the application of more refined observations or multivariate analysis of his original data does nothing to diminish the tremendous contribution of this landmark research.

Chapter 5

Hemenway, Hrdlička, and Hawikku: A Historical Perspective on Bioarchaeological Research in the American Southwest

Gordon F. M. Rakita

I. INTRODUCTION

> Dr. Matthews went to Los Muertos in the month of August, 1887. He found that no attention had been paid to the collection or preservation of human bones, which were extremely fragile, crumbling to dust upon a touch, and which had been thrown about and trampled under foot by curious visitors, so that but little remained of value from the work which had been previously done. (Billings, cited in Matthews *et al.*, 1893:141)

Dr. John S. Billings, surgeon with the United States Army Medical Museum, thus described how one of the first explorations of prehistoric remains in the American Southwest, Frank Hamilton Cushing's Hemenway Expedition (Matthews *et al.*, 1893), developed its bioarchaeological component (see Chapter 1). Then, as now, the analysis of prehistoric human burials was frequently an afterthought for many researchers working in the Southwest. Unfortunately, Buikstra's (1991:174) call for "[m]utually designed research strategies" between archaeologists and biological anthropologists still remains for the most part "an elusive goal" in this region.

There is reason for cautious optimism, however. Throughout the past century, a few intrepid researchers have maintained an interest in uniting archaeological

Bioarchaeology: The Contextual Analysis of Human Remains, Buikstra and Beck (eds.)

and biological data in the examination of indigenous groups of the desert west. Within recent decades, a number of integrated projects have greatly expanded our understanding of the prehistoric peoples and cultures of this important region. These projects have provided not only models for how such integration might be accomplished, but also examples of the rich intellectual rewards that result.

This chapter reviews key events and periods in the history of bioarchaeological work in the North American desert west. The focus is on research conducted within Arizona and New Mexico, occasionally extending to contiguous regions of the United States and Mexico. The chapter begins with the early history of regional research and then moves to a description of more recent developments. It ends with a discussion of possible future directions that are open to southwestern bioarchaeologists.

II. THE HEMENWAY EXPEDITION AND THE U.S. ARMY MEDICAL MUSEUM (1886–1888)

The Hemenway Southwestern Archaeological Expedition directed by Frank Hamilton Cushing represents the first organized research on southwestern prehistory that explicitly included physical anthropologists as part of a preconceived attempt at interdisciplinary research. The impetus for the project came from Cushing's research at the Pueblo of Zuni. During the course of his ethnographic work among the Zunis, Cushing's informants claimed that their ancestral roots lay to the Southwest of the region they currently inhabited. Consequentially, Cushing undertook excavations at several ruins in the Salt River valley near Tempe, Arizona, with the view of confirming these claims. In the fall of 1886, with financial support from Mary T. Hemenway, a Boston philanthropist, he assembled his multidisciplinary team (Hinsley and Wilcox, 1996; Haury, 1945). He arranged for the participation of a historian (Adolph F. Bandelier), an artist (Margaret Magill, his sister), a publicist (Sylvester Baxter), an expedition secretary (F. W. Hodge), a topographer (Charles Garlick, of the U.S. Geological Survey), an archaeologist–ethnologist–linguist (Cushing, himself), and a physical anthropologist (Dr. Herman F. C. ten Kate, of Holland).

In the winter of 1886 a small advance party of the Cushings, Magill, Hodge, Garlick, and several Zuni informants departed New England for Fort Wingate, New Mexico, from whence the expedition was launched. For the first months of 1887, the expedition conducted surveys and excavations along the Salt River in southcentral Arizona, having set up camp (Camp Augustus) near Tempe. Eventually a second camp was established in March near the site of Los Muertos where most of the remaining work was conducted. During these early stages, no one trained in anatomy or physical anthropology was present, as Dr. ten Kate was not scheduled to join the expedition until later (Brunson, 1989:20–24).

However, in August of 1887, Mrs. Hemenway, recently returned to Boston from a visit to the expedition in Arizona, consulted with Baxter on the health of Cushing. She was sufficiently concerned to request that Dr. Washington Matthews (Fig. 1), then a surgeon in charge of the United States Army Medical Museum (USAMM), join the expedition at Camp Hemenway. Matthews,[1] a professional friend of Cushing, arrived at Los Muertos in August of 1887. It was at this point that he made the observation quoted at the beginning of this chapter.

In order to salvage the skeletal remains, Matthews requested that Dr. John S. Billings, curator of the USAMM, immediately send out Dr. J. L. Wortman with the appropriate preservatives. At the time, Baxter characterized Wortman as one of the leading comparative anatomists, osteologists, and paleontologists in the United States, having previously been the assistant of the paleontologist Edward Cope (Baxter, 1888, 1889:10). Matthews then persuaded Cushing to take a rest from the expedition's work and travel to the west coast where he might be able to recover more suitably. Originally scheduled for a 3-week "vacation," Cushing's trip lasted 3 months. The Cushings, Magill, and Matthews left in late September and were not to return until December.

During his absence, Hodge was left in charge of the expedition's work. It was Cushing's desire, however, that while he was away the skeletal remains be exposed but left in situ until he could examine them upon his return. On November 18th of 1887, Dr. ten Kate arrived at the expedition's camp near Los Muertos. Hrdlička (1919:117) notes that ten Kate was a native of Holland who studied under Broca. He was a graduate of the University of Leyden with both a Ph.D. and a medical degree. During his original visit to the United States in 1883 sponsored by the Société d'Anthropologie de Paris, he collected data on the Iroquois, as well as tribes in southern California and the Southwest. In agreement with Cushing, ten Kate had arranged to compare the skeletons excavated by the expedition with the physical characteristics of the extant tribes living in southern Arizona (ten Kate, 1892). He had originally planned to begin anthropometric observations among the Pima immediately, but he set aside this project in order to attempt to salvage the Los Muertos skeletal remains. A week later, on November 25th, Dr. Wortman arrived at camp. Apparently, there was an initial controversy between the two surgeons (Brunson, 1989:23), perhaps regarding who had the mandate to carry on the skeletal preservation work. However, the issue was resolved quickly and the two prevailed upon Hodge to allow them to remove the uncovered remains in order to prevent further destruction.

---

[1]Matthews, along with his brother and father (an Irish surgeon who had settled in the United States in 1846), had served in the Union Army during the war between the States. He received his own medical degree in 1864 from the University of Iowa and was afterwards appointed assistant surgeon at the Rock Island Arsenal in Illinois. Matthews was quite familiar with the west and Southwest, having been posted both to Fort Union, Montana and Fort Wingate prior to his position with the USAMM in Washington (Schevill, 1948/1949).

Figure 1 Dr. Washington Matthews, 1843–1905 (Schevill, 1948/1949:2).

Baxter reported that after his return in December, Cushing initiated excavations at the site of Las Acequias in an attempt to locate well-preserved skeletal material. These indeed were discovered, and Baxter described the activities of the two surgeons:

> The two doctors [ten Kate and Wortman] are found grubbing in the pits, industriously at work over the skeletons, over whose anatomical characteristics their enthusiasm is aroused to a high pitch. They are intent on securing and saving every bone, and are regardless of personal discomfort, not only their clothes being covered with the dust, but their faces begrimed and their hair and beards thoroughly powdered, making them look like some strange burrowing animals. The result of their painstaking is one of the finest and most complete collections of ancient skeletons ever brought together, and the consequent discovery of certain anatomical characteristics that promise to be of high importance in the determination of racial distinctions. (Baxter, 1889:33)

Subsequently, between March and May of 1888, ten Kate was able to complete his intended work not only with the Pima, but also the Papago, Maricopa, and Yuman groups. Wortman remained with the expedition until June of 1888, as he cared for the skeletal material and oversaw their removal, curation, and eventual transportation to the USAMM (Matthews *et al.*, 1893).

In June, excavations in the Salt River valley stopped and the expedition moved on to the Zuni region, with excavations at Hàlona, Inscription Rock, and Hèshotaùthla. However, Cushing soon returned to the east due to illness, leaving Hodge in charge. Later, in 1889, Jesse W. Fewkes took over supervision of the expedition's work. No synthetic report of the expedition was ever completed. Cushing claimed that Fewkes misappropriated many of his field notes, limiting his ability to produce a final report (Brunson, 1989). Nor were relations between Cushing and Hodge (as both his secretary and his brother-in-law) always satisfactory. In fact, Hodge criticized Cushing's conduct in his introduction to Emil Haury's much more recent report (Haury, 1945). Despite the fact that Cushing never issued a final report, both ten Kate (1892) and Matthews, Billings, and Wortman (1893) were able to publish the results of their work. The report published by ten Kate in the *Journal of American Ethnology and Archaeology* compared 104 of the crania procured by the expedition (48 from the Salt valley, 56 from the Zuni region) with the observations he made on 445 living Zuni, Pima-Papago, and other southern Arizona groups. To the latter, he added data he had collected on 131 other individuals during his 1883 visit to the Southwest. After presenting data collected among the indigenous tribes, ten Kate provided a summary of the skeletal material, noting the great similarity between the Zuni and the Arizona materials. The overriding objective of his analysis, however, was determining the extant group that most closely resembled the prehistoric Salt River collections, the assumption being that physical similarity denoted genetic relatedness. His conclusion, based on the measurements of cranial breadth and length, as well as the cephalic index, was that the living Zuni displayed the greatest morphological similarity to the ancient Salt River sample. Moreover, he was quick to point out that this conclusion generally conformed to Cushing's determination regarding the ancestry of the Zuni peoples. This publication represents the first study of many in the Southwest that sought to draw conclusions regarding genetic similarity on the basis of cranial morphology.

Upon his return to Washington, Matthews examined and measured the materials sent to the USAMM from the excavations in both the Salt River valley and at the Zuni sites. The physical anthropological report was to have been published as part of the larger archaeological publication. However, as no report seemed forthcoming, Matthews, Wortman, and Billings' (1839:142) report was published independently. Other work was published by these medical doctors regarding southwestern collections or data considered tuberculosis among southwestern native groups (Matthews, 1887, 1888), Inca bone frequencies (Matthews, 1889), and observations of the hyoid bone (ten Kate and Wortman, 1888). Unfortunately, this was one of the few times researchers trained in human anatomy or skeletal biology would participate directly with an archaeological excavation in the region for the next 50 years.

III. THE U.S. NATIONAL MUSEUM, THE AMERICAN MUSEUM OF NATURAL HISTORY, AND THE MUSEUM OF THE AMERICAN INDIAN

The Hemenway Expedition contributed significantly to the collections of the United States Army Medical Museum. The USAMM was founded in 1863, with Dr. George A. Otis as curator from 1864 to 1881 (see the National Museum of Health & Medicine's "A Brief History of the Collecting of Anatomical Specimens by the Army Medical Museum"). Otis was quick to begin a symbiotic relationship with the 17-year-old United States National Museum–Smithsonian Institution (USNM-SI), under the direction of secretary Joseph Henry. In exchange for all human skeletal material at the USNM-SI, Otis and Henry agreed that the Army Museum would relinquish all ethnological and archaeological collections. Further, Hrdlička (1919:66) notes that Otis requested that Army and Navy medical personnel forward on to the USAMM skeletal materials of general interest. This resulted in the museum amassing a significant collection of osteological materials, which was later retransferred to the USNM-SI, beginning in 1898. This series included roughly 1500 crania and skeletons. While the USAMM retained pathological specimens, the contribution of over 3500 skeletal specimens by the USAMM no doubt provided the USNM-SI physical anthropology division with a firm foundation for subsequent research and publications in physical anthropology.

As discussed in Chapter 4, during 1903, Aleš Hrdlička became the first director of the newly formed Division of Physical Anthropology at the USNM-SI. Hrdlička was responsible for the collection of anthropometric data and osteological specimens from the North American desert west (Hrdlička, 1908a, 1909a, 1935a). During his earlier tenure at the American Museum of Natural History, he had collected somatological, medical, physiological, photographic, and skeletal data from indigenous tribes of the Sierra Madre in northern Mexico (including the states of Chihuahua, Sonora, Hidalgo, and Durango). Hrdlička also compiled data relating to tribal demography, stature, folk conceptions of illness, native diet, infanticide, and various pathological conditions, including influenza and smallpox. This trip augmented data collected by the Lumholtz Expedition (Hrdlička, 1919:98; Lumholtz, 1902). Hrdlička's work was so successful that he was subsequently able to arrange with F. W. Putnam of the AMNH and the Hyde Expedition to sponsor a similar trip to the American Southwest (Utah, Colorado, New Mexico, and Arizona) and additional regions in Mexico (Michoacan and Morelos). Hrdlička (1931:2) indicates that he was in charge of physical anthropology for the Hyde Expedition between 1898 and 1903, when he also examined Basketmaker remains from Utah. Subsequently, in the summer of 1909, Hrdlička collected and synthesized data on tuberculosis from a variety of indigenous groups in the western United States, including several from the Southwest (Apache, Hopi, Pima, and Navajo). While stressing that unsanitary living conditions among these groups was the most likely contributing factor to the high rates of respiratory disease, he did note that "... the Pueblos ... are among the tribes most free from tuberculosis."

The desert west skeletal collections housed in the USNM-SI were soon supplemented by the excavations conducted at various prehistoric Zuni ruins. Frederick Webb Hodge, former secretary to the Hemenway Expedition, initiated excavations supported by the Heye Foundation and the Museum of the American Indian in New York. Between 1917 and 1923, Hodge explored the protohistoric and historic sites of Hawikku and Kechiba:wa, approximately 11 miles Southwest of Zuni Pueblo. In the course of this work, roughly 950 burials were uncovered from Hawikku (Howell, 1994, 1995) and 266 from Kechiba:wa (Lahr and Bowman, 1992). Unfortunately, Hodge did not follow Cushing's example and engage anatomists or physical anthropologists in his research. Smith and colleagues (1966:192) report that in-field identifications of age and sex at Hawikku were made by individuals not trained in physical anthropology and are therefore suspect. They indicate that "[a]pparently no careful study of such details as tooth eruption, epiphysial union, or closure of cranial sutures was made." Skeletons sent to the USNM-SI and examined by Hrdlička were apparently only those that were among the best preserved (predominantly adults). Only 1 cremation out of a total of at least 317 excavated (Smith *et al.*, 1966:193, 203) was sent to the Museum of the American Indian and seems never to have been examined by a trained osteologist. Moreover, no systematic plan for the disposition or study of skeletons and associated artifacts was ever formulated. This resulted in miscellaneous remains being sent to both the USNM-SI and the Museum of Archaeology and Ethnology at Cambridge, England, leading to loss of contextual data for some of the burials (Smith *et al.*, 1966; Howell and Kintigh, 1998; Lahr and Bowman, 1992). A description of the mortuary practices at Hawikku was left unpublished until 1966 (Smith *et al.*, 1966), while an in-depth study was not completed until 1994 (Howell, 1994). Recent restudies of the skeletal materials from Hawikku have been plagued by concerns over the accuracy of the sex determinations (Corruccini, 1998; Howell and Kintigh, 1998). The human remains from Kechiba:wa sent to Cambridge, England, were studied in the early 1990s (Lahr and Bowman, 1992); however, the mortuary practices at this site have yet to be examined or even published systematically.

IV. 1930s AND 1940s: SELTZER, STEWART, AND SPUHLER

The 1930s saw a shift in southwestern bioarchaeological research, as in the rest of the country. The most obvious example of this shift is the publication of Hooton's *Indians of Pecos Pueblo* in 1930 (see Chapter 4). Previously, most osteological data had been collected by anatomists or surgeons. For example, R. W. Leigh, a dentist by training, published a comparison of dental pathologies in skeletal series from four North American indigenous groups, including the prehistoric Zuni crania available at the Smithsonian Institution (Leigh, 1925). Hrdlička had suggested this study (Leigh, 1925:179). Similarly, the neuroanatomist G. von Bonin studied skeletons from the Lowry ruin in southwestern Colorado (1936, 1937; see also Stewart, 1937a). Increasingly, however, collection, observation, and analysis of skeletal material were conducted by individuals trained in physical anthropology as a result of focused training programs in institutions of higher learning (Spencer, 1982a).

Some of the crania from Hawikku (referred to as the "Old Zuñi"), along with others from the Southwest, constituted the primary data sets for a series of craniometric studies published in the 1930s and 1940s (Brues, 1946d; Hrdlička, 1931; Seltzer, 1936, 1944; Stewart, 1940c) [see also Hooton (1930) and Chapter 4]. Several of these investigations returned to Cushing's fundamental question regarding the ancestry of the modern Zunis. Additionally, most attempted the general reconstruction of biological affinities throughout the prehistoric occupations of the greater Southwest.

Hrdlička (1931) reported the first southwestern biodistance study since ten Kate's initial observations on the Hemenway Expedition materials. He gathered cranial measurements from over 10 different locations throughout the Southwest, including southern Utah Basketmaker sites, Puyé in the Jemez mountains of New Mexico, Hodge's "Old Zuñi," Chaco Canyon materials, and the Salt River collection, as well as Hopi mesa, Chaves [*sic*] Pass, and Petrified Forest specimens. Using comparisons of various metric attributes of the crania, especially the cephalic index, Hrdlička reached several conclusions. Importantly, he claimed that the southwestern collections displayed two distinct morphological groups: one brachycephlic ("round-headed") and the other dolichocephalic ("long-headed"). Among the former were the Utah Basketmakers and the Hawikku and Salt River samples. The latter included the Puyé and Hopi. Additionally, some specimens (e.g., the non-Puyé Tewa) appeared to be intermediate between these two clusters. Hrdlička also noted that the geographic distribution of these two groups was unsystematic, which probably represented "considerable interpenetration."

A few years later, Carl Seltzer (1936, see 1944 for details) reanalyzed the collections examined by Hrdlička. In doing so he iteratively compared the mean and standard deviation of over 20 metric traits and 11 indices of the skull for each pairing of the Hawikku collection against each other sample. He agreed with Hrdlička's conclusion that Zuni crania were morphologically similar to the Salt River and Utah Basketmaker samples (cf. Corruccini, 1972). He further argued that the Zuni collection resembled those from Chaco Canyon, the Petrified Forest, and Chaves [*sic*] Pass. However, he did conclude that:

> The supposedly sudden appearance of large numbers of undeformed crania in the pre-Pueblo and the very earliest of Pueblo phases has caused the majority of archaeologists to believe that these deformed specimens marked the arrival of what they termed "a new race," "a round-headed invasion." The writer cannot be of the same opinion ... the writer is prone to believe that the deformed crania are more the expression of a change in fashion or ideals of beauty rather than in physical type. (Seltzer, 1944:25)

Stewart's (1940c) analysis of skeletons excavated by Frank H. H. Roberts (1939, 1940) in the Zuni region lent support to this conclusion, which challenged the traditional viewpoint of many archaeologists, including A. V. Kidder (1924), who had suggested that the Basketmaker–Pueblo transition was marked by the arrival of a genetically dissimilar people into the Southwest.

It is important to note that while both Hrdlička and Seltzer (as well as Hooton and most contemporary physical anthropologists) referred to portions of their collections as belonging to specific morphological "types," this was not an exercise in mindless, essentialist classification. Nor should it be seen as a glimpse into the racist attitudes of early 20th-century biological anthropologists. No doubt such attitudes existed (Brace, 1982). However, Seltzer was quick to point out that terms such as "dolichoid" or "round-headed" were descriptions of overall sample sets and often did not necessarily characterize significant variation within groups. "The impression conveyed by these statements is that all Basket Maker crania are dolichocephalic, that is, have indices below 75, and that all the Pueblo crania are brachycephalic with indices over 80. This is *not* true" (Seltzer, 1944:26, emphasis added). Physical anthropologists were, on the whole, cognizant of the variation exhibited in their collections. It was simply that their research interests did not lead them to explanations of that variability, a fact of their historical context. There were, of course, differences in the way in which the various typologies classified or grouped morphological variation (see Chapter 2).

While a few individuals continued to pursue craniometric studies into the 1950s (Spuhler, 1954), many of the techniques and assumptions of these research agendas were decried by both physical anthropologists (Stewart, 1954d) and archaeologists (Kraus, 1954) alike. Perhaps this was an outgrowth of increasing subfield specialization possible within large anthropology departments. Decreasing knowledge about developments in other subfields may have led to difficulties in integrated research.

V. DISINTEGRATION: 1940s THROUGH 1960s

Between the late 1940s and 1960s, few integrated bioarchaeological studies were conducted in the Southwest. In the 1940s, this was no doubt due to the wartime reduction in archaeological activity. Moreover, the period following the war was characterized by a shift from large site excavations to more modest salvage archaeology projects. These less intensive excavations, by their very nature, were unable to uncover the large skeletal samples that characterized previous research. While bioanthropologists continued to conduct research on excavated skeletal materials, often these analyses resulted in brief appendices in larger archaeological site reports (e.g., Brues, 1946d; Gabel, 1950; Kelly, 1943; Reed, 1953; Stewart, 1940c) or publications in strictly physical anthropological venues (e.g., Hanna, 1962; Hanna *et al.*, 1953; Miles, 1966; Zaino, 1968). Equally distressing is the amount of work from this period that is part of the gray (or difficult to acquire) literature (e.g., Reed, 1966, 1967; Snyder, 1959). Additionally, during the late 1950s and 1960s, archaeological research goals and methodologies underwent dramatic changes in response to an emerging processual paradigm. As demonstrated by the seminal work by Hill (1970) and Longacre (1970), interest focused on ceramic, *not* skeletal or mortuary, correlates of prehistoric social organization. Thus, a general hiatus in bioarchaeological studies occurred in the desert Southwest until the late 1970s and 1980s.

VI. A RESURGENCE: THE 1960s TO 1980s

Throughout the 1940s, 1950s, and 1960s, interest in the relative biological or phenotypic distance between southwestern skeletal collections was maintained.

This continuity with the early goals of southwestern bioarchaeology is exemplified by the work of Spuhler (1954) and Giles and Bleibtreu (1961). In the late 1960s and 1970s, a renaissance of genetic distance studies occurred across the Southwest. The phenotypic similarity of a skeletal collection from one site was often examined with those from multiple other locations. This fluorescence of studies included those by Benfer (1968) and Butler (1971) with the Casas Grandes (Paquimé) material of northwestern Chihuahua, Bennett's (1973a) study of the Point of Pines burials, El-Najjar's (1974) work with the remains from Canyon de Chelly, McWilliams' (1974) examination of the Gran Quivira sample, Heglar's (1974) Cochiti study, the dissertations by Birkby (1973) and Lumpkin (1976), and Corruccini's (1972) report. Some studies in the mid- to late-1970s, however, foreshadowed a growing interest in prehistoric health and disease (Brooks and Brooks, 1978; El-Najjar, 1976; El-Najjar *et al.*, 1976).

The 1980s saw a broadening of the research interests of southwestern bioarchaeologists. This expansion is exemplified by the symposium organized by Charles Merbs and Robert Miller entitled "Health and Disease in the Prehistoric Southwest" — Salud y Enfermidad en el Noroeste Prehistorico — held at Arizona State University in 1982 (Merbs and Miller, 1985). Conspicuously absent from the volume are any studies involving biodistance assessment. Instead, chapters focus on paleodemographic issues (see chapters by Berry, Palkovich), nonspecific indicators of nutritional stress (e.g., Martin *et al.*, Weaver, Walker), specific pathological conditions, including tuberculosis (e.g., Reinhard, Sumner, Miller), as well as methodological concerns (Alcauskas) and historic accounts of infectious disease (Russell). Moreover, the symposium was designed to be inclusive, incorporating research from Mexico and involving medical doctors as well as physical anthropologists.

The papers in the Merbs and Miller volume were largely an outgrowth of a resurgence in integrated archaeological projects. By the early 1970s, large burial samples were once again being recovered across the Southwest. Increasingly, archaeologists saw the advantages of once again consulting with biological anthropologists. Analysis of human skeletal material was appreciated as an important line of evidence for testing alternative hypotheses. For example, the 1972 and 1973 excavations at the late 13th-century site of Pueblo de los Muertos in westcentral New Mexico by Watson, LeBlanc, and Redman (1980) produced 26 burials. These remains were examined by R. Linda Wheeler (1985) in an attempt to test Watson, LeBlanc, and Redman's hypothesis regarding the abandonment of the area. Specifically, they suggested that abandonment was the result of two factors: local resource depletion and subsequent warfare with competing neighboring groups. Wheeler's paleopathological analysis described evidence for both nutritional stress and trauma. However, she did not feel that the pathologies exhibited by the Los Muertos sample deviated significantly from the usual southwestern pattern and thus concluded that skeletal data did not support the archaeologists' hypothesis. Subsequent work (Kintigh, 1996) has suggested that the 14th-century abandonment of such sites was the result of a lack of intracommunity social integration rather than intercommunity conflict, thus supporting Wheeler's conclusions.

This period also witnessed the development of a number of large-scale, multisite, interdisciplinary projects. Among others, these included the National Park Service's work in Chaco Canyon (Akins, 1986), the Black Mesa Archaeological Project (Martin *et al.*, 1991), the University of New Mexico field school at Tijeras Canyon (Cordell, 1980), the excavations of Pueblo Grande in Phoenix (Mitchell, 1992, 1994), the Dolores Archaeological Program in southwestern Colorado (Stodder, 1987), and the School of American Research's excavations at Arroyo Hondo (Palkovich, 1980). For the first time since the Hemenway Expedition at the turn of the century, biological anthropologists were an integral component of research design and implementation.

VII. NEW INTERESTS AND CURRENT APPROACHES (1990s–PRESENT)

Since this resurgence in southwestern bioarchaeology, numerous studies of previously excavated skeletal collections have been completed. In her 1990 dissertation, Ann Stodder compared 188 of the burials from Hawikku to a sample from the Galisteo Basin pueblo of San Cristobal. Stodder tested the proposition that the location of the San Cristobal population in the center of Spanish colonial activity resulted in greater overall nutritional and health stress compared to the more isolated Hawikku population. Her analysis suggested, however, that differences in health between the two populations were not straightforward. For example, the San Cristobal sample exhibited a higher mean life expectancy and a lower rate of juvenile mortality than the Hawikku series. Stodder suggested that different local economic and subsistence strategies may account for some of the variability observed. Her paleoepidemiological approach thus provided an excellent illustration of population-based bioarchaeological studies to southwestern prehistorians.

In 1992, Lahr and Bowman studied 54 remains excavated by Hodge when at Kechiba:wa. They documented widespread iron-deficiency anemia and arthritis in this collection. Lahr and Bowman proposed that people of Kechiba:wa suffered chronic ill heath. They also reported various other low-frequency pathological conditions, including button osteomas, osteomyelitis, and a possible case of venereal syphilis. Of special note is their diagnosis of three cases of possible tuberculosis in their sample. They thus disagreed with Morse's [1961, see Buikstra (1999) for an extended discussion of Morse] contention that New World population densities were too low to support infectious conditions such as tuberculosis. In particular, they suggested that aggregated communities such as Kechiba:wa may have exhibited population levels high enough to sustain such diseases.

However, the skeletal material analyzed by Danforth, Cook, and Knick (1994) from the relatively small community at the Carter Ranch site illustrates a "... relatively well-adapted population." Their sample of 34 displayed less evidence of malnutrition than those from more densely populated sites. Juveniles showed nutritional stress, as evidenced by linear enamel hypoplasias, Harris lines, cribra orbitalia, and porotic hyperostosis. However, adult stature estimates for the Carter Ranch individuals approximated those reported by Hrdlička (1935a) for living Puebloan groups and thus suggested that many individuals may have recovered from childhood malnutrition. Moreover, while dental lesions were common, they do not represent a significant deviation from expected frequencies for maize-dependent horticulturists. This study thus provided information on the health of a small-scale community that can be profitably compared to larger samples. Such comparisons can provide critical tests of alternative hypotheses regarding population aggregation and abandonment in the region.

In 1994, Howell reconstructed the mortuary treatment of 954 burials excavated by Hodge at Hawikku. In doing so, he compiled one of the largest databases of archaeologically recovered graves from a single community in the American Southwest. In collaboration with Keith Kintigh (Howell and Kintigh, 1996), Howell has been testing hypotheses about this protohistoric and historic Zuni community. Utilizing a statistical clustering technique and diversity measures, he has identified individuals who may have held community leadership roles. Moreover, he has been able to document a reduction in the number of females holding such positions in the immediate post-contact period (Howell, 1995). Howell proposes that this change was the result of colonial Spanish ideological and economic influences. He and Kintigh (Howell and Kintigh, 1996) have supplemented mortuary data with non-metric dental traits to compare biological affinity between and within spatially discrete clusters of burials. Their results suggest that leadership may have been hereditary in nature. While there has been criticism of their approach in an issue of *American Antiquity* (Corruccini, 1998; Howell and Kintigh, 1998), their study, as well as others (Schillaci and Stojanowski, 2000, 2002), illustrates both the value of integrating biological and archaeological data in the Southwest and that traditional concerns in southwestern bioarchaeology, such as genetic relationships, can be approached in novel ways that integrate both archaeological and biological data.

Bone chemistry — both trace elements and stable isotopes — has been used to infer diet in southwestern skeletal collections (e.g., Ezzo, 1993; Matson and Chisholm, 1991; Spielmann *et al.*, 1990). Price and colleagues (1994) have also used stable strontium ratios from east-central Arizona bones and teeth to evaluate hypotheses regarding prehistoric population mobility and residence rules. Such analyses provide greater geographic specificity for inferences of population movements.

Finally, and perhaps most dramatically, a number of scholars have been exploring evidence for both interpersonal violence and cannibalism in the prehistoric American Southwest [for brief reviews, see Hillson (2000) and Plog (2003)]. Indeed, reports of cannibalism have reached the popular press (Preston, 1998). While the earliest and most intensive investigations of possible cannibalism in the region include the works of the Turners (Turner and Turner, 1992, 1999) and White (1992), recent investigations have integrated cultural data while others have explored the utility of analyses of human myoglobin in ancient human feces in identifying instances of cannibalism (Billman *et al.*, 2000; Hurlbut, 2000; Ogilvie and Hilton, 2000). Case studies of osteological evidence for violence include those of Darling (1998), Kuckelman, Lightfoot, and Martin (2000, 2002), Martin and Akins (2001), and Walker (1998). Since the early 1990s, the perspective that prehistoric Southwesterners did indeed engage in warfare and other forms of interpersonal violence has been amply supported (Haas and Creamer, 1993; LeBlanc, 1999; Lekson, 2002; Wilcox and Haas, 1994), contrary to historically presented ideals of Puebloan society as harmonious and remarkably conflict free (e.g., Benedict, 1934). Of particular interest are so-called "extreme processing" events, which are characterized by human bone assemblages with indications of perimortem trauma, intentional disarticulation, burning, exposure, and cut or other processing marks (Kuckelman *et al.*, 2000; Lekson, 2002).

The cause and nature of violence or cannibalism in the Southwest are, however, sources of ongoing debate (Billman *et al.*, 2000; Dongoske *et al.*, 2000; Kantner, 1996; Lekson, 2002; Plog, 2003). Two alternative explanations for extreme processing events, cannibalism and witch persecution (or kratophany), dominate this debate. Alternative explanations include social or political intimidation, raiding, interpersonal strife, and small-scale warfare. The positive effect of this debate has been increased concern with the detailed examination of the spatial, cultural, and historical contexts of human remains, as well as the modern political milieu in which these scholarly reports are appearing (e.g., Dongoske *et al.*, 2000; Plog, 2003). Researchers have therefore been encouraged to engage their osteological study with contextualized archaeological observations to answer anthropological questions about human nature. Bioarchaeological research in the Southwest is thus carried to a higher level.

VIII. FUTURE DIRECTIONS

While it is true that bioarchaeologists throughout the United States now work in an era of NAGPRA regulations, this should be viewed as both a challenge and an opportunity. It is our responsibility to educate both the public and our colleagues about the possible contributions that integrated bioarchaeological research programs can make to both understanding the past and to issues of significance to Indian communities.

Furthermore, we must follow not only methodological but also theoretical advances in both archaeology and bioanthropology. In this respect, southwestern bioarchaeologists seem to be falling behind (Goldstein, 2001). For example, while archaeologists in other areas have been moving quickly beyond the Saxe–Binford (Saxe, 1970; Binford, 1971) approach to mortuary ritual (e.g., Beck, 1995), southwestern investigators continue to uncritically apply the assumption inherent in this program (e.g., Mitchell and Brunson-Hadley, 2001; Howell, 1995; Mitchell, 1994; Ravesloot, 1988). Similarly, southwestern skeletal biologists have been slow to consider how the osteological paradox (Wood *et al.*, 1992) might complicate their conclusion regarding prehistoric health and demography (e.g., Martin, 1994; Nelson *et al.*, 1994). We must not allow the rich empirical record of the Southwest to lull us into a false sense of confidence with our current methodologies and theoretical perspectives.

In fact, the uniquely rich archaeological record of the Southwest holds potential for resolving general issues that have long been a matter of concern to bioanthropologists and archaeologists. For example, the etiology of cribra orbitalia and porotic hyperostosis has been debated by bioanthropologists for close to 15 years (Holland and O'Brien, 1997). Are these conditions the result of maize dependency or intestinal parasites; are they symptoms of physiological imbalance or adaptive response? The prehistoric Southwest, where skeletal evidence of anemia is ubiquitous, represents a perfect laboratory to test alternative hypotheses.

Likewise, the Chavez Pass–Grasshopper Pueblo debate [see Wills (1994) and McGuire and Saitta (1996) for discussions] has highlighted southwestern archaeologists' interest in the nature of prehistoric social complexity. Were Puebloan communities egalitarian or were they controlled by an elite hierarchy? If the latter, how did elites obtain and maintain their status distinctions? Studies such as those conducted by Howell and Kintigh (1996) illustrate one novel approach to this issue. Unfortunately, such reanalysis and reevaluation of perennial anthropological questions require continuing and ongoing access to skeletal collections, particularly large skeletal series. Due to the long tradition of large-scale excavations, the American Southwest has been an excellent source of such collections. Buikstra and Gordon (1981) showed that large collections are often the focus of repeated reanalysis, often with novel methods and techniques. These restudies often lead to alterations in previously accepted research results. Reevaluations are the hallmark of scientific inquiry. While NAGPRA may act to limit such studies (see e.g., Turner, 2002), there are examples where bioarchaeological inquiries have been enriched through input from indigenous communities. In complementary fashion, issues of ancestral community, health, and heritage have been of

Figure 2 Erik Reed at Awatovi in 1939 (Courtesy of Museum of Northern Arizona Photo Archives, negative no. 72.578, photo by Marc Gaede).

special interest to living descendants. The Southwest, with its many traditional communities, holds great promise in this regard.

The trend in southwestern anthropology over the 20th century has been toward less and less integration of archaeology and biological research. Nevertheless, there has been a significant upturn in synthetic research in recent decades. Work such as Howell and Kintigh (1996), Spielman, Schoeninger, and Moore (1990), Schillaci and Stojanowski (2002), and the various research on violence and cannibalism in the Southwest represent innovative examples of scholarship that incorporate archaeological and physical anthropological data and methods. They also do justice to the exceptional example that Cushing's multidisciplinary Hemenway Expedition set over 100 years ago. The future is bright for bioarchaeology in the American Southwest, and we are on the threshold of realizing Buikstra's decade-old call for mutually designed research strategies.

Chapter 6

A New Deal for Human Osteology

George R. Milner and Keith P. Jacobi

I. INTRODUCTION

Franklin D. Roosevelt's New Deal relief programs in the 1930s to early 1940s decisively changed the practice of archaeology in the United States. The story of how this unprecedented funding and access to labor — most famously through the Works Progress Administration (WPA) and Civilian Conservation Corps (CCC) — transformed archaeology has been told a number of times (Baklanoff and Howington, 1989; Dye, 1991; Haag, 1985, 1986; Lyon, 1996; Milner and Smith, 1986; Schwartz, 1967; Seltzer and Strong, 1936; Seltzer, 1942, 1943). Excavation methods and field training were improved; institutional support was augmented; knowledge of prehistoric cultures was increased; and crippling rural unemployment was reduced. The last, of course, was the principal reason for federal involvement in this great endeavor.

The contribution of the New Deal excavations to physical anthropology, specifically the study of human skeletons, has received considerably less attention (Jacobi, 2002). This omission comes as something of a surprise because some of the largest and most heavily studied skeletal collections in the United States (indeed the world) were the direct result of the relief-work excavations, especially those in the Southeast. They include, among others, skeletons from the well-known Indian Knoll shell heap in Kentucky (Snow, 1948; Webb, 1946). Perhaps this lack of interest reflects the fact that studies of skeletons during the Great Depression and immediately afterward contributed little to the advancement of research questions and the development of new analytical methods. Nevertheless, this work resulted in one lasting contribution of considerable significance — the

Bioarchaeology: The Contextual Analysis of Human Remains, Buikstra and Beck (eds.)

Figure 1 Eleanor Roosevelt (third from left) visiting Alabama's WPA archaeological laboratory in Birmingham. She traveled widely during the Great Depression to promote her husband's various New Deal programs. Courtesy of the Alabama Museum of Natural History Photograph Archives, Mary Harmon Bryant Hall, 16CAL.

generation of large and generally well-documented skeletal collections (Fig. 1). These collections, which would be difficult or impossible to duplicate in today's financial and political climate, continue to be the subject of active research. Each year more is published on these skeletons, and their research potential is far from exhausted.

II. NEW DEAL PROJECTS

Archaeological work supported by relief-work projects started in 1933, but numerous large-scale excavations in several states did not get under way until early in the following year (Lyon, 1996; Milner and Smith, 1986). Among them were excavations at sites that would soon be covered by water rising behind newly built Tennessee Valley Authority (TVA) dams (Webb, 1938, 1939). William S. Webb, a physicist at the University of Kentucky with a deep interest in prehistory, was instrumental in getting the Tennessee Valley work funded, staffed, organized, and under way. Not only were the scale and number of excavations unprecedented, Webb had to contend with academic institutions that bickered over the ultimate disposition of the artifacts and skeletons, and separate agencies that controlled the labor needed to dig the TVA sites. A forceful personality with a no-nonsense approach to both field and laboratory work — Webb was commonly called "Major" and earned his nickname "Bullneck" — was essential for the successful initiation of the TVA projects and, later, the state-wide WPA and CCC archaeological program in Kentucky. In writing to a physical anthropologist about someone else's long-overdue report, Webb said that "unless he could make a report to you or to me in the course of the next two weeks, his investigations would serve no useful purpose for us" (Webb, 1940). He added that he "would be glad to be advised as to what you think of the situation and whether or not I owe him anything — money, courtesy, or anything else. I have no desire to be hard-boiled, but I dislike to be a sucker."

Over the next several years, archaeological projects in many states yielded vast numbers of artifacts and skeletons (Fig. 2). This huge effort ended early in 1942 when the last workers were drafted into the armed services or secured employment in burgeoning war industries. In the months immediately following the attack on Pearl Harbor, the archaeological projects that were still under way were abruptly ended. The WPA administrator for these projects noted that while the work was "suspended for the 'duration'" everything possible should be done

Figure 2 Impressions of the CCC camp at Moundville where many graves along with many houses and artifacts were excavated, as drawn in 1936 by Marshall Davis who participated in the work. Courtesy of Douglas Jones.

"to see that the closing is done in an orderly fashion and to the best possible advantage of our sponsors and of science in general" (Deignan, 1942). She added that "should better times come, it is my hope that this program may be reopened." The archaeological projects, of course, were never resumed, leaving a tremendous backlog of materials, some of which have yet to be systematically examined.

The suddenness of it all meant that unwashed materials were often left in their original field bags and boxes. Even the laboratories where artifacts and skeletons were cataloged and studied were used for other purposes, such as the one in Birmingham, Alabama, that closed in the spring of 1942 to make space for pressing mineral exploration needs (Griffin, 1978). In Kentucky, Webb was frustrated over the problems that arose when the laboratory "staff was taken for a WPA defense project (bus and truck Survey) and it [the laboratory] closed overnight" (Webb, 1942). He went on to explain what happened to the skeletons.

> The skeleton laboratory merely stopped where they had been working that afternoon when the order came through. Some skeletons were partially restored, some completely restored, some not yet attempted. In Dr. Snow's laboratory where the skeletons are being measured there are some 200 skeletons each lying on its own case, unwrapped and partly ready for storage. The process was merely stopped at a given hour, you see we had no previous notice of discontinuance. (Webb, 1942)

III. PROJECT PERSONNEL

There was a widespread feeling among archaeologists that laboratories provided with qualified personnel and adequate equipment should be established to handle all excavated materials, including skeletons. A need for physical anthropologists was one of the recommendations made by the National Research Council's Committee on Basic Needs in American Archaeology (1939), which were subsequently elaborated and published by Guthe (1939). The reality, however, was that funds for all analyses on New Deal projects were quite limited, regardless of whether village architectural remnants, mound strata, artifacts, or skeletons were the subject of study. Money for reports was similarly hard to find. After all, the purpose of the projects was to put people to work, not to study and write about what was found when sites were dug.

The people who handled the bones were a mixed lot. On many projects, such as those in Kentucky, the skeletons were excavated by skilled "trowel" men, as distinct from the "shovel" men who did the heavy work (Milner and Smith, 1986). Other laborers, both men and women, unpacked, cleaned, and labeled the bones after they arrived in the laboratories. Tight budgets and insufficient space meant that every effort had to be made to ensure that tasks went forward as quickly and smoothly as possible, resulting in a highly regimented workday. The staff of the Alabama Archaeological Laboratory, for example, was warned

Figure 3 Cleaning bones and artifacts in Alabama's WPA archaeological laboratory in Birmingham. Courtesy of the Alabama Museum of Natural History Photograph Archives, Mary Harmon Bryant Hall, 3CAL.

about "hanging around rest rooms," "unnecessary talking while trying to work," "preparing for rest periods before [the] bell rings," "using [the] telephone during work hours," and "unnecessary loitering in hallway" (Binning, 1941; Fig. 3).

Age and sex estimates for skeletons were made in the field by excavation supervisors with varied academic backgrounds, abilities, and practical experiences. Sometimes field assessments were published because there were not enough qualified laboratory personnel to look at the skeletons or the bones were considered too poorly preserved to warrant further study. Other skeletons were examined by researchers with more training in skeletal anatomy. One of the earliest to look at the skeletons was William D. Funkhouser, a zoologist and dean of the graduate school at the University of Kentucky (Funkhouser, 1938, 1939). Along with Webb, he was involved in some of the earliest projects in the Tennessee Valley, building on years of fieldwork in Kentucky (e.g., Funkhouser and Webb, 1928; Webb and Funkhouser, 1932). As Webb said to him in 1934 about the Norris Basin collection, "my guess is that this collection of skeletal material has many interesting features and should furnish the basis for an excellent report" (Webb, 1934). Other physical anthropologists who worked in one way or another with the skeletons included Marcus Goldstein, H. T. E. Hertzberg, Frederick S. Hulse, Madeline Kneberg Lewis, Georg K. Neumann, Marshall T. Newman, Ivar Skarland, and Charles E. Snow. All of them had professional careers in anthropology ahead of them, mostly in physical anthropology.

Snow, for example, continued his work with prehistoric skeletons long after World War II when he taught at the University of Kentucky and, for a time, continued to publish descriptions of bones from the New Deal projects (e.g., Snow, 1948; Webb and Snow, 1945).

Occasionally medical experts were asked to comment on unusual and typically rather extreme pathological cases. Shipping bones to get such opinions, however, was not without its difficulties. Snow once asked Webb about whether he had heard if bones shipped to Ohio had reached their destination (Snow, 1940b). The specimens did indeed arrive, and they were examined by Gustav C. Carlson (Department of Sociology, University of Cincinnati) and William McKee German (Department of Pathology, Good Samaritan Hospital in Cincinnati) (Snow, 1941d). Their opinions pleased Snow very much because he felt that "if we can rely upon the archaeological interpretation of the sites from which these specimens come . . . it may be possible to prove pre-Columbian occurrence of syphilis" (Snow, 1940c). Research on the origin and distribution of the treponemal infections, which include venereal syphilis as well as others such as yaws, continues to be a subject of great interest to paleopathologists.

IV. COLLECTIONS AND DOCUMENTATION

The archaeological excavations collectively yielded many thousands of skeletons—even now there is no complete count of them all. Most of the skeletons came from the southeastern states, especially Alabama, Kentucky, and Tennessee, for the simple reason that most of the excavations were conducted in that part of the country. Here archaeologists were particularly effective at securing funds and organizing large excavations, unemployment was widespread, there were few alternatives for federally supported projects in poverty-stricken rural areas, many large sites were to be destroyed by TVA dams, and mild winters allowed work to be conducted throughout the year. The prospect of many skeletons caused great enthusiasm among the people responsible for working on them.

> Some of our sites are enormous producers. The shell-mound area in West Kentucky has given out some 2,300 skeletons, 750 from one site, the latter in simply marvelous condition. Other sites aren't quite as good, but they approach this figure. And a new one, just started, may be even better. WOW. (Hertzberg, 1940)

The labor needed to catalog specimens was staggering, and the problems that accompanied this work were at times overwhelming. No sooner was a collection cleaned, numbered, and packed away than another arrived to take its place. As Snow (1941c) remarked to Webb about the Alabama laboratory late in the archaeological projects: "We have packed away approximately 2500 burials, leaving our shelves clear to receive the Moundville and other new material."

For the most part, the skeletons were well documented by the standards of that time. Field workers were sometimes provided with detailed instructions for excavating, recording, and removing skeletons. In Kentucky, they were warned that the "physical anthropologist is helpless if the archaeologist does not supply him with notes on stratification, intrusions, and cultural affiliations of a group of skeletons" (Anonymous n.d.a:12). Sometimes efforts were made to provide on-site training in proper excavation and recording methods, such as Hertzberg's visits to the Kentucky sites.

> About once a month or six weeks I like to get out to the various sites to see how things are going and to instruct the workers in exhumation and the rudiments of physical anthropology. It pays dividends in recovered zygomal [*sic*], nasals, face fragments, and the like. (Hertzberg, 1940)

Field methods were improved, and on many sites each skeleton received its own form to record body position, grave goods, and other pertinent information. Moreover, a genuine effort was made to come up with consistent terms for describing the burials and their archaeological contexts. They included suggestions written by Neumann and James B. Griffin—the latter was one of the Young Turks who were then shaking up the field of archaeology—for the Society for American Archaeology's Committee on Archaeological Terminology. They felt obliged to point out that "some effort has been expended in the selection of the words and objections should be made on the basis of indefiniteness, colloquialism, or reduplication [*sic*]" (Griffin and Neumann, 1940:1). Terms such as extended, fully flexed, semiflexed, and bundle burials, or others like them, were already in use on many of the New Deal excavations (Anonymous, n.d.a; Lewis and Kneberg, n.d.).

Generally the skeletons were photographed, and the results were astounding. In fact, the clarity of prints from large-format negatives is often as good as, if not better than, the field photographs taken today on 35-mm film. The many black-and-white prints provide excellent documentation that supplements the written descriptions of burials and drawings of them. Occasionally the photographs are the only surviving record of burial positions, and they are sometimes the only documentation of trauma and pathological lesions in skeletons that are no longer available for study. The photographs were often so good that recent investigators can sort out confusions in burial numbers by distinctive breaks or other features that show up on various bones.

Fortunately, the excavators removed the majority of the skeletons from the field for later study and permanent storage. For the most part all skeletal elements, not just skulls as was the widespread practice in earlier years, were transported to the laboratories, which often required lengthy trips over rough roads. The exceptions were usually poorly preserved skeletons that were thought to have little research significance. Badly decomposed bones could not be measured, and because osteologists were mostly interested in skeletal dimensions, they were not considered worth the effort of removing from the field. Nonetheless, attempts were often made to stabilize fragile bones in the field so they might arrive safely in the laboratory.

Researchers interested in bone chemistry should be aware that it is often difficult to determine exactly what preservatives were applied to particular skeletons in the field and laboratory. Alvar and acetone were generally favored for preservation purposes, although other materials were also used (Anonymous, 1938, 1940b, n.d.a). For example, some 4600 skeletons were treated with alvar as they passed through the Alabama Museum of Natural History's WPA archaeological laboratory (Snow, 1941g). Snow (1941g) called it "a most invaluable panacea for all archaeological ills . . . [it] is one of the most remarkable advances made in the preserving of precious specimens and artifacts of the past." First-rate preservatives, however, were not always available. Laboratory crews in Kentucky were forced to improvise as war approached and shortages of critical materials worsened. Alcohol was substituted for acetone, and it was found that old training aircraft windshields, when dissolved by lengthy immersion in acetone, produced an acceptable coating for the bones (Anonymous, 1941). The procedures used in Kentucky, from the field to the laboratory, indicate the range of materials that might be slathered on easily broken bones — the emphasis on measurable bones is clear.

> As a skeleton is removed from the grave in the field, it is packed in its own individual box, well supported with soft wrapping to insure its safe arrival in the laboratory. Most of the skeletal material upon exposure in the ground is found to be rather badly fractured from weight of earth and other causes. Steps are then taken, upon removal, to preserve the bones, especially the whole ones and the skull, by application of thin paper and shellac, to prevent further deterioration while awaiting study. When the box is opened in the laboratory, and the bones unpacked, the first step in their rather involved processing is their cleaning. This is accomplished by removing the dirt and by immersing the bones in alcohol to remove the shellac and paper, which now have served their purpose. When the bones have dried after their washing, they are soaked for fifteen minutes in a resinous solution which hardens and strengthens them. Two days are required for the preservative to set thoroughly, after which time the skeletons are ready for repairing. Skilled laboratory workers then begin the tedious work of assembling the fragments. Their purpose is to reconstruct all parts of the skeleton which have significance from the standpoint of measurement. (Anonymous n.d.b:3–4)

V. RESEARCH OBJECTIVES

During the 1930s, studies of human osteology mostly focused on bone measurements, and the work with skeletons from the newly excavated sites was no exception. Tables showing the dimensions of bones, especially crania, dominated the contributions of physical anthropologists to archaeological site reports. For example, in Snow's (1948) lengthy report on the much-studied Indian Knoll skeletons from Kentucky, over three-fourths of the text was devoted to quantitative and qualitative descriptions of bone morphology, mostly the former. The rest of the text covered pathological conditions, skeletal anomalies, cut marks on bones, and the like. The descriptive aspect of this work was very much in keeping with the research of that time. Articles on osteology that appeared in the *American Journal of Physical Anthropology* during the Great Depression were similarly descriptive, and they too focused on skeletal anatomy (Lovejoy *et al.*, 1982).

Most of the physical anthropologists involved in the WPA work had received Harvard training where they were heavily influenced by Earnest A. Hooton, one of the leading physical anthropologists of his day. Hooton even made available his osteometric equipment for the study of skeletons from TVA's Pickwick basin (Newman and Snow, 1942). It is not surprising that his interest in skeletal measurements and the classification of skulls provided direction to the New Deal project studies [see the contributions of Hooton's typological interests to physical anthropology in Armelagos *et al.* (1982)].

Considerable effort was spent fine-tuning the measurements that were taken with so much care. Neumann (1940) describes remeasuring the Indian Knoll series so he would have "a large number of detailed measurements of it." The comparability of data was a cause of concern because the measurements were to be used to identify the morphological characteristics of distinct groups of people. This interest in identifying discrete cranial types contributed to Neumann and Snow's frustration with data published by Aleš Hrdlička, one of the most influential physical anthropologists in the early 20th century. One of the problems they saw in his work was a tendency to use geographical and temporal categories that were far too coarse for the kinds of comparisons they thought were important. For example, Neumann (1940) complained that skeletons from Illinois, when simply lumped together, included "Hopewellian, Middle Mississippi, Upper Mississippi as well as late Woodland (probably Algonkin-speaking) skulls." It is for this reason that the relief-project excavators were cautioned to pay particular attention to the contexts of the skeletons they found: "If cultural or stratigraphic grouping is not made, the physical anthropologist is obliged to treat all of the skeletal material as a single group, a procedure that may give false averages" (Anonymous n.d.a:12).

Snow, Neumann, and their colleagues took it upon themselves to push for greater standardization in measurements intended for comparative purposes (Fig. 4). They found that even individuals trained by one man—it was, of course, Hooton—showed wide variation in exactly how the measurements were taken and the terms used for various skeletal dimensions (McKern, 1940; Neumann, 1940). Such concerns, and an interest in defining discrete cranial types

Figure 4 Charles E. Snow was one of the more prolific physical anthropologists who worked on the New Deal projects. Courtesy of the Alabama Museum of Natural History Photograph Archives, Mary Harmon Bryant Hall, 12CAL.

corresponding to different peoples, are evident in a number of letters the young osteologists circulated among themselves.

> As I visited the different institutions I further made it a point to find out how the different physical anthropologists felt toward the standardization of measurements. As you probably know that question was brought up at one of the AAPA meetings by Miss Tildesley of the Biometric Laboratory in London, and immediately squelched by Hrdlička and Pearl. The younger physical anthropologists who are working with American Indian material on the whole feel differently about it, and desire a uniform descriptive method for routine measurements so that their work is comparable. Among those who would like to do this are Shapiro, McCowen [*sic*], Stewart, Skarland, Snodgrasse, Krogman, Newman, myself and others. In all about fifteen. None of the physical anthropologists whom I have approached, however, feel that a committee should be formed as this would immediately arouse opposition, but would rather to begin to straighten out as much as possible by correspondence and quietly agree on a set of measurements and use them. Since you are working up American Indian material I would like you to be in on this too.
>
> The first step would be to get a list of measurements from every physical anthropologist to find out how many measurements that are being used are the same ones, that is, to find out just what we have in common and needn't quibble about. As examples of cases where measurements are not comparable I might cite that Hrdlička's head length is not the same as Hooton's. Hooton's external palatal length is not the same as Wilder's. Hrdlička's orbital breadth is not the same as Shapiro's. (Neumann, 1940)
>
> In my opinion, the meetings at Chicago were extremely worthwhile and many of us regretted your inability to attend. In close huddles with Neumann, von Ronin, Newman, and myself, much profit was forthcoming from our discussions of mutual

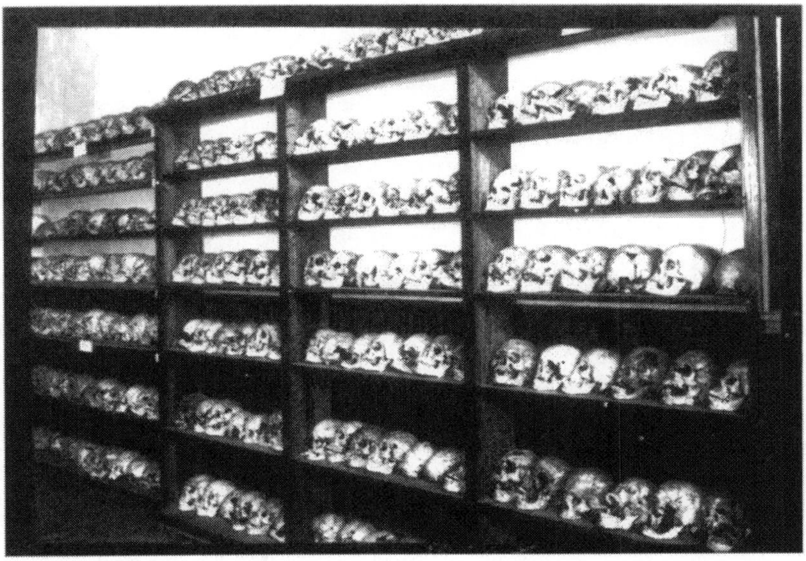

Figure 5 Skulls lined up in an effort to identify various cranial types in Alabama's WPA archaeological laboratory. Courtesy of the Alabama Museum of Natural History Photograph Archives, Mary Harmon Bryant Hall, 80CAL.

> problems concerning the physical types in the Southeast. Neumann's exhibit at Krogman's laboratory was particularly helpful since, for the first time, all of us could see crania exemplifying Neumann's physical types. (Snow, 1941h)

A key reason for measuring bones was to identify distinctive physical types linked to separate archaeologically defined cultures (Fig. 5). As noted in a WPA quarterly report on the Kentucky work, "the physical anthropologist is seeking to describe the several sub-varieties of American Indian who inhabited this area, and to classify them in their proper categories. Thus while the cultural anthropologist describes the materials [*sic*] aspects of a culture, the physical anthropologist provides an idea of the type of person who carried the culture" (Anonymous, 1940b:22). The Kentucky laboratory's brochure prepared for "This Work Pays Your Community Week," a nationwide effort to explain the WPA's projects to the public, explained that "we are interested in learning their [prehistoric Native American] physical appearance; and in comparing them with other groups which have produced similar cultural manifestations" (Anonymous n.d.b:4). In Kentucky, like other states, the objective was simply "to discover as much as possible of biological interest relating to the physical type of an early dweller in this state" (Anonymous, 1940b:26). Researchers firmly believed that with enough data it would be possible to reconstruct the temporal and geographical distribution of morphologically distinguishable groups of people associated with equally distinctive artifact inventories. Toward the end of the projects, Snow (1940a) wrote confidently to Hertzberg about the prospects of their work: "It seems that gradually the continuities and affinities of the various racial types are showing up more distinctly as time passes. It won't be long until we should have worked out the racial history of the United States."

Yet little was done, or indeed could be done, with the many measurements taken with so much effort. Adding machines used to calculate means and other summary figures were in short supply, even in laboratories such as the one in Alabama where thousands of skeletons were pouring in from the field (Snow, 1941e). The means of organizing vast amounts of metric data and using them in multivariate analyses would not be available for several more decades.

There was, however, a bigger problem — one that lay at the heart of the entire enterprise. The physical anthropologists were interested in the identification of ideal types, not the population-oriented analyses of morphological variation that are so common today. It was widely thought that different cultural baggage was carried by physically separable groups of people. As a result, the description of morphological characteristics, specifically bone dimensions, was of utmost importance. Because separate artifact inventories were thought to have been associated with equally distinctive groups of people, changes over time in ways of life were commonly attributed to the appearance of new populations identifiable by their long or broad heads, or some other distinguishing physical characteristic. It seemed reasonable to suppose that "the long-headed individuals" from one of the Norris Basin sites in Tennessee were from "an Iroquoian invasion," as there was a tendency toward "dolichocephalism in certain Iroquoian groups" (Funkhouser, 1938:248). Lacking effective multivariate techniques for characterizing cranial shape, but reflecting this emphasis on discrete morphological types, Madeline Kneberg Lewis directed her artistic talents to drawing busts of men, women, and children from the sites in Tennessee where she worked (Lewis and Kneberg, 1946; Lewis and Lewis, 1995). These portraits, based on considerable experience with newly excavated crania, were intended to capture the essence of the physical type typical of each time and place (Fig. 6).

Even as this work was being undertaken, there were disquieting signs that sorting crania into a few distinctive categories would ultimately prove unproductive. Snow expressed concerns about the cranial types and the variation that existed within skeletal collections from single sites to Neumann (1952) who spent much of his time trying to identify distinctive "varieties" associated with particular times and places.

> After measuring and observing most of the restorable material, I am now faced with a variation of cranial index which runs from sixty-five to an undeformed brachy [*sic*] of ninety, and a head form which varies from a narrow ellipse to a frankly spheroid shape. Can we consider these as simply variations within a homogeneous sub-racial stock?

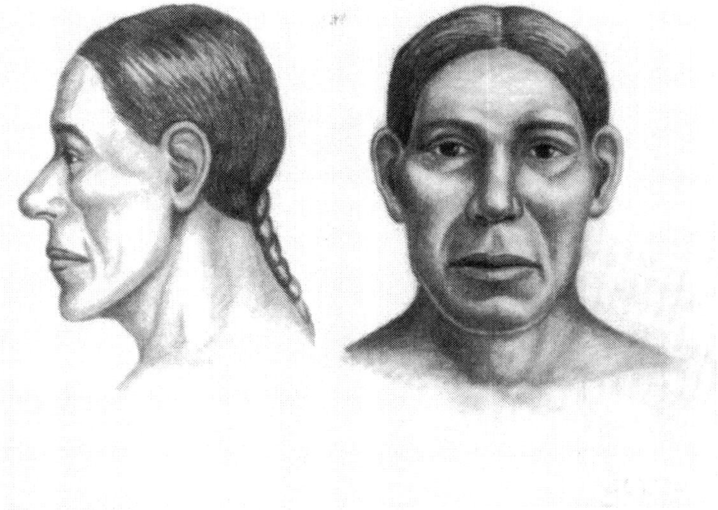

Figure 6 One of Madeline Kneberg Lewis' drawings based on crania found at archaeological sites in Tennessee. Courtesy of the Frank H. McClung Museum, the University of Tennessee.

> Just what are the limits of this variation? Some of these crania can actually be lost among series which Newman and I consider typically Shell Mound coming from horizons which long antidate [*sic*] middle Mississippian. At present I am inclined to think that the Middle Mississippi people must have mixed with the earlier longheads and that some of the genes must have been preserved so as to express themselves in a small percentage of the later population. In short, it seems to me that this variation is too great to be regarded as normal for a fairly homogeneous population. (Snow, 1941f)

Despite such concerns, Snow, like his contemporaries, continued to place great emphasis on his ability to sort out the physical features of the people who lived in different times and places. At the end of his influential work with Kentucky's Adena skeletons, he would write that "Professor Webb and I have recently looked at each Adena skull in the face once more before closing the book on the Adena Complex" (Snow, 1944). Such personal experience — difficult if not impossible to replicate — was the basis for the recognition of the supposedly discrete cranial types that were said to characterize separate populations associated with particular cultures.

In addition to describing the appearance of prehistoric people, the New Deal physical anthropologists were interested in how they lived. Snow (1943), for example, became fascinated with achondroplastic dwarves. They occur very rarely at archaeological sites, but he was fortunate in having two from

Moundville, a large late prehistoric mound center in Alabama. Of more importance was a concern with the health of these ancient people. In fact, physical anthropologists and archaeologists alike felt that studies of skeletons could make a "definite and important contribution" to an understanding of dietary adequacy in the past (Anonymous, 1940b:27). According to Webb (1945), bones were also important because through them "the ravages of such diseases as existed in prehistoric times where the disease was unhindered and the condition unameliorated by modern medicine" could be determined. He went on to say that "such skeletal material, because it is prehistoric is a great aid to medical and dental students as indicating the extent of damage when pathological conditions are unchecked."

It is indeed unfortunate that this interest was hardly ever followed by action. There were occasional reports on specific specimens, including those by various specialists, but there was no attempt to investigate the effect of living conditions on the health of people at various times and places. This omission comes as a bit of a surprise because Hooton, who had such a strong influence on the young physical anthropologists, had pioneered a population-based approach to the study of ancient diseases in his Pecos Pueblo work [see Hooton's contributions to paleopathology in Ubelaker (1982)].

Examinations of skeletons also contributed to the description of mortuary practices. Many of the skeletons found at the sites, often all of them, were included in long lists of burial numbers, age and sex estimates, burial offerings, and other information. Assessments of age and sex were provided by the physical anthropologists when they had an opportunity to examine the bones—otherwise, field identifications were used. Unfortunately, the archaeologists' interest in mortuary practices rarely extended beyond the identification of the usual ways bodies were handled at a particular site. Typical burial treatment was considered one of the traits that could be used to sort out various cultures. This approach, however, never proved particularly successful, and trait lists soon gained a poor reputation among the younger generation of archaeologists who would dominate the field after World War II.

The research on ancient health and mortuary practices—both receive much attention today—were frustrated for several reasons. First, time and money were lacking for anything more than descriptive site reports. Second, there was insufficient integration of the work of separate specialists. Osteological information appeared in reports as separate chapters or appendices, just like other specialized analyses. Here again money was an issue. Most of the funding went to alleviate the plight of the unemployed. After all, that was the reason the New Deal programs were put into place. Seltzer (1943) estimated that as much as 85% of all the money allocated to most of the archaeological projects went into the pockets of the poor. There simply was not much left over for the analysis and report-preparation aspects of the archaeological projects. Third, there were too few trained osteologists. This was a time when there were only a few anthropology graduate programs, and the suddenness and unprecedented scope of the New Deal archaeological projects took everyone by surprise. Fourth, the physical anthropologists were more interested in skeletal morphology and cranial classification than diseases and mortuary practices.

VI. LATER COLLECTIONS RESEARCH

The collections made at that time — excavated by archaeologists and initially described by physical anthropologists — continue to hold great research value. In fact, some of the most frequently studied skeletal collections in the United States, most notably Indian Knoll, came from these hectic years of intensive fieldwork.

Research questions, of course, have changed over the years since the original work was done. Included among these interests is a concern with the health of people in the past and how shifts in disease patterns were related to changes in basic ways of life. In particular, archaeologists and osteologists alike share a concern with the benefits and costs of the long shift to a more settled existence based largely on agriculture. This interest only gained prominence during the late 1960s and, especially, in the 1970s. At that time the so-called New Archaeologists were directing much of their attention toward how people once lived, specifically their subsistence and settlement practices, as an important component of their studies of the processes underlying cultural change. The impact of agricultural economies and urbanization on human health continue to be a major part of osteological work (Cohen, 1989; Cohen and Armelagos, 1984; Larsen, 1997).

Research methods also have changed. Even the ways the age and sex of skeletons are determined are not the same as they were back in the 1930s and 1940s. For example, many of the catalog cards filled out by Snow and co-workers in the Alabama archaeological laboratory indicate that ages were based on cranial sutures, and they were reported as unreasonably precise estimates (e.g., 27 years old). It did not take physical anthropologists long to recognize that this work needed revision (Johnston and Snow, 1961; Stewart, 1962e). We now know that studies of New Deal skeletons contributed to the tendency to classify too many individuals as males, as noted first by Kenneth Weiss (1973). This problem arose because the skull was emphasized over the pelvis when estimating the sex of skeletons, even though both cranial and postcranial features were said to have been used (e.g., Newman and Snow, 1942). It undoubtedly came about through an excessive fixation by earlier osteologists on skulls to the exclusion of other parts of the skeleton, including the pelvis. Reexaminations of New Deal project collections show that most discrepancies in the sex assigned to skeletons involves males being reclassified as females in the more recent studies (Milner and Jefferies, 1987; Powell, 1988). Thus despite Lucile St. Hoyme and

Mehmet İşcan's (1989) casual dismissal of Weiss' observation, studies of old collections support his conclusion that males tend to be overrepresented in skeletal reports published up through the 1960s. This literature, of course, includes the New Deal site reports with their long lists of skeletons and artifacts.

There is, however, another cause for concern: archaeologists have long used the published age and sex estimates in their studies of mortuary behavior (Pedde and Prufer, 2001; Rothschild, 1979; Shryock, 1987; Winters, 1968). While excusable back in the 1960s and perhaps in the 1970s, there is no reason to continue that practice. There is simply no shortcut that avoids the lengthy reexamination of skeletons. Fortunately, this work is now being done on a number of collections from the New Deal projects.

Recent studies of New Deal collections have also revealed hitherto unnoticed or poorly documented skeletal conditions. That comes as no surprise because the original work was done hurriedly and typically lacked any objective other than the description of bone shape and size. A good example of the value of examining skeletons is the discovery of many more victims of violence than was recognized previously, as is apparent in collections from Alabama, Kentucky, and Tennessee (Bridges, 1994a; Bridges *et al.*, 2000; Jacobi and Hill, 2001; Mensforth, 2001; Smith, 1995, 1997; our examinations of Kentucky and Alabama collections). This particular finding is important because these collections can add much to our knowledge of variation over time and space in conflicts among small-scale societies, an issue that has gained wide attention among archaeologists only since the late 1980s (Keeley, 1995; Lambert, 2002; Milner, 1999). The fact that these specimens have sat largely unrecognized on museum shelves is a large part of the reason why an overly romantic view of harmony in prehistory has dominated archaeological thought over the past half century.

VII. CONCLUSION

Too few researchers, too many skeletons, too little funding, too little time, and too narrow a research focus meant that most collections were studied incompletely in the Great Depression and immediately afterward. For the most part they still await comprehensive study. Perhaps in a strange way this state of affairs is a fitting tribute to the great efforts of the New Deal physical anthropologists. The collection of enormous numbers of skeletons was their most significant contribution to physical anthropology, not advancements in theory or method. By any measure, theirs was a remarkable achievement. Researchers today owe a large debt of gratitude to the men and women who ensured that the skeletons would be housed properly for future study. We must also be thankful for the great efforts of the many people — the shovel and trowel men, the trained supervisors, and the photographers — who labored under arduous conditions to collect large numbers of well-documented skeletons.

New research questions and methods require the further examination of all these collections. Fortunately, several institutions have recently redoubled efforts to organize and preserve these collections in order to enhance access to them. Nobody can know what lies in the future, but we can hope that the results of one of the most notable archaeological endeavors are not undone by politically expedient but short-sighted decisions over reburial that forever deny these invaluable collections to later generations of researchers. Here is one time where the dead can truly speak for themselves about what they ate, how healthy they were, their relations with members of their own societies, and their dealings with neighboring groups.

ACKNOWLEDGMENTS

Examination of the University of Kentucky notes was facilitated by a Kentucky Heritage Council grant for collections improvement and Virginia G. Smith's able assistance, both in the mid-1980s. For this chapter, Lynne Sullivan and Jennifer Barber provided the University of Tennessee's Frank H. McClung Museum figure; Douglas Jones and the Alabama Museum of Natural History also made available several figures. Bob Pasquill provided information on the CCC at Moundville.

Chapter 7

Invisible Hands: Women in Bioarchaeology

Mary Lucas Powell, Della Collins Cook, Georgieann Bogdan, Jane E. Buikstra, Mario M. Castro, Patrick D. Horne, David R. Hunt, Richard T. Koritzer, Sheila Ferraz Mendonça de Souza, Mary Kay Sandford, Laurie Saunders, Glaucia Aparecida Malerba Sene, Lynne Sullivan, and John J. Swetnam

I. INTRODUCTION; JANE E. BUIKSTRA

Frank Spencer's monumental compendium, *The History of Physical Anthropology: An Encyclopedia*, Volumes 1 and 2 (1997c), contains approximately 304 primary name entries. Of these, 296 are men and 8 are women. Among the women, only the anatomist and skeletal biologist Mildred Trotter is from the United States; one other was associated with physical anthropology in the Americas (Diaz Ungría, educated in Spain and a cofounder of the Department of Anthropology at the Central University of Venezuela). Have there truly been so few women's contributions? We beg to disagree.

There are many potential reasons for the invisible status of women in our discipline, including myriad social, economic, and political influences on women's career paths and choices made in decades past. Of the women who contributed to the study of archaeologically recovered human remains prior to or during the development of bioarchaeology, many abandoned promising careers following marriage, e.g., Boyle-Hamilton and Dillenius, or substantially tailored their careers to those of their husbands, e.g., Sawtell Wallis, Bullen, and Ericksen. Kneberg could not marry her long-term partner, Thomas Lewis, until they had both retired from the University of Tennessee due to nepotism rules. Her biography also notes that even though she was ABD at the University of Chicago and Lewis merely held a BA, she was commonly addressed in academic circles as

Bioarchaeology: The Contextual Analysis of Human Remains, Buikstra and Beck (eds.)

"Ms. Kneberg" and he as "Dr. Lewis." Only Brooks seems to have enjoyed a fully egalitarian professional marriage. Others chose or were redirected by their institutions into career paths such as education, collections curation, and conservation that were viewed as "more suitable" for women than field research, e.g., Barrett, Mosny, and Mello e Alvim. Sadly, several promising careers were cut short by illness and premature death, e.g., Studley and Sublett.

Other forces involving mentoring may also be at work. Powerful male mentors assisted several of these women in their careers, e.g., Frederic Ward Putman (Studley), Fay-Cooper Cole (Kneberg), Aleš Hrdlička and T. Dale Stewart (St. Hoyme), Earnest Hooton (Brues), Georg Neumann (Robbins), and James Anderson (Sublett). Although Studley's early death precluded reciprocal mentorship on her part, each of the others was known for her active involvement with students, repayment for the time-consuming assistance they had received in the formative stages of their careers.

But career choices and time-consuming mentoring are not the only variables. Trotter's entry in the Spencer volume is remarkably brief and incomplete. It occupies less than a full page and neglects to mention that Trotter was one of only two female founding members of the American Association of Physical Anthropologists and its first woman president. Also omitted is her status as the first woman to be awarded the prestigious Wenner-Gren Foundation's Viking Fund Medal. Her landmark efforts as a pioneering researcher in combining identification of war dead (WWII) with scientific research (Trotter and Gleser, 1952) are trivialized, their significance justified solely in terms of utility in T. D. Stewart's more recent study of remains from the Korean conflict. The second female president of the AAPA, Alice Brues, whose many, varied, and important contributions are described later, is not even among those listed in Spencer's volume. These surprising omissions suggest that one cannot exclude gender bias as a factor, not to mention more subtle underevaluations of women's contributions.

We therefore include within this volume a chapter dedicated to the invisible women pioneers in bioarchaeology, whose contributions are significant and, in general, underappreciated. Chapter 7 begins with the first woman with a single-authored publication in bioarchaeology (Studley, 1884); the first Canadian physical anthropologist (Boyle-Hamilton); and the first woman to receive a Ph.D. degree for research in physical anthropogist in the Americas (Dilenius). Others are singled out for the importance of their contributions to collaborative research and studies that helped counter arguments concerning the racial basis for intelligence and individual achievement (e.g., Wallis and Kneberg). Kneberg is also significant due to her prominence during the Works Progress Administration (WPA) era and her attention to the incomparable WPA collections at the University of Tennessee. She was concerned with the public dissemination of ancient lifeways long before this became a mandate from funding agencies. Both Bullen and Brues made careful, significant observations about the history of disease, specifically the treponematoses, in North America. Brues tended to emphasize functional, anatomically informed interpretation of her observations, whereas Bullen contextualized her remarks within archaeological and ethnohistorical literatures. Bullen attenuated her career as an observer of American Indian remains due to her concern for the attitude of the Indian community in 1975, long before such issues were perceived by most bioanthropologists. Similarly, Audrey Sublett—through her collaboration with the Seneca in the conduct of a historical cemetery removal—pioneered methods for field observations. One result of this project, conducted in 1964, was the widely cited Lane and Sublett (1972) study of residence patterning and inheritance during the historic period.

Alice Brues received the 14th Ph.D. degree supervised by Earnest Hooton (1940). She was broadly based in physical anthropology and anatomy, with direct experience in forensic anthropology, human variation, genetics, skeletal biology, and the study of ancient remains. The paper most influential within bioarchaeology did not, however, deal directly with empirical data from archaeological materials. "The Spearman and the Archer: An Essay on Selection in Body Build" (1959b) hypothesized that hunting technology could influence body proportions, thus linking biology with culture in a manner that encouraged ensuing studies of behavioral adaptations in the past.

As did Bullen, St. Hoyme made excellent use of ethnohistoric sources, especially evident in her 1962 paper (Hoyme and Bass, 1962) on understanding past lives of individuals whose remains were excavated in the course of the John Kerr reservoir project. Her work, like that of Kneberg and Bullen, emphasized the humanity of those ancients who were being studied. Similarly, Louise Robbins, whose early work followed the typological perspective of an earlier generation, shifted to a focus upon representing multiple facets of individual lives. Her 1977 contribution to the Blakely symposium volume epitomized this perspective.

In sum, there are numerous women who have contributed significantly to our scholarly heritage as bioarchaeologists, yet who remain largely or entirely invisible within the discipline. Their lives and their professional achievements are summarized in the following biographies.

II. CORDELIA A. STUDLEY (1855–1887); DELLA COLLINS COOK

A. Biography

The first woman to publish work in physical anthropology in North America was Cordelia Adelaide Studley. Her father, William Sprague Studley (1823–1893), was pastor of the Tremont Street Methodist church in Newton, Massachusetts.

The U.S. Census for 1880 presents a snapshot of the family: her father at 57, her mother at 47, Cordelia 24, a younger daughter and son, and an Irish servant. An obituary notice published by her mentor Frederic Ward Putnam in the *Proceedings of the Boston Society of Natural History* is the most complete account of her life (Putnam, 1888). She studied medicine at Boston University and the University of Michigan before becoming first, in 1881, Putnam's student and later his assistant at the Peabody Museum for the 5 years in which she was active. Putnam writes of health and financial limitations that prevented her from continuing her career. Discussing her death he says that she was "nervously prostrated" and that at Michigan "she overtaxed her strength and returned to Boston for treatment under one of our highest specialists" (Putnam, 1888:420). His language suggests mental illness.

B. Professional Career

The University of Michigan was an exceptional place in the late 19th century. Women were admitted to medical school there in 1870. In 1882, the first Ph.D. degree awarded to a woman in the United States went to Alice Freeman, a historian trained at Michigan and president of Wellesley College. In 1880, women constituted 19% of medical school graduates (McGuigan, 1970). Cordelia Studley must have attended in the late 1870s. While she did not complete her degree, Putnam notes her expertise and command of literature in several of his publications (Putnam, 1884). During her 5 years at the Peabody Museum, her name appears frequently in its annual reports: "To another assistant, Miss Studley, we are indebted for three Indian skeletons, from Marion, Massachusetts. A clay pipe and other European articles found with these skeletons prove that they were of Indians who were buried after contact with the whites" (Putnam, 1886:413). An additional skeleton and shell mound artifacts appear in the corresponding accession list. Many of the entries relate to her role as an excavator for the museum. She excavated Archaic shell mounds on the Damariscotta River in Maine in 1885 (Bourque, 2002), producing an unpublished 102-page manuscript on her work (Studley n.d.).

Studley published just one paper, "Notes upon Human Remains from the Caves of Coahuila, Mexico" (1884). This collection had been donated in 1880 by Dr. Edward J. Palmer, perhaps better known as a botanist, who explored four caves near San Antonio del Coyote in north Mexico, acquiring mummies and artifacts for the museum. Studley summarizes his notes on the caves. She describes the material minutely, noting posterior bridging of the atlas, perforated olecranon fossa of the humerus, "carinated" linea aspera of the femur, and flattening of the tibia. Stature is estimated and various indices of limb proportion are calculated. The 25 crania are described in detail, grouping by cranial index and emphasizing shape categories. Fusion of the internasal suture, auditory exostosis, and Wormian bones are described. Tables explore the relationship between cranial index and cranial capacity, distribution of sutural bones, auditory exostosis, gnathic index, nasal aperture shape, and orbit shape. Pathologies include an inflammatory lesion of a fibula, four skulls with superficial wounds, and a healed arrow wound to the face, as well as tooth wear, caries, abscesses, and antemortem tooth loss.

Studley was familiar with the work of French anthropologists Paul Broca and Paul Topinard, British scholars William Henry Flower and Charles Carter Blake, and Americans Jeffries Wyman and George A. Otis, although there is no formal citation of literature in her paper. She mentions *Crania Ethnica*, but does not name its authors, Hamy and Quatrefages. This is a fairly complete account of the literature of her day. She compares the mummy crania with her own measurements from 42 tribally identified skulls from the Southwest and previously published series, concluding that cave crania are more dolichocephalic than recent inhabitants of the region. Hrdlička (1914a) and Whitney (1886) cited her study favorably.

A paper titled "Description of the Human Remains Found in the 'Intrusive Pit' in the Large Mound of the Turner Group, Little Miami Valley, Ohio, during the Explorations of Messrs. Putnam and Metz" (Studley, 1885) suggests the direction her work might have taken, but the paper itself has apparently been lost. Earnest A. Hooton had access to it and published this summary: "This paper consists of a careful account of the pathological features of the skulls and a minute description of the perforations found in six of the crania, together with measurements and observations on the specimens" (Hooton, 1922:99). He compliments her skill at measurement and reconstruction. One notes that Hooton's description of the crania with drilled holes is rather perfunctory.

Putnam was an enthusiastic mentor to women, perhaps because his institution made heavy use of unpaid contributors, and he had constant difficulty attracting sufficient financial support. Cordelia Studley joined his staff in the same year as Alice Fletcher, who had a long career as an ethnomusicologist and Plains ethnologist (Brew, 1968; Mark, 1980). The ethnohistorian Zelia Nuttall was another of Putnam's protégés (Mark, 1980). What distinguished these women from Cordelia Studley — apart from their longevity — is that they were women of independent means. Putnam, the founder of section H of the American Association for the Advancement of Science, promoted these women as members (Mark, 1980). He was apparently in the process of doing the same for Studley at the end of her brief career.

III. SUSANNA PEEL BOYLE-HAMILTON, M.D., C.M. (1869–1947); DELLA COLLINS COOK AND PATRICK D. HORNE

A. Biography

Susanna Boyle was Canada's first female physical anthropologist. She was the eldest of five children, born in Elora, Ontario, in 1869. Her father, David Boyle (1842–1911), began life as a Scottish blacksmith. He immigrated to Canada and became a teacher in Elora, first at Middlebrook School, where he pioneered Pestalozzian methods, then at the Mechanics' Institute, where he was noteworthy for his interest in equal access to education for women. He was an amateur naturalist, incorporating field studies in natural history in his grammar school and secondary school teaching. In 1884 he joined the Canadian Institute in Toronto, where he served as field archaeologist and curator until his death. His professional contacts included Putnam and Boas. Gerald Killan (1983) details his founding role in Canadian archaeology and museology.

After Boyle's family moved to Toronto (Killan, 1983), Susanna attended Trinity Medical College, where she was awarded her M.D. degree in 1890 (Hafner, 1993). She became a professor at the Ontario Medical College for Women on her graduation and was demonstrator in anatomy, among other positions, from 1891 to 1898 (Killan, 1983; Sheinin and Bakes, 1987). In 1898, Susanna Boyle and two of her colleagues founded a dispensary for women that provided clinical training for women students (De la Coeur and Sheinin, 1990). The politics of medical education for women was complicated in Toronto. Women could sit for examination for degrees at Trinity Medical College, but they could not attend classes. The Ontario Medical College for Women was established in 1883 to prepare women students for exams. Anne Rochon Ford writes that "[t]he Ontario Medical College for Women graduated a total of 109 women in its twenty-two years of existence. The Faculty of Medicine at the University at last opened its doors to women in 1906 and the Ontario Medical College for Women closed" (Rochon Ford, 1985). However, this was a decidedly mixed blessing. De la Coeur and Sheinin (1990) argue that the demise of the OMCW had the unfortunate consequence of ending teaching positions for women. It is perhaps no coincidence that Susanna's youngest sibling Anne Anderson Perry was a prominent suffragist and journalist.

B. Professional Career

Susanna Boyle's published work in anthropology was limited to appendices to her father's reports. Hrdlička says of David Boyle that "he was not a somatologist, but his friendly attitude towards this branch of science is well seen in his detailed and well-illustrated 'archaeological reports,' many of which contain valuable notes on Indian ossuaries, other burials, on the collected skeletal material, and on other subjects of direct interest to physical anthropology" (1918:161). In support of this statement he cites Susanna Boyle's 44 page paper (Boyle, 1892). There is a brief discussion of method citing Morton, Broca, and Topinard, followed by cranioscopic and craniometric observations on 48 crania, each illustrated with an engraving. The crania are provenienced by county in Ontario, with the exception of four from British Columbia and two from Arkansas (Fig. 1). These last are included to illustrate artificial cranial deformation. The engravings emphasize suture closure, Wormian bones, and other minor variants. David Boyle's 1895 book summarizing his seven annual appendices to the Report of the Minister of Education for Ontario includes several of the skull drawings, but measurements and descriptions are omitted (Boyle, 1895). The Boyle osteology collection remains part of the patrimony of Canada (Anderson, 1961).

Surviving members of the Boyle family living in Canada had no further information about her when they were interviewed by historian Gerald Killan. He writes "She was a slight disappointment to her father in that she had put family before career and practiced medicine only on a part time basis" (Killan, 1983:220). We have pieced together her later history from archival sources.

From 1898 to 1904 she was women's physician at the State Hospital for the Insane, Independence, Iowa (Sheinin and Bakes, 1987). Immigration from Ontario to the northern United States in the late 19th century was quite common, the predominant pattern being that rural laborers who were unemployed as a result of mechanization of agriculture and falling wheat prices sought jobs in urban areas south of the border (Widdis, 1988). In Iowa she met a Pennsylvania-trained physician, Arthur S. Hamilton (1872–1940). They married in 1902 and moved in 1904 to Minneapolis, where he was both professor of nervous and mental diseases at University of Minnesota and a practicing physician (Anonymous, 1940a). The Hamiltons had one child, David A. Hamilton. The American Medical Association's files list Susanna as licensed in 1898, but not thereafter (Hafner, 1993), suggesting that she did not practice medicine after her marriage. However, University of Toronto alumni records indicate that she returned to serve as staff pathologist at the Hospital for the Insane, Independence, Iowa, during World War I, taking the place of physicians called into medical service (University of Toronto Archives). She practiced in Minneapolis during the war and was resident surgeon at the Massachusetts Hospital for Women, as well (Sheinin and Bakes, 1987). We have no dates for her residency, but it played a role in her interactions with Putnam. She died in Minneapolis in 1947.

Between 1896 and 1898, Susanna Boyle translated some eight lengthy articles from the Italian for the medical journal *Alienist and Neurologist* (St. Louis, MO). "Alienist" was the term used for physicians who devoted their practice to the care

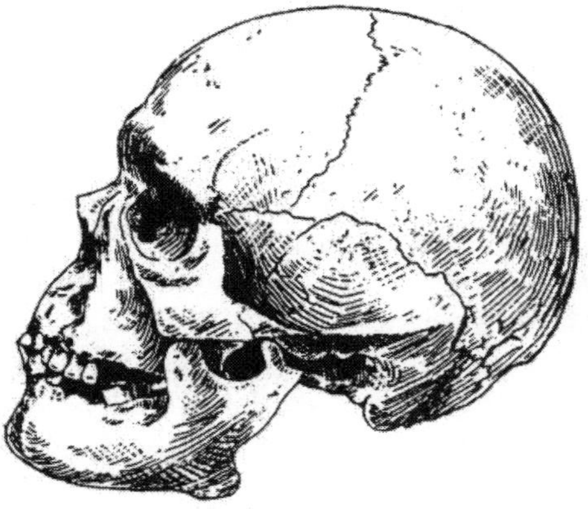

Fig. 101. No. 20.

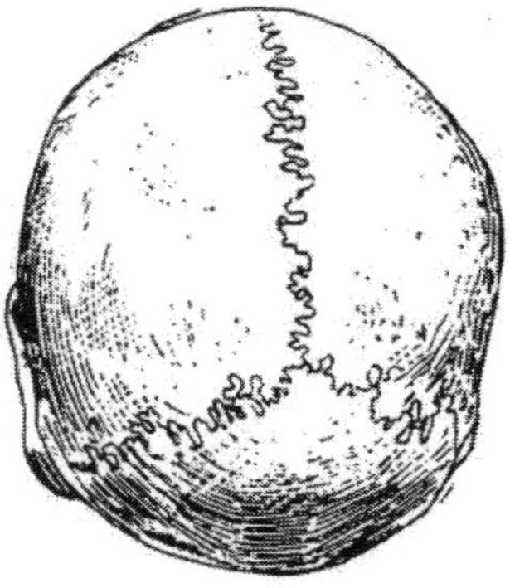

Fig. 102. No. 20.

Cranioscopy.	*Craniometry.*	
		Millimetres.
Age, 50	1	190
Sex, male	2	140
Sutures, deeply serrated	3	140
Pterion, H, bridge narrow	4	95
Inion, 4, rough	5	111
Artificially lengthened occipital foramen, I	6	110
Glabella, medium	7	15
Superciliary ridges, medium	8	111
Frontal eminences, absent	9	132
Forehead, retreating	10	76
Lateral parietes, slightly bulging	11	48
Probola, globular	12	295
Sub-iniac curve, extends below condyles	13	230
Mastoid process, small but very massive at root	14	289
Hollow at root of nose, deep	15	519
Inferior border of nares, curves wide, shallow	16	72.5
Wormian bones, three present; two in right side of lambdoid, and one in occipito-mastoid suture of left side. In front of each occipital condyle on basilar process is a distinct tubercle	17	136
	18	49
	19	28
	20	73.6
	21	67.8
	22	85.5
	23	73.6

See page 60 for Key to Measurements.

Figure 1 Susanna Boyle's descriptions of crania 102 and 103 from York County, Ontario (Boyle, 1892:78).

and treatment of the institutionalized insane until it was replaced by the term "psychiatrist" in the 20th century. A footnote to the first article (Vol. XVII # 1, 1896, p. 14) states that it was translated by Susanna P. Boyle, M.D., C.M., Professor of Normal and Pathological Histology, Ontario College for Women; Physician to the Girl's Home, Toronto. But why the *Alienist and Neurologist*? Why would a professor of histology and general practitioner opt to translate papers for a psychiatric journal, albeit a leading one? Here, perhaps, lies the connection to Dr. Joseph Workman.

Joseph Workman (1805–1894) is a hero in the annals of Canadian medical history. Internationally acclaimed for his reforms in the treatment and compassionate care of the institionalized insane, he was superintendent of Toronto's lunatic asylum from 1854 to 1875. According to the authorative *Dictionary of Canadian Biography*: "Unquestionably Canada's most prominent nineteenth-century alienist, Workman was much admired in his lifetime, being known in his last years as 'this Nestor of Canadian specialists,' in the apt words of the English alienist Daniel Hack Tuke" (Brown, 2000). Workman restored order to what was chaos at the Toronto asylum and as father of Canadian psychiatry ushered in a new era in the treatment of the insane.

As an accomplished linguist, upon his retirement, Workman spent much time translating Italian and Spanish for the leading medical journals in North America, especially the *Alienist and Neurologist*. Active and competent to the end, he translated numerous articles for the *Alienist and Neurologist* over nearly 30 years. It would seem that Susanna Boyle stepped into the void created by Workman's death regarding the translation of articles. That these two would have met was almost inevitable. Was it the influence and inspiration of this contact that led Susanna to take up a post in an asylum upon leaving Canada? There is one further piece of evidence that would appear to back this connection. In 1894, the year of Workman's death, Susanna Boyle's father David wrote the first and only contemporaneous biography of Joseph Workman. It would seem that the Boyles and Workmans were indeed both colleagues and friends.

IV. JULIANE A. DILLENIUS (1884–1949); DELLA COLLINS COOK

A. Biography

Juliane A. Dillenius was the first woman to earn a doctorate for her research in physical anthropology in the Americas. Information about her life is reported by Baffi and Torres (1997). Dillenius, an Argentine of German ancestry, was a doctoral student under the German physical anthropologist Robert Lehmann-Nitsche

(1872–1938) at University of Buenos Aires, completing her dissertation in 1911. She received the first Ph.D. degree awarded at UBA. After completing her degree, she spent 2 years in the Munich laboratory of the anatomist and anthropologist Johannes Ranke (1836–1916), with whom Lehmann-Nitsche had studied. "She returned to Buenos Aires in 1913 and married Lehmann-Nitsche, essentially bringing her promising career to an end" (Baffi and Torres, 1997:611). On her husband's retirement in 1930 they moved to Berlin. Following his death in 1938 she returned to Argentina, spending the last decade of her life in Buenos Aires.

B. Professional Career

Dillenius published five articles during her brief career (1909, 1910, 1912a,b, 1913). Her research focuses on the shape of the parietal in deformed skulls. She used a subtens caliper to describe the anterioposterior and lateral flexion of the parietal and calculated the resultant angles (Fig. 2). She noted increased complexity in the pars complicata of the coronal suture in deformed skulls and commented on its relationship to the temporal line. Her methods are modeled on those of Ranke and Damasus Aigner, his student. The definitive publication includes photographs and halftone drawings, as well as comparisons of means and ranges with Aigner's data on brachycranic and dolichocranic skulls (Dillenius, 1912b). Both Lehmann-Nitsche and Ranke published descriptive studies of South American crania that are cited in Dillenius's publications.

Lehmann-Nitsche's sojourn in Argentina coincided with a period of social liberalism; a military coup in 1930 signaled the beginning of a reactionary era characterized by the restriction of immigration and implementation of eugenic policies (Stepan, 1991). Conflict over national loyalties and German cultural imperialism was frequent within the university community (Garcia and Podgorny, 2000). One issue was faculty publication and instruction in Spanish rather than German, or other languages, together with changes in Argentina's international relations during World War I (Podgorny, 2001). One might ask how these events affected Dillenius in her later career decisions.

V. MARIAN KNIGHT STECKEL (1889–1982); DELLA COLLINS COOK

A. Biography

Marian Vera Knight was born in Fall River, Massachusetts, to John G. Knight and Edith Woodward. After her brief career at Smith College, she married Harvey S. Steckel of Allentown, Pennsylvania in 1917. There were two children.

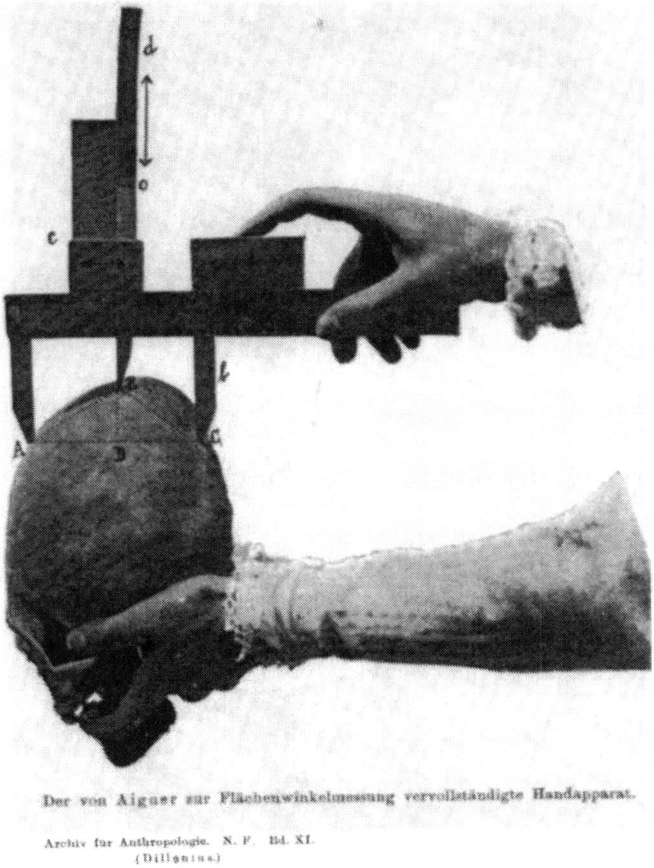

Figure 2 Method of measuring the parietal subtens and a Calchaqui' skull: perhaps these are Juliane Dillenius's hands (Dillenius, 1912: plate 3).

In 1953 she was honored by the Lehigh County Medical Society for her service as founder and president of the Public Health Nursing Association, as well as her work for the Red Cross and the Lehigh County TB Society. This award suggests that her later years were active, though not in physical anthropology.

B. Professional Career

At Smith College, Knight was a student of Harris Hawthorne Wilder (1864–1928), professor of biology. She earned her A.B. degree in 1912 and her M.A.

in 1914. She worked at Smith College as a demonstrator in zoology from 1912 to 1914 and as an assistant in zoology from 1914 to 1917. Knight published her master's thesis, a description of 93 Indian crania from New England (1915). Her study included collections from the Peabody Museum at Harvard, the Park Museum in Providence, Amherst College, Smith College, and private collections. Some of her specimens figure in Wilder's papers on facial feature reconstruction and mortuary archaeology (1912, 1923).

Knight's paper is a cutting-edge craniology for its day. Eight indices are given the most prominent treatment: each is presented as a histogram and as a table of three or five categories by sex. Several arcs and chords are compared, and cranial capacity is determined. Averages for 46 measurements and 22 indices are reported separately for males and females, and the typical skull of each sex is described using index categories.

Six skulls are illustrated in photos and two are presented as sagittal section drawings (Fig. 3). Individual data are presented in fold-out tables. Tooth loss

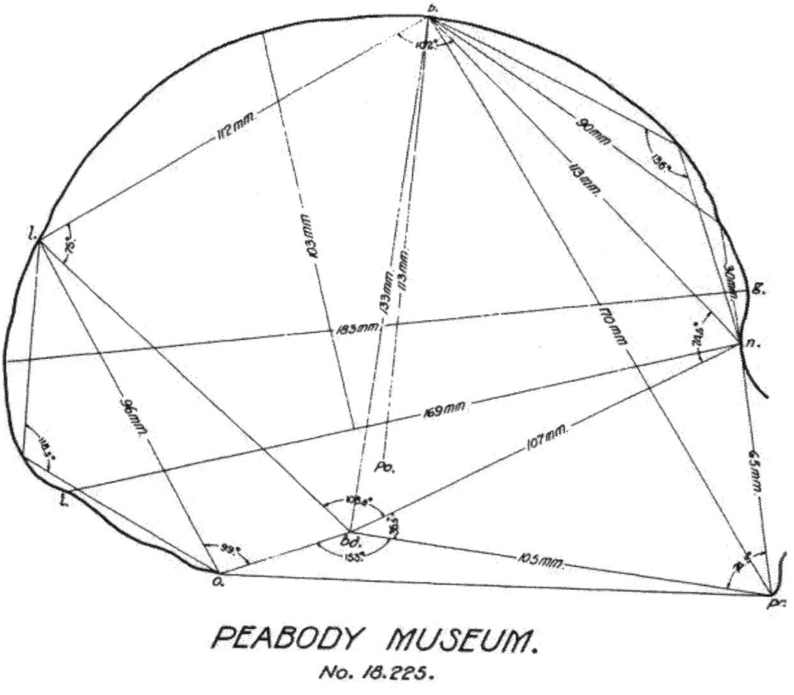

Figure 3 Marian Vera Knight's analysis of a skull from Swampscott, Massachusetts, using a cubic craniophor and Lissauer diagraph (Knight, 1915: plate X).

is discussed where it contributes to facial asymmetry, but pathological changes and dentition are given minimal attention otherwise. One skull is described as unusually large and is interpreted as a tribal leader.

In his essay on the history of physical anthropology, Hrdlička (1918) points out Knight's publication and tells us that Wilder began teaching anthropology at Smith in 1905. One of his earliest students must have been Inez Whipple (1871–1929), who, Hrdlička points out, later became Mrs. Wilder. She earned her bachelor's degree at Brown and her master's degree at Smith. She published her M.A., a long article on the comparative anatomy of the dermal ridges (Whipple, 1904), and several papers on amphibians. Dermatoglyphics was continuing research interest of her husband's (Wilder, 1902; Wentworth and Wilder, 1932). Inez Whipple Wilder also taught biology at Smith College. She survived her husband and wrote his biography.

Wilder was perforce a mentor of women. His textbook of anthropometry (1920) identifies two women as having completed M.A. degrees in his anthropological laboratory: Marian Knight and Margaret Washington. Tables of data from their theses appear in the appendix to his textbook. Margaret Washington Pfeiffer later published her anthropometric research (Wilder and Pfeiffer, 1924). Charlotte Day Gower Chapman (1902–1982), known for her later work on the ethnography of Sicily and the Antilles, was also Wilder's student. Her M.A. research on the morphology of the nasal aperture (Gower, 1923) seems to have escaped the attention of later scholars, perhaps because it was critical of the racial generalizations of the day. Comparison of portrait photos of Knight on file at Smith College and photos of Whipple (Cummins and Midlo, 1943) show that Whipple was the model used in several illustrations in Wilder's textbook (Fig. 4).

VI. RUTH SAWTELL WALLIS (1895–1978); DELLA COLLINS COOK

A. BIOGRAPHY

Ruth Sawtell Wallis is remembered primarily for her contributions to the anthropometry of children and, with her husband, Wilson D. Wallis, to the ethnography of the Micmac, Malecite, and Eastern Dakota. However, her early career was in bioarchaeology. Full accounts by June Collins (1979) and Patricia Case (1988) provide the basis for this sketch of her life.

Her father was a shopkeeper in Springfield, Massachusetts. She was the eldest child. After taking a B.A. in English, she earned her M.A. at Radcliffe in 1923, studying under E. A. Hooton. Fellowships from Radcliffe and the

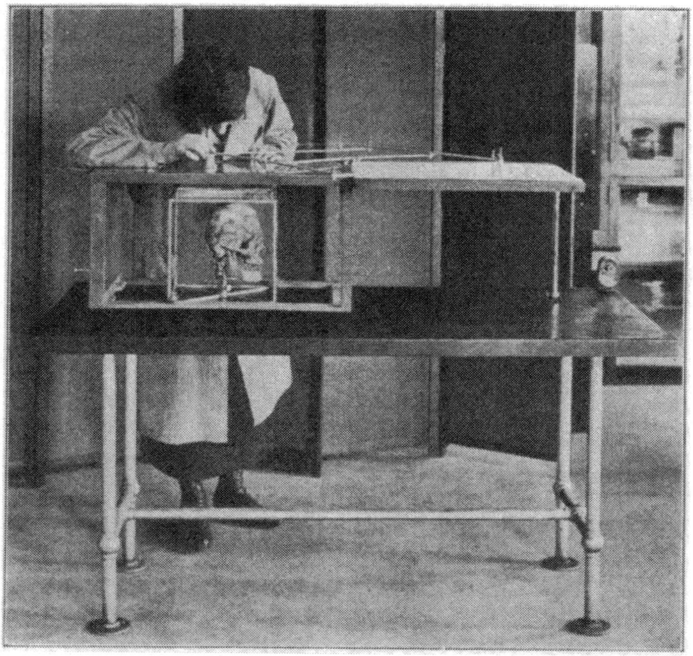

Figure 4 Miss Inez Whipple demonstrating the dioptograph of Lucae in the anthropological laboratory at Smith College (Wilder, 1920:24).

National Research Council allowed her to spend 1923 to 1925 in France, Germany, and Britain, where she collected data for the studies discussed later. On her return, she entered the graduate program at Columbia University, completing her Ph.D. in 1929 and assisting Franz Boas in his work on head form in immigrants. She held faculty positions at University of Iowa (1930–1931), Hamline University (1931–1935), and Amherst College (1956–1974). Case (1988:363) attributes the hiatus in her career to "lack of funding, professional jealousy, and the attitudes of day toward married women in academia." From her marriage in 1931 to her husband's retirement she collaborated on his ethnographic research, conducted several anthropometry projects for government agencies, and wrote mystery novels with an anthropological bent. Wilson Wallis was a widower with two children, but none were born to his second marriage. Both Case and Collins stress Ruth Sawtell Wallis' role as a mentor to students at University of Minnesota, where her husband spent his career.

B. Professional Career

Not the least of her accomplishments in bioarchaeology was her substantial role in Hooton's groundbreaking study, *Indians of Pecos Pueblo*. Hooton says of her, "I cannot express my sense of obligation and gratitude to the three young Radcliffe graduates who have been my principal assistants in the preparation of this report. Dr. Ruth Otis Sawtell, now of the Bureau of Educational Experiments, New York, was my sole helper from 1921 to 1925. During this period she repaired and catalogued the bulk of the Pecos collection, recorded my observations, measured the cranial capacities, and made substantial progress in the work of statistical reduction" (1930:viii).

Few members of our profession have served so thorough an apprenticeship. Her fellowship to Europe yielded three publications of interest here. She collaborated with Ida Treat and Paul Valliant-Couturier in excavating Trou Violet, a Mesolithic site in the French Pyrenees. June Collins (1979) reports that Sawtell later looked back on her work in Europe as one of the happiest periods of her life. Sawtell's monograph on two skeletons from these excavations is a model of careful description using Hooton's typological framework (Sawtell, 1931). Her attention to context and taphonomy is noteworthy. Comparisons are made to several ancient European series, including Cro-Magnon, Mentone, Chancelade, Kaufertsberg, Ofnet, and Mugem, as well as to published data on modern humans. Hrdlička's data on Munsee and U.S. Whites are oddly prominent in her analysis. Pathology is discussed at some length. She notes healed cranial trauma and dental abscesses, attributing osteitis in an ulna to the latter: "such a condition as has already been described in the molar region of the maxilla and mandible must have sent out a septic stream to all parts of the body which may well account for the state of the ulna" (Sawtell, 1931:233). Vertebral lesions are diagnosed as arthritis deformans, with some discussion of Bartels' evidence for Neolithic tuberculosis. Her analysis of the cranium led her to support a scenario of variability rather than a sequence of discrete types.

This question is pursued in two later publications. Her contribution to a 1932 conference on eugenics held at the American Museum of Natural History stands in sharp contrast to the remainder of the papers, most contributed by true believers in racial improvement. She uses a correlation analysis of cranial, facial, and nasal indices in Medieval Merovingian and Reihengraber crania to ask whether there ever was a Nordic race characterized by "harmonic" long heads, long faces, and long noses. She concludes that such harmonic types are uncommon, whether in Medieval Europe, Canary Islanders, or California Indians, the latter two data sets from Hooton's publications.

Her writing has a satirical flavor, as her lead sentence suggests: "There is a great desire for purity, when purity can be obtained painlessly through a mental remolding of ancestral contours. Pure race, tall stature, long head, long narrow face, high, narrow nose; thus a yearning for simple, clear-cut human origins, a sense of the aesthetic and a sense of superiority have clustered around the Nordic, parent of the people one asks to dinner" (Wallis, 1934a:99). She extended this study in a more rigorous paper in *Human Biology* the same year (Wallis, 1934b). The combined Medieval French and German series she had measured, as well as 19 other series from the literature, are examined for bivariate correlations of linear measures rather than indices. She shows that correlations are low with respect to those among body segments and that patterns of correlation differ strikingly from series to series. She ends with a critique of overreliance on correlation in the absence of other statistical techniques, a thoroughly modern point of view. One wishes her interest in the human skeleton had continued after 1934.

Wallis' craniological work is obviously consistent with Boas' programmatic efforts at undermining the race concept in physical anthropology, and it is puzzling that Sawtell is not more widely cited. She carried on a warm correspondence with Boas that is particularly poignant after her Iowa job was terminated because funding to the Iowa Child Welfare Station was cut in 1931. He attempted to help her find grant support and wrote several letters on her behalf. The correspondence ceases with her marriage to an anthropologist for whom Boas had little respect (Boas, 1972). Her last letter to Boas in 1936 congratulates him on his 78 birthday and is more formal in tone. It alludes to an illness that made her a "not efficient assistant" (Boas, 1972) when she was his student.

VII. MILDRED TROTTER (1899–1991); DELLA COLLINS COOK, MARY LUCAS POWELL, AND JANE E. BUIKSTRA

A. Biography

Mildred Trotter was born on February 3, 1899, in Monaca, Pennsylvania. Her parents, James R. Trotter and Jennie Zimmerman Trotter, were farmers; her father also served (for a time) as the director of the community school. She had two sisters, Sara Isabella and Jeanette Rebecca, and one brother, Robert James. After attending a one-room grammar school in her home town, she entered high school in nearby Beaver, Pennsylvania, where she encountered opposition from the principal for choosing geometry over home economics in her course schedule. She graduated from Mount Holyoke College in 1920 with a B.A. degree in zoology and promptly began graduate study in anatomy at Washington University

School of Medicine (WUSM) in St. Louis, Missouri, as a student of Robert J. Terry, earning her Ph.D. on hypertrichosis during 1924.

B. Professional Career

Trotter's research of relevance to bioarchaeology focused primarily upon the influence of sex, age, and race on skeletal development (Spencer, 1997c). After spending 14 months in 1948–1949 at the U.S. Armed Forces Central Identification Laboratory in Hawaii working on identification of skeletal remains from the Pacific theatre of World War II, she developed a series of formulae (with Goldine Gleser) for estimation of stature based on long bone lengths from identified individuals of known stature (Trotter and Gleser, 1952). This study was one of the first to employ data from military casualties for scientific research, and the formulae are widely used in forensic anthropology today. During her long and very productive career at WUSM she taught anatomy to more than 4000 students during five decades, and the WUSM alumni association honored her by establishing an endowed scholarship in her name in 1975. Despite her active schedule of research, teaching, and publication, she encountered numerous difficulties and discouragement in her academic advancement because of her sex. However, she did not submit willingly, and after 16 years at associate professor rank she demanded that her chairman either promote her or document her deficiencies; shortly afterward she because the first full professor at WUSM, in 1946. Her experience with sex discrimination prompted her to mentor her female students with particular care, although her male students also received attentive guidance. Trotter was a member of the American Association for the Advancement of Science, the American Anthropological Association, the Anatomy Societies of Great Britain and Ireland, and a founding member (and first female president) of the American Association of Physical Anthropologists.

Trotter published very little on human remains from archaeological contexts. While on a National Research Fellowship to the University of Oxford, England, she decided to conduct a comparative study of spines of Egyptian mummies and ancient Britons instead of continuing research on human hair (the topic of her doctoral dissertation). Her report (Trotter, 1937) appeared in *AJPA* in a series of articles on the human spine by students of R. J. Terry; it was the only one on ancient remains, with the remainder of papers reporting data from Terry's eponymous collection of modern American cadavers at WUSM. The preceding paper in this series is by Caroline Whitney, suggesting that Terry encouraged at least one other female student. Trotter discovered, during her time at Oxford, that she much preferred working on bones and continued to do so after returning to the United States. Her research on human hair is little cited today, but her other contribution to bioarchaeology was a descriptive paper on the hair of South American mummies that discusses hair morphology and taphonomy (Trotter, 1943).

VIII. MADELINE D. KNEBERG (1903–1996); LYNNE P. SULLIVAN

A. BIOGRAPHY

Madeline D. Kneberg was born in Moline, Illinois, on January 18, 1903, the youngest of three daughters of Charles E. and Anna Anderson Kneberg. Her parents were the children of Swedish immigrants, and the family maintained Swedish traditions through church and civic groups. Both parents were professional interior decorators, and the family was financially secure. Charles Kneberg encouraged Madeline's athletic abilities (giving her baseball gloves instead of dolls) and taught her to drive a car by the age of 13. Later in life, Kneberg reminisced "My father did his best to make a boy out of me." After her father was killed in a tragic accident in 1916, Kneberg left Moline at age 16 to attend preparatory school at Southern Seminary in Buena Vista, Virginia, and then enrolled in Martha Washington College in Fredricksburg for a year. During this time, she also coached a girls' basketball team and taught horseback riding. Kneberg's father also had encouraged her early artistic interests, and her talent is evident in her many drawings and paintings for publications and exhibits (e.g., Lewis and Kneberg, 1946). In 1924, at age 21, she went to Florence, Italy, to study opera singing for four years, but chronic sinus infections threatened her future vocal studies and she finally decided to seek a different career.

Kneberg returned to the United States and enrolled at Presbyterian Hospital in Chicago, graduating 3 years later as a nurse. In 1931 she enrolled at the University of Chicago, completing a bachelor of science degree the following year. She majored in sociology with a minor in psychology and courses in anthropology and history. The encouragement and support of Fay-Cooper Cole, head of the Anthropology Department at Chicago, encouraged Kneberg to make her ultimate career choice: physical anthropology.

B. PROFESSIONAL CAREER

Kneberg's first professional employment as a physical anthropologist came in 1932, when Cole employed her for $50 per month on a grant project studying possible racial differences in human hair, conducted in the anatomy laboratory of William Bloom at the University of Chicago with support from the

National Research Council. Her first publication, in the *American Journal of Physical Anthropology* (Kneberg, 1935), demonstrated considerable variation in the shape of the hair shafts from one person, illustrated by her photomicrographs of hair cross sections. Kneberg noted that a previously developed index of hair shaft diameters (Martin, 1914) did not reliably correlate with racial groups, and she proposed that the form of the hair shaft was related to the structure of the follicle, thus providing a physical basis for a hair's relative curliness or straightness. Kneberg's second set of publications (1936a,b) on human hair describes the procedure for creating scalp sections for studying hair follicles and the results of a study of hair weight. The latter study concluded that there is no correlation between hair weight or size and racial groups. Her work on the hair project eventually provided the basis for her master's thesis (Kneberg, 1936) and her doctoral research.

The four University of Chicago faculty members who had the most influence on Kneberg's training and intellectual development as an anthropologist were Franz Weidenreich, A. R. Radcliffe-Brown, Fay-Cooper Cole, and Thorne Deuel. Kneberg learned techniques for reconstructing human anatomy from fossil bones in Weidenreich's laboratory, and Radcliffe-Brown's view of society as a system of interrelated parts undoubtedly influenced her work with T. M. N. Lewis on the prehistory of the Chickamauga Basin in Tennessee (Lewis and Kneberg, 1946; Lewis *et al.*, 1995). Fay-Cooper Cole, her lifetime mentor and friend, was one of a now nearly extinct breed—a general anthropologist. Before coming to the University of Chicago in the late 1920s, he had trained primarily as a physical anthropologist, conducted research in the Philippines in both physical anthropology and ethnology, and worked at the Field Museum. His UC summer field schools "became famous for their system of horizontal and vertical control of archaeological excavations" of mounds and villages (Fowler, 1985:7). Although Kneberg did not attend his field school, she was clearly influenced by his emphasis on scientific data collection procedures. Deuel's collaborative publications with Cole examined systematically all of the excavated materials from an archaeological site with an eye toward interpreting the functional implications of the artifacts (Willey and Sabloff, 1977:135). In *Rediscovering Illinois* (Cole and Deuel, 1937), traits were classified by functional categories such as "agriculture and food-getting," an innovation that focused the investigators' attention on the activities that produced the artifacts rather than on the objects themselves (Willey and Sabloff, 1977:135). Cole was fond of saying that he and Deuel were developing techniques that would "make the dead past live again" (Fowler, 1985:9). Kneberg adopted this approach and added a strong focus on interpretation of the past for the general public as well as for academic scholars, one of her most significant contributions to southeastern archaeology.

At Chicago, Kneberg took premed classes, including gross anatomy, microscopic anatomy, and physical chemistry, as well as anthropology. However, caught in the middle of the Great Depression, her financial situation became increasingly difficult. To make ends meet, she lived with five nurses in a one-bedroom apartment and earned $5 per Sunday singing in the university choir. Eventually she was forced to abandon her dream of becoming a physician and decided instead to pursue a career in physical anthropology because she enjoyed the subject and realized that her studies in human anatomy would prove to be very useful.

After completing her M.A. degree in anthropology at Chicago in 1936, Kneberg took a temporary job at Beloit College in Wisconsin for 6 months, where she taught general anthropology and a class on the Pueblo Indian and worked in the college museum. In the spring of 1937, she arranged a collaborative project between Beloit College and the University of Chicago to excavate a conical mound near Shireland, Illinois, as a field experience for her Beloit students. The mound was the largest of a group of 16 that included effigy mounds. Thorne Deuel brought the field equipment and some student supervisors from Chicago. The 2-day dig was Madeline's only archaeological field supervisory experience before going to Tennessee.

In 1937, Kneberg returned to Chicago to continue work on her doctorate, resuming her study of human hair and passing her preliminary examinations. However, her work on her dissertation ended when Krogman and Cole recommended Kneberg to T. M. N. Lewis for the job as a physical anthropologist to analyze the skeletal material from Lewis' archaeological excavations for the WPA in Tennessee. The financial pressures of the Great Depression persuaded Kneberg to accept the job, and at age 35, she moved to Knoxville, accompanied by her mother and her sister.

Thomas M. N. Lewis was a graduate of Princeton, served in the Navy during World War I, and entered graduate school at the University of Wisconsin where he took anthropology classes. When his father became ill, Lewis helped with the family firm in Watertown, Wisconsin, and occasionally did archaeological fieldwork with William C. McKern at the Milwaukee Public Museum. In 1934, McKern recommended Lewis as a field supervisor on the large Tennessee Valley Authority reservoir projects in Tennessee. Lewis subsequently was appointed at the rank of Associate Professor in archaeology at the University of Tennessee, charged with supervising all archaeological work in the state. The Division of Anthropology was established as a section of the history department. Lewis set up a laboratory in Knoxville to catalog materials brought from the huge field projects, but it was not until 1938 that the WPA archaeological program was restructured to support laboratory analysis and promote publication.

By this time Kneberg arrived in Knoxville in June 1938; the laboratory work was 4 years behind the fieldwork (Lyon, 1996:149). As director of the laboratory (1938–1942) she oversaw half a dozen supervisors and 40 laboratory workers. Kneberg was responsible for the preparation, restoration, cataloging, and analysis of all archaeological material. Under her direction, the Knoxville laboratory developed an innovative attribute-based system for artifact classification, a technique for pottery vessel reconstruction, and numerous card files for analytical purposes and collections management. For the Chickamauga project alone, the laboratory staff classified over 360,000 pottery sherds and some 100,000 stone, bone, shell, and copper artifacts. They also reconstructed several hundred pottery vessels and examined all of the nearly 2000 recovered skeletons for age, sex, and skeletal pathology. She met her goal of clearing the laboratory's enormous backlog by the time the Chickamauga project ended and was remembered decades later by archaeologists active in Southeastern WPA projects as a dynamic force and the source of inspiration for much of the analytical work.

Lewis and Kneberg (n.d.) jointly developed a *manual of field and laboratory* techniques employed by the UT–Knoxville Division of Archaeology, based on methodologies developed at Fay-Cooper Cole's archaeological field schools and refined by Lewis' field experience and Kneberg's laboratory innovations. This manual provided detailed instructions for the organization of archaeological projects, the selection and deployment of crew members, methods of excavation, artifact and data recording, mapping, and laboratory analysis, including a series of carefully designed field and laboratory forms with directions for proper completion. Thanks to Lewis and Kneberg's rigorous attention to detail and insistence on consistency in data collection and recording, the vast systematic collections and associated documentation from the WPA archaeological projects that they supervised have provided a wealth of data for successive generations of scholars (including many of the authors of chapters in this edited volume).

In 1940, the University employed Kneberg to teach anthropology, including human evolution. Knoxville was only about 60 miles from Dayton, Tennessee, the scene of the infamous 1926 Scopes trial. The president of the university told her to teach what she thought she should, and the university would stand behind her; Fay-Cooper Cole offered to testify for her if she got arrested, as he had for Scopes. Within weeks after the bombing of Pearl Harbor in 1941, the federal government stopped all WPA fieldwork and disbanded most of the laboratory. By June of 1942, the program ended because the workers were needed for the war effort and several of the WPA supervisors were drafted. The end of the New Deal projects came so fast that Lewis and Kneberg personally had to pack up the laboratory. Federally supported archaeology ground to a halt during the war and never again had the funds or labor force for large projects. The New Deal era "golden age" of southeastern archaeology was over.

Kneberg entered physical anthropology when the field was just coming into its own. The analytical program for human osteology that she instituted for the WPA laboratory in Knoxville (Lewis and Kneberg, n.d.) incorporated detailed data collection methods with important innovations in interpretation. For example, she was clearly influenced by Hooton's landmark study of Pecos Pueblo (1930) in her methodology for collection of systematic cranial measurements for typological comparisons. However, instead of interpreting her data with reference to the prevalent scheme of racial groupings worldwide, she sought to construct specifically regional typologies based on the human remains recovered from successive Native American occupations at different archaeological sites, an approach far more useful to Southeastern archaeologists than, for example, descriptive categories such as Hooton's "Pseudo-Negroids" at Pecos. She compared archaic Native American postcranial morphology at the Eva site in eastern Tennessee with Archaic population samples from WPA archaeological excavations at Indian Knoll, Kentucky (Snow, 1948), and noted how the late prehistoric inhabitants of Hiwassee Island differed physically in some respects from historic accounts of 18th-century Creek Indians in eastern Tennessee [e.g., less sexually dimorphic than Bartram's descriptions of the "Muscogulge" would suggest (Bartram, 1928)]. Her documentation of specific biological changes through time in cranial shape and dental features in Native Americans of Tennessee stimulated later scholars' functional analyses linking biological to cultural changes in subsistence regimen and technology from Archaic to Late Prehistoric times.

Kneberg maintained a lifelong friendship with Wilton Marion Krogman, a leading forensic anthropologist and former classmate of hers at UC and regularly consulted him on matters of skeletal analysis. This connection may have inspired her reconstructions of the facial characteristics of prehistoric individuals, nicely illustrated by her own hand in the reports on Hiwassee Island (Lewis and Kneberg, 1946) and the Chickamauga basin (Lewis *et al.*, 1995). These reconstructions were among the first efforts to apply forensic techniques to archaeological material and served Kneberg's strong focus on humanizing the scientific study of the past.

Kneberg's reports both present detailed demographic data (presented in tabular form) for the different component samples, and for Eva she employed statistical analysis (χ^2 tests) to compare observed rates of death for males and females, suggesting that the dangers of pregnancy and childbirth were responsible for the high mortality among females aged 15–30 compared with males. She drew upon historic accounts of the devastating impact of newly introduced European infectious diseases such as smallpox and whooping cough to explain the sharp differences in subadult mortality between pre-contact and post-contact components at Hiwassee Island. Her descriptions of individual burials contain numerous observations on skeletal pathology, and she was quite interested in the effects of different subsistence regimens on dental health, such as the heavy tooth wear observed in the Archaic inhabitants of Eva and the lighter pattern of wear but heavier burden of caries in the Late Prehistoric components at Hiwassee Island. Her diagnoses of specific conditions include osteoarthritis, *osteoporosis symmetrica* and *cribra orbitalia,* trauma, and periostitis, but she does not attempt to identify specific infectious diseases except to note that "there is a suggestion that the [cranial lesions in a young child from Hiwassee Island] are of syphilitic origin. This would not be unexpected in a skull of the historic period" (Lewis and Kneberg, 1946:167).

During the years as laboratory director for the WPA projects, Kneberg gradually shifted her career focus away from physical anthropology; as she commented later in life, "I lost interest in physical anthropology to a large extent because archaeology was such a challenge." Her analyses on the population samples from Hiwassee Island and Eva are well known, but aside from these two reports there are no other osteological studies among her numerous archaeological publications. As a result, her contributions to American bioarchaeology have not received the recognition that they deserve in the scholarly community.

Kneberg reminisced that she and Lewis were constantly exchanging ideas and carried on a continuous conversation about interpretations and what they were trying to do; she noted that they "worked as one person" and she never knew "where she stopped and where Tom began." She referred to Tom as the "more practical" of the two, in the sense that she was more able to imagine past peoples' lives in a "popular" sense. She also felt that their work drew criticism at times because it was too popular, noting that the emphasis in the 1930s was on developing scientific techniques for data collection, not on reconstructing the past. It seemed to her that some people were overly obsessed with scientific techniques. Certainly one should have accurate information on which to base reconstructions, but, in her view, the point was to interpret data.

Both Tom and Madeline kept up their scholarly contacts through regular attendance at scientific conferences, hosting scholarly visitors, performing service activities, and helping to found chapters of Sigma Xi and Phi Beta Kappa at the University of Tennessee in Knoxville. Even though Kneberg had turned her interest toward archaeology, she still maintained scholarly ties to physical anthropology as evidenced by a visit to Knoxville in the spring of 1950 from the noted physical anthropologist T. Dale Stewart of the Smithsonian. She also fondly recalled discussions with archaeologists Robert Wauchope, William Haag, and James B. Griffin; Madeline related that she and Griffin "used to argue," as evidenced by a discussion of pottery typology at the 1959 meeting of the Southeastern Archaeological Conference (Williams, 1962).

In 1961, the University of Tennessee Press published the report on the Eva Site, one of the WPA excavations in the Kentucky reservoir (Lewis and Lewis, 1961). This now classic report on the archaic period in the midsouth was Kneberg and Lewis' final collaborative publication; later that year they decided to retire, to marry, and to move to Florida. Lewis died in 1974. In 1994, at 91 years of age, Madeline was still interested in sharing her ideas about archaeology with the public and agreed to be interviewed for television programs about the WPA/TVA archaeology projects in Tennessee and about her own career. In 1995, she was awarded the Southeastern Archaeological Conference's highest honor, the Distinguished Service Award. After suffering a stroke in 1995, Madeline never recovered her health and she died in her sleep on July 4, 1996, at 93 years of age.

Madeline Kneberg Lewis was a woman of diverse talents and interests, who followed (often of necessity) a winding, but always interesting, path through life. Her early work in physical anthropology and her emphasis on integration of biological with cultural information are worthy of recognition. Madeline was one of the few women to win professional recognition for her work on WPA archaeology projects of Roosevelt's New Deal. She was elected a fellow in the American Association for the Advancement of Science and was one of the first, if not the first, women to hold a full professorship in anthropology in the Southeast.

To Madeline Kneberg Lewis, the purpose of archaeology was straightforward: "to reconstruct the past." She believed strongly that archaeology has the potential to show the human race its pitfalls and past mistakes, and she was concerned that science today tends to emphasize the physical sciences at the expense of the social sciences, a dangerous trend because many of the modern world's problems — overpopulation and competition for space and natural resources — have social causes. She and Tom realized that they could never take advantage of the full potential of the vast WPA collections during their own lifetimes, but they were determined to secure this outstanding legacy for future scholars of the past.

IX. KATHARINE BARTLETT (1907–2001); DELLA COLLINS COOK

Katharine Bartlett, an anthropologist who devoted her working life to the archaeology of the southwestern United States, was born in Denver, Colorado, in 1907. She received her M.A. in anthropology from the University of Denver in 1929, under the direction of Dr. E. B. Renaud. In 1930, she took a summer job at the newly formed Museum of Northern Arizona in Flagstaff and joined the permanent staff later that year. Bartlett spent more than a half century (1930–1981) at the museum, working to develop research programs in archaeology, geology, ethnology, zoology, and botany in the Colorado plateau. As the first Curator of Anthropology (1931–1952), she created the cataloguing system for the museum's collections and, later, as Librarian and Curator of History (1953–1981), she acquired and organized library materials and archives for the most comprehensive collection of research material in northern Arizona. During these decades she also designed exhibits, edited museum publications, and produced more than 60 publications based on her research. In 1984 Bartlett was honored as the first Fellow of the Museum of Northern Arizona, and in 1986 took part in a conference at the Arizona State Museum on the work of early women anthropologists in the Southwest; this conference formed the basis for the Smithsonian Institution's popular traveling exhibit, "Daughters of the Desert" (Anonymous, 2001).

Bartlett's original area of interest was physical anthropology. Her mentor, Etienne Bernardeau Renaud (1880–1973), began his education in romance languages at the University of Paris. He converted, like Hooton, from humanities to anthropology after earning his doctorate at the University of Denver. He served as professor of anthropology there and helped build archaeological collections at the Denver Museum of Natural Sciences. The archaeologist Marie Wormington also studied with Renaud. Bartlett analyzed human skeletal material from many archaeological excavations directed by the Museum of Northern Arizona and the National Park Service. Her physical anthropology is to be found in the appendices of field reports, but it bears wider reading. In general, she describes crania in the context of Hooton's work at Pecos Pueblo, stressing, as did Hooton, variability and deformation (Bartlett, 1941, 1954). Bartlett's contribution to WPA excavations at Montezuma Castle (1954) includes an account of a woman who may have briefly survived three arrow wounds that begs for further study. A brief article on Pueblo foodstuffs touches on nutrition (Bartlett, 1931). The published catalog of the "Daughters of the Desert" exhibit includes a well-illustrated account of her contributions, stressing her publications on history and folk arts in the Southwest (Babcock and Parezo, 1988).

Ms. Bartlett never married. Her companion of 30 years was the archaeologist and artist Gene Field Foster (Anonymous, 2001).

X. ADELAIDE KENDALL BULLEN; MARY LUCAS POWELL

A. Biography

Adelaide Kendall (Bullen) was born on January 12, 1908, in Worchester, Massachusetts, to Grace Marble Kendall and Oliver Sawyer Kendall. She married Ripley Pierce Bullen, 6 years her senior, on July 25, 1929, and they had two sons, Dana Ripley Bullen II (born in 1931) and Pierce Kendall Bullen (born in 1934). R. P. Bullen's first professional career was in engineering research for General Electric, after earning a degree in mechanical engineering from Cornell in 1925. However, his true interest was in archaeology, and in 1939 he helped to found the Massachusetts Archaeological Society. The following year he left G.E. for a position at the Robert S. Peabody Foundation for Archaeology at Phillips Academy in Andover. When Ripley began graduate studies in Harvard in 1940, Adelaide entered Radcliffe and received her B.A. degree cum laude 3 years later, despite the responsibilities of raising two small children. She promptly began graduate work in cultural and physical anthropology at Harvard, working respectively with Clyde K. M. Kluckhohn and Earnest A. Hooton.

In 1948, Ripley was offered the post of assistant archaeologist with the Florida Board of Parks and Memorials, and the Bullens moved from Massachusetts to Gainesville, Florida. When Ripley joined the staff of the Florida State Museum (later the Florida Museum of Natural History) in 1949, Adelaide joined him there as a volunteer on archaeological projects, listed as an "associate" in anthropology, and they began their lifelong affiliation with that institution.

The Bullens were founding members of the Florida Anthropological Society in 1948 and served as officers and editors of *Florida Anthropologist.* Despite their many significant contributions to the Florida State Museum and to the development of Florida archaeology, neither of them held academic teaching positions in the Department of Anthropology at the University of Florida, as did most of their colleagues at the museum. According to Adelaide Bullen's biographer Rochelle A. Marrinan, this surprising omission was most likely due to the fact that neither of them had completed advanced academic degrees in anthropology (Marrinan, 1999).

B. Professional Career

Adelaide Bullen maintained a very active schedule in research and publication in different areas of physical anthropology throughout her lifetime. Her first published paper was a cross-cultural study of stuttering in Navajo, Oceanic, and U.S. White children (1945a). Other reports drew upon her research on somatotyping conducted while she was at Radcliffe (Bullen and Hardy, 1946; Bullen, 1948, 1952, 1953a, 1967a) and her work at Harvard as a civilian consultant to the Department of the Army and a member of the fatigue research staff studies on body fatigue (Bullen, 1956, 1967b). Her paper on qualitative and quantitative aspects of body build research, which appeared in the *Quarterly Journal of the Florida Academy of Sciences*, was awarded the Academy's Phipps-Bird award for the best article published that year. She was interested in anthropological perspectives on aging (Bullen, 1962a), penned a detailed biography of her husband after his death for the *Proceedings of the Seventh International Congress for the Study of Pre-Columbian Cultures of the Lesser Antilles* (Bullen, 1978) and, late in life, wrote a children's book, *Jim Tall and Count Small* (Bullen, 1975), about the friendship between a circus tall man and a midget (perhaps inspired by her early interest in body type studies). As contributing editor in physical anthropology for the *Handbook of Latin American Studies* (Bullen, 1969, 1971), Bullen provided annotated bibliographic entries for a broad range of recent publications (many of them in Spanish) on bioarchaeological analyses and studies of living populations not widely available to English-speaking scholars.

Adelaide and Ripley Bullen began their joint career in archaeology in 1941, when they attended the University of New Mexico summer field school at Chaco Canyon (A. K. Bullen and R. P. Bullen, 1942). In their first collaborative projects, Adelaide focused on archaeological materials, not human remains, as in, for example, their joint report on the excavation of the homestead of a freed slave, Lucy Foster (A. K. Bullen and R. P. Bullen, 1945), "an excellent early treatment of a domestic site related to slavery and the circumstances of free persons of color" (Marrinan, 1999:150). Her thoughtful essay in *American Antiquity* urged archaeologists working in the southwestern United States to familiarize themselves with the ethnohistoric record — to "catch our archaeology alive" — in order to interpret correctly materials recovered from their excavations (Bullen, 1947:133). Her first published skeletal report, a careful analysis of two historic period burials from Rhode Island (A. K. Bullen and R. P. Bullen, 1946), incorporated biological and historic data to support her suggestion that the two juveniles were of mixed Native American and African ancestry.

After their move to Florida, Adelaide was actively involved with many archaeological projects with her husband (A. K. Bullen and R. P. Bullen, 1950, 1953, 1954, 1961a,b, 1963, 1966a,b,c, 1970; R. P. Bullen and A. K. Bullen, 1956, 1963, 1967, 1968a,b,c, 1972, 1973a,b, 1974a–d, 1976a,b,c; R. P. Bullen, 1966; Bullen *et al.*, 1967, 1968, 1973a,b). When human burials were recovered, Adelaide conducted a careful analysis of the skeletal remains while Ripley interpreted the cultural materials. In some publications, Adelaide's analysis is credited to her in the table of contents but she is not listed as coauthor [e.g., her brief interpretation of the burial from Burtine Island, in R. P. Bullen (1966:11)]. In other reports, she is credited for coauthorship and her analysis appears at the very end under her name, as does her discussion of the very fragmentary human skeletal material "collected from the treasure seeker's spoil" at the Lemon Bay School mound (R. P. Bullen and A. K. Bullen, 1963:56). Her contributions to projects directed by other archaeologists, however, are often "invisible" in the bibliographic record; for example, her description of human skeletal remains recovered during an archaeological reconnaissance in the U.S. Virgin Islands appears on a single page within the text (Sleight, 1962:25) and she is credited by name but not listed as a coauthor. Adelaide's anthropological training is evident in even her briefest reports: she weighs both biological and cultural data in forming her conclusions, she clearly states the potential and the limitations of the archaeological record, and refuses to extend interpretation beyond evidence by identifying, for example, postmortem damage to bones with cannibalism or absence of body parts with trophy taking.

Adelaide Bullen's most significant contribution to American bioarchaeology was a lengthy survey article published in the *Florida Anthropologist* (Bullen, 1972) titled "Paleoepidemiology and Distribution of Prehistoric Treponemiasis (Syphilis) in Florida." She states her goal clearly on the first page: the search in the archaeological record for "incontestible evidence of the presence of syphilis in Florida in pre-Columbian times" (Bullen, 1971:133, her emphasis), followed by a review of the paleopathological literature on this topic, citing Aleš Hrdlička, Herbert U. Williams, and Aidan Cockburn. Her initial focus is on one skeletal individual, an adult female from the Palmer Mound radiocarbon dated around A.D. 850, with extensive pathological lesions. Adelaide consulted Ellis R. Kerley and Lent Johnson at the Armed Forces Institute of Pathology and T. Dale Stewart at the Smithsonian Institution, three physicians with considerable expertise in paleopathology and a strong interest in infectious diseases. Kerley's differential diagnosis considered yaws (another treponemal disease similar to, but not identical with, venereal syphilis) as well as other diseases known to produce lesions of similar form and distribution: tuberculosis, Paget's disease, osteomyelitis, and pulmonary osteoarthropy. His final diagnosis of venereal syphilis rather than yaws was based on multiple points of agreement with the clinical profile of that disease, as well as upon two points of "negative evidence:" the absence of palate and nasal lesions frequently associated with the latter disease and the lack of any historical record of yaws in Florida.

Bullen noted that additional skeletal material from the Palmer site was also diagnosed by Kerley as syphilitic, an important point in the epidemiology of treponemal disease, and that Charles Snow (1962) had diagnoses of syphilis (supported by Kerley) in a burial from the Bayshore Homes site, dated slightly later than the Palmer mound. She then carefully described pathological skeletal material from eight other sites in Florida tentatively identified as "syphilitic" by C. B. Moore, Williams, Kerley, and other scholars, as well as pathological specimens excavated by Moore from the great Mississippian site of Moundville in west-central Alabama and sent for evaluation to Dr. D. S. Lamb at the Army Medical Museum (Moore, 1907). Bullen concluded that, taken together, the skeletal evidence strongly suggested that "syphilis may have been present in a recognizable form as early as 3300 B.C. in Florida" (Bullen, 1972:166), but increased in prevalence (or, for various reasons, in visibility in the archaeological record) during Late Prehistoric times.

Bullen's article is copiously illustrated with skeletal pathology from all of the sites discussed and well referenced from both the clinical and the paleopathological literature, a very significant contribution to the documentation of the natural history of treponematosis in North America (Powell and Cook, 2005). Her choice of the term "treponemiasis" instead of the more synonymous "treponematosis" is unusual, and her reasons for using it are unknown. Her review was paired in the same issue of *Florida Anthropologist* with an article (Warren, 1972) devoted to epidemiological, clinical, and historical aspects of the modern treponemal syndromes — pinta, yaws, and endemic syphilis (bejel) — other than venereal syphilis. Reviewing Bullen's data and arguments, Warren hypothesized that the form of treponematosis identified in her Florida skeletal material was, in earlier times, "previously endemic as pinta or bejel" but "acquired a more extensively endemic or even epidemic form for certain groups by Weeden Island times" due to specific cultural changes [increased population size and density, a shift to agricultural-based subsistence, and broader contacts — "some of which may have been venereal" — with other regional groups (Warren, 1972:185)].

In 1975, after almost three decades spent analyzing Native American skeletal remains, Adelaide Bullen decided to end this aspect of her career. She informed William R. Maples, the chief osteologist at the Florida Museum of Natural History, that "... for reasons connected with Indian reactions to osteological studies," she had "decided not to do any detailed analyses of the FSM collections" and instead to conduct "... only general observations connected with publication of wider scope than osteology" (Marrinan, 1999:155). Her biographer, Rochelle Marrinan, noting that "[t]his reaction seems very early to what, in the succeeding decade, became a full-blown problem for osteological studies," (Marrinan, 1999:155) interpreted Adelaide's decision as "a passing of the guard" — her explicit handing over to Maples the role of FSM bioarchaeologist. Although Adelaide's decision did indeed precede federal legislation to protect Native American graves and sacred artifacts (the Native American Graves and Repatriation Act of 1990, known as NAGPRA; see Chapter 15) by some 15 years, Native American protests against archaeological excavation and analysis of burials had first gained a measure of public and professional attention in the late 1960s. Adelaide's great respect and deep sympathy for the first inhabitants of Florida are very evident in her extensive chapter titled "Florida Indians of Past and Present" (Bullen, 1965) in *Florida from Indian Trail to Space Age*, a general history of the state written for a nonprofessional audience a decade before her decision was made. In this chapter she combines archaeological and ethnohistoric data to tell the stories of six different groups, from the 16th-century Timucuans to the modern Seminoles, to search for answers why the Seminoles alone survive in Florida today. It seems reasonable (to me, at least, a fellow bioarchaeologist) that Adelaide's decision was perhaps motivated as much by a growing disjunction between the emotional detachment required by her professional study of the physical remains of Native Americans and her increasing sympathy with the objections of their modern descendants, as by a simple desire for retirement from a demanding area of research.

Bullen's 1972 article on treponematosis is her only single-authored publication cited in Hutchinson's recent monograph, *Bioarchaeology of the Florida Gulf Coast* (2004), although she appears as coauthor of five additional references (R. P. Bullen and A. K. Bullen, 1956, 1976; A. K. Bullen and R. P. Bullen, 1953, 1963; Bullen *et al.*, 1967). Her contributions are almost invisible in Larsen's 2001 edited volume, *Bioarchaeology of Spanish Florida*: only her appendix to the Goodman mound monograph (Bullen, 1963) is cited, in a table listing previous research in Florida, and her analysis of skeletal material from diverse archaeological projects at Amelia Island, curated at three different institutions, is not mentioned, although this site is otherwise well covered. Both Larsen's and Hutchinson's books appeared in the Florida Museum of Natural History's monograph series named in honor of Adelaide's husband, Ripley P. Bullen. The first half of Marrinan's title for her biography of Adelaide K. Bullen — "Best Supporting Actress?" — seems sadly appropriate.

XI. GRETE MOSTNY (1912–1991); PATRICK D. HORNE AND MARIO CASTRO

A. Biography

Grete Mostny was born in Linz, Austria, on September 11, 1912. During her school years she attended the Mädchen Royal Gymnasium, graduating in 1933. During the mid-1930s she started working as an ad-honorem teaching assistant at the Egyptology Department of the Künsthistorisches Museum in Vienna. Later, she entered the University of Vienna and graduated with honors in egyptology and African studies in 1937. By 1939 she had obtained her doctorate in philology and oriental history from Brussels Free University.

B. Professional Career

During her university years, she participated in numerous archaeological excavations throughout Europe. In 1939 she moved to Chile. Dr. Mostny was appointed curator of anthropology at the National Museum of Natural History in Santiago in 1943, a post she held until 1964 when she resigned to become director of the same museum. In 1946 she became a Chilean citizen.

From 1950 until 1972 she taught courses in anthropology and prehistory as a member of the Faculty of Philosophy and Education of the University of Chile. She created the *Monthly Newsletter* of the National Museum of Natural History in 1956, and in 1968 she founded the National Center for Museology. From 1964 until her retirement in 1982, she was a member of the National Monuments Council.

At the international level, from 1954 until the late 1980s Dr. Mostny was a member of the Permanent Council of the International Congress of Prehistoric and Protohistoric Sciences. As president of ICOM-Chile, she was a member of ICOM's executive committee. She also held memberships in the anthropology section of the Pan American Institute of History and Geography and the Museums Association of Great Britain. Throughout these years, she also represented Chile at numerous international scientific meetings and was frequently invited to give lectures at various institutions abroad.

It is thanks to Dr. Mostny's long, illustrious, and productive career that professional archaeology was established in Chile. Her complete bibliography lists more than 178 professional publications, the majority of them dealing with the archaeology and ethnography of Chile (e.g., Mostny, 1954). One of her earliest reports on biological anthropology dealt with the collection of Egyptian mummies at the National Museum of Natural History in Santiago (Mostny, 1940), and she subsequently published studies of mortuary site archaeology (Mostny, 1947, 1952, among others) and various aspects of the biological anthropology of the indigenous inhabitants of Terra del Fuego (Mostny, 1964; Lipschutz *et al.*, 1946a,b, 1947).

Mostny is perhaps best known to North American scholars for her numerous publications (Mostny, 1955, 1956, 1957a–e, 1967a,b) on the mummy known as "The Prince of El Plomo," a young Inca boy sacrificed on a mountaintop in the Peruvian Andes. Her edited monograph (1957a) constitutes the most professional and complete mummy examination to that date, a significant contribution to the bioarchaeology of high-altitude Inca Andean sites and a benchmark example in mummy studies. The various chapters cover the discovery of the mummy and its conservation, a suite of histological, morphological, pathological, and radiological examinations of the very well-preserved body, and detailed descriptions of the artifacts, particularly the boy's elaborate clothing. Published in Spanish, this work is, unfortunately, not well known to English-speaking anthropologists, except through summaries in edited anthologies on South American archaeology (1967a,b).

Dr. Mostny died in Santiago, Chile, on December 12, 1991.

XII. ALICE MOSSIE BRUES (1913–); MARY KAY SANDFORD AND GEORGIEANN BOGDAN

A. Biography

Alice Mossie Brues was born on October 9, 1913, in Boston, the second of two children. Her choice of an academic career and her approach to the natural world were profoundly influenced by her parents. She credited her father, Harvard professor and renowned entomologist Charles Thomas Brues, with teaching her to "think biologically at a very early age." Her mother, Beirne Barrett Brues, was an amateur field botanist and collector of Native American baskets who published with her husband. Alice participated in her parents' travel and fieldwork in natural history from an early age (Dufour, 1988).

Brues completed her undergraduate work in 1933 at Bryn Mawr with double majors in philosophy and psychology. Her decision to pursue graduate work in anthropology resulted from an introductory anthropology course at Harvard that she took the summer following her graduation from Bryn Mawr. She entered graduate study at Radcliffe later that year, working under the direction of Earnest A. Hooton, who was already an established leader in the relatively new field of physical anthropology, and earned her Ph.D. in 1940 along with two other Hooton students, Sherwood L. Washburn and C. Wesley Dupertuis (Giles, 1997:500). As was true for virtually all of Hooton's students, her doctoral dissertation, "Sibling Resemblances as Evidence for the Genetic Determination of Traits of the Eye, Skin and Hair in Man," focused on aspects of contemporary human variation, and publications resulting from it (Brues, 1946a, 1950) are still cited as basic sources on inheritance of these features.

B. Professional Career

As a pioneer in the development of physical anthropology, Brues' contributions to the discipline are vast in number and diverse in breadth. As one of only five women who have held the office of President of the American Association of Physical Anthropologists, Brues' impact on the discipline extended far beyond her research, which is perhaps most familiar to contemporary human biologists who focus on the genetic basis of modern human variation and to physical anthropologists who apply their knowledge of skeletal biology in forensic situations. In both domains, Brues' contributions were seminal. Her dissertation study on the linkage of body build, eye color, freckling, and sex (Brues, 1940) helped introduce the science of human genetics to the anthropological study of human variation (Brues, 1946a; Dufour, 1988:23). Also of great importance is Brues' early use of computers to simulate the processes of population genetics—work that culminated in her classic paper, "Selection and Polymorphism in the ABO Blood Group System," in which she suggested that the A-B-O blood groups were maintained in human populations by natural selection favoring heterozygote genotypes (Brues, 1960, 1963a, 1964; see also Weiss and Chakraborty, 1982:382; Dufour, 1988:24; Spencer, 1986:343).

Brues can also be regarded as one of the founders of forensic anthropology in the United States. She forged solid ties with local law enforcement agencies while a faculty member at the University of Oklahoma School of Medicine and authored one of the first guides to the identification of skeletal material (Brues, 1958a) written for members of the law enforcement community and published in the *Journal of Criminal Law, Criminology and Police Science*. The article summarized key methods for determining age, sex, race, stature, and individual identification while integrating examples from her own case files.

While the breadth of Brues' scholarship may seem unusual today in our era of increasingly specialized scholarships, it was not unusual among the members of the founding cohort of the discipline, many of whom were Hooton's students. Indeed, during the infancy of this field, there were many interconnections among diverse modes of exploration of past and present human variation and evolution. Examples of this phenomenon abound: Brues' guide to skeletal identification, while written for a specialized professional audience, was later recommended in an early physical anthropology text (Montagu, 1960) as a basic reference for information on the methods and techniques of osteological analysis. This observation, together with Brues' own words, speaks to the shared roots of forensic anthropology, bioarchaeology, and human osteology during this era. Stated another way, at this phase in the history of our field, seminal contributions in one arena were also fundamental in establishing the foundations of other related areas. Thus, it is from within this broader historical context — the founding of the larger subdiscipline of physical anthropology — that Brues' contributions must be considered. From this vantage point, an understanding of the historical roots of physical anthropology in natural history and clinical anatomy are particularly relevant to understanding the initial perspectives and approaches of bioarchaeology.

During the period of infancy of physical anthropology, the field of practicioners was very small. In 1993, when Brues presented the American Association of Physical Anthropologists Lifetime Achievement Award to T. Dale Stewart at the annual meeting of the AAPA, she remarked.

> Physical anthropologists, such as they were, were kind of odd-ball anatomists... I remember the meeting in 1937 well.... This was the 8th meeting [of the American Association of Physical Anthropologists], and it was held at the Harvard Faculty Club, in a room which perhaps could have held 50 people, but was by no means crowded. The treasurer's report stated that the "expenses in connection with the 7th meeting had been $13.93." But the association had big names by then — Hrdlička and Stewart, and Hooton and Schultz, and Krogman, to name only a few — and we students sat in awe of these great people while we listened to a total of 29 papers. (Brues, 1993:556)

Hooton's writing provides insight into Brues' own perspectives on human variation as well as a historical context for her emphasis on anatomical observation and description. Description as scientific activity in this era often has been criticized as being inherently typological, "pre-Darwinian," or reflective of racist thinking (Armelagos *et al.*, 1982; Lovejoy *et al.*, 1982). For example, Wolpoff and Caspari assert that "[o]ne of the drawbacks of skeletal biology is its emphasis on description, identification and classification. In other words it is typological. While description will always be important in skeletal biology, its preponderance is a part of the discipline's pre-Darwinian legacy" (1997:142). However, a missing component of such evaluations is the recognition that observation and description are essential initial stages in natural history, whether executed by pre-Darwinians like John Ray and Linneaus or by scholars operating from an evolutionary perspective like Darwin himself or his advocate Huxley. Moreover, Brues made it crystal clear in her own writings that these essential initial stages were neither inherently typological nor at odds with a revised concept of biological race, one grounded in statistical, populational, and evolutionary theory. Brues later fully developed this modern conceptualization of biological race, making clear its essential differences and incompatibility with typological categorizations of human variation (Brues, 1990a).

Brues' career in bioarchaeological research began in 1940 at the Peabody Museum at Harvard when she served as a research assistant in charge of the skeletal collections (Brues, 1990a:2). She analyzed skeletal material excavated from Alkali Ridge in southeastern Utah, excavated by John Otis Brew of the Peabody Museum in 1931–1933 (Brew, 1946; Brues, 1946b,c). Brew's investigation of cultural evolution in the Mesa Verde region had led him to conclude that the transition from Basketmaker to Pueblo represented cultural continuity rather than invasion and population replacement (Brew, 1946:ix). In fact, the Alkali Ridge volume includes a rather lengthy critique of the use of craniomorphometry in connection with studies of cultural evolution in the region, thus representing a departure from previous work (see especially Brew, 1946:67).

Brues' analyses of this material, consisting of 16 burials from five sites, appeared in two appendices at the end of the volume (as was typical for the time). The quantity of specific information is striking, presented in far more detail than the norm. Morphometric data were presented in tabular form (1946b:316), and her findings pertaining to age, sex, and pathology were summarized in a short but thorough narrative (1946c:327). Photographic plates illustrated cranial and postcranial pathology, including cranial deformation in the form of lambdoidal flattening, as well as signs of traumatic injury and interpersonal violence (Brew, 1946c). Brues described examples of reactive bone lesions (e.g., periostitis), anomalies (e.g., bifurcated ribs), neoplasm (e.g., button osteoma), metabolic/nutritional disorders (e.g., osteoporosis), and neuromechanical (e.g., osteoarthritis). Based on the evidence, Brues concluded that Alkali Ridge populations suffered from "... malnutrition and hostile attack" (1946c:329).

Brues' investigations of 17 crania from several sites in the San Simon Valley in southeastern Arizona (Brues, 1946d) focused on morphometric analyses for purposes of elucidating the transition from the preceramic Cochise culture to the San Simon culture. In addition to listing craniometric data in tabular form, Brues compiled, in even greater detail than in the Alkali Ridge study, a written narrative of each skull, including observations about age, sex, anomalies, and dental pathology.

The entry of the United States into World War II prompted Brues to change focus briefly to the collection and statistical analysis of anthropometric data collected from the U.S. Air Force, an early application of physical anthropology to the field of ergonomics aimed at improved design in military equipment and uniforms. Brues strongly supported Hooton's vigorous acceptance of statistical methodology at a time when physical anthropology was dominated by Aleš, who, in Brues' words, suffered from "math anxiety" (Brues, 1990b:5). Hrdlička was unfamiliar with some of the most elementary statistical terms and concepts (e.g., standard error and probable error), and Brues later enjoyed relating to her students and colleagues one anecdote in particular in which "Hrdlička ... is reputed to have said, commenting on a paper presented by a nervous graduate student, that 'Miss So-and-So herself admits that there are probable errors in her work.'" (Brues, 1990b:5). In 1942, Brues moved to Wright Field in Ohio as an assistant statistician to work on similar projects for the military (Dufour, 1988:24). After the war, she consulted for a short time at Massachusetts Institute of Technology on a project pertaining to gas mask design. Her participation in statistical research would prove to be pivotal in influencing her later choice to learn computer programming and apply that knowledge to problems in population genetics.

In 1946, Brues began her teaching career at the University of Oklahoma School of Medicine, where she taught gross anatomy. Throughout her career, Brues often commented on the need for more coverage of anatomy in medical school curricula (see Brues, 1966:107). During almost two decades spent at this institution, her scholarly contributions encompass the fields of bioarchaeology, forensic anthropology, paleopathology, and modern human variation. Brues, like Hooton, viewed the study of human variation as a kind of common denominator for all of physical anthropology, uniting the study of prehistoric and modern human variation and, above all, defining physical anthropology as a unique field of study. In describing the field in an article written for law enforcement officials, Brues explained these interconnections in this way:

> Physical anthropologists begin their course of study with human anatomy, giving special attention to the skeleton which ... furnishes information about prehistoric as well as modern peoples. Also, they must study ... human variations ... with reference to sex, age, race and environmental influences. Lastly, they should have direct experience in handling human skeletons in adequate numbers .... either in medical school collections or ... in collections recovered from archaeological excavations .... Having completed their training, they are generally employed either as anatomists in medical schools or as teachers and researchers in universities or museums, where they are concerned with the study of prehistoric human remains and the significant comparisons that may be made between them and the skeletons of living or contemporary peoples. (Brues, 1958a:551)

During her years at the University of Oklahoma, Brues expanded her research program involving bioarchaeological analyses. While progressing through the academic ranks in the Department of Anatomy, Brues was appointed as curator of physical anthropology at the Stovall Museum, a position she held from 1956 to 1965. There she completed studies of prehistoric skeletal materials from the Nagel (Brues, 1957), Horton (1958), Morris (1959a), McLemore (Brues, 1962), and Spiro Mound (Brues, 1996) sites in Oklahoma. These analyses presented both detailed descriptions of individual skeletons and summaries of the basic patterns of demography, health, disease, and genetic affinity. Brues' descriptions of skeletal pathology from each site employed broad categories such as trauma, arthritis, dental pathology, periostitis, and osteoporosis, and she also compared the different site patterns and noted interesting variations among them. The 52 skeletons she examined from the McLemore site, for example, did not evince the marked signs of infectious disease and nutritional disorder that Brues found for the Morris, Horton, and Nagle sites, situated in the eastern part of the state.

In 1965, Brues moved to the University of Colorado at Boulder, and her decades there (1965 to 1984) concentrated on institution building. She developed an astonishing variety of "courses in human gross anatomy, quantitative methods, human osteology, human variation, primate neuroanatomy, nutritional anthropology, and human growth" (Dufour, 1988:26). The graduate program at Colorado produced many professional physical anthropologists during her tenure, and her textbook, *People and Races* (1977), has influenced generations of students through its notable integration of skeletal and dental variation with soft tissue variation. Brues continued to be a prolific contributor to the literature of physical anthropology, until limited by poor health in her late 80s.

Brues' most important contributions to bioarchaeology lie in the realms of paleopathology and in biomechanical studies that examine associations between cultural and biological features in human groups. Her early reports (Brues, 1946b,c,d) on skeletal pathology from archaeological sites in California and Arizona were characterized by an unusual breadth of detail and carefully considered interpretations. Her identification of skeletal material from Oklahoma sites as representative of some form of treponemal disease, resembling (but not identical to) yaws or syphilis, reflected a diagnostic caution based on her knowledge of clinical literature. Brues described the typical appearance of these lesions, particularly common at Morris and Horton, on long bones as longitudinal streaks, periosteal plaques, and gross proliferation of new bone (Brues, 1959:67) and categorized such lesions as "slight," "moderate," and "severe," noting that the severity of expression was variable among different bones within individual skeletons or, at times, on the same bone. She then compared the patterns with similar pathology from other prehistoric collections and from contemporary clinical cases, intrigued by the apparent variation of the skeletal pathology over time and space and speculating about factors such as nutritional disorders that might contribute to this variation (Brues, 1959:67). In the decades since Brues' reports were published, her work continues to be cited as an example of prehistoric treponemal disease in the New World (Baker and Armelagos, 1988; Powell and Cook, 2005), and she maintained her interest in the evolutionary nature of human disease (Brues, 1967, 1975).

In January of 1965, Brues was invited to speak on her pioneering research in bioarchaeology, osteology, and paleopathology as a discussant in a symposium organized by Saul Jarcho, sponsored by the Subcommittee on Geographic Pathology of the National Academy of Sciences–National Research Council. In her commentary on papers by Douglas Osborne and James Miles on the paleopathology of the Mesa Verde region, Brues called for more collaboration and communication between clinicians and physical anthropologists (1966:107). Her point of departure was a question raised by Miles over the excessive degree of anteversion of the femoral neck displayed by the prehistoric materials. Recalling an early paper by Hooton on the relevance of physical anthropology to biomedicine, Brues pointed out that physical anthropologists are uniquely poised to recognize such variation as normal due to their knowledge of skeletal anatomy. She also spoke directly to findings related to New World treponemal disease, urging caution in referring to prehistoric diseases by contemporary clinical names such as syphilis or yaws and reminding her listeners that past diseases have been modified by evolutionary processes in ways that are unknown to us. In offering these comments, she anticipated some of the most important precepts of contemporary paleoepidemiology (Buikstra and Cook, 1980), particularly important in the study of prehistoric treponemal disease:

> . . . it is a very bad idea to give a disease a name. . . . I have been very fortunate in dealing with material from southeastern Oklahoma in which a similar type of disease . . . was extremely prevalent. . . . We ought not to discuss it in terms of whether it is syphilis. We want instead to establish a syndrome of related symptoms. We should establish the fact that they are correlated with one another and let the syndrome be unnamed at present. . . . I believe that diseases may change, they may evolve or mutate, and diseases also come in groups. . . . In the past there may have existed pathological agents which were related to both (syphilis and yaws) and perhaps in a strict sense were neither. (Brues, 1966:108)

Brues' speculations about the impact of cultural behavior on the physical form of human groups helped set the stage for future biocultural studies. Brues' (1959b) paper, "The Spearman and the Archer: An Essay on Selection in Body Build," moved beyond the restrictive realm of typology and somatotyping to focus on the effects of basic biomechanical concepts and principles—ideas that Brues explicitly defines and illustrates in her essay—upon the human body. She reasoned that populations with more linear body proportions (e.g., longer arms) would have been more efficient at throwing spears than more compactly built populations. Following the same logic, populations with lateral builds would have been more efficient at using the bow and arrow (see also Brues, 1977:171). In this theoretical essay, Brues hypothesized an association between body shape and size and weapon use, while urging readers to consider other factors, including climate, that may have had an impact on the evolution of body build. The impact of "The Spearman and the Archer" lies less in the specific nature of the hypothesized relationships and more in its general message about the potential for dynamic interactions between biology and culture. At the time it was written, there is little doubt that it offered a significant departure from previous work on body build, which had been dominated by typological approaches, including somatotyping (discussed in Brues, 1990b). Brues' work clearly demonstrated her conviction that description of biological variation was not an end in itself but instead provided the necessary prologue to essential questions about human evolution. Following her lead, some researchers focused on assessing the impact of habitual activities on skeletons of individuals (e.g., see Angel, 1966a), whereas others applied studies of cross-sectional geometry to prehistoric populations (Ruff and Hayes, 1983a,b). All of these subsequent studies (e.g., Brace and Montagu, 1977) recognize the importance of "The Spearman and the Archer," with its emphasis on the effect of biomechanical variables as a potential field of inquiry.

XIII. SHEILAGH THOMPSON BROOKS (1923–); MARY LUCAS POWELL, JOHN J. SWETNAM, AND DELLA COLLINS COOK

A. Biography

Sheilagh Thompson was born on December 19, 1923, in Tampico, Tamaulipas, Mexico, the first child of Robert Thompson and Lea Levine Thompson. After growing up in Belfast, Ireland, where his family owned a restaurant, and enduring the horrors of trench warfare in World War I, he moved to Mexico to work as an accountant for the Huasteca Oil Company, a local subsidiary of the Dutch company, Shell Oil. Sheilagh's brother Charles was born in 1926. The family enjoyed living in Tampico, but after both children contracted malaria, their parents decided to move to Los Angeles, California, where Lea's sister lived with her husband, a geologist with Shell Oil. Robert Thompson was subsequently employed as an accountant by Van Camps and Lea taught mathematics at a local school.

As a child, Sheilagh had no special interest in science or natural history, but she was a voracious reader (in her own words) on many different subjects (including poetry and literature) and recalls hiding on the roof of their small house so that she could read undisturbed. She was an excellent student in high school, impelled by her intense curiosity, and her mother insisted that she apply to the University of California at Berkeley for her undergraduate education rather than attending a small local college as her father had suggested. She recalled that her attention was captured by the very first anthropology course that she took at UC–Berkeley, taught by the archaeologist William H. Olsen, and she resolved to pursue a career in this fast-growing and exciting discipline. Another childhood hobby was hiking, a skill that was very useful to her in archaeological reconnaissance and fieldwork.

She met her future husband, Richard Brooks, in 1950, when he visited UC–B on a field trip with his major professor from San Francisco State University. When they decided to marry in 1951, Sheilagh asked her Episcopalian minister to perform the ceremony; however, scheduling conflicts arose and so they were married by a friend and colleague from the UC–B Museum of Paleontology, who was a Buddhist priest. Sheilagh and Richard had two daughters, who accompanied their parents to fieldwork locations in Mexico and the southwestern United States during their childhood. Neither of them, however, chose anthropology as a profession. Kathleen Mary Brooks is a counselor at a shelter for battered women in Las Vegas and her younger sister, Caroline Nora Brooks Harris, earned her Ph.D. in film and communications at the University of Wisconsin–Madison and is an assistant dean for human resources at the University of Hawaii.

B. Professional Career

Sheilagh Brooks received her B.A. degree (with honors) in anthropology from the University of California–Berkeley in 1944. During her undergraduate studies, she discovered that she preferred studying ancient cultures rather than collecting ethnographic data. Her M.A. degree (1947) was in paleontology and anthropology, with a thesis titled "A Comparative Study of Mammal Bone Recovered from Archaeological Sites in Marin County, California." Four years later, she was the first woman to earn a Ph.D. in physical anthropology from UC–Berkeley. Her 1951 dissertation topic, "A Comparison of the Criteria of Age Determination of Human Skeletons by Cranial and Pelvic Morphology," reflected a major research interest of her dissertation advisor, Dr. Theodore C. McCown (Brooks, 1970; Kennedy and Brooks, 1984), who had become interested in forensics through his service in identification of war dead. Afterward, he applied forensic methods in several historically important cases from California mission cemeteries.

After completing her Ph.D., Sheilagh spent a year (1953) as curator at the Paleontology Museum at UC–Berkeley. From 1958 until 1963, she was employed as a researcher in the Physiology Department at that institution, working under Sherbourne F. Cook. During the years from 1959 to 1963, she held a number of part-time teaching positions at various universities in the western United States, including Arizona State University, Nevada Southern University, and Pasadena City College.

From 1963 to 1966, Sheilagh taught anthropology at the University of Colorado (Boulder). She served as editorial manager for the AAPA *Yearbook of Physical Anthropology* (1963–1964) and as assistant editor for volumes 11–13

(1965–1967) of the *Yearbook*. In 1964–1965, Richard was Acting Curator of North American Archaeology at the Museum at Southern Illinois, and Sheilagh worked for a short time in 1965 as a research assistant at the museum, analyzing skeletal data from archaeological projects. In 1966 they both joined the Anthropology–Sociology Department at Nevada Southern University.

In 1961, Sheilagh began teaching anthropology courses during the summer session at the University of Nevada–Las Vegas, repeated in 1963 and 1965. The following year she became the first full-time anthropologist in UNLV's Department of Sociology and Anthropology. At that time, the university numbered only a few thousand students, and she realized that to build an independent anthropology program would require strong relationships with other departments within the university and with the wider community. She was active in developing team-taught courses with both the Department of Biology and Geoscience, even though this meant teaching overloads for many semesters. Cooperative relations in the community included not only service on museum boards and state anthropological associations, but also a continuing relationship with the metropolitan police department, making use of her skill at determining the age and sex of skeletal remains. By building the number of majors and developing extradepartmental alliances, she was able to justify a rapid expansion in the number of anthropologists hired, and by 1972, a separate Department of Anthropology and Ethnic Studies was established.

From the beginning, Sheilagh's interest in the interdependence of cultural and physical causes of human behavior was apparent. The department was established with a strong four-field approach. Instead of labeling one introductory course, "physical anthropology" and the other course "cultural anthropology," the courses were identified as "Introduction to Anthropology: Cultural Factors" and "Introduction to Anthropology: Physical Factors." Working closely with the biology department, Sheilagh created courses designed to educate students on the abuse of the concept of race. As the department expanded, Sheilagh's was a persistent and effective voice in promoting the expansion of the department to include diverse anthropological perspectives. Her strong commitment to issues of social justice was especially evident in her support of a close relationship between anthropology and ethnic studies programs, which, at the insistence of the department, was housed with anthropology instead of being set adrift as another interdisciplinary program.

One element in the success of the department was the close cooperation between the department, headed by Sheilagh, and the Archaeological Survey (and later the Museum of Natural History) headed by Richard Brooks. The survey provided needed employment for anthropology majors at a time when resources were scarce in the university system, and the Brooks were able to give students a chance to work closely with professionals in the field.

In 1985 Sheilagh Brooks received the UNLV Distinguished Faculty Award. In 1987, the Board of Regents created the rank of Distinguished Professor with the proviso that only one individual a year from the university could be elevated to that rank. Sheilagh was the first scholar at UNLV to be chosen for this honor. In 1989 she was honored with the Barrick Distinguished Scholar Award. Her years of community service (local and regional) were recognized in 1990 when she was awarded the Crystal Flame and named the 1990 Woman of Achievement in Science by the Greater Las Vegas Chamber of Commerce Women's Council. In 1991 she was included in the book, *Distinguished Women in Southern Nevada*. She was invited to write the obituaries or tributes for several well-known colleagues, including her mentor T. C. McCown (Brooks, 1970), Sherbourne F. Cook (Brooks, 1976), Audrey Sublett (Brooks, 1979), and Louise Marie Robbins (Brooks, 1988).

Sheilagh T. Brooks' interest in skeletal sexual dimorphism was evident early on in her choice of dissertation topic, and her search for improved skeletal criteria for age and sex determination remained a prime research interest throughout her long, active career in physical anthropology, applied in both archaeological and forensic contexts (Brooks, 1951, 1955; Brooks and Suchey, 1990; Suchey *et al.*, 1988). Her fellow student at UC–B, Thomas McKern, was also influenced by McCown's interest in age-related skeletal changes, and he worked during the 1940s with T. Dale Stewart of the Smithsonian Institution on the development of improved methodological criteria for the identification of U.S. military personnel who died during World War II and the Korean war (McKern and Stewart, 1957; McKern, 1970). She was particularly concerned that the widely used pelvic aging standards developed from study of male skeletons might not be appropriate for aging female skeletal individuals, given the sex-specific physiological demands of pregnancy and childbirth. During the 1980s, Brooks collaborated with Judy Myers Suchey on this line of research, focusing on the implications of essential differences between the aging process in male and female pelves for individual identification in forensic contexts (Brooks and Suchey, 1990). Together Suchey and Brooks developed a new set of age-determination pelvic criteria based on more than 1000 sets of pubic symphyses of documented sex, age, and parity aimed at refining the aging of skeletal individuals beyond the capability of existing systems of pubic aging criteria (Suchey *et al.*, 1988).

Sheilagh's interest in skeletal tissue analysis encompassed fossilized bones (Cook *et al.*, 1961, 1962), mammal bone from archaeological sites in California (Brooks, 1947), and human remains (Brooks *et al.*, 1966; Brooks, 1969). She worked closely with her husband, Richard, and other scholars throughout her career, publishing numerous reports on the bioarchaeology of Native Americans and Euro-Americans throughout the Great Basin area. Some of these studies focused specifically on diet (Reinhard *et al.*, 1989), whereas others covered a broad range of physical features (Brooks and Brooks, 1974; Brooks *et al.*, 1977, 1981, 1984, 1985, 1988, 1990; Brooks and Brooks, 1985, 1988, 1990).

The definition and interpretation of distinctive biological characteristics of specific population groups were other main themes in Brooks' career, and her numerous publications in this area covered both multiple traits and specific features such as paleoserology (Brooks and Heglar, 1972; Brooks *et al.*, 1979) and dental non-metric traits (Kobori *et al.*, 1980; Perizonius *et al.*, 1991). Her thorough grounding in "four-field" anthropology at UC–Berkeley and her close collaboration with her husband's archaeological research prompted her interest in the cultural interpretation of artificially manipulated physical features, such as cranial deformation as a sign of ethnicity (Brooks and Brooks, 1980), and in mortuary analysis (Brooks and Brooks, 1969). They collaborated on several important excavations, most notably the Stillwater Marsh in Nevada (Brooks *et al.*, 1988, 1990; Brooks and Brooks, 1990), Zape Chico in Mexico (Brooks and Brooks, 1978, 1980; Reinhard *et al.*, 1989), and Niah Cave in Sarawak (Brooks and Brooks, 1969; Brooks and Helgar, 1972; Brooks *et al.*, 1979). Theirs was an exemplary marriage of professionals, and Sheilagh (unlike the great majority of female colleagues in her cohort of physical anthropologists) was accorded equal professional recognition by her peers rather than being regarded as a "useful assistant" to her archaeologist husband.

A third important aspect of Sheilagh Brooks' investigations of past populations focused on aspects of health and disease: the physical toll exacted by mobility and subsistence patterns of Native Americans in the Great Basin (Brooks *et al.*, 1988, 1990; Haldeman and Brooks, 1987), the potential impact of infectious disease on mortuary patterns (Brooks and Brooks, 1978), the identification of specific infectious diseases (Brooks, 1989; Brooks *et al.*, 1994), metabolic disorders (Brooks and Melbye, 1967), or developmental defects (Brooks and Hohenthal, 1963; Hohenthal and Brooks, 1960; Brooks and Brooks, 1991), and paleopathological "profiles" of single individuals (Brooks and Brooks, 1991) or population samples (Brooks and Brooks, 1987). Her overviews of Great Basin paleopathology (Stark and Brooks, 1984, 1985) covered several millennia of human occupation from Archaic hunter-gatherers to Historic settlers in that harsh yet beautiful landscape.

As noted earlier, one inspiration for Sheilagh's research interest in skeletal aging criteria was the work by her Ph.D. advisor, Theodore D. McCown, in one particular practical application of physical anthropology: individual identification of military war dead (Kennedy and Brooks, 1984). Other aspects of forensic anthropology that caught her interest included identification of human vs nonhuman bone (Brooks, 1975), the teaching of forensic anthropology in the United States (Brooks, 1981), the importance of scientific archaeological methods in the exhumation of buried individuals (Brooks, 1984), stature estimation from incomplete remains (Brooks *et al.*, 1990), and accurate descriptions of soft tissue features for analysis of wounds (Rawson and Brooks, 1984). In 1978,

Sheilagh Brooks became a diplomate of the American Academy of Forensic Science. She was awarded the T. Dale Stewart Award by the academy in 1993 for her contributions to forensic anthropology.

XIV. MARY FRANCES ERICKSEN (1925–); DELLA COLLINS COOK

A. Biography

Mary Frances Ericksen was born in Fortville, Indiana, in 1925 (Anonymous, 2003). She was one of the first two anthropology undergraduates at Indiana University, earning her B.A. in 1947. She recalls that Hermann Wells made special arrangements so that she and the other student could get degrees a semester before the major was formally on the books. They participated in Glenn Black's excavations at Angel Mounds, and Ericksen remembers being given the run of his field library. He drove the crew to their respective churches on Sunday and picked them up afterward! She has no recollection of classes with physical anthropologist Georg Neumann while she was at Indiana University. In 1948, she married George E. Ericksen (1920–1996), who completed his M.A. in geology at Indiana University in 1949.

B. Professional Career

George Erickson's career with the U.S. Geological Survey took them to Latin America for many years (Evans, 1997). This experience deepened her interest in applications of physical anthropology to understanding prehistoric cultures. From 1949 to 1952 she was a guest researcher at Museo Nacional de Arqueologia y Antropologia, Lima, Peru. From 1955 to 1962 she was associated with Museo Arqueologico de La Serena and Museo Nacional de Historia Natural de Chile, and several of her early publications report her research on South American skeletal series (Ericksen, 1962a,b,c, 1977/1978).

She earned her M.A. from Columbia University in 1957 and her Ph.D. in anatomy from George Washington University in 1973 using collections at the U.S. National Museum, Smithsonian Institution. Her dissertation compared Eskimo, Pueblo, and Arikara femur histology (Ericksen, 1973, 1976a, 1980). She joined the anatomy faculty at GWU as a special lecturer in 1973, becoming an assistant professor in 1978 and adjunct professor in 1992. She authored a series of papers on gross and histological age changes in the skeleton using the Terry collection at the USMN (Ericksen, 1976b, 1978a,b, 1979, 1982a,b, 1991) and has made important contributions to histological study of ancient populations (Ericksen, 1995, 1997; Ericksen and Stix, 1991; Ortner and Ericksen, 1997).

XV. LOUISE ROBBINS (1928–1987); MARY LUCAS POWELL, DELLA COLLINS COOK, MARY KAY SANDFORD, AND GEORGIEANN BOGDAN

A. Biography

When Louise Marie Robbins passed away from a brain tumor in 1987, Sheilagh Brooks noted in Robbins' obituary in the *American Journal of Physical Anthropology* that she was "associated with footprint analysis in the minds of most physical anthropologists" because of her numerous contributions in this field of research over the preceding decade aimed at both forensic and paleoanthropological applications. However, Robbins' early interests were much broader, both within anthropology and within the natural sciences at large.

Robbins was born in Chicago. Her parents were Harry S. and Gladys Robbins. She grew up in rural Indiana, graduating from Clark's Hill High School in Tippecanoe County. After service in the Navy from 1950 to 1956, she began her studies at Indiana University, earning her B.A. in 1960, her M.A. in 1964, and her Ph.D. in 1968. Her undergraduate degree was in chemistry with minors in physics and zoology, a background that may have stimulated her later interest in soil trace elements (Robbins, 1977). Her graduate work at Indiana University reflects an early interest in somatology (Robbins, 1962, 1963, 1965), but a 1962 fieldwork experience with Georg Neumann at the Dan Morse site in Illinois seems to have changed the direction of her career. Her M.A. and Ph.D. projects built on Neumann's interests in craniology of archaeological and recent American Indians (Robbins, 1964, 1968; Robbins and Neumann, 1969, 1972; see Chapter 2). Robbins served as a lecturer in anthropology at University of Nebraska in 1964–1965, participating in research with the dental anthropologist Sam Weinstein and the anthropometrist Edward I. Fry.

B. Professional Career

In 1965 Robbins' career took a decisive turn toward bioarchaeology when she began work as a lecturer at University of Kentucky, serving first as a temporary and later as the permanent replacement for Charles E. Snow. She was the first of a long line of osteologists who failed to achieve tenure in that position. However, she approached her job with enthusiasm for the long, if unsung, history of anthropology in Kentucky (1970). Her time at Kentucky is reflected in several preliminary reports on at-that-time largely unpublished collections that have since served as foundations for the careers of several of our colleagues. A review paper (Robbins, 1977) summarizes her otherwise unpublished work on several field and laboratory projects while at University of Kentucky. She comments on underreporting of deaths in pregnancy and childbirth at Indian Knoll (Oh2) because excavators failed to recognize fetal bones and on biases resulting from discarding poorly preserved bone at Carlson Annis (Bt5), both WPA-era Archaic sites. She discusses Archaic diet and dental disease, reflecting her field work with Patty Jo Watson at Carlson Annis. She suggests endemic goiter as an explanation for high rates of bone pathology in Fort Ancient skeletons at the Buckner site (B12) and calls for trace element studies of sites from the goiter belt. She makes a convincing case for infanticide at the Fort Ancient Incinerator site in Ohio (now Sun Watch 33My57), where she participated in excavations. She first touches on the question of prehistoric syphilis in a discussion of a young female skull from the Barrett site (1974a).

Her most substantial publications from her time at Kentucky arise from her collaborations with Patty Jo Watson's program of cave research. Robbins' case study of the "Little Al" mummy from Salts Cave, Kentucky, is remarkable for its careful use of radiography, autopsy, coprolite analysis, and blood typing (Robbins, 1971). She reviewed mummy finds from western Kentucky and Tennessee and described fragmentary human remains from the Salts Cave Vestibule in a subsequent publication (Robbins, 1974b). The most unusual aspect of this project is her argument for cannibalism as an explanation for the high frequency of burned and broken human bones in the latter collection.

Following her move to Mississippi State University and thence to University of North Carolina at Greensboro, Robbins addressed the question of specific infectious agents, reexamining the question of tuberculosis in prehistoric Kentucky (1978a) and developing a very convincing diagnosis of a "yawslike" disease process in a Native American skeletal series from southern Louisiana (1976, 1977b, 1978b). Her analysis of some 24,900 bone fragments representing 275 individuals from the Early Mississippian (Coastal Coles Creek, AD 900–1000) community at Morton Shell Mound in Iberia Parish, near the Louisiana coast, is an outstanding model of painstaking bioarchaeological analysis. The extreme fragmentation of all skeletal elements (resulting from a mortuary program that apparently required that even the smallest bones be fractured before final interment in shell mounds) brought one unexpected benefit amidst the myriad disadvantages: Robbins was able to examine closely the endosteal surface of each long bone fragment and relate any pathological alterations observed there to other lesions visible on the outer surface. Robbins drew upon a fairly extensive literature on the nonvenereal treponematoses that affected bone and to combine relevant clinical and epidemiological information in her interpretation of skeletal pathology at

Morton Shell Mound. After careful consideration of the progressive nature of the disease entity responsible for this suite of lesions, Robbins wrote: "When the pathology of the Morton Shell Mound people is scrutinized closely with regard to its overt expression, degree of severity, pattern of dispersion through the skeletal system, and its predisposition for adult individuals, all factors point to a particular causal agent, i.e., the treponemal infection or disease called yaws" (Robbins, 1976:70).

During her time at North Carolina Greensboro, Robbins' attention turned to forensic anthropology. She pioneered research on individuation and on reconstruction of stature from footprints and shoe prints (Robbins, 1978c, 1982, 1984, 1985, 1987). This research has been the focus of a great deal of criticism (Buikstra and Gordon, 1992). Robbins' companion in this phase of her life and her executor, Margaret Bushnell, elected to send only the footprint portion of Robbins' papers to the National Anthropological Archives, an unfortunate choice from the perspective of those of us who are more interested in her contributions to bioarchaeology. Her data from SunWatch and other Fort Ancient sites are on file at the Boonschoft Museum, Dayton, Ohio. While Robbins indicated that some of her research materials were placed on file at the Museum of Anthropology, University of Michigan (Robbins and Neumann, 1972), this does not appear to be the case (personal communication A. R. Nelson). Neumann's metric forms are on file at Indiana University, but we are unaware of the present location of Robbins' manuscript material related to his data.

Robbins joined the UNCG faculty as an associate professor in 1974. One year prior to her arrival, UNCG formally established the Department of Anthropology apart from the Sociology Department. Recruited by the new department head, Dr. Harriet Kupferer, Robbins was given the task of building the physical anthropology component of the program. By the end of her first year, Robbins had added three new physical anthropology courses — Human Variation, Human Evolution, and Human Identification — to the anthropology curriculum. Her "hands-on" approach to teaching was reflected in her insistence on including laboratory components as corequisites to certain courses, including Human Identification. Also during her first year, she created a laboratory section for the foundational course in introductory physical anthropology and — in an innovative and astute move — worked with Kupferer to gain approval for the course as a natural science elective in the General Education Curriculum. While designing new courses, and attracting a large student following through her infectious enthusiasm for her subject matter, Robbins worked tirelessly to acquire study collections — comparative osteological materials, fossil casts, and study skeletons — for the teaching and research laboratories.

Robbins moved the physical anthropology laboratory spaces several times during her tenure at UNCG, including a move from the basement of the Foust Building — the oldest structure on campus — to more modern environs in the

Graham Building, where the anthropology department is housed today. She would have been pleasantly surprised to learn that when plans for the New Science Building were drawn up on the UNCG campus in the late 1990s, a teaching laboratory was included in the blueprints. Students who learn physical anthropology today at UNCG do so in a state-of-the art facility where Robbins' strong emphases on osteology and comparative anatomy are still maintained.

Shortly before her death, Robbins was promoted to full professor. She provided many UNCG students with their introduction to physical anthropology. Those who studied under her close tutelage remember her as an avid teacher, a devoted mentor, and a passionate scientist.

XVI. LUCILE E. ST. HOYME (1924–2001); DAVID R. HUNT, RICHARD T. KORITZER, AND MARY LUCAS POWELL

A. BIOGRAPHY

Lucile Eleanor Hoyme was born at the Garfield Memorial Hospital in Washington, DC, on September 8, 1924. She was the only child of Guy L. Hoyme, a U.S. government architect. Hoyme was 55 years old when he married Helen Bailey, aged 35. The Hoymes lived all their lives in northeast Washington, DC. Throughout her life, Lucile felt a strong responsibility to her parents, never separating herself from them and often making sacrifices on their behalf in both her professional career and personal life. In their later lives, Lucile provided home health care for both her parents and taught additional classes at the surrounding universities in the city to help pay for this medical service. Her mother suffered with Alzheimer's disease, and her father was physically debilitated for a number of years before his death in September 1967.

As was characteristic of many professional women's careers of her era, the caretaking of her parent's medical and emotional needs was often quite stressful for St. Hoyme and impacted her professional responsibilities. Teaching at universities in the Washington area allowed her to both share her expertise and earn additional income to help with her parent's medical costs. During the academic year of 1966–1967, she taught part time at the University of Pennsylvania to help defray her father's high medical bills during his hospitalization and meet the cost of his funeral. After the passing of her parents, instead of being free to proceed with her life, St. Hoyme was tragically plagued with health problems of her own (e.g., a colon resection, the onset of adult diabetes, and gynecological symptomatology causing metabolic imbalances, ultimately resulting in gynecological surgery with discovery of cancerous ovaries). These debilitating ailments often prevented her from focusing on her work, and at times even left her bedridden at home.

St. Hoyme's life was strongly grounded in a deep religious faith, with an abiding interest in bible study, and close social ties with her lifelong church home, Wallace Presbyterian Church in Hyattsville, Maryland. She was not dogmatic but a practitioner of the golden rule. She often struggled with the doctrines of the Christian faith and the tenets of scientific inquiry, and more specifically with the dichotomy of conservative Christian beliefs and evolutionary theory. Many afternoons were spent with her close friend, Richard Koritzer, at the Newman Bookstore at Catholic University, discussing all aspects of intellectual dilemmas in pursuit of scientific endeavor, always remaining objective but never doubting that there was a God and debating "how (not if) He did it."

Because of Lucile's nonconformist appearance and eccentricities, she was often wrongly judged. This sometimes pained her, but to a certain degree she utilized this "cover" to play the "dumb broad" or the "frumpy eccentric" that allowed her to ask questions or lead conversations in paths that would never have progressed down certain avenues or have been permitted in other circumstances. Her skill as an ethnographer was often applied to the unwitting "interviewed." She was expertly versed in many subjects, particularly in botany and herbal medicine, comparative zoology, and was fluent in French, Italian, German, Spanish, and Greek, was familiar with Cyrillic, and could read several other languages. She was an avid seamstress and knitter, and in her college days had received medals and trophies in archery.

Lucile St. Hoyme died on November 15, 2001, after several successive strokes and complications from diabetes. She is remembered by her friends for her quick wit, dry sense of humor, and unpretentious manner in dealing with students and professionals alike. Many people also remember her for her compassionate nature, as she never passed up the opportunity to be generous to the homeless and to friends in need.

B. Professional Career

Hoyme began her 40 years of employment with the Smithsonian Institution on April 6, 1942, as assistant clerk–stenographer to Dr. Aleš Hrdlička; she was not yet 18 years old. The position was opened by the Civil Service Commission to assist Dr. Hrdlička. Her work primarily included typing manuscripts and tables and running statistical computations on the adding machine. Although her hire was only "to extend for the duration of the present war and for not more than six months thereafter," fortunately the personnel office did not pay heed to the initial hiring memo. Supported by Dr. Hrdlička and T. Dale Stewart, she was hired permanently as a clerk–typist in September 1943. With the passing of

Dr. Hrdlička in September 1943, her new supervisors were T. Dale Stewart in the Physical Anthropology Division and Matthew Sterling in the Bureau of American Ethnology. During her years of service, she was promoted from the clerk–typist position to Museum Aid in 1955, to Museum Anthropological Aid in 1956, to Museum Specialist in 1961, and finally to Associate Curator in 1963.

While working full-time at the museum, Hoyme began taking courses at George Washington University, completing the Bachelor of Science degree in zoology in 1950. In 1953 she received her Master of Science in Biology degree, with a thesis titled "The Role of Saliva in Inheritance by the Ability to Taste Phenyl-thio-carbamide" (Hoyme, 1950, 1954, 1955). In 1956, with the support and encouragement of T. Dale Stewart, she began a research project on the problem of sex determination on the skeleton, a topic that she pursued throughout her career. She realized her lifelong avocation as an educator when she coordinated and judged the District of Columbia science fair exhibition at the Smithsonian in December of that year. She continued supporting science fairs in the Washington, DC schools for another decade and published a short article on biology projects in science fairs (Hoyme, 1964b). In her collections care duties, Lucile received recognition in 1957 from the Smithsonian incentive awards committee for her design and implementation of a new accessioning and cataloging card and information format for the Physical Anthropology Division.

Hoyme received a National Science Foundation research grant in 1957 for study at Oxford University, England, to pursue a Ph.D. degree under the direction of J. S. Weiner and Sir W. E. Le Gros Clark. In 1964, she received the degree of Doctor of Philosophy in Anthropology, Oxford University; her dissertation was titled "Variation in Human Skeletal Characteristics." While at Oxford, she also worked as a laboratory assistant and part-time anatomical demonstrator. The great mental and physical stress incurred by taking extra coursework in addition to her normal curriculum, the environmental and social change that she encountered, and the inability to provide day-to-day assistance to her parents forced her to request an additional 6 months of leave from the Smithsonian, which she received. As a respite from purely academic pursuits, she took some time to investigate her family roots, pursuing the early history of the Hoymes in England. She discovered that the original family surname was "St. Hoyme," which had been shortened when her forbears immigrated to the United States. Lucile later legally adopted the original surname, although out of respect for her father she waited until close to the time of his death to do so. (Editor's note: In the bibliography for this chapter, all entries for "Hoyme, Lucile E." after 1966 are listed as "St. Hoyme, Lucile E.")

Upon returning to Washington in 1960, Hoyme resumed her position in the Department of Anthropology. In addition to her usual duties, she became the new radiographer for the department, providing radiographic images of skeletal and other anthropological materials using the newly purchased X-ray machine.

She was sent to the main offices of the Eastman Kodak Company in Rochester for training and received outstanding accolades by Kodak staff. Thus motivated, she vigorously implemented the new procedure at the NMNH and received a Smithsonian Superior Performance Award for her radiographic work in November 1960.

Hoyme's professional abilities as an osteologist and physical anthropologist soon began to win recognition. Bureau of American ethnology head Frank H. H. Roberts commended Lucile with a formal letter of acknowledgment and a cash award for her significant contribution resulting in the John Kerr Reservoir (Buggs Island) report, published in BAE Bulletin 182 (RBS Paper No. 25) (Hoyme and Bass, 1962). Given her accomplishments and contributions, T. Dale Stewart recommended her promotion to the Museum Curatorial Series as an associate curator in the Department of Anthropology. Hoyme's duties as Associate Curator primarily encompassed the actual curation and care of the human skeletal collections: organizing the pending physical anthropology accessions and collating the background information, field records, and laboratory records for the site of collection. Her involvement with review and analysis of the collections included supervising the inventory, cataloging of specimens, and providing skeletal inspection and analysis of the specimens. This analysis, along with the documentation, provided the basis for the reports and outlines for the various collections. She also continued her personal research investigations of specimens, provided materials for training and exhibitions, and upgraded the collections storage and cataloging data as necessary.

During her curatorship, St. Hoyme took the opportunity to collaborate with J. Lawrence Angel on various FBI forensic cases during the 1960s and penned his obituary for *AJPA* in 1988. She also enjoyed collegial interactions with other physical anthropologists, among them staff members Marshall Newman and Donald Ortner, as well as a host of students and academicians who visited the collections to conduct their own research. She would often query the students about their projects and offer advice on improving their data collection procedures, a different approach to the problem, or even suggest other collections that might be conducive to their research question.

In the 1960s, Lucile became acquainted with Richard T. Koritzer, one of the authors of this biographical sketch, through their mutual colleague, Larry Angel. A shared passion for dental research led to collaboration on numerous presentations at the annual meetings of the American Association of Physical Anthropologists and other professional societies (Koritzer and St. Hoyme, 1970a, 1971a, 1974, 1979a, 1984, 1985, 1986, 1988, 1989, 1993, 1997, 1999; Koritzer *et al.*, 1987), as well as collaboration on a teaching film (1971b) for Georgetown University. Their enthusiasm for dental anthropology research led them to organize the Dental Anthropology Group, which hosted several sessions at the American Association of Physical Anthropology annual meetings and was seminal in the development of the dental anthropology specialty. Outstanding members of the Dental Anthropology Group who advanced this field of research included Steven Molnar, C. Loring Brace, and the eminent Albert Dahlberg, an early friend, mentor, and a great scientist. A statistician in another department who became a great friend and mentor was Neil Roth. Although Lucile was never quite sure that some of Neil's statistical manipulations might not be "black magic," she used them and they worked.

Koritzer and St. Hoyme's joint publications covered such topics as the first penetration of the dental enamel to its fullest depth with fluoride, the first evidence of an intrinsic, metabolic effect on caries, a description of masticator muscle anatomy discussing motion and implication in the human chewing complex, trace element studies in prehistoric population dentition, a study of dental pathology (i.e., TMJ), and descriptions of skeletal landmarks of function and determination of age and sex from ossuary fragments to an accuracy at the 80% level (Koritzer and St. Hoyme, 1970b, 1971c,d,e, 1979b, 1980, 1992; Koritzer *et al.*, 1982, 1992; Hoyme, 1982; St. Hoyme and Koritzer, 1971, 1976).

During her long and productive career, St. Hoyme published numerous reviews of books in English or other languages (Hoyme, 1956b, 1958, 1968a,b, 1969b,c) and presented many papers and posters at the annual meetings of the American Association of Physical Anthropologists (Hoyme, 1951, 1954, 1956a, 1957a, 1963, 1964a, 1965, 1966; St. Hoyme, 1972, 1976, 1984a). She was a fellow of the American Anthropological Association and an elected fellow of the American Association for the Advancement of Science and a long-standing member of the American Association of Physical Anthropologists, the Royal Anthropological Institute of Great Britain and Ireland, the Anthropological Society of Washington, the New York Academy of Sciences, Sigma Xi, the Society of Women Geographers, the Society for Systematic Zoology, and the Omicron Kappa Upsilon Dental Honor Society.

St. Hoyme frequently engaged in fieldwork opportunities. In 1966, she spent several months in Jamaica with Dr. Jane Phillips from Howard University, gathering anthroposcopic data of value to her research with genetic and climatic influences on human phenotypes. To continue her study of climatic influences expressed in the human skeleton, she traveled to Poland in 1969 to measure crania at the University of Warsaw in the laboratory of Andrzej Weircinski, as well as at the University of Jagiellonskiego (Krakow) and the Polish Academy of Sciences in Warsaw. In the 1970s St. Hoyme traveled again with Dr. Phillips to Ethiopia to study population dynamics and environmental effects on those populations. She also visited the Bishop Museum in the 1970s to study the skeletal collections there, investigating dental wear and dental pathology.

Students enlivened St. Hoyme's existence and kept her intellectual passion alive. Some of her best times were spent guiding young minds through the process of inquiry and research. From 1964 to 2001, Lucile taught anthropology courses as a professorial lecturer or associate professor at American University and George Washington University and she was an adjunct professor at Georgetown University and at Howard University. She collaborated in research at Baltimore College of Dental Surgery and the Dental School of the University of Maryland, Baltimore, and was a valuable contributor to the educational outreach programs of the Smithsonian Naturalist Center, serving their advisory committee from 1981 until her death in 2001. During her emeritus years, Lucile served as a professional mentor for American, Howard, and George Washington Universities, and numerous intern students at the Smithsonian had the fortunate experience to receive individualized study with her.

After retirement, Lucile continued in some of her curatorial activities, revisiting some of her early publications on anthropometric tools and their history (Hoyme, 1951, 1953). She produced an inventory of all the historic anthropometric instruments in the NMNH anthropology department, including those collected and used by Hrdlička, Stewart, and Angel. Many of these instruments are the early prototypes of the standard measuring instruments used in physical anthropology today. Others in the collection are interesting historic apparatus from our typologically oriented past and are the only examples of these early French and German instruments in the United States.

St. Hoyme made significant contributions to the development of American bioarchaeology in two different areas of research: her noteworthy investigations of morphological features in the human skeleton, particularly on the sex characteristics of the pelvis (St. Hoyme, 1957b, 1959, 1984b), and her thoughtful approach to interpretation of skeletal pathology. She approached the study of the human skeleton as a bioanthropologist, i.e., she was always aware of cultural factors that influenced an individual's development. This method is well demonstrated in her study of variation in the cranial base and morphological variation due to climatic and genetic controls (Hoyme, 1964b). She recognized the importance of taphonomic factors (both anthropogenic and natural) in the analysis of human remains recovered from archaeological or forensic contexts, as her research on key features in long bones useful for associating isolated elements demonstrates (St. Hoyme, 1980). In 1989, she coauthored (with M. Y. İşcan) a significant review of the methods in skeletal biology for the determination of sex and race, which assessed the accuracy of these accepted methods and warned of the many often-overlooked assumptions underlying them.

In the early 1960s, Hoyme collaborated with William Bass on the publication of the analysis of two skeletal series from archaeological sites in the Kerr Reservoir in Virginia. Bass conducted the majority of the data collection and basic analysis during his tenure at the Smithsonian during the summers of 1956–1958. Hoyme utilized Bass's skeletal analysis for the report, providing a brief description of each burial and a detailed comparison of the two series with respect to their very different demographic profiles, patterns of skeletal and dental pathology, archaeological evidence for subsistence regimens, and mortuary treatments (including defleshing marks). Hoyme added to the report by providing the proper cultural perspective, integrating both biological and cultural information (from archaeological, historical, and clinical sources) to create a biocultural interpretation of two chronologically successive Native American societies.

In the Kerr report, a careful approach to interpreting observed skeletal pathology is followed. Four distinct categories of "inflammatory changes" are described, noting that the first stage probably represented "simply a normal variation; or a relatively minor injury, such as a severe bruise, which irritated the periosteum, resulting in a temporary increase in the blood supply, but which would normally heal without further complication." The other three stages represented successively severe pathological involvement, with both macroscopic and radiological evidence of cortical and endosteal alterations in the most frequently affected long bones. It is warned by the authors that "the diagnosis must remain tentative.... The 'swollen, bowed tibia syndrome' does not seem to show any clear correlation with the age or sex of the affected persons, although it appears primarily in adults. The radiographs gave no indication of the increased density typical of syphilis" (Hoyme and Bass, 1962:374).

A historical context to the pathological observations in this study was enhanced greatly by Hoyme's searching historic accounts of Native American diseases in the mid-Atlantic coastal region. She discovered a description by the Englishman John Lawson (1709) of a particular ailment of the Carolina Indians: "... they have a sort of rheumatism or burning of the limbs which tortures them grievously, at which time their legs are so hot, that they employ the young people continually to pour water down them." Hoyme conjectured, "[i]t is tempting to equate this 'rheumatism' with the swollen, bowed tibiae described above, but the necessary evidence is lacking.... Eventually, with sufficiently detailed descriptions, comparisons with clinical reports may make differential diagnosis possible" (Hoyme and Bass, 1962:378).

A few years later, Hoyme (now St. Hoyme), in an essay on the origins of New World paleopathology in the *American Journal of Physical Anthropology* (St. Hoyme, 1969a), noted that although many of the infectious diseases that afflicted the Old World were apparently unknown in the Western Hemisphere before 15th-century European contact, many pathogens endogenous to the human species, such as staphylococci, would certainly have entered the New World with the earliest human groups and that Native Americans were at risk from zoonotic infections, trauma, nutritional, metabolic, developmental disorders, and other natural perils, many of these were not detectable in skeletal remains. She discussed features of Native American medicine (such as the nonisolation of infectious individuals) that would have promoted the spread of introduced infectious diseases and drew upon her knowledge of clinical literature to warn against overhasty diagnosis of bone lesions, as different causes (e.g., tumors and tuberculosis) may produce lesions similar in appearance. It provided a measured assessment of the possibilities and pitfalls for evaluating pre-Columbian Native American health status, written to encourage further carefully informed research in this important little-known realm.

To understand the complexity of a life such as Lucile St. Hoyme's is much like viewing a prism. Of the many facets, one reflects the life of a young girl of 17 starting out her career as a stenographic assistant to the physical anthropology giant, Aleš Hrdlička. A parallel facet reveals her rise in the ranks at the National Museum of Natural History, Smithsonian Institution, ultimately to curatorial status. Yet another shows a scholar who earned academic degrees in biology, zoology, and anthropology and excelled in languages and mathematics. A fourth facet reveals her continual commitment and attendance to her parents and her struggles as a woman professional in mid- to latter 20th-century physical anthropology. In summary, the varied sides of this prism reflect a resultant rainbow of this complex woman's life. Through her contributions, our intellectual reserves have grown and our understanding of human variation and skeletal biology has advanced. She enriched the community with the brightness of an active and sharp mind, and to a chosen few, she extended the privilege of seeing her as she really was. Although she lived in penury during her life, she was richly blessed and her legacy is precious to us today.

XVII. MARÍLIA CARVALHO DE MELLO E ALVIM (1931–1995); SHEILA FERRAZ MENDONÇA DE SOUZA AND DELLA COLLINS COOK

A. Biography

Marília Carvalho was born in Rio de Janeiro on May 19, 1931 and died on January 2, 1995. Her father, Manoel Fontoura de Carvalho, was a prominent civil engineer who was active in state government. Her mother, Zelia Fontoura de Carvalho, was a housewife. Marília was the younger of two daughters; the older was named Zelia after her mother. Although he died when she was still a young adult, her father had encouraged her intellectual interests, and she remembered him as a model parent. She did her undergraduate work in geography and history at Instituto Lafayette (Bacharelado — 1952) and in the Faculty of Philosophy of the University of Distrito Federal (Licenciatura — 1953), now renamed the Universidade do Estado Da Guanabara. Her mentors were Professor Heloísa Alberto Torres, an ethnographer and for many years director of the Museu Nacional, and Professor Pedro Ribeiro, a specialist in history and classics.

To expand her academic preparation she studied anatomy in 1962 with Dr. Vinelli Baptista and Dr. Alvaro Froes da Fonseca, the founder of the Sociedade Brasileira de Anatomia.

Marília Carvalho began her academic career as assistant to the Chair of History of Antiquities, with Pedro Ribeiro, at the University of Distrito Federal (later the Federal University of Rio de Janeiro) and Assistant in Anthropology and Ethnography in Universidade do Estado da Guanabara (1954–1961). In 1957 she began her long association with the Museu Nacional as an assistant naturalist to Torres for whom she catalogued ethnographic specimens. In following years she had the opportunity to attend short courses on anthropology offered in Brazil in 1958 by Juan Comas (México) and Almeirindo Lessa (Portugal). These contacts developed her interest in physical anthropology. In 1960 she was promoted to the position of anthropologist/technician at the Museu Nacional, and in 1963 she presented her *Livre Docencia* to the Chair of Anthropology and Ethnography in Universidade do Estado da Guanabara. This research was the basis of her first published papers (Messias and Mello e Alvim, 1962; Mello e Alvim, 1963b). In the same year she became Doctor of Sciences. Finally, in 1966, she was promoted to Research Anthropologist in the Division of Anthropology of the Museu Nacional.

She married Walmir Mello y Alvim, an engineer who worked with her father. The marriage endured until his death just a few years before her own. They had two sons, Mauricio, a physician, and Ricardo. Marília was devoted to her husband and children, and her family was a priority in her life. She seldom traveled to international meetings as a result.

B. Professional Career

Marília Carvalho de Mello e Alvim was associated throughout her long career with the Museu Nacional in Rio de Janeiro. She also taught undergraduate students at the Universidade do Estado do Rio de Janeiro, former Universidade do Estado da Guanabara, for decades, even in retirement. Like many of the museum staff, she taught specialization and graduate courses at Universidade Federal de Rio de Janeiro. Many graduate and undergraduate students took her labs and classes. A whole generation of archaeologists and physical anthropologists in Brazil learned from her expertise and many of them subsequently coauthored publications with her: Dorath Pinto Unchoa (Mello e Alvim and Uchoa, 1972, 1973, 1975, 1976, 1978, 1979, 1980, 1990, 1993, 1995, 1995/1996a,b, 1996; Mello e Alvim *et al.*, 1991), Denizart Mello Filho (Mello e Alvim and Mello Filho, 1965, 1967/1968), Giralda Seiferth (Mello e Alvim and Seiferth, 1967, 1969a,b, 1971a,b), Edson Medeiros de Araújo, Lilia Maria Cheuiche Machado (Mello e Alvim and Machado, 1987), Nanci Vieira de Almeida, Margareth Carvalho Soares

(Mello e Alvim and Soares, 1981, 1981/1982, 1984), João Carlos de Oliveira Gomes (Mello e Alvim and Gomes, 1989; Mello e Alvim *et al.*, 1987), and the senior author of this entry (Mello e Alvim and Mendonça de Souza, 1984, 1986, 1990; Mendonça de Souza and Mello e Alvim, 1992, 1992/1993), among others.

She served as curator of the museum's skeletal collections and facilitated the work of many visiting scholars, among them Wesley Hurt, Christy Turner, Anna Curtenius Roosevelt, Annette Laming-Emperaire, and the junior author. For much of the late 20th century she was the only prominent physical anthropologist working with ancient human remains in Brazil. Throughout her long and productive career she published extensively on this subject (Bertolazzo *et al.*, 1983/1984, 1984, 1985; Mello e Alvim, 1963a, 1966, 1971b, 1972a,b, 1977a,b,c, 1978, 1991, 1996), as well as producing numerous articles on specific and general aspects of the physical anthropology of living populations in Brazil (Mello e Alvim, 1962; Mello e Alvim and Pessoa de Barros, 1971/1972) and other anthropological topics (Mello e Alvim, 1971a, 1986; Pourchet and Mello e Alvim, 1975). She did no fieldwork with archaeologists, but she contributed descriptions to many field reports (Cunha and Mello e Alvim, 1969, 1971; Mello e Alvim *et al.*, 1987; Mello e Alvim and Dias, 1972; Mello e Alvim *et al.*, 1973/1974a,b; Mello e Alvim and Ferreira, 1984; Mello e Alvim and Seyferth, 1971; Mello e Alvim *et al.*, 1975; Roosevelt, 1991).

The first phase of her research is almost exclusively confined to descriptive craniology, following the conventions of French research of the 1960s and using Martin and Saller's *Lehrbuch der Anthropologie* (1957) as her standard. Her doctoral project (1963b) addressed the central issue in South American physical anthropology: the contrast between the earliest human remains from the continent and its later indigenous populations. She compared the Lagoa Santa materials at the Museu Nacional with recent crania, the so-called Botocudo, and with crania from coastal shell mounds, the sambaquis. Many of her publications expand on this theme or extend it to additional samples (e.g., Mello e Alvim, 1963c, 1977d,e, 1992; Mello e Alvim and Mello Filho, 1965). Her concept of the Lagoa Santa material as a homogeneous race has been critiqued recently in new research on the Lagoa Santa remains (Neves and Atui, 2004). In 1979 she published the first craniometric manual for Brazilian students, including selected craniometric points and measurements based on Martin and Saller (1957) and added some cranioscopic information, age, and sexing parameters and radiologic craniometry (Pereira and Mello e Alvim, 1979). This very simple and inexpensive book is still the most detailed textbook available in Brazilian Portuguese for this purpose. A second edition, revised with added postcranial data, was unfortunately never completed. After the 1980s, she turned to non-metric traits as a method of investigating the question of biological distance between different Brazilian Indian and prehistoric populations. A noteworthy article with a less typological focus is a study titled "Non-metric Traits . . ." (Mello e Alvim *et al.*, 1983/1984,

1984), in which the same groups are used to demonstrate that auditory exostosis, mandibular torus, maxillary torus, and palatine torus vary independently in these populations.

Mello e Alvim was very meticulous in the laboratory, and approached her research with great discipline. She stressed careful methodology and was not concerned with innovations or speculation. Her writing was carefully crafted in both language and the structure, and her analyses typically follow a regularized format. She was reticent about engaging in controversy and seldom criticized the work of others. Throughout the last half of the 20th century, hers was the face that Brazilian physical anthropology presented to the international community, and her influence on this discipline in South America was profound.

XVIII. LILIA MARIA CHEUICHE MACHADO (1936–2005); GLAUCIA APARECIDA MALERBA SENE AND SHEILA FERRAZ MENDONÇA DE SOUZA

A. Biography

The Brazilian anthropologist Lilia Maria Cheuiche Machado was born in Rio Grande do Sul State on October 17, 1936. She received her undergraduate degree in history in 1962 from Pelotas University, in the same state, and then spent several years teaching second grade in a local elementary school. However, archaeology was her real passion, and for that reason she moved to Rio de Janeiro to improve her knowledge of this subject. Betweeen 1971 and 1976 she took classes in archaeology at the Instituto de Arqueologia Brasileira (IAB), where she began her career as a researcher. During this same period, she worked in the laboratory of the Department of Anthropology of the Museu Nacional, supervised by Dr. Marília Carvalho de Mello e Alvim and Tarcísio Torres Messias. These contacts gave her the opportunity to begin her career in physical anthropology, to which she dedicated the rest of her life.

B. Professional Career

In 1976, Lilia (as she prefered to be called) had the opportunity to go to the United States with her husband, Luiz Renato Dantas Machado, who represented Brazil in the Organization of American States (Organização dos Estados Americanos). While living in Washington, DC, she took courses in biological anthropology and forensic osteology at the Smithsonian Institution and was awarded a Smithsonian Predoctoral Fellowship in Biological Anthropology and Archaeology, which gave her the opportunity to work with Douglas Ubelaker, Jane Buikstra, and Christy Turner II, among others. Her time spent at the Smithsonian Institution strengthened the ties of friendship and cooperation between the IAB in Brazil and the Smithsonian in the United States, thanks to her contacts with SI archaeologists Betty J. Meggers and Clifford Evans.

Returning to Brazil at the end of 1980, Lilia entered the doctorate program in anthropology at the Universidade de São Paulo, under the supervision of Luciana Pallestrine, a Brazilian archaeologist. In 1984, she obtained her Ph.D. degree, writing a thesis titled "Analysis of Human Remains from the Corondo Site, Rio de Janeiro State, Brazil: Biological and Cultural Aspects" (Análise de Remanescentes Ósseos Humanos do Sítio Corondó, RJ: Aspectos Biológicos e Culturais, Machado, 1984a). Her thesis was the first detailed study of this subject in Brazil, describing different aspects of bioarchaeology and funerary archaeology of the individuals buried at this archaeological site. It became a guidebook for a whole generation of students of bioarchaeology in Brazil interested in associating field and funerary archaeology with physical anthropology. Dr. Machado became Director of the Laboratory of Biological Anthropology at the Instituto de Arqueologia Brasileira and also the president of that institution from 1985 until her retirement in 2000. She was very proud of IAB, a private institution that has been a major pioneer in archaeological research in Brazil.

Dr. Machado took part in 33 field projects in archaeology, conducted by the Museu Nacional or IAB, and she supervised the work of many undergraduate students in biological anthropology and archaeology. She conducted research in the States of Rio de Janeiro and Minas Gerais in Brazil, sometimes as a researcher and sometimes as project director. Between 1987 and 1997, she held the position of researcher at the Conselho Nacional de Desenvolvimento Científico e Tecnológico, receiving grants to fund her research on skeletal remains recovered from archaeological sites of the archaic period at coastal and inland locations. In 1998, she returned to Washington, DC, on a visiting scholar grant from the Smithsonian Institution.

During her long and productive career, Dr. Machado published some 65 papers and reports. Some of her papers dealt with nonbiological topics (Machado, 1975/1976, 1991a, 1995a; Mello e Alvim *et al.*, 1974; Dias Jr. *et al.*, 1975, 1976a,b; Carvalho and Machado, 1975; Carvalho *et al.*, 1973; Kneip *et al.*, 1991; Barbosa *et al.*, 2003) or obituaries of colleagues (Machado and Crancio, 2003). Her explicitly biocultural approach reflected her strong conviction that the biological anthropology of a population could not be understood adequately outside of its cultural and ecological context. This focus appeared early in her professional career (e.g., Machado, 1977; Machado and Silva, 1989b) and remained a major theme in her research (Machado, 1986, 1999a,b, among others).

She published numerous descriptive analyses of skeletal series from archaeological sites (Machado, 1984b, 1988, 1995b), and other publications focused on specific aspects of biological anthropology, including paleopathology (Machado, 1990b, 1991b; Ferreira *et al.*, 1982, 1984), experimental archaeology (Machado and Almeida, 2001), demography (Machado, 1985c,d, 1992a), and dental anthropology (Machado, 1981/1982, 1992b, 1995d; Machado and Kneip, 1993b, 1994; Machado and Pereira, 1985; Machado and Tsaknis, 1985; Turner II and Machado, 1983; Turner *et al.*, 1992).

Within the realm of mortuary studies, Machado examined both modes of body disposal, such as cremation (Machado, 1990a; Machado and Kneip, 1991, 1992), funerary structures (Machado *et al.*, 1987a, 1989a), taphonomy (Machado, 1997), worked human bones (Machado *et al.*, 1995, 1996), and associations between biological and social identities (Machado, 1995c,e,f, 2001a,b; Machado and Alves, 1995, 1995/1996; Machado and Kneip, 1993a; Machado and Sene, 1999, 1999/2000, 2001; Machado *et al.*, 1992, 1994).

Lilia Maria Cheuiche Machado died on July 20, 2005, from complications of diabetes and other health problems.

XIX. AUDREY J. SUBLETT (1937–1977); LORRAINE P. SAUNDERS

A. Biography

The only child of Burkett J. and Sissy Sublett, Audrey Jane was born April 27, 1937, in Ingram, Texas, a small town in the hill country just west of Kerrville. While her father's family had been long-term residents of the area, her mother had been born in Houston and was considered a "big city" woman by the local inhabitants. Rather than discourage this perception, Sissy Sublett cultivated her image as a refined, well-educated woman with a distinctive personal style. Burkett Sublett, however, clung to his provincial roots, brusque and distant in his contacts with others.

The marriage was Sissy's second; she had relocated from Houston to Ingram with her first husband, who died from tuberculosis shortly after their move. Although in his youth Burkett Sublett had worked in the oil fields, at the time of their marriage he owned a restaurant/gas station in Ingram. When Audrey Jane was born to them, Sissy was in her early 40s and Burkett was a decade older. By the time Sublett graduated from high school, her father had sold his business and was comfortably retired — due in part to an income from their pecan orchard.

The differences in her parents' backgrounds, personalities, and dispositions were reflected in their interactions with their daughter. Sissy was a nurturing, supportive mother whose attentions were perhaps exacerbated by Audrey's bout with polio as a young child. To Sissy, who had high expectations for her daughter, there was no question that Audrey would attend college and earn an advanced degree. While Audrey regarded her father as being indifferent to her future prospects, it has been suggested that this was a misconception based on their very diverse personalities. Nevertheless, it is likely that her intense involvement in outdoor activities was an attempt to gain his attention and approval.

Sublett's early education, although in the mid-1940s, was more characteristic of a bygone era. She rode horseback to school, as did many of her classmates, some of whom lived on the far-flung cattle ranches in the hill country. In fact, it was common for ranch families to also own houses in town, with mother and children residing in town during the school year, returning to the ranches on weekends and for the summer break. As her small town had no high school, Sublett attended the institution in nearby Kerrville. These years were eventful for her, involving not only scholarly activities but also rodeo competitions, specifically barrel racing, at which she won several awards.

While it was expected that Sublett would attend the University of Texas at Austin, she procrastinated in submitting her application, as this was not the school of her choice. She had earlier applied to the University of Arizona and, upon acceptance by that institution, her parents acceded to her wishes. Sublett's undergraduate years at Arizona introduced her to the field of physical anthropology. In fact, her specific interest in skeletal biology was fostered there and led her to pursue advanced studies in that area.

B. Professional Career

Audrey Sublett dated the beginning of her professional life to her initial contact in 1959 with Dr. James E. Anderson at the University of Toronto, where she earned her master's degree. Anderson was a physician who required his students to be trained thoroughly in the composition and structure of the human body. The course of study included gross anatomy, osteology, dentition, and growth and development, and students were required to acquire practical experience in these areas in addition to their academic work. As part of her program at Toronto, Sublett assisted Anderson in his longitudinal growth study initiated in the late 1950s. This project involved measurements of stature, weight, and fat deposition, as well as assessments of dental calcification and radiography to evaluate skeletal development.

The most significant aspect of the curriculum at Toronto was Anderson's tutelage in human osteology, specifically his development of new analytical methods for non-metric skeletal traits. Anderson's research in human growth and development prompted his interest in these traits, as many of them represent defects in skeletal development. He employed this method of analysis in his studies of the skeletal remains of Native American populations in both Canada and the

United States. Sublett was an enthusiastic adherent of this methodology (as were fellow students such as Nancy Ossenberg), as she had viewed metric analysis — with its goal of assigning individuals to predefined "types" — as inherently racist. She also adopted Anderson's pioneering analytical approach, which favored documentation of a total population profile (including data on metric and non-metric features, dentition, pathology, and maturation) in osteological studies.

Completion of Sublett's M.A. program at Toronto coincided with Anderson's acceptance of an academic position at the State University of New York at Buffalo (SUNY/Buffalo), and he invited her to enter the doctoral program there as his student. Despite her strong interest in skeletal research, she had decided against pursuing doctoral studies and declined Anderson's offer. Shortly thereafter, while on a driving tour of Mexico, she and several friends were involved in a serious accident in which all were injured and one was killed. This caused Sublett to reconsider her decision to cut short her studies, although she could never put into words exactly why the experience affected her in that way.

Her years at SUNY/Buffalo were eventful, not only in terms of her doctoral studies but also her involvement in a very important project related to the Reservation Period of Seneca Iroquois history. The Flood Control Act of 1936 had as its goal the protection of cities in Pennsylvania from destructive flooding, and pursuant to this, in 1961 ground was broken for a dam (Kinzua) on the Allegheny River near Warren, Pennsylvania. It was predicted that, when the dam was completed, the backwater would completely inundate the Cornplanter Grant in Pennsylvania and flood the southern one-third of the Allegany Reservation. This would affect a number of Seneca cemeteries dating from the late 1700s to the 1960s. While the Seneca had not prevailed in the legal battle to prevent the Kinzua dam construction, they did succeed in requiring the Army Corps of Engineers to relocate all of the cemeteries threatened by the flooding. At this point, SUNY/Buffalo entered the picture, proposing a study of archaeological contexts, genealogical research, and skeletal analysis. This last feature was the brainchild of Audrey Sublett, who saw this as an opportunity to test the accuracy of aging and sexing standards established for Caucasian and African populations for analysis of Native American skeletal remains. This was an ideal sample, as the age and sex of many of the individuals in the affected cemeteries were documented. The next of kin of a number of these individuals consented to the study and, in 1964, the Cornplanter Grant phase of the project was completed; the following year the Allegany cemeteries were investigated.

Sublett's work was supported by a National Parks Service grant in 1964 and, in 1965, by the American Philosophical Society. As she was limited to the amount of time the gravediggers required to transfer the remains from the graves to the burial containers (10 to 15 min in most cases), Sublett often could not complete a comprehensive analysis. A complementary genealogical study was carried out by fellow SUNY/Buffalo student George Abrams. While these data sets were to be included in the dissertations of Sublett and Abrams, future researchers would also come to benefit from this project.

In 1966, Sublett completed her Ph.D. degree at SUNY/Buffalo. Her dissertation, titled "Seneca Physical Type and Changes through Time," was based on the study of 550 skeletal remains from sites that spanned a period of approximately 800 years in the Iroquois region of New York. The most recent burials were those from the Kinzua dam cemetery relocations, and the earliest were from prehistoric Late Woodland (ca. 1100 AD) occupation. Both metric and non-metric data were utilized in testing the hypothesis that a distinctive Seneca physical type could be defined, and she also traced temporal trends in morphology. Included in this study were comparisons with other Iroquois groups. The major conclusion resulting from the dissertation research was that there was indeed a generalized Iroquois physical type and that the Seneca, while distinctive, clearly fit into this pattern.

During the last year of her doctoral studies, Sublett had taken advantage of all opportunities to assess the job market. The mid-1960s was a favorable period for employment in physical anthropology, and she had a number of interviews and job offers. Her choice of Florida Atlantic University (F.A.U.) in Boca Raton, Florida, was due in large part to the ongoing Fort Center Site excavation; the proximity of the ocean and opportunities for scuba diving and deep-sea fishing were also considerations. Fort Center was a large ceremonial complex in central Florida just south of Lake Okeechobee, which included mounds, causeways, and a charnel house platform over an artificial pond. The latter contained a large number of skeletal remains that had spilled from the platform when a lightning strike had set it ablaze. The excavation, which had begun in 1961, was directed by Dr. William H. Sears. Sublett's involvement in the field began in 1967 and ended when the site was closed in 1969, but analysis of the recovered skeletal remains continued for several years thereafter. Both her field and laboratory work were supported by a 3-year NSF grant awarded in 1968.

The work at Fort Center was complicated by Sublett's commitment to a salvage project near Binghamton, New York. Situated in a gravel pit, the Engelbert Site was a multicomponent occupation area in use from the Archaic Period (ca. 2000 BC) through the Late Woodland (until the early 1500s). Student archaeological crews from SUNY/Buffalo (directed by Dr. Marian E. White) carried out the initial excavations in 1967, prior to the formation of the Engelbert Site Project directed by Dolores Elliott. Sublett was assisted in her work by Joyce Sirianni and Rebecca Lane. In addition to 600 pit features, 135 burials were salvaged, with excavation barely staying ahead of the contractors who were mining the gravel; by the end of the 1968 field season, the site was completely obliterated. The Fort Center and Engelbert excavations were ongoing simultaneously during the summers of 1967 and 1968, and Sublett was able to offer field and laboratory experience to F.A.U. students at both sites, in addition to collecting a considerable amount of skeletal data.

From 1969 through 1977, Sublett divided her time between teaching and research, occasionally becoming involved in small-scale burial excavations such as the Onondaga Bloody Hill site (directed by Dr. James Tuck). Having collected considerable skeletal data from two significant sites (in addition to Allegany), much of her research time involved processing this information and preparing it for publication. She included students in this work, providing them with research and publication opportunities. During this time, Sublett collaborated with non-professionals in small-scale projects, coauthoring research reports with them. She also presented scholarly papers at national and international professional meetings. Her accomplishments and contributions in the fields of both archaeology and physical anthropology were acknowledged when she was named a Fellow of the New York State Archaeological Association.

Sublett's untimely death in 1977 at the age of 40 prevented her completion of the large-scale projects in which she was the principal researcher. However, the Seneca cemetery relocations (Cornplanter, Allegany) resulted in a considerable body of data that has been used by other researchers and will continue to be a valuable resource in the future. The same is true of the Engelbert and Fort Center sites and her data on the nearly 400 Contact Period Seneca skeletal remains studied for her dissertation and provided by Charles F. Wray. In the case of the Cornplanter Grant Project, which was in fact a pilot study undertaken to establish the protocols for the Allegany phase of the project, Sublett and George Abrams published their results in the *Pennsylvania Archaeologist* in 1965. While both Sublett and Abram produced manuscripts reporting the Allegany excavations and analysis, neither published the results as separate publications, although Sublett included data in her doctoral dissertation.

Perhaps the most significant work to result from the Allegany research was an article Sublett coauthored with her student, Rebecca Lane (Lane and Sublett, 1972). In this report—published in *American Antiquity*—the utility of non-metric osteological data in describing cultural components of life in earlier times was demonstrated. Based on the biological affinities calculated for males and females within and between the neighborhood cemeteries, it was determined that the traditionally matrilocal Seneca people had adopted a patrilocal residence pattern prior to or during the Reservation Period. The work of both Sublett and Abrams at Allegany was the basis for Rebecca Lane's 1977 doctoral dissertation. Employing genealogical data as a control, Lane developed an improved statistic for estimating biological affinities based on non-metric osteological trait frequencies. Designated the "standard effective divergence" (SED), this statistic rectifies some of the problems with measures of biological distance such as the "mean measure of divergence" (MMD). The SED is a direct measure of biological difference, while the MMD *may* be an *indication* of this. Also, unlike the MMD, the SED is accurate with small samples and when comparing samples of different sizes.

In 1986, Lorraine P. Saunders, another Sublett student, completed a doctoral dissertation which employed Lane's SED in testing the accuracy of the Contact Period (1550–1686 AD) Seneca village site sequence; also, approximately one-half of the osteological data had been collected by Sublett. Thus, even long after her death, Sublett continued to contribute to advancements in skeletal studies and provide opportunities for her students.

Another feature of Sublett's professional life was her encouragement of students and nonprofessionals, and her frequent coauthorship with them of scholarly works. As was typical in decades past, Sublett's osteological analysis was at times included as an appendix in archaeological Site reports. An example would be Marian E. White's Kleis Site report. She also shared her research with colleagues in the form of oral presentations at professional conferences.

Sublett often expressed concern about the future of human skeletal research, and she accurately anticipated the shift in attitudes that has come to pass. As early as 1970, she predicted that in the near future burial excavation would no longer be considered an acceptable research strategy. In anticipation of this, she asserted that the focus of research must be to develop new analytical methods that would maximize the recovery of information from the skeletal collections that were already in existence. It is ironic that, while she did anticipate the trend, she did not forsee enactment of a Federal law such as NAGPRA, which would eliminate many curated skeletal assemblages. The professional career of Audrey Sublett was brief but noteworthy. The considerable body of data that she generated, her dedication to improving and refining method of analysis, and the influence that she had upon colleagues and students serve to secure her place in the field of physical anthropology.

ACKNOWLEDGMENTS

Section V

For details of Marian Knight's later life we are indebted to Nanci Young, Smith College archivist.

Section X

I am greatly indebted to Rochelle A. Marrinan's valuable biography of Adelaide K. Bullen in the volume *Grit-Tempered: Early Women Archaeologists in the Southeastern United States*, edited by Nancy M. White, Lynne P. Sullivan, and Rochelle A. Marrinan (1999), for the information about Adelaide's life and career.

Section XIII

We thank Kathleen Brooks, Diane France, Stan Rhine, and Alan Simmons for providing essential information for this biographical entry.

Section II

Emerging Specialties

Introduction by Jane E. Buikstra

I. INTRODUCTION

This section turns to the themes that figure prominently in today's bioarchaeological research. The first of these centers on ritual studies, mortuary theories, and archaeological definitions of burial programs. This theme remains highly visible in archaeological inquiries and has been conspicuous in recent book length treatments (Arnold and Wicker, 2001; Chung and Wegers, 2005; Insoll, 2004; Parker Pearson, 2001; Rakita *et al.*, 2005; Sprague, 2005; Thorpe, 2002; Williams, 2003). As emphasized by Goldstein in Chapter 14, such theoretical developments are (too) frequently underappreciated by bioarchaeologists.

The second segment of this introduction emphasizes the history of age and sex assessments as background information essential to appreciating developments in paleodemography (Chapter 9). This discussion is followed by a summary of Chapters 8–12.

II. THEORIES OF MORTUARY BEHAVIOR

As emphasized in Chapter 1, archaeological contexts are fundamental to the interpretation of excavated human remains. They link individual remains to a specific community that existed in time and space and—as Cushing so aptly noted over a century ago—provide essential information concerning life histories, including social distinctions and ethnicity (Cushing, 1890; Hinsley and Wilcox, 1996, 2002). The definition of context is therefore an essential baseline for addressing other bioarchaeological issues, such as behavioral interpretations, division of labor, or health (Buikstra, 1977).

As Binford (1971) and others have emphasized, burial domains are frequently partitioned by personal attributes. If age, sex, circumstances of death, or health status affect the distribution of graves, potential for bias in the excavated sample is enormous. For example, hunter–foragers from Illinois were initially characterized in terms of extreme ill health (Neumann, 1967) before spatial distinctions in their Middle Archaic burial program were understood (Buikstra, 1981b). In this ~6 millennium BP example, people who were unable, either by age or infirmity, to perform a full round of activities were buried in the village, separate from the bluff crest community cemeteries. Excavations from midden deposits thus recovered disproportionate numbers of the young, the elderly, and individuals with activity-limiting pathology. In contrast, contemporary bluff crest cemeteries were sites where many young to middle-aged adults without signs of activity-limiting pathology were interred. Thus, these spatially distinct burial areas are best considered together when attempting to estimate health status during Middle Archaic times.

Cushing's emphasis on archaeological contexts led to his comparison of cremated remains to primary interments, which conflated chronological and cultural distinctions. Even so, his approach was remarkably nuanced and prescient in emphasizing nonmaterial aspects of mortuary rituals, which are frequently invisible archaeologically. Cushing's interest in religion, death, and the soul reflected issues also being considered by Tylor (1871) and Frazer (1890)[1] in their seminal 19th-century ethnographic research on ritual.

During the early 20th century, members of the French sociological school, Durkheim (1965), Hertz (1960), and Van Gennep (1960), published studies of ritual that would influence more recent scholarship on the archaeology of death, especially that of the "processual school." Importantly, Hertz and Van Gennep emphasized mortuary ritual as process, implying that its final stage is merely a single frame or snapshot of a much longer drama. Hertz chose to use the single case of secondary burial to illustrate general constructs and, in so doing,

[1] See Bell (1997) for a history of the scholarly study of ritual.

emphasized a triadic relationship among the corpse, the mourners, and the soul. Van Gennep's discussion of liminality and the dynamic, sequential stages of rituals, including mortuary rituals, have anchored powerful models for interpreting death rituals (Huntington and Metcalf, 1979; Metcalf and Huntington, 1991). Their work encouraged processual archaeologists, such as Brown (1981), to emphasize that archaeological contexts reflect only the final stage of a funerary sequence or burial program.

Before the rise of renewed archaeological interest in mortuary studies during the 1970s, many American archaeologists accepted Kroeber's (1927) interpretation of mortuary practices as being matters of fashion and style and therefore socially uninformative. As recently as 1969, Ucko echoed this cautionary tale (Rakita and Buikstra, 2005).

As part of the "New Archaeology" (Binford, 1962)[2], researchers again sought meaning in cemetery sites. A processual "Saxe–Binford–Brown" approach lent credibility to the study of mortuary practices (Binford, 1971; Brown, 1971, 1995; Saxe, 1970, 1971). Saxe, for example, emphasized the relationship between spatially bounded cemeteries and control of restricted resources. This association, reflected in his ethnographically tested Hypothesis 8, has been refined (Goldstein, 1980), extended (Charles and Buikstra, 1983), and remains useful (Morris, 1991). It may also be considered one stimulus for landscape archaeological studies that became common currency during the final decade of the 20th century (Ashmore and Knapp, 1999; Bowser, 2004). While the original arguments by Saxe, Binford, and Brown were subtle, recognizing both the significance of mourners and that the grave is only the final stage in complex interment rituals, many subsequent studies have assumed that there is a direct relationship between tomb elaboration and social rank (e.g., Tainter, 1978; Whittlesey and Reid, 2001).

The Saxe–Binford–Brown approach to mortuary studies received pointed critical review during the 1980s (Braun, 1981; Hodder, 1980, 1982a,b; McGuire, 1988; Miller and Tilley, 1984; Parker Pearson, 1982; Shanks and Tilley, 1982; Tilley, 1984). The most visible critiques were developed within symbolic, structuralist, and interpretative theoretical responses to processual archaeology (Rakita and Buikstra, 2005). Such "post-processual" perspectives argue that since the mourners and not the dead conduct mortuary rituals, the relative ostentation of burial programs frequently reflects political or social relations within the living community. Mortuary rituals and grave elaboration could therefore misrepresent or mask the social persona of the deceased. As Cannon (1989) has emphasized, mortuary ostentation may also exhibit cyclical trends. North American archaeologists, especially those working in the Southwest, have tended to ignore these critiques and many continue to assume a direct relationship between grave

[2]As with all paradigm shifts, there were early precursors; notable among these for the "New Archaeology" was Walter Taylor's conjunctive approach (Taylor, 1948).

elaboration, body treatment, and individual identities (see Goldstein, 2001). On the other hand, bioarchaeologists consistently fail to appreciate the significance of archaeological theories for their interpretations. Larsen (2002, Chapter 13, this volume) correctly emphasizes that bioarchaeology has become increasingly inter- and multi-disciplinary, but as Goldstein (Chapter 14, this volume) underscores, the associated discipline is seldom archaeology.

Several highly visible late 20th-century archaeological foci would seem to naturally link to the study of human remains. These include individualized study of the body, gender, and ethnicity. Gender studies, however, require an appreciation of the social differences between sex and gender (Walker and Cook, 1998). Bioarchaeologists have blurred this distinction (Buikstra and Ubelaker, 1994; Larsen, 2002; Steele and Bramblett, 1988), thus ignoring an important potential source of social and behavioral information. A survey of collected works on gender in archaeology (Arnold and Wicker, 2001; Claasen and Joyce, 1997; Gero and Conkey, 1991) enumerated only two bioarchaeological approaches among 38 articles from this period.

While popular subjects such as the individual and the body appear ideal for inter-(sub)disciplinary perspectives, including biological anthropology, as Goldstein (Chapter 14) points out, this is not necessarily the case, e.g., Meskell, 2000. In addition, some skeletal biologists who use complex, self-referential social terms such as ethnicity (Barth, 1969) have ignored social processes and thus confuse ethnicity with biological heritage, e.g., Howells (1995). Bioarchaeological studies of ethnicity and cultural modifications to the body, such as cranial deformation and dental modifications, whereby identity is permanently and physically inscribed, are more productive, e.g., Milner and Larsen, 1991 and Reycraft, 2004.

In sum, there is much for bioarchaeologists to gain through an appreciation of archaeological theorizing and the richness of ethnographic approaches to ritual. As detailed below, bioarchaeological approaches to skeletal analysis became increasingly technically sophisticated during the late 20th century, reflecting the appropriation of methods from the biological and physical sciences. Potentially productive venues developed in conjunction with the other social sciences remain a challenge for bioarchaeologists of the 21st century.

III. PALEODEMOGRAPHY: IN THE BEGINNING

Paleodemography is defined as the study of past population dynamics. Within bioarchaeological inquiry, paleodemographic inferences are based upon estimates of age-at-death and sex from skeletal remains. The transformation of these data facilitates comparisons designed to reconstruct population structure and health status or to detect census error. Vital rates, such as birth or death, may be estimated in the course of such investigations (see Buikstra, 1997, Frankenberg and Konigsberg, Chapter 9, and Hoppa, 2003 for other recent reviews of paleodemography). Another aspect of paleodemography involves the assessment of population density parameters or site occupation length based upon the investigation of cemeteries, either locally or regionally.

Morton, focusing upon adult skulls, did not partition the observations presented in his *Crania Americana* (1839) by sex. However, in a later work he laid to rest the assertion that a "pigmy race" once inhabited the Mississippi valley. Upon examination, a prospective short adult was found to have many deciduous teeth, with only the first molars and incisors of the permanent dentition present, which "as every anatomist knows" appear at about age seven. Morton concluded that these and other examples provide "convincing proof of what he had never doubted—viz. that the so-called Pigmies of the western country were merely children . . ." (Morton, 1841:126).

Later in the 19th century, as problem sets broadened, researchers—primarily medical doctors—regularly "sexed" skeletal remains and also separated the immature from the mature, upon occasion remarking upon older adult individuals. In attributing sex, pelvic observations were frequently privileged. When describing the Madisonville remains, Langdon (1881:237) remarked that "sex has been determined, so far as practicable, from the general skeletal development and the shape of the pelvis . . . ." With refreshing candor that predates the heated debates of the late 20th century concerning accuracy, he went on to caution that "it is hardly necessary to add that due allowance should be made here for possible errors" (Langdon, 1881:237).

Whitney, in his survey of the Peabody collection for signs of pathology, opined that the collection was especially valuable:

> The remains have been dug up with particular care for the preservation of the bones of the body as well as those of the head. The importance of this cannot be overestimated, for not only can the sex and age be more accurately determined, but also it can be more easily settled whether any pathological changes are the results of a local affection or of a general (constitutional) disease. (Whitney, 1886:433)

Thus, Whitney also looked well beyond the skull when estimating sex and age-at-death.

In pursuit of rigor, Matthews and co-workers (1893:220) extended their observations to quantification, designed for comparison between the sexes and with other groups.[3] They partitioned 18 pelves by morphological sex (8 males,

[3]The authors emphasize that none of the measurements originated with them, but are instead borrowed from the European "Pelvimetry" of Garson (1881–1882) and Verneau (1875).

10 females) and then assessed the performance of four indices,[4] concluding that "especially the indices of pubo-ischiatic depth and that of sacral length, show very prettily the natural grouping of the sexes" (Matthews, 1893:221). An immature pelvis is reported, but researchers declined to offer a sex assignment.

Hence, by the end of the 19th century, there were ample precedents for focusing sex diagnosis on the bony pelvis. Medical doctors, well aware of developmental skeletal and dental anatomy, were satisfied to assign individuals to general age-at-death categories.

Hrdlička's (1920a) handbook, *Anthropometry*, was heavily influenced by his instructor Manouvrier, to whom he dedicated the volume. In his discussion of sexing, he first focused on the skull, turned to the pelvis, the major long bones, and then the sternum, scapulae, ribs, spine, patella, calcaneus, and first phalanx of the great toe (Hrdlička, 1920a:91). While today's standards for morphological observation would explicitly privilege the pelvis, many of the cranial *and* postcranial attributes recommended in *Anthropometry* for sex diagnosis remain in handbooks from the late 20th century (Bass, 1987; Buikstra and Ubelaker, 1994; Steele and Bramblett, 1988; White and Folkens, 1991). The only truly significant addition has been Phenice's (1969b) three ischio-pubic features that have proved valuable in a variety of modern and ancient contexts. Rather optimistically, Hrdlička (1920a:92) asserted that an experienced observer could, with the cranium alone, diagnose sex correctly 80% of the time—in his opinion, adding the mandible raised the percentage to 90% and a full skeleton assured near perfection of 96%.

One trend prominent in late 20th-century sex diagnosis was the statistical evaluation of accuracy. The bias observed by Weiss (1972), whereby more males than females are recorded in archaeological remains, has been attributed to several possible causes: that (1) there are relatively few positive skeletal attributes in female skeletons compared to males (Phenice, 1969b); (2) in situations of marginal preservation, older female skeletons may be more friable than those of males (Walker *et al.*, 1988); and (3) the skulls of older females assume features associated with males, such as relatively rugose areas of muscle attachment (Walker, 1995). While selective archaeological recovery and nonrandom cemetery organization may also be invoked, the trend for more males than females to be classified correctly through morphological assessment of the skull was confirmed (Konigsberg and Hens, 1998). Konigsberg and Hens (1998), emphasing parametric approaches, e.g., probit analysis, in morphological evaluations, demonstrated the robusticity of their approach when faced with fragmentary remains and the unbalanced sex ratios anticipated in archaeological samples. Their research, including summary data from other sources, reported correct classification rates of ~80% for cranial features (including chin form),

[4]These were (1) breadth–height index, (2) index of superior strait, (3) index of the pubo-ischiatic depth, and (4) index of sacral length.

with much higher rates presented by other workers when pelvic attributes are available, e.g., 96–97% (Meindl *et al.*, 1985; Phenice, 1969b). The correct classificatiaon rate of 90–92% claimed by both Hrdlička (1920a) and Meindl *et al.* (1985) for complete skulls is perhaps an unrealistic expectation in archaeological samples, where post-depositional destruction is a key limiting factor.

In reference to age-at-death, Hrdlička (1920a:96) opined that "[f]or the anthropologist himself it generally suffices to determine whether the skull or skeleton is subadult, adult, or senile. . . ." Criteria include dental and epiphyseal development among juveniles. The newly formalized pubic symphysis method (Todd, 1920) is mentioned in one sentence as showing "important changes with age" (Hrdlička, 1920a:98).[5] Dental wear, cranial suture obliteration, and diminished bone weight are also considered. After nearly a century of critical evaluation and refinement, late 20th-century bioarchaeologists follow Hrdlička in recommending observations of multiple features of the developing dentition, epiphyseal union, and long bone length in juveniles, whereas grossly observable features of both pelvic articular surfaces and cranial suture closure continue to be favored in estimating age-at-death in adults (Buikstra and Ubelaker, 1994; Hoppa and Vaupel, 2002; Meindl and Lovejoy, 1985, 1989; Scheuer and Black, 2000; Smith, 1991). Researchers have also developed standards based on dental and bone histology (Fitzgerald and Rose, 2000; Robling and Stout, 2000). These are the attributes that serve as the basis for bioarchaeological characterizations of individuals and for paleodemographic comparisons.

IV. CHAPTERS 8–12: THEMES IN BIOARCHAEOLOGY

Chapter 8, entitled "Behavior and the Bones," begins by underscoring the multiplicity of ways in which bioarchaeologists have inferred behavior, both habitual and extreme, from skeletal and dental remains. J. Lawrence Angel's seminal work, beginning during the mid-20th century, is also recognized. Angel's work follows a tradition that can be traced to late 19th-century studies of skeletal plasticity, including those of Rudolf Virchow. Such approaches, developed in the German tradition, stand in marked contrast to the typological perspectives favored by Americans such as Hrdlička. Hooton, however, had been exposed to the German perspective while studying in the United Kingdom.

As underscored in Chapter 1, American studies of human remains from archaeological contexts were conducted during the 19th century by workers such as Matthews. Following the early 20th-century focus on typology, interest again

[5]Interestingly, Hrdlička's revised handbook, "Practical Anthropometry" (1939), provides no expanded discussion of Todd's methods.

developed in the wake of the "New Physical Anthropology". Biomechancial approaches became popular, following Ruff and Hayes' (1983a,b) highly visible study of Pecos Pueblo. Bridges' studies of osteoarthritis and cross-sectional geometry, enriched by her nuanced approach to the archaeological record, were also influential during the 1980s and 1990s. Other methods used for behavioral reconstruction include musculoskeletal markers (MSMs), evidence of violence, and nonmasticatory use of teeth as tools. The current status of each type of indicator is reported here, along with critical review. As Pearson and Buikstra emphasize, while bioarchaeologists have been unable to find signatures for specific activities, aggregate level information defining sexual division of labor and subsistence strategies has been productive.

Paleodemography is the subject of Chapter 9, "A Brief History of Paleodemography from Hooton to Hazards Analysis," by Sue Frankenberg and Lyle Konigsberg. They offer both a history of paleodemographic research in bioarchaeology and an evaluation of recent critiques. Hooton's attempts to estimate community sizes from cemetery data[6] are cited as the first significant contributions to paleodemography, well ahead of their time. Frankenberg and Konigsberg offer a sophisticated critique of Hooton's inferences about living population size and then rework his data through both life table and hazards approaches.

J. Lawrence Angel, a student of Hooton (see Chapter 4), knew of formal demographic methods such as life table construction, but eschewed them in favor of his own original methods for reconstructing population parameters. By the 1980s, however, life table construction had become standard practice. During this period, critiques such as those mounted by Bocquet-Appel and Masset (1982) targeted paleodemography. Challengers argued that parameters estimated in paleodemographic reconstructions were unrealistic, that life tables represented fertility better than mortality, and that age estimation methods were too imprecise, especially for adults, to permit accurate reconstructions.

Frankenberg and Konigsberg also report recent responses to such critiques, including the development of more sophisticated modeling methods, such as hazards analysis. Sampling issues have been addressed, and new methods for age estimation have been developed. The authors close by predicting a bright

[6]In addition to demographic reconstructions, regional population distributions can also be based on cemetery density, best gained through systematic survey. One of the best examples of this work is that of Charles (1992), who used transect survey data for Woodland mounds from the lower Illinois river region to generate a chronological sequence that charts initial Middle Woodland repopulation of the region from the north approximately 2000 years ago. Three subsequent Middle Woodland mound types reflect the following 350-year segment, followed by two distinctive, temporally sequential Late Woodland tumulus forms. The Late Woodland sites saturate the region over the subsequent 750-year period. Approaches such as this are possible only in contexts where census data for cemetery distribution are known and temporal assignments can be made with accuracy.

future for paleodemography, provided ongoing attention is paid to refining both measurement and analytical methods. They also stress the need to engage in evaluative processes for age estimation techniques, as well as the importance of generating realistic models of paleodemographic processes and identifying appropriate statistical approaches.

Konigsberg, in Chapter 10, "A Post-Neumann History of Biological and Genetic Distance Studies in Bioarchaeology," offers a sequel to Cook's (Chapter 2) discussion of craniology. He first considers the events surrounding the shift from Neumann's "varietal" thinking to population genetic modeling. Long's (1966) mathematically sophisticated critique of Neumann's typology and the intellectual climate associated with midcentury paradigm shifts in physical anthropology and archaeology are cited as key factors stimulating a shift to population-based, multivariate approaches to inferring ancestral relationships.

The second, more extensive focus of Chapter 10 is a history of biodistance studies conducted during the second half of the 20th century, with examples drawn heavily from Konigsberg's extensive experience in eastern U.S. bioarchaeology. He first notes that most of the regional studies dating to the 1970s that examined evidence for large-scale migrations reported strong evidence for genetic continuity. He finds this uniformity puzzling and speculates that it may be a methodological or analytical artifact. Small-scale processes such as residence patterns, although incompletely informed by population genetics, became the subject of study during the 1970s. In an appendix, Konigsberg provides a population genetic model for the effects of differential migration on genetic variance within the sexes that corrects his earlier derivation (Konigsberg, 1988).

Konigsberg then traces the history of population and quantitative genetic modeling in biodistance research, beginning with Rebecca Lane's (1977) prescient dissertation and culminating in the Relethford-Blangero model, which is designed to test for long-range gene flow. He also reaffirms the need to develop more sophisticated theoretical models that link biodistances to time and space simultaneously. In closing, Konigsberg calls attention to two developments that may transform 21st-century biodistance study: the "new morphometry" and ancient DNA analysis.

Cook and Powell, in Chapter 11, "The Evolution of American Paleopathology," begin their review with 19th-century studies that set the pattern for the first part of the 20th century. Both Warren and Morton's early contributions are noted, as are key late 19th-century reports by scholars such as Wyman, Jones, and Harrison Allen. Although the authors emphasize that the twin themes of artificial cranial modification and syphilis dominated the century, a broad range of conditions—congenital, infectious, and traumatic—was also recognized and is reported here.

During the early 20th century, Cook and Powell argue (as does Jarcho, 1966a) that the field of paleopathology benefited primarily from scholarship representing nonanthropological fields, including paleontology, anatomy, medical, and dental science. The authors cite key contributions to the study of syphilis by Herbert U. Williams, a physician who applied a range of contemporary medical methods in his research on archaeological bone lesions. Studies of ancient disease by Hrdlička and by Hooton are also evaluated, with Hooton's *Indians of Pecos Pueblo* cited as a landmark effort.

Hooton's population-based legacy is visible in the approaches taken by Stewart and Angel at the Smithsonian Institution, but three decades passed before this perspective truly flourished, beginning with the work of George Armelagos in Nubia, as emphasized by Cook and Powell. A series of papers by Armelagos and his students, beginning in the 1970s, address the political and economic impact of agriculture on various aspects of health in prehistoric Native Americans in Illinois. This extension of Hooton's population-based approach, based primarily on nonspecific markers of developmental stress,[7] has recently influenced global health projects initiated by anthropologist Jerome Rose and economist Richard Steckel in 1990. Such large-scale projects necessarily summarize multiple data sets from different regional and temporal units and incorporate the work of researchers whose data collection standards may not be identical. In this manner they depart from the unified, contextually focused research that bioarchaeology emphasizes, but they have revitalized interest in the study of health of global populations from a multidimensional perspective.

Cook and Powell also report productive collaborative efforts in the study of ancient health, begun by William Bass in the Great Plains and by Clark Larsen and David Hurst Thomas in the northern portion of La Florida. The latter project has extended the pre-contact perspective on health into the historic period and should therefore be considered significant on several levels. The authors include a review of recent texts and other compendia on paleopathology, as well as the history of professional associations and international congresses that focus on paleopathology and mummy studies. They close by considering new biomolecular and imaging methods that hold great promise for 21st-century bioarchaeology.

In Chapter 12, "The Dentist and the Archaeologist: The Role of Dental Anthropology in North American Bioarchaeology," Rose and Burke explore the manner in which the study of human teeth has contributed to bioarchaeology. Explicitly linking paradigm shifts in dental anthropology

---

[7] A penetrating critique of this approach was published in 1992 by Wood and colleagues, who pointed out that in a sense the presence of nonspecific indicators was an indication of health sufficiently good to survive the insult and register it. This critique has stimulated both negative (Cohen, 1994; Goodman, 1993) and thoughtful (Saunders *et al.*, 1995; Storey, 1997; Wright and Chew, 1998; Wright and Yoder, 2003) responses.

to temporal divisions defined in American archaeology and paleopathology, they identify four chronological units: (1) Classificatory–Descriptive (1840–1914), (2) Classificatory–Historical (1914–1940), (3) Contextual–Functional (1940–1960), and (4) Modern (1960+). In each of these periods, they consider four data categories: caries, dental wear, developmental defects, and dental size/shape — and three interpretative themes: dietary reconstructions, analysis of childhood disease and stress patterns, and genetic relationships.

Beginning their discussion by considering the vigorous 19th-century debates concerning dental health at meetings of the Odontological Society of Great Britain, Rose and Burke report scholars' early attempts to explain why caries rates were higher among the developed countries than in ancient times. They highlight the problem-oriented work of Mummery on dental caries, dental wear, and diet. Despite the innovative, problem-oriented research of Mummery and others,[8] Hrdlička tended to simply describe caries rates and use dental wear to estimate age-at-death rather than using it as a source of dietary information. Hrdlička did, however, report the relatively high rate of shovel-shaped incisors in North American Indians.

During the Classificatory–Historical period, the tempo of work increased. Leigh (1925) published a "classic" comparative study of diet and dental health. Caries became firmly linked to dental decay and dental wear continued to be studied, both as an age indicator and in relationship to diet. After World War II, Rose and Burke see little advancement in the study of caries and wear. Dental histological methods did advance, however, and, under the influence of Dahlberg, studies of dental morphology became more rigorous and systematic.

More recent studies (falling in the Modern, 1960+ category) have focused explicitly upon the relationship among dental caries, wear (including microwear), and diet, concerned especially with changes associated with the transition to agriculture. Dental enamel defects and microdefects of the enamel and dentine have been used as measures of childhood health and adaptation. Studies of dental morphology and measurement have not been so visible during this period, although Turner's landmark work in standardizing morphological observations is of singular importance. Rose and Burke also underscore the key texts by Brothwell (1963a,b), which have been immensely influential within dental anthropology, human osteology, and archaeology.

---

[8]Matthews, Wortman, and Billings (1893) present an early example of comparative studies of caries and diet across groups. This predates Leigh's (1925b) widely cited study by over three decades.

Chapter 8

Behavior and the Bones

Osbjorn M. Pearson and Jane E. Buikstra

I. INTRODUCTION

The reconstruction of the behaviors and lifestyles of prehistoric peoples from their skeletal remains and archaeological contexts constitute primary goals of bioarchaeology. Today bioarchaeologists attempt to meet these goals through a combination of biomechanical analyses, studies of osteoarthritis and trauma, and other observations (Larsen, 1997; Bridges, 1992, 1994b, 1996; Ruff, 1992, 2000; Hawkey and Merbs, 1995). The effort to use such data to produce an impression of prehistoric lifeways has become increasingly visible over the last three decades, owing its popularity to the influential work of a host of earlier researchers. J. Lawrence Angel was one of the earliest advocates of what has become the current approach, as illustrated by his description of three 9000-year-old skeletons from Hotu Cave, Iran:

> Femoral neck torsion, tibial head tilting, gluteal crest development, platymeria, platycnemia, and stressed extensor and rotator muscle insertions form a complex [cf. Wagner 1927 (1926):115–117] called the bent-knee gait, often misinterpreted. This applies to the use of the legs flexibly, like a skier, and *not* a posture. Stress on the ilio-tibial band, iliac crest, and lower and upper lumbar areas (possible herniation of lowest nucleus pulposus in number 2) suggests further that the Hotu women may have done some standing and working with braced legs (as pulling on a fish net) as well as much climbing in rough country, carrying, and digging. The injuries to the thumb-wrist joints and little finger of number 3 suggest possible fighting but more plausibly hard manual work perhaps more specialized than digging for roots: flint chipping, plaiting baskets, net-making, or possibly midwifery or shaminism. The pelves of the two women show enough bone reaction at ligament attachments and insertion of the abdominal wall muscles (rectus and external oblique aponecosis [*sic*]) to hint that pregnancy may have been frequent and without rest period. (Angel, 1952:265)

Bioarchaeology: The Contextual Analysis of Human Remains, Buikstra and Beck (eds.)

Angel's work on other skeletal samples such as the Archaic period remains from Tranquillity, California, further exemplified this holistic approach to behavioral reconstruction and allowed Angel to paint a detailed portrait of at least some of these people's activities and to advance informed speculation about others:

> The Tranquillity people show other postural specializations in the frequency of flexion facets at the ankle (80 percent) and in retroversion of the tibia in two out of four cases. Together with the marked femoral pilaster and platymeria, these suggest active running in rough terrain. In five out of nine cases the olecranon fossa floor is perforated, a condition linked with elbow hyperextensibility. As expected, four out of these five cases are female. This may relate to the general "economy of bone" which the Tranquillity people show: the shafts of all long bones are flattened about to the degree seen in Old World Paleolithic and other hunting populations and often show a sinuosity and extra sharpness of muscle attachments which approach the bowing of sabre shin seen in actual malnutrition. (Angel, 1966a:3)

II. ROOTS

Much of the recent work by bioarchaeologists to reconstruct the activity patterns and lifestyles of prehistoric peoples has followed Angel's lead, but with attempts to incorporate improved methods, new approaches, and a wider comparative framework of populations for which homologous data are available. It should not be forgotten, however, that Angel also stood upon the shoulders of giants, and the roots of behavioral reconstruction are to be found much earlier in time.

Functional and behavioral interpretations of skeletal remains ultimately arose from anatomy and those trained in it, whether in England, Germany, or the United States. By the late 1800s, European physicians and anatomists followed one of two traditions: a traditional one that emphasized typology and classification and a relatively new one that focused on the plasticity and adaptability of the body over a lifetime. The second approach became almost synonymous with the name Rudolf Virchow, whose profound influence led him to be regarded as the father of the medical study of pathology.

By the late 1800s, Virchow—and, by extension, nearly the entire German anatomical and medical establishment—placed great emphasis on the plasticity of the body, including the skeletal system, in response to external forces. At the time, German academia also led the world in technological innovation and engineering, and the exuberance and vigor of this field of inquiry also influenced German anatomists. The most visible product of this intellectual cross-fertilization was the work of Julius Wolff on the structure and development of trabecular bone, research that formed one of the bases of what 20th-century researchers came to regard as Wolff's "law" (for a historical summary, see Martin *et al.*, 1998). Wolff originally formulated his proposition as a means of understanding how trabecular bone adopted an architecture that allowed it to resist mechanical stresses with a minimum amount of material. This "law" formed a homologue to models that mechanical engineers of the time were developing for iron trellis systems that could bear great loads with a minimum of material (Martin *et al.*, 1998). This emphasis on plasticity and adaptation, characteristic of the German anatomists from Virchow's day onward, greatly influenced the work of a number of important figures in anthropology, including Franz Boas, Rudolph Martin, Franz Weidenreich, and, more recently, Friedrich Pauwels, Adolph Schultz, and Holger Preuschoft.

Most contemporary anatomists in other European nations, Great Britain, and America could read German and were at least aware of the German emphasis on functional adaptation. Some, including Sir Arthur Keith, adopted a perspective heavily influenced by functional considerations (Keith, 1940); most, however, remained committed to more traditional, typological approaches to anatomy and, by extension, the nascent science of anthropology. In America, Aleš Hrdlička embodied and greatly advanced the traditional, typological approach toward morphology. Trained as a physician in the Czech Republic, Hrdlička was clearly cognizant of contemporary German anatomical studies, but his approach to anthropology was to remain firmly typological (see Chapters 1–3). Hrdlička's work on the shapes of the femur and tibia included a typological categorization of shapes of the shafts (Hrdlička, 1898, 1934a,b), but his later work also included a perspective on the development of distinctive shapes of femoral shafts (Hrdlička, 1934a,b), a study of comparative shapes of homologous primate femora (Hrdlička, 1934d), and a comprehensive treatment of femoral third trochanters and hypotrochanteric fossae (Hrdlička, 1934c, 1937b).

Earnest Hooton exerted a strong influence on the development of bioarchaeology and functional interpretations of human remains, in part due to his detailed descriptions of the remains from Pecos Pueblo (Hooton, 1930; see also Chapters 1, 2, and 4). Hooton had trained in anthropology in England, where he was exposed to a broad range of research methods, including the new German focus on somatic plasticity, the early developments in biometry, and statistical descriptions of populations pioneered by Francis Galton and Karl Pearson, as well as the classic, typological approaches to morphology that still dominated British anatomy and were to form the basis of much, but not all, of Hooton's work.

The first work in North America on behavioral interpretations of human remains preceded the more influential, later work on the topic by Hrdlička, Hooton, and others. Some of the earliest investigators realized what has become a dominant paradigm today: a comprehensive bioarchaeological approach to inferring behavior—individual or group—requires consideration of both archaeological contexts *and* human remains. The 19th-century Hemenway Expedition discussed earlier (see Chapters 1 and 5) serves as an early North American example. One of Cushing's goals, influenced by his prior ethnological and archaeological experiences, was to study grave accompaniments in order to know the sex, the condition of life, and other facts about the individual. As mentioned previously, he believed this information would lead to "vivid, even historic knowledge of the people" interred at Los Muertos (Hinsley and Wilcox, 2002:200). In complementary fashion, Washington Matthews and colleagues (1893) were quite eager to infer behavior through the study of human bones. Noting that neither septal apertures of the humeri nor platycnemia occur in children, these authors argued that both conditions arose due to specific activities. For example, they inferred that grinding corn led to the development of septal apertures among women. They also took issue with Manouvrier's (1888) deduction that platycnemia necessarily developed through hyperactivity of the *tibialis posterior* muscle and was necessarily or even frequently associated with hunting lifestyles on rough terrain (see also Kennedy, 1989; Ruff, 2000). They argued instead that behavioral interpretations of platycnemia should be based on a more broadly based consideration of biomechanical principles.

> When the *tibialis posticus* assumes the inverse action, the tibia becomes a lever of the second class, with the fulcrum at the ankle joint, the power at the insertion of the muscle, and the weight (which in ordinary cases is but the weight of the body and the clothing) at the knee joint. There are three ways (besides frequency of impulse) in which the distance through which the lever moves, as in climbing hills; second, by diminishing the time in which it moves, as in running and jumping; third, by increasing the weight, as in lifting and carrying heavy loads. Largely to the third way we are inclined to attribute the prevalence of platycnemia among various American races, including the Saladoans. (Matthews *et al.*, 1893:224)

Following such 19th- and earlier 20th-century scholarship, research that focused on functional and behavioral interpretation of human remains experienced a great acceleration from the 1970s onward. This increased interest has its roots in Washburn's (1951, 1953) "New Physical Anthropology" and in the holistic conception of anthropology imparted by Hooton upon his students. With respect to bioarchaeology, the contributions of J. Lawrence Angel, Sherwood Washburn, and T. Dale Stewart loom large. At the close of the 20th century, interpretations of prehistoric people's patterns of activity have been based on four primary forms of data: cross-sectional geometry, osteoarthritis and trauma, and muscle markings in addition to an assortment of other traces of behavior left on bones or teeth.

III. INTERPRETING PREHISTORIC PATTERNS OF ACTIVITY

A. Cross-Sectional Geometry

1. History and Application

Following a period of near invisibility, midcentury, biomechanical approaches once more assumed significance in late 20th-century interpretations (Bridges, 1985, 1989a; Bridges *et al.*, 2000; Larsen, 1995; Larsen *et al.*, 1995, 1996; Ruff, 1991, 1992, 1994, 1999, 2000; Ruff and Hayes, 1983a,b; Ruff *et al.*, 1984; see also Chapter 13). The cross-sectional geometry of long bones of an animal are commonly upheld as one of the best indicators of the mechanical forces that the animal had adapted to resist in life, and thus a reasonable reflection of habitual activities (Ruff, 2000). Stimulated by the structural analysis of platycnemia (Lovejoy *et al.*, 1976),[1] researchers investigated topics such as mobility patterns and sexual division of labor across time and space in a variety of archaeological skeletal samples, basing their inferences on bone shape.

The thickness of limb bones of animals has long been of interest in functional morphology, from Galileo's observations of allometric changes in animal limb bones to the present (Preuschoft, 1971; Wainright *et al.*, 1976; Alexander, 1977; Pauwels, 1980; McMahon and Bonner, 1983; Currey, 1984; Schmidt-Nielsen, 1984; Currey and Alexander, 1985; Martin and Burr, 1989). Anthropological interest in the relationships between bone cross-sectional geometry and function largely grew out of the broader fields of biomechanics and functional anatomy as reflected in the work of Pauwels (1980). Early applications of beam mechanics to model the strength of human long bones were made by Pauwels (1980), Endo and Kimura (1970), Kimura (1974), Lovejoy *et al.* (1976), and Lovejoy and Trinkaus (1980), among others. The development of technology to digitize the cross sections of long bones and of computer programs such as SLICE (Nagurka and Hayes, 1980) that could rapidly calculate second moments of area from bone sections allowed the proliferation of studies of cross-sectional geometry during the 1980s and 1990s.

For bioarchaeologists, Ruff and Hayes' (1983a,b) study of the cross-sectional geometry of the Pecos Pueblo femora and tibiae proved to be an influential landmark. The study was quickly followed by investigations of changes in limb bone

---

[1] It was Lovejoy and colleagues (1976) who called Wolff's "law" and modern derivative biomechanical principles to the attention of the physical anthropological community in America; the concept had been well known for many years to functional anatomists. Their interpretation of the behavioral correlates of platycnemia, although tentative ["this hypothesis does not seem improbable" (Lovejoy *et al.*, 1976:505)], was reminiscent of Manouvrier's (1888) of nearly a century before in its emphasis on active locomotion on uneven substrates.

cross-sectional geometry that had accompanied the shift to agriculture on the Georgia Coast (Ruff *et al.*, 1984). This study corroborated Larsen's (1981) earlier findings that a decline in femoral strength, a decrease in the development of the femoral pilaster, and an overall decrease in size accompanied the transition to agriculture in the same region. In the late 1980s and early 1990s, the notion that hunter–gatherers were taller, healthier, and led more physically demanding lives than later horticultural or more intensive agricultural populations became a widely accepted paradigm (e.g., Cohen and Armelagos, 1984; Larsen, 1982; Ruff *et al.*, 1993). It is significant, therefore, that Bridges (1989a) described an instance from northern Alabama in which the transition to Mississippian agriculture failed to produce the expected pattern and instead found that the Mississippian males had stronger legs than their Archaic predecessors and that Mississippian females had both stronger legs and considerably stronger humeri than their Archaic counterparts, a change that was accompanied by a decrease in upper limb asymmetry.

Bridges (1989a) pointed to a variety of other studies (Pickering, 1984; Goodman *et al.*, 1984; Lallo, 1973; Hamilton, 1982) that had suggested that bone size, muscle marks, or arthritis incidence — all of which tended to be treated at the time as nearly equivalent indicators of activity — provided additional evidence that changes in subsistence with agricultural intensification had required increasing amounts of labor and activity rather than the reverse. Bridges (1991a) soon reported that the comparison between frequencies of osteoarthritis in Archaic and Mississippian people from northern Alabama produced the opposite pattern of what the cross-sectional geometry indicated: the foragers had more osteoarthritis in their joints. For the time, Bridges showed a great sensitivity to such contradictions (see also Bridges, 1989b, 1990, 1991b, 1992, 1994b, 1996). Toward the end of her career, Bridges (1997) began to test the relationships between various traits taken to be indicators of activity, a research direction that foreshadowed one of the current forefronts of research and to which we return at the end of this chapter.

Additional studies of cross-sectional geometry of the long bones of prehistoric populations continued to appear at a rapid pace in the late 1980s and early 1990s. Prominent examples include Brock and Ruff's (1988) study of changes in cross-sectional geometry in the American Southwest; Robbins and co-workers' (1989) study of Late Woodland limb bones from Delaware; Fresia and colleagues' (1990) documentation of the decline in the bilateral asymmetry of the humerus on the Georgia Coast; Larsen and colleagues' (1995) report on the rugged skeletons from Stillwater marsh and other Great Basin sites (see also Ruff, 1999); Ruff's (1994) description of extraordinary development of the femoral pilaster in femora from the southern Plains; Ledger and co-workers' (2000) analysis of the limbs of 18th-century slaves from Cape Town, South Africa; and Stock and Pfeiffer's (2001) documentation of substantial variation in limb bone structure between two groups of hunter–gatherers, Andaman Islanders and Precontact Khoisan from the Cape of South Africa. By no means is this list exhaustive and it reflects the visibility that cross-sectional geometry has achieved as the most highly regarded measure of activity patterns. In addition to work on recent populations, a large number of studies were devoted to the cross-sectional geometry of Upper Paleolithic people, Neanderthals, and still earlier hominins (Senut, 1985; Grine *et al.*, 1995; Churchill *et al.*, 1996; Holliday, 1997a; Pearson and Grine, 1996, 1997; Churchill and Formicola, 1997; Ruff *et al.*, 1994, 1999; Ruff, 1995; Trinkaus and Ruff, 1999a,b; Trinkaus *et al.*, 1991, 1994, 1999; Pearson, 2000; Holt, 2003).

Data for such studies were initially digitized from photographs of sectioned bones or from CT scans (Ruff and Leo, 1986), but Runestad *et al.* (1993) developed a method of molding the external contour of a bone, taking biplanar, orthogonal X-ray films of the bone and using the endosteal surface visible in the X-rays to approximate the endosteal contour of the section. The contour mould and X-ray method has been subsequently used in a large number of other studies (Churchill, 1996; Churchill and Formicola, 1997; Holliday, 1997a,b; Holt, 2003).

Likewise, the biomechanics of primate and human mandibles have been reproduced utilizing a beam model, with the cross section of the corpus acting as the beam section (Hylander, 1988; Daegling, 1989; Daegling and Grine, 1991; Dobson and Trinkaus, 2002). Furthermore, some studies of mandibular cross-sectional geometry have been able to compare their results to experimentally determined strains acting on the mandible (Hylander and Johnson, 1994; Chen and Chen, 1998; Daegling and Hylander, 1998, 2000; Daegling and Hotzman, 2003). Given the amount of data available for the bony structure of the mandible, the direction and magnitudes of the muscles that act upon it, and the amount of bite force that can be generated, it has also been possible to construct finite-element models of how human and primate mandibles and crania deform during mastication (Korioth *et al.*, 1992; Richmond *et al.*, 2005; Strait *et al.*, 2005; Ross *et al.*, 2005).

2. Criticisms of Cross-Sectional Geometry

During the late 20th century and into the 21st century, skeletal biologists began to question certain fundamental assumptions of biomechanical approaches to behavioral reconstructions. Concerns were expressed concerning uncritical acceptance of fundamental, 19th-century assumptions (Wolff's "law") and failure to recognize recent research in bone biology, especially "mechanobiology." Tendencies to interpret nonsignificant results and to dismiss confounding variables were also cited (Bice, 2003). Lovejoy and colleagues expressed this concern:

> A common assumption that has long pervaded interpretations of the hominid postcranium is that the distribution of bone, in both its cortical and cancellous forms, can be viewed as an uncomplicated "record" of the bone's loading history. However, during the past decade, highly aggressive research protocols, together with their

> continual reintegration into novel theoretical approaches, have cast strong doubt on this presumption. It can no longer be used as a perfunctory basis for the direct interpretation of skeletal form. Too many data have accumulated which negate so simplistic an approach. (Lovejoy *et al.*, 2002:97)

A variety of experimental studies have found that bones are not actually loaded or bent in the directions that anthropologists initially expected they would be and that the axis of bending does not pass through the centroid of area of the section as analytical programs such as SLICE (Nagurka and Hayes, 1980) assume it does (Gross *et al.*, 1992; Demes *et al.*, 1998, 2001; Lieberman *et al.*, 2004). Both kinds of exceptions to expected functional patterns in bones constitute sobering findings for those who wish to use cross-sectional geometry in their reconstructions of the lives of prehistoric people. Jurmain (1999) also questioned the current utility of studies of the cross-sectional geometry of prehistoric people's long bones to shed light on their patterns of activities because very few clinical studies have actually documented the effects of specific activities on the cross-sectional geometry of human long bones. Without such data from living subjects, interpretations from studies of ancient bones will likely remain only interesting speculation, regardless of whether such inferences seem plausible or not.

With regard to the problems created by the fact that bones can be loaded in directions we might not predict and that the neutral axis of bending may not pass through the centroid of area of a section (as we generally assume), Lieberman and colleagues (2004) found that the section modulus of a bone is likely to contain more error than other variables such as the torsional second moment of area (J). The section modulus (Z) of a cross section of a beam or bone is defined as the section's second moment of area divided by the perpendicular distance from the bending axis to outermost point of bone mass in the section (Martin *et al.*, 1998). The section modulus is currently (Ruff, 2000) considered the most useful—and most biomechanically meaningful (Martin *et al.*, 1998)—cross-sectional property to analyze in skeletal material. In light of Lieberman and co-workers' (2004) finding, anthropologists might be well advised to emphasize analyses of J standardized for body size, which was popular from the early 1990s until 2000 (Ruff *et al.*, 1993, 1994; Larsen and Ruff, 1991; Larsen *et al.*, 1995).

A final, recent development in the study of cross-sectional geometry that affects its utility in making behavioral inferences about prehistoric populations is the growing realization that bones may not model in response to exercise or habitual activity in the same way across the life span. Ruff and co-workers (1994) pioneered some of the recent interest in the ontogeny of cross-sectional geometry with a model of the ontogeny of femoral cross sections that hypothesized that activity during childhood would produce extra subperiosetal apposition and decrease the rate of endosteal resorption, whereas strenuous exercise during adulthood could produce endosteal stenosis but only a modest amount of additional subperiosteal deposition. A variety of recent studies have suggested that activity during childhood, especially during the adolescent growth period, appears to exert a more substantial influence on the size and shape of adult bones than exercise later in life (Kannus *et al.*, 1995; Khan *et al.*, 2000; Kontulainen *et al.*, 2001; for a review, see Pearson and Lieberman, 2004). The implications of these findings have yet to be fully explored by bioarchaeologists, but the great variety of subsistence practices employed in prehistory offer a promising area of inquiry for studies of the ontogeny of bone shape and strength.

B. Osteoarthritis and Trauma

1. Arthritis

While reports of arthritic change appeared during the 19th century, emphasis was frequently on describing the most extreme cases. For example, Langdon (1881:249) discussed the fusion of all thoracic and lumbar vertebrae in ancient remains from the Madisonville, Ohio, cemetery site, attributing the condition to *arthritis deformans*. Similarly, Whitney (1886:444) described remains of an older man recovered from a stone box grave near Brentwood, Tennessee: "both elbow joints are roughened and irregular and the surface in spots looks like ivory. His joints must have grated like a rusty hinge when he attempted to move them, and the stiffness and restricted motion must have been the same as is seen in the rheumatic cripple of to-day." Thus, 19th-century observers emphasized description, diagnosis, and the degree to which behavior had been limited by the arthritic condition, not on the behaviors that might have caused the condition.

Comparative, population-based descriptive studies are found throughout the 20th century, e.g., Hrdlička (1914a), Stewart (1947, 1966), and Jurmain (1977a,b, 1980, 1990, 1991). Stewart's research included age-related patterning as well as population comparisons for Native American skeletal series. He reported extreme arthritis in the lumbar vertebrae of Inuit peoples when compared to Pueblo Indians. Extending this comparison to the knee, hip, elbow, and other joints, Jurmain (1977a) described more severe arthritic changes along with earlier onset for Alaskan Eskimos. Ortner (1968), concentrating upon the elbow, also concluded that arthritic change in Inuit remains was more extreme than that observed in Peruvians.

Angel's (1966a) study of 35 Archaic period skeletons from the Tranquillity site, mentioned earlier, provides a vivid example of one of the best of the early studies of the pattern of arthritic degeneration to draw inferences about living conditions and labor. Angel wrote:

> There is plenty of evidence that they lead strenuous lives. All four preserved vertebral columns show fully developed hypertrophic arthritis in cervical and lumbar regions and one shows a healed fracture at waist level plus herniation of the disk nucleus into the body of the fourth lumbar vertebra. This degree of wear and tear of the disks and

> ligaments at the age of 25–40 is typical of hardworking populations (Gejvall, 1960, ch. VIII; Nathan, 1962; Stewart, 1958a) and one or two decades ahead of our vertebral column aging. (Angel, 1966a:3)

Several additional points of interest are illustrated by the preceding quotation: Angel's profound knowledge of human anatomy; his close, collegial association with T. Dale Stewart, whose deft studies of vertebral anomalies and pathologies influenced both Angel and the subsequent adoption of the entire field of paleopathology; and Angel's familiarity with contemporary studies in Europe.

Angel's work on the Tranquillity remains became an oft-cited landmark study that proposed an explicit link between osteoarthritis (OA) and specific behaviors (see also Chapter 11). In this report, Angel noted "6 of 13 people have arthritis in the elbow joint, usually including eburnation after friction removal of cartilage over the capitulum" (Angel, 1966a:3). Consideration of possible causes for the high frequency of this pathology led Angel to the idea that throwing darts from a spear-thrower (or atlatl, to use the Aztec word) might be the cause. He wrote:

> The spear thrower, of course, puts extra stress on the arm muscles and elbow. Hence it seems logical to describe this special pathological change as "atlatl elbow." Laughlin (1963), Stewart, Merbs, and others have noted it among the Alaskan Eskimo and Aleut. It is less frequent in female skeletons. But it does occur in two out of four Tranquillity females even though the arthritic lipping is slight. Possibly seed-grinding has some effect. It is equally likely that a genetic weakness or avascularity of the joint plays a part in small and isolated populations. This is given point by the frequency of a similar elbow avascular necrosis in baseball playing Japanese, as opposed to Westerners (Nagura, 1960). (Angel, 1966a:3)

Angel noted that other throwing actions should also cause shoulder and clavicular stresses, not observed in this sample. Angel's term, "atlatl elbow," for the condition proved to be influential in many subsequent studies (e.g., Jurmain, 1977a; Bridges, 1990). Angel attributed the pathology in females to seed grinding using a mano and metate, inspiring Merbs (1980) to coin the term "metate elbow" for it. Recognizing that genetic factors can also influence patterning, Angel's research combined both focused observations within joints and considerations of overall patterning.

As noted by Jurmain (1999), the use of OA to infer behavior became less popular during the final decade of the 20th century. This may in part be attributable to critical evaluations, such as those of Bridges (1992:80): "while arthritis is undoubtedly related in part to forces placed on the joint, it is not a straightforward indicator of the level or type of normal activities." Jurmain (1990, 1991, 1999) concurs, concluding that the most productive approaches appear to be those that investigate patterning in multiple joints, using the total available skeletal sample as the database (see also Rothschild, 1995; Waldron, 1994). Other matters of concern are nonstandard data-recording protocols and the absence of statistical testing (Bridges, 1992).

Jurmain's (1999) review of the clinical and epidemiological literature on living people showed that such studies provide only ambiguous and contradictory evidence for the link between activity and OA. Instead, injury to joints, which may or may not be a predictable consequence of certain activities, emerges as the most important risk factor for the development of OA later in life. As Jurmain (1999) notes, it is clear from clinical studies that many joints are able to sustain vigorous, long-term loading from distance running and other activities without developing osteoarthritis (Hoffman, 1993; Panush and Lane, 1994; Lane *et al.*, 1993). Jurmain (1977b) described distinct patterns of age of onset of OA in various joints in different populations, and today it appears that developmental age and activity interact in complex ways to produce OA:

> Some joints (elbow and hip particularly, as compared to other joints) appear to be under differential risk, given the age of the onset of mechanical loading; early injury and/or modification of joint mechanics can produce OA changes later in life. (Jurmain, 1999:105)

Likewise, a variety of studies suggest that different joints may develop OA in response to dissimilar stimuli: "the knee appears to be most pone to activities involving repetitive bending, while the hip and spine appear to be more at risk as the result of heavy lifting" (Jurmain, 1999:105). If reinforced by future findings, such results will mean that the observation of OA in different locations in a skeleton may reveal different types of information about the physical activities of the person rather than providing a gauge of overall levels of activity. The clinical literature is replete with contradictory and complex findings about the associations between osteoarthritis and activity, however, and Jurmain's words of caution are worth repeating:

> The association of OA with *specific* activities is not clearly supported in contemporary contexts by either the occupational or sports literature. Further, the implications for and limitations on osteological interpretations are obvious. (Jurmain, 1999:105)

Part of the difficulty in applying clinical studies of risk factors and OA to bioarchaeological studies arises from the fact that clinicians usually define OA in a different way than anthropologists. In clinical settings, erosion of the cartilage in joints, damage to subchondral bone, and narrowing of the joint capsule are used to diagnose OA, whereas many osteologists' definitions have included the development of osteophytes around the joint capsule, a phenomenon that is not of clinical relevance unless the osteophytes interfere with the joint's function (Jurmain, 1999). Osteologists should score the two phenomena separately (Buikstra and Ubelaker, 1994). It remains likely that some—and perhaps much—of the OA that anthropologists have attributed to "activity" is in fact due to activities across the life span, but some is almost certainly due to injuries, and more research and more caution in drawing conclusions from traces of OA are both clearly warranted.

2. Trauma, Including Spondylolysis

During the 19th century, studies of paleopathology in ancient Native American remains typically included "injuries" as one of three categories, with the others being "anomalies" and "diseases" (Matthews *et al.*, 1893; Whitney, 1886). Some, such as Whitney (1886:436), distinguished between fractures and dislocations, although fractures were the most commonly described injury. In contrast to researchers' early preoccupation with the impact of arthritis upon activity, fractures, especially cranial fractures, were both described in exquisite detail and attributed causally to aggressive behaviors. Observed within the Madisonville sample, for example, was a partially healed, extensive fracture that retained a depression "just above the ear which nicely fits one of the round-headed stone hammers found in the cemetery" (Langdon, 1881:252). From a small sample drawn from across North America, Whitney (1886:439) felt that in three examples of cranial fractures "there was a strong presumption in favor of their being due to intentional violence. The seat, the left side of the head, especially favors this view, as it presupposes that the persons who gave the blows were right-handed." Arrow wounds were also reported (Langdon, 1881).

Twentieth-century bioarchaeological inquiry continued to report evidence of both intentional and accidental trauma (see also Ortner and Powell, 2006). While descriptive reports occurred throughout this period, comprehensive comparative studies appeared relatively late in the century, most postdating Lovejoy and Heiple's (1981) influential attempt to establish age-specific fracture patterns in the late prehistoric Libben site (Ohio) skeletal sample.

A 2001 summary of the history of violence by Walker described considerable variation across time and space in the Americas, as also reported by Ortner and Powell (2006). Walker and Lambert's extensive, contextualized studies of trauma, for example, identified increased violence during the Middle Period for the Santa Barbara Channel islands, a time of resource stress (Lambert, 1994, 1997; Walker, 1996). Late prehistory saw little violence in some locations, while chronic warfare apparently caused the death of at least one-third of the adults interred at the Norris Farms (Illinois) Oneota site (~AD 1300; Milner, 1995; Milner *et al.*, 1991). The roughly contemporaneous Crow Creek Massacre site (South Dakota) provides ample evidence of traumatic death and violent, postmortem treatment of nearly 500 individuals, including scalping and dismemberment (Willey, 1990; Willey and Emerson, 1993; Zimmerman *et al.*, 1981). Ortner and Powell (2006) emphasize that scalping clearly predates European contact, documented as early as Middle Archaic times (Mensforth, 2001; Smith, 1995, 1997).

Additional traces of human behavior have been described from other portions of the body. The nonmasticatory use of teeth as tools received scholarly attention during the early 20th century (Leigh, 1925a), an interest that has been maintained since that time (Milner and Larsen, 1991; Larsen, 1997). Transversely oriented occlusal grooves were noted in anterior teeth of several hunter–forager groups from Texas, the Great Basin, California, and British Columbia (Bement, 1994; Larsen, 1985; Schulz, 1977). The Great Basin samples were studied through scanning electron microscopy, which revealed multiple fine scratches following the main axis of the groove. It has been suggested (Larsen, 1985, 1997) that some form of flexible material, such as sinew or plant fibers, was passed repeatedly over the teeth. Notching and lingual surface wear associated with extramasticatory functions have also been reported for groups from Texas (Hartnady and Rose, 1991), Tennessee (Blakely and Beck, 1984), and the Georgia Coast (Larsen, 1982).

As noted in the first section of this chapter, 19th-century scholars such as Wyman (1875) considered cannibalism a likely explanation for the archaeological recovery of fragmented human bone that had been treated in the same manner as faunal remains. This subject again assumed marked visibility through the work of White (1992) and Turner (1983; Turner and Turner, 1999), who focused on evidence from the Greater Southwest. Both scholars concentrated on developing detailed protocols for identifying evidence of cannibalism, including evidence of burning, cut marks, pot polish (smoothed surfaces due to boiling), and fragmentation patterns. While alternative explanations were proposed, including the destruction of social deviants such as witches (Darling, 1993, 1998; Dongoske *et al.*, 2000; Ogilvie and Hilton, 1993; Martin, 2000), the recovery of human myoglobin from a human coprolite recovered archaeologically from the Cowboy Wash site (Colorado) demonstrated that at least one person consumed human flesh in the ancient Southwest, ca. AD 1150 (Marlar *et al.*, 2000).

Merbs' (1983, 1995, 1996a) work on degenerative changes among the Inuit was both influential and showed the potential of careful study of trauma to elucidate patterns of prehistoric activity. Merbs' work included a careful consideration of spondylolysis, including sacral spondylolysis (Merbs, 1996a), and, ultimately, a rigorous exploration of the etiology of spondylolysis (Merbs, 1996b). This work allowed Merbs to make interesting interpretations:

> Sacral spondylolysis was a relatively common phenomenon in Alaskan and Canadian males during late adolescence and early adulthood but . . . the condition would correct itself, leaving a permanent record only in those unlucky enough to die young. Although the unusually vigorous activity patterns of these males appear to have been a major cause of the stress fracturing that produced the spondylolysis, specific (but largely unspecified) anatomical variations and delayed vertebral maturation may also have been significant contributors. (Merbs, 1996a:365)

An explicit focus on the reconstruction of habitual behavior led Merbs (1969, 1983) to also investigate osteoarthritis, along with osteophytosis, compression fractures, spondylolysis, and anterior tooth loss. Working with historic period Canadian Inuit (Sadlerimiut) remains from Southampton Island,

Northwest Territories, Merbs formulated explicit behavioral expectations based on ethnohistoric accounts. These expectations guided his behavioral reconstructions, an approach that received widespread recognition and approval, (e.g., Bridges, 1992; Jurmain, 1999).

C. Musculoskeletal Stress Markers

We close this review with a discussion of musculoskeletal stress markers (MSMs), which are also commonly called enthesopathies, entheses, or, more coloquially, muscle markings. Along with cross-sectional geometry, many anthropologists consider MSMs and OA in joints and the vertebral column to be indicators of activity patterns and a reflection of skeletal responses to its mechanical environment (Jurmain, 1977a, 1980; Kennedy, 1989; Larsen, 1995; Hawkey and Merbs, 1995). The expression of both OA and MSMs tends to become more common and more pronounced with age (Jurmain, 1977a, 1980, 1999; Dutour, 1992; Hawkey and Merbs, 1995; Wilczak, 1998; Wilczak and Kennedy, 1998; Weiss, 2003a,b, 2004).

The early history of observation of MSMs by no means achieved the degree of precision and specificity that researchers have sought to achieve since the early 1990s, but less formalized or systematic observations of muscle markings constituted part of the examination of skeletal remains from the early days of American physical anthropology. Hrdlička (1937b), for example, penned a detailed account of structural variants associated with insertion of the gluteus maximus and offered a comprehensive summary of the etiologies that had hitherto been proposed for the development of those features. Perhaps the most influential observer and interpreter of the significance of muscle markings was J. Lawrence Angel. From his early work (e.g., Angel, 1946a), Angel displayed an acute sensitivity to what variations in both overall skeletal morphology *and* areas of tendon attachment might reveal about prehistoric lifeways and activity patterns. An early example of this sensibility may be found in the following passage from his description of the three Epipaleolithic indviduals from Hotu Cave, Iran:

> The upper surfaces of the tibiae are tilted more than usual and the laterally compressed shafts of the shinbones have a diamond-shape cross section. Fibulae are deeply fluted. The femora are distinctly platymeric or thickened transversely in the upper shaft as if to take stress from strong abductor and lateral rotator muscles. The deep gluteal fossae adjacent to marked crests, the strong adductor tubercles, the stressed origin areas for gastrocnemius, and on the tibiae the increased origin area for deep muscles supporting the arches of the feet confirm the suggestion that muscles involved in rough-country travel were well-developed. (Angel, 1952:259)

Likewise, Charles Snow (1974) paid careful attention to the development of muscle markings in his description of pre-contact Hawaiian skeletons from

Mokapu, Oahu. Snow's text paints an evocative portrait, as for example his summary of femoral muscle markings:

> Almost all of these bones show well-developed pilastering of the linea aspera. This buttressing, reinforced bony ridge was strong evidence for well-developed flexor and extensor muscles. . . . The bone relief of the trochanteric region was bold and showed extensive muscular areas. Likewise, in the popliteal region at the back of the knee, the adductor tubercle was very well developed. (Snow, 1974:47)

Snow was fortunate to have detailed accounts of the daily habits, work, recreation, and other physical activities of Hawaiians from the period of contact that he could use to draw links between behaviors and the osteological traces of heavy musculature that he observed. Such close attention to ethnographic accounts of labor and activity patterns have informed some of the best analyses of other ostensible markers of activity, including Ruff and Hayes' (1983a,b) analyses of the cross-sectional geometry of the Pecos Pueblo limb bones, Merbs' (1983, 1996a) work on trauma and degenerative disease among the Inuit, and Bridges' (1989a) account of changes in the cross-sectional geometry of limb bones in Indians from Northern Alabama during the transition from foraging to agriculture. However, while Bridges' (1989a) work illustrates the judicious use of ethnographic accounts, it also illustrates another problem with interpreting patterns of prehistoric activities: everything about the activities of the ancient foragers in Alabama must be inferred and thus are not "known." This problem becomes exacerbated in progressively more ancient societies and may be particularly problematic in Paleolithic societies, which experienced living conditions, including surprisingly low population densities (Stiner *et al.*, 1999, 2000), that may not have a close historical analog.

Returning to the present, other studies of MSMs have made use of accounts of labor conditions to enrich interpretations of the pattern of observed muscle markings. In an influential article on the life stresses of slavery, Kelley and Angel (1987) combined observations of muscle markings, patterns of arthritis, and historical information about living conditions and diet to interpret the pattern of morphology in skeletal remains. Their study constitutes an early, systematized attempt to quantify and compare the development of muscle insertions of enslaved ironworkers from Catoctin Furnace, Virgina. The resultant picture of life and activity patterns could be painted with broad strokes:

> Our best evidence for occupation and related pathology is from the Catoctin site. The muscle crests we compare are the deltoid, pectoral, teres, and supinator (see Figs. 3–5). The former are, of course, involved in the lifting of heavy objects. Their development in teenagers or young adults females indicates heavy work of a type not common to twentieth-century females. In combination with shoulder or vertebral breakdown, including separated L5 arch, and schmorl herniation, the picture of hard, heavy labor is substantiated. (Kelley and Angel, 1987:207)

While reports of osteoarthritis in behavioral reconstructions declined in the late 1990s, attention turned to MSMs. Use of the atlatl and similar behaviors were inferred by Kennedy (1983) to be related to hypertrophy of the ulnar crest to which the supinator muscle attaches. Studying the relative development of such "enthesopathies" (tendinous insertions or ligamentous attachments) became increasingly popular for behavioral inferences during the 1990s (Jurmain, 1999).[2] Hawkey and Merbs (1995:325) caution that such markers are ideal "for a study of activity-induced changes in a population" only in large, well-preserved skeletal series, preferably those dating to a relatively narrow time span where cultural and genetic isolation and a limited number of specialized, known activities exist. Another concern in the use of MSMs for behavioral inferences is that there is little scientific evidence that directly links enthesopathies to specific activities (Jurmain, 1999; Robb, 1994; Ruff, 2000).

Spurred by the examples presented by Angel (1952, 1966a; Angel and Kelley, 1986; Angel *et al.*, 1987; Kelley and Angel, 1987), Kennedy (1983, 1984, 1989), and others (e.g., Dutour, 1986, 1992) of the power of MSMs to provide grist for the mill of interpretation of prehistoric lifeways, work on more rigorous methods for quantifying and comparing MSMs began in earnest in the late 1980s and continued vigorously through the 1990s. Hawkey and Merbs (1995) produced an influential study of MSMs in Hudson Bay Inuit from two time periods, the "Early Thule (Classic Period)" and "Later Thule (Transitional/Historic)." Based in part on Hawkey's master's thesis (Hawkey, 1988), Hawkey and Merbs' study of this population has become a landmark in the study of MSMs. The methodology they used to quantify MSMs has been widely adopted—with and without modifications—by many subsequent studies (e.g., Steen and Lane, 1998; Weiss, 2004). Key aspects of the method include assessing each muscle origin or insertion site for three features: robusticity markers, stress lesions, and ossification exostosis. Each is scored along an ordinal scale with photographs and descriptions to guide the researcher in making allocations (Hawkey and Merbs, 1995). In this protocol, "robusticity" generally refers to the overall size and prominence[3] of the origin or insertion area, "stress lesions" usually refer to resorptive pitting in an attachment site, and "ossification exostoses" denote small spurs of ossified ligaments or aponeuroses protruding from the attachment site. These three features were then combined into an overall ranked score of expression that placed the least weight on "robusticity" and the most weight on the degree of development of "stress lesions" (Hawkey and Merbs, 1995).

Although Hawkey and Merbs' (1995) methodology proved highly influential, a large variety of other methods, many of them less clearly or precisely defined,

---

[2]See Kennedy (1989) for an inclusive listing of enthesopathies, osteoarthritic changes, and fractures, as used for behavioral inferences.

[3]"Modeling" in Frost's terminology (Frost, 1986; Martin *et al.*, 1998; Lieberman *et al.*, 2003).

have also been proposed for the quantification of MSMs. Among the best-defined methods are those of Wilczak (1998), who quantified attachment areas by digitizing chalk outlines of insertion areas, and of Robb (1998), who advocated a system of seriation of MSMs from least to most pronounced. Most studies find more pronounced muscle marks in males than in females, even when controlling for age.[4] The different methodologies for scoring MSMs have also produced some interesting, conflicting results that suggest that additional work is needed to clarify how closely they correspond and under what circumstances they will tend to produce differing results. For example, Hawkey and Merbs (1995:326) reported very little correlation with age among adults, noting that "[a]lthough a gradual increase in attachment robusticity was noted from young to middle to adult, the differences were not significant statistically, and all adult samples were pooled." Using the same methodology for scoring MSMs, Elizabeth Weiss (2003b, 2004) found significant correlations with age and bone length in both the humerus and the lower limb. Likewise, Wilczak (1998) reported a complex set of correlations between insertion size and age.

D. Criticisms of Enthesopathies, Osteophytes, and Osteoarthritis

Since 1995, there has been a large increase in publications on MSMs and an even larger number of presentations on muscle markings at the annual meetings of the American Association of Physical Anthropologists (e.g., Munson Chapman, 1997; Steen and Lane, 1998; Churchill and Morris, 1998; Peterson, 1998; Lovell and Dublenko, 1999; Molnar, 2003; Pany *et al.*, 2003; Toyne, 2003). Despite the surge in interest in MSMs, there are very few clinical studies that have actually linked MSMs, and their degree of development, with specific activities (Jurmain, 1999), largely because the osteophytes interpreted as MSMs by osteologists generally do not cause discomfort to living people and are thus not of clinical significance. However, a few researchers are now focusing on the problem of how activities in life correlate with the development of pits and osteophytes in MSM development (Zumwalt *et al.*, 2000; Zumwalt, 2004).

Many questions about MSMs still remain to be answered. Do repetitive activities or overuse injuries cause MSMs? Do occasional, high-stress activities produce MSMs and are such infrequent, high-magnitude strains more likely to produce MSMs than more repetitive but lower-strain activities? Are there individual differences in the risk of developing MSMs after performing specific activities? Are there population-level differences in the probability of developing

[4]With the exception of work by E. Weiss (2004), however, these comparisons between the sexes have not attempted to control for body size or muscularity.

rugged muscle insertion sites in response to performing specific amounts of given activities? Are there age effects so that activities performed at a young or old age have dissimilar probabilities of influencing the expression of MSMs? Until more is known about the etiology of MSMs, interpretations of what they show about prehistoric activities will necessarily remain speculative, however logical that speculation may seem.

So far, there have been very few ontogenetic studies of the development of MSMs from childhood into adulthood. The literature contains more ontogenetic studies of OA, and these show low frequencies in early adulthood followed by increasing frequencies later (Jurmain, 1999). Jurmain (1999) has urged anthropologists to pay special attention to the age of onset of OA in specific joints in comparisons between sexes and populations. A problem with studying age of onset arises from the fact that in clinical studies, injury to a joint, particularly injury in childhood, repeatedly emerges as a major risk factor for OA later in adulthood (Micheli and Klein, 1991; Jurmain, 1999). Most people survive from mid-childhood to early adulthood (Wood *et al.*, 2002). As a result, most osteological series contain very few skeletons of juveniles older than about 5 years of age, making it very difficult to accurately assess the probability of injury to joints. The upshot for osteologists is that the best way to solve the problem of etiology of OA will be via more clinical research on living people. Studies of archaeological populations may also prove invaluable, but it is doubtful that they will ever be able to match the diagnostic ability of clinical studies in which many more factors such as body mass, actual activity patterns, diet, history of injuries, and the like can be accurately measured and taken into account.

IV. CONCLUSIONS

The reconstruction of prehistoric lifeways and activity patterns from skeletal remains has been one goal of physical anthropologists from the very origin of the discipline in the United States. Key early influences on American physical anthropologists and anatomists primarily included contemporary British, French, and German anatomists, who often worked under differing research paradigms, yet were also generally mutually aware of each other's work. In particular, under the direction of Virchow from the 1850s onward, the "German school" of anatomy placed great emphasis on the plasticity of tissues, including muscle and bone, to environmental factors, including work and activity. Many other anatomists, importantly including Hrdlička, remained firmly rooted in the older tradition of typology, which today has much less importance in the functional interpretation of skeletons than the paradigm championed by Virchow and his students.

The approach to reconstructing behavior espoused by most modern bioarchaeologists perhaps owes its origin to the combination of the holistic approach to skeletal anatomy fostered by J. Lawrence Angel, a student of Earnest Hooton, and the development and subsequent surge in popularity of studies of the cross-sectional geometry of bones. Modern methods include cross-sectional geometry, patterns of osteoarthritis, musculoskeletal stress markers, trauma, and other observations. Virtually all of these data sets are problematic, as thoughtfully critiqued by Jurmain (1999), and considerable work remains to clarify which of these features provide the best indications of activity and how the various forms of data are interrelated.

Bridges' (1989a, 1991b, 1997) work highlighted the fact that cross-sectional geometry, osteoarthritis, and the development of muscle markings might not be closely correlated and might, in fact, not be interchangeable indicators of activity. Rather, her work suggested that these aspects of skeletal morphology might arise from differing influences. More work on skeletal and living populations is clearly needed to elucidate how the various forms of data are intercorrelated as well as what activities are responsible for the development in life of the features that we can observe in skeletal populations. Encouragingly, some workers have already taken additional steps in this direction, including Churchill's (1996) factor analysis of upper limbs, which included measures of cross-sectional geometry, muscle lever and load arms, and other dimensions in Neanderthals, early modern humans, and a series of recent comparative populations. Likewise, E. Weiss' (2004) work on the intercorrelations among MSMs scored via the Hawkey-Merbs' method, cross-sectional geometry, body size, sex, and age stands as a very useful study of how these properties are interrelated. More studies of living people and the factors that we assume have generated the patterns of cross-sectional geometry, osteoarthritis, and MSMs in prehistoric populations are badly needed. Physical anthropologists have cause to feel optimistic at this juncture: all of these studies are feasible and will undoubtedly serve to enrich our understanding of the lives of our ancestors.

In sum, although bioarchaeological studies of behavior have failed to establish signatures for specific activities, group-level inferences have compared and contrasted groups with different lifeways. Sexual division of labor has also been addressed, as have topics such as cannibalism and the extramasticatory use of teeth as tools. While the goal of behavioral reconstruction is central to 21st-century bioarchaeology, researchers must also pay attention to the need for rigor in their studies and not fall into the "activity-only myopia" decried by Jurmain (1999).

Chapter 9

A Brief History of Paleodemography from Hooton to Hazards Analysis

Susan R. Frankenberg and Lyle W. Konigsberg

I. INTRODUCTION

Paleodemography, or the study of past population dynamics, should be and often has been an important component of bioarchaeology. Studies of population structure provide ways to evaluate the contributions and impacts of past behaviors, social structure, economics, and environment on human life and well being, with the relationships between past behavior and biology being a principal concern of bioarchaeology. The application of paleodemography in bioarchaeological studies has not been consistent through time, however, for a number of reasons. These reasons include the facts that many demographic methods have not been easy to translate to archaeological skeletal samples and that many bioarchaeologists have been ignorant of available methodologies. This chapter traces the history of paleodemography within bioarchaeology, identifying first and continued uses of various demographic techniques and tracing the underlying mathematical threads common to diverse approaches. This history also seeks to evaluate criticisms of and continuing problems in paleodemography from both outside and within the field.

The first clear use of paleodemographic concepts and methods on the American scene began in the 1920s–1930s with Hooton's work on Madisonville and Pecos Pueblo. Hooton was concerned with estimating living population size and evaluating survivorship from cemetery samples, but apparently was unaware of then-common formal demographic methods that could assist in these endeavors.

Bioarchaeology: The Contextual Analysis of Human Remains, Buikstra and Beck (eds.)

Little new was published in paleodemography after Hooton until Angel's work in the late 1940s through 1960s. In contrast to Hooton, Angel was aware of work in formal demography, but rejected life table methods in favor of his own obtuse calculations. The number of paleodemographic methods and applications exploded in the 1970s, beginning with use of formal life table methods and the development of life tables for anthropological populations. By the 1980s, when calculation of life tables was standard procedure in skeletal studies, paleodemography came under attack both within and outside the field. The principal criticisms of paleodemographic methods at that time were that demographic reconstructions implied unrealistic rates of survival and other parameters, that paleodemographic life tables often were more informative about fertility than about mortality, and that methods of age estimation were too imprecise to allow meaningful demographic analyses.

The mid-1980s to the present time has been a period of mixed results and varied successes. Dissatisfied with the discrete age nature of life tables and the limitations mentioned earlier, numerous scholars have moved on to hazards analysis as a means of modeling paleodemographic processes more realistically. These types of studies have addressed interval-censored data and nonzero growth rates and are beginning to build uncertainty of age estimation into the models. Some scholars also have moved on to identify the diverse sampling, measurement, and analytical issues in paleodemography that must be addressed in order to understand past population processes successfully (for examples, see Milner *et al.*, 2000). While a small number of researchers continue to reject paleodemography as viable and others continue to follow methodologies now shown to be unrealistic and inaccurate, we believe we can expect more from paleodemography in the future. This chapter presents a brief history of the accomplishments and pitfalls of paleodemography beginning with Hooton and ending with current and future directions of study.

II. THE HOOTONIAN PALEODEMOGRAPHIC LEGACY

Hooton's (1920) paleodemographic analysis in his monograph on the "Indian Village Site and Cemetery Near Madisonville, Ohio," like much of his bioarchaeological work, was decades ahead of its time. In a few brief pages, Hooton (1920:20) compared rather deficient skeletal age-at-death data from Madisonville to age-at-death data from European populations in order to estimate a reasonable crude death rate. This step presaged Weiss' (1973) far more elaborate development of model life tables for anthropology and Coale and Demeny's (1966) models for broader applications. Hooton also applied the central relationship between death rate, length of cemetery use, and number of burials in order to estimate living population. He wrote "assuming the total number of burials in the cemetery to have been about 1350 and the annual death rate to have been about 3 per hundred, a village of 450 to 500 inhabitants would have been sufficient to fill this cemetery in a century" (Hooton, 1920:27). Symbolically, we have

$$\frac{N}{T \times d} = P$$
$$\frac{1350}{100 \times 0.03} = 450, \tag{1}$$

where N is the number of skeletons, T is the length of time for which a cemetery is used, and d is the death rate per annum. Many years later Ubelaker (1974) used this same relationship to estimate population size from an ossuary sample. In a similar vein, Konigsberg (1985) treated N and d as known in order to logarithmically plot population size against length of cemetery use. A semilogarithmic plot for estimating T and P using Hooton's data from Madisonville (Hooton, 1920) is shown in Fig. 1.

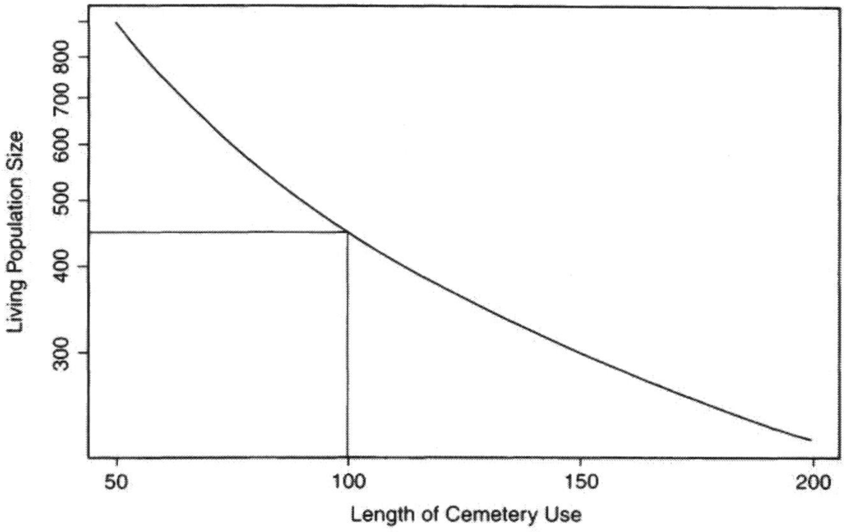

Figure 1 Semilogarithmic plot of living population size (P) and length of cemetery use (T) for the Madisonville site based on the number of skeletons (N) and annual death rate (d). Straight lines show that a living population of roughly 450 individuals could generate the observed cemetery size over a period of 100 years.

Ten years after his work on the Madisonville site, Hooton (1930) published a monograph on *The Indians of Pecos Pueblo* that also included a paleodemographic analysis ahead of its time in bioarchaeology. Compared to formal demography, however, his analysis of the Pecos Pueblo population was quite primitive. For example, Lotka (1922) had published the underpinnings of stable population theory 8 years earlier, and the use of life tables in mortality analysis had occurred considerably earlier [see Newell (1988) and Smith and Keyfitz (1977) for reviews and collected papers]. Hooton used a rather ad hoc survivorship graph (Hooton, 1930:335) to drive his analysis and attempted to find the number of skeletons that would be produced by a stationary population with a particular population size over a 100-year period. We have reproduced his figure here using a slightly different format (see Fig. 2). Hooton's graph is discussed in detail, as it allows us to point out the refinements that come from (later) use of formal life tables.

The logic underlying Hooton's calculation of the number of deaths in a 100-year period was as follows. Suppose that the living population size is constant at 1000 individuals, with survivorship such that 0.5 of the population survives until age 33.33, 0.1 survives until age 66.67, and all are dead by age 100. Hooton divided a 100-year span into three cohorts of 33.33 years each. The first cohort, which is represented with vertical hatching in Fig. 2, consists of 1000 individuals, all of whom die within the 100-year interval, generating 1000 deaths.

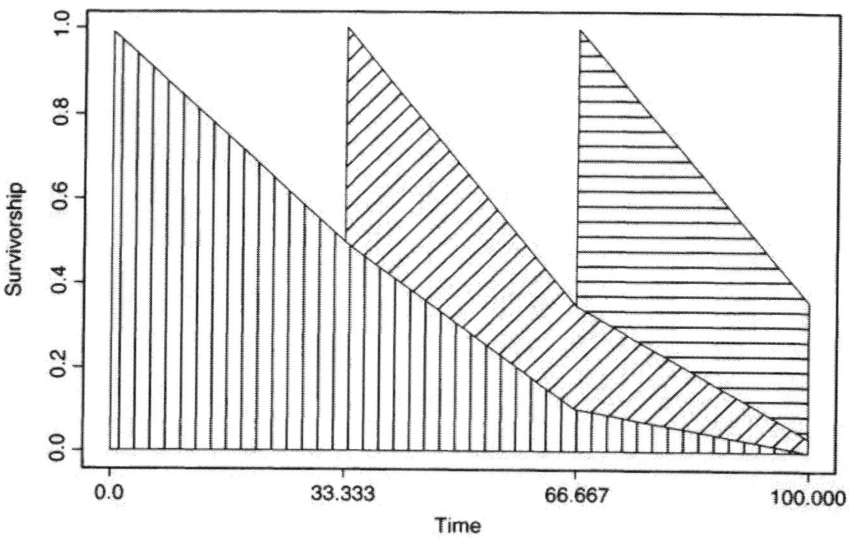

Figure 2 Survivorship of three cohorts over 100 years at Pecos Pueblo based on Hooton's (1930) Figure XI-2.

By 33.33 years into the 100 span, a second cohort (diagonal hatching) would have been born. This second cohort includes 500 individuals, which, with the remaining 500 individuals from the first cohort, totals a population of 1000 individuals. Of these 500 individuals in the second cohort, 90% would have died within the following 66.67 years, generating 450 deaths. The final cohort (horizontal hatching) consists of 650 individuals, again bringing the total population to 1000 individuals alive at time 66.67 years. These 650 individuals augment the 100 individuals still alive from the first cohort and the 250 individuals still alive from the second cohort. Of the third cohort, 50% will have died by the end of the 100-year span, yielding 325 deaths. Summing the deaths from the three cohorts we have 1000 + 450 + 325 = 1775 deaths, which is the value Hooton gave near the top of his page 336. Hooton (1930:336) recognized that his graph was a bit unrealistic, noting that he had assumed "that all of the population at the beginning of the century are young, whereas at least 10 percent are very old persons left over from the preceding century." He consequently suggested adding an additional 100 (10% of 1000) individuals, bringing the total number of deaths to 1875. This was a completely ad hoc, and in fact, logically inconsistent adjustment. Based on this adjustment, there should have been 1100 people alive at the beginning of the century, except that Hooton took great pains in the rest of his graph to assure that births accrued such that the population size stayed constant at 1000.

There are a number of unsupported assumptions that Hooton used to draw his figure and that conspire to lead us to the wrong answer. As Hooton noted, there should be some individuals who enter the century at age 66.67 years, but he curiously neglected to mention that there also should be individuals who enter the century at age 33.33. If there are three cohorts, then they all should be represented at any one point in time. This is not the case in Hooton's figure and our reproduction because the population starts far from its stable age distribution (i.e., the characteristic age distribution that follows from a fixed regime of age specific mortality and a fixed growth rate). The concept of a stable age distribution was well known to demographers in Hooton's time (e.g., Sharpe and Lotka, 1911), although the available mathematical form was better suited to continuous time problems than to the three age-class model Hooton used. Had Hooton understood the mathematics behind life table analysis he could have found the stable age distribution directly, and because he assumed that the population was stationary, the stable age distribution would have remained constant in both proportions and raw numbers. One way to find the stationary age distribution is to iterate through Hooton's figure many times, something that was not particularly feasible in the 1920s with paper and pencil. Doing this we find that the population stabilizes at 625 0–33.33 year olds, 312.5 33.34–66.66 year olds, and 62.5 66.67–100 year olds. Applying Hooton's logic, we start the century with 1000 individuals, all of whom will be dead by the end of the century, at 33.33 years into the century we pick up 625 new individuals by births, of whom 90% will die by the end of the century, and at 66.67 years we pick up an additional 625 individuals, 50% of whom will die within the next 33.33 years. Consequently, there should be $1000 + 1.4 \times 625 = 1875$ skeletons, the number Hooton actually estimated, but for the wrong reasons!

If we calculate a life table using Hooton's Pecos Pueblo data, we find that the number of skeletons that should accrue in one century is actually much greater than Hooton's estimate (Hooton, 1930), totaling about 2727 individuals instead of 1875. The reason for Hooton's underestimate is that although his graph implies a linear decline in survivorship (the usual assumption in a life table), his calculations treat survivorship as a step function. In other words, his mathematical modeling implies that everyone who is born survives until 33.33 years, at which age half of the cohort immediately dies. Then the remainder of the cohort lives until exactly 66.67 years, at which age 80% of the cohort promptly dies. The remaining 10% then live until 100 years, at which age they all die. In fact, what Hooton clearly intended from his graph was a continuous birth and death process, for which he needed to apply a life table. Table I gives a rather abbreviated life table to show these calculations. The first column (X) gives the age, the second column [$l(x)$] gives the survivorship from Hooton, and the third column [$d(x)$], although not in Hooton (1930), is calculated directly from survivorship. For example, $d(33.33) = l(33.33) - l(66.67) = 0.4$. The next three columns [$L(x)$, $T(x)$, and $c(x)$] are calculated first assuming a linear decrease in $l(x)$ across the age category and then assuming a step function for survivorship. The fourth (and seventh) column [$L(x)$] is one of the least simple to understand and is, in fact, the source of Hooton's underestimate. We consequently explain $L(x)$ in some detail.

$L(x)$ is a column that is specific to life tables, and that has no analog in the hazard models presented here. Technically, $L(x)$ represents the person-years lived within an age interval. If we make the age intervals infinitesimally small,

Table I

Pecos Pueblo Life Table Calculated from Hooton's (1930) Data[a]

	Usual life table					Hooton's method		
X	$l(x)$	$d(x)$	$L(x)$	$T(x)$	$c(x)$	$L(x)$	$T(x)$	$c(x)$
0.00	1.00	0.50	25.00	36.67	0.6818	33.33	53.33	0.6250
33.33	0.50	0.40	10.00	11.67	0.2727	16.67	20.00	0.3125
66.67	0.10	0.10	1.67	1.67	0.0455	3.33	3.33	0.0625
100.00	0	0	0	0	0	0	0	0

[a]The "usual life table" $L(x)$, $T(x)$, and $c(x)$ are calculated assuming a linear decrease in $l(x)$ across the age category, whereas "Hooton's method" values are calculated assuming a step function for survivorship.

as is the case in hazard models, then $L(x)$ simply becomes $l(x)$. However, in life tables the width of an age category can be rather large, as are Hooton's age categories, which are 33.33 years long. To find $L(x)$, we might proceed by adding together the $l(x)$ values year by year or, in the continuous case, we would use calculus to integrate $l(x)$ across the age category. The usual assumption in a life table is that $l(x)$ decreases linearly across the age category, as in Hooton's graph but not in his math. Following standard procedures, we use the trapezoid rule to find that $L(x)$ is the average of $l(x)$ at the beginning and end of the age category multiplied by the width of the age category. Thus, $L(33.33) = (0.5 \times l(33.33) + 0.5 \times l(66.67)) \times 33.33$ as shown under "usual life table" in Table I. In contrast, adopting Hooton's assumption of a step function for $l(x)$, we integrate across a rectangle so that $L(x)$ is simply $l(x)$ times the age width. This is shown in Table I under the section labeled "Hooton's method."

The fifth and eighth columns in Table I, $T(x)$, represent the people years to be lived in age category X and in all subsequent age categories. Since the value of $T(x)$ is calculated by subtracting the $L(x)$ of the previous category from the $T(x)$ of that category, the overestimation of $L(x)$ using Hooton's method compared to usual life table calculations also results in overestimation of $T(x)$. The sixth and final columns, $c(x)$, show the living age distribution, which is calculated as $L(x)/T(x)$. The final column in Table I gives the age distribution implied by Hooton's method, while the sixth column gives the slightly different age distribution that arises from a traditional life table analysis. Differences in the stable age distributions between the usual life table and Hooton's method do not appear great enough to have much effect on the predicted number of skeletons. Instead, Hooton's substantial underestimation of total population size is a product of the discrete nature of his step function survivorship. Because of the step function in survivorship, the birth of 625 individuals in Hooton's model only occurs every 33.33 years at a single point in time. Consequently, he does not account for a substantial number of individuals who were moving into the population by birth and out of the population by death. We need a continuous time birth and death model, which the life table can provide, albeit with linear survivorship through age intervals. In a stationary population, $T(x)/l(x)$ is the average age-at-death, and its inverse is equal to the crude birth and death rates. The crude death rate is the proportion of the population that dies per annum. Using these facts, we find that the number of deaths from a stationary population of 1000 individuals across a century is $1/36.67 \times 1000 \times 100 = 2727$ using the correct life table approach and $1/53.33 \times 1000 \times 100 = 1875$ using Hooton's step function survivorship.

There are additional complications that we could address here, including the fact that Pecos Pueblo is known to have been in a population decline so that a stationary model is inappropriate. In the interest of following a historical thread, we will forestall discussion of nonstationary models until later in this chapter, but for the moment we move from "Hooton to hazards" within the context of the Pecos Pueblo example. We argue that the linear decline of survivorship is an unappealing aspect of the usual life table approach. Indeed, it is common to make different assumptions about the shape of survivorship within the very young (Coale and Demeny, 1966:20). In hazards analysis we replace the piecewise linear survivorship function with a smooth curve. For our example here we will fit a Gompertz model to the survivorship values that Hooton used. The Gompertz model survivorship is

$$l(t) = \exp\left(\frac{a_3}{b_3}\left(1 - \exp\left(b_3 \cdot t\right)\right)\right). \qquad (2)$$

The numbering of the parameters in this equation keeps this model in line with the Siler model (Gage, 1988). We estimate the parameters a_3 and b_3 using the method of maximum likelihood (see Appendix) and find for Hooton's Pecos Pueblo example that $a_3 = 0.01265$ and $b_3 = 0.02693$. Figure 3 compares the linear survivorship from a life table approach to the Gompertz model.

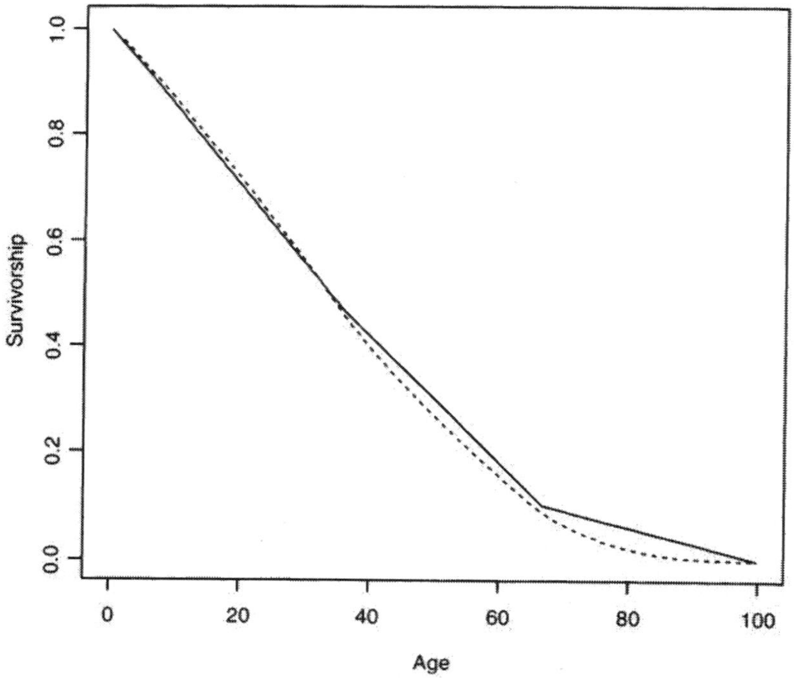

Figure 3 Linear survivorship calculated from a life table (solid line) and from a Gompertz model (dashed line) using Hooton's Peco Pueblo data (see Table I).

We can integrate Eq. (2) (see Appendix) to arrive at the stationary population mean age-at-death of 35.5 years, which is close to the value we found from the life table. As shown later, hazards analysis has a number of advantages over life table approaches. For now, we simply take solace in the fact that hazards analysis is not giving us a radically different answer from the life table approach.

III. THE DOLDRUMS PRIOR TO THE ADOPTION OF LIFE TABLES IN PALEODEMOGRAPHY

While Hooton may or may not have known how to calculate a life table, his student Larry Angel, whose 1969 article (Angel, 1969) is frequently cited as a cornerstone of paleodemographic research, rejected the use of life tables in bioarchaeology. Angel's (1947) first major publication in paleodemography on "the length of life in ancient Greece" cites a number of then standard demographic and life table works (e.g., Dublin and Lotka, 1936), but he chose not to present life tables for ancient Greece. Instead, Angel presented a frequency analysis of crania in broad age classes (his Table 1), a frequency analysis based on a more refined categorization of suture closure (his Table 2), and average ages at death from the first two tables and from a finer 5-year categorization (his Table 3). In his Figure 1, Angel presented a graph of life expectancy against age, which was calculated by forming successive samples prior to averaging ages at death, and then subtracting the floor of the interval. For example, life expectancy at age 35 would be the average ages at death for those who die past age 35 years, minus 35 years. While this gives results identical to those that would come from a life table, the life table provides additional measures (e.g., age-specific probability of death) that apparently were of no interest to Angel.

Angel's 1947 article is not the work for which he is best known in paleodemographic circles. Instead, his 1969 article "The Bases of Paleodemography" is widely cited in historical accounts. In this latter work Angel took an outwardly hostile approach toward life tables, writing that "one can construct a model life table (including life expectancy) from ancient cemetery data (Angel, 1947, 1953), but this falsifies biological fact to a greater or lesser degree . . ." (Angel, 1969:428). In lieu of calculating life tables, Angel presented a rather bizarre analysis that, among other errors, assumed that the age distribution of the living was equal to the age distribution of the dead. We can see examples of this in such statements as "There are 74 'living' children per 62+ adults in the 18 to 34 age range = 2.4 per couple" (Angel, 1969:433). The count of 74 "living" children comes from a method we have never been able to understand, as it involves a prorating of deaths and Angel never gave any instruction on how deaths were supposed to be apportioned. The 62+ adults in the 18 to 34 age range is also a bit mysterious, although we know that it is based on the observed distribution of deaths.

Angel closed his article with the statement that:

> The ultimate bases for paleodemography are accurate identification of each individual (sex, age, disease, fecundity, fertility, family relationship) from his or her skeleton and accurate counting of an adequate sample of such individuals collected by meticulous excavation techniques. The key is close collaboration between archaeologist and physical anthropologist. (Angel, 1969:434)

His work is probably best remembered for this message of collaboration between archaeology and biological anthropology, the battle cry of the bioarchaeologist. While Angel's work frequently is cited in paleodemographic contexts, it is really his mentor Hooton who deserves the greater credit.

Other physical anthropologists and bioarchaeologists contemporaneous with Angel also attempted to evaluate past population structure without formally using life table analysis. For example, Howells (1960) judged past population sizes for various archaeological groups based on the age structure of skeletal samples informally combined with information on settlement patterns and other sociocultural information. In the same volume, Vallois (1960) tabulated the age structure of archaeological and fossil skeletal samples and compared these groups both among themselves and with early historical (i.e., Roman period) documents in an effort to assess both the representativeness of various archaeological samples and temporal trends in mortality. Similarly, in a reassessment of age and sex at Indian Knoll, Johnston and Snow (1961) graphically compared the age structure of this archaeological sample with both fossil and more recent archaeological samples worldwide. Some bioarchaeologists continued to follow into the next decade the same approach of tabulating skeletal samples by age classes, taking percents and then graphing the "mortality profiles." For example, Blakely (1971) compared Indian Knoll to archaeological samples from Illinois Archaic, Hopewell, and Middle Mississippian sites in this way, although he did attempt to formalize his comparisons of mortality using χ^2.

IV. LIONIZATION OF THE LIFE TABLE

It is not entirely clear to us who should claim credit for the first application of life tables in paleodemography. In their 1976 treatment of paleodemography, Swedlund and Armelagos cited Hooton and Angel's work as the beginning of paleodemography in physical anthropology. However, they went on to note that "the application of demographic principles has not been an important aspect of skeletal studies until the last decade" (Swedlund and Armelagos, 1976:34). Acsádi and Nemeskéri's (1970) monograph on the "History of Human Life Span and Mortality" was certainly one of the first major treatments in paleodemography to fully exploit life table calculations. Swedlund and Armelagos' book contained an entire chapter on paleodemography, and the publication of Ken Weiss' (1973) "Demographic Models for Anthropology" brought life tables to the attention of both archaeologists and biological anthropologists. Ubelaker's (1974) study of ossuary paleodemography provided detailed information on how to construct life tables, as did Buikstra's (1976) demographic analysis of Illinois Middle Woodland. David Asch's monograph, again on Illinois Middle Woodland, brought a level of mathematical sophistication to paleodemography that was decades ahead of its time, as it required an intimate knowledge of stable population theory (Asch, 1976). By the mid-1970s, life tables clearly had become fairly commonplace in bioarchaeology. This is all the more remarkable when we consider that Angel, in 1969, wrote his "bases of paleodemography" without ever using life tables per se.

As there have been so many descriptions in the literature of how to construct life tables we will not give a description here (for reviews, see Meindl and Russell, 1998; Milner *et al.*, 2000). The all-too familiar columns are age, d_x or proportions of deaths in age classes, l_x or survivorship, q_x or age-specific probability of death, L_x or people-years lived in age intervals, T_x or people-years to be lived, e_x or life expectancy, and c_x or proportion of the living population in the age class. Only the last column may be unfamiliar; it is equal to L_x/e_x. One of the best accounts for constructing paleodemographic life tables is given by Moore and colleagues (1975) who give a detailed description of how to adjust a life table for populations with nonzero growth rates and also indicate the effects of underenumeration on various life table columns. We will hold off on the discussion of nonzero growth rates until the following section. Concerning the effects of underenumeration, Moore and colleagues (1975) note that a deficit of infants, a common occurrence in archaeological samples, will affect d_x and l_x throughout, but will have no effect on q_x and e_x beyond the first age interval. This is true because these latter two parameters are conditioned on survival past infancy for values beyond q_x and e_x.

V. THE PROBLEM OF NONZERO GROWTH RATES

If human populations grow exponentially, then the population size at time t (N_t) is a function of the population size at time 0 (N_0), as follows:

$$N_t = N_0 e^{rt}. \qquad (3)$$

Here r is the population growth rate, which is 0 for a stationary population, negative for a declining population, and positive for an increasing population. Moore and colleagues (1975) used the slightly different equation:

$$N_t = N_0 (1 + r)^t, \qquad (4)$$

but Eq. (3) is generally preferred to Eq. (4) because the former uses continuous compounding, whereas the latter compounds annually.

It is quite rare to have information on growth rate in a paleodemographic setting. In the case of Pecos Pueblo, some estimates of population size were available from the historic period, although the site extended well into the prehistoric period from which there are no population estimates. Hooton (1930:332) provided what was known about historic population sizes for Pecos Pueblo, which we have plotted as a semilogarithmic plot in Fig. 4. If the growth rate were constant, then the points should fall along a straight line, which they obviously do not. We have drawn in the least-squares regression line for didactic purposes; this line has a slope of about –0.015. In the following calculations, it is assumed that the growth rate for Pecos Pueblo throughout its history and prehistory was –0.015. This is a value that is clearly unrealistic, as Pecos Pueblo was inhabited for about 1000 years, and to reach its historically known population size after 1000 years of population decline (at $r = -0.015$) the founding population size would have to have been well into the millions. In all likelihood, the population was stationary through the prehistoric period and only entered population decline in the historic period. Furthermore, the rate of decline accelerated through the historic period, as seen in the nonlinearity in Fig. 4.

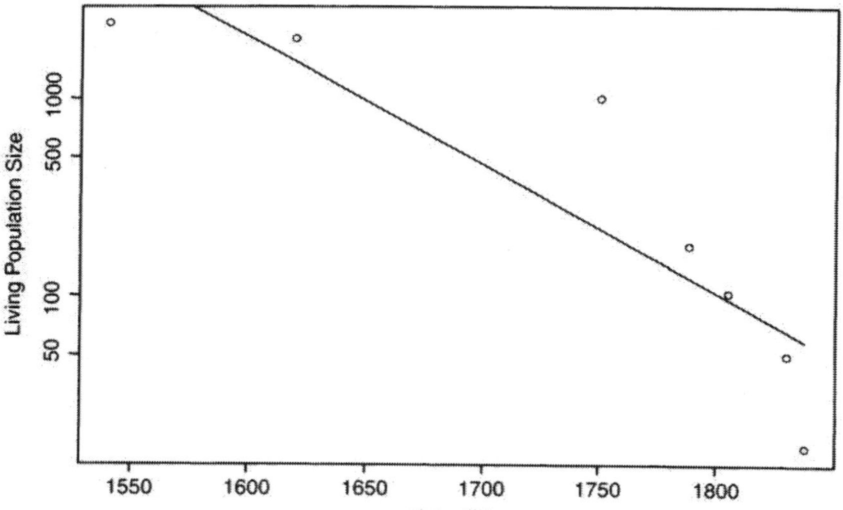

Figure 4 Semilogarithmic plot of living population sizes at specific dates reported by Hooton (1930:332). Points represent recorded population sizes, whereas the straight line is a least-squares regression line with a slope of roughly −0.015.

The method for calculating a life table from skeletal samples derived from growing or declining populations has been given in a number of different paleodemographic sources (Asch, 1976; Bennett, 1973b; Moore *et al.*, 1975; Weiss, 1973). The logic is as follows. For our current example with a growth rate of −0.015, individuals who die at exactly age 50 come from a birth cohort that was larger than the current birth cohort. Specifically, the birth cohort for those who die at age 50 years was $exp(0.015 \times 50) = 2.117$ times larger than the current birth cohort. Note that this expression is $exp(-rt)$, where t is the exact age-at-death and r is the growth rate. As a consequence, the number of deaths that occur at exactly age 50 need to be adjusted down by dividing by 2.117 in order to set them to the same cohort size as those who die at exactly age zero. Dividing the number of deaths by $exp(-rt)$ is the same as multiplying the deaths by $exp(rt)$, which is the correction given in the aforementioned sources and is equivalent to the annualized correction of $(1 + r)^t$.

As an example of calculating a life table corrected for growth we will continue with the Pecos Pueblo example, using the more detailed tabulation of ages from Palkovich (1983). There has been considerable discussion of Hooton's original age assessments on which Palkovich's figures are based (Mobley, 1980; Palkovich, 1983; Ruff, 1981). While Ruff has reexamined many of Hooton's age assessments, any new paleodemographic analysis should proceed from scorings of age indicators in both a reference sample and the archaeological sample of interest, as noted later. As the Pecos Pueblo skeletons have now been reburied (Tarpy, 2000), any such reanalysis will have to rely on archival data. Table II lists Palkovich's original life table with $r = 0$ and our recalculated life table with $r = -0.015$. The adjustments to number of deaths shown here are found using

Table II

Pecos Pueblo Life Table Under Zero Population Growth ($r = 0$) and Population Decline ($r = -0.015$)[a]

	$r = 0$				$r = -0.015$					
x	$D(x)$	$l(x)$	$q(x)$	$e(x)$	$D(x)$	$l(x)$	$q(x)$	$L(x)$	$L'(x)$	$e(x)$
0	322.0	1.0000	0.1765	28.18	319.6	1.0000	0.2548	0.8726	0.8792	21.84
1	117.0	0.8235	0.0779	33.12	113.5	0.7452	0.1214	1.3999	1.4425	28.13
3	120.0	0.7593	0.0866	33.83	108.9	0.6547	0.1326	4.2789	4.7171	29.88
10	145.0	0.6935	0.1146	29.71	115.8	0.5679	0.1626	5.2169	6.5332	26.92
20	113.2	0.6140	0.1011	22.90	77.8	0.4755	0.1304	4.4451	6.4675	21.17
30	808.8	0.5520	0.8033	14.92	443.9	0.4135	0.8560	4.7305	8.6195	13.60
50	198.0	0.1086	1.0000	15.00	74.7	0.0596	1.0000	0.8934	2.3685	15.00
80										

[a]Age data are from Palkovich (1983). $L(x)$ are years lived in the age interval, and $L'(x)$ is $L(x)$ adjusted by e^{-ra}.

the midpoint of each age interval, as is common practice. In Palkovich's original table the life expectancy at birth assuming a stationary population is 28.18 years so the crude birth and death rates equal 1/28.18 = 0.035. In our recalculation under an unreasonable growth rate of −0.015, the birth rate is 0.032 and the death rate is 0.047. The calculation of birth rate from a nonstationary life table is rather complicated, involving an adjusted column L'_x equal to $e^{-ra}L_x$. We refer the interested reader to Asch (1976:39), where the calculations are given. The death rate is then simply $b - r$.

VI. ADIEU TO PALEODEMOGRAPHY?

By the mid-1970s the scope of paleodemography came to cover the routine calculation of life tables, usually under the assumption of zero population growth (Lallo *et al.*, 1980; Lovejoy *et al.*, 1977; Owsley and Bass, 1979; Owsley *et al.*, 1977; Ubelaker, 1974). During this period of numerous publications in paleodemography, there were occasional lapses that demonstrated that authors did not understand the basics of stable population theory. For example, Blakely (1988a:22), in commenting on his graph of the percentage dead in age categories (his Figure 2), noted that "[t]he Dickson Mounds curve is pyramidal, indicating age stability." His comments are based on a misreading of Weiss (1973:65), where Weiss notes that the *living* age distribution from a census should be pyramidal if the population was stable. In skeletal samples, the living age distribution must be calculated from survivorship, and as a consequence it is always pyramidal if age intervals are of equal width. The reason Blakely (1988a:22) found that the King site did not have a pyramidal distribution is because he was using d_x instead of c_x. With this and a few other minor exceptions, bioarchaeologists by the early 1980s had become fairly facile at calculating and interpreting life tables.

Also by the early 1980s the status quo in paleodemography had come under serious attack. The first scathing critique of paleodemography came from William Petersen (1975). Petersen came to his discussion as an outsider, a social demographer, and his critique was largely leveled at older literature in archaeological demography. In fact, his only references to specific paleodemographic studies that we have cited so far are Hooton, Acsádi and Nemeskéri, and Weiss (Acsádi and Nemeskéri, 1970; Hooton, 1930; Weiss, 1973). Petersen criticized Hooton (1930) for his comparison of Pecos Pueblo mortality to modern European national data, while he mentioned Acsádi and Nemeskéri (1970) only in passing in his discussion of the accuracy of sex determination. Weiss (1973) was the recipient of the greatest amount of abuse by far:

> . . . Weiss's contribution shows the typical faults of a pioneer. In a paper addressed to an audience generally poorly versed in mathematics, he uses an unnecessarily

> cumbersome notation. More important, he displays an ignorance of fundamentals (or, at best, a carelessness in presenting them) that contrasts sharply with his technical pretentiousness. Weiss's work differs from the norm in archaeology and anthropology mainly in that it makes some genuine effort to assimilate the elements of demography. (Petersen, 1975:228)

Weiss [commenting in Petersen (1975:240)] responded: "a great deal of needless confusion comes from the author's fixation upon the basic traditions and concepts of his own field along with an inevitably spotty knowledge of the literature of the other." As an outsider, and one clearly unfamiliar with then current work in paleodemography, Petersen's critique did not stick.

The same cannot be said for comments that were published in 1982 and 1983 (Bocquet-Appel and Masset, 1982; Howell, 1982; Sattenspiel and Harpending, 1983). The comments by Howell and by Sattenspiel and Harpending came from anthropologists who were quite familiar with anthropological demography. Bocquet-Appel and Masset's "Farewell to Paleodemography" was written by practitioners who had actively worked with prehistoric skeletal material. These criticisms from within the field are summarized briefly here as they relate to historical developments in paleodemography. For fuller treatment and reponses to these criticisms, we refer the reader to the original exchanges (e.g., Bocquet-Appel, 1986; Bocquet-Appel and Masset, 1985, 1996; Buikstra and Konigsberg, 1985; Greene *et al.*, 1986; Konigsberg and Frankenberg, 1992, 1994; Masset, 1993; Masset and Parzysz, 1985; Piontek and Weber, 1990; Van Gerven and Armelagos, 1983). In a nutshell, Howell (1982) critiqued Lovejoy and colleagues' (1977) demographic reconstruction for the Libben site, noting that it implied an unrealistically low rate of survival into mid- and old-adulthood. Sattenspiel and Harpending (1983) showed that the inverse of the mean age-at-death is approximately equal to the crude birth rate in a nonstationary population (when the growth rate is unknown), but is relatively unrelated to the crude death rate. They (Sattenspiel and Harpending, 1983) consequently suggested that paleodemographic data were more informative about fertility than about mortality, a viewpoint that was rather foreign to a generation of paleodemographers who had grown up with life table analysis as an indicator of mortality. Finally, Bocquet-Appel and Masset (1982) argued that methods of age estimation were too imprecise and biased to produce usable results for demographic analyses.

VII. HAZARDS ANALYSIS: THE DEATH OF THE LIFE TABLE IN PALEODEMOGRAPHY?

By the mid-1980s there was growing dissatisfaction with the discrete age nature of the life table, as well as with the limitations summarized earlier.

A number of authors (Gage, 1988; Gage and Dyke, 1986; Wood *et al.*, 1992) suggested the application of hazard models as a logical alternative to life tables, and as shown in subsequent sections, there are many practical advantages to hazards analysis. The most detailed account to date of hazard models for anthropology is by Wood and colleagues (1992). In hazards analysis, the little $l(x)$ column of the life table is replaced with a survivorship function, usually written as $S(a)$ or $S(t)$, which is the probability of survival to exact age a or t. The little $d(x)$ is replaced with a smooth function, usually written as $f(a)$, which is the probability density function for age-at-death. The age-specific probability of death, $q(x)$, is replaced with the hazard rate, $h(a)$. Wood and colleagues (1992) give the relationships among the hazard rate, probability density function for age-at-death, and survivorship in their equations 5–8.

One practical benefit of hazards analysis is that it allows us to deal with the different age ranges that might be assigned to individual skeletons. For example, in the Pecos Pueblo sample there were 51 skeletons that could be aged no more precisely than 20 to 80 years old at time of death. While Palkovich (1983) apportioned these skeletons to the adult age classes based on the age distribution of the other adult skeletons, hazards analysis allows us to treat these individuals as interval censored. In fact, all skeletons could be treated as interval censored data, with the interval lengths varying from skeleton to skeleton. Thus, the arbitrary binning into traditional age classes is unnecessary. We fit survivorship to Pecos Pueblo data using a four-parameter model:

$$l(t) = \exp\left(-\frac{a_1}{b_1}\left(1 - \exp\left(-b_1 \cdot t\right)\right)\right) \exp\left(\frac{a_3}{b_3}\left(1 - \exp\left(b_3 \cdot t\right)\right)\right), \quad (5)$$

where the first exponential is a negative Gompertz function representing juvenile survivorship and the second exponential is the Gompertz function from Eq. (2) that represents adult survivorship (Gage, 1988). Figure 5a compares the survivorship fit by this four-parameter hazard model to the survivorship values that Palkovich calculated from a life table. Figure 5b compares Palkovich's $d(x)$ to the probability density function for age-at-death calculated from the hazard model. Details of the fitting procedure are given in the Appendix, and values for the hazard parameters are given in Table III.

If there is some estimate of the growth rate, then the hazards analysis can be adjusted for growth rate. We define a new survivorship

$$S'(a) = \frac{\int_a^\omega h(a) S(a)\, e^{-ra} da}{\int_0^\omega h(a) S(a)\, e^{-ra} da}, \quad (6)$$

which can be obtained from Asch (1976:72) or from Milner and colleagues' (2000) with an additional integration in the numerator. This adjusted survivorship is used to fit against the unadjusted deaths, as shown in the Appendix. Figure 6a gives the fitted survivorship curve at $r = -0.015$ for Pecos Pueblo, the hazard model survivorship curve at $r = 0$, and plotted interval-wise survivorships

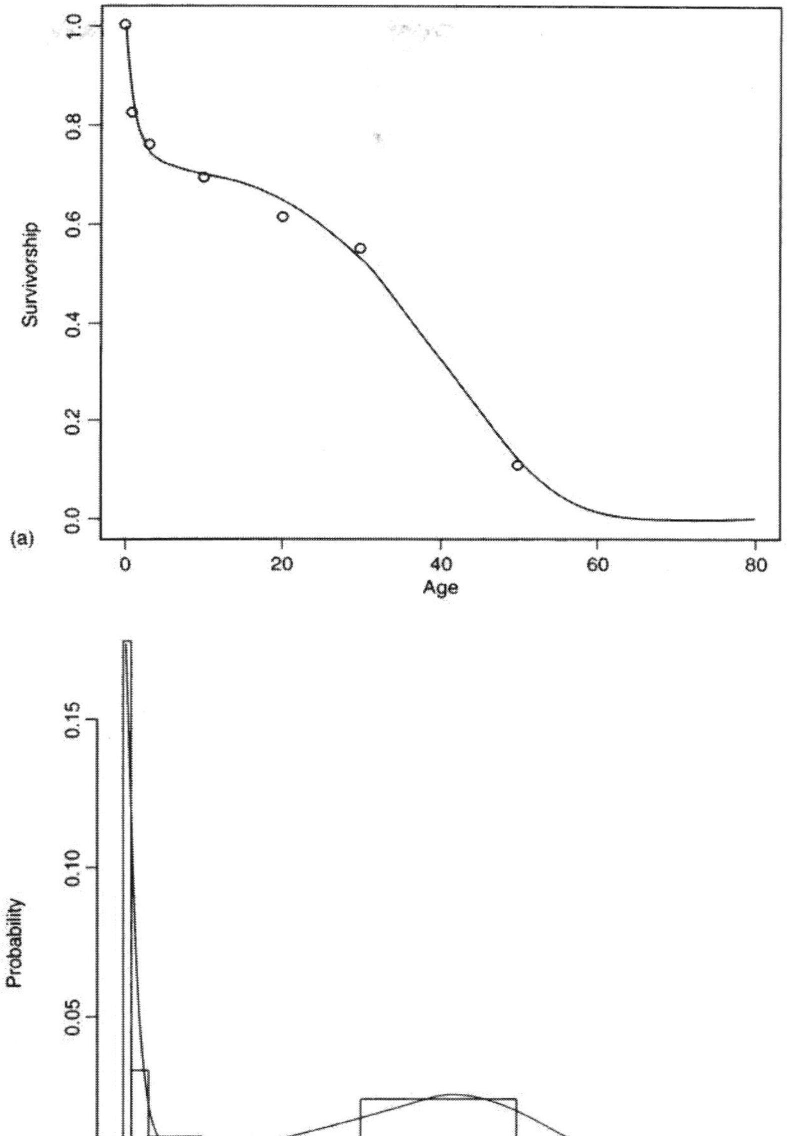

Figure 5 A comparison of Palkovich's life table data with a four-parameter hazard model for Pecos Pueblo. (a) Life table survivorship values (open circles) compared with survivorship fit by the hazard model (solid line). (b) Palkovich's $d(x)$ (open bars) compared with the probability density function for age-at-death calculated from the hazard model (solid line).

Table III

Hazard Parameters Estimated for Pecos Pueblo under Zero Growth, a High Rate of Population Decline ($r = -0.015$), and a Low Rate of Population Decline ($r = -0.003$)[a]

	$r = 0$	$r = -0.015$	$r = -0.003$
a_1	0.2398	0.3529	0.2628
b_1	0.7519	0.7156	0.7564
a_3	0.0024	0.0039	0.0028
b_3	0.0832	0.0746	0.0804
Birth rate	0.0366	0.0342	0.0362
Death rate	0.0366	0.0492	0.0392
MAL	20.61	22.22	20.85
MAD	27.29	27.09	27.10

[a]Also shown for each model are the crude birth and death rates, mean age in the living (MAL), and mean age-at-death (MAD).

calculated from the life table at $r = -0.015$ (from Table II). Figure 6b shows the probability of death functions and Fig. 6c shows the conditional probability of death functions for both stationary and nonzero growth rate hazard models. As seen from Fig. 6a, a negative growth rate lowers the survivorship from what we would calculate if we assumed a stationary population. Similarly, under a positive growth rate the survivorship would be elevated over what we would calculate if we assumed a stationary population. In a figure caption, Milner and colleagues (2000: Figure 16.1) appear to say the opposite, when they write that "positive values of r make it appear as if survival is lower at each age, whereas negative values have the opposite effect." However, they are talking about the effect of fitting a stationary model to a positive growth population, whereas we are talking about fitting a growth model to a stationary population. From the hazard model fit with $r = -0.015$, we find crude birth and death rates of 0.0342 and 0.0492 (see Appendix), which agree well with the values of 0.032 and 0.047 from the nonstationary life table.

We also can generalize Eq. (1) to give the size of a founding population that would generate the observed number of skeletons in a nonstationary setting. Following Asch's (1976) equation B-5 on his page 72, the size of the founding population (P_0) is estimated as

$$P_0 = \frac{N}{b \int_0^T e^{rt} dt \int_0^{\omega} h(a) S(a)\, e^{-ra} da} \tag{7}$$

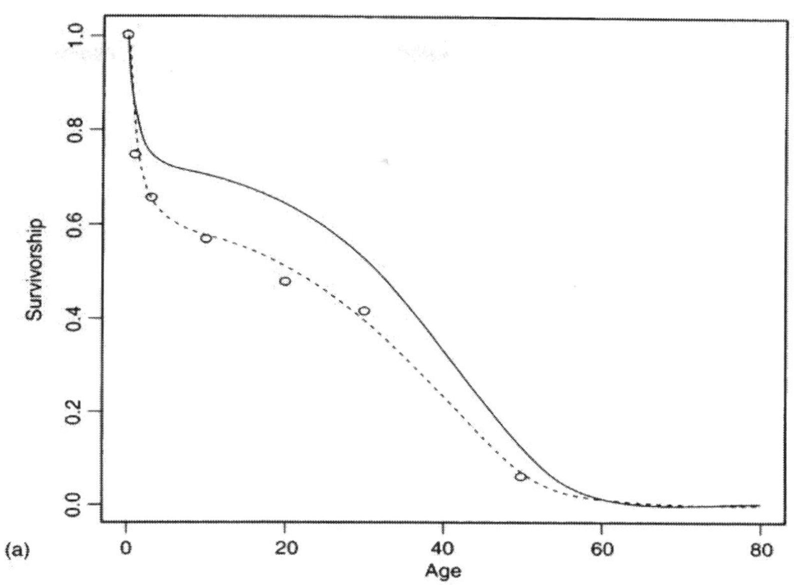

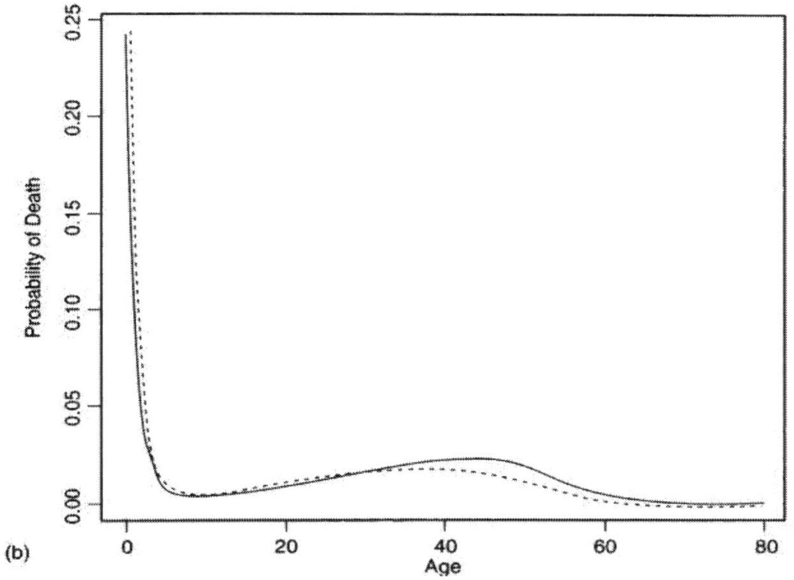

Figure 6 A comparison of hazard models for Pecos Pueblo assuming stationarity ($r = 0.0$, solid line) and population decline ($r = -0.015$, dashed line). (a) Survivorships fit by both hazard models and life table survivorship values (open circles) calculated for $r = -0.015$. (b) Probability of death functions for the two hazard models. (c) Conditional probability of death functions under both stationary and nonzero growth rates.

(c)

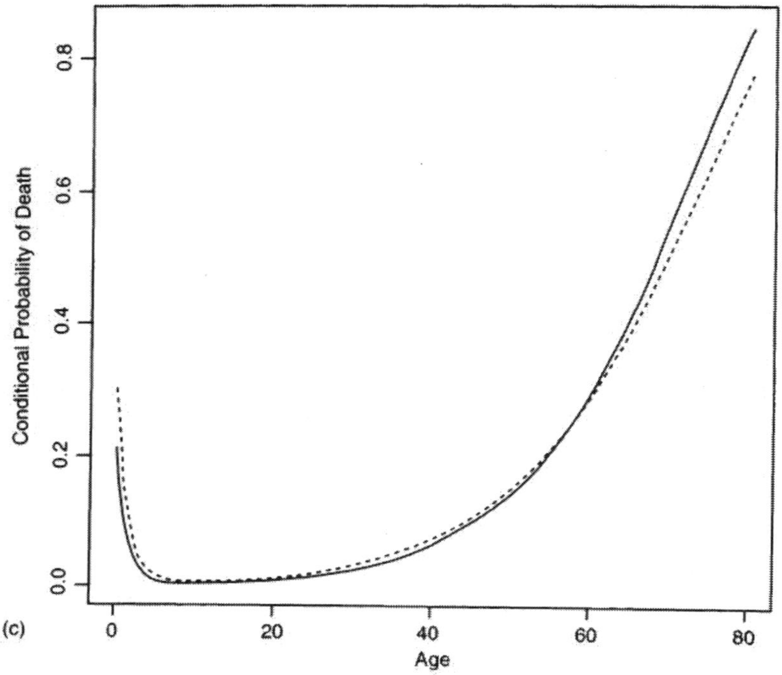

Figure 6 *(Continued)*

where T is the length of the time interval and N is the number of skeletons. In the stationary case, $d = b$, the first integral equals T, the second integral equals 1.0, and P_0 is simply P. Equation (1) is consequently a special case of Eq. (7). We use Eq. (7) and information from Hooton (1930) to try to estimate the number of deaths that should have occurred between 1500 and 1700 CE at Pecos Pueblo. Hooton (1930:336) used historical accounts and interpolation to suggest that the population sizes for 1500, 1533, 1566, 1600, 1633, 1666, and 1700 were 2600, 2440, 2280, 2120, 1920, 1640, and 1400 individuals, respectively. These population sizes imply a growth rate of about –0.003 per annum. Using this growth rate, assuming a "founding population" of 2600 people, and reestimating the hazard model over a 200-year time span (see Table III), we find that there should have been slightly more than 15,000 deaths during this 200-year span. This number of deaths is almost twice what Hooton (1930:337) estimated. The primary reason for his underestimate was that he did not accrue deaths continuously [as the integrals in Eq. (7) do], but instead treated Pecos as a series of 33-year-long birth cohorts.

VIII. IS MEAN AGE-AT-DEATH A MEASURE OF MORTALITY IN NONSTATIONARY POPULATIONS?

In their 1983 article, Sattenspiel and Harpending raised the near heretical notion that when a paleodemographic life table from a nonstationary population is treated as if it were stationary, it will yield a mean age-at-death that is nearly the inverse of the crude birth rate. They then noted that the inverse of the mean age-at-death in such a setting is not a particularly good indicator of the crude death rate (Sattenspiel and Harpending, 1983). As their argument is based on continuous age rather than the discrete ages approximated by life tables, hazards analysis is an appropriate mechanism for exploring their argument. From the Pecos Pueblo example, we found a mean age-at-death of 27.29 when we fit a stationary model. If the population truly were stationary, then this figure is also the life expectancy at birth and the inverse of the crude birth and death rates, which would equal 0.037. In the nonstationary hazard model we found a mean age-at-death of 27.10, a mean age in the living of 22.22 years, a birth rate of 0.034, and a death rate of 0.049. Note that if we use a stationary model when the population was actually in a decline with $r = -0.015$, we would overestimate the crude death rate by about 24.5%, but we would underestimate the crude birth rate by only 8.8%. In the appendix to their paper, Sattenspiel and Harpending (1983:497) derived the relationship between mean age-at-death ($\bar{a}_D$), mean age in the living ($\bar{a}_L$), growth rate (r), and crude death rate (d) as

$$\bar{a}_D = \frac{1 - r \times \bar{a}_L}{d} \tag{8}$$

In a later publication, Horowitz and Armelagos (1988:191) note after a rather lengthy derivation riddled with typos that "on close reading of the appendix to Sattenspiel and Harpending (1983) one does find a similar formula ..." to the one Horowitz and Armelagos present. The equation given by Horowitz and Armelagos (their equation 9) is

$$\begin{aligned} \bar{a}_D &= \frac{1 - r \times \bar{a}_L}{b\left(1 - \frac{r}{b}\right)} \\ &= \frac{1 - r \times \bar{a}_L}{b - r} \\ &= \frac{1 - r \times \bar{a}_L}{d}, \end{aligned} \tag{9}$$

which actually is identical with Sattenspiel and Harpending's equation. Equation (8) can be rewritten as

$$b = \frac{1}{\bar{a}_D} + r\left(1 - \frac{\bar{a}_L}{\bar{a}_D}\right)$$
$$d = \frac{1}{\bar{a}_D} - r\left(\frac{\bar{a}_L}{\bar{a}_D}\right). \quad (10)$$

Equation (10) shows that when the growth rate is zero, the inverse of the mean age-at-death equals the crude birth and death rates. In a nonstationary population the ratio of the mean age of the living to the mean age-at-death is a critical element in determining whether crude birth rate or crude death rate would be better estimated by the inverse of the mean age-at-death. As the ratio of mean age in the living to mean age-at-death is generally in the vicinity of 1, we can see that the inverse of the mean age-at-death is typically a better estimator for the birth rate than for the death rate. When the mean age in the living and the mean age-at-death are equal, the birth rate will be equal to the inverse of the mean age-at-death, while the crude death rate will be misestimated by a factor of r.

An issue related to the relationship between birth rate and mean age-at-death is the claim that the error in fitting stationary models to growing populations can be isolated in certain more complex models. Gage (1988) suggests that in the Siler model [a model identical to Eq. (5) but with the addition of a constant baseline hazard parameter called a_2] the error involved in fitting a stationary model to nonstationary data is entirely subsumed under one parameter (a_2). The Appendix shows that there are different ways to apply the associative law to rewrite the Siler model and that the growth rate affects different parameters depending on how the model is rewritten. We also show by example that changes in growth rate affect all the parameter estimates (see Table IV) and that Gage's claim consequently is incorrect. Hazard models, in and of themselves, do not allow us to circumvent the nonstationarity problem.

Table IV

Example of the Effect of Various Growth Rates on Estimation of the Five Parameters in a Siler Model (see Appendix)

r	a_1	b_1	a_2	a_3	b_3
−0.010	0.2749	1.5716	0.0091	2.3578E-08	0.3359
−0.005	0.3084	1.4616	0.0103	2.5403E-08	0.3344
0.000	0.3343	1.3207	0.0115	3.4808E-07	0.2786
0.005	0.3916	1.3221	0.0131	2.6210E-07	0.2848
0.010	0.4343	1.2540	0.0148	2.0782E-07	0.2897

IX. CAN WE EXPECT MORE FROM PALEODEMOGRAPHY?

The central criticism from Bocquet-Appel and Masset (1982) was that age determination methods and methods of paleodemographic reconstruction are so crude and inexact that paleodemography simply is not worth studying. They have continued to write prolifically about this topic (Bocquet-Appel, 1986, 1994; Bocquet-Appel and Masset, 1985, 1996; Masset, 1989, 1993), usually to maintain their critique, but on rare occasions to revisit that which they dismissed previously. Obviously, if raw data on which paleodemographers base their analyses are so tragically flawed and unfixable, then we must end our history in 1982 with the farewell from these authors. However, we believe that since 1982 there have been very hopeful developments in paleodemography, and we intend to end our history on this more upbeat note. A large number of salvos followed quickly on the heels of Bocquet-Appel and Masset's critique and continued for a few years (Buikstra and Konigsberg, 1985; Greene *et al.*, 1986; Lanphear, 1989; Piontek and Weber, 1990; Van Gerven and Armelagos, 1983). Although one of us was involved in this initial response, it was not until 10 years after the original Bocquet-Appel and Masset "farewell" that we published what we thought might end the debate (Konigsberg and Frankenberg, 1992). We were wrong in thinking this, and consequently we need to examine the full history of events over the last two decades.

We started working earnestly on the statistical use of age estimators in paleodemography in 1988. A publication in the journal *Biometrics* during the previous year (Kimura and Chikuni, 1987) entirely drove our way of thinking at that time. The authors of that article, both of whom worked in fisheries departments, had iteratively applied what is known as an "age-length key" in the fisheries literature. The age-length key is a tabulation of age classes against fish length developed using fish of known age as determined by counting otolith rings. As Kimura and Chikuni pointed out (1987), iterative application of the age-length key constitutes what is known as an EM algorithm (Dempster *et al.*, 1977) and consequently is a maximum likelihood method. We also found in looking through the fisheries literature that the problem of "age mimicry" (Mensforth, 1990) that Bocquet-Appel and Masset described in 1982 had been described about 5 years earlier in the fisheries literature (Kimura, 1977; Westrheim and Ricker, 1978). Kimura (1977:318) wrote "the age-length key will give biased results if applied to a population where the age composition differs from that of the population from which the age-length key was drawn."

The age-length key without iteration is a Bayesian method, and as such it uses age distribution of the reference sample as an informative prior when determining ages in a target sample. We knew this, fisheries workers certainly knew this, and Bocquet-Appel and Masset must have known this since they said as much in their 1982 article. Where we differed from Bocquet-Appel and Masset was that we felt, as did people in fisheries research, that this was not an insurmountable problem and that maximum likelihood estimation was the answer. It has continued to befuddle us through the years that Bocquet-Appel and Masset have so tenaciously argued against almost everything we have tried to do, when at the same time we have seen no "Farewell to Fisheries Demography" in the fisheries literature. We suspect that as fisheries science has a much greater economic impact than paleodemography, any abandonment of fish demography would have to come after considerable scientific and emotional expense. Granted, some of the methodological problems in paleodemography may be more difficult than those faced in fisheries science, but at least incorrect answers to paleodemographic questions are unlikely to have the drastic management effects that could arise from misestimates of fish stock.

We will not make a point-by-point response to Bocquet-Appel and Masset's (1996) most recent critique, as this is not the appropriate place. Some of our differences seem to be simply based on misreading each other's work. For example, in their 1996 paper, Bocquet-Appel and Masset discuss the idea of conditioning the reference sample on age, which has the effect of making age in the reference sample distributed uniformly. Bocquet-Apel and Masset (1996:573) then wrote: "This is where the idea to construct a uniform reference sample comes from, which was unfortunately interpreted as discarding data by Konigsberg and Frankenberg!" The reference they make to us is in regard to the following quote:

> Bocquet-Appel (1986) suggested that if the reference sample has a uniform age distribution, then the target sample age distribution will be estimated independent of the reference. . . . While this solution does consequently remove the problem of dependence between the target and the reference age distributions, it is not in general a useful way to proceed. The chief problem with selecting a reference sample with a uniform age distribution is that this requires discarding data, which certainly cannot be an efficient way to proceed. (Konigsberg and Frankenberg, 1992:239)

The passage we were referring to from Bocquet-Appel was the following:

> The only acceptable strategy for avoiding the influence of a particular reference population is to use a reference population in which the distribution is truly randomly distributed over the ages and, in this particular case an a priori uniform distribution . . . . (Bocquet-Appel, 1986:127)

A decade later Bocquet-Appel and Masset (1996:573) appear to have reinterpreted this passage to imply that one should condition on age in order to get the probability of being in a particular indicator state (their "simple technical trick"). If this was the intended message from Bocquet-Appel's earlier article (Bocquet-Appel, 1986), then it is curious that he never indicated in that article how to use these sets of conditional probabilities in order to estimate the age-at-death structure for the target. That would await the 1996 publication.

What we presented in 1992 was essentially the use of maximum likelihood methods in order to estimate the age-at-death structure for a paleodemographic sample using aging information from a reference sample (Konigsberg and Frankenberg, 1992). This was not an especially novel concept, as Boldsen (1988), Paine (1989), and Siven (1991) had already discussed likelihood applications in paleodemography. Bocquet-Appel and Masset's 1996 article was a claim for historical priority, as well as an argument that only the mean age-at-death can be estimated reliably using what we would call contingency table paleodemography (see the Appendix; Konigsberg and Frankenberg, 2002). They did not feel that the actual age structure (i.e., distribution of age-at-death within categories) could be determined accurately. Bocquet-Appel and Masset (1996) tried to demonstrate their point using simulation studies, arguing that their method, which they refer to as iterative proportional fitting, differed from what we presented in 1992 (Konigsberg and Frankenberg, 1992), which they refer to as iterative Bayesian. We are disinclined to trust their simulations because they managed to demonstrate differences between two methods that are identical (i.e., if you start both methods with the same data, each steps through the parameter space in the same way and thus gives identical results, as shown in the Appendix).

There have been other suggested methods for determining adult age-at-death within paleodemography. Jackes (1985) suggested using normal distributions of age within pubic symphyseal phases in order to get smooth distributions of age for target samples and has continued to apply this method (Jackes, 2000). The chief problem with this method is that the resultant age distributions for the target are in part dependent on the reference sample, a problem that was specifically noted in Bocquet-Appel and Masset's (1982) original critique. Jackes (2000:435) also has tried using the contingency table paleodemographic approach, finding that the method "is shown to be completely ineffective in replicating the real age-at-death distribution." However, she was attempting to fit a life table with 17 age categories using a six-phase indicator, and her solution has many age categories estimated with zero frequencies. This represents a solution on a boundary of the parameter space, and as such there is no unique likelihood solution (Fienberg, 1977:132). In other words, as stated by Clark (1981:299), "If $I < J$ (i.e., the number of length intervals is less than the number of age-groups with distinct length distributions), there will usually be a multiplicity of algebraic solutions and therefore no useful estimates."

An additional topic from Bocquet-Appel and Masset (1996) is the use of what they have called a "juvenility index" (or JI). The JI is a ratio of "the number of dead from 5 to 14 and the number of dead after 20 ($D_{5-14}/D_{20-\omega}$)" (Bocquet-Appel and Masset, 1996:580). Bocquet-Appel and Masset first introduced this index in 1977 in order to estimate mean age-at-death for samples that had under-enumeration of 0–5 year olds and where age-at-death estimates might be highly questionable (Bocquet and Masset, 1977). They then used 40 paleodemographic life tables and regressed mean age-at-death on the JI (Bocquet-Appel and Masset, 1982). By 1996 they had elaborated these regressions to include nonstationary growth rates (Bocquet-Appel and Masset, 1996). Buikstra and colleagues (1986) took a similar approach, but calculated the proportion of deaths over age 30 out of deaths over age 5 ($D_{30-\omega}/D_{5-\omega}$). Proportions have slightly simpler statistical properties than ratios [compare Buikstra *et al.* (1986) to Masset and Paryzysz (1985)], which is the reason Buikstra and colleagues used the former. Buikstra *et al.* (1986) then used regressions of crude birth rate and crude death rate on their death proportion in nonstationary models drawn from Coale and Demeny (1966). Using these regressions, they showed [following Sattenspiel and Harpending's (1983) suggestion] that the birth rate was more highly correlated with the death proportion than the crude death rate. However, Buikstra and colleagues (1986) cautioned against using the regressions to estimate crude birth rate and instead used a direct comparison of death proportions across time in west central Illinois to suggest an increased birth rate with the development of Mississippian culture. Storey (1992:174) has since applied such an analysis to Tlajinga-33, and there has been extensive discussion of using death ratios and proportions in the literature (Corruccini *et al.*, 1989; Hoppa, 1996; Konigsberg *et al.*, 1989; Paine and Harpending, 1998).

Near the end of our 1992 paper we spelled out future directions for paleodemography that included both reworking of then-current approaches and development of new methods. For example, we suggested switching to appropriate methods for age structure estimation, understanding reference samples, evaluating the efficiency of parameter estimators, developing methods for comparing different anthropological or paleodemographic samples, and incorporating uncertainty of age estimates into reduced parameterizations of life table functions (Konigsberg and Frankenberg, 1992). The incorporation of age estimation within hazard models is one direction that others and we have taken (Konigsberg and Holman, 1999; Milner *et al.*, 2000; Müller *et al.*, 2002; O'Connor, 1995) and is a cornerstone of the "Rostock Manifesto" (Hoppa and Vaupel, 2002). An additional area we did not consider at the time was the use of ordinal parametric models such as logistic and probit regression to describe the development of age indicators that are phase- or stage-based. Boldsen (Skytthe and Boldsen, 1993) pioneered this approach in paleodemography, referring to it as transition analysis because it models the age-to-transition between phases (see also Milner *et al.*, 2000; Boldsen *et al.*, 2002). As a consequence of the developments since 1992 we have almost completely abandoned contingency table approaches to paleodemography because they do not make good use of the ordinal nature of stage data or the continuous nature of age-at-death. Contingency table type approaches are still found within paleodemographic analyses (Gowland and Chamberlain, 2002; Jackes, 2003), but we suspect that they will eventually decrease in popularity.

Another direction since our 1992 paper that we did not anticipate was the growth of Bayesian methods in age estimation (Di Bacco *et al.*, 1999; Lucy, 1997; Lucy *et al.*, 1996). Although we (Konigsberg and Frankenberg, 1994; Konigsberg *et al.*, 1998) and others (e.g., Milner *et al.*, 2000) have found Bayesian logic and terminology useful, we are uncomfortable with the wholesale implementation of Bayesian methods in paleodemography. Lucy (1997) and Lucy *et al.* (1996) used the reference sample age distribution for their informative prior, which returns us to the original Bocquet-Appel and Masset (1982) critique. Di Bacco *et al.* (1999) have presented a Bayesian solution where they take vague priors for hazard parameters and then update these with information from the reference and target samples. This is a more hopeful method, although we have not yet seen an implementation or example of this type of analysis in the literature.

It seems clear to us that the statistical issues currently swirling around paleodemographic analysis will settle in the near future. A good sign that some consensus is being reached is the publication of "Palaeodemography: Age Distributions from Skeletal Samples" coedited by Rob Hoppa and Jim Vaupel (2002). That volume shows a fairly united front shared by both North American and European researchers and codified in the so-called "Rostock Manifesto." Consensus does not, however, mean the solution to all our problems. Cultural and archaeological sampling issues remain a considerable problem (Hoppa, 1999), as does the issue of Howell's (1976) "uniformitarian assumption" regarding rates of aging. Hoppa's (2000) publication of different rates of aging for the pubic symphysis is a disturbing message, although we suspect that the differences between reported samples reflect interobserver differences, not intersample differences in aging. If this is indeed the case, then there is a strong argument for standardization of observation methods and for interobserver error studies conducted on the same samples. Ultimately, the "uniformitarian assumption" is just that, an untestable assumption. However, the calibration literature (Brown, 1993; Brown and Sundberg, 1987, 1989; Konigsberg *et al.*, 1998), which has been rather widely ignored in paleodemography (exceptions are Aykroyd *et al.*, 1996, 1997) could be applied in this context. In multivariate calibration there is a "consistency diagnostic" that can be calculated for samples of unknown age-at-death. This diagnostic tests for whether aging rates in the unknown age sample are discordant when compared to the reference sample.

The history of paleodemography is far from over, as researchers continue to refine both measurement and analytical methods and to address the issues described earlier (e.g., Chamberlain, 2000; Meindl and Russell, 1998; Milner *et al.*, 2000; Paine, 2000). It is hoped that paleodemographic studies are now moving into an evaluation phase. By evaluation we mean systematically critiquing the strength, applicability, and reliability of age estimation techniques and developing models that realistically reflect paleodemographic processes and measure the impacts upon them. Evaluation also means using statistically sound, anthropologically pertinent measures to assess what is being measured and to enable researchers to assign confidence limits to their results. The refinement of paleodemographic methods, the solution of measurement and analytical issues, and the development of consensus among some paleodemographers will play only minor roles in the future of paleodemography, however, unless bioarchaeology as a field adopts current methods and contributes to this evaluation.

ACKNOWLEDGMENT

Order of authorship was determined by a draw of straws.

Appendix

I. INTRODUCTION

This appendix makes extensive use of the general statistical, mathematical, and graphics package known as "R." "R" is an S-like free software package initially developed by Robert Gentleman and Ross Ihaka in the Statistics Department at the University of Auckland and added to by numerous members of a working group. It is available under Free Software Foundation's GNU General Public License for a number of computing platforms and operating systems, UNIX, FreeBSD, Linux, and Windows 9x/NT/2000. We strongly recommend that readers of this chapter download "R" so that they can work through some of the examples given. Additional information on "R" is available on the Internet at http://www.r-project.org/.

II. MAXIMUM LIKELIHOOD ESTIMATION OF HAZARD MODEL PARAMETERS

In the maximum likelihood estimation method we form a parametric model to describe observed data. The parametric model is characterized by its parameters, and once we establish particular values for the parameters, we can find the log probability (up to an additive constant) of obtaining observed data. If we maximize this log probability by searching through the parameter space (i.e., trying different values for the parameters), then we will have found the most likely parameter values to have generated observed data; hence, the name maximum likelihood estimate. In a hazard model we have a parametric description for survivorship to exact age t. The probability of death between two exact ages is the difference between survivorship at the beginning and at the end of the age interval. The log-likelihood is defined as the sum across intervals of the products of the observed count of deaths in each interval with the estimated log probability of death in that interval. The "R" function calculates the log likelihood for the Gompertz model using the Pecos Pueblo example where 0.5 of the deaths fall between 0 and 33.33 years of age, 0.4 fall between 33.33 and 66.67 years of age, and 0.1 fall between 66.67 and 100.0 years of age.

```
function(x) {
    a <- x[1]                      # put x vector in a & b
    b <- x[2]

    t<-c(0,100/3,200/3,100)        # ages are 0, 33.33, 66.67,
                                     & 100
    l<-exp(a/b*(1-exp(b*t)))       # Gompertz survivorship
                                     at ages
    d<-l[1:3]-l[2:4]               # difference l(t) to get d(x)
    obs<-c(.5,.4,.1)               # observed d(x) for Pecos
                                     Pueblo
    lnlk<-obs%*% log(d)            # form the log-likelihood
    return(lnlk)                   # return log-likelihood
}
```

Now the "optim" function in "R" can be called in order to maximize the log likelihood. We call "optim" with starting values of 10^{-6} for a and b, assume that the aforementioned function is stored as "Gompertz," and set "fnscale" to a negative value in order to maximize, instead of minimize, the log likelihood.

```
optim(c(1E-6,1E-6),Gompertz,control=list (fnscale=-1))
```

We can also fit more complicated hazard models by maximum likelihood. For example, we fit a four-parameter model to Pecos Pueblo data. The model is

$$l(t) = \exp\left(-\frac{a_1}{b_1}\left(1 - \exp\left(-b_1 \cdot t\right)\right)\right) \exp\left(\frac{a_3}{b_3}\left(1 - \exp\left(b_3 \cdot t\right)\right)\right). \quad (1)$$

Data are in a file called "pecos," which looks like the following:

```
     [,1] [,2] [,3]
[1,]    0    1  322
[2,]    1    3  117
[3,]    3   10  120
[4,]   10   20  145
[5,]   20   30  108
[6,]   30   50  772
[7,]   50   80  189
[8,]   20   80   51
```

The first column ([,1]) is the beginning of an age interval, the second column ([,2]) is the end of that age interval, and the third column ([,3]) is the number of skeletons in the interval (from Palkovich, 1983). The eighth row ([8,]) contains

51 individuals who can be aged no more precisely than as adults between 20 and 80 years old. The likelihood function can be calculated with the following "R" function:

```
function(x,deaths=pecos) {
    a1 <- x[1]
    b1 <- x[2]
    a3 <- x[3]
    b3 <- x[4]

    t<-deaths[,1:2]
    l<-exp(-a1/b1*(1-exp(-b1*t)))*exp(a3/b3*(1-exp(b3*t)))
    d<-l[,1]-l[,2]
    obs<-deaths[,3]
    lnlk<-crossprod(obs,log(d))
    return(lnlk)
}
```

As with the Gompertz model, this function also can be maximized using "optim."

III. EXTRACTING SUMMARY MEASURES FROM HAZARD MODELS

Most summary measures from hazard models, such as mean age-at-death, birth rate, and death rate, require numerical integration. This can be done using the add-on library "integrate" in "R." For example, to find the mean age-at-death in a Gompertz model we would integrate survivorship from age 0 to the maximum (we assume 100 years), which in "R" could be written as

```
integrate(function(t,a=.01265,b=.02693)
    {exp(a/b*(1-exp(b*t)))},0,100)
```

This call would return the value 35.52587, which is rounded to the nearest first decimal place in the text of the chapter.

IV. FITTING A HAZARD MODEL WITH A KNOWN NONZERO GROWTH RATE

To fit a nonzero growth rate hazard model we have to include integration within the likelihood equation [see Equation (6) in the text of the chapter]. An "R" function for calculating the likelihood is given. As with the preceding functions, this function can be called by "optim" in order to maximize the function, but this call should be made only after fitting the stationary model to get good starting values.

```
function(x,r=-.015,deaths=pecos) {
     a1 <- x[1]
     b1 <- x[2]
     a3 <- x[3]
     b3 <- x[4]

     t<-deaths[1:7,1]

     ipdf<- function(t) {
      l<-exp(-a1/b1*(1-exp(-b1*t)))*exp(a3/b3*(1-exp(b3*t)))
            *exp(-r*t)*(a1*exp(-b1*t)+a3*exp(b3*t))
         }
      for(i in 1:7){
         L[i]<-integrate(ipdf,t[i],80)$value
         }

    L<-L/L[1]
    d<-L[1:6]-L[2:7]
    d<-c(d,L[7],1-d[1]-d[2]-d[3])
    obs<-deaths[,3]
    lnlk<-crossprod(obs,log(d))
    return(lnlk)
}
```

V. EFFECT OF AN UNKNOWN GROWTH RATE ON ESTIMATION OF HAZARD PARAMETERS

Gage has suggested that for a five-parameter Siler model:

> ... the effect of applying the stationary model to a population that is growing or declining will cause misestimation of a_2 equal to the magnitude of the intrinsic rate of increase. On the other hand, the remaining parameters are unaffected and can be compared across populations without restriction.... (Gage, 1988:440)

However, his statement is based on an erroneous derivation. He notes that when fitting survivorship to a stable age distribution, the frequency of individuals exactly age t years old is

$$\frac{N_t}{B} = \exp(-rt)\exp\left(-\frac{a_1}{b_1}(1-\exp(-b_1 \cdot t))\right)\exp(-a_2 t)$$
$$\exp\left(\frac{a_3}{b_3}(1-\exp(b_3 \cdot t))\right), \tag{2}$$

where N_t is the number of individuals t years old and B is the number of births. He then uses the associative property of multiplication to rewrite Eq. (2) as

$$\frac{N_t}{B} = \exp\left(-\frac{a_1}{b_1}(1-\exp(-b_1 \cdot t))\right)\exp(-(r+a_2)t)$$
$$\exp\left(\frac{a_3}{b_3}(1-\exp(b_3 \cdot t))\right), \tag{3}$$

which gives the appearance that the growth rate is "absorbed" onto the a_2 parameter. However, there are many other ways to apply the associative law, such as

$$\frac{N_t}{B} = \exp\left(-\frac{a_1}{b_1}(1-\exp(-b_1 \cdot t)) - rt\right)\exp(-a_2 t)$$
$$\exp\left(\frac{a_3}{b_3}(1-\exp(b_3 \cdot t))\right), \tag{4}$$

which gives the impression that the growth rate only affects estimation of the two parameters in the juvenile component of mortality.

Table IV tabulated estimates of the five parameters in a Siler model starting with a stationary model, adjusting the deaths to follow a nonstationary model, and then estimating the Siler model assuming a stationary population. It is clear from Table IV that, at least for this mortality pattern, changes in growth rate affect all of the parameter estimates.

VI. CONTINGENCY TABLE PALEODEMOGRAPHY

Konigsberg and Frankenberg (1992) and Bocquet-Appel and Masset (1996) have presented what we would refer to as "contingency table paleodemography." In contingency table demography, a reference sample is cross-tabulated by age and an age indicator, and tabulation of the indicator in a target sample is used to estimate the (unobserved) marginal distribution of age in the target.

As an example of such an approach we use reference and target data taken from Bocquet-Appel and Bacro (1997), as follows:

	Stage					
	I	II	III	IV	V	VI
Age						
23-29	8	19	30	7	1	0
30-39	2	18	43	25	1	0
40-49	0	6	29	27	5	1
50-59	0	2	26	37	13	0
60-69	0	0	9	28	9	1
70-79	0	0	7	28	10	3
80-89	0	0	2	10	10	5
Target	2	8	31.5	40.5	12	2

We have written two "R" functions, the first after Konigsberg and Frankenberg (1992) following Kimura and Chikuni (1987), and the second after Bocquet-Appel and Masset (1996). In each function "n1" is the reference sample data given as age by indicator (as shown earlier), and "n2" is the target vector. These are rather Spartan routines (e.g., they have the number of iterations "hard-wired," with no convergence check) that show the simplicity of the algorithms.

```
function (n1=nij1,n2=nj2,niter=200)   # Konigsberg and
                                        Frankenberg
{
      nr<-NROW(n1)
      nc<-NCOL(n1)
      pia<-n1/apply(n1,1,sum)          # pia from reference
      N<-sum(n2)
      da<-rep(N*1/nr,nr)               # Start from uniform da

      for(iter in 1:niter) {
      pai<-(da%o%rep(1,nc)*pia)/       # K&F eqn. 9
            (rep(1,nr)%o%
            (apply(da%o%rep(1,nc)
            *pia,2,sum)))
      da<-as.vector(t(n2)%*%t(pai))   # K&F eqn. 10
        }
      return(da)
}
```

```
function (n1=nij1,n2=nj2,niter=200)     # Bocquet-Appel and
                                           Masset
{
      nr<-NROW(n1)
      nc<-NCOL(n1)
      ma<-rep(sum(nj2)/nr,nr)
      fia<-(n1)/((n1%*% rep(1,nc))       # fia from reference
       %*%rep(1,nc))
      fi<-apply(fia,2,sum)               # fi from reference
      for (i in 1:niter) {
      fai<-fia/(rep(1,nr)%*%t(fi))       # B-A&P eqn 1.
      ma2<-fai%*%n2                      # B-A&P eqn 2.
      fia<-fia*(ma2/ma)%*%rep(1,nc)      # B-A&P eqn 3.
      fi<-apply(fia,2,sum)               # B-A&P eqn 4.
      ma<-ma2
      }
      return(as.vector(ma))
}
```

After 200 iterations the estimated age distribution from either function is 13.556024, 11.238048, 16.889118, 20.163258, 18.448143, 12.991268, and 2.714141, while after 20,000 iterations the age distribution is 15.851453, 2.182010, 27.973884, 17.821325, 14.883364, 15.727492, and 1.560473. We also have fit the age distribution by numerically maximizing the log-likelihood (in "optim"), which gives an age distribution of 17.761260, 2.902435, 30.032820, 22.258262, 12.926188, 20.902711, and 1.216324. While the mle converged properly, the standard errors on the estimated parameters are enormous. None of the standard errors is less than 100, whereas the parameters are all proportions between 0 and 1. This is a clear indication that there is insufficient information in the data, a result of fitting more age categories than there are indicator states. Other examples we have tried [e.g., Hoenig and Heisey's (1987) example at the bottom of their page 242] do converge properly with reasonable standard errors.

Chapter 10

A Post-Neumann History of Biological and Genetic Distance Studies in Bioarchaeology

Lyle W. Konigsberg

I. INTRODUCTION

This chapter reviews briefly the circumstances surrounding the shift from "varietal thinking" to population genetic-based approaches to archaeological skeletal samples. It then turns to the meat of the chapter—a history of biological distance analysis within bioarchaeology in the last half of the previous century. Before starting this adventure, I should lay some preliminary ground rules and sketch the focus and a broad outline for this chapter.

As concerns rules, I will, whenever possible, cite only published works, avoiding the many dissertations, theses, and other unpublished documents that are difficult for many readers to obtain. I make no claim that the history I give here is uniform in focus and degree of coverage. My personal bias (through experience) has been toward eastern U.S. bioarchaeology and, as a consequence, my coverage of other areas often borders on the paltry. I can only apologize at the outset for what may appear as a slight to researchers who work in other areas. As one of these researchers (Jantz, 1973) is just a few doors down the hall from me, I can assure him and the many others that any exclusions are born of my own ignorance. They are not comments on the quality of the work. While on the subject of focus, I should also make a few comments about the level of mathematical sophistication assumed for this chapter. History and mathematics are not often

Bioarchaeology: The Contextual Analysis of Human Remains, Buikstra and Beck (eds.)

a comfortable mix, and I have tried to avoid the use of mathematical concepts here. That said, I must admit that biological distance analysis and paleodemography are probably the two most mathematical branches of bioarchaeology, and it simply is not possible to discuss the history of either without some recourse to equations. I have learned over the years that equations are best left to "boxed text" that the reader can choose to ignore. This is not a textbook, and as such the "box" is not an option. I do provide a very brief appendix both to correct my own previous errors (of which, sadly, there are more than I might care to admit) and to provide in one place some information that may be of use to those who do biodistance analysis using archaeological skeletal material.

II. THE END OF THE BEGINNING

As described elsewhere, Della Collins Cook discussed the development of the "biological distance" concept within American bioarchaeology up through the 1950s and 1960s. As she points out, the period following World War II was epitomized by the work of Georg K. Neumann on "varieties" among American Indians. Although his seminal work on American Indian variation (Neumann, 1952) was titled "Archaeology and Race in the American Indian," it is clear that Neumann viewed Native American prehistoric biological variation as reflecting genetic "types" rather than "races." While Neumann's understanding of the works of Fisher, Haldane, and Wright on population genetic theory was clearly extremely limited, Neumann did attempt to objectively classify prehistoric Native American skeletal remains. Neumann's background in multivariate statistical analysis was also limited to nonexistent, and it was on this front that his work would ultimately fall out of favor. The seeds of discontent were also sowed by the development of the "New Archaeology," especially the "New Physical Anthropology" (Washburn, 1951).

A mere 14 years after the publication of Neumann's seminal work, Joseph K. Long (1966) published a devastating statistical critique of Neumann's typological views on Native American prehistoric biological variation. Specifically, Long applied a multiple discriminant analysis to craniometric data from 151 male adults from archaeological contexts. Long attempted to reclassify the crania into Neumann's "Lenapid," "Walcolid," "Iswanid," and "Otamid" types, but found that overall the classification was rather poor and was in part driven by whether crania bore artificial cranial "deformation." To quote from Long (1966:462): "Nothing here supports Neumann's (1952) explanation of subgroups based on large-scale migrations into the area." Long (1966:463) argued for using discriminant function analysis on a more regional level, stating that "this approach assumes a rejection of the simplistic picture of race and culture in North America."

Neumann and his students generally reacted to Long's (1966) article by either dismissing its results in a sentence or two or by outright ignoring its existence. In her monograph on the Oneota, Elizabeth Glenn (1974) wrote that "one of the lines of procedure in this analysis will be to test the cranial data of the different Oneota foci to determine to what extent they coincide with Neumann's varietal groupings of the area." One hundred and forty pages later, on the last page of her text, Glenn (1974:141) notes: "It might be added that one might wonder about the utility of the varietal approach, as such, with this population. Not only do the varieties not 'describe' all segments of the Oneota population, but also, the key traits associated with the differentiation of these varieties have not been the critical discriminants in these analyses." But she then continues that "... the value of the populations on which the varieties are based ... are unquestionably valuable to the consideration of this [the Oneota] population." Curiously, Glenn chose not to cite Long (1966), despite the fact that her 1974 monograph is based on her 1971 doctoral dissertation written under Neumann at a time when Long's work was well known and easily available. Robbins and Neumann (1972), in a lengthy tome (the bulk of which is taken up by 285 tables) on the Fort Ancient Culture dispatch with Long's work in one sentence that ignores the results of his work: "In testing the validity of Neumann's varietal typology, Long (1966:235) uses multiple discriminant analyses of specific measurements and indices to establish a metrical classification of the varietal groupings." This is the sole reference to Long's article in Robbins and Neumann (1972).

While it is not on the surface clear how Neumann and his students could simply choose to ignore Long's article, it is important to realize that Neumann was an established researcher, while Long was just beginning his own anthropological career. Long's publication was based on his M.A. thesis from the University of Kentucky, written in 1964 under Dr. Charles E. Snow. Snow himself, as a student of Hooton's, initially subscribed to a "varietal" classification of prehistoric human skeletal remains from the eastern U.S. By the 1970s, when Snow published his last monograph (Snow, 1974) he had clearly rejected a Neumann-type approach, but at the time that Long was writing his thesis there is no particular reason to suspect that Snow had departed from typological thinking. Long continued his career in anthropology after his discriminant function analyses, but he obtained his Ph.D. in medical anthropology at UNC–Chapel Hill (Long, 1973) and never returned to skeletal biology/bioarchaeology. He is probably best remembered as the founder of the Society for the Anthropology of Consciousness. Regardless of Neumann and his student's reaction (or lack thereof) to Long's publication, with the advent of the "New Archaeology" the time was ripe for change at the end of the 1960s. By the beginning of the next decade there were a host of dissertations that used biological distance analysis within archaeological contexts to address issues of past population structure. These studies were radical departures from their predecessors, which had addressed issues of mass migration and definition of "varietal types" and "races." Buikstra (1979:226) summarized this shift specifically within Hopewell (U.S. midwestern Middle Woodland) archaeology in a section she titled "Georg Neumann/Charles Snow: The End of an Era." Although our discussion until this point has focused on the eastern United States, and consequently on Georg Neumann, this shift occurred for other regions of the world [see, e.g., Van Gerven *et al.* (1973) for Nubia and Rightmire (1970) for South Africa]. In a curious attempt to rewrite history, Armelagos and Van Gerven (2003:58) argued that within skeletal biology "typology continues despite our understanding of adaptation and the processes of morphological change." They thus appear to reject the idea that the legacy of Neumann has ended. As should be clear from the remainder of this chapter, I disagree with Armelagos and Van Gerven's view of the recent history of biological distance studies.

III. THE RISE OF MULTIVARIATE ANALYSIS: *IN SITU* DEVELOPMENT OR MIGRATION OF PEOPLE?

Following on the heels of "varietal" and "racial" approaches, many biodistance studies were undertaken in the 1970s to determine whether archaeologically or ethnohistorically defined cultures arose by *in situ* development or by external migration into the region of interest. Unlike the earlier "varietal" studies, which had used only rudimentary statistical approaches, these more modern studies tended to use multivariate approaches—either discriminant function analysis (i.e., canonical variates or Mahalanobis distance analysis) for metric traits or (typically) Smith's Mean Measure of Divergence (MMD) for discrete traits. Overwhelmingly, the osteological evidence presented was in favor of local continuity of archaeologically defined cultures (Bennett, 1973a; Buikstra, 1976, 1977; Droessler, 1981; El-Najjar, 1978; Jacobs, 1993; Mackey, 1977; Molto, 1983; Reichs, 1984; Sciulli and Mahaney, 1986; Suchey, 1975; Van Gerven *et al.*, 1973; Wolf, 1977). This was not, however, an absolutely universal finding. Turner (1980:26) noted that the results of his analysis of discrete traits "clearly support the notion that the development of the Mississippian culture period in northern Alabama involved the movement of people into the area, either replacing, displacing, or hybridizing the descendants of previous inhabitants." Berryman (1980:12) similarly cited craniometric evidence for a "migration of Middle Tennessee people into the eastern Tennessee area" at or prior to the emergence of Mississippian. And in areas where there was well-documented archaeological or ethnohistoric evidence for migration, such as the Tiwanaku "colonization" in Peru (Blom *et al.*, 1998), the spread of Zapotecans during the formative period (Christensen, 1998), and the intrusion of the Oneota into central Illinois (Steadman, 1998), bioarchaeologists have cited biodistance evidence to support these large-scale migrations. Further, as Adams and colleagues (1978) note in "a world tour of migration theories" (which curiously lacks Oceania) "migration theory obviously has a higher probability value in island environments than elsewhere, for here diffusion over any distance must necessarily also involve migration." As a consequence, biodistance studies focused on Oceania [amply reviewed in Pietrusewsky's (2000) chapter] have always been framed around various migration hypotheses.

From a historical standpoint, it is particularly interesting to look at the accumulation of biodistance evidence against large-scale migrations during the 1970s and 1980s. As Adams *et al.* (1978) note, migration "theories" were largely on the wane by the 1970s. They view this development as a logical result of, among other things, the rise of positivist thought (and decline of specific historicity) in archaeology that came with the advent of the "New Archaeology." They go on to write (p. 516) "archaeology's retreat from migrationism has had a profound impact on current trends in Amerindian craniometry" and (p. 523) "whether one views physical anthropology as a 'handmaiden to human history' or worries that 'we have somehow drifted farther and farther away from prehistory,' our views of history continue to effect the methods and goals of our research." At the end of their review Adams and co-workers (1978) sketch "the beginnings of scientific migrationism," but in truth interest in any form of a migration resurrection has remained tepid, at best, until Anthony's (1990) sounding of the waters a decade ago. It seems almost inconceivable that the biodistance studies of the 1970s and 1980s could, as a group, have come to wrong answers to the migration versus *in situ* development question. However, aside from any platitudes about "statistics never telling lies," what is very curious about these many studies is that they never established any particular methodology or guidelines for using biodistance data to answer questions about migration. Although the studies often appealed to population genetic analyses of living groups, there were no methodological or theoretical developments along the lines now presented in the modern human origins debate (Cole, 1996; Konigsberg *et al.*, 1994; Relethford, 2001; Relethford and Harpending, 1994; Rogers, 1995; Waddle, 1994a,b), and the studies were often unclear on whether they had provided necessary (or simply sufficient) evidence for biological continuity of archaeological cultures. With the advent of direct genetic assays for prehistoric skeletal remains [see O'Rourke *et al.* (2000), Stone (2000), and later], we are beginning to return to the old migration questions, and there is likely to be an increase in such studies over the foreseeable future. Clearly, from a modeling/statistical standpoint there is a need to return to the question of *in situ* development versus long-range migration. This is made quite clear in Alan Rogers' (1995) article "How Much Can Fossils Tell Us about Regional Continuity?" Rogers surmises that it would be very difficult in a statistical sense to provide evidence from fossil material for regional continuity. If this is the case, then why were bioarchaeological studies from the 1970s and 1980s so successful in demonstrating (repeatedly) that there was regional continuity? Sampling issues alone, a hoary issue in paleoanthropology, do not appear to explain this divergence.

IV. EXAMINING LOCAL MIGRATION

While the discussion over *in situ* development versus migration was playing out in the literature, a new bioarchaeological focus arose during the early 1970s. With the publication of Lane and Sublett's (1972) "Osteology of Social Organization: Residence Pattern" article in *American Antiquity*, the study of short-distance migration (Anthony, 1990), more specifically mating/residence practices, became a legitimate domain of study. Lane and Sublett argued that cranial discrete traits could be used to test for post-"marital" residence pattern by comparing biological distances across cemeteries within males and within females. Specifically, they stated that if residence patterns were female–female based (i.e., uxorilocal), they would expect cemeteries to be heterogeneous in female comparisons and homogeneous in male comparisons. Alternatively, if residence patterns were male–male based (i.e., virilocal), they would expect cemeteries to be heterogeneous in male comparisons and homogeneous in female comparisons. In either case (uxorilocal or virilocal), they predicted that within cemetery comparisons of males with females would demonstrate heterogeneity. Although these postulates could have been motivated through population genetic theory and models [and indeed were in Lane's (1977) dissertation], Lane and Sublett chose to support their suggested patterns by analogy to expected material culture distributions in archaeological data [see, e.g., Longacre (1964) and Deetz (1968)]. As they were publishing in *American Antiquity* this was a fairly logical choice for citation. The Lane and Sublett model was ultimately based on the idea that if one looked only at skeletons from adults, then if males had a higher migration rate than females this would homogenize males across sites (because any cemetery would include migrant males). As an interesting historical sidebar, it should be mentioned that Corruccini (1972), in the same year as Lane and Sublett, published an analysis of prehistoric and historic Pueblo skeletal material that used the same logic as the Lane–Sublett model. To quote from Corruccini:

> . . . the Pueblo female samples are more tightly bound by a considerable margin than the males. The Pueblo sex differences . . . point to a proportionality between amount of genetic differentiation and amount of mating and residence flexibility, the latter factor being connected with the matrilocal patterns mentioned earlier. (Corruccini, 1972:386)

Following quickly on Lane and Sublett's publication there were additional extensions given in the literature. Spence (1974a) extended the Lane–Sublett model by also considering the effect of migration on within-sample variation in discrete traits by sex. He suggested that within a single cemetery male homogeneity would be expected if residence were based virilocally, whereas the opposite (female homogeneity) would be expected if residence were based uxorilocally. Like Lane and Sublett, Spence also motivated his model by analogy to the distribution of archaeological artifacts. In the same year, he (Spence, 1974b) also published a test case using within-sample variances by sex for craniometric data. Buikstra (1980) drew explicitly on the Lane–Sublett model, but also added an additional layer. She looked at discrete trait biodistances between mounds by sex within a single site (Pete Klunk mounds). Addition of the between mound component allowed her to contrast expectations depending on whether the mounds were used contemporaneously or in serial fashion. Other authors (Birkby, 1982; Bondioli *et al.*, 1986; Droessler, 1981; Kennedy, 1981) also drew on the Lane–Sublett model, or its logic, to make statements about past residential patterns.

It is interesting to note that the Lane–Sublett model grew out of analogy to the archaeological record. Like the antimigrationist paradigm of the 1970s and 1980s in bioarchaeology, which came fairly directly from the "New Archaeology," the focus of the Lane–Sublett model on short-distance migration was also born of the "New Archaeology." Like the antimigrationist paradigm, the Lane–Sublett model could also be critiqued on the grounds that it had little supporting theory. It was, as well, belabored by a host of implicit but unstated assumptions. Brenda Kennedy (1981) first pointed out a potential shortcoming of the Lane–Sublett model in that the predictions were ambiguous in certain settings. For example, in discussing a patrilocal (really, virilocal) pattern Kennedy wrote that:

> It may be assumed that the hypothetical "first group of men to engage in these marriage practices" formed a relatively homogenous group, since up to this point no exogamic unions had occurred. However, what of the second, third, and fourth generations of males taking part in these customs? Given the input into the male gene pool of the genes of their mothers who have come from a variety of sources, do the males remain a relatively homogenous group? (Kennedy, 1981:28)

The answer to this ambiguity lay in reformulating the predictions of the Lane–Sublett model on the basis of population genetic models (Konigsberg, 1988), a subject taken up in the Appendix. The conundrum in the Lane–Sublett model is that the effects of differential migration by sex are only displayed in the current postmigration generation because autosomal alleles are assigned randomly to the sexes in the next generation. The stability of the Lane–Sublett model was also challenged by an influential article by Cadien *et al.* (1974) in which they argued that the time depth represented within skeletal series could affect interpretations adversely. More recently, the simple-minded approaches I took in the late 1980s to the Lane–Sublett model have been rightly critiqued by others (Aguiar and Neves, 1991; Williams-Blangero and Blangero, 1990). Among these critiques, I do not include Tyrrell's (2000:299), who has referred to my dissertation as "a self-defeating exercise." He objected to my eliminating discrete traits that appeared to be dependent on sex of the individual as "artificially skewing the frequency differences." He apparently did not read my dissertation particularly closely, as I wrote:

> Clearly, only traits which are genetically or physiologically dependent on age or sex should be eliminated, and it may therefore be desirable in some instances to form grand samples across populations in order to reduce the effect of extraneous variation due to residential practices. (Konigsberg, 1987:113)

V. EXAMINING TEMPORAL VARIATION

Many studies have attempted to correlate biological variation with the passage of time in archaeological contexts. By and large, these studies have started from the premise that the local temporal sequences represent *in situ* development. Consequently, long-range migration is assumed to have occurred at a trivial or nonexistent rate. Often this assumption is tested in some way, but in other analyses it stands as a tacit assumption. For example, Larsen (1982) has compared bone biomechanical properties for preagricultural and postagricultural groups from the Georgia Bight and writes (Larsen and Ruff, 1991) that "given the well-established record of cultural continuity in prehistoric Georgia coastal populations, especially during prehistory (see discussion in Larsen, 1982), it is appropriate to suggest that biological change in this region did not likely result from population replacement." This is taken as an untested assumption (to my knowledge, there are no biological distance or ancient DNA studies for the prehistoric Georgia Bight), and the skeletal material is then used to document what are generally viewed as plastic responses to environmental change rather than the result of directional evolutionary forces (such as gene flow or selection). Larsen (1997) gives numerous examples of bioarchaeological studies that have suggested plastic responses of bones through temporal sequences, with the prime mover usually being the adoption of agriculture. I will not comment on the history of such studies here, as they are outside the direct purview of biological distance analysis.

There are two research topics where temporal variation has, pretty much by necessity, been explained in evolutionary terms. First, in geographic areas such as parts of the American Great Plains, where there is no clear evidence for short-term environmental change, gene flow or other evolutionary arguments have taken priority (Jantz, 1973; Jantz and Willey, 1983; Key, 1983; Key and Jantz, 1981). Second, changes in dental size (typically dental reduction) usually must be explained using some form of evolutionary mechanism, as tooth size is unlikely to exhibit plastic responses to environmental change. Guagliardo (1982) suggested that tooth size could be modulated by interuterine stress and that differential mortality associated with the stress could then lead to different average tooth size across age classes. As a consequence, differences in adult tooth size across age cohorts could be interpreted either as a result of selection (Perzigian, 1975) or, following Guagliardo, as evidence for interuterine stress followed by differential mortality. Although Guagliardo's argument could be extended to suggest some form of plastic response in dental size, if there is a nonzero genetic correlation between stress-induced mortality and stress-induced interuterine dental size reduction, then the appropriate quantitative genetic model would be one of evolution by natural selection on correlated characters. In this case, the direct selection operates through stress-induced mortality, whereas selection for reduced dental size comes as a correlated response. Less complicated selection models have been given to explain the almost universal reduction in dental size over the course of human history and prehistory; these are reviewed briefly in Larsen (1997:245). One of these models, the "probable mutation effect" (Brace, 1964), has been a source of long-term debate, which is discussed next.

VI. POPULATION GENETICS AND BIODISTANCE

It is exceedingly difficult to pinpoint when population and quantitative genetic theory first made their entrada into bioarchaeology. As sketched out earlier, many of the methods and debates in biodistance analysis of the 1970s and 1980s were motivated by developments from the "New Archaeology." As such, they did not make explicit use of genetic theory. Much as I might like to claim some historical priority from my dissertation work (Konigsberg, 1987), the roots of genetic theory in biodistance analysis run much deeper. While it was resoundingly criticized in some quarters, McKee's (1984) deterministic computer simulation of dental size reduction via the probable mutation effect represents an early explicit use of models from the great population geneticist of the last century, Sewall Wright. But there are earlier threads than this.

I would trace the incorporation of population genetic theory to three influences from the literature. First, the publication of Cadien *et al.'s* (1974) influential article on "Biological Lineages, Skeletal Populations, and Microevolution" set the stage for incorporation of population genetic theory. In truth, they did little other than critique skeletal biologists for treating samples (or "lineages") with considerable time depth as if they were single snapshots of a biological population. However, Cadien and colleagues (1974) at least explicitly referred to and cited the relevant evolutionary works of Fisher, Haldane, and Wright and focused attention on the diachronic nature of skeletal samples. The lemons that Cadien and colleagues saw would eventually become lemonade for others (e.g., Konigsberg, 1990b; Owsley and Jantz, 1978; Owsley *et al.*, 1982), but their article stands as an important historical piece and one that was almost always cited before embarking on a biodistance study using archaeological human skeletal material. A second historical influence, although sadly one with relatively little impact, was the completion of Rebecca Lane's dissertation in 1977. While Lane and Sublett's (1972) *American Antiquity* article greatly overshadowed Lane's later dissertation, Rebecca Lane's dissertation is a remarkable study that unfortunately was never published. As she was working with historic Allegany Seneca Indian cemeteries that were relocated, she was able to collect both osteological information and genealogical information. Individuals within cemeteries were not identified, but she was still able to calculate kinship coefficients between cemeteries from the genealogical information associated with cemeteries and to compare this with biological distances from cranial discrete traits. Based on her empirical regression work, she suggested that a biological distance measure she derived to measure between cemetery divergence (the "standard effective divergence") was a hyperbolic function of the average kinship between cemeteries. So far as I can tell from the literature, this was the first (and for many years the only) attempt to directly relate biodistance to genetic kinship in archaeological samples. The final historical thread was an increasing emphasis on "The Use of Quantitative Traits in the Study of Human Population Structure," to quote directly from the title of Relethford and Lees' (1982) seminal paper. This emphasis led to a resurgence of interest in quantitative traits [see, e.g., the brief review in Williams-Blangero *et al.* (1990)], an area that had been pushed aside for many years. The near abandonment had come as a result of increasing interest in physical anthropology on single locus genetic markers and the dissatisfaction with metric approaches that appeared to be "nongenetic" or, worse, racist [see Washburn's (1951) review of the "old physical anthropology"].

As Rebecca Lane's (1977) dissertation does represent the earliest use of explicit population genetic theory in bioarchaeology, I will start this brief history in 1977. As mentioned earlier, Lane suggested that a measure of biological distance she derived was inversely related to the average genetic kinship between groups. As genetic kinship is a measure of similarity, it makes sense that distance measures should have some form of monotonic decrease with increasing genetic kinship. Indeed, Morton (1975) had given the relationship in the literature, and the relationship between Mahalanobis squared distances (or any squared Euclidean distance measure) and average kinship is now well known [see the Appendix, Eq. (7)]. In this regard, it is unfortunate that Smith's MMD, a nonlinear distance measure often used for discrete skeletal traits (see review in Tyrell, 2000), has persisted. John Blangero derived a threshold trait distance that is analogous to the Mahalanobis distance commonly used with metric traits (see Pietrusewsky, 2000), and consequently its relationship to genetic kinship is known. A number of authors have now used Blangero's generalization of the Mahalanobis distance

(Ishida and Dodo, 1997; Konigsberg, 1990b; Konigsberg *et al.*, 1993) in analyses of cranial discrete traits.

The ability to frame skeletal biodistance analyses within population genetic frameworks depended on advancements in quantitative trait theory. Although most of the foundations for these advancements were laid initially outside of anthropology (Crow and Denniston, 1974; Jacquard, 1974; Morton, 1973), the publication of Crawford and Workman's (1973) edited volume on "Methods and Theories of Anthropological Genetics" brought these foundations to the anthropological forefront. Relethford and Lees' (1982) review of quantitative trait analyses began to spark interest among skeletal biologists, who had a very long history of analyzing quantitative traits and a short to nonexistent history of incorporating quantitative and population genetic models in their analyses. At the time I was writing my dissertation in the mid-1980s there was already a considerable number of dissertations on biodistance analysis using archaeological human skeletal samples, and many of these appealed at least indirectly to population genetic theory and models. In this sense, many of these earlier works could be considered to fit within Relethford and Lees' category of "model-free" analyses. There were, however, few previous works that could be categorized as falling into Relethford and Lees' category of "model-bound" analyses. Rebecca Lane's dissertation was probably the only example of an explicit use of population genetic theory in skeletal analysis. In truth, while I attempted to use population and quantitative genetic models in my dissertation, most of my work there should be classified as "model-free." Although I used population and quantitative genetic models, I did not attempt to directly estimate population genetic parameters.

An interesting area of analysis that has grown directly out of the quantitative genetics literature (Lande, 1976; Lande and Arnold, 1983; Lofsvold, 1988) is the assessment of natural selection as versus genetic drift to explain temporal sequences (Sciulli and Mahaney, 1991). While this may sound like an esoteric area that could little inform us about our prehistoric past, the magnitude of genetic drift is fairly directly related to population size and migration. Consequently, if it is the case that natural selection does not account for some short-term changes, whereas drift does, then we may be in a position to make estimates of past population sizes. Conversely, the drift explanation may require population sizes that are so small that only natural selection remains as a probable explanation for the observed temporal pattern. In Sciulli and Mahaney's (1991) study of tooth size reduction between Late Archaic and Hopewell samples from Ohio they found that the population sizes necessary to get the observed amount of change by drift were so small that natural selection provides the best explanation.

Another area that has seen expanding application is the Relethford–Blangero model (Relethford and Blangero, 1990). Relethford and Blangero extended the Harpending–Ward (Harpending and Ward, 1982) model, which was given for allele frequencies to cover the case of multivariate quantitative traits. In the

Relethford–Blangero model an **R** matrix (see Appendix) is estimated for a number of populations using quantitative traits. The diagonal of the matrix gives a standardized distance for each population to the centroid (the hypothetical group that would exist if the populations were not divided from one another). In the Harpending–Ward model each population has an observed level of heterozygosity, which is replaced in the Relethford–Blangero model with a summary measure of additive genetic variance. In either model, the variance (heterozygosity is a measure of variance) is related to the distance to centroid in a negative fashion. Populations that are near the centroid tend to have considerable internal variation, whereas populations far from the centroid tend to have very little internal variation. This is true because drift and low migration in isolated populations move the populations away from the centroid and homogenize them. If all populations receive long-range migrants (migrants external to the considered populations) at the same rate, then the regression of within-group variance on distance to the centroid should be negative. Populations that have greater long-range gene flow than the average should fall above the regression line, whereas populations with less long-range gene flow than expected should fall below the regression line. Steadman (1998) has applied such an analysis for west-central Illinois. It is possible to apply the Relethford–Blangero model to published distance analyses (see Appendix), although this requires within-group variance–covariance matrices, which are rarely published. These matrices are necessary to obtain the observed within-group variances, while only a distance matrix is necessary to obtain the **R** matrix.

The theoretical relationship of prehistoric biological distances to time and space (simultaneously) is an area that is rather poorly developed. This is at the core of the Cadien *et al.* (1974) article, but has received relatively little attention. Konigsberg (1990a,b) provided some simple rudimentary models, while Epperson (1993) has given much more sophisticated models. Interestingly, space–time models for biological distances have begun to appear in the modern human origins debate (Relethford, 1999). This shows that these old problems with which we have dealt (migration versus *in situ* development) are now common fodder on the more global scale of the origins debate. As Relethford has pointed out repeatedly, the issue of documenting continuity versus replacement is complicated greatly by the fact that unequal population sizes can distort our interpretations. As mentioned previously, Rogers (1995), however, asks the question "how much can fossils tell us about regional continuity?" and answers "not much." If Rogers is correct in this statement, then the numerous "demonstrations" of regional continuity cited earlier for 1970s and 1980s biodistance analyses must fall by the wayside. However, Rogers takes as requisite evidence for regional continuity the demonstration of a nonzero temporal correlation within regions. This discards the between region synchronic relationships, as well as the cross time–space relationships. While there is certainly no agreement on how these relationships should be analyzed (Cole, 1996; Konigsberg, 1997; Konigsberg *et al.*, 1994; Sokal *et al.*, 1997; Waddle, 1994a,b), most authors do support looking beyond the within-region sequences.

VII. THE RESHAPING OF MULTIVARIATE ANALYSIS AND THE DAWN OF ANCIENT DNA

Any history of biodistance analysis at the turn of the millennium would be incomplete without at least briefly mentioning two fairly recent developments. First, there has been "a revolution in morphometrics" (Rohlf and Marcus, 1993) within the previous decade based on analyzing three-dimensional coordinate data. Benfer (1975) first described a caliper-based method for "digitizing" the human skull, but because of the awkwardness (and high error rate?), routine analysis of three-dimensional coordinate data had to await the development of relatively inexpensive, reliable, and transportable three-dimensional digitizers. To date, the "new morphometry" has been applied to problems in the analysis of human cranial sexual dimorphism and growth, but there has been only one study whose focus was biodistance analysis among archaeological human skeletal samples (McKeown, 2000). The other development that stands to radically transform biodistance analysis, and possibly replace it with genetic distance analysis, in the future is the new area of ancient DNA (aDNA) analysis. There have been three excellent reviews of aDNA (Kaestle and Horsburgh, 2003; O'Rourke *et al.*, 2000; Stone, 2000). As I have "successfully" dropped an open Eppendorf containing DNA samples into a buffer reservoir on more than one occasion, I refer the interested reader to the literature for a historical review of the expanding field of aDNA analysis.

Appendix

I. FROM MARITAL MIGRATION TO UNEQUAL VARIANCES BY SEX

Konigsberg (1988) gave a population genetic model for the effects of differential migration on genetic variances within the sexes (both across sites and between sites). While the basic results in this article were correct (e.g., equation 7 from that article), the logic of the derivation was not. Consequently, I correct here that derivation.

In the island model (Wright, 1969) there are an infinite number of subpopulations, all of size N, that exchange migrants with rate m. A standardized genetic variance between the islands, symbolized as F_{st}, can be used to characterize the effects of genetic drift and migration. Wright (1969:294) defined F_{st} as "the correlation between random gametes within subdivisions, relative to gametes of the total population," but I will use an equivalent definition as the probability of identity by descent within subdivisions (see Hartl and Clark, 1997). The probability of identity by descent within subdivisions is just the probability on sampling two alleles (with replacement) that they will be the same allele because they derive from a common ancestor. The recurrence relationship for F_{st}^t (F_{st} in the tth generation) is

$$F_{st}^t = \left[\frac{1}{2N} + \left(1 - \frac{1}{2N}\right) F_{st}^{t-1}\right](1 - m)^2 . \qquad (1)$$

Assuming an equal sex ratio but different migration rates by sex, we can write F_{st} statistics for males, females, and males with females. The F_{st} for males is the probability of sampling identical by descent alleles within males within subdivisions (similarly for females), whereas the F_{st} for males with females is the probability of sampling identical by descent alleles from a male paired randomly with a female from within subdivisions. These F_{st} values and the total F_{st} are

$$F^t_{st,\triangle} = 0.25\left[\frac{1}{2N} + \left(1 - \frac{1}{2N}\right)F^{t-1}_{st}\right](1 - m_\triangle)^2$$

$$F^t_{st,\bigcirc} = 0.25\left[\frac{1}{2N} + \left(1 - \frac{1}{2N}\right)F^{t-1}_{st}\right](1 - m_\bigcirc)^2 \qquad (2)$$

$$F^t_{st,\triangle\bigcirc} = 0.5\left[\frac{1}{2N} + \left(1 - \frac{1}{2N}\right)F^{t-1}_{st}\right](1 - m_\bigcirc)(1 - m_\triangle)$$

$$F^t_{st} = F^t_{st,\triangle} + F^t_{st,\bigcirc} + F^t_{st,\triangle\bigcirc},$$

where triangles and circles are used to represent male and female. The ratio of male to female F_{st} values is then

$$\frac{F^t_{st,\triangle}}{F^t_{st,\bigcirc}} = \frac{(1 - m_\triangle)^2}{(1 - m_\bigcirc)^2}, \qquad (3)$$

as in Konigsberg (1988). Equation (3) can be used with a quantitative genetic model to specify the ratio of male to female within-group genetic variances or the ratio of between-group variances (see Konigsberg, 1988). Wood (1986) gives a migration matrix method for calculating the equilibrium **R** matrix (see next section for definition of the **R** matrix) that could be used to find separate male and female matrices.

II. FROM D² TO "R" AND BACK AGAIN

We will first look at how to use a Mahalanobis (or any squared Euclidean) distance matrix to estimate an **R** matrix. The **R** matrix is a standardized variance–covariance matrix. As Relethford and Harpending (1994) note: "**R** matrices have certain properties that make them useful for studying genetic relationships among populations." They enumerate a number of important properties of **R** matrices, some of which are exploited in the Relethford–Blangero (1990) model that Steadman (1998) used in a bioarchaeological context. As the literature often provides $\mathbf{D}^2$ matrices for archaeological skeletal samples, it is useful to have a way to convert these to **R** matrices without requiring raw data.

To find the **R** matrix we first need to calculate what is known as a "codivergence" matrix, usually written as **C**. The codivergence matrix measures the variance around the centroid. A population's codivergence with itself is just the squared distance from a population's multivariate means to the centroid. Codivergences between two populations can be zero (if they are both at the centroid), positive if they are both "on the same side" of the centroid, or negative if they are on "opposite sides." The **C** matrix is given as

$$\mathbf{C} = -0.5\left(\mathbf{I} - \mathbf{1w}'\right)\mathbf{D}^2\left(\mathbf{I} - \mathbf{1w}'\right)', \tag{4}$$

where **I** is a g by g identity matrix (g being the number of groups), **1** is a g by one column vector of ones, $\mathbf{D}^2$ is the matrix of squared distances (calculated on t traits), and **w** is a g by one column vector of relative census sizes. By definition we have $\mathbf{1w}' = 1$. Using results from Relethford and Harpending (1994), we can calculate F_{st} (see earlier) as

$$F_{st} = \frac{\mathbf{w}'\mathrm{diag}\,\{\mathbf{C}\}}{2t + \mathbf{w}'\mathrm{diag}\,\{\mathbf{C}\}}, \tag{5}$$

where diag $\{\mathbf{C}\}$ is an operator that places the diagonal of a matrix into a column vector. The **R** matrix is then

$$\mathbf{R} = \mathbf{C}\,(1 - F_{st})/2t. \tag{6}$$

From Williams-Blangero and Blangero (1989) (their equation 7) we can write the original $\mathbf{D}^2$ matrix as

$$\mathbf{D}^2 = \left(\frac{2t}{1 - F_{st}}\right)\left((\mathbf{R} \odot \mathbf{I})\,\mathbf{J} + \mathbf{J}\,(\mathbf{R} \odot \mathbf{I}) - 2\mathbf{R}\right), \tag{7}$$

where **J** is a g by g matrix of ones and $\odot$ represents a Hadamard product.

Written without the first parenthetical term of $2t/(1 - F_{st})$, Eq. (7) represents a standardized distance often used in human population genetics studies. As another definition for the **R** matrix is as a matrix of average kinship coefficients between populations (off the diagonal) and within populations (on the diagonal), Eq. (7) shows that there is a direct relationship between biological distance and average kinship. However, this relationship only holds for Euclidean distance measures (such as Mahalanobis distances). Consequently, it is not possible using Eqs. (2) and (5) to estimate F_{st} or the **R** matrix from nonlinear distances, such as Smith's MMD. The **R** matrix method given earlier can be applied to discrete traits using Blangero's generalization of the Mahalanobis distance for threshold traits. Tyrell and Chamberlain (1998) attempted to estimate F_{st} and "effective genetic distances" from cranial discrete traits, but they used a model appropriate for diploid genetic markers, not for threshold traits. Where they refer to the "heterogeneity" of a trait they are actually using the formula for heterozygosity at a biallelic locus.

Chapter 11

The Evolution of American Paleopathology

Della Collins Cook and Mary Lucas Powell

In Memoriam: Patricia S. Bridges; Sit tibi terra levis

I. INTRODUCTION

Paleopathology, the study of disease in past populations, lies at the intersection of several disciplines, among them, paleontology, medicine, dentistry, and anthropology. Perhaps because it is so interdisciplinary, it has resisted professionalization until recently. Much of that professionalization has occurred in North America.

Sir Marc Armand Ruffer, a British physician known today primarily for his extensive and innovative research on Egyptian mummies, is widely credited with the invention of the term "palaeopathology." However, this term was actually coined by an American physician and ornithologist, R. W. Shufeldt, in an article titled "Notes on Palaeopathology" that appeared in 1892 in the journal *Popular Science Monthly*. He wrote: "Palaeopathology: (Greek *palaios*, ancient, and *pathos*, a suffering), the word used in the title of this paper, is a term here proposed under which may be described all diseases or pathological conditions found fossilized in the remains of extinct or fossil animals" (1892:679). Shufeldt's choice of the word *pathos* explicitly links this area of inquiry to the specialized study of pathology in current medical science. The term appeared in the 1895 edition of *Funk and Wagnall's Standard Dictionary* and was popularized by Ruffer in his 1921 treatise, *Studies in Palaeopathology of Egypt*, but it

Bioarchaeology: The Contextual Analysis of Human Remains, Buikstra and Beck (eds.)

was not included in the *Oxford English Dictionary* until 1985 with the expanded definition, "the study of pathological conditions found in ancient human and animal remains" (Cockburn, 1997, in Elerick and Tyson, 1997; Aufderheide and Rodríguez-Martín, 1998).

The evolution of paleopathology from a minor interest of Renaissance antiquarians and pastime of Victorian physicians to a modern scientific discipline in the 20th century has been reviewed in detail by a number of scholars, beginning with the American anatomist Roy L. Moodie (1880–1934). Moodie's "Studies in Paleopathology. I. A General Consideration of the Evidences of Pathological Conditions Found among Fossil Animals" was published in *Annals of Medical History* (1917). His *Paleopathology: An Introduction to the Study of Ancient Evidences of Disease*, the first comprehensive review, soon followed in 1923, and a popular summary, *The Antiquity of Disease*, was published the same year. This slim book, published in the University of Chicago Science Series and aimed at "the educated layman," devoted more than half its pages to evidence for disease in fossils, because Moodie's purpose was to challenge Henry Fairfield Osborn's thesis that disease was an important factor in extinction. His discussion of human paleopathology is concerned with trepanation and other evidence for "primitive surgery" as much as with evidence for disease. A more anthropological focus is found in the work of Herbert Upham Williams (1866–1938), who was a physician and professor of pathology at University of Buffalo. His lengthy review article titled "Human Paleopathology" in *Archives of Pathology* (1929) focused exclusively on examination and interpretation of human remains, a departure from the older, more comparative approach defined by Shufeldt.

In 1965, Saul Jarcho organized an international symposium titled *Human Palaeopathology*, supported by the National Academy of Sciences/National Research Council. The papers had a strongly Americanist focus. New studies were presented on skeletal remains from Mesa Verde in New Mexico and a Middle Horizon site in California, and there was critical reassessment of earlier work in the Southwest, the Arctic, and Peru. In addition to anthropologists, the participants included a virtual who's who of bone biology at midcentury, Walter Putschar, Henry Jaffe, Lent Johnson, and Harold Frost among them. Innovations in radiology and histology, as well as the need for greater sophistication in diagnosis, were stressed by several contributors. The symposium papers were published the following year (Jarcho, 1966b).

This volume is particularly notable for Jarcho's essay *The Development and Present Condition of Human Palaeopathology in the United States*. He presented brief biographies of the heroes of what he regarded as the Golden Age: John Collins Warren, Samuel George Morton, Joseph Jones, Jeffries Wyman, Frederic Putnam and William Whitney, Aleš Hrdlička, Marc Armand Ruffer, Roy L. Moodie, Herbert U. Williams, and Earnest A. Hooton. He situated them in an account of the institutional origins of our field in Boston and Washington.

Jarcho called for "a revival of palaeopathology in the United States that should counteract the doldrums of the last three decades" (1966b:28), emphasizing the discovery of cranial deformation and syphilis among the ancient inhabitants of North America as questions that motivated these pioneers. Jarcho was critical of the relative isolation of paleopathology from medical sciences and of its marginal position in archaeology. He stressed the potential for innovations in method and the need for systematic data collection. He was particularly critical of publication practices in American paleopathology, pointing out that pathology journals did not recognize it as a specialty and that archaeological publications were slow, secondary, and lacking indices. He stressed the need for cross-disciplinary bibliography. In 2006, we can report that the "Renaissance and Revolution" (1966b:27) that Jarcho called for have come to pass.

However, the renaissance or revival had already begun when Jarcho wrote his essay. The first evidence of renewed life was a review essay by Erwin H. Ackerknecht (1953) included in the influential graduate text book *Anthropology Today.* In the 1960s, several book-length overviews of paleopathology were published by European scholars eminent in the field: Calvin Wells' (1964a) *Bones, Bodies, and Disease* (1964), Ackerknecht's (1965) *History and Geography of the Most Important Diseases*, and Paul A. Janssens' (1970) *Palaeopathology: Diseases and Injuries of Prehistoric Man.* These authors were generally optimistic about the future of the discipline and encouraged collaborations with medically trained scholars. A substantial edited volume, Don R. Brothwell and A. T. Sandison's (1967), *Diseases in Antiquity*, collected together what are now the classic studies in the field. Some researchers, however, expressed fears that paleopathology would become excessively self-referential, and some, such as Jarcho, lamented the lack of theoretical and methodological advances over the previous three decades.

The revival of paleopathology addressed a wide scientific public. Articles published by American paleopathologists during this period in *Science*, the first by Saul Jarcho (1965b), the second by J. Lawrence Angel on paleodemography and anemia in the Mediterranean (1966b), a third by Ellis Kerley and William Bass (1967) on the history of the discipline, and a fourth by George Armelagos (1969) on studies of health and disease in ancient Nubia, brought modern paleopathology to the attention of the larger scientific community and emphasized the necessity for interdisciplinary collaborations.

One sign of vigorous development in a scientific discipline is the steady proliferation of literature published by scholars all over the world. By that measure, paleopathology enjoyed a booming economy during the latter half of the 20th century. In 1971, George J. Armelagos and colleagues published the first comprehensive *Bibliography of Human Paleopathology*, which included 1778 individual international contributions. In the same year, Jay B. Crain published a substantially complementary bibliographic list of 1222 sources

(Crain, 1971). In 1980, Michael R. Zimmerman raised the ante regarding quality with a carefully abstracted and indexed bibliography of 628 sources (Zimmerman, 1980). The work he began continues to grow through the contributions of many of our colleagues to the annotated bibliography section of the *Paleopathology Newsletter.* Most recently, in 1997, the San Diego Museum of Man issued a massive volume titled *Human Paleopathology and Related Subjects, An International Bibliography*, edited by Elerick and Tyson, which includes more than 18,000 individual entries. Six supplements to this reference work, compiled by a vast array of international scholars working in concert with the original editors, are now available in electronic form, with more additions planned in the future. Jarcho (1966b) identified systematic bibliography comparable to the *Index Medicus* as a critical need for our discipline. We have not yet achieved that level, but great strides have been made.

The more recent evolution of paleopathology in the United States has been addressed in three review articles, which appeared almost simultaneously in the early 1980s: "Palaeopathology: An American Account" by Jane E. Buikstra and Della C. Cook (1980) in *Annual Review of Anthropology*, "History and Development of Paleopathology," by J. Lawrence Angel (1981b) in the jubilee issue of *American Journal of Physical Anthropology*, and "The Development of American Paleopathology," by Douglas H. Ubelaker (1982), in Frank Spencer's edited volume, *A History of American Physical Anthropology, 1930–1980*. These authors noted a new focus in the decade following Jarcho's essay on detailed differential diagnosis (based on carefully constructed models of disease processes) and on explicit integration of biocultural context, dietary reconstruction, analyses of growth and development, and paleodemographical analysis into interpretations of skeletal pathology. Because the early history of American paleopathology has already been discussed in considerable detail in these reviews, we begin with a brief review of the first century and a half and then focus our attention primarily on major theoretical and methodological developments of the last quarter of the 20th century.

II. DEFORMED CRANIA AND ANCIENT SYPHILIS: THE BEGINNINGS OF AMERICAN PALEOPATHOLOGY

The discovery of intentionally modified crania in burials ranging from Egypt to Chile attracted the attention of North American anatomists and physicians such as John Collins Warren (1822, 1838) and Samuel George Morton (1839, 1844b). Their interest lay primarily in human cranial morphology rather than paleopathology per se, but they built large collections that remain useful to the present day. Their treatises (Warren, 1822; Morton, 1839) present deformed crania as extreme examples (albeit deliberately produced) of metrical and morphological variability, and their omission of other pathology has been noted (Ubelaker, 1982). Jarcho (1966b) correctly points out that Morton illustrated cranial lesions without noting them in his text and suggests that Morton's interest in paleopathology was minimal. However, Samuel Morton's enormous output includes several gems of paleopathology apart from his interest in cranial deformation. For example, he debunked claims for an extinct pygmy race in North America, pointing out that its proponents had mistaken the skeletons of children for very short adults (1841), and he described anomalies and pathologies ranging from atlanto-occipital fusion (1847), to bullet (1839:167) and axe wounds (Morton, 1839:131) (Fig. 1), the latter perhaps associated with unrecognized trepanation.

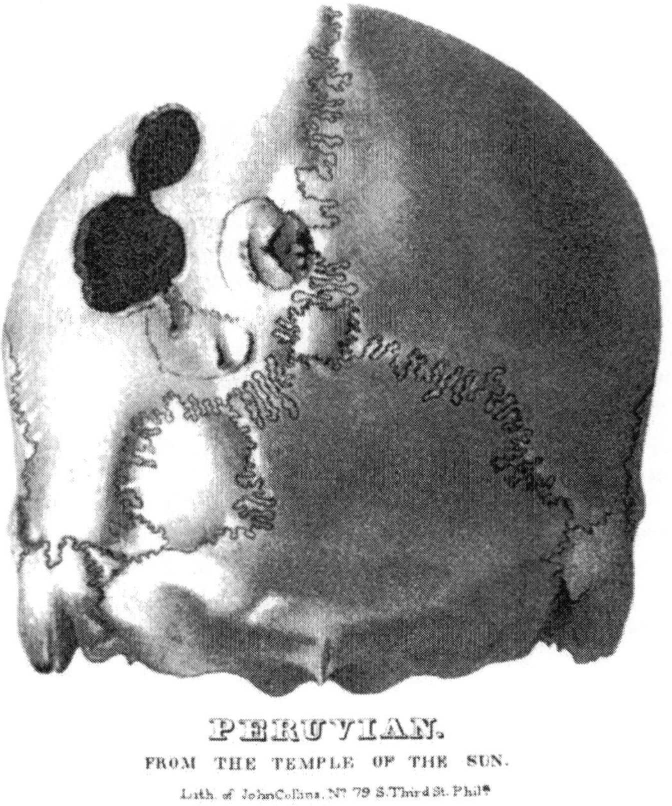

Figure 1 Axe wounds in a Peruvian skull (Morton, 1839).

The generation that followed Morton remained chiefly concerned with skeletal morphology and anthropometry, primarily of the skull, aimed at documentation of worldwide human variation. In the words of Ubelaker (1982:341), "to some extent, disease processes and cultural modifications presented unwanted 'noise' in the system" because they distorted the "pure" biological evidence of population affinities. We will touch on several exceptions to this largely accurate generalization.

Jeffries Wyman (1814–1874), a physician who served as first curator of the Peabody Museum of American Archaeology and Ethnology at Harvard University, was Warren's student and successor. His work focused primarily on descriptions of normal and artificially modified cranial morphology in archaeological specimens from Oceania, the Pacific Coast of South America, and the southeastern United States (1866, 1871, 1874, 1875), but he supplemented his craniology with extensive notes on anomalies and pathologies. Ubelaker (1982) points out that Wyman's comparative study of "auditory nodules" in prehistoric Peruvian and Polynesian crania (1868) was the first American study to take an explicitly comparative population approach to understanding lesions in ancient populations. Wyman's work foreshadowed Hrdlička's (1935b) monograph on auditory exostosis by nearly 70 years. Wyman's brief contribution to the Museum's fourth annual report (1871) compares 18 crania from his excavations in Florida to crania from Kentucky and Peru, touching on deformation, cranial capacity and variations, but the most interesting observations for our purposes concern the postcranial skeleton. Flattening of the tibia is discussed with reference to Gillman's observations in Michigan Indians and Broca's studies of Cro-Magnon. Septal aperture and pelvic diameters are also explored, and there is a brief discussion of pathological changes that includes what may be the first report of spondylolysis. Jarcho notes that "lesions that we might class as presumably syphilitic are described clearly and concisely but are attributed merely to 'periosteal inflammations'" (Jarcho, 1966b:11, citing Wyman, 1875), thereby initiating the controversy that has occupied much of the attention of American paleopathologists to the present day. Wyman's (1874) last report is a brief but lurid account of evidence of cannibalism from his Florida excavations that might be revisited in the light of today's controversies about taphonomy and human remains.

The first treatise that focused primarily on skeletal evidence of disease in ancient North America was Joseph Jones' detailed examinations of pathology in Late Prehistoric stone box burials in the Nashville Basin of Tennessee (1876). In his *Explorations of the Aboriginal Remains of Tennessee*, Jones sounded a second theme that dominated American paleopathology throughout the late 19th and early 20th centuries: the antiquity of syphilis. Jones (1833–1896), a physician who had served as a medical officer in the Confederate Army during the recent Civil War, was the son of the pioneering southern archaeologist,

Charles Colcock Jones (Schnell, 1999). At the time of his Tennessee studies, Joseph Jones held the position of professor of chemistry and clinical medicine in the medical department of the University of Louisiana in New Orleans. In his paleopathological diagnoses, he drew extensively upon his own biological training and clinical experience for his diagnosis of "syphilis" and employed both macroscopic and histological examination of lesions (Fig. 2). He applied chemistry—hydrochloric acid digestion—to the question of the antiquity of the Tennessee remains, concluding that they were so old that contact with the Spanish could not account for the disease. Ten of 171 pages in Jones' monograph are devoted to the syphilis question. The remainder is largely devoted to archaeological context. However, mortuary practices, cranial deformation, fading of hair color in mummies, misinterpretation of children's skeletons as pygmies, and cranial capacity are explored at length. Jones might be credited as the first scholar to focus on weeding out pseudopathology. Jones cites both Morton and Wyman, as well as several European sources. Two years later he described similar

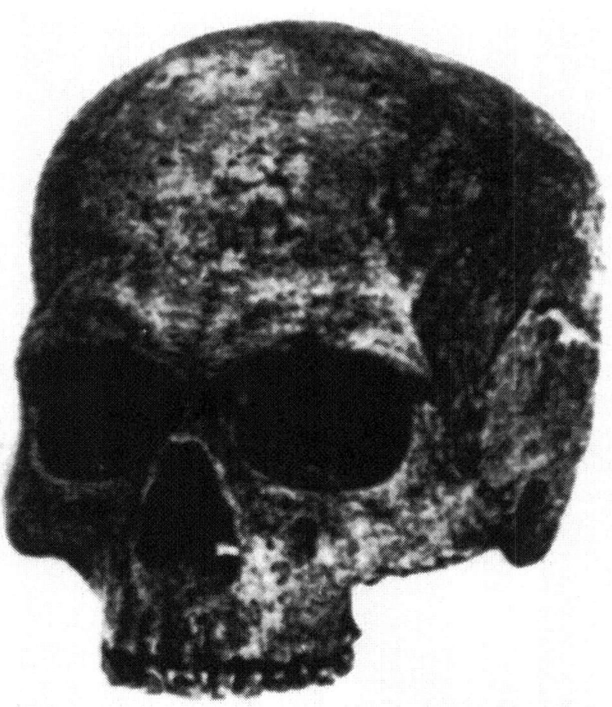

Figure 2 Syphilitic cranium from Big Harpeth River (Jones, 1876; Williams, 1932).

lesions in shell mound remains from Louisiana and reviewed historical accounts of disease among American Indians (Jones, 1878). Jones' careful approach to diagnosing past diseases by reference to current medical knowledge provided a powerful model for the subsequent theoretical and methodological development of American paleopathology, as well as an informed examination of one of its enduring themes.

Harrison Allen (1841–1897) was a physician and professor of physiology, zoology, and anatomy at the University of Pennsylvania. His contributions to paleopathology begin in 1867 with his detailed description of the Moulin Quignon jaw. He compared the French fossil with 300 mandibles from Samuel Morton's collection in order to rebut claims that it was morphologically distinct from modern humans, describing the effects of tooth loss in the process (Allen, 1867). He also produced one of the first careful case studies, a diagnosis of cleft palate versus trauma in a Seminole cranium from Morton's collection (Allen, 1898), unfortunately not figured in his publication and clearly not one of the three Seminole crania described by Morton in *Crania Americana* (1839). Allen's two monographs on collections from excavations, *Crania from the St. John's River, Florida* (1895) and *The Study of Skulls from the Hawaiian Islands* (1898), are descriptive craniologies in the mainstream of the typological anthropology of the late 19th century. However, both include detailed studies of discrete traits, reflecting Allen's deep interest in functional and comparative anatomy of the skull, and in the borderland between normal and pathological variation. An example is the discussion of the relationship of metopism to interorbital distance, citing Lombroso's criminal typology, in the Florida monograph (Allen, 1895a). This aspect of Allen's work is discussed at length by Hrdlička (1918).

Allen's (1898) study of Hawaiian crania is quite modern in its comparison of noble with commoner and pre-contact with post-contact specimens. Allen thus precedes Hooton in pioneering the comparative approach by more than 30 years, although his comparisons are less quantitative than Hooton's. Osteoporosis, osteitis attributed to syphilis, hyperostosis of the mandibular condyle, and alveolar atrophy following mortuary ablation of the anterior teeth are described. His most interesting diagnosis attributes vault and nasal lesions, enamel hypoplasia, and small maxillae in a 13-year-old to measles (1898; see Fig. 3). Ubelaker (1982) notes that both Harrison Allen and Joseph Jones studied with Joseph Leidy (1823–1891) at the University of Pennsylvania. Leidy had been a student of Samuel Morton's, and while Leidy made no contributions to paleopathology himself, he was an innovative contributor to paleontology, pathology, particularly to parasitology (Warren, 1998). Allen and Jones thus belong to an academic genealogy of Philadelphia physicians beginning with Morton.

Two physical anthropologists in the Department of Anthropology at the National Museum of Natural History, the successor to the United States

TRANSACTIONS WAGNER FREE INSTITUTE OF SCIENCE

PLATE X.

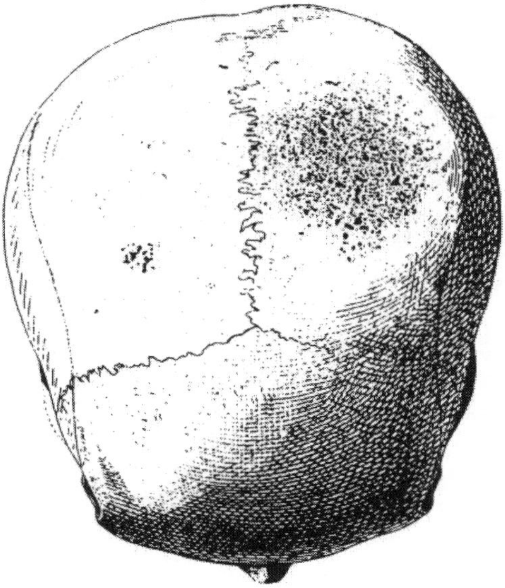

No. 1756, A. N. S.

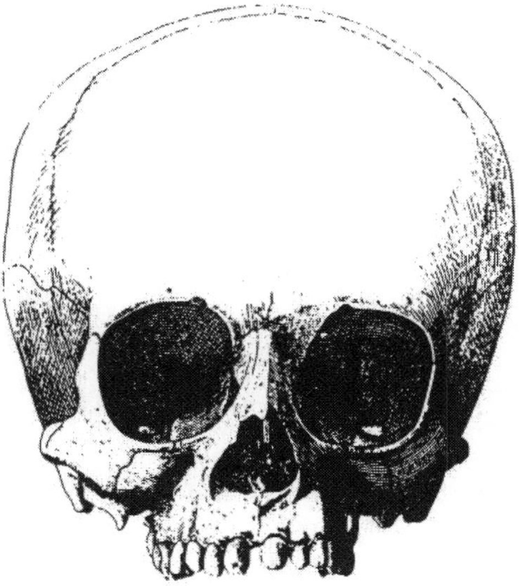

No. 1124, H. U.

Figure 3 Hawaiian child with dental and facial lesions of measles (Allen, 1898).

National Museum in the Smithsonian Institution, were strongly influenced by Hrdlička's interests in skeletal pathology. The *International Bibliography on Human Paleopathology and Related Subjects* (Elerick and Tyson, 1997) lists 85 single-authored and 9 coauthored publications by T. Dale Stewart and 108 single-authored and 15 coauthored publications by J. Lawrence Angel. Stewart published numerous studies of the effects of diet and cultural practices on skeletal and dental structures (e.g., Stewart, 1931a, 1939a, 1941a,b, 1942; Stewart and Groome, 1968), the history of premodern surgery (1937b, 1958b), forensic anthropology (1948, 1951b, 1954c, 1962d, 1979a), skeletal pathology (1931b, 1935, 1956a,b, 1966, 1974), and Native American populations of the New World (1960a, 1970b, 1973b, 1979b, n.d.).

The Harvard genealogy was even more prolific. William F. Whitney (1850–1921), a physician who served as curator of the Warren Anatomical Museum at Harvard University, contributed two articles to our literature, the first a short essay adding to the literature on pre-Columbian syphilis (1883) and the second (1886) a lengthy discussion of evidence for diseases of bone in ancient Indian remains in the collections of the Royal College of Surgeons in London, the Société d'Anthropologie in Paris, the Army Medical Museum in Washington, the Peabody Museum, and the Warren Anatomical Museum. In the latter paper Whitney discusses, among many topics, the association between Wormian bones and cranial deformation, the high prevalence of auditory exostoses, and specific lesions diagnosed as healed trauma, syphilis, and tuberculosis. He notes the rarity of several conditions, e.g., osteoporotic fracture: "it is remarkable that no case of impacted fracture of the neck of the femur has been found, which is of such frequent occurrence in old people" (Whitney, 1886:440). In all, he discusses 176 cases by catalogue number and includes skeletal material from Arkansas, California, Colorado, Kentucky, Iowa, Ohio, Rhode Island, Tennessee, Vancouver, Quebec, and Mexico. Writing of Whitney's relationship to F. W. Putnam, Jarcho infers that "paleopathology was now a separate and almost segregated area of research" (Jarcho, 1966b:12), but it is an equally reasonable inference that Putnam, who lacked an advanced degree, deferred to physicians, including Whitney and Dr. Frank W. Langdon (1881), in the study of ancient diseases.

Dentistry professionalized independently of medicine in the United States with its own associations and journals. This isolation is reflected in relatively limited attention from historians and bibliographers of medicine, including those interested in paleopathology. The earliest American paper of which we are aware is a notice of remains from excavations by the Kansas City Academy of Science in Clay County, Missouri, that remarks on heavy dental wear without caries: "as the enamel of the crowns of the teeth of the Mound-Builders was absent for the greater portion of their lives, and yet the teeth remained sound, it follows that when a portion of the enamel is removed decay of the rest of the

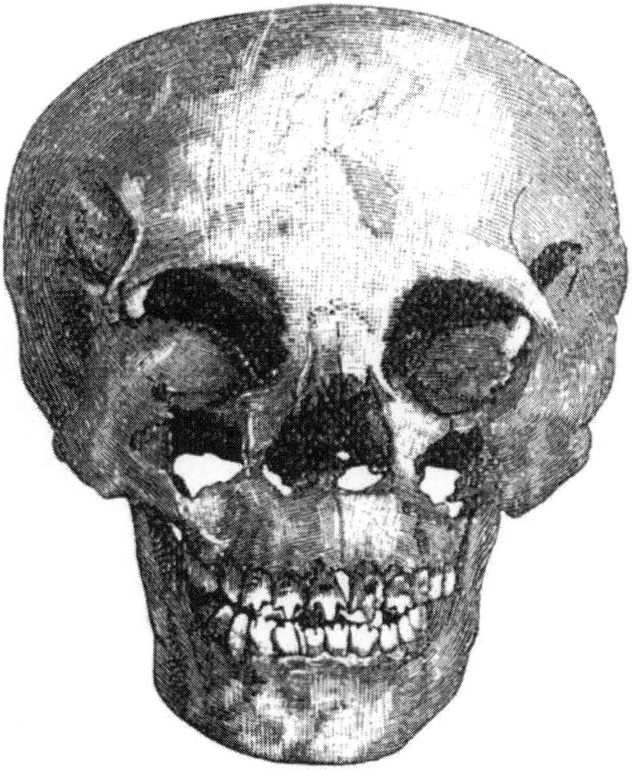

Figure 4 Dental filing in a skull from Labna, Yucatan (Andrews, 1893).

tooth does not necessarily follow" (Sozinskey, 1878:498). Dental science retained this open attitude toward drawing such general inferences from paleopathology. At the end of the 19th century, Robert R. Andrews (1893) described crania with filed and inlayed teeth from Labna in Yucatan (Fig. 4) and Copan in Honduras, noting heavy calculus deposits and a stone implant, and marveling at the skill of the ancient practitioners. Andrews' paper reports on material recovered for the Peabody Museum's Hemenway Expedition. It appears in an issue of *The Dental Practitioner and Advertiser* that celebrates the association's meeting in conjunction with the World Columbian Congress in Chicago that year. This case was published again 2 years later in *Dental Cosmos* amidst a survey of ancient teeth in museum collections in the United States; the author also presents data from his own excavations at Cahokia (Patrick, 1895).

III. THE EARLY 20TH CENTURY

The early 20th century roots of American paleopathology were well established in disciplines other than anthropology, as Jarcho has shown (1966b). Roy L. Moodie (1884–1934) was professor of anatomy at University of Illinois at Chicago and later at the College of Dentistry of the University of Southern California. He had studied under the University of Chicago paleontologist Samuel Wendell Williston (1851–1918). By far the largest part of Moodie's work in paleopathology concerns fossil animals, and many of his ideas, e.g., his interpretation of hyperextension of the spine as evidence for tetanus, are now the province of taphonomy rather than paleopathology. Chicago became the locus of an active community of paleopathologists in the first decades of the 20th century. In addition to his review publications (Moodie, 1917, 1923a,b), Moodie conducted an extensive radiographic study of North American, Egyptian, and Peruvian mummies in the collections of the Field Museum of Natural History, presenting the results in a remarkably detailed atlas (Moodie, 1931). Another Chicago physician encouraged by Williston was Charles A. Parker, who described osteoarthritis in the knees of Lansing Man (1904). Henri Stearnes Denninger (b. 1904), a physician affiliated with the Fay-Cooper Cole's Illinois archaeological survey during his student years at Illinois Medical School, published a series of careful case studies of Illinois and southwestern remains (Denninger, 1931, 1933, 1935, 1938; Cook, 1980b; see Fig. 5). H. U. Williams' (1932, 1936) magisterial papers on the origins of syphilis include specimens and information sent to him by both Moodie and Denninger.

Other noteworthy paleontologists who found paleopathology relevant to their research include Franz Weidenreich (1873–1948), an Alsatian who spent the last years of his career at the American Museum of Natural History (Weidenreich, 1939), and William L. Straus (1900–1981), an anatomist at Johns Hopkins University (Straus and Cave, 1957). Both explored the utility of pathological changes in making behavioral inferences about early humans.

Herbert U. Williams (1866–1938), a physician, was professor of bacteriology and pathology at the University of Buffalo. Jarcho (1966b) points out his innovative use of radiology, microscopy, and serology in study of ancient bone. Williams set out to review the literature, but added new critical and analytical material throughout his review. His careful study of porotic hyperostosis is a classic in its use of sections and radiographic correlation (1929). His principal contribution to paleopathology is his extensive review of Old and New World evidence regarding the origins of syphilis (1932, 1936). These papers are noteworthy for their rigorous examination of both documentary and skeletal evidence, and for their novel arguments for paleoepidemiology: "Where the seeds of corn could be carried, the seeds of syphilis might also be carried. What now takes place in a few days may have required centuries, but one is dealing with centuries"

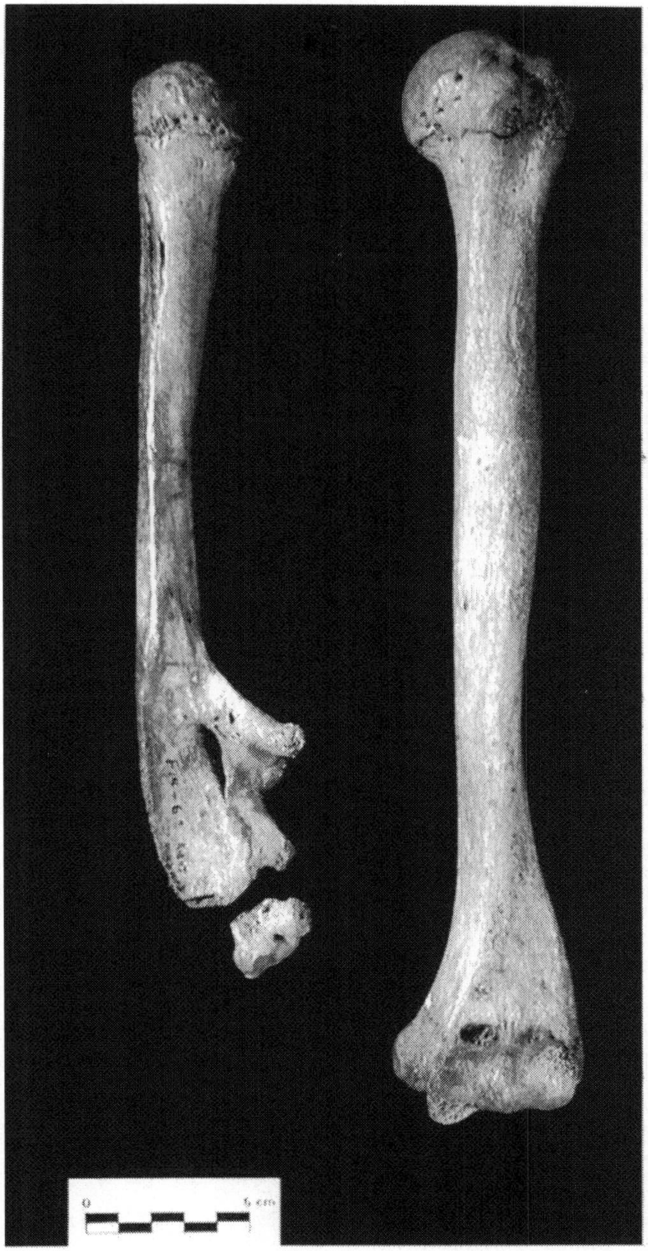

Figure 5 Denninger's hemimelia case from Fulton County, Illinois (1931).

(Williams, 1932:980). Williams reexamined an enormous quantity of skeletal material ranging from that reported by Jones (see Fig. 2) and Whitney to that published early in the century by the Peruvian physician and archaeologist Julio C. Tello (1880–1947). Tello wrote a controversial monograph, his dissertation, on cranial and ceramic evidence for syphilis, to which Williams lent his stature (Tello, 1909; Tello and Williams, 1930; Stewart, 1943c). Williams was active in the Buffalo Museum of Science and contributed collections and articles on fossil fishes and local archaeology (Goodyear, 1994). An accomplished amateur musician, he wrote the University's surprisingly Darwinian fight song, *The Bison Is King*!

Several archaeologists were early contributors to paleopathology. Perhaps the most prominent was George Grant MacCurdy (1863–1947), who was a curator and professor at Yale University. He is best remembered for his extensive excavation and collecting in Peru with Bingham at Machu Picchu, as well as elsewhere in the Americas. He was one of the founding members of the American Association of Physical Anthropologists. His publications concern trepanation and other surgical procedures as well as a wide range of pathologies in Peruvian skeletal material (1905, 1918, 1923). The cranial osteosarcoma from Paucarcancha, Peru, that has served in several publications as an icon of ancient disease is from MacCurdy's work (1923: see Fig. 6).

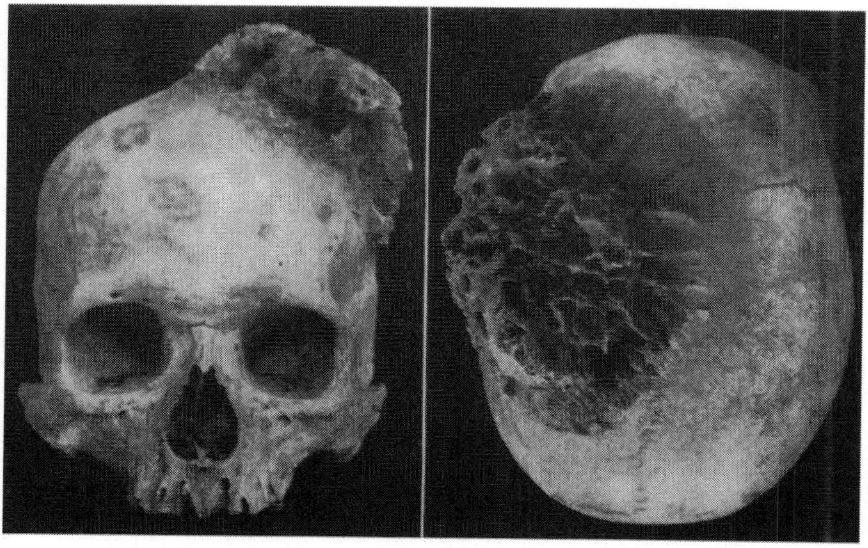

Figure 6 Cranial osteosarcoma from Paucarcancha, Peru (MacCurdy, 1923).

Dental paleopathology continued its somewhat separate history. Jarcho points out that Moodie's later career led him to dental paleopathology and to an interest in testing the then-current focal infection theory linking dental disease with arthritis (Jarcho, 1966b; Moodie, 1928). One of Moodie's papers on this subject appeared in a health magazine for the general public (1930). However, the true pioneer in this field was Rufus Wood Leigh (1884–1964). Leigh was a D.D.S. who also earned an M.A. in anthropology. Long associated with the Army Medical Museum, he taught at Georgetown University and later at the University of Utah. His systematic studies of caries and dental wear in prehistoric Native Americans and in ancient Egyptians explored the effects of way of life on oral health (1925a,b, 1928, 1929, 1930, 1934, 1937). The first of these papers compared Indian Knoll, Sioux, Arikara, and Havikuh Zuni collections at the Smithsonian. Leigh believed the Indian Knoll collection to be maize farmers — Ritchie's concept of Archaic was still in the future — but found this group to have the most severe attrition and Sioux the least (1925a). He notes in this study and his later study of Peruvian crania (1937) that caries were common in these ancient peoples, but that the age of onset was much later than in modern Americans, a quite modern epidemiological insight. Leigh's parallel publications in anthropological and dental journals reflect a commitment to the value of each discipline for the other, and they are still widely cited. The dentist Samuel Rabkin's studies of several sites in Northern Alabama (1942) and Indian Knoll (1943) are similar in scope. In the later 20th century Albert A. Dahlberg (1908–1993), professor of dentistry and anthropology at the University of Chicago, continued this emphasis on the common ground between his disciplines (1960; Mann and Murphy, 1990).

Physical anthropology developed as a profession under the strong influence of the Czech-American physician Aleš Hrdlička (1869–1943), who was curator of the Division of Physical Anthropology of the United States National Museum, Smithsonian Institution, founder of the *American Journal of Physical Anthropology*, and energetic collector of skeletons. As Ubelaker points out, paleopathology was secondary to his major focus on racial typology and the antiquity of human occupation in the Americas. Much of his paleopathology concerns cultural practices that alter the morphology of the skeleton, an interest that, along with his interest in anomalies, can be seen as "a direct outgrowth of his career interest in documenting human variation" (1982:340). In 1913, Hrdlička traveled to Peru to collect specimens of skeletal pathology and surgical treatment from a broad range of Native American archaeological sites for the purpose of developing an exhibit on physical anthropology for the Panama–California Exposition to be held in San Diego in 1915. This collection, comprising more than 1000 pathological specimens, is now curated at the San Diego Museum of Man. A photographic catalogue with descriptions by Charles F. Merbs places the collection in modern scientific context (Tyson and Alcauskas, 1980). His equally vigorous efforts for the U.S. National Museum to collect large, documented skeletal series continue to provide samples for generations of researchers, albeit shaped by Hrdlička's theories of the peopling of the New World (Hunt, 2002; Keenleyside, 1998, 2003; Keenleyside and Mann, 1991).

Hrdlička published numerous studies of prehistoric Native American skeletal pathology, many of them dealing with cultural modifications of the teeth and skull (Ubelaker, 1982; Elerick and Tyson, 1997). His studies on trepanation in Mexico (1897) and Peru (1914b, 1939b), on Arctic tooth ablation (1940b), and on ethnomedicine (Hrdlička, 1932; Lumholtz and Hrdlička, 1897) are among his many contributions to paleopathology that remain important to modern anthropologists. He contributed the term *symmetrical hyperostosis* to the lively international discussion of skeletal signs of anemia (1914b) and documented the extensive variability in the frequency of auditory exostosis among Native Americans (1935b). His early papers on congenital anomalies in ancient skeletons (1899a, 1933) are thoroughly articulated with the medical literature of his day. The earliest of these, a description of a Mexican skeleton with cervical ribs, a bicipital rib, low mandibular condyles, and numerous features he calls "anthropoid" (1899a:103), is interesting in its weighing of individual versus "ethnic" features. Hrdlička's ideas about the antiquity of humans in the New World had not yet crystallized; he cites Morton's work as science rather than history, and he is open to the notion of important ethnic differentiation. Among his last publications is a detailed description of a tiny skull from Peru, which he attributes to a "midget . . . without . . . any detectable pathological condition" (1943b:81), citing just his own work on normal variation. Ortner (2003) has revised this diagnosis to congenital idiocy, finding evidence for hypoplasia of the frontal lobes. It is remarkable that a physician who began his career working in an insane asylum could construe this skull as normal.

Hrdlička's larger concerns color his paleopathology in other ways. On the one hand, Hrdlička's view of infectious disease is closely tied to his view of the profound isolation of New World populations from the Old World and its diseases. An early interest in, or open-mindedness toward, the diagnosis of syphilis (1908b) and tuberculosis (1911) gave way to a radical vision of a New World Eden. Late in life he argued for the absence of infectious pathogens, except for infant diarrhea and pneumonia, in the pre-Columbian Americas (1932). On the other hand, his view of Indian health is grounded in his early experience in health surveys of reservation populations, socially conscious studies that linked infectious diseases to human misery (1908c, 1909a).

In 1930, the classicist convert to physical anthropology Earnest Albert Hooton (1887–1954) of Harvard University produced the first comprehensive study of a specific pre-contact Native American population, *The Indians of Pecos Pueblo* (see Chapter 4). Hooton took a surprisingly modern approach in this analysis: he not only collected the usual craniometric and morphological data but also systematically recorded pathological lesions and carefully evaluated them within the Pecos cultural, behavioral, and temporal contexts. Thanks to the careful archaeological excavation and analysis of this large site, which had been occupied for several centuries, Hooton was able to divide the large series of burials into temporally distinct subsamples. This degree of chronological control, which permitted detailed diachronic comparisons of specific skeletal features, represented a significant advance over earlier analyses that often barely distinguished pre-Columbian from post-Columbian contexts. Hooton satisfied established anthropological expectations by publishing detailed metrical and morphological analyses of the Pecos Pueblo crania (with some postcranial data included), but he also investigated associations between patterns of diet, food preparation, and habitual activities (farming, hunting, warfare, etc.) and patterns of observed skeletal pathology. He tabulated frequency data on specific conditions, including osteoarthritis, trauma, inflammatory lesions, and porotic hyperostosis (which he called "osteoporosis symmetrica"), suggested that syphilis had been present at the site, and evaluated health for the different time periods. Hooton recruited six physicians to collaborate in the analysis of the Pecos Pueblo remains, the most prominent among them being Herbert U. Williams. Ubelaker (1982) calls Hooton's approach *epidemiological*. Hooton's student J. Lawrence Angel (1981b:510) credits Hooton with ending the early 20th-century eclipse of paleopathology through "his insistent stress on the population as a unit of study," and Ackerknecht (1953) viewed Hooton's concept of *integration* as the essential step in making paleopathology meaningful within anthropology.

Despite the integration of relevant biological and cultural data apparent in Hooton's landmark analysis, more than three decades would pass before the widespread adoption of his approach. A large number of prehistoric Native American skeletal series were excavated by various New Deal archaeological projects funded by federal and state relief agencies during the later 1930s and early 1940s. (See Chapter 6.) Some of these series, such as those from Indian Knoll and other sites in Kentucky and Moundville in central Alabama, both analyzed by Hooton's student, Charles E. Snow (1941a,b, 1942, 1943, 1945a,b, 1948, 1951; see Fig. 7), included several hundred well-preserved individuals and would have been ideal material for diachronic and comparative studies that built upon Hooton's foundation. However, the physical anthropologists, including Snow, Marshall Newman (Snow and Newman, 1942; Newman, 1951), Ivar Skarland (1939), and Fred S. Hulse (1941), who analyzed these and other series for the New Deal archaeological reports, were primarily concerned with traditional typological analysis for the purpose of delineating biological relationships between Native American populations throughout North America, particularly in the eastern United States. The reporting of skeletal and dental pathology is typically nonsystematic, even anecdotal, although the diagnosis of specific infectious diseases (particularly syphilis) was often attempted and associations

(a) (b) (c) (d) (e)

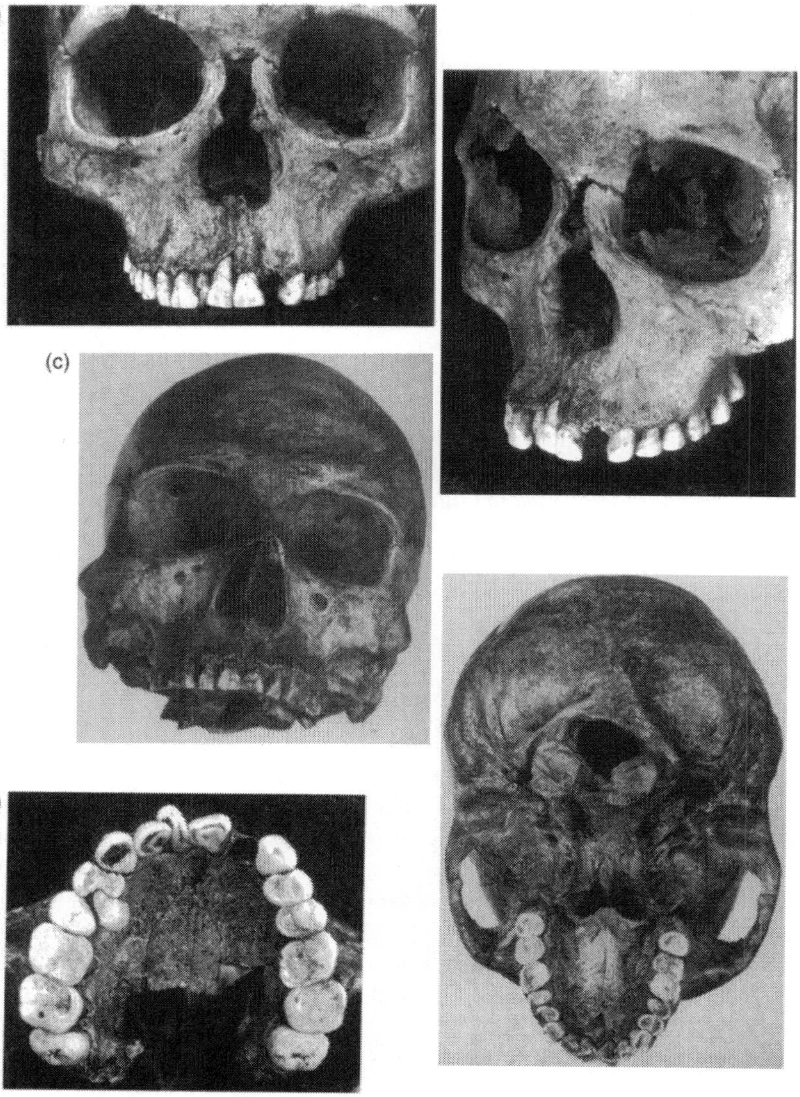

Figure 7 Photographs of skeletal pathology from the Indian Knoll site, taken by Charles Snow. (a and b) Adult female (Burial 9), Indian Knoll, remodeled nasal aperture. Reproduced from Snow (1948) with permission of the William S. Webb Museum of Anthropology (WSWMA), University of Kentucky, Lexington. (c and d) Adult male (Burial 105), Indian Knoll, atlanto-occipital fusion and facial asymmetry. (e) Adult female (Burial 349), Indian Knoll, with severe anterior malocclusion. Previously unpublished photographs curated at WSWMA, used by permission.

among diet, food processing technology (e.g., the grinding stones characteristic of the Archaic period sites), and dental pathology were noted (see Rabkin, 1942, 1943). Nevertheless, Charles E. Snow (1910–1967) did produce fine case studies of scalping (1941b) and achondroplasia (1943) at Moundville that continue to be cited.

The focus on typological anthropometry was continued in the published analyses of skeletal series excavated under the auspices of numerous River Basin Surveys and other federally funded archaeological projects over the next two decades. An important exception to this generalization is the work of Alice Brues, who was Hooton's student. While her descriptions of skeletal remains include typological analysis of crania, she devoted equal attention to evidence of disease, and her reports are excellent examples of careful diagnosis (Brues, 1946b,c, 1957, 1958b, 1959a).

IV. LATER 20TH CENTURY

Two physical anthropologists in the Department of Anthropology at the National Museum of Natural History, the successor to the United States National Museum in the Smithsonian Institution, were strongly influenced by Hrdlička's interests in skeletal pathology. T. Dale Stewart (1901–1997) was trained as a physician, although he had begun work at the Smithsonian before beginning his medical studies at Johns Hopkins. During his long and productive career he published on the effects of diet on skeletal and dental structures (e.g., Stewart, 1931a, 1963b), ancient surgery (1937b, 1943c, 1958b), skeletal pathology (1931b, 1935, 1943c, 1956a,b, 1966, 1974; Mann *et al.*, 1990), and Native American populations of the New World (1960a, 1963a,b, 1970b, 1973b, 1979b). His studies on cranial and dental deformation constitute a large proportion of the important work on these topics (1939a, 1941a,b, 1942, 1943c, 1963a; Stewart and Groome, 1968). Stewart's interests in population differences in arthritis patterning (1947), spondylolysis (1931b, 1935, 1956a), and neural tube defects (1975) have been particularly influential on recent research in paleopathology. Many of Stewart's papers correct errors in the literature of paleopathology (1937b, 1966, 1969, 1975, 1976).

J. Lawrence Angel (1915–1986) studied with E. A. Hooton at Harvard, earning his Ph.D. in anthropology in 1942. His career included fieldwork in WPA projects and teaching at Jefferson Medical College, in addition to his position at the NMNH. He focused his attentions on patterns of health and disease in the ancient eastern Mediterranean and their causal relationships with temporal change in diet, activity patterns, population affinities and migrations, and demography (Angel, 1966b, 1971, 1972, 1973b,c, 1974a, 1978, 1979, 1982, 1984; Angel and Bisel, 1985). While his earliest papers are largely typological in focus, they include brief discussions of pathology, touching on subjects as diverse as trepanation (1943), ankylosing spondylitis (1946b), and sex differences in frequencies of skull wounds (1943). Many of his papers contain stimulating insights not amenable to testing in the small and disparate samples available for study. An example is *The Cultural Ecology of General versus Dental Health* (1974a), a paper exploring the association of longevity, subsistence, and caries and abscesses. These ideas recur through his work; they harken back to Moodie's interests and have stimulated more recent scholarship with more powerful methods (Meiklejohn *et al.*, 1992). Angel's central focus in paleopathology was surely his effort at teasing apart malaria and thalassemia as contributors to anemia in Mediterranean remains, as well as distinguishing them from health consequences of agriculture and urban life (1966b, 1969, 1971, 1972, 1973b,c, 1978), a problem that remains unsolved. These questions led him to attempt to infer fecundity from changes in the pubic symphysis (1969), a project that, albeit mistaken, has stimulated much useful research.

During the last decade of his life, Angel became interested in the analysis of biological and cultural determinants of health and mortality among free and enslaved African-Americans in the Middle Atlantic region of the United States (Angel, 1981a; Angel *et al.*, 1987). Angel shared with his mentor Hooton a focus on the effects of technology and diet on skeletal morphology (Angel, 1943, 1946b, 1973b, 1982; see Ackerknecht, 1953). An innovative feature of Angel's work was his lifelong interest in activity-related modifications of the skeleton. He coined the term "atl-atl elbow" to describe patterned arthritis in California foragers (Angel, 1966a) and contributed extensively to the literature on activity markers (Kennedy, 1989). The breadth of his interests is strongly emphasized in the chapters of *A Life in Science: Papers in Honor of J. Lawrence Angel* (Buikstra, 1990).

The internist, medical historian, and editor Saul Jarcho (1906–2000) was a close associate of T. Dale Stewart's. In addition to the review publications discussed earlier, he produced a series of elegant case studies (Jarcho *et al.*, 1963, 1965; Jarcho, 1964, 1965a) that serve as models for curing the malaise he identified in midcentury paleopathology. Among them, his use of bone chemistry to demonstrate lead poisoning in Pueblo potters (1964) is a landmark in joining laboratory sciences, archaeology, and physical anthropology. Another influential medical historian was Erwin H. Ackerknecht (1906–1988). His 1953 review argued that "the pathology of a society reflects its general conditions and growth and offers, therefore, valuable clues to an understanding of the total society" (1953:120), an inclusive view consistent with his desire to integrate the study of medicine with social anthropology. Like Jarcho, Ackerknecht was an M.D., but, fleeing Germany for political reasons in 1933, he studied ethnology under the French anthropologist Paul Rivet and later jointed the Boasian community at the American Museum of Natural History. He was professor of history of medicine at Wisconsin and Zurich. His sojourn in the United States (1941–1957) generated important studies of ethnomedicine that should be read by anyone interested in ancient surgery (Ackerknecht, 1971).

Several archaeologists made important contributions to paleopathology in the mid-20th century. William A. Ritchie (1903–1995) was curator at the New York State Museum and is best known for defining the Archaic period in eastern North America. His excavation reports include observations of pathologies, and he authored excellent case studies of tuberculosis and multiple myeloma (Ritchie and Warren, 1932; Ritchie, 1952). Kenneth E. Kidd (1906–1994), who was curator of ethnology at the Royal Ontario Museum and professor at Trent University, contributed case studies of ankylosing spondylitis and torticollis (1954).

In Canada, James E. Anderson (b. 1926) performed much the same role as Stewart and Angel in the United States. Anderson was a student of the anatomist J. C. Boileau Grant (1886–1973) and taught anatomy and anthropology at University of Toronto, SUNY Buffalo, and McMaster University (Jerkic, 2001). He was particularly interested in development, variability, and the borderlands of pathology (Anderson, 1960). His contributions to paleopathology include diagnosing treponematosis in some of the earliest remains from the Western Hemisphere (1965) and extraction of meaningful information about ancient disease from ossuary materials (1964). In Chile, Juan Munizaga (1934–1996), who trained with T. D. Stewart, was an institution builder and important contributor to paleopathology (Aspillaga, 1996; Munizaga, 1965, 1991; Munizaga *et al.*, 1978a,b). In Mexico, Juan Comas (1900–1979) was similarly a pioneer in paleopathology. Comas was a Mallorcan who had studied with the Swiss physical anthropologist Eugène Pittard (Spencer, 1997c). His Americanist interests included population-appropriate age, sex, and stature standards, cranial deformation, and cranial anomalies (1942a,b, 1943, 1965, 1976). Both Comas and Munizaga were interested in the history of anthropology (Munizaga, 1993; Comas, 1968). Luis Fernando Ferreira similarly pioneered a remarkable program in paleoparasitology at Fundação Oswaldo Cruz in Brazil (Ferreira *et al.*, 1980, 1984; Araújo *et al.*, 1980, 2003).

In the 1960s and early 1970s, paleopathological analyses, which harkened back to Hooton's population approach, increased dramatically. These studies explicitly integrated archaeological, ethnographic, and historical data with detailed observations of skeletal pathology to explain differences in demography and skeletal evidence of changes in nutrition, trauma, infectious disease, iron deficiency anemia, and dental health in temporally sequential population samples in ancient Sudanese Nubia by George J. Armelagos (1968) and, in the New World, prehistoric Virginia by Lucille E. Hoyme and William M. Bass (1962), Illinois by Armelagos' students John W. Lallo (1973) and Jerome C. Rose (1973), and Kentucky by Claire M. Cassidy (1972). These studies shared (not by accident) several important theoretical and methodological features: they emphasized the necessity of the population approach, focusing on population samples rather than unrelated "interesting" specimens in order to obtain a valid demographic "cross section" of the populations in question; they examined a broad range of skeletal and dental indicators of health; they employed methods of analysis borrowed from other disciplines, such as radiography (a long tradition in paleopathology); and they all tested the proposition long held by historians and anthropologists that the change from nomadic hunting–gathering to horticulture- or agriculture-based sedentism invariably improved "the quality of life" of human populations as reflected in patterns of growth, development, nutrition, workload, infectious disease experience, and longevity. Finally, they all concluded that increasing reliance upon cereal-based agriculture (maize in the New World, millet in ancient Nubia) was at best a mixed blessing. The earliest of them (Hoyme and Bass, 1962) was conducted as part of a federally funded dam construction project in Virginia, but the other four were doctoral dissertation projects in the Departments of Anthropology at (respectively) the University of Colorado, the University of Massachusetts, and the University of Wisconsin.

These multifaceted investigations interpreted biological data within appropriate cultural contexts for the explicit purpose of shedding light on a theoretical issue of growing importance within American archaeology: the myth of the "unmixed blessings" of the Neolithic revolution (i.e., plant and animal domestication and the rise of sedentary village life). They heralded an era of interaction between paleopathology and mainstream American archaeology, which was just entering the first stage of its own revolution: the "New Archaeology", explicitly scientific and eager to question all previously received wisdom about the human past. In hindsight, that interaction has been more unidirectional than one would like. These studies and the large literature that followed them are relatively little cited among archaeologists, a problem that is not helped by our colleagues' persistence in characterizing Late Woodland farmers in the Midwest as hunter–gatherers (Goodman and Armelagos, 1985).

However, the human cost of agriculture was not the only question that motivated the late 20th-century renaissance in paleopathology. A strong focus on cultural context animated new work on activity-related pathology in the Arctic by Charles F. Merbs (1969, 1983) and the Southwest (Jurmain, 1975, 1977a). By the 1970s, programmatic research integrating paleopathology as a means of understanding ancient ways of life became the rule rather than the exception for several anthropologists whose regional interests have led them to a broad focus on disease conditions. The work of Marvin Allison and colleagues in Chile (Allison, 1973, 1974a,b,c, 1979; Elzay *et al.*, 1977; Munizaga *et al.*, 1978a,b) and Frank and Julie Saul in Mesoamerica (Saul, 1972, 1976; Saul and Saul, 1991) are excellent examples.

V. AMERICAN PALEOPATHOLOGY AT THE END OF THE 20TH CENTURY

The last three decades have seen a welcome proliferation of interdisciplinary projects involving archaeologists, biological anthropologists (including paleopathologists), historians, ethnographers, and other scholars aimed at broad-scale reconstructions of specific lifeways in the past. As Jane E. Buikstra predicted in the title of her chapter in the 1991 volume, *What Mean These Bones? Studies in Southeastern Archaeology* (Powell, Bridges, and Mires, editors), paleopathology had finally come "Out of the Appendix and Into the Dirt," i.e., become an active partner in the formulation of research objectives in such projects from the initial planning stages onward.

The longest running of these collaborations is surely William M. Bass' career-long focus on understanding the culture history and adaptations of Plains Indians. Bill Bass was Snow's student at the University of Kentucky, completing a craniological master's thesis on the Moundville series in 1956. His subsequent career at the University of Kansas and University of Tennessee is grounded in his participation in the River Basin Survey excavations on the Plains in cooperation with many archaeologists, historians, and other specialists, notably Waldo Wedel and Donald J. Lehmer. Many of his contributions to paleopathology reflect a long and fruitful collaboration with the physician and professor of otolaryngology, John B. Gregg, who had begun to apply his specialty to paleopathology in the 1960s (Holzhueter *et al.*, 1965; Gregg *et al.*, 1965, 1982; Bass *et al.*, 1974). Among Bass' numerous students, Douglas Ubelaker, Douglas Owsley, David Hunt, Ted Rathbun, and Richard Jantz have been important contributors to paleopathology, and they continue to recruit colleagues from other disciplines (e.g., Logan *et al.*, 2003). A summary account of his influence can be gleaned from a volume dedicated to Bass, *Skeletal Biology in the Great Plains* (Owsley and Jantz, 1994).

Another outstanding example of such interdisciplinary cooperation is the *La Florida* bioarchaeological project, coordinated by biological anthropologist Clark S. Larsen and archaeologist David Hurst Thomas over the past two decades. Larsen's investigation of diachronic changes in health, disease, mortality, and osteological markers of habitual activity patterns in Native American populations of the northern portion of *La Florida* (the Georgia Coast) began with his dissertation research in the late 1970s at the University of Michigan (ironically, in a graduate program that offered no systematic training in paleopathology). After completion of the dissertation in 1980, he expanded his research to include several Spanish colonial period sites on coastal islands or the adjacent mainland in the region (Larsen, 1982, 1990). So far, two generations of students in bioarchaeology, history, ethnology, and archaeology from four institutions (including Northern Illinois University, University of

North Carolina, Chapel Hill, and Ohio State) have been actively involved under Larsen's tutelage with *La Florida* investigations covering some 3000 years of Native American life in the extreme southeastern United States. Larsen's latest book, titled *Bioarchaeology of Spanish Florida: The Impact of Colonialism* (2001), summarizes this landmark project. Larsen and several of his long-time colleagues on the Georgia Bight project have also applied their multidisciplinary approach in a collaborative salvage project with archaeologist Robert L. Kelly (Larsen and Kelly, 1995) to examine dimensions of health, disease, and skeletal morphology among hunter-gatherers of the Great Basin in the western United States.

VI. TEXTS AND PROFESSIONALIZATION

The last quarter of the 20th century was inaugurated nationally in 1976 by celebrations of America's bicentennial and in paleopathological circles by the publication of R. Ted Steinbock's *Paleopathological Diagnosis and Interpretation* (1976), based on his dissertation in medicine and anthropology at Harvard and written specifically to guide paleopathologists in performing differential diagnosis of specific diseases. It opens with a discussion of bone as a biological organ system, emphasizing that pathological skeletal morphology can only be recognized and interpreted by reference to normal patterns of growth and development. Subsequent chapters focus on specific categories of disease affecting bone: trauma, hematological disorders, metabolic bone disease, arthritis, and tumors, and selected specific (treponematosis, tuberculosis, leprosy) and nonspecific (pyogenic osteomyelitis) infections. Steinbock emphasized the importance of detailed comparisons between firmly contextualized archaeological specimens and medical examples from both modern and preantibiotic-era medical collections.

This landmark text was soon joined by Donald J. Ortner and Walter G. J. Putschar's *Identification of Pathological Conditions in Human Skeletal Remains* (1981; see also Ortner, 2003), which greatly expanded the range of conditions discussed by Steinbock and featured almost 800 photographic illustrations of pathological specimens from medical and museum collections around the world. Some publications adopted the atlas format: the *Atlas of Human Paleopathology*, by Michael R. Zimmerman and Marc A. Kelley (1982), and the *Regional Atlas of Bone Disease*, by Robert W. Mann and Sean P. Murphy (1990). Others were explicitly comprehensive in nature, such as *The Cambridge Encyclopedia of Human Paleopathology*, by Arthur C. Aufderheide and Conrado Rodríguez-Mártin (1998), an excellent companion volume to *The Cambridge World History of Human Disease* (Kiple, 1993) published 5 years earlier. Some texts from this period were devoted to the analysis of naturally or artificially desiccated bodies, such as Aidan and Eve Cockburn's (1980) edited volume, *Mummies, Disease, and Ancient Cultures*, which was expanded in a second edition (Cockburn *et al.*, 1998). New noninvasive (CAT scanning and MRI) and minimally invasive (endoscopy) methods developed for use in clinical medicine now not only reveal more detailed information than the older styles of autopsies, popular since the late 19th century, but are far more acceptable to museum curators concerned about the destruction of valuable specimens.

Other recent advances in American paleopathology include molecular investigations of ancient microbial DNA and metabolites of earlier forms of infectious diseases (e.g., tuberculosis in the pre-contact Americas), significant critical reevaluations of long-accepted methods of interpretation of skeletal evidence of morbidity, the incorporation of a broader range of relevant disease models considered in epidemiological perspective (e.g., nonvenereal forms of treponemal disease as well as venereal syphilis), the growth of scholarly societies aimed at promoting the professionalization of the discipline and strengthening international collaborations, isotopic evaluations of dietary regimens that contribute to multifocal studies of noninfectious diseases (e.g., acquired iron-deficiency anemia), and a new pedagogical emphasis upon intensive study of *normal* biological processes which shape bone as well as the *abnormal* processes that deform it.

VII. DIFFERENTIAL DIAGNOSIS AND INTERPRETATION IN EPIDEMIOLOGICAL PERSPECTIVE

In his three-part review of paleopathology, H. U. Williams devoted 92 pages to syphilis and just 62 pages to everything else (Williams, 1929, 1932, 1936). While we have done our part to maintain this imbalance (Cook, 1980b; Powell, 1992, 2000; Powell and Cook, 2005; Jacobi *et al.*, 1992), our colleagues have rectified it to a large extent in the last two decades. A number of interesting infectious diseases are now visible to paleopathologists, among them smallpox (Jackes, 1983), coccidioidomycosis (Harrison *et al.*, 1991), Chagas' disease (Guhl *et al.*, 1999; Reinhard *et al.*, 2003; Aufderheide *et al.*, 2004), and perhaps Lyme disease (Lewis, 1998). The interior zoo of ancient intestinal and parasites has been enlarged greatly (Araújo *et al.*, 2003), and there is renewed interest in external parasites (Reinhard and Buikstra, 2003). Our field has tended to view one of the most ubiquitous of infectious diseases — streptococccal infection in the form of dental caries — as unimportant because it is a trivial condition in our age of readily available antibiotics. Osteomyelitis secondary to caries has been discussed by Ortner (2003:197), and dental caries has been evaluated carefully as a heath hazard in several recent studies (Sciulli and Schneider, 1986; Sledzik and Moore-Jansen, 1991; Saunders *et al.*, 1997). Philip Walker has ventured into pathology among the living to explore social status and caries (Walker and Hewlett, 1990), and status differences in caries rates have been demonstrated in some ancient populations (Cucina and Tiesler, 2003).

The variety and sophistication of diagnoses of arthritis, orthopedic, and degenerative diseases has increased as well. Instructive case studies of inflammatory arthritis (Ortner and Utermohle, 1981), juvenile rheumatoid arthritis (Buikstra and Poznanski *et al.*, 1990; Lewis, 1998), DISH (Crubézy and Trinkhaus, 1992), and Scheuermann's disease (Merbs, 1983; Cook *et al.*, 1983) have appeared in recent years. Claims about the evolutionary history of rheumatoid arthritis in the Americas (Rothschild *et al.*, 1988) remain to be substantiated with careful presentation of evidence. Nevertheless, it is still true that these conditions remain understudied in American paleopathology (Bridges, 1992).

Age-related bone loss is a topic for which paleopathology offers perplexing, and perhaps useful, comparative perspectives to medical science. Several studies have noted the scarcity of osteoporosis-related or fragility fractures in most ancient populations. It has been suggested that few people survived to advanced ages (Lovejoy and Heipel, 1981) or that activity and diet promoted better bone maintenance in the past (Agarwal and Grynpas, 1996). The claim that osteoporotic fracture is absent in the past (Agarwal and Grynpas, 1996) reflects quite stringent criteria and an incomplete reading of the literature in paleopathology in that vertebral compression fractures are seen among older people in some groups (Merbs, 1983:116; Cybulski, 1992). A recent symposium volume explores these complexities in detail (Agarwal and Stout, 2003). Other degenerative diseases have received relatively little attention, but conditions as diverse as Morgagni syndrome (Armelagos and Chrisman, 1988) and atherosclerosis (Cybulski, 1992; Zimmerman and Trinkaus *et al.*, 1981) are now represented in paleopathology.

In contrast, there has been a remarkable growth in interest in congenital defects and genetic syndromes in the last two decades. There is now a comprehensive atlas of anomalies of the axial skeleton illustrated largely with material from ancient North America (Barnes, 1994a); a companion volume on the appendicular skeleton would be highly desirable. Some landmark new discoveries include severe neural tube defects (Dickel and Doran, 1989), congenital scoliosis (Cybulski, 1992), probable Apert syndrome with hydrocephalus (Pedersen and Anton, 1998), Rubenstein–Taybi syndrome (Wilbur, 2000), and a variety of anomalies of the extremities (Cybulski, 1988; Barnes, 1994b; Keeleyside and Mann, 1991; Murphy, 1999). More extensive studies of aural atresia (Hodges *et al.*, 1990) and Klippel–Feil syndrome (Merbs and Euler, 1985; Danforth *et al.*, 1994) revisit topics explored by Hrdlička and Jarcho and raise interesting issues about relatively high frequency of these conditions among certain Native

American isolates. Genetic syndromes expressed in the teeth have a smaller literature than their diversity and ease of recognition in skeletal material might suggest (Cook, 1980a; Mann *et al.*, 1990), but the discovery of ancient anomalies with little or no visibility in the clinical literature holds out promise of interaction between fields (Skinner and Hung, 1989; Lukacs, 1991). It is a measure of the potential of the paleopathology of congenital defects for useful application in medical research that our colleagues in clinical medicine seek out our ancient case studies (e.g., Berg, 2003).

VIII. TUBERCULOSIS: A FERTILE FIELD FOR MOLECULAR PALEOPATHOLOGY

Late 20th-century paleopathology overturned the received wisdom of the midcentury regarding the absence of tuberculosis in the New World, and it is perhaps difficult to recall how controversial the discoveries made by Ritchie (1952), Allison *et al.* (1973), and others (Buikstra, 1981c) once were. Recent advances in molecular analysis of ancient amplified DNA and other organic materials have provided independent verification of previous pathological diagnoses of tuberculosis in human remains from New and Old World archaeological sites. Refined methods of DNA "fingerprinting" had been used successfully to identify index cases in modern localized outbreaks of antibiotic-resistant tuberculosis, by comparing different strains of related pathogens, and protocols for the recovery of DNA from ancient human tissues had been standardized. Comparisons of ancient mycobacterial DNA recovered from skeletal individuals from archaeological sites in Peru (Salo *et al.*, 1994), Germany (Baron *et al.*, 1996), England (Roberts and Dixon, 1993; Taylor *et al.*, 1996), Hungary (Haas *et al.*, 2000), Scotland, Turkey (Spigelman and Lemma, 1993), Illinois, Canada (Braun *et al.*, 1998), and Egypt (Zink and Nerlich, 2003) with DNA "profiles" of members of the modern *Mycobacterium tuberculosis* complex (principally *M. tuberculosis* and *M. bovis*, the two species that infect mammalian hosts most commonly) suggest that tuberculosis has an exceedingly ancient "pedigree" as a human disease.

The majority of these molecular studies have focused on a 123-bp segment of DNA unique to the *M. tuberculosis* complex, known as *IS6110* (Salo *et al.*, 1994). At the 1997 International Congress on The Evolution and Palaeoepidemiology of Tuberculosis in Szeged, Hungary (Pálfi *et al.*, 1999), discussions of mycobacterial DNA and RNA were featured in 17 of the 86 papers presented, a rate of almost 20%. Nine papers presented molecular aspects of modern mycobacteria in the United States, Guadeloupe, France, Hungary, Slovakia, and Austria. Eight other presentations summarized analyses of ancient mycobacterial DNA in human remains from Egypt, the Middle East, Borneo, Moravia, Lithuania, the midwestern and southwestern United States, Ontario, Peru, and Chile.

Another new molecular technique is based on the detection of mycolic acids produced by pathogenic mycobacteria inside infected human hosts. This has been applied successfully to bone samples from identified tuberculous patients in a historic hospital cemetery (Newcastle Infirmary) as well as to samples from pathological and nonpathological skeletons from two archaeological sites in England (Gernaey *et al.*, 1999). These fatty acids in the cell walls of the mycobacteria can be readily detected by liquid chromatography (Ramos, 1994), and this method has been employed for some time in clinical settings for the verification of *M. tuberculosis* complex infection in living patients. Because mycolic acids may be found in tissues distant from the site of frank tuberculous lesions in ill individuals and are also detectable in individuals who have been exposed to the pathogenic mycobacteria but had not developed symptoms of clinical disease, the exciting application of this clinical technique to archaeological specimens opens the way for investigations of true prevalence of prehistoric tuberculosis, i.e., the proportion of infected to uninfected individuals in a skeletal sample rather than the mere identification of those few individuals who developed recognizable bone lesions before death. Because modern tuberculosis affects bone in fewer than 10% of clinically ill patients (Schlossberg, 1994), lesions identifiable by paleopathological criteria represent merely "the tip of the tip of the iceberg" of tuberculous infection in past populations. A critical review of the current state of paleopathological research on tuberculosis is presented in Roberts and Buikstra's (2003) comprehensive volume, *Tuberculosis: Old Disease, New Awakening*.

This new technology has spurred a revival of comparative paleopathology, a relatively dormant field since Moodie's time. *M. tuberculosis* complex DNA has been recovered from the skeleton of a historic Neutral Indian dog from Ontario that showed hypertrophic osteoarthropathy, a characteristic lesion of end-stage tuberculosis in that species (Bathurst and Barta, 2004). Positive spoligotyping results from a Pleistocene bison with less characteristic lesions have also been reported (Rothschild *et al.*, 2001). This intriguing finding awaits confirmation with other samples of comparable antiquity.

IX. SYPHILIS: A BROADER VIEW OF DIAGNOSTIC ALTERNATIVES

While the emphasis placed by Wood *et al.* (1992) on interindividual differences in host immune response and developmental integrity in determining the form and degree of biological response to pathological insults is both apposite and timely, another quite different "osteological paradox" also deserves careful consideration. Differences in pathogenicity and synergistic interactions among species and strains of pathogenic organisms also play key roles in determining patterns of skeletal response to infection seen in skeletal series (Powell, 2000). American paleopathology at the end of the 20th century incorporates an increasingly sophisticated range of diagnostic models both for identification of specific ancient infectious diseases and for evaluation of the burden of morbidity and mortality upon their human hosts. Not all infectious diseases, even those variant forms caused by closely related pathogens, exact the same toll on human populations either individually or in the aggregate effect. A disease such as tuberculosis, identified in numerous pre-Columbian Native American skeletal populations and of particular interest to many American paleopathologists, may be nearly invisible in dry bone examinations of skeletal series, simply because most infected individuals follow one of three common trajectories: (a) they never develop clinical disease, (b) they recover with latent infection maintained for decades, or (c) they die before bone involvement develops. Only a few follow a fourth pathway: (d) the development of diagnostic bone lesions. As a result, mortality from this disease potentially far outstrips skeletal morbidity, but just the reverse is true for treponematosis, another major infectious disease frequently identified at pre-Columbian sites. While venereal syphilis had a high potential for severely impacting mortality at all ages (including prenatal life), two of its three modern "cousins" (yaws and endemic syphilis, but not pinta) can produce high population levels of skeletal morbidity but only very rarely cause death. These two diseases are in effect "mirror images" of one another in their capacities for morbid and mortal impact and, when present simultaneously in an individual or a population, may each exacerbate the effects of the other.

Arguments pro and contra the pre-Columbian New World presence of another major infectious disease, venereal syphilis, have been debated for some five centuries, beginning soon after the dramatic "outbreak" of this apparently new disease in 1493 in southern Europe following the so-called siege of Naples. At the first International Congress on the Evolution and Paleoepidemiology of Infectious Diseases (ICEPID), held in Toulon, France, in 1993 (Dutour *et al.*, 1994), this question was reexamined by a broad range of scholars employing the most current evidence from historical, medical, archaeological, and paleopathological research conducted on a large number of New and Old World population samples. No clear consensus was reached on the origin(s) of venereal treponemal disease, but the pre-Columbian presence of some form(s) of treponemal disease is (at the present time) more clearly apparent in the American bioarchaeological record than in its European counterpart. Molecular investigations of treponemal disease have lagged behind similar studies of ancient tuberculosis, but the first successful identification of prehistoric treponemal antibody in a pre-Columbian Native American skeleton from Virginia (Ortner *et al.*, 1992) will surely stimulate further research in this area. It seems possible that the introduction of New World treponematosis, fundamentally endemic in its native form and acquired through variable means from Native American contacts (Powell, 1994), into the quite different epidemiological context of late 15th-century European populations may have resulted in a *venereally spread* outbreak of a "new" disease. Furthermore, if Old World forms of treponematosis were indeed as rare in western European populations as they appear in the bioarchaeological record, perhaps because they had been very recently introduced from tropical Africa or the Near East, it is possible that these populations' immunological vulnerability may have contributed to the savage virulence of the new disease as reported in contemporary medical accounts of the first decades of its appearance (Quetel, 1990).

A comprehensive review of the current state of research on treponematosis is presented in Powell and Cook (2005).

X. DEVELOPMENTAL STRESS MARKERS AND THE "OSTEOLOGICAL PARADOX": CRITICAL REEVALUATIONS OF ACCEPTED INTERPRETATIONS

During the last quarter of the 20th century, advancements in theoretical and methodological aspects of analyzing a broad range of physiological "stress markers" have led to increasingly sophisticated interpretations of the relative contributions of poor diets, widespread infectious disease, heavy workloads, and significant parasite infestations to patterns of health and disease, particularly for subadults. Diets deficient in essential nutrients and/or calories fail to provide strong resistance to opportunistic infectious diseases (e.g., tuberculosis) and promote multiple specific signs of biological stress, including reduced adult tooth and body size, fluctuating asymmetry of dental and skeletal epigenetic traits and metric features, dental enamel defects, decreased neural canal diameter and cranial base height, increased risk of spina bifida, altered pelvic inlet shape, and markers of specific metabolic malfunction such as iron deficiency, anemia, rickets, and scurvy (Angel *et al.*, 1987; Cook, 1979, 1981; Buikstra and Cook, 1980; Clark, 1988; Clark *et al.*, 1986; Goodman and Armelagos, 1989; Huss-Ashmore *et al.*, 1982; Cohen and Armelagos, 1984; Larsen, 1987, 1997; Stewart-Macadam and Kent, 1992; Ortner *et al.*, 2001). Because diet plays such an important role in the development of many forms of pathology, American paleopathologists have long been interested in efficient methodologies for biochemical analysis of bone and dental tissue samples aimed at paleodietary reconstruction (Buikstra, 1992; Larsen, 1997; Sobilik, 1994). During the 1970s, paleonutrition emerged as a specialty increasingly independent of paleopathology, initially focusing on trace elements, but moving in the early 1980s to stable isotopes of specific elements (primarily carbon and nitrogen; Price, 1989; Sandford, 1993; White, 1999). Interpretation of pathological change in bone can now be augmented by independent investigations into diet at the individual and population levels.

These advances have prompted a series of critical reevaluations of current paradigms for the evaluation of levels of health in past human populations. In an article in *Current Anthropology* titled "The Osteological Paradox: Problems of Inferring Prehistoric Health from Skeletal Samples" (Wood *et al.*, 1992), the authors challenged conventional assumptions concerning morbidity (i.e., that high frequencies of observed skeletal lesions, interpreted without reference to specific etiologies or age-at-death distributions, could directly reflect low levels of population health) and mortality (i.e., that low mean age-at-death invariably signaled demographic decline), by positing significant "intervening variables": demographic nonstationarity and selective mortality due to intrapopulation heterogeneity in risk of death ("individual frailty"). Drawing upon clinical and epidemiological reviews of associations between morbidity and age, sex, social status, and other contextual variables, they argue that changing patterns of skeletal pathology and mortality reported worldwide for population samples undergoing the transition from hunting/gathering lifeways to reliance on cereal-based agriculture could arguably be interpreted as reflecting either improved or worsened levels of health if widely used criteria (e.g., skeletal inflammatory response) were employed without reference to demographic contexts.

A point that is equally important in such evaluations, although not addressed by Wood and colleagues (1992), is critical consideration of the different potentials for skeletal morbidity and for mortality associated with different infectious diseases. For example, tuberculosis produces recognizable skeletal pathology in relatively few of its victims yet carries a very high risk of death, while the endemic treponematoses (yaws and nonvenereal syphilis) produce some degree of skeletal pathology during the advanced stages of illness in a relatively high proportion (20–50%) of infected individuals, yet carry a negligible direct risk of death because they do not affect internal organ systems as does venereal syphilis (Powell, 1992). Overemphasis by researchers on isolated aspects of undifferentiated nonspecific skeletal pathology, e.g., periostitis on tibia shafts, without simultaneous consideration of other, more useful indicators of specific etiologies, as well as careful analysis of the mortality profile of the sample in question, can result in erroneous conclusions as to the major risks of early death in that population.

A second important reevaluation challenged the biological reality of "penalty-free adaptation" to suboptimal diets and living conditions, as set forth in the "small but healthy" hypothesis formulated by Seckler (1980, 1982). Seckler argued that "smallness may *not* be associated with functional impairments . . . the mild to moderately malnourished people in the deprivation theory are simply 'small but healthy' people in the homeostatis theory" (emphasis in original, Seckler, 1982:129). Goodman (1994) disagreed, citing numerous recent studies of living populations by researchers in biomedicine, epidemiology, nutrition, and political economy that link poor nutrition with small adult stature *and* a host of pathological conditions, including decreased energy expenditure, reduced cognitive skills, and premature mortality, as well as the biological markers of poor development of skeletal and dental tissues familiar to paleopathologists. For example, minor reductions in adult tooth size, apparently harmless in themselves, are correlated positively in many archaeological populations with another seemingly harmless developmental defect, linear enamel hypoplasia; the important point is that both are correlated positively with shortened life span (Simpson *et al.*, 1990). As Larsen notes in his review of the role of stress markers in bioarchaeological investigations, "clearly, there are negative consequences of small body size in disadvantaged settings, indicating that this reduction is maladaptive" (Larsen, 1997:62). Goodman and colleagues have joined forces with other proactive anthropologists who study the biological impact of political economies in modern disadvantaged populations (Crooks, 1995; Leatherman and Goodman, 1997; Stinson, 1992) to decry what they see as "Cartesian reductionism and vulgar adaptationism" (the title of his 1994 paper) in the benign interpretation of malignant consequences of politically determined inequalities of essential resources. A particularly satisfying application of this thinking is a meta-analysis of data from the American Southwest that includes social status, density, and ecological factors; substantial evidence for cultural buffering of crises emerges from this effort (Nelson *et al.*, 1994). A recent review puts these issues in a context stressing political complexity and gender as important variables in archaeological studies (Danforth, 1999).

The role of dietary iron intake in the production of iron-deficiency anemia and the consequences of chronic anemia for healthy biological development and maintenance was the focus of a third particularly heated debate, also led by Goodman (1994). Kent (1986; Kent *et al.*, 1990, 1994) and Stuart-Macadam (1992a,b), citing clinical studies of the negative impact of low levels of circulating iron in a human host (achieved through sequestration of iron in the liver) upon multiplication rates of invading microbes (e.g., Weinberg, 1974, 1992), argued that mild to moderate chronic iron-deficiency anemia may be interpreted as a positive adaptation to prevalent parasite and microbial loads. Goodman countered that the chronic iron deficiency typical in many women and children in disadvantaged living populations (and signaled by the skeletal lesions of porotic hyperostosis and cribra orbitalia in archaeological samples) carries profound negative impacts on normal growth and development, oxygen transport (and therefore work performance), and cognitive levels, therefore depressing effective behavioral skills necessary for survival and reproduction. Therefore, he concludes, it cannot be seen as a "neutral" adaptation but must be considered a dangerous "adjustment" to iron-poor diets and heavy parasite and microbial loads, a devil's bargain, as it were, that exacts a heavy price in the form of increased morbidity and shortened life span.

XI. MULTIPLE MARKERS AND AN INDEX OF HEALTH IN THE PAST

In 1990, economic historian Richard H. Steckel and physical anthropologist Jerome C. Rose organized a conference at Ohio State University to inaugurate a new project focused on the status of health in New World populations from 7000 years BCE to the early 20th century. *The Backbone of History* project incorporated data from more than 12,500 skeletal individuals from 230 sites in North, Central, and South America. The site samples were combined to provide 65 regional samples large enough to permit statistical analysis. Each individual was coded for a set of specifically defined variables: cultural (time period, subsistence mode, social status, community type), biological (genetic background, age, and sex), and ecological (elevation, maritime vs inland, etc.). Then each one was scored for a set of seven biological variables: stature, enamel hypoplasia, dental caries/antemortem tooth loss, and skeletal evidence of anemia, infectious disease, degenerative joint disease, and trauma. The individuals were assigned a "health index" ranking, based on a composite of the feature scores, and this process was then applied to each regional population sample. The volume that presented the results of this project (Steckel and Rose, 2002) focused on large-scale trends, e.g., the gradual decline in health for Native Americans from 7000 BP to the period of European contact, but also reported variations in different variable scores at different times and places. A second stage of the project, inaugurated in 2002, will examine variation in the seven features within more specific contextual analyses.

Steckel and Rose's project builds on a long tradition in North American paleopathology that begins with Lawrence Angel's (1966a) contributions at midcentury. Far more than elsewhere, we have been interested in diachronic comparisons of disease frequencies as a means of assessing the adaptedness of ancient peoples. The marked increase in such studies through the 1970s coincided with the emphasis on ecological and processual models that characterized the "New Archaeology" movement. The volume *Paleopathology at the Origins of Agriculture* (Cohen and Armelagos, 1984) serves as a high-water mark for this era, summarizing a number of research projects in which archaeological and osteological data are integrated. Unfortunately, as archaeologists have moved to new theoretical issues, the work of paleopathologists has become less central to their interests. *The Backbone of History* is a revitalization of this tradition in a new context that derives its rigor and scope from econometrics and economic history. It requires considerable glossing over of local factors in reaching its big picture view, and those of us with strong investments in local and regional explanations look forward to the second, contextually focused, stage of this project.

XII. TRAUMA AND ITS CULTURAL MEANINGS

Until quite recently, most reports of trauma in paleopathology consisted of brief mentions of healed fractures. To be sure, trepanation, scalping, dental mutilation, and tooth ablation have generated lively debate, but fractures and bruises have been seen as less anthropologically meaningful. Trauma has emerged as a fascinating and controversial locus for interdisciplinary study in the past few decades. Lovejoy and Heiple (1981) found a remarkably high frequency of healed fractures in a Late Woodland site in Ohio, but demonstrated using survivorship curves that most were accidental and that treatment was effective. They argue that child abuse and interpersonal violence were not important in this population. Buikstra (1981b) found evidence for social bias against persons with healed but handicapping injuries in Archaic sites from Illinois. Walker (1989) showed strong evidence for interpersonal violence in healed cranial injuries in a large series from California. These studies set the stage for research into the fate of persons who were not survivors.

It is now difficult to imagine that in the mid-20th century there was a consensus that warfare and interpersonal violence were largely confined to state-level societies in the ancient Americas and that scalping was probably introduced during the colonial period. The discovery of the Crow Creek site in South Dakota with its evidence for slaughter of 500 people (Willey, 1990) and many smaller scale examples has produced a picture of endemic warfare in the societies of the Great Plains, both before and after European contact (Owsley and Jantz, 1994). Chiefdom-level societies in eastern North America have yielded similar evidence for raiding and between-group conflict (Smith, 2000, 2003; Milner *et al.*, 1991). A fascinating collection of careful interpretive studies, *Troubled Times* (Martin and Frayer, 1997; see also Standen and Arriaza, 2000; Verano *et al.*, 2000), explores gender, status, deviance, and ritual activities as factors in within-group patterns of trauma in both Old and New World contexts. The paleopathology of several historic battlefields has also been explored in detail (Scott *et al.*, 2002; Williamson *et al.*, 2003). These studies represent a fruitful synthesis of forensic pathology and paleopathology. For a fuller review, see Phillip Walker's (2001) account of current research.

Two projects that employ both forensic and archaeological techniques for tool or cut mark analysis have generated considerable controversy in the Southwest. Evidence for systematic butchering and burning and boiling of at least 29 individuals at the Mancos in Colorado has been interpreted as cannibalism (White, 1992). Rigorous comparison to faunal assemblages is important in this study. Evidence from this site and 75 others is shown to be regionally and chronologically bounded (Turner and Turner, 1999). Because cannibalism was contemporary with contact with Aztec Mexico, Christy Turner argues for state-level terrorism and for cult-drive social pathology in interpreting these findings. A lively debate about these findings continues among archaeologists, ethnologists, and members of descendant groups. Melbye and Fairgrieve (1994) have presented similar evidence for cannibalism related to ethnic conflict from the Canadian Arctic.

The long dormant application of paleopathology to understanding the life ways of fossil hominids has been revived recently. Survived trauma and conditions associated with substantial disability have been seen as evidence for care giving, arguments with antecedents in the work of early 20th-century paleontologists. It is hardly surprising that these speculations are controversial. Kathy Dettwyler (1991) has provided an ethnographically grounded critique of this literature, and David DeGusta (2003) has convincingly demolished one of the more recent claims for human-like caregiving, and the question of whether disability is evidence for altruism or resilience among modern human hunters has been explored (Keenleyside, 2003). Comparative pathology lies beyond the scope of this chapter, but we note a fascinating revival of comparative paleopathology in Blaire van Valkenburgh's (van Valkenburgh and Hertel, 1993) studies of fractures in Pleistocene carnivores.

One of the features that distinguishes paleopathology from pathology as a medical specialty is our focus on the relationship between osteoarthritis and activity patterns. Much of the literature in paleopathology conceptualizes joint disease as the result of chronic, patterned, microtrauma susceptible to interpretation in terms of movement and force. Pat Bridges (1992) has critiqued these assumptions and called for a focus on duration and intensity of activities on the one hand and on arthritis syndromes on the other. Angel's (1966a) atl-atl elbow concept has faired badly in the half-century since he proposed it. Patricia S. Bridges showed conclusively that Angel's inferences about laterality and sex differences were based on insufficient samples and that only frequent activities — weaving or grain grinding rather than weapon use — leave evidence in the skeleton (Bridges, 1990, 1991b, 1994b). Bridges used to say that she decided to compare activity-related muscle markers in Archaic and Mississippian skeletons because she was "sick and tired of hearing archaeologists yammer on about how 'men were men in the Archaic, much bigger and stronger than those little puny Mississippian wimps.'" It is hardly a coincidence that the waning of the hunting hypothesis as an explanation for various aspects of human evolution has accompanied the entry of large numbers of women into our profession!

In contrast to the fate of Angel's speculation, Stewart's (1931b, 1956a) inference that population differences in the frequency of spondylolysis might be related to way of life has been amply born out. Charles F. Merbs has demonstrated that spondylolysis is a fracture of the neural arch of the vertebrae, not a congenital defect, and has explored its relationship to activity patterns and material culture (Merbs, 1995, 1996a, 2002a). His work is an excellent example of research in paleopathology that advances medical science.

Recent interesting studies of groups with well-characterized, unusual activity patterns include Canadian voyageurs (Lai and Lovell, 1992; Lovell and Dublenko, 1999), prehistoric and Colonial Pecos Pueblo farmers (Munson Chapman, 1997), and Eskimo groups with differing technologies (Hawkey and Merbs, 1995; Steen and Lane, 1998). Two thorough reviews (Jurmain, 1999; Kennedy, 1989) attempt to put other inferences about activity on solid ground.

XIII. QUINCENTENNIAL

One important stimulus to the "golden age" of paleopathology at the turn of the last century was the Columbus Quadricentennial. The 400th anniversary of Columbus' landfall in the Caribbean was the occasion of institution building and collection building, particularly in New York, Chicago, and Mexico City. Documentary and archaeological research on the conquest was important in Hrdlička's development of his radical concept of a New World free of infectious disease and Stewart's articulation of cold-filter hypothesis. The quincentennial has been the occasion for reflection on the role of introduced pathogens in the demographic catastrophe that followed in the New World and for regional syntheses. Review articles (Ortner *et al.*, 1992; Merbs, 1992), regional syntheses (Blakely, 1988b; Jantz and Owsley, 1994; Larsen and Kelly, 1995; Whittington and Reed, 1997), and explicitly comparative volumes of collected papers (Verano and Ubelaker, 1992; Larsen and Milner, 1994) attest to the rich and varied picture of health consequences of contact between Old and New Worlds. Medical historians have been similarly active (Guerra, 1993; Settipane, 1995; Boyd, 1999; Fenn, 2001), while the historian Alfred W. Crosby, who gave us the concept "Columbian exchange," has moved from what he calls his "role as a biological determinist" (1997: x) to stress quantification as the key to European colonial expansion. The picture that emerges from paleopathology is decided nonuniformitarian. Social factors, including workload, living conditions, and competition for resources, were as salient as infectious diseases in population decline.

The quincentennial was observed nationally in the United States with the passage of the Native American Grave Protection and Repatriation Act (NAGPRA) following more than a decade of state regulations. A great part of the interesting literature of the last decade reflects our profession's efforts to document collections that are threatened with destruction. A series of volumes from the Archaeological Survey is noteworthy (e.g., Owsley and Rose, 1997) for the integration of bioarchaeology and paleopathology with archaeology. However, the long-term consequences of NAGPRA for our discipline may be visible in the relative silence of several contributions to recent archaeological literature on the Conquest regarding information from human remains (e.g., Thomas, 1989, 1990, 1991).

Perhaps in response to the repatriation movement and certainly in response to opportunities arising from cultural resources management, there has been a striking increase in applications of paleopathology to historic groups. Two recent collections of papers offer an introduction to this literature (Saunders and Herring, 1995; Herring and Swedlund, 2003), and a slightly older symposium christens this field "biohistory" (Rose and Rathbun, 1987). Paleopathology has been particularly useful in this context in shedding light on the aspects of lives of persons who are largely invisible to historians.

XIV. AMERICAN PALEOPATHOLOGY AT THE MILLENNIUM: PROFESSIONAL AND INTERNATIONAL IN NATURE

In 1985, the first "Short Course in Paleopathology," organized by Donald J. Ortner at the Department of Anthropology, National Museum of Natural History (Smithsonian Institution), drew participants (both teachers and students) from all over the world for 3 weeks of intensive focus of workshops and lectures on a wide range of topics and methodologies, incorporating hands-on examination of pathological specimens from the department's extensive skeletal collections. In 1988, Ortner and Arthur C. Aufderheide continued this international coverage by organizing a symposium titled "Human Paleopathology: Current Syntheses and Future Options" for the International Congress of Anthropological and Ethnological Sciences in Zagreb, Yugoslavia (Ortner and Aufderheide, 1991). This symposium included numerous participants in Ortner's 1986 "Smithsonian Short Course," as well as other paleopathologists from all over the world. The presenters were charged with summarizing important advances in specific aspects of human health and disease from ancient times to the present. Ortner's research and teaching efforts (including subsequent Short Courses in Paleopathology offered at the Smithsonian and the University of Bradford, UK) have spearheaded the Smithsonian Institution's strong international presence in paleopathology.

The strong interest in the natural history of infectious diseases that has always characterized American paleopathology was evident in the high level of participation by American scholars in the series of International Congresses on Evolution and Paleoepidemiology of Infectious Diseases (ICEPID) held in various European venues between 1993 and 2001. The first two congresses focused on diseases of particular interest to paleopathologists studying health in the pre-Columbian New World: syphilis, at the 1st ICEPID in Toulon, France, in 1993, and tuberculosis at the 2nd ICEPID in Budapest and Szeged, Hungary. However, American paleopathological research was also well represented at the 3rd ICEPID held in 1997 at the University of Bradford, England (Roberts *et al.*, 2002), and the 4th ICEPID held in 2001in Marseille, France, although the "disease in question" for both of those conferences (respectively, Hansen's disease, also known as leprosy, and plague) were apparently both absent from the Western Hemisphere before the 16th century AD. Greenblatt and Spigelman (2003) provide a broader overview of new clinical, molecular, epidemiological, and paleopathological perspectives on this topic in their recent edited volume, *Emerging Pathogens, the Archaeology, Ecology, and Evolution of Infectious Diseases.*

Although artificially or naturally mummified human remains are not so common in North American as South America, Europe, Egypt, or Asia, an interest in their scientific analysis stimulated the formation in 1973 of a scholarly society, the Paleopathology Association (PPA), "an informal group of scientists from many disciplines, whose common link is that they are interested in disease in ancient times" (Cockburn, 1994:135). The Paleopathology Association (known for its first year or so as the "Paleopathology Club") had its origin in a working group of U.S. and Canadian scholars (joined by a colleague from Czechoslovakia) who convened in Detroit, Michigan, for a symposium cosponsored by Wayne State University Medical School, the Smithsonian Institution, and the Detroit Institute of Arts. The focus of the symposium was the carefully conducted multidisciplinary autopsy of an Egyptian mummy (PUM II) on loan from the Pennsylvania University Museum (Cockburn, 1994). The five founders were Aidan and Eve Cockburn, Theodore A. Reyman, Robin A. Barraco, and William H. Peck, who had conducted two previous autopsies of mummies (DIA I in 1971 and PUM I in 1972). The Cockburns and Reyman subsequently published two editions of a worldwide paleopathological study of mummies (Cockburn and Cockburn, 1980; Cockburn *et al.*, 1998).

The success of the first edition stimulated further interest in the scientific analysis of mummified remains, leading eventually to the organization of five World Congresses on Human Mummies: in 1992 in Tenerife, Canary Islands (Proceedings 1995). In 1995, Congress II met in Cartagena de Indias, Colombia (Cardenas-Arroyo and Rodríguez-Martín, 2001). In 1998, Arica, Chile hosted

Congress III (Santoro, 2001), and in 2001, Congress IV was organized in Nuuk, Greenland (Lynnerup *et al.*, 2003). In 2004, Congress V was held in Torino, Italy. Arthur C. Aufderheide (2003), founder of the International Mummy Registry, recently published a monumental and detailed worldwide survey of scientific studies of mummies, covering more than 200 years of investigations of naturally and artificially mummified bodies. Perhaps the most unusual scholarly publication on this topic appeared in 1998 by PPA member Christine Quigley, *Modern Mummies*, a serious treatise on methods of preservation of the human body in the 20th century.

Both the size and the interests of the PPA members have expanded in the past quarter of a century to include the entire universe of human health and disease. The Paleopathology Association has held annual meetings since 1974 (cojointly with the American Association of Physical Anthropology since 1980), and the European members have sponsored biennial meetings since 1976, which include practical workshops, roundtable discussions, and both focused symposia and general sessions of podium and poster presentations. Since 1973, PPA has published a quarterly newsletter containing research articles and reports, notices of upcoming professional conferences of interest, comments and inquiries about specific topics and cases, abstracts of recent theses and dissertations on paleopathological topics, and a regular series of annotated bibliographic references (including journal articles, books, films, and, most recently, web sites).

The motto of the association ("mortui viventes docent," the dead teach the living) celebrates not only the ancient classical world interests of many members, but also the unbroken tradition of scholarly interest in past human societies reaching back more than seven millennia (Cockburn and Reyman, 1982; Cockburn, 1994). However, it most definitely looks forward to the future as well, sponsoring the first Aidan and Eve Cockburn Student Award for an outstanding presentation at the annual North American meeting in 2000. The first Bioanthropology Foundation, now Institute for Bioarchaeology, award was made for a poster presentation at the XIII Biennial European Members meeting in Chieti, Italy, that same year. The institute fosters mentoring partnerships and provides editorial assistance to nonnative English language speakers for presentations and publications. The membership of the Paleopathology Association has grown steadily both in numbers and in diversity since its founding in 1973. By 1978, it included more than 300 members representing 22 countries. In 2003, it included some 500 anthropologists, historians of medicine, physicians, and others (including museums, libraries, and research institutions) representing 40 countries on six continents (only Antarctica lacks a member, sadly). Nearly half of the current members reside in the United States or Canada. The membership during its first two decades was heavily weighted toward senior professionals, but at the present time nearly 30% (168/600) are undergraduate or students, primarily in anthropology. Male members outnumbered female members 30 (88%) to 4 (12%) in the "brief biographies" published in Nos. 3 and 4 of the PPA newsletter in 1973, but by 1999 the two sexes were almost equally represented in the membership database: 245 women (47%) and 273 men (53%).

Beginning in the early 1980s, a number of PPA members began seriously considering the loss of scientific data from collections of human remains threatened by the growing calls for repatriation by Native Americans and Australian Aborigines. A working group, the Database Committee, was formed in 1988 to develop effective formats for systematic detailed collection of osteological data, and in 1991 the Paleopathology Association published an extensive set of guidelines, *Skeletal Database Committee Recommendations* (Rose *et al.*, 1991). That same year, Jonathan Hass organized a seminar/workshop at the Field Museum of Natural History for the purpose of further development of effective standardized protocols. The participants included invited scholars with significant expertise in specific varied topics in skeletal biology, as well as representatives of major training programs in bioarchaeology in the United States. The result of their labors was *Standards for Data Collection from Human Skeletal Remains* (Buikstra and Ubelaker, 1994), a concise manual of detailed methodologies for collecting a broad range of skeletal and dental data covering demography, metric and non-metric observations, taphonomic alterations, paleopathology, cultural modifications of skeletal tissues in living individuals (e.g., cranial shaping), and recovery and analysis of samples for biochemical and microstructural analysis. Detailed instructions are included for compiling verbal, numeric, and photographic databases. This reference is today used widely throughout the United States and abroad to develop and implement effective data recovery protocols in museums and other institutions that curate human skeletal collections (Quigley, 2001). However, it represents a minimum standard. Substantially more sophisticated protocols have been developed at the Smithsonian's Repatriation Office, although not in published form.

Another scholarly society, the Paleopathology Club, was formed in the late 1970s as an affiliate of the International Academy of Pathologists. This society sponsors annual meetings with presented papers and symposia on specific topics, held in conjunction with the International Academy of Pathology, United States and Canadian division. Its focus has differed from the beginning from that of the Paleopathology Association, although both share the same interest on health and disease in past human populations. It serves as a network for exchanging information among its members about specific pathological cases and disorders, and each issue of its newsletter contains one or two slides illustrating a problematical case on which members are invited to comment. The case study series is now available on the web and in CD form. The Paleopathology Club has been sponsored since its formation by the pathologists Marvin J. Allison and Enrique Gerszten of the Medical College of Virginia, Virginia Commonwealth University. Their efforts at publicizing the importance of paleopathological research among physicians and other medical scientists has served to encourage closer interdisciplinary collaborations between medicine and anthropology.

Professional venues in North America for presentation and publication of research in paleopathology increased dramatically during the last quarter of the 20th century. Articles on paleopathology had regularly appeared in the *American Journal of Physical Anthropology* since its first volume in 1918. A review of the mean number of articles per year on this topic (taken at 5-year intervals) from 1960 to 2000 showed a steady rate in the first two decades: 1.25 in 1960, 0.75 in 1965, 1.25 in 1970, and 0.33 in 1975. However, in 1976 one complete issue (vol. 46:3 Part II) was devoted to a symposium honoring T. Dale Stewart; 9 of the 25 papers covered topics in paleopathology. Beginning in 1980, the average numbers of articles on paleopathology per volume of *AJPA* steadily increased, from 1.00 in that year to 1.25 in 1985, 1.75 in 1990, 1.60 in 1995, and 2.1 in 2000. The discipline was also well represented in major professional journals of archaeology during these decades, including *American Antiquity, Midcontinental Journal of Archaeology, Southeastern Archaeology*, and *Journal of Archaeological Science,* as well as *Archaeology*, the magazine published by the American Institute of Archaeology for its general readership. Two new journals devoted totally or primarily to paleopathology have appeared in Europe in the last two decades: the *Journal of Paleopathology* in 1989, published in Italy, and the *International Journal of Osteoarchaeology* in 1991, published in England. Both journals include several North American paleopathologists on their editorial staffs and regularly feature articles, research reports, and book reviews by North American contributors, although the relatively high cost of individual and institutional subscriptions for these journals has unfortunately limited their availability in the United States.

XV. SUMMARY AND CONCLUSIONS

Over the past century and a half, American paleopathology has steadily expanded its original focus from specific individual cases diagnosed almost exclusively by reference to current medical knowledge of diseases toward a broad interdisciplinary interpretation of health status of individuals and populations based on simultaneous interpretation of a wide range of biological, cultural, and environmental data. It is now a well-recognized research focus within the discipline of bioarchaeology and draws upon the latest advances in biomedicine, nutrition, and genetic analysis. "Molecular medicine" applied to ancient samples provides new insights into the evolution of infectious diseases, even those that leave no discernible trace of the skeleton. New techniques in medical imaging technology, unavailable even in the 1970s, are now widely employed, and actual autopsies of mummified human remains have been largely replaced by much less destructive examinations of internal structures and tissues through the use of computer axial tomography, endoscopic examination, and even magnetic resonance imaging (although this latter technique is less effective because of the extremely low moisture content of mummified tissues). We may rephrase slightly the concluding statement by Buikstra and Cook (1980:461): "Though we are *still* far from an exact science, it is clear that paleopathologic study has made notable advances within the past decades, and that the laments of the 1960's are now best viewed as part of the history of the discipline."

The dark ages preceding the midcentury Renaissance of paleopathology were perhaps not as dark as they have been portrayed. They were a consequence of professionalization in anthropology and in medicine. On the one hand, physicians became highly specialized and were no longer broadly educated in classics and social sciences; on the other, physical anthropologists became obsessed with craniology and osteometry. To the extent that paleopathologists have now professionalized in our turn, we run the risk of turning inward. Specialty journals are highly desirable, but it is important that we continue to publish in medical journals and general anthropology journals in order to avoid isolation. As a field we do little writing for the general public, and the attention paid to paleopathology in prestigious general journals such as *Science* in the 1960s (Anderson, 1965; Angel, 1966b; Jarcho, 1965b; Kerley and Bass, 1967; Armelagos, 1968) has no recent parallels. Ackerknecht's goal of integration with other aspects of anthropology and medicine is still unrealized. We are still excessively self-referential in the sense that paleopathologists coin new terms — from Hrdlička's symmetrical hyperostosis to LSMAT — or invest old terms with new meanings — periostitis as equivalent to infection — and attach explanatory scenarios without grounding them in medical science. Some of us are excessively self-referential in citing only paleopathology or, like the latter-day Hrdlička, just ourselves. The cure to this malaise is integration in Ackerknecht's sense, not only with archaeology but with the several disciplines that share a stake in the health of ancient populations. Paleopathology remains resolutely multidisciplinary and international in its scope as we begin the 21st century.

Chapter 12

The Dentist and the Archaeologist: The Role of Dental Anthropology in North American Bioarchaeology

Jerome C. Rose and Dolores L. Burke

I. INTRODUCTION

In the increasingly interdisciplinary and interdependent academic world, it has become clear that no discipline is an island—that, in fact, formerly rigid boundaries are displaying a considerable amount of elasticity that encourages interaction. An example of the new approaches may be found in the relationship of the classic field of archaeology, dating to the 19th century, and the newer study of bioarchaeology, originating in the last third of the 20th century. This chapter examines an activity that contributes to the broader world of archaeology via the channel of bioarchaeology: dental anthropology, the study of teeth in humans. This historical review concentrates on the development of dental anthropology with respect to American archaeology and osteology (after British beginnings), and the history would be somewhat different if we had included the extensive literature from other parts of the world.

Teeth have been studied for thousands of years. The physicians of the pharoahs of ancient Egypt examined and ministered to the teeth of their ruler gods, and in so doing kept records of their research and treatment that would assist their successors. Indeed, the study of teeth has always been a case of inquiry

Bioarchaeology: The Contextual Analysis of Human Remains, Buikstra and Beck (eds.)

and application: What causes the condition of this tooth? What can be done about it? In modern times the process has been expanded, and rather than simply describing anomalies found in old skeletons, dental researchers have used available skeletal collections to answer questions about the present. This unique orientation made the results of their research relevant to other dental researchers and to archeologists interested in reconstructing the way of life of past peoples.

We know that intellectual history has demonstrated that there is a common progress in the development of science and that even diverse fields have followed a common historical path, with leaps of progress occurring at about the same time. Thus it was no surprise for us to find that major paradigm shifts and methodological innovations in the development of dental anthropology coincided, for example, with temporal schemes proposed by both Willey and Sabloff (1993) for American archaeology and Jarcho (1966b) for paleopathology. As the temporal divisions in these schemes do not differ by more than 5 years and we are relating dental anthropology and archaeology, we employ the schema proposed by Willey and Sabloff (1993) to organize our history of dental anthropology. Our divisions are Classificatory–Descriptive (1840–1914), Classificatory–Historical (1914–1940), Contextual–Functional (1940–1960), and Modern (1960–). Using these temporal divisions to discuss the development of dental anthropology permits us to better understand its contribution to archaeology and the birth of bioarchaeology. Table I shows a comparison of activity among archeologists, osteologists, and paleopathologists within these periods.

Within our discussion along temporal lines, we use data categories and recognize a standard research sequence. There are four data categories that encompass the vast majority of dental anthropology research: dental disease (specifically caries), dental wear, developmental defects of the enamel and dentin, and dental morphology and measurement. These data categories have been combined in various ways to make contributions to three interpretive themes: dietary reconstruction, analysis of childhood disease and stress patterns, and reconstruction of genetic relationships. The body of research, encompassing data categories, reflects a standard research sequence in stages of plausibility, methodology, and application, as follows.

A. Plausibility

The researcher believes that an interesting idea could work. This is represented by a publication that demonstrates, for example, that microscopic scratches on teeth could be used to reconstruct the physical consistency of food eaten in the past. Generally, the Classificatory–Descriptive division in our review illustrates a stage of plausibility.

Table I

Temporal Divisions of Archeology, Osteology, and Paleopathology, 1840 to Present

Period	Archeology	Osteology	Paleopathology
Classificatory–Descriptive (1840–1914)	Describing architecture and artifacts while developing typological systems	Developing racial categories from cranial types	Discovering that pathological lesions found in ancient skeletal material can be diagnosed
Classificatory–Historical (1914–1940)	Developing methods for ordering sites and artifacts in time	Cranial typology continues to dominate	Identifying first and oldest cases of specific diseases, applying new methods from clinical medicine to analysis of ancient materials
Contextual–Functional (1940–1960)	Focusing on context and function	Focus on cranial typology attacked within profession; "New Physical Anthropology" (Washburn, 1951)	Becoming interested in problems of entire populations (Jarcho, 1966b)
Modern (1960–)	Shifting to problem-oriented, expalanatory research	Publication of *Dental Anthropology* (Brothwell, 1963b); birth of bioarchaeology (Blakely, 1977)	Beginning shift to paleoepidemiology

B. Methodology

Numerous studies focus on developing and refining methods of data collection and analysis that would lead to reliable diet reconstruction from microscopic scratches. In the study of the past this stage must also incorporate what in archaeology would be called enthnographic analogy. Here clinical research, studying living nonindustrial peoples, and even animal experimentation and observation in the wild are used to establish the relationship between the specific feature or condition of the teeth and its causal agent, e.g., looking at the relationship between specific and known foods and the scratches produced on the teeth. This interpretive stage is a necessary one before data from the past can be interpreted because in the study of the past we must infer the presence of the causal agent from the evidence or impact that it has left behind on the skeleton. The stage of methodology is best expressed in the Classificatory–Historical period, with some further evidence in the Contextual–Functional period.

C. Application

This final stage is the development of research methods where collected data can be analyzed in such a way that we can progress to a higher level of providing explanation and understanding process. Here we have reached the stage where data can be routinely employed in bioarcheological analysis where we are interested in problems such as the origin of agriculture. This routine application of methods comes in two forms. First, we have the routine application of the method in the analysis of a single-site, specific skeletal sample, where all possible data are collected from the skeletons. These usually descriptive reports, often added as chapters or appendices in archeological publications, are the building blocks of synthetic bioarchaeology research. Second, we have the application of the methodology to the solving of a particular problem. Here one or more methods are employed to collect data from many skeletal series to look at one phenomenon, such as determining the dates of earliest agriculture in the American midwest. This final stage is most clearly shown in the Modern period.

Examining the history of dental anthropology within the temporal scheme borrowed from Willey and Sabloff (1993) and the proposed sequence of methods development, we demonstrate the unique contributions that dental anthropology has made to the development of bioarchaeology. We contend that, unlike other medical professionals and osteologists, dentists and dental researchers have a long history of problem-oriented research using ancient human skeletal remains. The problems being addressed often may have been simple ones, but because the research focused on basic biological processes, dental research has for more than a century made substantive contributions to our understanding of ancient biology and culture, using human skeletons recovered by archeologists.

In this short chapter it is not possible for us to provide a complete review of the literature for each of our methodology themes; instead we employ more selective citations to document the development of dental anthropology, specifically focusing on its role as a major contributor to the development of bioarchaeology. Fortunately, comprehensive literature reviews have been published for each of the dental anthropology data categories. For an extensive bibliography and a more detailed history of dental caries research and its use by dental anthropologists, the reader is referred to Powell (1985) and Larsen (1997), and also to Caselitz (1998), who provides a survey of the worldwide distribution of caries over time. For a history of dental wear studies, there are the contributions of Molnar (1972), Powell (1985), and Rose and Ungar (1998). A history of developmental dental defects can be found in Goodman and Rose (1990), while for a more detailed history of the development of morphological studies the reader is referred to Scott and Turner (1988, 1997).

II. CLASSIFICATORY–DESCRIPTIVE 1840–1914

While archeologists, osteologists, and paleopathologists were describing materials, developing typologies and categories, and determining the possibility of analysis, dental researchers were engaged in their own specialized attention to human teeth. Meetings of the Odontological Society of Great Britain were alive with discussion and great debates and one of the most frequently asked questions was "Why are diseases of the teeth more common now in civilized life then they formerly were?" (Mummery, 1870:73). Even at this early date dental researchers sought answers to this question using problem-oriented research to study excavated skeletal remains of past peoples. Is this not by definition what we call bioarchaeology today?

Mummery examined variation in dental decay between groups that varied by diet and culture using a sample of teeth from the skulls of 203 ancient Britains, 143 Romano-Britains, 76 Anglo-Saxons, and 36 ancient Egyptians, in addition to another 1175 skulls from 18 modern groups on other continents. He demonstrated that dental decay increased with civilization and decreased when more primitive conditions returned. For example, when comparing the teeth of Anglo-Saxons to the Romano-British he stated that "the simpler habits of the Anglo-Saxons, together with their nourishing food" produced higher quality teeth and less disease (Mummery, 1870:33). It was not too long after Mummery's work that Miller (1883a) proposed that acids produced by the fermentation of carbohydrates by bacteria are the primary cause of dental decay. Miller (1883b) did later temper this causal explanation by stating that imperfections in the structure of teeth permitted the acids to cause decay. It was only two decades later that it was confirmed that indeed it was the fermentation of sugar and starches that cause the destruction (Turner and Bennett, 1913).

It is true that some of Mummery's explanations for the cause of caries were far wide of the mark, such as attributing the poor quality of British children's teeth, hence prone to caries, to too much cerebral activity (i.e., studying) among the little tykes that was depriving their teeth of resources necessary for sound development. Here, although wrong, he was among the first to postulate that structural imperfections made modern teeth more prone to decay. This supposition led to 80 years of research on ancient and modern dental defects that laid the ground work for using them to reconstruct childhood stress patterns. There were a few strange interpretations in Mummery's papers, but other conclusions were sound and have been supported repeatedly by more recent studies using larger samples with better temporal control. Hardwick (1960) employed data on 13,450 ancient British teeth to reproduce the same variation in caries frequencies over time as did Mummery, although he more correctly identified sugars and refined flour in the diet as the major variable, while still including defective dental development as a major contributor to increased decay. Using even larger samples and more sophisticated methods, Moore and Corbett demonstrated the same trends by time and culture while attributing increased decay to increased consumption of sugar and refined flours (e.g., Moore and Corbett, 1971, 1973, 1975; Corbett and Moore, 1976).

Mummery (1870:42) used modern skulls from around the world to demonstrate that the amount of sand and grit introduced into food during processing led to variation in the degree of tooth wear observed in both ancient and "primitive" teeth that contrasted greatly with the minimal wear on modern British teeth. Although his knowledge of what these nonwestern peoples were eating derived from the sometimes fanciful reports of explorers, this is in effect the use of ethnographic analogy. Some decades had to pass before methodologically sound studies of living hunter–gatherers and horticulturalists were conducted to verify his conclusions. Again Mummery led the way by concluding that the amount of dental wear seen on teeth is directly related to the types and consistency of the foods being eaten. Not content with looking at degree of wear, Mummery (1870:42) also noted oblique wear (when the occlusal surface becomes angled to the cheeks on the lower molars) and attributed it to the chewing of tough, hard foods. It was not until a century later that both Hinton (1981) and B. Holly Smith (1984) used variation in the angle of the occlusal wear plane to document the transition to agriculture with its attendant shift from hard to softer foods.

Mummery touched on other areas, such as the effect of disease, minerals in foods, and even medicines on dental development and increased susceptibility of teeth to decay. It was, however, not long before Berten (1895, cited in Sheldon *et al.*, 1945) suggested that physiological disturbances are reflected in the dental microstructure. Black (1906) compared the occurrence of various surface and histological defects of the enamel to the occurrence of various local and systemic disturbances.

The fact that archeological teeth are amenable to histological/microscopic analysis was demonstrated by Professor C. S. Tomes (1892). While taking a short cut through a cemetery on his way to a meeting, Professor Tomes, still known for his contributions to dental development and histology (e.g., Tomes process of ameloblasts), found a tooth and took it home for analysis. He subsequently delivered a paper to the Odontological Society on the histology and postmortem changes that he observed in this tooth, laying the ground work for the study of ancient dental microstructure and taphonomy. Thus, when subsequent dental researchers pursued Mummery's idea of the relationship of caries frequency and dental defects, they knew that ancient teeth could be sectioned and examined microscopically.

This long discussion of Mummery's contributions to dental research and the development of bioarchaeology is not presented to establish his genius or far-sightedness; on the contrary, we wish to point out that he was not unusual for his time, and all of these topics were discussed again and again at Odontological Society meetings, in Britain and elsewhere. We further contend that the study of ancient teeth was an integral part of this research enterprise of finding out why modern industrialized people suffered from extensive dental disease. We do, though, find that three of the four major data categories or research themes that we think characterize dental anthropology and bioarchaeology are to be found in Mummery's article. First, variations in decay frequencies are related to variation in diet, providing the basis for using caries to document dietary change and offering a major tool for studying one of the great human revolutions, the transition to agriculture. Second, increased rates of decay might be due to defects or imperfect dental development, a notion that led to searching for variations in the quality of dental development among living peoples, as well as those from antiquity. This focus on developmental irregularities provided data needed to employ defects in dental development (e.g., hypoplasias, enamel microdefects, interglobular dentin) to reconstruct the patterns of childhood stress. Third, variation in dental wear was due to variation in diet, and analysis of both living and ancient teeth led to the use of dental attrition, at both gross and microscopic levels, for dietary reconstruction, with its first significant use being documentation of the transition to agriculture.

The research of Mummery and colleagues provided the foundation for using teeth as one source of data for the interpretation of ancient skeletons (i.e., the plausibility stage of research). There is no better place to see if indeed this research had any impact on the analysis of skeletons being excavated by archeologists than in the work of Aleš Hrdlička, curator at the Smithsonian Institution, father of American physical anthropology and the founder of the *American Journal of Physical Anthropology* in 1918. Hrdlička (1908b,c; 1909c,d; 1910; 1912d; 1913) produced osteological appendices on skeletons excavated by C. B. Moore in Arkansas and Louisiana, frequently reporting the presence and frequency of dental decay but never drawing any conclusions or inferences from these observations. In contrast, Turner and Bennett (1913), in examining ancient Egyptians (26–30 dynasties) excavated by Sir Flinders Petrie for Karl Pearson (the well-known statistician), compared their decay rate to that of modern people and attributed the ancient rate, which was three times lower than the modern, to the ancient Egyptian diet of rough bread with an absence of the refined carbohydrates that characterize the modern diet.

Mummery (1870) used only qualitative statements (e.g., extensive, more than, less than) to compare dental wear on his various skeletal samples, but soon afterward Broca (1879) produced a dental wear scoring technique that became employed extensively. Hrdlička (1908b, 1909c) used and reported scores based on Broca's five-level scale, without citation, in his osteological appendices.

Despite the previously discussed innovative work on the relationship of wear and diet, Hrdlička (1908b:563) made, at best, minimal attempts at dietary reconstruction. For example, when he was reporting on crania from Arkansas excavated by C. B. Moore, he noted that dental wear was less than usual and stated that "their food was not coarse." Rather than using variation in wear for dietary inferences as did Mummery, Turner, and Bennett, he employed the scheme as one of his methods for determining age-at-death. With the later publication of this method in his laboratory guide (*Practical Anthropometry*, Hrdlička, 1939), dental wear became, and remained, one of the primary means of determining age-at-death for American archeologists and physical anthropologists alike. As there were few reliable indicators of age available to osteologists and the dental wear method was easy to use, it acquired an emphasis for age determination and was seldom employed for dietary reconstruction. In these same osteological appendices, Hrdlička (1908b; 1909c) noted the high frequency of shovel-shaped incisors, stating that this is a characteristic of American Indians. His only other comment on these Native American teeth was that the number and morphology of the molar cusps resembled those found on people of European ancestry. These general observations were the first stirrings of dental morphology studies and eventually led to analysis of large comparative collections that developed into the seminal work on recording standards and the distribution of shovel-shaped incisors (Hrdlička, 1920c).

Thus, the Classificatory–Descriptive period came to a close with considerable progress having been made in establishing that teeth can be used effectively in the study of ancient skeletons. Mummery demonstrated that both dental disease and wear varied between two different diets, showing that it was possible to use these two data sets to reconstruct diets (i.e., plausibility stage). Also, the hypothesis that the high frequency of dental disease in modern populations was due to a decline in the quality of dental structure was firmly established. The testing of this hypothesis was to be the motivation for studying ancient teeth for more than 50 years.

In similar fashion, the first wear scoring system was developed and applied to determining the age-of-death, and this application guided the use of dental wear methods in osteological analysis for many decades to come.

Less well recognized as a useful tool was the modern understanding introduced by clinical dental research that dental decay resulted from the fermentation of carbohydrates by bacteria and that physiological disturbances could result in dental developmental defects. Skeletal analyses did not use any of these concepts for the reconstruction of ancient diets; there simply were not enough comparative data to make interpretations possible.

Finally, high frequencies of shovel-shaped incisors among Native Americans were recognized as a significant distinguishing characteristic, but the popularity of cranial morphology hindered any further advancements in dental morphology research.

III. CLASSIFICATORY–HISTORICAL 1914–1940

This period between the beginning of the First and Second World Wars saw a settling-in process — a development of chronology and reinforcement of existing knowledge. In dental research there were numerous studies conducted by dentists that firmly established the relationship between increased sugar and refined carbohydrates in the diet and increased dental decay (e.g., Price, 1933). As was the case earlier, this understanding was applied to the interpretation of caries frequencies observed in ancient skeletal samples. The presence of dental decay in Paleolithic skeletons was reported by early researchers (Praeger, 1925; Vallois, 1936; Krogman, 1938). Ruffer (1920) discussed the dietary implications of the low decay and high wear rates among the predynastic Egyptians that he was studying. However, to go beyond simple statements to more detailed interpretation of ancient decay and wear rates required knowledge of variation among living nonindustrialized peoples, i.e., "ethnographic analogy." These data were soon available from many parts of the world, including Africa, Australia, and the Arctic (e.g., Nicholls, 1914; Campbell, 1925; Orr and Gilkes, 1931; Waugh, 1928, 1931, 1933; Staz, 1938; Oranje *et al.*, 1935; Schwartz, 1946). Campbell (1925), studying the native Australians, pointed out various features in the environment and diet that could have contributed to the advanced wear and minimal incidence of diseased teeth that he was recording. Buxton (1920) took special note of the soft diet of the African pastoralists and pointed out that they had considerably less dental wear than their agricultural neighbors. These associations of known diets and food processing technologies could then be used as analogies by the osteologists, enabling them to associate a pattern of wear and frequency of decay with specific information provided by archeologists about food remains and processing technologies, ultimately providing a reconstruction of the ancient diet (i.e., methodology stage).

The first significant study of skeletons (i.e., application stage) using this clinical and ethnographic knowledge base examined four different prehistoric Native American groups with different subsistence patterns (Leigh, 1925b). Leigh demonstrated that agriculturalists have more decay than hunter–gatherers, and the association of maize agriculture and frequent dental decay is firmly planted in the literature. In this classic study of the Sioux, Arikara, Indian Knoll, and Zuni, Leigh concluded that the Sioux had less wear than the others because they did not use stone grinders to prepare their maize for consumption. Leigh's 1925 article had a major impact on the use of caries and wear for reconstructing ancient diets and is still one of the most frequently cited articles in the literature. This causal relationship of diet with wear and decay is exploited to establish that specific peoples were agricultural. For example, Christopherson and Pedersen (1939) examined caries differences between Neolithic and Bronze Age skeletons from Denmark in the investigation of agricultural origins.

It is true that Leigh's (and others') study of caries and subsistence can be classified as good problem-oriented research in bioarchaeology, but the studies also represent the methodology stage where specialized studies of a small number of data categories focus on refining methods and establishing the validity of the interpretations. Mummery and others provided evidence that caries and wear could be used to interpret past diet. Ethnographic studies provided the details necessary for interpretation and use. We contend that the use of decay rates for subsistence reconstruction must also extend to the "field" portion of osteology where skeletal samples are subjected to comprehensive analysis. The use of caries and wear to reconstruct diet must also be found in the osteological appendices and monographs that are likely to be read and used by archeologists before this line of research can be considered to have had a role in the development of bioarchaeology. One seminal work in the early 20th century that in some respects represents an initial attempt at bioarchaeology is Hooton's (1930) *The Indians of the Pecos Pueblo*. In the dental appendix to this monograph, Rihan attributed the 47.9% decay rate to a poor diet that resulted in imperfect structure, thus making the teeth susceptible to caries. Here the author employed the theme of inferior structure rather than an agricultural diet high in carbohydrates. Despite numerous studies linking decay and dietary carbohydrates, blaming substandard dental development and structure as postulated by Mummery (1870) and Miller (1883b) was simply not abandoned. Bodecker (1930) made histological sections of ancient Pueblo teeth and found that, despite many developmental imperfections, the teeth were free of decay. This same interpretation is present well into our next chronological period when Moorrees (1957:144, 150) in his study of living Aleuts stated that the variation in dental decay is due to the poor structure of teeth that resulted from the consumption of a modern diet rather than associating the increase in decay with the consumption of foods with high proportions of fermentable carbohydrates.

Lux (1935, 1936, 1937) was fairly typical of the writer of osteological reports and appendices when he reported the presence of dental disease in detail (here for skeletons from central Texas), but does not draw inferences concerning diet. Others of this era working in North America, such as Goldstein (1932) and Moodie (1929), also made inferences about diet from dental wear observed on the teeth of the skeletal remains. One constant theme was the presence or absence of stone food-processing utensils following the interpretation of Leigh; another theme was the confirmation of the inverse relationship between increased wear and decreased caries. All noted the cleansing effect of the coarse diets. Colquitt and Webb (1940), publishing in the *Tri-State Medical Journal*, associated matter-of-factly the high decay rates among the prehistoric Caddo with increased consumption of maize. Later Webb (1944) published an article in the *American Journal of Orthodontics and Oral Surgery* to focus the attention of the dental profession on the study of ancient skeletons, stating that his purpose was to dispel the mistaken belief that dental disease, especially caries, was not just a product of civilization but existed in the past.

As suggested earlier, we think that the reason that dental wear was not used extensively for dietary reconstruction in most early skeletal studies was that dental wear had another more immediate use: the determination of age-at-death. Simply put, there were few reliable indicators of age available for use, and dental wear systems such as Hrdlička's were available and an easily used option. As an illustration, we can return to Hooton's (1930:18) monograph on Pecos Pueblo where he listed dental wear as one of his aging techniques along with cranial sutures and general texture of the bone. It seems that in 1927, when the analysis was well advanced, Wingate Todd visited and assisted in aging 594 of the skeletons using his newly developed technique of pubic symphysis aging. It is primarily due to the lack of easily used options for aging that dental wear was retained as a widely used technique for age determination. Consequently, further advances in recording dental wear and in the use of dental wear for dietary reconstruction had to wait until the 1960s.

Hrdlička's 1920 work on shovel-shaped incisors established solid recording standards and provided distribution data on the trait. He went on to make the link between Native Americans and Asians and suggested an Asian origin for Native Americans. His later study (Hrdlička, 1921b) addressed other morphological variations but had little impact due to the absence of rigorous recording standards. Research on morphological variation, among ancient peoples does not progress much beyond here. It is possible that Gregory's (1922) comment that there was little morphological variation among races discouraged further application of dental morphology analysis to the study of ancient peoples and hence delayed the development of morphology methodology. In contrast, morphological analysis of fossil teeth blossomed and prospered. Similarly, the measurement of teeth had been standardized and some data were recorded (e.g., Campbell, 1925), but little was done with them. The reason for this lack of interest in dental morphology and metrics is abundantly clear when we realize that cranial types and measurements had established themselves as the quintessential means of establishing racial affiliations and reconstructing the migration histories of ancient populations (e.g., Hooton, 1930).

Researchers during this period between the World Wars made substantial contributions to two of our four research themes in dental anthropology. Clinical and ethnographic research clearly linked diets high in sugar and carbohydrates to increased decay. Ethnographic research showed solid correlations between the physical consistency of the diet and rates of dental wear. Significant problem-oriented research was conducted on ancient skeletons, which established that the consistency and content of ancient diets could be reconstructed from dental disease and wear rates. However, much progress was still needed in developing consistent methods for recording and analyzing data. Some of the site-specific osteological analyses were using dental disease to make inferences about diet, while the vast majority that recorded decay frequencies made no inferences. Dental wear observations remained primarily subjective, but were often used to postulate the presence or absence of stone grinders. One explanation for interest in the latter is that the presence of stone grinders could at least be confirmed by archeological excavation. Interest in the structural quality of the tooth and its relationship to dental disease remained strong, but little work on ancient teeth was conducted, probably due to the difficulty in making histological sections of ancient teeth, which were often relatively delicate. Dental morphology research progressed little other than using incisor shoveling to determine that the skeletons were Native Americans, which was self-evident. Further, cranial morphology and metrics held sway in the determination of genetic affiliations and documenting migrations.

As a result, the Classificatory–Historical period saw major advances being made for using both dental disease and wear for reconstructing ancient diets (i.e., methodology stage). Data collected from living nonindustrial peoples demonstrated that increased decay was associated with higher amounts of sugar and carbohydrates in the diet, while wear decreased with increased food processing. This use of ethnographic analogy was necessary to establish the association of specific diets with specific levels of dental disease and wear because without this knowledge, reconstructions of ancient diets was not possible. It then had to be demonstrated that this knowledge could be used effectively for establishing ancient dietary patterns, especially in differences between hunter–gatherers and agriculturalists.

By the end of the period, some osteologists were using dental disease and wear to establish the presence or absence of an agricultural subsistence adaptation (i.e., application stage). However, dental wear made a limited contribution because of its typical use in determining age; the availability of a simple system for associating specific stages of wear with specific age groups made this inevitable.

During this period, very little progress was made in the study of dental defects and morphology.

IV. CONTEXTUAL–FUNCTIONAL 1940–1960

The years during and immediately after World War II made up a period that was a time of both quiescence and transition in dental anthropology and osteology research as a whole. Great advances were being made in some areas of dental clinical research while little occurred in the application of this new knowledge to dental anthropology until the 1960s. Washburn's (1951) article titled the "New Physical Anthropology" promoted evolution and its associated concept of adaptation as the theoretical backbone of the field, and he stated that physical anthropology research must focus on process and explanation. This article was extremely critical of osteology as being totally descriptive, and it probably had a negative effect on the amount of research being done with human skeletons by both professionals and graduate students. The article eventually had a significant impact on osteology and dental anthropology, but not until almost a decade later. During these same years archaeology was undergoing a spurt of fieldwork, with increased excavation of human skeletons necessitated by U.S. Army Corps of Engineers reservoir construction.

The authors of many osteological reports and appendices reported on dental caries, but made no inferences concerning the diets of the ancient people that they were studying. For example, neither Alice Brues (1957, 1958b, 1959a, 1962, 1963b), who would soon become a well-known physical anthropologist, nor Aaron Elkins (1959), developer of a clever physical anthropologist murder mystery character, made any inferences from caries data they reported from Oklahoma. Other researchers reported the frequencies of dental decay and from their data drew conclusions about diets, especially the presence or absence of agriculture in the ancient cultures (Cran, 1959; Goldstein, 1948, 1957; Newman, 1951; Snow, 1945a; Stewart, 1943a; Webb, 1959). These references are not exhaustive as they concern skeletal collections deriving primarily from the states of Arkansas, Louisiana, Missouri, Oklahoma, and Texas, but they do show trends in using caries as an indicator of dietary change, particularly the advent of agriculture. Similar trends undoubtedly can be found concerning the study of ancient skeletons from throughout ancient North America.

The use of dental wear for dietary reconstruction went into decline after its great start with Leigh's work because of the overriding need for simple age determination systems. The publication of Hrdlička's *Practical Anthropometry* (1939) provided both archeologists and osteologists with a simple and easily used age determination method, which included a stage of wear and a corresponding age category. Both studies of ancient skeletons and living peoples made it patently clear that rates of wear varied between different diets. For example, Hrdlička (1939a:45) provided a clear warning, frequently unheeded by those who used his system, that his scheme was only applicable to "American aborigines" and that it must be recalibrated before it could be used for other groups of people. In an attempt to control for dietary variation between ancient peoples, subsequent researchers sought a way to calculate rates of wear that could then be used to determine age. Zuhrt (1955) used age intervals between the eruption of the first, second, and third molars (roughly 6 years between each) and the differences in wear scores between these teeth to calculate a wear rate and determine age-at-death for his historic German skeletal series. Miles (1958) devised a similar system, and its 1963 publication in Brothwell's widely read *Dental Anthropology* made it well known. At the end of the period being considered here we had the first methodological improvement in wear scoring methodology when Murphy (1959a) proposed an eight-point scheme for scoring dentin exposure in the molars and a wear gradient system (Murphy, 1959b) that could be used for age determination.

In contrast to the scant progress seen in caries and wear research, considerable progress was made in the study of dental defects (i.e., methodology stage). Various clinical studies had been exploring the relationship of physiological growth disturbances, dental defects, and caries following the lines of research begun during Mummery's time (e.g., Mellanby, 1929; Day, 1944). Between 1932 and 1944, University of Illinois dental researchers produced a number of studies relating the time of tooth development to histological structures (Massler *et al.*, 1941; Sarnat and Schour, 1941; Schour and Massler, 1941; Schour and Van Dyke, 1932), while others examined the relationship of various diseases and physiological disturbances and the development of dental defects (e.g., Kreshover, 1940, 1960; Kreshover and Clough, 1953). The reader is referred to Goodman and Rose (1990) for a more complete discussion of the history of clinical and epidemiological dental defect studies. The ultimate conclusion that would be drawn from this research was that various physiological upsets produced dental defects seen both on the surface (enamel hypoplasia) and in histological sections (Wilson bands) and that these defects are products of childhood stress and poor nutrition.

Although Tomes (1892) had shown that ancient teeth could be sectioned and studied microscopically, it was not until a half century later that any significant work was done. Despite the work going on in Illinois, this increased interest in ancient dental structure was still being driven by searching for the cause of high frequencies of modern dental decay. Sognnaes (1955) sectioned 233 archeological teeth and carefully documented the postmortem changes. He went on to demonstrate that prehistoric teeth are rarely superior in structure when compared to modern teeth and yet they always exhibit less dental disease (Sognnaes, 1956). Following the trends seen in all aspects of dental anthropology studies, hypotheses were also tested among living, but nonwestern/nonindustrial, peoples. Moorrees (1957) attributed the decline in Aleut enamel and dentin microstructure to a decline in the quality of the diet. Cran (1960) used dental histology to demonstrate that the low dental decay among native Australians was due to the absence of refined carbohydrates rather than poor microstructure. Falin (1961) demonstrated the same relationship in his comparison of dental enamel microdefects among teeth from the "Stone Age," Bronze Age, and modern eastern Europeans. These studies put an end to the search for superior dental structure in the past and paved the way for using dental defects to reconstruct stress levels and dietary quality in past peoples (i.e., methodology stage).

After Hrdlička the first major contribution to the study of Native American dental morphology was made by Dahlberg (1945), whose study of morphological variation of Native Americans, especially the Pima, spanned decades. This work became more broadly known among biological anthropologists with Dahlberg's (1951) publication of "The Dentition of the American Indian" in the Laughlin edited volume, *Papers on the Physical Anthropology of the American Indian*, published by the Viking Fund (the parent of the Wenner-Gren Foundation). The critical contribution was the establishment of standards for recording morphological variation (i.e., methodology stage). Beginning in 1956 and continuing through the 1970s, Dahlberg distributed plaster plaques for use in standardizing the scoring of morphological variation. These plaques are familiar to many biological anthropology students and are still in use today. Shortly thereafter, the study of Aleut teeth by Moorrees (1957) included extensive morphological data and comparison to other groups. However, nothing occurs in the routine study of ancient Native American skeletons except the occasional referencing of some morphological variation. The research role of morphology is still filled by reliance upon variation in cranial shape and dimensions. The potential for use of morphology is there; Scott and Turner (1988) credited Klatsky and Fisher (1953) as the first to use dental morphology to discriminate among human races. While the morphological analysis of the teeth of excavated skeletons is particularly sparse, Scott and Turner (1997) saw the 1950s as a period of basic research where Kraus, Garn, and Lasker, among others, were establishing the genetic foundations of morphological variation.

The period during and after World War II produced relatively little advancement in the use of dental anthropology in the study of ancient skeletons. Increased dental decay was being used as the signature of an agricultural economy, but little use was being made of the decay frequencies being reported in the osteological reports and appendices. The use of dental wear was still confined primarily to the determination of age-at-death, but major methodological improvements were made in controlling for dietary variation and recording degrees of wear. The notion that a decline in the quality of enamel structure was the cause of the modern epidemic of dental disease was finally laid to rest, making it possible for new dental development standards to be used by researchers in the next period to reconstruct childhood stress patterns from dental defects. Pioneering work in dental morphology studies was under way, and the inclusion of some of this work in the influential publication of *The Physical Anthropology of the American Indian* (Laughlin, 1951) set the stage for its use in the study of excavated skeletal remains.

V. MODERN 1960–TODAY

The decade of the sixties was an exciting time for archaeology and physical anthropology that had a great impact on the growth of dental anthropology as a specialty and its application to the study of human skeletons — bioarchaeology. One of the most important was the great growth of anthropology departments that provided jobs for the young innovators in both archaeology and physical anthropology. More importantly, this growth had an impact on the size of the graduate student body, and it is this group that produced many of the methodological innovations in dental anthropology, as well as their immediate application to testing the hypotheses generated by the theoreticians of the "New Archeology." It is only natural that archeologists interested in understanding the reasons for culture change should focus their attention on one of the great human revolutions — the advent of agriculture. Dental anthropology had been preparing itself, and here was the ideal application for its one solidly developed method — dietary reconstruction.

We can see some changes in caries research among ancient peoples, with Dahlberg (1960) pondering the anomalously low rates of decay observed among the Neolithic agriculturalists at Jarmo and Carbonell (1966) examining the teeth from ancient Kish. As influential as was the work of Leigh earlier in the century, it was Don Brothwell's (1963a) chapter on "Macroscopic Dental Pathology of Some Earlier Human Populations" in his edited volume *Dental Anthropology* (Brothwell, 1963b) that included a significant synthesis of the literature on ancient caries and made it clear that dental disease could be used for the reconstruction of ancient diets. *Dental Anthropology* (Brothwell, 1963b) was the first textbook on the study of ancient teeth, and it is here that so many of the advances in dental research were brought to the attention of archeologists and osteologists alike. In fact, our recently purchased copy of this book was once owned by a well-known Midwestern archeologist. Further extending Brothwell's influence on archeologists were the graphs and discussion of dental decay and its relationship to diet found in his even more widely read and used *Digging Up Bones* (Brothwell, 1963c). On a worldwide basis this is the most frequently cited book in the bibliographies of archeological reports that include a discussion of excavated skeletons. Here he provided a detailed system for recording dental disease and wear when studying ancient skeletons, and the two books together led to the extensive use of teeth for reconstructing ancient diets and establishing the presence of agriculture. However, it was still some years before inferences about diet became a standard part of routine osteology. There are many reports and appendices referring to dental disease, but with no inferences concerning the diet (e.g., Bass and Rhule, 1976; Bennett, 1973a; Buikstra *et al.*, 1971; Buikstra and Fowler, 1975; Egnatz, 1983; Ford, 1963; Maples, 1962; McWilliams, 1965, 1968; Phenice, 1969a). However, some researchers reported the frequencies of dental decay and, from their data, made inferences about diets, especially the presence of agriculture (Black, 1979; Hoyme and Bass, 1962; Keith, 1973; Klepinger, 1972; Mehta and Sensenig, 1966; Scott and Birkedal, 1972).

One reason for the lack of dietary inference in studies of specific skeletal collections was that a question still remained as to what frequency of diseased teeth was necessary to support an assertion that these people were agriculturalists. There was no established threshold; researchers had to compare their data to other skeletal series and then qualify their reconstructions with the words possible or probable. Christy Turner II (1979) faced this problem in trying to establish that the Jomon people of central Japan were agriculturalists. He thus compiled published caries data on 64 skeletal samples representing hunter–gatherers, mixed economy, and agriculturalists and found that hunter–gatherers never exceeded a rate of 1 to 2% diseased teeth. This provided a benchmark for comparison when the researcher was working with only one or two skeletal series.

Between the late 1960s and middle of the 1970s, archeologists promoted the study of cultural process, ecological relationships, and subsistence pattern reconstructions (Willey and Sabloff, 1993). These changes encouraged the integration of archaeology and osteology in a way that had not existed previously. The commingling of "New Archeology" (Willey and Sabloff, 1993) with "New Physical Anthropology" (Washburn, 1951) motivated osteologists to focus on reconstructing ancient lifeways and asking analytical questions about when and why subsistence shifts were occurring; i.e., the beginnings of bioarchaeology and the study of dental disease were in the forefront. This paradigm shift and integration of the two fields through the 1970s saw a major increase in problem-oriented research on ancient skeletons such that when Cohen and Armelagos' volume *Paleopathology and the Origins of Agriculture* was published in 1984, 68% of the chapters discussed the trend in decay frequencies in relation to the adoption of an agricultural subsistence economy. At the same time there began the production of numerous skeletal analysis methods volumes, starting with Wing and Brown (1979) and followed by Gilbert and Mielke's (1985) *The Analysis of Prehistoric Diets*, that included the use of dental decay to reconstruct the proportion of sugar and carbohydrates in the diets of ancient peoples (see also İşcan and Kennedy, 1989; Katzenberg and Saunders, 2000).

Dental wear, mired in its application to age determination, had a much more tortuous path to achieve its routine application to dietary reconstruction. Miles (1958, 1963) devised a system of age determination that took into account the different diets of ancient peoples, and its reintroduction in Brothwell's widely read *Dental Anthropology* made it widely known after 1963. A review of numerous site-specific osteological articles, appendices, and monographs turned up only one that employed the Miles technique (Black, 1979). The reason for this lack of use in standard skeletal analyses is very simple: this technique requires the user to determine the wear rate for all the skeletons in the sample and then, once calibrated, data are used to age the individual skeletons. Lack of use is also true for more recently developed and similarly complex systems such as that by Lovejoy (1985). Most archeologists and osteologists did not use the more complex population-based systems because they wanted one that was simple and that could be used for aging one skeleton at a time—the forensic approach. This simple system was provided by Brothwell (1963c) in *Digging Up Bones*, where he provided a diagram of dentin exposures on molars that were divided into four adult age periods (replacing Hrdlička's scheme of 1939). Although he admonished the reader that these age groupings by wear stage were only applicable to premedieval British skeletons, we doubt if the many hundreds of users of this book ever adjusted the age-wear scoring for their particular skeletal samples before using it to age skeletons. At least none ever said that they had done so! Harn (1971) reproduced, almost exactly, Brothwell's schema in his monograph on the Dickson Mounds Illinois skeletons, and although in the text he warned of a culture-specific limitation, he did not change Brothwell's chart to use it for his western Illinois skeletal series. The problem is simply that the users of Brothwell's book who have found a nicely visual aging system seldom searched through the text looking for caveats and cautions. Brothwell's dental wear chart is also reproduced in the most widely used American osteology manual, authored by Bill Bass (1971). Here again Ubelaker, who wrote the section on teeth for the second edition, clearly stated that wear is diet specific and that the system cannot be used on populations whose diets differ from those for whom it was developed. Despite these admonitions, examination of hundreds of skeletal analyses shows that Brothwell's dental wear aging system is one of the most frequently used means of age determination (e.g., Bennett, 1973a).

The situation began to change in the 1960s, when more reports were concerned with dietary reconstruction, and dental wear slowly assumed an important role. One example of an early osteological chapter from a salvage excavation project will serve to make this point. In the analysis of skeletons salvaged from the John H. Kerr Reservoir Basin in Virginia, published in the *River Basin Survey Papers*, Hoyme and Bass (1962) used decay and wear rates to explain the differences in diet between the groups—one classified as preagricultural and the other as agricultural. They talked about how "the coarse diet produced rapid wear of teeth" (Hoyme and Bass, 1962:351) and later that "it is safe to assume that the diet . . . was varied and well cooked, soft and fairly starchy" (Hoyme and Bass, 1962:354) while "the well worn teeth confirm the suggestion of a diet containing far more coarse, fibrous foods, than soft, starchy mushes and puddings" (Hoyme and Bass, 1962:355). This is just one example in the paradigm shift that brought wear back as an important tool for dietary reconstruction.

The paradigm shift resulting from the cross-influence of "New Physical Anthropology" and "New Archeology" upon the young physical anthropology faculty and their graduate students can be seen easily in the study of dental wear as these students developed new techniques to elucidate the mechanics of wear. The most prominent problem being addressed by archeologists was when and why agriculture developed; once the graduate students in physical anthropology had a problem to attack using data obtained from excavated skeletons, methodological innovations became bountiful. In one early study, Patricia Smith (1972) examined the archeological hypothesis that the Natufians of the Jordan River valley were protohorticulturalists. She modified the Broca/Hrdlička wear scoring system to demonstrate that the Natufians exhibited the increased wear that should be associated with increased grain consumption. Lunt (1978) used the Murphy system to document refinements in diet between prehistoric and medieval Danish skeletal samples.

Mummery (1870:42) was one of the first to observe that oblique wear (the occlusal surface becomes angled to the cheeks on the lower molars) can be attributed to the chewing of tough hard foods. Similar observations were made by Brace (1962), Leek (1972), Murphy (1959a), and Taylor (1963). Building on these ideas, Molnar (1971) devised a system for measuring the angle of wear and classifying the shape of the worn occlusal tooth surface. He tested his methods on preagricultural and agricultural skeletal samples. Hinton (1981) made improvements to Molnar's system and tested the improved methods by comparing samples of hunter–gatherers and agriculturalists. B. Holly Smith (1984) contended that flat wear should be associated with tough, fibrous diets and could identify the earliest stages of the agricultural revolution. She developed a technique for measuring occlusal wear angles and demonstrated that a change in angle within any given degree of wear would identify the earliest stages of subsistence change. Further improvements in scoring methods were subsequently offered by Scott (1979a,b) and Walker (1978) to name two, while improvements in statistical analysis of data were offered by Conover and Iman (1980) and Benfer and Edwards (1991) as examples. Applications of wear analysis to routine skeletal studies increased as researchers focused on the agricultural revolution. Again as an example of widespread use we have Cohen and Armelagos' (1984) *Paleopathology at the Origins of Agriculture*, where 37% of the 19 substantive chapters used dental wear data to document the transition to food production.

A significant methodological revolution in the study of wear occurred when Dahlberg and Kinsey (1963) examined the scratches on teeth with a light microscope and suggested that the analysis of microwear could be useful in dietary reconstruction. The switch to using the scanning electron microscope (SEM) began with Walker *et al.* (1978) when they demonstrated differences in microwear in hyraxes that were either browsers or grazers [see Rose and Ungar (1998) for a detailed history and discussion of microwear studies]. The real push to develop microwear methods derived from controversies concerning the dietary reconstructions of fossil hominids (e.g., Grine, 1977; Ryan, 1980; Walker, 1981). Development of methods occurred rapidly as researchers turned to experimental studies in the laboratory (e.g., Covert and Kay, 1981; Peters, 1982), the development of methods for quantifying the microwear (e.g., Gordon, 1982, 1984;

Kay, 1987), and the collection of quantified data from living primates (e.g., Teaford, 1985, 1986; Teaford and Walker, 1984). The earliest applications of microwear analysis to the study of excavated skeletal samples were entirely qualitative, but focused on dietary reconstruction and, more specifically, the transition to agriculture (e.g., Harmon and Rose, 1988; Hojo, 1989; Puech *et al.*, 1983; Rose and Marks, 1985). These studies, especially their routine use in skeletal analyses, were hindered by the expense and time needed to extend qualitative observation to quantified analyses. A major breakthrough for widespread application of microwear analysis was the introduction of a semiautomated system for collecting data from SEM micrographs scanned into a computer (Ungar, 1995). Like the analysis of gross dental wear, microwear studies quickly focused on the transition to agriculture for their problem orientation. Examples include Bullington (1991) and Teaford (1991) working on North American problems, Pastor (1993) on the Indian subcontinent, and Molleson and colleagues (1993) on Neolithic sites in Syria.

The seventies witnessed the greatest revolution in research methodology in the area of dental defects. The research of those doing animal experimentation and clinical studies came together neatly with the work of those examining ancient teeth looking for the elusive evidence of superior structure and an explanation for the modern plague of dental decay. Once again Brothwell's (1963b) edited *Dental Anthropology* played a pivotal role in promoting the use of the latest methods by the large corps of graduate students working on dissertations and interested in problems posed by archeologists. It is not so much that each of these chapters was truly innovative, but that enough research had been published that the authors could securely establish how dental anthropology could make a significant contribution to the study of the past. Second, the book was widely read by physical anthropologists and archeologists, thus taking these studies from the relative obscurity of dental and medical journals to a broad spectrum of anthropologists and archeologists alike. In his chapter, Clement (1963) reviewed the previous studies integrating the knowledge gained from clinical and animal studies with that of living nonindustrial people and ultimately the study of ancient teeth. His comparison of prehistoric and modern dental microstructure clearly established that ancient peoples also had dental microdefects and that variation in enamel structure was not the leading cause of the modern increase in dental disease. We think that this chapter had a direct—and through the lectures of faculty an indirect—influence on a new generation of dental anthropologists.

Swärdstedt (1966) conducted the first systematic study of enamel hypoplasias in an archeological population, demonstrating that hypoplasias were more common among the lower social classes of a medieval population in Westerhus, Sweden. His work was followed in the 1970s by a steady stream of studies applying the analysis of enamel hypoplasias to reconstruction of childhood stress in ancient skeletal populations. Here the guiding concept is that the frequency and age distribution of dental developmental defects could be used to reconstruct the amount and age pattern of childhood stress, and these patterns then could be compared among samples to establish trends in childhood stress associated with such phenomena as the adoption of agriculture and the decline of cultures. These included studies of the decline of the Maya (Saul, 1972; Saul and Hammond, 1974); prehistoric California Indians (Schulz and McHenry, 1975); the shift to agriculture in the prehistoric Ohio Valley (Sciulli, 1978); the temporal trends in the rise of Egyptian civilization (Hillson, 1979); and the adoption of agriculture in the lower Illinois River Valley (Cook, 1979, 1981, 1984; Cook and Buikstra, 1979). All of the studies focused on testing hypotheses derived from the new problem-oriented archeological research concerning the human response to culture change. Unlike caries and wear, hypoplasias had been noted occasionally in skeletal studies and appendices, but virtually no interpretations were made. Knowledge of childhood stress episodes was of little interest until the researchers were engaged in problem-oriented research. This is demonstrated by 10 of the 19 chapters in *Paleopathology at the Origins of Agriculture* (Cohen and Armelagos, 1984) using hypoplasias extensively to document the increased stress caused by the transition to agriculture.

It took slightly longer for the analysis of dental microdefects to be used for evaluating the adaptation of ancient cultures. First, it had to be demonstrated more broadly to American researchers that the analysis of dental defects observed in thin sections was possible (Molnar and Ward, 1975). Once the technique was shown to be plausible, it then had to be demonstrated that the frequency of microdefects did correspond with the expected change in childhood stress levels. Further, methods for recording and interpreting the age of occurrence had to be standardized. These studies came very rapidly, fueled by the abundance of young researchers (Clarke, 1978; Condon, 1981; Cook, 1981; Jablonski, 1981; Rose *et al.*, 1978; Rudney, 1981; Wright, 1987, 1990). Methodological innovations are still occurring, but the routine application of microdefects to testing archeologically derived hypotheses is standard and is included in standard works on dental anthropology (e.g., Hillson, 1996; Kelley and Larsen, 1991; Alt *et al.*, 1998) and bioarchaeology (e.g., Larsen, 1997).

Once again Brothwell's (1963b) *Dental Anthropology* makes a significant contribution to the study of dental morphology because it is here that Dahlberg (1962) promoted the use of dental morphology to assess population affiliations and to document migrations among ancient peoples. Another pivotal study in the use of dental morphology in skeletal analyses is David Green's (1965) dissertation and subsequent publications (Green, 1966, 1967) on the dental morphology of an ancient Nubian skeletal series. He used dental morphology to counter the assertions derived from cranial analysis in establishing that a temporal series of skeletal collections are genetically homogeneous and are the product of evolution *in situ*. This alternative to craniometrics for establishing genetic homogeneity, a crucial first-step analysis for osteologists attempting to interpret changes in pathological lesions over time, is unfortunately not adopted for routine skeletal analyses. There are a number of reasons for this indifference, but it is most likely due to the amount of time such data collection requires, the complexity of the statistical analysis, and the lack of standards for collecting comparable morphological data. Dahlberg's dental models were excellent but included only a small number of the available morphological variants.

During the post-1960 period, Turner single-handedly altered the use of dental morphology in the study of ancient skeletons from the Americas. Beginning in 1983, Turner produced a stream of publications that used dental morphology to reconstruct the peopling of the Americas (Turner, 1983a,b, 1984, 1985, 1986). Further, his series of plaster plaques illustrating the scoring of the morphological variations was made available to researchers at nominal cost and are now widely used (Turner *et al.*, 1991). This series, known as the *Arizona State University Dental Anthropology System*, provided an easily learned and standardized scoring system. This is not to say that Turner was the only researcher using dental morphology to reconstruct population affiliations. Lukacs (1983, 1987, 1988) used dental morphology to establish genetic affiliations for his ancient skeletal series from the Indian subcontinent. However, the use of dental morphology in American bioarchaeology has been very limited and has produced few studies of note other than those by Turner and his students (e.g., Sciulli, 1978). However, this may change as Turner's ASU dental morphology system has been incorporated into the widely used [especially for Native American Graves Protection and Repatriation Act (NAGPRA)-related skeletal analyses] "Standards" volume edited by Buikstra and Ubelaker (1994).

Just as Brothwell's book promoted all the other areas of dental research, dental metrics were covered by Goose (1963). He mentioned Carr's (1960) measurements of Middle Minoan teeth, Lundstrom and Lysell's (1953) study of medieval Danish skeletons, and Nelson's (1938) measurements of the Pecos teeth, but concluded that little more than reporting had been accomplished. In contrast, he discussed the application of multivariate techniques of dental measurement analysis to fossil hominids, all of which was very exciting but of little interest to osteologists. However, his chapter did lead to the development of a dental sexing technique developed by Ditch and Rose (1972) and further enhanced by Black (1978), among others, for deciduous teeth. However, the technique has never been widely used as formulae must be developed for each skeletal series. Perzigian (1975) used measurements of the teeth to assess fluctuating dental asymmetry for the interpretation of childhood stress among excavated skeletal samples (i.e., the Arikara), but the technique was heavily criticized on methodological/analytical grounds (e.g., Green, 1984). Although methods for measuring teeth are provided in the skeletal "Standards" edited by Buikstra and Ubelaker, the use of dental metrics in routine North American skeletal analyses remains minimal at best. Kelley and Larsen's (1991) *Advances in Dental Anthropology* contains only one chapter on dental measurements concerning ethnic variation in tooth size, while Alt *et al.* (1998) have three chapters mostly devoted to sex determination.

VI. CONCLUSIONS

From the 19th century onward dentists wondered why dental decay was so rampant in industrialized western populations. To answer this question they turned to the past for answers and instituted problem-oriented research using ancient skulls. Thus, caries research progressed using both living groups and ancient teeth. From the earliest research using ancient North American skeletal collections, changes in ancient diets were associated with changes in the frequency of dental decay, and by the second decade of the 20th century increased dental decay had been associated with the switch to agriculture. Unfortunately, osteologists analyzing excavated collections and producing numerous site-specific appendices, chapters, and reports were sporadic, until the 1960s, in their use of decay to reconstruct ancient diets. The use of dental wear for dietary reconstruction was also established during the earliest periods, but the great need for some means of determining age-at-death diverted applications of dental wear from dietary reconstruction. It was not until the 1950s and 1960s that the development of reliable age indicators enabled dental wear studies once again to focus on dietary reconstruction. Once archeologists began to focus their attention on documenting the transition to agriculture, methods for studying caries and wear were available for testing their hypotheses and quickly saw widespread application. During this same time methods for analyzing enamel defects and morphology were developing, but had not reached the stage where they were readily applied to the interpretation of ancient human remains.

The turning point in the application of dental research methods to the analysis of skeletons was the publication of Brothwell's *Dental Anthropology* and *Digging Up Bones*. The publication of these volumes in 1963 made the results of decades of dental research widely known to both osteologists and archeologists. The routine analysis of caries and wear increased rapidly from this date onward, and the stage was set for rapid progress in the use of dental defects, tooth morphology, and metrics in analyzing ancient skeletons. At the same time archeologists were developing their own problem-oriented research strategies with an emphasis on subsistence reconstruction. Dental anthropology was ready with the tools necessary to document the switch to an agricultural economy using excavated human skeletons. Anthropology departments were growing at a rapid rate, and the young "New Archeologists" on the faculties were interacting with the growing population of physical anthropology graduate students interested in osteological research. Under the influence of the "New Physical Anthropology" paradigm requiring problem-oriented research, these graduate students (needing acceptable dissertation topics) began testing the hypotheses and subsistence reconstructions developed by the archeologists using the accumulated knowledge of dental anthropology made widely known by Brothwell's publications. Major methodological advancements were made in dissertation after dissertation during the 1970s and early 1980s. It would not be too much of an exaggeration to say that dental anthropology led the way in the development of the new field of bioarchaeology because it was here that the study of human skeletons could directly address the problems of greatest interest to archeologists — the development of agriculture.

The last 15 years of dental anthropology research have not been covered extensively in this chapter because although research and methods development has continued at a rapid pace, the place of dental studies in bioarchaeology had already been firmly established. The focus on documenting and explaining the advent of agriculture by both archeologists and osteologists motivated a great amount of significant research in dental anthropology. Other events promoted the collection of dental data from skeletons during the routine analysis of collections produced by archeological excavation. The most significant was the passage of NAGPRA, which made it clear to all that excavated skeletons had to be analyzed soon after excavation or not at all. Thus, funding for osteological analysis was available more readily in cultural resource management contracts than ever before. The potential loss of museum collections to repatriation encouraged the analysis of long-neglected skeletons and motivated bioarcheologists to develop a standard suite of data collection standards. The *Standards for Data Collection from Human Skeletons* edited by Buikstra and Ubelaker (1994) includes methods for scoring dental pathology, wear, dental defects, measurements, and morphological variation. This protocol was adopted by numerous museums for collecting data from their curated skeletons, and so these dental data have been collected from thousands of individuals. The U.S. Army Corps of Engineers has adopted this protocol for documenting the extensive Federal skeletal collections. Further, scopes of work being issued for the analysis of newly excavated skeletons from CRM mitigation projects, which once commonly listed the collection of age, sex, stature, and pathology data, now cite the "Standards" volume as the specified protocol and require the collection of extensive dental data. As a result, the collection of dental data is now a routine activity in bioarchaeology in all of its various research forms, from problem-oriented research to site-specific analyses.

Section III

On to the 21st Century

Introduction by Jane E. Buikstra

I. INTRODUCTION

In this ultimate section we begin by considering other influential 20th-century American approaches to the study of human skeletal remains from archaeological contexts that serve to anchor 21st-century "bioarchaeologies." These include J. Lawrence Angel's "social biology," "Frank Saul's osteobiography," and the "biocultural" method championed by workers such as Robert Blakely, Alan Goodman, Thomas Leatherman, and Michael Blakey. We note differences in scope and emphasis, arguing that such diversity should be considered a measure of the vitality within this developing field. Our discussion of the "bioarchaeologies" is followed by an introduction to the final chapters of the volume.

II. THE BIOARCHAEOLOGIES

As Goldstein emphasizes in Chapter 14, "bioarchaeology" carries different meanings, depending upon who is using the term. For Larsen (Chapter 13), bioarchaeology is an interdisciplinary endeavor focused primarily on questions of quality of life, behavior and lifestyle, biological relatedness, and population history. Goldstein also underscores that Larsen's tremendously productive research program has led many scholars to follow his definition of the term (Armelagos, 2003), although this departs from the original usage (Buikstra, 1977). Buikstra considered many of the same topics, but placed more emphasis on social theories and an equal partnership between archaeology and bioanthropology. In this manner, Buikstra's bioarchaeology is more "biocultural" both in the sense explicitly stated in the Blakely (1977) volume and in the manner defined by researchers such as Goodman and Leatherman (1998) and Blakey (2001), who emphasize models drawn from political economy and critical theory. As definitional distinctions are nontrivial in interpreting the history of bioarchaeology, this section begins by considering the varied 20th-century labels that American scholars have applied to bioarchaeological approaches for studying past peoples. We also briefly consider recent critiques of bioarchaeology (Armelagos and Van Gerven, 2003; Armelagos, 2003).

While a label was not explicitly defined, Wilton Krogman emphasized the integration of physical anthropology with archaeology in his 1935 article "Life Histories Recorded in Skeletons," published in the *American Anthropologist*. Broadly trained in anthropology at the University of Chicago,[1] Krogman's goal in this publication was to draw attention to the breadth of information available in archaeologically recovered human skeletons, including those of children. In this example, Krogman investigated age-at-death, growth patterning, and health as reviewed through the study of radiographically visible indicators of growth arrest, or "Harris Lines," in two sets of immature remains attributed to Euro-American pioneers interred near Hartsburg, Missouri. In closing, he reinforced the significance of skeletal material for archaeological inquiry: "No matter how fragmentary the skeleton, how incompletely it is present, each part tells its own story in the recording of the age and health and physical history of the individual" (Krogman, 1935:103). This *life history* focus appears to have emerged as Krogman's interest in forensic anthropology was developing (Krogman, 1939, 1962). It is an

[1] After 1926, under the guidance of Fay-Cooper Cole, excavations at nearby Illinois sites became a regular feature of the University of Chicago graduate curriculum (Stocking, 1979). One of the students who received field training and was placed in charge of excavations was Wilton Marion Krogman (Haviland, 1994). Krogman (Ph.D. 1929) recalled his experiences in Fulton County, Illinois, with clarity and clearly enjoyed sharing accounts with the new generation of bioarchaeologists during annual meetings of the American Association of Physical Anthropologists even after his retirement in 1983.

apt precursor to Saul's more recent osteobiographic approach, which is also influenced by forensic anthropology.

The first explicitly labeled strategy-conjoined archaeological-human osteological study of the past was the *social biology* promoted by J. Lawrence Angel.[2] As Cook and Powell note in Chapter 11, Angel's approach was heavily influenced by Hooton's population perspective. Drawing heavily upon his dissertation research, Angel's *social biology* was defined in an article published in the *American Anthropologist* (Angel, 1946a). Like Krogman, Angel wished to bring the significance of his approach to a broad anthropological readership.

As with many of Hooton's students, Angel's focus was upon testing his mentor's ideas concerning the vitality of biological heterosis and culture change. Hooton argued that biocultural adaptation was positively associated with "mixing"—both biological and cultural. Angel, knowing that both pathology and heritage could be investigated in past populations, thus set out to study diachronic changes in skeletal series drawn from the eastern Mediterranean to see if those that were more morphologically heterogeneous were also healthier (Ortner and Kelley, 1990).

Angel's social biology was rooted in several contextual lines of evidence, archaeological, environmental, ecological, and historical. While he indeed focused on population-based approaches in the past, his model also emphasized the individual:

> But since cultural tradition is a product of interaction between contrasting individuals rather than an average of the qualities, thoughts, and acts of all the people in a society, individual differences have dynamic importance quantitatively. Hence vital and scarcely seen aspects of human processes in Greek culture growth can be brought out by social biology. (Angel, 1946a:494)

It is tempting to also see this focus on the individual in the works of his mentor Hooton, whose research focus was shifting to constitutional studies, "the anthropology of the individual" (Giles, 1997:499), a subject that Angel also studied early in his career (Angel, 1946a, 1947; Angel and Wagner, 1945).

Social biology was thus theoretically driven, contextually grounded, and population based. As Angel's career progressed, so did his regional research in the eastern Mediterranean, later extended to North American forensic and archaeological contexts. He studied historic period remains from North America, both African-American and Euro-American (Angel, 1976b; Angel *et al.*, 1987). Among his many contributions were significant advances in clarifying the relationship between the environment and disease, paleodemography, behavior reconstructions, and microevoluion (Buikstra and Hoshower, 1990; Ortner and

[2]Angel was the 19th of Hooton's 28 Ph.D. students (Giles, 1997).

Kelley, 1990). His was thus a broadly based study of past peoples, with many parallels in more recent, contextually sensitive bioarchaeology.

In 1972, another Harvard Ph.D., Frank Saul, published a method he labeled an *osteobiographic* analysis.[3] Like "bioarchaeology" (Buikstra, 1977), this approach is also explicitly problem oriented, driven by questions about paleodemography, ancestry, behavior, and health. Saul explicitly wished to reconstruct the lives of his study series from Atlar de Sacrificios "as individuals and as a population" (Saul, 1972:8). His focus on the Maya world led him to emphasize health-related questions about the Maya past, especially any biological clues concerning the Maya collapse. Questions of origins and migrations were also addressed. Saul then discussed his rationale for creating a new term.

> A newly coined term, "osteobiography," has been used in the title of this report in order to indicate that just such a comprehensive and reconstructive approach was being applied to the study of the recovered skeletons of the ancient inhabitants of Altar de Sacrificios. Rather than talk about measuring (sic) "sexing," "ageing," "sickening" (?!), and so on, the term osteobiography has been used to indicate in a single word that this study is concerned with all of the foregoing aspects of skeletal analysis. This study has, in fact, attempted to interpret the Altarians' life histories as recorded in their bones, hence the creation and use of the term osteobiography.
>
> In so doing, emphasis is being placed upon the meaningful and comprehensive use of skeletal materials in an archaeological context, an approach best exemplified to date by the studies of E. A. Hooton, Pecos Pueblo (1930); W. W. Howells, the early Christian Irish (1941); J. L. Angel, ancient Greece (1946–1959); and J. E. Anderson, Fairty Ossuary (1964). (Saul, 1972:8)

Saul, like Krogman, had forensic experience, which influenced his focus on the individual and life history construction. In fact, Frank and Julie Mather Saul (1989:300) have explicitly stated that they "use much the same approach in our efforts to reconstruct the lives of individuals whose remains are brought to us by the police." They go on to reinforce a synergism between forensic and archaeological studies of human remains. "One sphere of activity enhances and can learn from the other" (Saul and Saul, 1989:301). Archaeological contexts and historical questions figure heavily in Saul's osteobiography.

In their more recent work the Sauls (1989) also credited J. Lawrence Angel's influence upon their research, as well as that of the British physical anthropologist, Calvin Wells (see Chapter 16). They note with appreciation Wells' research relating bone pathology to the manner in which individuals and groups "actually functioned in life" (Saul and Saul, 1989).

Thus, the Sauls' osteobiography resembles other contextually sensitive research programs for studying the past. Explicitly problem-oriented, their

[3]In 1967, Saul had presented this approach in papers given at the annual meetings of the American Association of Physical Anthropologists and the American Anthropological Association (Saul, 1972:80). Saul and Saul (1989) indicate that Frank Saul first coined the term in 1961.

approach recognizes a broad range of possible analytical methods that may be used in individual and population reconstruction. While the individual is emphasized on occasion, especially when encountered in unusual archaeological contexts, e.g., Saul and Saul (1989:291), ultimate goals center primarily on population-based questions about the Maya ranging from the status of women to health.

The 1977a Blakely volume, in which the term bioarchaeology was defined, in fact focused primarily on "biocultural" adaptation. "Humans survive not through cultural adaptation nor through biological adaptation, but through *biocultural* adaptation" (Blakely, 1977b:1). Holism was championed, followed by clearly defined volume goals:

> (1) to document specific ways in which biological anthropologists can contribute to studies of cultural processes; (2) to illustrate the interrelationships between the biological, cultural, and environmental variables that affect the adaptedness or maladaptedness of prehistoric populations; and (3) to demonstrate the need for cooperation among biological anthropologists, archaeologists, ethnologists, and other expert investigators toward problem-solving in behavioral anthropology. (Blakely, 1977b:3)

This holistic, biocultural approach was emphasized in several contributions, especially those of Blakely, Robbins, Buikstra, and Perzigian. New methods, such as bone chemistry in dietary reconstruction, were also illustrated (Robbins, Buikstra, Gilbert). Blakely (1977c) related demographic patterning to flexible social adaptations at Etowah. Health and agricultural intensification were addressed in three studies. Robbins and Buikstra concluded that certain measures of health status argue for a poorer quality of life among agriculturalists, whereas Perzigian's perspective, drawn from dental anthropology, concluded the reverse. Thus, holism, interdisciplinary study, complex systems-based approaches, and new methodologies drawn from other disciplines were all visible aspects of the bioculutral approach advocated in this influential volume.

During the 1970s the question of biocultural adaptation with agricultural intensification was also driving another North American research program, guided by George Armelagos at the University of Massachusetts. As noted by Cook and Powell in Chapter 11, the ambitious research agenda, initially centered on remains from the University of Colorado Nubian expeditions, was later extended to the central Illinois River valley. Armelagos' focus in the Nubian project was skeletal and dental pathology, leading to his 1968 dissertation, which presented a paleoepidemiological model integrating the host, disease, and the environment. The Nubian sample was divided into three successive chronological units: Meroitic, X-Group, and Christian. Armelagos (1968) reported mortality patterns, activity-related stress, and specific conditions, including infectious disease and congenital anomalies. In parallel, David Greene's (1965) dissertation established genetic continuity for the temporally sequential skeletal samples (but see Chapter 10). Armelagos and colleagues explicitly emphasized a population-based perspective, in reaction to older, typological craniology and case approaches to diagnosing disease (Armelagos, 1968; Van Gerven *et al.*, 1973), critiques that would reappear in later work (Armelagos *et al.*, 1982; Armelagos and Van Gerven, 2003; Armelagos, 2003). The Nubian samples served as the basis for numerous other studies, frequently on health-related subjects (for a summary, see Martin *et al.*, 1984). This research is contextualized archaeologically, with a growing trend toward considering political and economic factors that may have affected food availability, nutrition, and health (e.g., Martin *et al.*, 1984).

The biocultural program developed in the Nubian context served to anchor Armelagos' students' research on the Dickson Mounds excavated collection, which was excavated and transferred to the University of Massachusetts during the 1970s. Once again, a tripartite temporal division was considered: Late Woodland (hunter–gatherer), Mississippian Acculturated Late Woodland (transitional), and Mississippian (agricultural). Unfortunately, as discussed in Buikstra and Milner (1989), the archaeological cultural assignments were not finalized at the time of the University of Massachusetts studies. Because most of the original osteological data recording forms are not now available, skeletal observations cannot be recast into the newer archaeological model.

As emphasized by Cook and Powell in Chapter 11, the University of Massachusetts program and similar studies served to revitalize paleopathology and bring it new scholarly (Cohen and Armelagos, 1984; Cohen, 1989) and popular (Goodman and Armelagos, 1985) visibility. Again, although restudy according to current archaeological cultural assignments would be desirable, as would the integration of data collected from the now closed *in situ* Dickson Mounds skeleton display, this research continues to exert considerable influence today.

Political-economic and critical approaches ground the biocultural approach recently advocated by researchers such as Goodman and Leatherman (1998). They argue for an engaged, action-oriented biocultural perspective that supersedes the earlier adaptationist paradigm; also underscoring the importance of considering social relations, especially power relations, when interpreting the remains of past peoples. This biocultural approach is thus contextualized both culturally and socially, holding promise for rendering studies of the past more theoretically sophisticated.

As illustrated in Chapter 15, engaging descendant communities has become crucial in 21st-century bioarchaeology. One example of a biocultural bioarchaeology is that of Michael Blakey and the African Burial Ground Project (Blakey, 1998a,b, 2001). As Goldstein points out (Chapter 14), Blakey's bioarchaeology of the African diaspora is contextualized archaeologically, historically, and in terms of oral traditions. The present-day African-American community helped set a research agenda that ranged from issues of ancestry to reconstructing health, diet, and behavior. As Blakey (2001:414) emphasizes, of "extraordinary interest" to the descendant community were the studies attempting to link the skeletons to specific African societies. However, as Goldstein (Chapter 14) also notes, Blakey's critique of forensic approaches would seem overdrawn.

Forensic anthropology is also singled out for criticism in two recent reviews of bioarchaeology (Armelagos and Van Gerven, 2003; Armelagos, 2003). These articles argue for population-based research that is theoretically sophisticated, endorsing the type of health assessments Armelagos and his students have developed in Africa and in North America. Forensic anthropology is devalued for being descriptive, inaccurate, and racist. Similarly, disease diagnosis and studies of disease distributions are criticized, as are historical approaches generally. Finally, ancient DNA studies are characterized as not having produced important new knowledge, and all studies of inherited features appear to be glossed as typological and racist. Several chapters in this volume suggest that such arguments are somewhat extreme. As emphasized by Konigsberg (Chapter 10) and Stojanowski and Buikstra (2004), biological distance studies today are not typological and are increasingly grounded in population genetics. Cook and Powell (Chapter 11) argue that a balance should be struck between the identification of specific diseases, tracing their histories, and the study of general health in the past. In that epidemics of infectious diseases, for example, have been potent forces in human history, their identity and distribution are not trivial issues. Race is a contentious issue, both within the discipline of anthropology and outside. Most forensic anthropologists practicing today consider race a social construct, and ancestry determinations are a very small part of what forensic anthropologists do. If knowledge of bones can assist in medico-legal contexts, then this application of skeletal biological knowledge would seem socially quite significant. While Armelagos and Van Gerven (2003) argue that we have learned very little from the study of mtDNA, it is just these studies that the descendant communities prioritized in their heritage quest for those interred at the African Burial Ground. All investigations of heritage are not by their very nature racist. Establishing genetic relationships is also an important issue within the context of repatriation initiatives (see Chapter 15). Chapter 13 also provides compelling arguments for the significance of new analytical techniques, including ancient DNA.

III. BIOARCHAEOLOGY IN THE 21ST CENTURY

Chapter 13 first defines the questions and topics that data from archaeological skeletons can usefully address, focusing on (1) quality of life, (2) behavior and lifestyle, and (3) biological relatedness (biodistance) and population history.

Second, important recent technological and methodological advances, especially those adapted from other disciplines, are reviewed.

In assessing quality of life, Larsen emphasizes dietary reconstruction and nutritional inference, disease, and growth. He highlights the significance of new methodologies drawn from bone chemistry for inferring diet. For example, stable carbon isotope analyses have virtually revolutionized our perspective on ancient agriculture in the Americas. In assessing disease patterning, Larsen advocates the paleoepidemiological approach pioneered by Hooton and Angel (see also Chapter 11). Recognizing the importance of disease diagnosis and the ambiguity of gross skeletal lesion morphology, he underscores the significance of microscopic and ancient pathogen DNA approaches. Developmental insults are also advocated as measures of health and nutritional stress, including linear growth of long bones and dental defects that are grossly and microscopically accessible (see also Chapter 11). In discussing behavioral reconstructions, Larsen emphasizes biomechanical approaches, for the postcranial skeleton, and dental wear, including microwear. For reconstructing population histories, Larsen reviews recent studies of ancient DNA for tracing ancestry and stable isotope ratios (strontium) for tracing patterns of mobility. In closing, Larsen reaffirms the importance of bioarchaeologists' engagement in multidisciplinary archaeological research from the beginning in order to advance our knowledge of quality of life, behavior, and population histories. He also considers the contributions that bioarchaeologists can make to studies of gender and political complexity.

Chapter 14 addresses the degree to which bioarchaeology has achieved its goals, as stated in the Blakely (1977) symposium volume. Her methodology involved examining eight journals, six archaeological and two anthropological, over the years 1995–2000 for published evidence of the collaborative integration of archaeology and biological anthropology yearned for by the contributors to the 1977 volume. She also integrates recent reviews from the physical anthropological literature (Larsen, 1997, 2002), emphasizing that, as an archaeologist, her perspective may differ from others represented in this volume.

Goldstein concludes that despite the hopeful projections of productive collaborations voiced in 1977, physical anthropology and archaeology have proceeded along very different trajectories over the intervening half-century. Larsen's influential definition of bioarchaeology emphasizes multidisciplinary skeletal research anchored by new technologies for studying quality of life, behavior, and population histories. It thus develops a skeletal biology of the past and does not closely link the bodies of people to their archaeological contexts.

Goldstein also argues that as the study of archaeological skeletons has become more laboratory oriented, it may have developed a sense of false precision and failed to note recent developments in bone biology. She contends that treating the archaeological record simplistically and ignoring recent developments in archaeological theories and methods is a dangerous path for bioarchaeology.

Post-1977 advances in the archaeology of mortuary sites are then summarized. Goldstein cites gender studies, landscape archaeology, and research centered on the individual as venues important to a contemporary bioarchaeology. The impact of NAGPRA upon potential collaborations between physical anthropologists and archaeologists is also insightfully considered. Three articles that utilized isotopic approaches to inferring diet, residence, and relationships between weaning behavior and fertility are singled out as being unusually integrative of archaeological context and human osteological data. She closes her chapter by considering ways in which bioarchaeology today would benefit from closer integration with archaeology and again underscores the significance of a contextualized bioarchaeology. She draws parallels between this approach and biocultural studies of the past, as defined in the work of Blakey (2001).

Buikstra (Chapter 15) treats a topic of central importance in framing bioarchaeology of the 21st century. She traces repatriation initiatives and legislation in the United States and Canada. The two North American political contexts provide an intriguing contrast in that the United States has federal legislation mandating reburial whereas Canada does not; it appears that relations between bioarchaeologists and First Nations are generally better in Canada than in the United States.

Buikstra considers the impact of NAGPRA and related legislation upon bioarchaeological research and the professions of archaeology and bioanthropology. The Kennewick decision is discussed in the context of NAGPRA, with emphasis on the tensions between scientific–archaeological/humanistic–traditional approaches to heritage issues. The role of the Society for American Archaeology in framing the discourse about repatriation is considered, including the society's position on the repatriation of culturally unaffiliated materials.

Case studies that underscore the tensions inherent in repatriation are reported, as are productive, positive collaborations from both the United States and Canada. Repatriation issues have undoubtedly complicated bioarchaeology, as practiced during the late 20th century, and will continue to do so. Problem-oriented excavations are no longer feasible unless sites are endangered. Deaccessioned collections cannot be restudied to verify earlier observations or to apply new methods. However, conversations and partnerships between Native American people and bioarchaeologists do hold promise for enriching the interpretations of past histories and making contemporary studies more meaningful to descendant communities in a socially responsible manner. There are indeed both challenges and opportunities inherent in repatriation initiatives. It is hoped that as mutual respect develops in the course of mutual ventures, the bioarchaeology of the 21st century will emerge as an even more robust approach to the peopling of the past.

In Chapter 16, the closing chapter of this volume, Roberts compares and contrasts the development of bioarchaeology in North America and Britain.

In so doing, she underscores the fact that in general, Britain has lagged behind the United States in the development of a richly contextualized bioarchaeology.

Roberts begins her discussion with a consideration of various English terms that have been applied to the study of human and other remains from archaeological sites, including "bioarchaeology" and "osteoarchaeology." As noted in the preface to this volume, "osteoarchaeology" was initially coined by Vilhelm Møller-Christensen (1973, 1978) to denote a careful examination and excavation of burial places conducted by someone familiar with human remains and the manner in which they are studied. This observational emphasis has been extended by French anthropologists in a technique known as "l'anthropologie de terrain" (field anthropology[4]), a complement to funerary archaeology, paleodemography, and paleopathology (Leroi-Gourhan *et al.*, 1962; Masset, 1972; Duday, 1978; Duday and Masset, 1987). Developed to facilitate the interpretation of complex Neolithic sepultures, the method specifies precise field observations and related analyses that facilitate reconstructing extended burial programs, circumstances of death, and life histories. The dynamic process of cadaver decomposition and decay is emphasized in relationship to depositional contexts. While the approach has been brought to the attention of American forensic scientists as a taphonomic method (Roksandic, 2002), it is seldom referenced by archaeologists or bioarchaeologists working in North America or Britain.[5]

Following her brief review of the history of physical anthropology in Europe, Roberts focuses upon 19th- and early 20th-century British "racial" studies as a background for the manner in which scholars undertook the study of skeletal remains. The development of the profession of physical anthropology is considered, along with the history of archaeological and anatomical museum collections. She notes the absence of large collections of complete skeletons from known individuals in Britain, contrasting with the situation in North America and in Portugal.

Turning to the second half of the 20th century, Roberts first summarized seminal contributions by European scholars such as the medical doctor Møller-Christensen (1903–1988), who pioneered in both excavation methods and in skeletal observations, especially of infectious diseases such as leprosy. In Britain, another physician, Calvin Wells (1908–1978), is considered to have been among the pioneers in combining archaeological and biological evidence in reconstructing ancient lives. Beginning to publish in the mid-1950s, Wells proved a remarkable prolific writer and thus his influence extended well beyond Britain's boundaries. His approach was clearly "bioarchaeological," as the term is used in

[4]Given that "anthropology" denotes physical anthropology in France, l'anthropologie de terrain could also be glossed as "field physical anthropology."

[5]However, see Malvido *et al.* (1997), Scarre (2004), Tiesler Blos (2004), and Pereira (1999).

North America. Contributions of other medical doctors, Keith Manchester, Cecil Hackett, and Eric Hudson to bioarchaeological investigations are also cited.

Another productive British scholar, Don Brothwell (b 1933), is perhaps best known for his handbook *Digging Up Bones.* Trained in geology and zoology, as well as anthropology and archaeology, Brothwell's approach to the archaeological record is eclectic, with his contributions including primary research on human skeletal and mummified tissues, including hair. His current work on zoonoses focuses on human animal transmission in an evolutionary framework, holding promise for yet another series of innovative contributions.

Turning to recent decades, Roberts describes the variety of contributions being made by the current contributing generation of bioarchaeologists and the training/research programs from which they have emerged. Britain has assumed a prominent role in advancing biomolecular approaches for the study of heritage, health, and residence histories. The recent development of a set of standards for data recording should also advance British bioarchaeology, which has yet to meet the challenge of developing a systematic national database concerning collections locations and composition—an initiative developed in the United States following NAGPRA's mandates. Standards for field data recording are variable in both countries.

In closing, Roberts considers factors that may affect future bioarchaeological research in Britain. Limiting attributes include repatriation, damage to existing collections, and funding for other than biomolecular studies. Enhancing variables include scholarly interest in the history of disease, the presence of large skeletal collections, historical records beginning in the medieval period, and a strong tradition of rigorous, scientific study of the past.

Chapter 13

The Changing Face of Bioarchaeology: An Interdisciplinary Science

Clark Spencer Larsen

I. INTRODUCTION

Bioarchaeology has developed into a distinctive discipline due in part to growth in interest in the role of human remains in understanding the history of the human condition. Previous generations of bioarchaeologists typically studied archaeological skeletons without ever having seen the context of recovery. Typically, the remains were excavated by an archaeologist and then transported to the laboratory, where a sole worker — the bioarchaeologist — studied them. Thus, collaborative research was limited to the interaction between the individual who excavated the skeletons and the individual who studied them. Oftentimes, the results of the investigation of the remains ended up in an obscure archaeological publication or report as an appendix.

Although the disconnection between archaeological context and bioarchaeological study continues to be an all too frequent practice, the on-site presence of skeletal specialists and supervision of excavation by bioarchaeologists is becoming commonplace. My sense of the field of bioarchaeology is that the results of study of skeletons are now incorporated into the body of research reports versus the earlier practice of relegation to unread appendices (and see Buikstra, 1991). Moreover, in contrast to the earlier practice of the lone osteologist attempting to report on all aspects of skeletal variation, bioarchaeological research increasingly involves teams of scientists — often drawn from nonanthropological disciplines. These teams draw from a range of areas of expertise that join together to address common problems and questions.

Bioarchaeology: The Contextual Analysis of Human Remains, Buikstra and Beck (eds.)

The purpose of this chapter is twofold. First, it discusses the questions and topics that archaeological skeletons are especially useful at addressing. Second, it discusses important technological and methodological advancements made in the field in the last several decades. Bioarchaeologists have been successful at drawing from research protocols developed in other sciences, and this chapter identifies some tools and skills that have been especially useful in helping us develop a more informed understanding of the human past, at least as it is based on the human biological component of that record.

II. THE SCOPE OF BIOARCHAEOLOGY

The changing scope of bioarchaeology, especially its interdisciplinary orientation, reflects what has happened in most other areas of science in the later 20th century and into the 21st century. Bioarchaeologists have been adept at identifying and coopting developments in other disciplines in addressing problem-oriented research. As discussed in this chapter, the disciplines that have contributed to bioarchaeology include those housed in a diversity of fields, including the biological, geological, chemical, physical, and engineering sciences — the so-called hard sciences — and the social and behavioral sciences. More than anything else, this diversity of fields that bioarchaeology draws from reflects the fact that human beings are highly complex. Unlike other animals, *Homo sapiens* and their ancestors involve a complex interaction among biology, culture, and the environment. This interaction is expressed in multiple ways in biological tissues that are often difficult to interpret. It is the bioarchaeologist's primary task to identify and interpret this complex interaction from the remote (and sometimes not so remote) past, relying on preserved biological tissues. Understanding this complex interaction helps us understand past populations as though they were alive today — as living, breathing, and functioning human beings.

There are some limitations in what we can learn about past human biology. Bioarchaeologists are almost always restricted to the study of bones and teeth (rarely partial or whole bodies are available for study), whereas human biologists who study the living are able to look at a range of behavioral and anatomical parameters that either cease to exist once a person dies or disappear altogether once the soft tissues have decayed away. Moreover, skeletons from a cemetery may not always be representative of the living population from which they are drawn. For example, a person who dies from an infection may do so well before the osteological signature has had time to develop. However, a person who displays an indication of a past perturbation or stress, such as hypoplasia or periostitis, may have had relatively robust health — enough so that the person survived the stress that caused the osteological modification. Thus, it is quite possible that there is a positive correlation between pathology and health in individuals drawn from a cemetery population (see Wood *et al.*, 1992; Cohen, 1998; Buikstra, 1997). Key in understanding the representation of archaeological cemeteries and the skeletons drawn from them is that they are aggregations of samples of populations, usually covering multiple generations.

III. AREAS OF BIOARCHAEOLOGICAL INQUIRY

Despite the limitations inherent in the study of ancient remains, there is much to learn about the past from them. This section focuses on the following three areas that archaeological human remains are informative about when drawing inferences about the past: (1) quality of life, (2) behavior and lifestyle, and (3) biological relatedness (biodistance) and population history.

A. Quality of Life: The Living and the Dead

The measurement of quality of a person's life is highly subjective and reflects a wide variety of circumstances—social, cultural, and biological. Because of this subjectivity, quality of life is difficult to measure and can represent many different things to different people (Bennett and Phillips, 1999). For example, the number of labor-saving devices a person owns might be one measure. Most measures of quality of life include health as the chief component, especially regarding disease and its consequences for the individual or population (Ware, 1987; Guyatt *et al.*, 1993; Allison *et al.*, 1997).

The central role that health plays in measuring quality of life in the living indicates that documentation of health indicators in ancient skeletons can provide bioarchaeologists with a means of assessing quality of life in the past. This chapter focuses on dietary reconstruction and nutritional inference, disease, and growth, especially new tools developed in other sciences that have helped refine our understanding of the past.

1. Diet and Nutrition: You Are What You Eat

An understanding of diet (the foods eaten) is fundamental to understanding health, mainly because diet provides the nutrition (the nutrients that these foods provide) one needs for a healthy life. For most of the history of anthropology, diet in past populations was based largely on the study of plant and animal remains. Traditionally, these areas of investigation—paleoethnobotany and zooarchaeology, respectively—were outside the purview of bioarchaeology. Skeletal biologists did not involve themselves in the reconstruction of diet or implications that these diets had for nutrition in past populations.

With the development of stable isotope analysis of archaeological human bone for dietary reconstruction, bioarchaeologists became deeply involved in the growing discussion in anthropology of issues relating to food use, dietary reconstruction, and nutritional inference. Stable isotope analysis was coopted by bioarchaeology from other disciplines—the theory is based in physics and it was first applied in geological sciences, especially geochemistry (Schoeninger, 1995). Elements that comprise tissues of plants and animals occur in various forms called isotopes, including, for example, carbon (C), nitrogen (N), hydrogen (H), oxygen (O), and strontium (Sr), which differ according to the number of neutrons found in their nuclei. Plants use either one of three photosynthetic pathways, which, because of the differences in how carbon is acquired from atmospheric carbon dioxide, express differences in the ratios of the stable forms of the element (^{13}C to ^{12}C). Because these differences in ratios of ^{13}C to ^{12}C—expressed as $\delta^{13}C$ values—are passed up the food chain to the consumer (animals and humans), researchers realized the tremendous potential for paleodietary study. Using instrumentation developed in chemistry called mass spectrometry, Vogel and van der Merwe (1977; van der Merwe and Vogel, 1978) measured amounts of stable carbon isotopes ^{13}C and ^{12}C and their ratios in archaeological bone samples from New York state. They hypothesized that because maize was the primary economically important plant consumed by prehistoric humans in this region of North America, it should be possible to identify the timing and importance of its use by examining the stable isotope ratios in a temporal succession covering the transition from foraging to farming. That is, maize has a C_4 photosynthetic pathway, in contrast to most other plants eaten in this region, which are almost exclusively C_3 plants. Because of differences in the way that carbon is acquired between C_3 and C_4 plants, plants of the former variety have lower (more negative) $\delta^{13}C$ values than plants of the latter variety. This means that the bone tissue (collagen) should express higher (less negative) $\delta^{13}C$ values when the shift to maize agriculture occurred, and the higher the value, the greater the importance of maize in diet. Their pilot study provided compelling evidence that contrary to the assertions of many archaeologists, maize did not become an important part of diet until late in North American prehistory, mostly after about AD 800 or so.

Since the mid-1980s, archaeologists and bioarchaeologists have analyzed thousands of human bone samples from around the world to address the shifting patterns of human diet based on stable carbon isotopes. Other stable isotopes and trace elements have also provided an enormously important perspective on diet and nutrition, such as for identifying relative use of terrestrial vs marine resources in coastal settings [see Schoeninger (1995)].

The knowledge that maize was grown and harvested later in prehistory provided a new and fresh perspective on other changes that took place in prehistoric societies in a number of regions. In eastern North America, at about the same time that maize became a key part of economy, there were widespread changes in sociopolitical complexity and settlement. Populations in the final centuries of prehistory became more complex and more sedentary than their predecessors and were more dependent on agriculture. The increasingly sedentary nature of these groups was also accompanied by an aggregation of population in villages and towns. Most authorities are convinced that the cultural florescence that took place during this time, known as the "Mississippian," was fueled by an economy reliant on production of domesticated plants, specifically focused on maize (Smith, 1989).

The focus on maize has important implications for health and nutrition in prehistoric societies. For example, maize is a poor source for protein in that it is lacking or deficient in several essential amino acids necessary for growth and development. This and other factors suggest that a change in diet led to a decline in nutritional quality (Larsen, 1995).

2. Disease

Disease is also an important component of quality of life. The representation of disease in ancient remains has been the focus of study by bioarchaeologists and others since at least the 18th century, and inferences about quality of life have been made (Ubelaker, 1982; Buikstra and Cook, 1980; Larsen, 1997). Unfortunately, due to the overlap in bony responses to specific pathogens, it is oftentimes difficult, if not impossible, to diagnose specific diseases. Beginning with the publication of Earnest Albert Hooton's classic monograph, *The Indians of Pecos Pueblo, a Study of Their Skeletal Remains* in 1930, a new approach to the study of ancient disease commenced. Hooton presented the frequency of a variety of skeletal pathological conditions present in the Pecos series, such as for osteoarthritis, infection and inflammatory conditions, and trauma. His inchoate paleoepidemiological study set the stage for population-oriented research undertaken by J. Lawrence Angel in the eastern Mediterranean region (Angel, 1966b, 1984) and later workers (e.g., Cook, 1984; Larsen, 1982; Ubelaker, 1994; and many others). Especially important about Hooton's approach is the shift in focus from mostly diagnosis of specific diseases to the development of a greater understanding of the importance of context and biocultural setting in interpreting disease prevalence and pattern. That is, bioarchaeologists seek to identify the environmental, social, cultural, and other factors that best explain the presence of a disease or set of diseases in the past and circumstances for changing frequency.

Diagnosis is still an important element of understanding disease and health history in an ancient population. Skeletal lesions are notoriously difficult to match with a particular disease that caused them. Some general patterns have emerged that are consistent with uniformitarian notions of how pathogens operate in a living population. For example, bioarchaeologists have documented in many settings a general pattern of increase in frequency of periostitis and bone infection in later prehistory where populations have increased in size and are concentrated in settled communities (Larsen, 1995). Epidemiological theory tells us that infection increases when human populations become denser and sedentary. Indeed, this is the general pattern that we see in the prehistoric past, especially in North America where the bioarchaeological record is most complete.

This pattern change in periostitis is interesting, but still leaves open the question of what diseases were present in the past. The identification of specific diseases is not just an "academic" question. Rather, their documentation offers an important avenue for reconstructing the evolutionary history of infectious diseases and for understanding their interaction with humans and other organisms. This knowledge provides an important tool for control of the disease in the living.

The osteological signatures of disease are somewhat clear for several chronic infectious diseases, including tuberculosis, treponematosis, and leprosy (see Ortner and Putschar, 1985; Larsen, 1997). However, the overlap in skeletal manifestations between these and other disease syndromes greatly constrains our ability to identify specific infectious disease in ancient remains. Tools developed in other sciences are offering important — and potentially revolutionary — insight into the history of disease. First, histology, the microscopic study of tissue structure, offers perspective on disease diagnosis that is informative in ways not possible with gross inspection of pathological bone. One application in particular is the histological analysis of cranial bone exhibiting cribra orbitalia and porotic hyperostosis. There has been a growing consensus in bioarchaeology and paleopathology that these two pathological conditions represent iron deficiency anemia. However, detailed examination of thin sections of bone lesions reveals that although diagnosis of anemia is correct in many cases, scurvy, inflammation, and other agents or circumstances can be involved (see Kreutz, 1997; Carli-Thiele, 1996; Schultz, 1993; Schultz *et al.*, 2001).

Second, DNA extracted from ancient bone is beginning to confirm the presence of specific diseases in osteological remains. Like the human host, the pathogens that cause disease are living organisms. Therefore, just as human remains retain nucleic acids with potentially amplifiable DNA, so too should the pathogens that the human host was carrying at the time of death. Spigelman and Lemma (1993; see also Rafi *et al.*, 1994) completed one of the earliest studies on DNA in order to detect the presence of the pathogenic organism that causes tuberculosis, *Mycobacterium tuberculosis*. They applied polymerase chain reaction (PCR) analysis — a technological breakthrough in molecular biology that allows identification and amplification of DNA (Mullis and Faloona, 1987) — to bone samples from skeletons that have skeletal lesions diagnosed as tuberculosis. PCR analysis revealed the presence of *M. tuberculosis* DNA. Similarly, extracted DNA from soft tissue (lung) tuberculosis in a 1000-year-old mummy from Chiribaya Alta in southern Peru was analyzed by PCR, identifying a segment of DNA that is unique to *M. tuberculosis* (Salo *et al.*, 1994). This evidence for presence of tuberculosis in a pre-Columbian New World setting is enlightening because it presents unequivocal evidence for the disease in the New World well before the arrival of Europeans. This finding runs counter to the traditional argument that tuberculosis was introduced by early Spaniards. This powerful tool also now reveals the presence of tuberculosis in a range of settings around the world, both before and after the late 15th century (e.g., Spigelman and Donoghue, 1999; Faerman *et al.*, 1999). Moreover, unlike human DNA extracted from archaeological skeletons, pathogen DNA is relatively contamination free, making it a highly promising material for future investigation (see Kolman and Tuross, 2000).

3. Growth and Development

Growth and development of skeletal and dental hard tissues offer important perspectives on health and nutritional status, perhaps better than any other indicators. The current understanding of human skeletal growth as it is documented in ancient skeletons is backed by an extensive literature, especially following Stewart's (1954) study of Eskimos and Johnston's (1962) study of Indian Knoll, Kentucky (and see review in Hoppa and FitzGerald, 1999). Most bioarchaeological studies focus on linear growth of long bones and dental development. The resulting growth profiles from bones and teeth suggest that environmentally disadvantaged populations have retarded growth.

Studies of enamel defects called pathological striae or lines of Retzius (or Wilson bands) from ground thin sections of teeth using histological techniques provide an important retrospective picture on growth history for skeletal individuals. Jerome Rose (1977, 1979; Rose *et al.*, 1978) was among the first to show the important value of histological analysis in understanding growth history and to infer quality of life in archaeological populations. Based on his study of dentitions from central Illinois (Dickson Mounds) in a temporal succession of populations, he was able to identify a pattern of increasing physiological stress based on a higher frequency of microdefects in later than in earlier teeth. Rose argued that the increase in stress was caused by a decline in nutritional quality with the increased emphasis on maize, along with population aggregation and increased disease stress.

In historic settings where stress can be documented through written sources, the context for study of microdefects is provided. In Spanish Florida, Roman Catholic missions were established among native populations in the 16th and 17th centuries. For this region of North America, there is a range of evidence—bioarchaeological and documentary—indicating deteriorating quality of life in native populations (Hann, 1988; Larsen, 2000). During this time, native populations increased maize production and consumption, relocated to crowded mission communities, and increased labor generally. Simpson (1999, 2001) analyzed longitudinal thin-sections of anterior teeth (incisors, canines) from prehistoric and mission-era native populations in northern Florida. Whereas Rose's analysis involved light microscopy, Simpson applied scanning electron microscopy (SEM) for observation and analysis. Scanning electron microscopy is now the preferred tool in histological study, mainly because it offers much greater power of magnification, greater depth of focus, and greater resolution in detail of structure than standard light microscopy (see discussion in Teaford, 1991). SEM uses an electronic beam that scans the tooth section, causing an emission of electrons. These secondary electrons are amplified, resulting in an image reflecting very slight variations in brightness created by surface features in the tooth section. Like the image seen on a standard television, the image of the section represents an electronic composite of many points of light rather than the object itself.

Normal Retzius lines are slightly darkened and radiate outward from the dentine–enamel junction to the tooth's surface. The regular spacing of the lines reflects periodic growth deposition of enamel lasting anywhere form 6 to 9 days of growth. Pathological Retzius lines are abnormally dark bands reflecting acute stress episodes lasting from several hours to several days. Simpson's comparison of mission teeth with pre-mission (mostly prehistoric) teeth revealed a marked increase in pathological Retzius lines (from 48 to 83% of teeth). This finding is consistent with the notion that mission peoples experienced more stress than their pre-mission ancestors. Pathological Retzius lines are a nonspecific indicator of health, but Simpson (2001) believes that for this setting the lines are likely caused by dehydration from infantile diarrhea. Under these circumstances, severe dehydration results in dysfunction of ameloblasts caused by intracellular fluids moving into intercellular spaces in the developing enamel. His interpretation is consistent with the fact that the mission setting was highly unsanitary and conducive to conditions that would cause widespread infantile diarrhea (see Larsen *et al.*, 1992; Larsen and Sering, 2000).

B. Behavior and Lifestyle

1. Bones and Beams: Behavioral Reconstruction and Interpretation below the Neck

Physical activity is a defining characteristic of humans generally and shows a high degree of patterned variability across the world. Workload, for example, varies considerably across the spectrum of different subsistence strategies, ranging from heavy to light. There are some human groups that spend much of their day in highly demanding physical activities, whereas there are others that are involved in relatively little physical activity (e.g., most Americans). In order to reconstruct and interpret behavioral patterns in past humans, bioarchaeologists commonly rely on the study of osteoarthritis, primarily because the disorder is caused in large part by wear and tear on the joints of the skeleton. Because other factors also influence osteoarthritis, such as environment, climate, body weight, and genetic predisposition, it is not possible to equate frequency or type of osteoarthritis with frequency or type of workload or physical activity. Generally speaking, however, populations that had demanding physical lifestyles had more osteoarthritis than populations living under less demanding circumstances (Larsen, 1997).

Biomechanical analysis of long bones (e.g., femur, humerus) is much more revealing about activity level and type. Just like building materials that go into the construction of buildings or bridges, long bones are structured so as to be able to withstand breakage due to excessive mechanical loading, such as from bending or twisting (torsion). The ability of building materials and long bones to resist mechanical loads is called "strength" or "rigidity" (see Ruff, 1992; Larsen, 1997). "Beam theory" from engineering science provides an important framework for present-day bioarchaeologists for drawing inferences about activity and workload in past humans. In this regard, long bones can be modeled as hollow beams. When viewed in cross-section, the magnitude of physical stresses in these hollow beams is directly proportional to the distance from a central or "neutral" axis running down the midline of the bone. Mathematically, the stresses are equal to zero at this midline, but the further one is from the midline, the greater the magnitude of mechanical stress. Thus, a bone that is the strongest is one in which the material (cortical bone) is placed furthest from the midline.

Engineers have developed standard formulas for calculating cross-sectional geometric properties that measure the strength of a cross-section. Unlike building materials analyzed by engineers, human bone tissue is highly dynamic, and the strength of the cross-section changes over the course of an individual's lifetime. For example, when viewed in cross-section, the diaphyses of the long bones of the skeleton continues to expand throughout life. This continued expansion appears to maintain the mechanical integrity of the element, especially as bone tissue is lost after about age 40 (see Ruff and Hayes, 1982).

Two key cross-sectional geometric properties analyzed by bioarchaeologists include values called "I," which measures the ability of the bone to resist bending, and "J," which measures resistance to torsion. J represents the sum of the strength values of I_x and I_y (resistance to bending in the "x" plane and "y" plane, respectively) and is a good overall indicator of bone strength.

Cross-sectional geometric properties can be measured either invasively or noninvasively. In the former, the bone is cut with a fine-tooth saw at the section location (e.g., femur midshaft), the section is photographed, the photograph is projected onto a digitizer screen, and the outlines of the endosteal and periosteal surfaces are digitized manually. With computer software developed by engineers and modified for study of human bone, the cross-sectional geometric properties are calculated automatically from the digitized section images (see Ruff, 1992). Alternatively, images can be generated via computed axial tomography (commonly known as CT scans). Computed axial tomography was developed in medical science as a noninvasive means to observe body tissues in living persons. Bones offer excellent material for observation of surfaces not visible (i.e., the endosteal surface). Due to advances in CT technology, it is now possible to create high-quality images that are as accurate as photographic images of the actual bone cross-section.

Long bones from a range of archaeological populations, mostly in North America, have been analyzed by various workers in an effort to characterize type and level of physical activity (Ruff, 1992; Larsen, 1997). For example, comparison of I and J values in the aforementioned prehistoric and mission-era populations from Spanish Florida reveals patterns of behavioral change. In particular, there is an increase in bone strength in comparison of late prehistoric and mission Indians that reveals an increase in workload once Spanish arrive in the region (Larsen and Ruff, 1994; Larsen *et al.*, 1996; Ruff and Larsen, 2001). Historic records indicate that the mission Indians were heavily exploited for labor, including food production, transport of heavy materials, and construction projects. Thus, the increase in bone strength reflects an adaptation to increased labor demands during the 16th and 17th centuries. Biomechanical evidence provides clear biological evidence for changing patterns of lifestyle and behavior not possible from other sources.

2. Masticatory Function: Activity above the Neck

Prior to their consumption, most foods have to be processed in some manner in order to make them chewable, enhance their taste, or provide key nutrients that might otherwise be unavailable once it passes to the digestive tract. Foods most humans eat today are highly processed, so much so that the amount of chewing has been minimized. Still, nearly all food requires some amount of chewing before it is passed on to the digestive tract. Study of gross wear on the occlusal surfaces of teeth reveals different patterns and severity of wear, reflecting the kinds of foods being eaten or the type of processing (e.g., with grinding stones) before they enter the mouth (Smith, 1991). The identification of wear patterns on the teeth allows the bioarchaeologist to draw inferences about foods that members of a particular population ate. Moreover, unusual patterns of wear, such as heavy wear on the lingual surfaces of maxillary incisors (e.g., Irish and Turner, 1987; Larsen *et al.*, 1998; for review, see Milner and Larsen, 1991), indicates the use of teeth in extramasticatory functions.

Scanning electron microscopic study of occlusal surface tooth wear has been an important technological breakthrough in refining our understanding of tooth use and masticatory adaptation. A large body of experimental research on animals and humans fed different types and textures (e.g., hard vs soft) of foods reveals key patterns of variation. For example, subjects fed soft or otherwise nonabrasive foods show a strong tendency for having fewer microwear features (e.g., scratch width, pit size) than subjects fed hard or abrasive foods (Teaford, 1991; Teaford and Lytle, 1996).

Bullington (1991) examined occlusal surfaces of deciduous teeth from part-time agriculturalists and later intensive agriculturalists from west-central Illinois. Ethnobotanical research indicated that the earlier group ate wild plants and animals, along with various domesticated starchy seeds having hard seed coats. In contrast, the later group replaced (at least partially) these plants with maize. Analysis of ceramic technology indicates that plants in the later period were likely boiled for long periods of time, which would soften the food into a mushy consistency. Her comparisons of microwear using SEM revealed that the deciduous teeth of two prehistoric groups had similar types and frequencies of microwear on their occlusal surfaces. However, comparison of microwear in the youngest age cohort (ca. 0.5–1.0 years) indicated that the later intensive agriculturalists have a lower frequency of microwear features than the earlier less intensive agriculturalists. This difference indicates the strong possibility that the later infants were eating softer foods than the earlier infants.

In contrast to the setting from west-central Illinois, microwear appears to have changed dramatically in other areas of North America and elsewhere in major subsistence shifts. Teaford and colleagues (Teaford, 1991; Teaford *et al.*, 2001) found a general reduction in frequency and severity of microwear features with the shift from foraging to farming on the southeastern U.S. Atlantic coast. In mission populations during the period of more intensive agriculture, the occlusal surfaces of molars contain fewer pits and scratches, reflecting both the change in foods consumed (more maize) and how they were prepared (e.g., more prolonged boiling of food). In contrast, Molleson and co-workers (1993) and Pastor (1992) documented an *increase* in frequency of microwear features in western and southern Asia, respectively. In these settings, the change appears to be related to the adoption of the use of grinding stones, used in preparing grains into flour. This new technology added grit to the foods being eaten, and hence more microwear.

C. Population History

1. Identifying Biological Relationships

The identification of relatedness between human groups has been a major area of discussion in anthropology since the 18th century, when measurement of skulls began to be used to identify biological/anatomical differences and to infer biological history and population relationships. Bioarchaeologists today use biodistance to identify temporal and spatial relationships between and within past groups based on the study of polygenic skeletal and dental traits (Buikstra *et al.*, 1990). The approach assumes that sharing of skeletal and dental attributes indicates affinity (e.g., presence of a persistent metopic suture or accessory cusps on molars). These traits include both discrete (non-metric) and metric features, which for the most part do not bear a one-to-one relationship with a person's genome. However, biodistance analysis has provided an important tool for providing information on population structure and relatedness.

Recent advances in the last few years in extraction and amplification of DNA from archaeological bone are beginning to make possible the identification of genetic distance, a development that was thought to be unlikely a decade ago. Although application of the PCR to archaeological bone is still very much in its infancy in bioarchaeology, the situation is changing rapidly as protocols are established and reliable results begin to emerge.

Hypotheses about population movements and relationships in North America have generated a great deal of debate among archaeologists and linguists. In the American Great Basin (Nevada and parts of surrounding states), Sidney Lamb (1958) suggested that based on glottochronological evidence the ancestry of present-day Numic speakers living throughout the region today could be traced to a founding population in southeastern California. He argued that the founding group spread throughout the Great Basin from this homeland at about AD 1000. Some archaeologists believe that changes in material culture, settlement pattern, and subsistence seen at this time were caused by the Numic expansion and replacement (e.g., Bettinger, 1994), whereas others see no record of cultural discontinuity (e.g., Raven, 1994).

Studies of mitochondrial DNA (mtDNA) from living Native Americans reveal that at least four distinct founding matrilines (haplogroups A, B, C, and D) account for most groups (Kaestle *et al.*, 1999; Smith *et al.*, 1999; Crawford, 1998). Comparison with Asian populations reveals the strong likelihood of a northeastern Asian origin for Native Americans. Moreover, different language groups today possess different frequencies of haplogroups, including Numic and non-Numic speakers. Theoretically, the identification of haplogroups from mtDNA extracted from bone samples from archaeological contexts in the Great Basin should reveal either a discontinuity or a continuity between prehistoric and living peoples in the Great Basin.

Kaestle and co-workers (Kaestle, 1995; Kaestle *et al.*, 1999) extracted and amplified DNA from a sample of skeletons from the Stillwater Marsh region of western Nevada in order to identify ancestral–descendant relationships. Their findings indicate that the most parsimonious explanation appears to be some degree of admixture between Numic and pre-Numic peoples. This appears to be the case because the presence of mtDNA and albumin phenotypes for *both* Numic and non-Numic speakers are present in the Stillwater series. Kaestle and co-workers suggest that if Numic speakers did move into the Great Basin, they did not replace the earlier pre-Numic population. Thus, while the native languages spoken at the time of European contact were derived from the end of the first millennium AD, the biological composition involved both pre-AD 1000 and post-AD 1000 groups. This conclusion jibes with the observations made by Smith and co-workers (1995) from their analysis of serum protein albumin extracted from skeletal samples from the Stillwater Marsh remains from the Great Basin. That is, the combined absence of Al^{Na} and the presence of Al^{Me} phenotypes indicate that the ancestry of the prehistoric populations from the region are neither Athapascan nor Algonkian, which is consistent with most linguistic reconstructions for the region (see Smith *et al.*, 1995). The mtDNA evidence makes clear that the prehistoric and historic populations in the region are likely biologically related.

At the other side of the Great Basin, from prehistoric skeletons recovered from near the Great Salt Lake in northern Utah, O'Rourke and colleagues (1996, 1999; Parr *et al.*, 1996) have provided another context for testing hypotheses about the so-called Numic expansion. Like the evidence derived from the study of skeletons from Stillwater Marsh, DNA evidence from the Great Salt Lake region indicates a probable continuity between pre-AD 1000 and post-AD 1000 populations in the eastern Great Basin. Interestingly, the genetic marker associated with haplogroup B, the 9-bp deletion, is present in the earliest and latest samples in this region, providing additional support for biological continuity in the Great Basin.

2. Where Did They Come From? Residence Shifts in the Prehistoric Past

The documentation of a person's residence history through the analysis of stable isotopes in earlier and later forming tissues offers an important means of identifying patterns of mobility. Strontium isotope ratios ($^{87}Sr/^{86}Sr$) in skeletal and dental tissues are useful for identifying the amount of marine and terrestrial foods eaten by a person. In the South African Cape region, strontium ratios differ between people who live in the coastal setting vs those who live in a terrestrial setting in the interior mainland. Ratios determined from bone samples for prehistoric people living in the interior are higher than for prehistoric people living on the coast. These differences reflect local geochemistry such that people living in the interior are exposed to—the geochemical isotope ratios are passed directly through the food chain without undergoing fractionation (Sealy *et al.*, 1991). Indeed, comparison of earlier formed teeth with later formed teeth in South Africa reveals isotopic differences reflecting different location of the individual when the teeth were forming. Similarly, Price and co-workers (1994a,b; Grupe, 1995) analyzed strontium isotope ratios in teeth (earlier forming tissues) and mature bone (later forming tissues) in the American Southwest and Southern Bavaria. In the Southwestern setting, only some of the individuals displayed strontium isotope ratios that matched local geochemistry, indicating that these individuals likely spent their entire lives at their natal residence (or close to it). In Bavaria, comparison of isotope ratios in earlier and later populations from the Bell Beaker period (2500–2000 BC) reveals a decrease in variation, which can be interpreted as representing a decline in mobility of the population in general. This interpretation is consistent with archaeological settlement analysis showing a shift to increasing sedentism.

IV. THE STATE OF THE SCIENCE OF BIOARCHAEOLOGY

My reading of the record of the changes seen in the field of bioarchaeology since the mid-1980s is one of vitality, innovation, and increasing sophistication. Long past are the days when the sole bioarchaeologist working on an archaeological skeletal series would be expected to develop a comprehensive analysis of a series of archaeological skeletons. More commonly, the bioarchaeologist today is involved from the start of an archaeological project and where the skeletons are excavated in order to address questions that will be used to improve our understanding of quality of life, behavior, and population history in past societies. The bioarchaeologist called upon to study the skeletons will likely call upon other experts, such as those who study ancient DNA, bone geometry, or tooth microwear.

Bioarchaeologists have recognized the strength of the tools developed in other sciences that can help address key issues about the human biological past. The newly evolving instrumentation and techniques discussed in this chapter—scanning electron microscopy, computed axial tomography, mass spectrometry, and so forth—are not just "bells and whistles." Rather, they offer a means to address issues about human biological history. Based on recent history of bioarchaeology, we can expect to see continued growth of the discipline fueled by new methodological and theoretical developments. New opportunities to learn about the human past will continue to help us.

Any discussion of the history of bioarchaeology will almost certainly highlight difficulties relating to comparability of data sets generated by different bioarchaeologists. The development of common standards for data collection is helping to address this problem (e.g., Buikstra and Ubelaker, 1994). A broader concern is the need to increase even more the level of collaboration between archaeologists and bioarchaeologists. I believe that further refinements in this arena are necessary for the placement of bioarchaeology within the larger context of anthropological and behavioral sciences generally.

There is often the misperception that archaeological bones and teeth are not especially informative about human social or political behavior. For example, it is often assumed that the study of gender is inaccessible in past settings. Some argue that gender attribution—and therefore the issue of gender overall—is too ambiguous in archaeological settings to be able to reconstruct and interpret past human behavior. Wylie (1991:31) noted that "the very identification of women subjects and women's activities is inherently problematic; they must be reconstructed from highly enigmatic data."

Contrary to Wylie's statement, gender is a highly visible part of the past. Nowhere is the potential for elucidating human social behavior where gender is concerned than in human skeletal remains. Indeed, human remains provide the *only* direct means of identifying the sex of a person in archaeological contexts, and arguably sex identity—female or male—provides a key window into gender. Indeed, this point is underscored with the publication of an entire volume devoted to the discussion of sex and gender in relation to disease and health in the past (see Grauer and Stuart-Macadam, 1998). Clearly, this biocultural approach to the study of gender has important meaning for the emerging studies in archaeology, other areas of anthropology, and other disciplines in general and for health and behavior in particular.

The link between quality of life and gender also speaks to the larger issue of the relationship between health and political complexity. The political structure of a population is strongly integrated with its subsistence base. In this regard, access to food (and, by extension, nutrition) is influenced by the political system: access differs according to age, gender, status, and other cultural identities. In her extensive overview, Danforth (1999) documents clear links between quality of life (especially nutrition) and political complexity, finding that most members of egalitarian societies have good nutritional health and that low-status individuals in state-level societies have poorer health than high-status individuals. This is a pattern that is very much the same as seen in societies around the world today (e.g., see various studies in Strickland and Shetty, 1998). The discussion of political complexity and implications for health and quality of life is an issue that is of enormous concern to a range of disciplines, and bioarchaeology has much to offer, especially with regard to the past.

V. SUMMARY AND CONCLUSIONS

Bioarchaeology is enjoying a period of robust growth. This growth relates to the increasing recognition that human remains offer valuable insight into human behavior, health, and quality of life in the past. More importantly, the growth of the discipline reflects the strengths that it brings to the table in addressing issues about the past as well as the success of the discipline in adopting and applying developments — technological, methodological, and theoretical — from other sciences in new, innovative, and highly creative ways.

ACKNOWLEDGMENTS

Much of the discussion in this chapter reflects my own education in bioarchaeology. I especially thank my colleagues and collaborators who have contributed in so many ways to the advancement of the field in general and who have contributed to my own research in particular — Christopher Ruff, Margaret Schoeninger, Mark Teaford, Dale Hutchinson, Katherine Russell, Mark Griffin, Frederika Kaestle, David Smith, and Scott Simpson. I am fortunate to have been involved in the kind of archaeological–bioarchaeological collaboration that I espouse in this chapter. I especially acknowledge the projects that I have coordinated with David Hurst Thomas, Bonnie McEwan, Jerry Milanich, Rebecca Saunders, Robert Kelly, and Joseph Craig. I am grateful to the National Science Foundation, the National Endowment for the Humanities, and the St. Catherines Island Foundation, the primary agencies that have funded my research.

Chapter 14

Mortuary Analysis and Bioarchaeology

Lynne Goldstein

I. INTRODUCTION

In her preface to the volume *Biocultural Adaptation in Prehistoric America*, Gwen Kennedy Neville, editor of the Southern Anthropological Society Series in which the volume was published, noted that the papers in the volume represented an important attempt to combine and share knowledge (1977):

> (T)he participants converge in their genuine interest in applying multiple perspectives and in their commitment to the creative sharing of knowledge. . . . (I)n the search for answers they are not afraid to look beyond their own disciplinary boundaries into the promising territory of the holistic study of human beings. (Neville, 1977)

The question or issue that I try to address in this chapter is whether we have made progress since the early 1980s in bioarchaeology in the creative sharing of knowledge and holistic study of human beings. The six authors whose work is represented in the 1977 volume attempted to demonstrate the value of physical anthropology and archaeology working together to address problems of the past, and although this may have been a somewhat ambitious and perhaps even naïve goal, it was certainly a worthy one.

My own perspective is as an archaeologist, so my bias is perhaps different than others in this volume. In order to address the current state of knowledge and to assess the range of research being done, I examined eight journals over the period 1995–2000, as well as some more recent issues of a few other journals. The eight journals were *American Antiquity, American Anthropologist, Current Anthropology, Journal of Anthropological Archaeology, Antiquity, Journal of*

Bioarchaeology: The Contextual Analysis of Human Remains, Buikstra and Beck (eds.)

Archaeological Research, *Historical Archaeology*, and *World Archaeology*. With a few exceptions, I did not systematically examine the *Journal of Physical Anthropology* or other specifically physical anthropology journals because Larsen (1997, 2002) has provided relatively recent reviews of much of this work that could be incorporated into this analysis.

The simplest way to summarize what I found is to say that archaeology and physical anthropology[1] have followed very different trajectories since the early 1980s, without extensive or creative sharing of knowledge. There were a total of 87 articles that had some discussion of mortuary practices, although this may not have been the primary focus of the article. The work in both areas has been extensive, interesting, and innovative, but it has not necessarily been collaborative or interactive to the degree that the participants in the 1977 Blakely volume had hoped. There are a variety of reasons for this state of affairs, and this divergence is worth exploration and examination.

II. CURRENT BIOARCHAEOLOGICAL RESEARCH DIRECTIONS IN PHYSICAL ANTHROPOLOGY

The simplest explanation for the lack of shared and combined research between archaeology and physical anthropology is that research in each of these areas has taken very different directions. In physical anthropology, researchers realized that they could use the tools of skeletal biology, DNA research, and chemistry to answer new questions that had never before been considered. Research became more and more science and laboratory oriented (Larsen, 1997). To be a physical anthropologist focused on skeletal remains from archaeological sites meant that one had to know much more than how to determine age, sex, trauma, and obvious signs of disease. As Larsen notes in a 2002 article documenting current directions in bioarchaeology, the field is now focused on using human bone to study such topics as dietary reconstruction from bone chemistry, infectious disease and health, physiological stress and disruption of growth, violence and trauma, masticatory function and tooth use, lifestyle reconstruction and interpretation, population history and biological relatedness, and paleodemography (Larsen, 2002:120). In this framework, the physical anthropological researcher does not necessarily need the archaeologist once the archaeologist has excavated the

[1]I use the terms "physical anthropology" and "biological anthropology" to refer to those researchers who are anthropologists who study human remains because of the way in which Larsen has appropriated the term "bioarchaeology." Because I use the term bioarchaeology differently and more broadly [and more like the way in which Buikstra (1977) used the term], I feel the need to distinguish it here. I choose not to use the term osteology, as an osteologist may not necessarily be an anthropologist.

bones — research can be accomplished by the physical anthropologist working alone or with other scientists. Indeed, Larsen has redefined bioarchaeology as something exclusive to physical anthropology — he sees it solely as the study of human remains recovered from archaeological settings (Larsen, 1997, 2002). This is in dramatic contrast to the definition and interpretation of several other scholars [such as Buikstra's (1977:69) "active participation of both archeologists and physical anthropologists in all phases of research design"], but because of Larsen's impressive and prolific publication output, his definition has *de facto* become the most common, or at least the most ubiquitous, definition of bioarchaeology.

Larsen emphasizes the point that skeletons are especially important in making statements about the past because they are the physical remains of the people themselves and "are the most direct evidence of the biology of past populations" (Larsen, 2002:145). While this is an important point and one that is difficult to argue, he then goes on to discuss the issue of determination of gender and social inequality, noting that the sex of an individual is nearly always revealing about their gender. "Indeed, the jump from sex identification to social identity and behavioral inference is not a big one" (Larsen, 2002:145). Unfortunately, while perhaps it may not be a big leap in some instances, it is a huge leap at other times, and is precisely the reason that one should never rely on biological data alone — sex and gender are *not* the same. Context is everything, and while physical anthropology can do a lot with the bones once they are out of the ground, there is much that cannot be done without context and other nonosteological data. Indeed, there are several articles in the literature review conducted for this chapter that focus on the topics of gender, status, identity, and the changing role of gender over time; in each case, context is key in interpretation (e.g., Crown and Fish, 1996; Sofaer Derevenski, 1997, 2000; several articles in Arnold and Wicker, 2001; see also Parker Pearson, 1999).

It is possible, and even likely, that some physical anthropologists have started to ignore archaeological data because physical anthropology has gotten complicated and requires such specialized training that researchers do not have time to do everything they have to do for their specific analysis, plus work with the archaeological data as well. It is also possible that archaeological data are different enough that they are hard to interweave into the osteological analysis. Archaeological data are often "messy," requiring more interpretation and more work than osteological data.

A survey of the physical anthropological literature on bioarchaeology (Larsen, 2002) makes it clear that the field has become much more scientific or laboratory oriented and, with that move, has come the idea that the field is more absolute in what it can do and in how accurate it is. This idea has not been without its critics. Jurmain (1999) has perhaps been among the most vocal of these, demonstrating that some applications of work taken from biology and anatomy have not been carefully done and can be called into question for their conclusions.

Most recently, a dissertation by Bice (2003) critically analyzes the biomechanical model for reconstructing behavior from archaeological skeletal remains. She examines the use of cross-section geometry as representative of physical activity, sexual division of labor, and differential usage of the upper limb. All of these have been used extensively in the bioarchaeological literature since the 1980s. Bice demonstrates that ongoing nonanthropological theoretical, experimental, and clinical research on the adaptation of bone to mechanical loading has shown that the assumptions routinely made in bioarchaeology do not hold, and a number of alternative explanations are possible. Bice also provides evidence that the conclusions reached in bioarchaeological publications about kinds of activities conducted are likely false and that data presented in published papers often do not support the conclusions presented; behavioral inferences are sometimes based on statistically insignificant group differences in the cross-sectional geometric properties analyzed. In other words, Bice's work demonstrates that for one area of physical anthropology, conclusions about past human behavior are based on simplistic and unsupported assumptions. The researchers have borrowed work from another discipline, but have failed to keep up in that discipline and did not realize that more recent research has demonstrated that their measurements and analyses can have alternative explanations. Science is rarely absolute; alternative explanations are important to consider, and multiple lines of evidence are critical.

Larsen and many other scholars have developed bioarchaeology in exciting and interesting new directions. However, many have not taken note of what was said back in 1977—that to do biocultural or bioarchaeological work well requires a sharing and combining of information and questions. Biological data may appear more scientific and more precise, and it may be more precisely measurable, but it is not inherently better data—it is just different, and we need to learn how to work with different kinds of data. As the field has become more seemingly scientific, some bioarchaeologists have decided that they no longer need to work as closely with archaeological data, which is a huge mistake. When we work with data from the past we are working with parts of the whole so we cannot afford to dismiss any of it.

Perhaps worse or more dangerous than totally ignoring archaeological data is treating these data simplistically. Archaeology has changed significantly since the early 1980s and it is an even greater mistake to reduce archaeological mortuary theory to such simplistic assumptions as "there is always a one-to-one relationship between grave wealth and social status," or the sex and gender assumption cited earlier. While these assumptions may hold true sometimes, they do a serious disservice to theoretical developments in archaeology, they are culturally insensitive, and they fail to recognize the complex social fabric of the past and the present.

In the set of literature examined for this chapter, I found few examples in which physical anthropologists acknowledged that there were important theoretical developments in anthropological archaeology that need to be taken into account in their work, which, in fact, could bias bioarchaeological interpretations if ignored. One example is examined in a later section of this chapter, and another example is Buikstra (2000:15) in her discussion and critique of historical bioarchaeology, where she notes that, especially for historic contexts, there is a "complex interplay between social, economic, and ideational factors" (Buikstra, 2000:17). As Parker Pearson (1982) and Shanks and Tilley (1982) noted a number of years earlier, many different factors can enhance, distort, or mask relationships among social standing, burial treatment, and cemetery structure.

III. CURRENT MORTUARY RESEARCH DIRECTIONS IN ARCHAEOLOGY

At the other end of the spectrum are archaeologists. What happened to them since 1977? In mortuary studies, Hodder (1982, 1984, 1986) and others (Parker Pearson, 1982; Pader, 1982) argued that the focus on social ranking and social organization was far too narrow and inappropriate and was also too mechanistic as applied by some analysts. Further, others argued that areas such as gender and symbolism were largely ignored (Conkey and Spector, 1984; Ehrenberg, 1989; Shanks and Tilley, 1982; Wylie, 1991). In terms of what was missing from mortuary analyses, such critiques were certainly accurate, and the 1980s and 1990s saw a shift away from the social ranking—social organization orientation of mortuary analysis. Unfortunately, instead of supplementing or enriching mortuary analysis with additional approaches, the analysis of social organization was abandoned by many, only to be supplanted by topics such as emotive analyses (e.g., Tarlow, 1999), which is fine, but also limited. There is no reason why one cannot focus on both or at least develop aspects from each, creating a richer and more complex picture of the past, but this tended not to happen, with a number of notable exceptions, such as Chesson (1999), Kuijt (1996), and Kus and Raharijaona (1998). With this shift away from social ranking and organization was a shift away from most statistical analyses and work that appeared to be scientific or positivist; research tended to be more humanistic and symbolic [see a number of examples cited and discussed in Parker Pearson (1999)]. Interestingly, one of the few calls for an integrated approach to mortuary analysis or biocultural study was by Ian Morris in his book on how he thought burials in the ancient Graeco-Roman world should be analyzed (Morris, 1992). Morris thoughtfully and carefully put together an integrated plan for analysis of the skeletons and the archaeological materials.

As noted earlier, two important areas of study that had been largely ignored in the 1970s and 1980s were the study of gender and the study of symbolism.

Although mortuary analysts certainly noted the sex of burials, there was little discussion of the meaning and importance of gender in any way. At most, there might be some surprise when a woman was buried with lots of grave goods or was placed in a high status position. The topic of gender as a focus of study was largely ignored, as was the idea of a gendered past. As far as symbolic issues were concerned, archaeologists had long considered it an area of interest, but one that was beyond what they could extract from the archaeological record. Work such as James A. Brown's early research at Spiro (e.g., Phillips and Brown, 1983) was considered a notable exception. In the 1990s, these two areas developed both theoretically and practically, with a number of researchers demonstrating that excellent research could be accomplished in different ways and from different directions. Arnold and Wicker's (2001) edited volume on gender and mortuary practices presents a wide range of some more recent examples on approaches to gender analysis in mortuary sites, and people have also taken a variety of approaches to symbolism, religion, and ideology (e.g., Bradley, 1998a; Brown, 1996, 1997; Burley, 1995; Hill, 1998; Pollex, 1999; Siegel, 1996).

Another interesting direction of recent mortuary studies has been to focus on the individual and on the emotive (Tarlow, 1999; Meskell, 2000). While the archaeologist may see such a study as most likely limited to those instances in which one has documents or inscriptions or evidence about specific individuals, it is possible that one could propose or develop an emotive analysis based on individual or idiosyncratic treatments at a particular mortuary site. Sets of behavior could lead to such an interpretation, and Tarlow (1999) in particular outlines some of the ways in which an emotional archaeology might be possible further back in the past, through the analysis, for example, of metaphors and meanings. Most interesting here, however, is that nowhere in these published papers or examples did anyone suggest that such studies could be improved if they were done in conjunction with a physical anthropologist. Who better to assess and determine details about an individual? Working with a physical anthropologist who has additional information about the person, this area seems a natural one for collaboration, yet it rarely happens unless the archaeologist uses some already published skeletal data. One can only imagine how much more might be gained if Meskell's (2000) rich and impressive narrative analysis of Deir el Medina could be interwoven with a detailed osteological analysis.

Perhaps some of the most exciting of the developments in the archaeological mortuary analysis arena have been in the area of landscape studies. In the edited volume *Regional Approaches to Mortuary Analysis* (Beck, 1995), a variety of authors attempted to place mortuary analysis in a larger perspective, providing a regional context for mortuary sites and allowing archaeologists to place their work in a broader setting as part of the overall landscape in which people live. Creating a sense of place and developing the notion of social memory have been critically important research directions because researchers can connect disparate and messy data from a region and link it in ways that had not been possible previously. Notable in this regard is Bradley's work on votive deposits in the Thames River (Bradley, 1990), as well as his work on megaliths and their relationship to the natural landscape (Bradley, 1993, 1998b). Equally impressive and influential has been Alcock's (1993) work on Roman Greece. Papers in the Beck (1995) volume have provided a variety of examples of the kinds of regional mortuary analyses and landscape research that are possible.

In the literature review for this chapter, there were at least four papers that focused on landscape issues and mortuary sites in a way that highlights one of the reasons why collaboration and context are so critical. Coming out of Bradley's (1993) seminal work on megaliths and landscape, a number of archaeologists have begun to examine monuments as a continuing part of the landscape rather than as an artifact built at one time and abandoned. We know these monuments were not abandoned, but were used again and again in patterned ways, and are still being used today. This behavior toward monuments comes from cultural memory. "Ancient monuments represent the past in the landscape and cultural memory gives them meaning and cultural significance" (Holtorf, 1998:24). Holtorf goes on to note that practices such as secondary burial are examples of a closely related phenomenon that he terms "history culture." History culture includes all of the appearances of the past in everyday social life (Holtorf, 1998:24). Holtorf (1998:24) notes: "Ancient monuments in the landscape influenced cultural memories of subsequent societies whose history cultures, in turn, transformed the monuments." These monuments not only link the present with the past, but they also link the present with the future. While Holtorf focused on megaliths, Semple (1998) examined how Anglo-Saxons reused Bronze Age burial mounds and Neolithic long barrows. In these cases, the reasons were different than those developed by Holtorf, but Semple was able to outline a strong case for the Anglo-Saxon perception of the landscape and its meaning. Hingley (1996), in his focus on Atlantic Scotland, makes a strong case that Neolithic peoples were deliberately reinventing monumental aspects of the past as a strategy related to developing a contemporary identity. The fourth case of reuse of the landscape is examined separately later.

The question to address here is why are these studies important to the development of a proper bioarchaeology or biocultural approach? Precisely because they demonstrate the importance of context and the fact that human bone is not simply placed in the ground or in a tomb and forgotten. Mortuary sites are active places and they are used and changed. If the analyst does not understand this, it is unlikely that the analysis will properly reflect what actually happened. There is not a single or simple answer or interpretation at a mortuary site, but possibly many.

The fourth example of a monument serving as a continuing part of the landscape is documented in an article by Moss and Wasson (1998); this is perhaps the most remarkable case of all. The site is known as the Pistol River Site and is located on the Oregon coast. The site was the main village of the *Chet-less-chun-dunn-dunne*, an Athapaskan group. In 1856, a party of 34 Euro-American men burned the village. There are materials at the site from the AD 1600s through the 1800s. The site was of interest to 19th-century archaeologists, as well as to amateur archaeologists. The Oregon Archaeological Society excavated it, as did some professionals during highway construction. Later, professionals from the University of Oregon tried to return and do more work, but some locals also dug here. Among the excavators are local residents who are Native Americans and descended from those killed at the site. These people are friends of another family who are descended from one of the vigilantes who burned the village in 1856. The site yielded many artifacts as well as burials, and the artifacts have attracted many collectors. The story of the site is too complicated to repeat here, but is intertwined with Native ties and affiliation, archaeological significance, local interest, materials not curated or reported, a site claimed by everyone, and a place that today is important for locals, archaeologists, and wind surfers. It is definitely a story of use and reuse that cannot be understood without context.

I would be remiss at this point if I did not raise the possibility of politics as a reason why archaeologists may have moved further away from physical anthropology. As the call for new laws increased, and in particular, as the Native American Graves Protection and Repatriation Act (NAGPRA) was implemented, physical anthropology was seen in a negative light by many native peoples and was cast as a colonial endeavor. In part this was a misunderstanding, in part this was because few physical anthropologists had relationships with tribes, and in part this was because of some past misdeeds. Physical anthropologists were also not quick to publicly stand up and defend themselves and their work. Many archaeologists found it easier to avoid burials, and even physical archaeologists began to look at other portions of the subfield rather than continue to focus on bioarchaeology.

As interesting and important as the different kinds of landscape work have been, it has moved archaeologists further and further away from the bones themselves. Rejection of the processual approach to mortuary analysis with its ranking and focus on social status apparently accompanied rejection of integration of the physical anthropological data and physical anthropologists, perhaps in part as a rejection of science and statistics. I also have no doubt that politics played a role in the avoidance of physical anthropology as well. Whatever the basis, it is relatively rare to find an article written by an archaeologist about a mortuary site that incorporates skeletal data or is coauthored by a physical anthropologist. Instead of making progress at integration of archaeology and physical anthropology since 1977, a review of the literature reveals that physical anthropology and archaeology have diverged even more since the early 1980s. In the nearly 90 articles reviewed for this chapter, there are few articles that are biocultural or bioarchaeological in the sense that was meant by Buikstra in her original 1977 definition.

The one bright spot of cooperation and integration of physical anthropology and archaeology is perhaps less an integration than it initially appears, and that is the work that uses bone chemistry to address issues of diet and residence patterning. This area of research brings together archaeologists and physical anthropologists to examine patterns of dietary change and possible measures of agricultural intensification, as changes happen in bone chemistry when maize becomes available in the diet. This kind of research has been ongoing since the 1980s, but in this review of the literature, at least three articles—Price and co-workers (1998), Schurr (1997), and Schurr and Schoeninger (1995)—demonstrate an integrated approach to data, attempting to employ a range of data, examining context, and exploring a variety of alternative explanations. Price *et al.* (1998) asked new questions about prehistoric migrations in central Europe, concluding that migration was substantial during the Bell Beaker period. Schurr (1997) compared demographic measures of fertility with age-related changes in stable nitrogen-isotope ratios and concluded that the relationship between fertility and weaning behavior is a complex one. I found the Schurr and Schoeninger (1995) article most impressive, as it examines the association between social complexity and agricultural intensification for late tribal and chiefdom level societies, but uses many, many different lines of evidence and examines alternative explanations. Of course, the irony of these analyses is that this time the archaeologists can use the bone samples without doing much more than noting that they are human bone.

IV. TOWARD AN INTEGRATIVE APPROACH

Integrative bioarchaeological work is not impossible, but it can be somewhat impractical and difficult. There is no question that for both archaeology and physical anthropology the work being done today is time-consuming and complex. To add another layer to that work adds more difficulties, more time, more cost, and so on. However, even though the overall benefit should be great, the individual researcher often has to worry about publications, raises, grants, tenure, and often calculates that maybe this is something that can be done at a later time. Perhaps only scholars who have infrastructures that support such integrative work can afford to do this research, but if this were true, that would be even more distressing. In this volume, Larsen discusses the increasing interdisciplinary work of bioarchaeology, but unfortunately, he does not include archaeology in that interdisciplinary or multidisciplinary umbrella, nor does he really include archaeology under the rubric of bioarchaeology, except as a source for data.

The lack of integration means that both archaeologists and physical anthropologist are ignoring data. Almost by definition, archaeological data are messy data — it is incomplete, it is imperfect, and it can be interpreted in multiple ways. By ignoring one or more aspects of data, the researcher is likely ignoring context or multiple lines of evidence or, at the very least, is not exploring alternative explanations. In other words, while there is nothing inherently wrong with physical anthropologists or archaeologists working on only one aspect or subset of data, that subset can never provide a full or complete picture of what happened in the past. Further, until those data are incorporated with other information, one cannot be sure of one's interpretations. As one example, some researchers have realized that the addition of geologists or geomorphologists is also important for mortuary analysis for these very reasons; their ability to understand site formation processes and construction can often provide critical information for overall interpretation (e.g., Buikstra *et al.*, 1998).

In a book focused on the anthropology and culture history of death, Elizabeth Hallam and Jenny Hockey (2001) explore the relationship among death, memory, and material culture. Their work is primarily cultural anthropology, but they explicitly explore memory through material objects that acquire meanings through practice, ranging from items such as mourning clothing to objects that represent the body, such as photographs or effigies, to memorials. Although the majority of the book focuses on Western death rituals, the authors also examine some non-Western ethnographic accounts of funerary ritual. In summarizing the ways that people bring material culture into death rituals, Hallam and Hockey comment on the real diversity in how different peoples sustain materialized relationships between the living and the dead; they note that "the body and its material surroundings become significant in the orientation of persons, both deceased and alive, in relation to their past, present and future" (Hallam and Hockey, 2001:190). One of their fundamental points is that the "processes of memory making, in relation to death and the dead, are not confined to institutionalized, public rituals" (Hallam and Hockey, 2001:201), but can be a factor of everyday life.

In 2000, Hendon wrote an article bringing together several different lines of evidence to talk about the past, and specifically about storage. She argues that storage is a situated practice "through which groups construct identity, remember, and control knowledge as part of a moral economy" (Hendon, 2000:42). Using this model, she discusses the social meaning embodied in other kinds of storage, such as burials. Burials are spots on the landscape that are remembered so "their presence informs a locale with meaning" (Hendon, 2000:47). Hendon (2000:50) goes on to discuss the rich symbolism that such burials may contain, arguing that groups, using both the natural and created landscape, construct a setting that gives permanent expression to both identities and relationships. At the same time, because not everyone knows about all of these places, some of this mutual knowledge is imagined, and burials are one set of the features that are imagined, remembered, and discussed. Her paper is a very powerful one and would have been even more so if her examples had included osteological, as well as archaeological and architectural, data.

Perhaps because so much of my own work has focused on the importance of the spatial dimension of mortuary practices (e.g., Goldstein, 1980, 1981, 1995, 2000), I find myself particularly concerned that much of the bioarchaeology practiced today has ignored the context and the spatial location of the human remains. If the relative placement of a burial has anything whatever to do with that person's life, ties to others, and social memory, it does not appear that that information will ever be determined from much of the modern bioarchaeological research done today, or at least it seems that independent lines of evidence verifying that relationship will be ignored. Similarly, archaeologists have focused so intently on the structural, social, symbolic, and landscape aspects of sites that they have tended to ignore the individuals themselves, except when those individuals left behind items of personal adornment or inscriptions. Still, archaeologists have not tended to ignore the context of the materials they have found and they have examined the relationships of things to each other.

Michael Blakey has prepared a very thoughtful review of the bioarchaeology of the African diaspora in the Americas (Blakey, 2001). In this review, he notes the well-known early work of scholars such as Boas and Herskovits and also the significant work conducted during the first part of the 20th century by African-American researchers such as W.E.B. DuBois, Jean Price Mars, Cheikh Anta Diop, Katherine Dunham, W. Montague Cobb, Fernando Ortiz, and Irene Diggs (Blakey, 2001:390). In contrasting the work of these early black scholars with white scholars, Blakey comments on the different intellectual trajectories and traditions. He points out that at certain institutions and under certain anthropologists, there was intellectual cross-fertilization with black scholars and collective scholarship, whereas at other institutions, this was not the case (Blakey, 2001:391). While it is beyond the scope of this chapter to summarize Blakey's arguments, he is trying to draw a parallel between what happened in black scholarship generally and what has happened in African diasporic bioarchaeology. He notes that "(a)rchaeology and physical anthropology have experienced even less interaction with the black intellectual traditions than did American sociocultural anthropology" (Blakey, 2001:394).

Blakey (2001) provides a history of "physical anthropology and the Negro" and, not surprisingly, there is a dominant racial deterministic trend to many of the studies that were done. Blakey rightfully criticizes many studies for focusing on race rather than on people and notes that this focus really does not change until the National Historic Preservation Act of 1966 and the emergence of cultural resources management (CRM). Once CRM took off and projects happened across the country and in cities everywhere, many firms began to focus more and more on historical archaeology and on African-American sites.

Blakey (2001) makes an interesting distinction between what he calls the "biocultural" approach and the "forensic" approach in physical anthropology. Indeed, his point is very similar to the one made earlier in this chapter; his forensic approach is what I am concerned about with some of the current trends in bioarchaeology, and his biocultural approach is what I would call good bioarchaeology. He also makes a historical linkage or connection, intellectually associating biocultural work with the University of Massachusetts–Amherst and forensic work with the Smithsonian. In fact, I think this distinction is somewhat of an overstatement, and there are people trained by other individuals and institutions in the country who can be said to practice each approach, but his point is worthy of note in terms of the history of the discipline. The University of Chicago, for example, was training physical anthropologists in the biocultural approach at the same time. Blakey did not himself coin the terms "biocultural" and "forensic" approaches.

More significantly, I am uncomfortable with such labeling of an approach. I certainly agree with Blakey (2001) that there is an approach in physical anthropology that focuses on bones and tends to ignore context, history, and the people themselves. However, I do not agree that it is useful to call that approach "forensic," as it suggests that all forensic work is also problematic. At one level one could argue that this is true, but there are instances when one does not have much context or history and one only has the human remains. At that point, a forensic analysis is what one does. Even though the point being made about lack of context and focus is absolutely correct, there is no reason to make the term "forensic" derogatory. Blakey's success with the New York African Burial Ground project provides an excellent example of why context and background are so important to good biocultural analysis.

A final irony in the comparison of archaeologists and physical anthropologists vis-à-vis mortuary sites is that, largely as a result of politics and changes in the culture of archaeology, archaeologists are conducting fewer excavations and analyses of mortuary sites, but physical anthropologists are probably doing more analyses of skeletons. Implementation of the NAGPRA and other laws in the United States (as well as similar laws in other countries) has meant that archaeologists excavate fewer burials and mortuary sites; when they do such excavations, the work is generally limited in extent and scope. While one might surmise that the same would be true for physical anthropologists, this has not necessarily been the case, as so many skeletons in museums and other institutions have had to be examined before they could be reported or returned to tribes or affiliated to specific tribes. Indeed, since the implementation of NAGPRA, a considerable amount of physical anthropology has been required, at least at the level of basic recording of standardized data (Buikstra and Ubelaker, 1994). At least for the United States, there is now a large set of detailed, roughly comparable, data on human skeletons that has never before been available. We should be able to now ask and answer questions that have never before been possible, as well as make comparisons across a number of different regions.

As exciting or interesting as these directions in physical anthropology may be, there are several things that dampen one's enthusiasm: (1) the bones themselves may no longer be available and (2) archaeologists have not done an equivalent job of recording context and detail on the artifacts and other information from these sites; in some instances, sites were excavated many years ago and context is poor at best. While we may hope that someone has field notes and records of how the site was excavated, as well as maps of what was done, archaeologists as a group or as a discipline have not carefully or systematically considered how to record artifacts and items that may be repatriated in the same way that physical anthropologists have done. Of course physical anthropologists have been able to focus on recording the skeleton, while archaeologists have to worry about every possible kind of artifact, but it would be helpful if there were at least some set of standards considered or some consistency in recording applied.

This chapter began with a quote from the Blakely 1977 volume on the importance and value of combining and sharing knowledge in the development of biocultural anthropology. A truly integrated project that incorporates the active participation of archaeology, physical anthropology, other scientists, and indigenous people in all phases of research design would still be a new form of research and an innovative way to proceed — while it was perhaps a naïve concept in 1977, it remains difficult today, but it is still a worthwhile goal.

Chapter 15

Repatriation and Bioarchaeology: Challenges and Opportunities

Jane E. Buikstra

I. NATIVE AMERICANS, BURIALS, AND COLLECTIONS BEFORE 1989

The latter half of the 20th century witnessed not only increased visibility of bioarchaeological research but also enhanced concern by Native Americans/First Nations[1] about excavation of ancient burial sites and collections of archaeologically recovered remains and associated funerary items held in museums and universities. Such concerns excited public support and led to state and federal repatriation legislation in the United States. At the federal level, the National Museum of the American Indian Act (NMAIA) was passed by Congress in 1989[2] and the Native American Graves Protection and Repatriation Act (NAGPRA) in 1990.[3] State laws vary widely (Ubelaker and Grant, 1989), as discussed later. Similarly variable laws have been promulgated in Canada at corporate, municipal, and provincial/territorial levels, although there was no encompassing federal legislation in place by the end of 2004 (Burley, 1994; Ferris, 2003; Nicholas and Andrews, 1997; Watkins, 2003).

[1]In this chapter, "Native American" will be used to refer to Indians, Native Alaskans, and Native Hawaiians. "First Nations" are the indigenous people of Canada.

[2]Public Law 101-185; 20 U.S.C. §§80q–80q-15.

[3]Public Law 101-601; 25 U.S.C. §§3001 et seq.

Bioarchaeology: The Contextual Analysis of Human Remains, Buikstra and Beck (eds.)

Tensions derived from burial excavations have deep roots in North America. For example, in 1654, Roger Williams reported the following incident:

> I have had a Sollemne debate with Nenekunat and the rest of the Nariganset Sachims in a late great Meeting at Warwick whether [whither] they came downe with 4 Score armed men to demaund Satisfaction for the robbing of Pesiccush his Sisters grave and mangling of her flesh agst John Garriard a Dutchman, whose crue, and it is feard himselfe Committed that gastly and stincking vilanie agst them. (LaFantasie, 1988:425)

Although the charges against the Dutch trader, Jan Gereardy, were dismissed when the plaintiffs failed to appear at court [Rhode Island, Court of Trials 1920:9–10 (1655:13)], a few days later the Warwick commissioners enacted strict sanctions for grave robbing (LaFantasie, 1988:430).

More recent, 19th-century tensions developed in the course of expeditions promoted by institutions such as Chicago's Field Museum and the Army Medical Museum (Bieder, 1990; Riding In, 1990, 1992). Collecting remains of the recently dead provoked angry responses from descendant communities, which brought no effective relief (Bieder, 1990). In the face of the devastation wrought by advancing waves of Euro-Americans, 19th-century policies anticipated Indian assimilation, which seemed inevitable as the 1890 census reported fewer than a quarter of a million Native Americans (McGuire, 1992). Peripheralized socially and regionally, Indian communities had little political power.

Late 19th-century archaeologists helped restore heritage to American Indians by laying to rest the myth of the Moundbuilders in eastern North America and similar attempts to argue that the Southwestern Indians were newcomers to their land of towering pueblos (McGuire, 1992; Silverberg, 1968). Yet these archaeological approaches reinforced theories that either kept Indians forever "primitive" or argued that they would only advance through acculturation (McGuire, 1992; Trigger, 1980, 1989). As the assimilation model persisted into the early 20th century, archaeologists and physical anthropologists became increasingly peripheralized not only from Indians but also from the ethnologists who, led by Boas (United States) and Jenness (Canada), argued that it was more important to record living Indians before their demise than to explore their heritage through excavations (Trigger, 1997).

After 1914, archaeologists turned to focus on historical sequences and chronology. Artifact analyses became increasingly formal and removed from functional concerns (Willey and Sabloff, 1993). Physical anthropologists sometimes contributed to archaeological interpretations, especially those of origins and migrations, but both specialists became progressively more isolated from ethnologists and the living communities they studied. While the "New Archaeology" of the 1960s and 1970s did restore concern for lifestyle, the search for law-like behaviors did not address the concerns of living communities. Much as they had been in the previous century, Native Americans were considered objects of study (Ferguson, 1997; McGuire, 1992; Trigger, 1980, 1989, 1997).

Indian activism would soon change the balance of power, however. Small skirmishes occurred as early as the 1940s, when a confrontation over a Woodland burial displayed at the W.H. Over Museum in Vermillion, South Dakota, led to a hostage situation at the adjacent university (Rhodd, 1990). Activism on both sides of the reburial issue became prominent in California during the 1960s (Meighan, 1984; Ubelaker and Grant, 1989), soon to be followed by additional involvement during the 1970s, both in Canada and in the United States.

In the province of Ontario, the disposition of any buried human remains is covered under the Cemeteries Act. Landowners negotiate with the designated representative of the deceased, who is the First Nations' government nearest the property. Archaeologists have no defined standing; in fact, studies other than to determine that the remains are Indian are prohibited under the law (Ferris, 2003). During the late 1970s, archaeologist Walter Kenyon was subject to civil suit by the Union of Ontario Indians, thus interrupting his excavations at the Grimsby site. The activist Canadian branch of the American Indian Movement (AIM) became involved and asked to rebury the materials, a move that was not popular with the Iroquois Six Nations Reserve who had unofficially approved Kenyon's excavations. After protracted negotiations, Kenyon completed his excavations and reburied the bones (Kenyon, 1977, 1982; Kristmanson, 1997).

Meanwhile, in the United States, incidents at three different Iowa sites led to the promulgation of an influential 1975 state law. In 1971, excavations at the multicomponent Kulh (13ML126) site recovered 26 Euro-American remains and one of a historic Indian woman.[4] The Euro-Americans were reburied immediately and the Indian bones were removed as archaeological specimens. After acrimonious debate, which included not only the state archaeologist but also the governor of Iowa, the remains were reburied without study. At a second site, the Siouxland Sand and Gravel Quarry (13WD402), human remains had been subject to ongoing destruction. Interaction between AIM and the director of Sioux City Public Museum became violent. Ultimately, an agreement was forged to excavate the site, study all the bones, and rebury them nearby. Such excavation and study did not occur, however, because AIM removed the bones to the Rosebud reservation and reburied them. The site was ultimately destroyed without study after quarrying resumed. In the third incident, the Lewis site (13PW5), an Archaic ossuary, was being destroyed during construction of a school. In this case, archaeologists and Indians forged an agreement and the bones were carefully excavated and

---

[4]Rose and colleagues (1996) report two Native American remains, although this is not reflected in the Anderson *et al.* (1978) discussion.

reburied after study. It was this collaboration that led to the Iowa reburial law (Anderson *et al.*, 1978; Tiffany, 1990).

The Iowa law accomplishes several goals. Importantly, it provides support for the excavation and study of threatened human remains. It also places the state archaeologist in charge of any excavation, study, and report preparation. State-owned cemeteries are established for reburial of remains following study, although no timetable is mandated (Anderson *et al.*, 1978; Tiffany, 1990). Although pre-1975 collections are not covered by this legislation, common practice is to also rebury them, after scientific reporting. Reflecting upon the law after over a decade of implementation, Joseph Tiffany (1990), associate director of the office of the state archaeologist (OSA), reported that one of the most positive aspects had been the trusting relationship developed between the OSA and the Indian community.

Other state laws differ in content and the degree to which the needs of the archaeological and Indian communities were met in constructing legislation. Beck and Teague (2001), for example, report that lobbyists representing developers and artifact collectors heavily influenced Arizona laws governing burials encountered on private property and state land. Thus, a maximum of only 10 days is allowed for burial removal. In addition, landowners can retain artifacts, including sacred items and those of cultural patrimony. Following a period of acrimonious encounters, the implementation of Arizona's law has moved from the political arena to that of bureaucracy. Beck and Teague (2001) report increasingly productive collaboration between archaeologists and claimant groups.

Summarizing the status of state laws the year before NAGPRA was enacted, Ubelaker and Grant (1989) reported that only Delaware's reburial law applies both proactively *and* retroactively. Many laws provide procedures for distinguishing cases of medico-legal importance from earlier interments. With the exception of California, state laws permit study, with varying time limits and different policies concerning destructive analyses.

Other 1970s federal laws were an important part of the mix. These included the Indian Self-Determination Act of 1973, the Indian Education Act of 1973, and the Indian Religious Freedom Act of 1978. The 1978 legislation was the first to give Native Peoples legal rights to burials and sacred sites (McGuire, 1992). These laws tend to contradict and supplant the 1906 Antiquities Act. The 1906 act, designed to preserve sites from looters, permitted excavation and retention of burials for only those holding permits for archaeological excavation and effectively separated Indians from the archaeological record. Similarly, the Archaeological Resources Protection Act (ARPA) of 1979 designated burials over 100 years old as "archaeological resources," a term distasteful to Indian people as it denies their humanity (Trigger, 1997; Watkins, 2003; Winski, 1992).

As the government's policy of assimilation waned, urban Indians moved back to the reservations. Initially, there were tensions between traditionalists and returning activists, but during the 1970s common interests emerged (McGuire, 1992). One of these involved national repatriation legislation. Rhetoric flowed from both sides of the issue.

Activists gained ascendancy by asserting that Euro-American ancestors were not excavated for scientific study and were never displayed. They further argued that osteological analyses had never benefited living Indian people and that museums were simply "collecting," gaining status through increasingly larger holdings (Pearson, 1990). One of the most vocal activist spokespersons was Vine Deloria (1969, 1995), who argued for the primacy of Indian traditions and discounted skeletal studies as being "only mildly interesting and, in any case, speculative in the extreme" [Deloria (1989), cited in Lippert (1992:18)].

Traditional Indian people focused on religious beliefs. Although the following examples postdate NAGPRA, they describe traditionalist perspectives well. For example, Rose Kluth and Kathy Munnell of the Leech Lake Reservation in north-central Minnesota expressed their views on life cycles in the following manner:

> We are taught that when we die, we are intended to go back into the earth, and that Mother Earth will take care of us. Burials feed the underground spirits and small creatures through the natural decomposition of the body. This is part of the natural order — we all depend on each other to survive. We must not disturb this cycle. (Kluth and Munnell, 1997:117)

The Zuni belief system also specifies the completion of a cycle.

> . . . traditional Zuni beliefs are that each person's life passes through four stages. The first stage is life as we know it. Little is known of the three other stages. It is essential that each person pass through each of the four stages of his or her life cycle before it is complete. All human burials with which the Zuni Tribe has a cultural affiliation are at some point in their journey through the latter stages of the life cycle. To disturb burials while on their life cycle journey is not the Zuni way. The ramifications of disturbing burials cannot be determined. How disturbance affects the life cycle journey, a journey that must be completed, is unknown, but it may well have detrimental results. (Ferguson *et al.*, 1996:268)

Although individual physical anthropologists such as Audrey Sublett had worked closely with Indians during removals and reburials of historic period (19th to 20th century) graves as early as 1965 (Abrams, 1965; Cybulski, 2001; Lane and Sublett, 1972), the initial response from professional organizations to Native American concerns was not accommodating. The American Association of Physical Anthropologists (AAPA) passed a strongly worded resolution in 1982 emphasizing that the organization "deplored reburial" except when direct lineal descendants were involved. They further resolved that no remains should be reburied without study. It was argued that all remains should be treated in the same manner, however, and "close and effective communication with appropriate ethnic groups" was encouraged. In the same year, the forensic anthropology section of the American Academy of Forensic Sciences endorsed a similar motion. Communication was encouraged, although not active collaboration. Physical anthropologists, including bioarchaeologists, based justifications of burial excavation and collection of remains on knowledge gained through scientific study. A few bioarchaeologists and medical scientists attempted to address directly the impact of reburial and the importance of the study of skeletal remains (Buikstra, 1981a, 1982; Buikstra and Gordon, 1981; Neiburger, 1988; Owsley, 1984; Owsley and O'Brien, 1982; Gregg and Gregg, 1987), but their arguments were not regarded as compelling by Native American communities, either activist or traditional.

The American Association of Museums (AAM) in 1973 encouraged sensitivity to the issues surrounding repatriation. Then in 1988, AAM developed a more extended statement, recommending case-by-case decision making. Museums were urged to acknowledge that the ethics of today should supersede those of yesterday. Living descendants were to be privileged and remains acquired illegally were to be repatriated upon request. Remains obtained in a manner legal at the time but today unethical or illegal should be considered on that basis. Groups requesting remains where there were no living descendants had to demonstrate compelling religious or cultural values that outweigh scientific interests (Ubelaker and Grant, 1989).

The Society for American Archaeology (SAA) increasingly assumed the role of spokesperson for the scientific community during the 1980s. The SAA executive committee passed an initial resolution in 1983 (Adams, 1984), which closely resembled that adopted by the AAPA and AAFS. This was followed by a revised version in 1986, reaffirmed in 1999. The four cornerstones of the most recent SAA statements include (1) the legitimacy of both native and scientific viewpoints; (2) careful consideration of scientific importance vs strength of affinity; (3) case-by-case implementation; and (4) communication and consultation. The SAA has argued that from 1986 on it recognized the legitimacy of repatriation and that it worked actively with Congress to create balanced legislation (Lovis *et al.*, 2004). Similarly, the Society of Professional Archaeologists' (SOPA) code of ethics urged sensitivity to and respect for legitimate concerns of groups whose culture histories are the subject of archaeological investigations (SOPA, 1981).

During the early phases of activism, a few U.S. archaeologists were persuaded by the Indians' arguments. Elden Johnson (1973), a professor of anthropology at the University of Minnesota, gently reminded the readership of *American Antiquity* that he believed archaeologists have an ethnographer-like responsibility to the American Indians who are the biological and cultural descendants of those whose sites we excavate. He noted that Indians protest the excavation of any burial and that they resented being objects of scientific study. Native peoples desired consultation and some feared that archaeologists were in league with looters

(Johnson, 1973:129). Similar concerns were voiced by Joseph Winter (1980) based on his experience in California. He emphasized the advantages of developing collaborations and trying to "see the sites from the Indians' perspective, which can be radically different from ours, yet just as legitimate" (Winter, 1980:125).

Archaeological support for reburial and repatriation efforts increased during the 1980s. For example, archaeologist Larry Zimmerman (1989, 1997) attended the 1982 SAA board meeting with Jan Hammil, a member of the American Indians against Desecration (AID). At that meeting, Hammil and Zimmerman unsuccessfully attempted to persuade the SAA board not to consider adopting an antireburial stance. Previously, Zimmerman had actively encouraged the 1979 reburial[5] of remains from the Great Plains, including those from the Crow Creek Massacre site (39BF11) in central South Dakota (Zimmerman *et al.*, 1981). He subsequently became active in the World Archaeological Congress and was central to the first inter-Congress on archaeological ethics and treatment of the dead held during 1989 in Vermillion, South Dakota. Although the discourse at the inter-Congress was sometimes acrimonious, the six points of the "Vermillion Accord" all emphasize respect—for mortal remains, for wishes of the dead, for local communities, for scientific research value, for final dispositions reflecting a balance between community and scientific/educational interests, and for concerns of various ethnic groups and those of scientists (Zimmerman, 2002:92). Zimmerman (1990:419) also argued that there is "no middle ground" on the issue of reburial.

The National Historic Preservation Act of 1966[6] (NHPA) has encouraged archaeological research and management of traditional cultural properties. While many tribes now have preservation and heritage programs, those of the Zuni, the Hopi, and the Navajo are among the oldest (Ferguson, 1997). Archaeologists associated with these initiatives have been some of the most vocal in support of reburial and full partnerships between Indian people and archaeologists in archaeological research. They became increasingly visible during the late 1980s and as NMAIA and NAGPRA were implemented (Anyon, 1991; Anyon and Ferguson, 1995; Anyon and Zunie, 1989; Ferguson *et al.*, 1993; Klesert, 1992; Klesert and Andrews, 1988; Klesert and Downer, 1990).

Another influence on the rapprochement between archaeologists and traditional peoples has been the impact of post-processual theories in archaeology. Humanistic and contextual, post-processual theorists argue for multivocality in interpreting the past (Hodder, 1984, 1985; Shanks and Tilley, 1987, 1989). This approach, which began to influence American archaeology during the 1980s,

---

[5]Bass (1981) reports that the site was discovered in the spring of 1978. Thus, the nearly 500 individuals had to be excavated and analyzed within a time frame insufficient for thorough skeletal analysis.

[6]Public Law 74-292; 16 U.S.C. §470 et seq.

appears better suited to the resolution of repatriation issues than processual perspectives (Zimmerman, 2002). Post-processual approaches, however, and their implications for archaeological study and repatriation issues have not been universally applauded (Clark, 1996; Mason, 1997, 2000).

II. FEDERAL REPATRIATION LEGISLATION IN THE UNITED STATES: NMAIA AND NAGPRA

The turbulent 1980s culminated in two federal laws concerning repatriation. The first, the National Museum of the American Indian Act (NMAIA, 1989), established a new national museum dedicated to the history, music, and art of American Indian cultures.[7] Important for bioarchaeological research, this act also affects collections of human remains and funerary items held by the Smithsonian Institution. All are to be inventoried and, if possible, tribal or descendant origins determined. Appropriate tribes or descendants may then request that the remains are returned (Winski, 1992). At the time of the law, the Smithsonian held 18,400 sets of Native American remains (Killion and Molloy, 2000). As of mid-October, 2004, an estimated 3323 individuals have been repatriated, with 500 more awaiting tribal decisions. Approximately 3000 more are under claim by tribes and additional claims are anticipated (Billeck, personal communication, October 15, 2004). Thus, it appears that at least 37% of the Native American remains within the Smithsonian's collections will be returned.

One positive aspect of the NMAIA is that systematic study of remains prior to deaccession is a priority. Destructive analysis is possible during this period. As Killion and Molloy (2000) emphasize, an immense amount of contextual knowledge has been gained due to this law, as well as a systematic catalogue. Repatriation from the Smithsonian's collections has not been without its tensions, however. Initiating their request in 1987, the Larsen Bay Tribal Council voted to request the return of all remains excavated by Hrdlička from the Uyak site, on Kodiak Island. Hrdlička had excavated there during four field seasons: 1932, 1934, 1935, and 1936 (Hrdlička, 1944). Negotiations were protracted, with the Smithsonian arguing against cultural affiliation, which was offensive to the Larsen Bay community. The Larsen Bay Tribal Council's claims were bolstered by legal assistance from the Native American Rights Fund (NARF). Following passage of the NMAIA, communication continued and expert outside advice was sought. Finally, secretary Adams decided in favor of repatriation.[8]

[7]This National Museum of the American Indian opened its public space on Washington, DC's mall during September of 2004.

[8]Although Donald Ortner, acting director of the National Museum of Natural History's repatriation program, pressed for convening the Repatriation Review Committee (RRC) to consider the

Ultimately, the remains and 144 associated funerary items were returned to Larsen Bay in September of 1991 and January of 1992, respectively (Bray and Killion, 1994:xiii). The published report of this repatriation experience contains contributions by Native Americans, archaeologists, physical anthropologists, and museum and legal experts whose views diverge substantially. By exposing the sometime fractious process fully, however, as well as perspectives on Hrdlička as a person and a professional, the Smithsonian paints a vivid picture of the complex issues repatriation initiatives raise. A well-articulated statement about the Smithsonian's newly formed Repatriation Office is also included in the Bray and Killion volume (Zeder, 1994).

Killion and Malloy (2000) also examined the proposition that research on Native American remains was becoming less visible over time due to repatriation. They noted that the number of scientists visiting the Smithsonian Institution's North American Native American collections decreased during the final decade of the 20th century. A survey of two relevant professional journals, the *American Journal of Physical Anthropology* (AJPA) and *American Antiquity* (AA), for the periods 1985–1989 and 1990–1996 saw no decrease in the numbers of articles based on Native American remains. For AJPA, the proportions are 6 and 7%, respectively. The figures for AA are 4 and 7% (Killion and Malloy, 2000:114). Given the small percentages, the change is probably not statistically significant. A further survey of the same journals for the years 1999–2003 conducted during preparation of this chapter yielded figures of 6.3% (AJPA) and 8% (AA), a small but apparently stable proportion. There is no support in these data for the inference that during the period from 1990 to 2003 published osteological studies of Native American remains decreased in the AJPA and AA. Katzenberg (2001), however, inferred that this might not be the case for presentations at professional meetings, referring to both Canadian and U.S. examples.

On November 16, 1990, the 101st Congress passed NAGPRA.[9] The provisions of this act are far reaching, including establishing tribal ownership

---

Larsen Bay repatriation request, secretary Adams made this decision without consulting the RRC (Bray and Killion, 1994:xiv).

[9]McKeown (2002:108) describes the legal complexity of NAGPRA as follows: "Congress sought to reconcile four major areas of federal law. As civil rights legislation, Congress wished to acknowledge that over the nation's history, Native American human remains and funerary objects suffered from differential treatment as compared with the human remains and funerary objects of other groups. They also wanted to recognize that the loss of sacred objects by Indian tribes and Native Hawaiian organizations to unscrupulous collectors negatively impacted on Native American religious practices. As Indian law, Congress founded their efforts on an explicit recognition of tribal sovereignty and the government-to-government relationship between the United States and Indian tribes. As property law, the Congress wanted to clarify the unique status of the dead as well as highlight the failure of American law to adequately recognize traditional concepts of communal property still in use by some Indian tribes. Lastly, as administrative law, Congress would direct the Department of the Interior to implement Congress' mandate, including the promulgation of regulations to ensure due process,

of Native American remains and cultural items recovered from federal lands[10] after the date the legislation was passed. Provisions also established parameters for considering trafficking in Native American human remains and "cultural items," specifically funerary objects, sacred objects, and objects of cultural patrimony, a crime. Most important for bioarchaeologists, NAGPRA, in its extension of American Common Law to Native American remains, required all federal agencies, except the Smithsonian Institution, and museums (including state agencies and universities) receiving federal funds to inventory their collections of American Indian remains and cultural items, as defined earlier. The National Park Service (NPS), as well as appropriate lineal descendants and culturally affiliated Indian tribes and Native Hawaiian organizations, was to be notified by these institutions of their holdings. While remains and all forms of cultural items could be requested by Indian tribes, Native Hawaiian organizations, and lineal descendants, only Indian tribes and Native Hawaiian organizations could request objects of cultural patrimony (McKeown, 2002). Nonfederally recognized Indian groups did not have standing to make a request, a provision that has been criticized (Walker, 2002).

The deadline for such notifications was established as November 16, 1995, 5 years after the passage of the law. It quickly became clear, however, that implementation was not proceeding smoothly. The NPS did not meet its 12-month deadline for development of regulations. Part of the problem was a lack of the funding necessary for Frank McManamon, consulting archaeologist (for NAGPRA), to assemble a staff and develop proposed regulations (Walker, 1992). The legislatively mandated funds to assist museums and Indian tribes during the repatriation process were not available until 1994 (McKeown, 2002). Ultimately, Congress appropriated perhaps one-tenth of the necessary resources (Rose *et al.*, 1996), although this may be an overestimate. Issac (2002) discusses the complexity of the problem for Harvard's Peabody Museum, with its large, geographically diverse, and early collections whose accession records long predate computers. For the Peabody, just the consultation process itself was estimated to require multiple contacts with approximately 500 of the 758 recognized tribes in the United States. Isaac (2002) estimated that the overall cost of inventory completion for Harvard at seven million dollars, only a portion of which had been subvented by grants from the National Park Service. Ferguson *et al.* (1996:271) reported that the issue is equally daunting for the tribes, for whom decision making on these unfamiliar procedures is remarkably complex and time-consuming. The "sheer volume of work anticipated as a result of NAGPRA is staggering for both the

awarding of grants and assessment of civil penalties." Legal background issues are also discussed in Trope and Echo-Hawk (1992).

[10]Approximately one-third of the United States is either federally or tribally controlled. In other circumstances, state laws apply (Rose *et al.*, 1996).

tribes and the museums. . . ." A third factor complicating this vastly under funded mandate was a delay in the Department of the Interior's appointment of the authorized seven-person NAGPRA Review Committee composed of three individuals nominated by Native American groups, three proposed by museums and scientific organizations, and one additional person selected by the committee (Walker, 1992). The first meeting of the Review Committee occurred on April 29–May 1, 1992, in Washington, DC (National Park Service, U.S. Department of the Interior, NAGPRA web page).

Simply completing and systematizing the required inventory has occupied considerable effort upon the part of the NPS. Winski (1992) had estimated that as many as 600,000 Native American remains were held in museums and private collections. Other estimates reached one and a half to two million (Trope and Echo-Hawk, 1992). The minutes of the July 19, 2004, meeting of the NAGPRA Review Committee reported that the inventory of culturally unidentifiable remains was 99% complete (NAGPRA Review Committee, 2004:10), "including 111,000 human remains, including 2,627 human remains that have been transferred as recommended by the Review Committee and 489 that have been affiliated" (National Park Service, U.S. Department of the Interior, NAGPRA web page). If the 18,400 from the Smithsonian Institution were added to this figure, along with an estimated but unknown number in private hands and others repatriated without reporting, it appears that the Congressional Budget Office's 1990 estimate of ~200,000 remains is consistent with data available in 2004.

NAGPRA's contentious issues include the definition of "cultural affiliation" and the disposition of "culturally unaffiliated" materials. A group with standing may demonstrate one of five relationships with human remains or cultural items: "(1) lineal descent, (2) tribal land ownership, (3) cultural affiliation, (4) other cultural relationship, and (5) aboriginal occupation" (McKeown, 2002:120). While all five categories are relevant to remains or objects recovered from federal or tribal lands after November 16, 1990, only lineal descendants and recognized Indian tribes and Native Hawaiian organizations may request collections held by federal agencies or museums on November 16, 1990 (McKeown, 2002). This makes the definition of cultural affiliation crucial to the repatriation process.

NAGPRA regulations define cultural affiliation in terms of relationships of shared group identity that can reasonably be traced historically or prehistorically between members of a present-day Native American group and an identifiable earlier group. Timothy McKeown, NAGPRA program officer, whose responsibilities in 2004 included regulations, reports that a

> wide variety of evidence can be introduced to document such a relationship, including geographic, kinship, biological, archaeological, linguistic, folklore, oral tradition, historic evidence and other information or expert opinion. Unlike claims of lineal descent in which the relationship between the claimant and the individual whose remains or objects are [*sic*] be claimed must be direct and without interruption, determination

> of cultural affiliation should be based on an overall evaluation of the totality of the circumstances and evidence and should not be precluded solely because of some gaps in the record. (McKeown, 2002:120)

A preponderance of evidence, i.e., 51% certainly, is considered legal proof rather than more stringent definitions centering on "reasonable doubt" or "scientific certainty" (Rose *et al.*, 1996:91; Thomas, 2001:230).

Thus, both humanistic and the scientific perspectives must be weighed in determinations of cultural affiliation. Especially problematic is the evaluation of oral traditions in relationship to biological and archaeological data.[11] Traditional Indians express surprise at the positivist discussion of "extinct cultures" (Binford, 1962) and resent their oral traditions being characterized as myths and legends (Anyon *et al.*, 1997). Archaeologists speak of a division between history and prehistory, while no prehistory exists in Navajo traditions. "Our history begins with the creation of life and the creation of this world" (Begay, 1997:163).

Biological determinations of cultural identity, including those involving modern and ancient DNA, have been characterized by TallBear (2000:1) as "racist ideology." Even so, the Western Mohegan, an unrecognized tribe, have presented DNA evidence to support recognition petitions (Kaestle and Horsburgh, 2002; TallBear, 2000). Similarly, inherited skeletal morphology has been used to identify Euro-Americans and African-Americans held within "Native American" collections (Jantz and Owsley, 1994; Ousley *et al.*, 2000; Owsley, 1999). Resolution of these apparently conflicting weightings of attributes legally accepted as evidence of "cultural affinity" requires considerable insight concerning rigor in both qualitative and quantitative measures.

Ethnologists and archaeologists have demonstrated convincingly that oral traditions of Indian people may reflect historical events, some occurring before 1492 (Echo-Hawk, 1994; Eggan, 1967; Pendergast and Meighan, 1959; Teague, 1993; Wiget, 1982). More equivocal are assertions that even late Pleistocene or very early Holocene events may persist across the millennia. Arguments have been made, however, for oral traditions about landscapes, including volcanic events and floods, being particularly conservative and accurately represented (Cruikshank, 1981; Echo-Hawk, 1994; Hanks, 1997). Validation methods for oral traditions have also been developed (Vansina, 1985; Whiteley, 2002).

As emphasized by Anyon and colleagues (1997), concepts of time and space differ for archaeologists and traditional Indian people. Archaeologists

---

[11]Attempts to link oral traditions and archaeological inquiry have considerable time depth. As noted in Chapter 1 of this volume, Cushing earnestly solicited Zuni opinion concerning symbolic meanings of pictographic and other art in relationship to his archaeological "test" of the myth of the Lost Others (Hinsley and Wilcox 2002:89, 101). Similarly, McGregor (1943:295) solicited opinions from traditional Hopi concerning the ceremonial nature of an elaborate interment discovered within a Pueblo III ruin approximately 20 miles east of Flagstaff, Arizona.

conceptualize the archaeological record in terms of linear time and known space, while for many traditional Native Americans, time is not linear nor are events firmly lodged in space. Thus, Anyon *et al.* (1997) argue that while oral traditions may inform archaeological interpretations and archaeology may be significant in issues such as land claims, archaeology has little to offer oral traditions. The two do, however, converge in certain broad themes, "such as migrations, warfare, residential mobility, land use, and ethnic coresidence" (Anyon *et al.*, 1997:80). Archaeological study at the site of the Battle of Little Big Horn has, for example, also been used to support Native American rather than western historical tradition (Anyon *et al.*, 1997).

Roger Echo-Hawk (1994, 1997) has illustrated a rigorous approach to conjoining information from oral traditions and archaeology in determining whether the Pawnee were culturally affiliated with the prehistoric Central Plains Tradition and thus had rights to repatriation. First identifying ancient, nonfictional oral traditions, he cross-validated information about earthlodge form from narratives about animal ceremonialism with the archaeological record. The Smithsonian's Native American Repatriation Committee has accepted his interpretations, and human remains from the prehistoric Steed-Kisker site have therefore been transferred to the Pawnee (Echo-Hawk, 1997).

Echo-Hawk (1997, 2000), like some other indigenous intellectuals today, e.g., Pullar (2001), argues not for the primacy of oral traditions, including origin narratives, but for respect, discounting faith-based literal interpretations that privilege oral traditions, e.g., Deloria (1995). Archaeologists and bioarchaeologists are also embracing the concept, voiced by Thomas (2001:231), that archaeology and oral traditions are "separate ways of knowing the past." Before archaeologists attempt to use oral traditions, however, they should be made aware that some tribes, such as the Hopi and Navajo, encourage archaeological reporting of oral traditions. Others, such as the Hualapai and Zuni, do not (Anyon *et al.*, 1997, 2000).

However thorny the issue of identifying "culturally affiliated" remains, even more daunting are repatriation issues surrounding "culturally unaffiliated" materials. As noted earlier, in 2004 most of the ~111,000 inventoried remains were classified as unaffiliated. The 1990 NAGPRA legislation does not specify a procedure for the repatriation of culturally unaffiliated remains and therefore decisions have been made by the Repatriation Committee on a case-by-case basis. During this process, for example, remains have been returned to unaffiliated tribes. A section of the NAGPRA regulations (Department of the Interior, Office of the Secretary, 1995:62167) has, however, been reserved to establish continuing responsibilities of federal agencies and museums (McKeown, 2002). At its June 8, 2000 meeting, the Review Committee developed a set of general recommendations concerning the repatriation of culturally unaffiliated remains. They noted that human remains may be culturally unaffiliated due to one of three factors: (1) they are affiliated with an unrecognized tribe, (2) they belong to a defined group for whom there are no living representatives, and (3) there is insufficient evidence to identify the earlier group. Regional solutions are recommended, with emphasis on consultation and consensus. In its advisory role, the Review Committee recommended that the NPS develop a draft proposed rule to be considered at the next committee meetings and then to be published for additional public comment in the *Federal Register* (Department of the Interior, Office of the Secretary, 2000:36463). The National Park Service has provided the Review Committee with draft regulations, which were initially approved and then rejected by the committee (Lovis *et al.*, 2004). As of the end of 2004, no further drafts were presented or published in the *Federal Register*.

The development of regional consensus may not be a solution well received by all stakeholders. The SAA, for example, does not believe that the secretary of the interior has the authority to issue regulations concerning culturally unidentifiable human remains and that new legislation is required (Lovis *et al.*, 2004). Several tribes, as well, have expressed concern about the repatriation of inappropriately attributed remains. The Wind River Shoshone rejected a proposed repatriation because they felt museum records may have been flawed. The Blackfeet, concerned about the affiliation of 15 skulls sent to the Army Medical Museum in 1892, requested scientific study by Smithsonian staff. Since they were at war with neighboring tribes in 1892, they were afraid that they might be repatriating their enemies rather than their ancestors. Following study, only those remains biologically identified as Blackfeet were repatriated (Thomas, 2000). Shortly before NAGPRA became law, the Museum of New Mexico inquired of the Zuni tribal council concerning repatriation of remains in the museum's collection excavated from Zuni lands. In response, the tribal council passed a resolution that applies to all ancestral remains and associated grave goods in museums. It stated that removal from their graves had so desecrated the materials that there was no way to reverse the process. Thus, Zuni remains in museums as of ~1989 are not to be repatriated but instead curated with respect (Ferguson *et al.*, 1996).

The issue of cultural affiliation and the differing interpretations of the Indian and scientific communities is writ large in the recent example of "The Ancient One" (Kennewick Man). This example has relevance for the legality of the National Park Service developing guidelines without new legislation and the form those guidelines might take (McLaughlin, 2004).

Reports on Kennewick Man abound, e.g., Chatters, 2000, 2001; Owsley and Jantz, 2001; Swedlund and Anderson, 1999; Thomas, 2000, 2004; Watkins, 2003; Tri-City Herald, Kennewick Man Virtual Interpretive Center: Legal Documents 2005; McManamon, 2004, and are not presented in detail here. Briefly, the case involves remains exposed by erosion in 1996 along the course of the Columbia River in Washington. Referred initially to anthropologist James Chatters as a potential forensic case, the remains contained an ancient projectile point and were

C-14 dated to 8000–8500 BP. During the course of his investigation, Chatters (1997) initially referred to the remains as showing more European features[12] than those of recent American Indians and was quoted in *The New York Times* as having thought the remains were those of a "white guy" (Egan, 1996, cited in Thomas, 2004). Chatters (2000) later published systematic observations that paralleled earlier conclusions of a team of experts consulting for the NPS. Both studies concluded that the Kennewick remains more closely resembled East Asian and Polynesia peoples rather than recent American Indians. His features were not, however, unlike the few available Archaic remains from North American. This sequence of assessments ultimately attracted considerable attention and led to repatriation requests from not only a coalition of nearby Indian tribes, but also the Asutru Folk Assembly, a traditional European pagan religion, and the Polynesian heritage activist Faumuina (Asatru Folk Assembly, 1997; Jelderks, 2001; Walker, 2002, 2004).

In September 1996, after a month of study by Chatters and the establishment of antiquity, the Army Corps of Engineers (COE) decided to repatriate the remains to the Confederated Tribes of the Umatilla Indian Reservation (Chatters, 2000; Owsley and Jantz, 2001; Watkins, 2003). Before repatriation could occur, however, eight anthropologists[13] filed suit that halted the return. As Watkins (2003) emphasizes, there were three main points at issue: (1) the equation of pre-1492 antiquity with "Native American," (2) the assertion that the study of these ancient remains would be of major benefit to the United States, and (3) the allegation that the scientists' civil rights were being denied by the Corps' actions.

Litigation proceeded, with attention focused on the issue of cultural affiliation and whether the Kennewick remains were "Native American." In March 1998, the NPS/COE commissioned a series of studies by 18 experts who evaluated the remains for cultural affinity through the study of archaeological, biological, historical, and traditional information used for determining the cultural affiliation of remains under NAGPRA. Sections of the report were submitted during 1999 and 2000 (McManamon, 2004). The report well illustrates the difficulty in gaining consensus from humanistic and scientific interpretative traditions in establishing cultural affinity for Native Americans. Studies of the physical remains, as no aDNA was recovered (Kaestle, 2000; Merriwether *et al.*, 2000; Smith, 2000), are cited earlier, and suggest that the skeleton is distinctly different from recent

---

[12]Cook (Chapter 2) reports that in 1790, Blumenbach remarked upon the Caucasian features of an archaeologically recovered skull from Illinois. Similarly, John Collins Warren (1822:135) reports that a skull from Ohio found in a cavity of a rocky bank 60 miles from Marietta, Ohio, appeared upon initial inspection to be a "young female of European origin."

[13]Robson Bonnischsen, C. Loring Brace, George W. Gill, C. Vance Haynes, Jr., Richard L. Jantz, Douglas W. Owsley, Dennis J. Stanford, and D. Gentry Steele.

Indian groups; the archaeological evidence (Ames, 2000) was equivocal and could not definitively support models for continuity or discontinuity. Boxberger (Boxberger and Rasmus, 2000), in his analysis of traditional historical and ethnographic information, concluded "that the ethnographic and historic data specifically place the Yakama, Wanapum, Palouse, Walla Walla, Umatilla, Cayuse and Nez Perce in this area. The oral traditions place these tribes in this area since the beginning of time." Linguistic evidence could not definitively establish continuity or replacement, although the ethnobiological terms of the traditional regional language core suggest considerable time depth (Hunn, 2000). Turning to oral traditions, however, Hunn (2000) notes a legend (*Lalíik*) that specifies a summit above an ancient flood that may reflect a Pleistocene event linking the contemporary language group (Sahaptin) to a very ancient, local past. Radiocarbon dates confirmed the ~8000–8500 BP antiquity of the Kennewick remains (Fiedel, 1999).

Although the report had focused on issues of cultural affiliation, the acceptance of Kennewick as a "Native American" also loomed large in the opinions presented to the court by the NPS. The remains were argued to be Native American, primarily based on the antiquity of the materials (McManamon, 2000). Secretary of the Interior Bruce Babbitt's (2000) letter to Louis Caldera, then secretary of the Army, concluded that the remains were Native American and that the preponderance of evidence supported cultural affiliation with the Confederated Tribes of the Umatilla Indian Reservation. In reaching his conclusions concerning cultural affiliation, Babbitt privileged geographical evidence and oral traditions. His letter also emphasized that due to the special relationship between Indian tribes and the federal government, "any ambiguities in the language of the statute must be resolved liberally in favor of Indian interests." As Ferris (2003:165) notes, one of the important lessons from the Kennewick case is "that federal agencies will be inclined to interpret the affiliation provisions in the act [NAGPRA] as meaning that *any* Native claim, no matter how tenuous or strong the 'science,' justifies repatriation."

The SAA has taken issue with the Department of Interior (DOI) conclusions, arguing that Congress intended that the "earlier group" should be "analogous to a modern tribe in terms of its composition and scale" (Lovis *et al.*, 2004:177). The SAA is also against the repatriation of remains to diverse sets of modern tribes under the guise of joint affiliations on the theory that if the set is sufficiently large, an appropriate relationship must be there somewhere (Lovis *et al.*, 2004:178). For these reasons and others, the SAA filed an *amicus curiae* brief in support of the plaintiffs. Significantly, however, the SAA has never contested the identity of the remains as "Native American" (SAA Kennewick Legal Briefs web page; see also Kelly, 2004).

On August 30, 2002, Judge John Jelderks, United States Magistrate Judge for the District of Oregon, found secretary Babbitt's conclusions unwarranted and overturned his decision [Bonnichsen v. United States, 217 F. Supp. 2d 1116 (D. Ore. 2002), Jelderks, 2002]. He noted that the coalition, which includes one unrecognized tribe, does not conform to the definition of "tribe" in the singular, as specified by NAGPRA. The judge also largely echoed the SAA's concern for clear definition of an earlier group, which was lacking in Babbitt's letter. He explicitly emphasized the difficulty posed by the antiquity of the remains and the absence of associated artifacts. Jelderks discounted the oral tradition evidence, concluding "narratives are of limited reliability in attempting to determine truly ancient events" (p. 53). He also concluded that even if oral traditions "could be relied upon to establish that the ancestors of the Tribal Claimants have resided in this region for more than 9,000 years, the narratives cited by the Secretary do not establish a relationship of shared group identity between those ancestors and the Kennewick Man's unidentified group" (p. 55).

Following an appeal by a coalition of four Plateau tribes and tribal groups, the SAA filed a second *amicus curiae* brief (SAA Kennewick Legal Briefs web page) supporting Judge Jelderks' finding that that the DOI had failed to convincingly argue its case for the definition of Native American and for cultural affiliation. Similarly, the SAA supported the district court's determination that the tribal claim was constructed improperly, in that the tribes were not a single cultural entity and that a federally unrecognized tribe was included. Concerns about levels of proof and the special relationship of the agency (DOI) to the Indian tribes were also raised.

The decision had been appealed to the Ninth Circuit Court of Appeals, who in February and April of 2004 denied both the appeal and a subsequent petition for a rehearing [Bonnichsen v. United States, 357 F.3d 962 (9th Cir. 2004); superseded and amended by Bonnichsen v. United States, 367 F.3d 864 (9th Cir. 2004)]. In a final statement, the court concluded that "[n]o recognizable link exists between Kennewick Man and modern Columbia Plateau Indians," and, in fact "the record does not contain substantial evidence that Kennewick Man's remains are Native American within NAGPRA's meaning" (Gould, 2004:1604). The opinion further emphasized that these "remains are so old and the information about his era is so limited, the record does not permit the Secretary of the Interior to conclude reasonably that Kennewick Man shares specialized and significant genetic or cultural features with presently existing indigenous tribes, people or cultures" (Gould, 2004:1608). Thus, while Judge Jelderks had raised several other issues in overturning Babbitt's determinations, the final appeal was denied largely because the court was not convinced that the remains conformed to NAGPRA's definition of "Native American."

As Watkins (2003) observes, this decision implies that Indian people will now have to focus on defining, perhaps at an evidentiary level, when immigrants became Native Americans. Or perhaps scientists can identify the threshold between the two? The real tragedy is that an estimated one to three million dollars were spent on this case (Schneider, 2004), when it could have been directed toward tribal or archaeological programs (Watkins, 2003).

The Kennewick case will doubtless be precedent setting, unfortunately so in that the archaeological context was poorly defined and quickly reburied by the COE. The absence of associated cultural materials also makes the issue of tribal associations difficult. However, reverberations may be profound, requiring, for example, that the DOI consider a fourth subcategory of culturally unaffiliated remains: remains of sufficient temporal and cultural distance from contemporary Native American tribes that claims of cultural affiliation cannot be supported under NAGPRA. Judge Jelderks' decision also reinforces the need for further legislation in reference to the repatriation of culturally unaffiliated remains (McLaughlin, 2004).

The clash between processual "scientific" archaeology and traditional Indian perspectives is frequently glossed as one of ethics. Processualist archaeology is said to be less ethical because it does not unconditionally support the perspectives of living traditional communities. The American Anthropological Association's ethics statement clearly states that if conflict occurs, the anthropologist must privilege the lives of the people studied. For archaeologists and bioarchaeologists that has traditionally meant the subject of their inquiry, a person not now living. Any shift to focus on living descendants largely postdates 1980. Rather than arguing about ethics and determining degrees of ethicality, a more productive middle ground may be that suggested by Goldstein and Kintigh (1990), who assert that the ethical dispute is better considered a conflict in cultural values. Ethics, as a code of behavior that derives from cultural values, is thus a cultural construct. The preferred behavior, they argue, in situations of culture conflict, is one of tolerance and respect, meaning that compromise is possible. That their proposal may represent a substantive middle ground is reflected in the fact that soon after their 1990 article appeared, it was critiqued by both conservative archaeological interests (Meighan, 1992) and those favoring reburial (Klesert and Powell, 1993). It appears that desirable tolerance is building, following increased communication and the emergence of collaborative community-based heritage programs (Dongoske *et al.*, 2000; Ferguson, 1996; Watkins *et al.*, 1995).

While such collaborations are moving apace in situations where archaeologists are simply mandated to avoid other than endangered interments, there are fewer cases of active involvement of U.S. bioarchaeologists directly in collaborative research efforts. Some examples do exist, however. One of these is the productive alliance forged by Phillip Walker, a physical anthropologist, and the Chumash Indians of southern California. In 2004, Walker had worked with the tribe for 25 years. During this period, the tribe and Walker created a specially designed subterranean ossuary located on Walker's campus in Santa Barbara, California. This location is near the center of the region where the Chumash lived during historic times and was specially designed in consultation with spiritual leaders to ensure it met their needs. It also protects the remains in a context where researchers may continue to provide information about the Chumash past, supervised by the living descendants. Walker has carefully developed this collaboration, sharing information, involving the Chumash in research projects, and assisting the tribe in other preservation initiatives (Walker, 2002).

Another productive collaboration occurred in the midcontinent, involving the Omaha Tribe and physical anthropologist Karl Reinhard. As a result of Nebraska's 1989 reburial law, the Omaha tribe asked the University of Nebraska for the return of historic remains. At the request of the tribe, Reinhard and colleagues collaborated with the tribe in developing research designed as follows:

> (1) to provide an idea of what Omaha life was like during 18th–19th century; (2) to correct some misinterpretations of Omaha culture and history, especially past archaeological studies that suggested the Omaha were warlike; (3) to address past and modern health issues: diet, diabetes, cancer and other diseases; (4) to explain the science of the analysis so that young Omaha people might become interested in pursing careers in science and technology; (5) to define the contributions of Omaha culture to Native American society as a whole and the world at large. (Reinhard, 2000:515)

The full scientific study has been reported in a book chapter (Reinhard *et al.*, 1994). The results have also been communicated to the Omaha, who have been particularly interested in dietary and activity reconstructions indicating that the current high level of diabetes among the Omaha result from recent changes in lifestyle. Reinhard and colleagues also attribute the 19th-century population decrease to epidemic disease rather than warfare, thus confirming Omaha oral traditions. Other points of interest concerning historical figures and the contributions of the Omaha to the history of the Great Plains were also of interest to both the tribe and the scientists. (Reinhard, 2000)

In the Southwest, as another example, the Hopi — who claim cultural affiliation with all prehistoric remains in the American Southwest (Anyon and Thornton, 2002) — appear open to the (nondestructive) study of remains. Importantly, the Hopi cultural advisors are willing to consider professional research designs that address specific problems using specified sets of data, which mutually benefit the Hopi and the anthropologists. Even though reburial is specified, the Hopi are interested in inherited features that may document migrations, as well as life history indicators that define age-at-death, sex, and health status. Funerary items as indicators of social status are also of interest (Ferguson *et al.*, 2000).

Other collaborations involve museums and Indian people. For example, during the 1990s, the University of Missouri opened the Museum Support Center, a new state-of-the-art, 25,000 square-foot curation facility. One portion has been dedicated as a mausoleum, a place that is secure and not open to most visitors. Several Indian groups have visited the mausoleum and blessed it in various ways, satisfied with the quality of curation and seclusion (Michael O'Brien, personal communication, November 18, 2004).

There are several important points raised by these collaborations. First is that there is no global solution. Walker's collaboration works because it is based in Chumash culture history. The issue of diabetes and warfare were of interest to the Omaha, but may not be relevant to many other Indian peoples. The Omaha desire that more young Indian people embark on science careers, which may not be the vocation of preference among other groups. The Hopi are concerned about migrations, which might not interest other tribes. Those visiting the University of Missouri facility are satisfied with the security of the mausoleum and the level of respect accorded the human remains stored there. This may not be an ideal solution in other regions. However, the need for openness, for communication, for mutual respect, and for initiatives that are of interest to all collaborating parties *is* global.

III. CANADA: COLLABORATION AND COMMUNITY CONTROL

Canada has no federal legislation comparative to NMAIA and NAGPRA. Some find this situation undesirable in that national heritage legislation could provide uniform policy for federal lands and it is perceived that funding levels would increase (Syms, 1997). Yet, as emphasized by Watkins (2003), relationships between First Nations and archaeologists are strong and becoming stronger. He attributes this to "archaeologists directly taking into consideration the wishes of the indigenous populations in the research arena rather than performing through a regulatory or legal framework" (Watkins, 2003:277). It would appear that relationships between First Nations and physical anthropologists/bioarchaeologists are also quite strong and to be envied by many colleagues in the United States. Why might this be?

One factor may lie in divergent historical paths taken by archaeologists and physical anthropologists working in Canada and in the United States. One clear difference is the timing in which the leadership in both professions began voicing sympathy with concerns expressed by First Nations about burial excavations and the curation of remains and funerary items.[14]

In 1976, The Royal Society of Canada sponsored a symposium on New Perspectives in Canadian Archaeology. One of five sessions was entitled "Archaeology: New Motivations and Attitudes." The introductory paper, by William Taylor, then director of the National Museum of Man (NMM, now the National

[14]While it is tempting to trace the history of liberal attitudes to Sir Daniel Wilson's attitude toward Native Americans, Wilson left no direct anthropological legacy (Trigger, 1966).

Museum of Civilization, NMC) in Ottawa, emphasized the growing engagement of indigenous people.

> Another major consideration in new motivations and attitudes in Canadian archaeology is, of course, the rapidly developing involvement of Canadian native groups, based on a long and lasting concern in Canadian archaeology. It is generally recognized now that it is no longer sufficient to explain that white archaeologists also dig up the bones of white men and put them on display. The profession seems to be adjusting with reasonable comfort to a very different situation as native peoples of Canada become more involved at both the local and national level. We have yet a considerable distance to travel. . . . It seems that we have not, in fact, seriously attempted to provide, by popular publication or museum display or by local teaching, the kind of direct return to the Indians and the Inuit of Canada to which they are entitled — and which we are capable of providing. (Taylor, 1976:154)

Speaking from his experience with Native peoples of Quebec, Laurent Girouard also emphasized the need for engagement.

> These days, whether we like it or not, the archaeologist who studies the ancestors of the Amerindians must bear the burden of the colonial past which was forced upon this continent's first inhabitants by the whites. He must choose one of two alternatives. Either he continues to study the Amerindians' past as something which has no political meaning, no relationship with the present and therefore with the life of the Indian and Inuit communities today, or he can study this past by consciously placing it in a historical continuum. If he does this, he must take into account the situation of the Amerindian today. (Girouard, 1976:159)

Two additional papers spoke of the need to engage communities as equal partners (Swinton, 1976) and the dissatisfaction of Indians over the excavation and display of burials (Johnston, 1976). Next, Jerome Cybulski, a leading physical anthropologist, located at the NMM spoke about skeletal analyses. Cybulski (1976) emphasized that in response to First Nations concerns, *in situ* analyses had occurred in Canada during the 1970s and that reburials were occurring in the course of truly collaborative efforts in British Columbia, projects initiated by the tribes. These interactions appeared to lack the antagonisms seen during the 1970s in Iowa, for example.

Other evidence of sensitivity to First Nations concerns is evident in Cybulski, Ossenberg, and Wade's "Committee Report: Statement on the Excavation, Treatment, Analysis and Disposition of Human Skeletal Remains from Archaeological Sites in Canada" (1979). Prepared for the Canadian Association of Physical Anthropologists, the report was a response to public concern over the "nature and purposes of scientific study of human skeletal remains from archaeological sites in Canada" (Cybulski *et al.*, 1979:32). A significant portion of the report is dedicated to "the concerns of living native peoples in that certain archaeological sites, particularly those of the late prehistoric period and those of the protohistoric period, have direct bearing on their cultural and biological heritage" (Cybulski *et al.*, 1979:32).

In response to such concerns, the committee emphasized that there were many close working relationships among archaeologists, physical anthropologists, and local native groups. Their recommendations included increased interaction and collaborations with local communities where biological and cultural heritage is demonstrated in archaeological initiative.

> It is recommended, therefore, that communication and consultation with local communities, on the part of both individual researchers and the provincial or federal agencies responsible for archaeological sites, becoming a working rule uniformly applied throughout the country. The Canadian Association for Physical Anthropology urges individual researchers — archaeologists and physical anthropologists — to consult with local native band councils about their projects and to keep local communities informed of the progress of those projects. The Association also encourages individual researchers to return information to the communities in the form of unpublished and published reports, and by means of formal lectures and informal/presentations before, during and after field work. (Cybulski *et al.*, 1979:35)

The authors close with an emphasis on equal treatment of all human remains, no matter what the heritage, including pioneer graves (Cybulski *et al.*, 1979). This statement contrasts markedly with the stand taken by the AJPA in 1982.

Thus, it appears that the Canadian bioarchaeologists were developing a pattern of community partnership and consultation that would visibly appear in the United States over a decade later and usually only after consultations were legally mandated. The longer history of collaboration with First Nations has yielded a number of exemplary collaborative case studies, including several projects by Cybulski (1978, 1992; Cybulski *et al.*, 2004), who speaks of his 35 year career in the following terms: "I've done field and laboratory osteology in several different regions of the British Columbia portion, ever with the co-operation and active participation of First Nations and their members" (Cybulski, 2001).

Another impressive project is represented in Williamson and Pfeiffer's *Bones of the Ancestors: The Archaeology and Osteobiography of the Moatfield Ossuary* (2003), developed in collaboration with the Six Nations Council of Oshweken, Ontario. Multivocal in presentation, it includes indigenous intellectual, physical anthropological, and archaeological perspectives. Destructive analyses to estimate diet were conducted, authorized by the Six Nations Council.

Two additional recent studies of ancient individuals also illustrate close collaborations between communities and bioarchaeologists. In August of 1999, hunters discovered the remains of an ancient body in the mountains of northwestern British Columbia, not far form the Alaska border. After 550–600 years, it was eroding from a glacier. At the request of the local Indian community — the Champagne and Aishihik First Nations (CAFN) — the remains, termed Kwäday Dän Ts'inchí or "Long Ago Person Found," were carefully recovered, kept in a relatively sterile and chilled condition, flown to Whitehorse, Yukon Territory, and studied prior to reburial. Portions of the associated hat and cloak were radiocarbon dated. The collaborative agreement developed between the Province and the local First Nation emphasized, among other points, the need for respectful treatment and a desire for state-of-the-art analyses of both the body and the artifacts. Thus, destructive analyses for various determinations, including diet, were approved (Beattie *et al.*, 2000). Such studies were informative in that they indicated that this young man had lived near the sea most of his life. Following scientific study and the recovery of samples, the ashed remains were buried near the site of discovery (Beattie *et al.*, 2000; Dickson *et al.*, 2004; Lundberg, 2001).

Another recent collaboration developed as a result of a burial eroding from the shoreline of Southern Indian Lake on the Churchill River, Manitoba, first noted during June 1993. Due to fluctuating lake levels, full recovery did not occur until the summer of 1994. These remains of a young woman were considered an ancestor of local Cree and were referred to as *kayasochi kikawenow* or "our mother from long ago." The elders concluded that she had permitted herself to be recovered so that she could share her knowledge with present and future generations. Her remains were studied in nondestructive ways, including X-rays of the tibia. Artifacts were photographed, cast, and the originals united with the remains, reburied near the original gravesite. A few artifacts were sacrificed for radiocarbon dating, indicating that she died approximately 330 years ago. In addition to the technical studies, Cree archaeologist Kevin Brownlee authored a book for general audiences in collaboration with Leigh Syms (Brownlee and Syms, 1999). Brownlee also created a permanent interpretative display for the community school at South Indian Lake.

These Canadian examples both show differences and similarities to those reported for the United States. The Kwäday Dän Ts'inchí project promoted state-of-the-art study, including destructive analyses. Many of the other cases also included destructive analyses within their research designs, usually directed toward research questions of mutual interest to the physical anthropologists, archaeologists, and the First Nation community. While Reinhard and colleagues also engaged in destructive bone chemistry analysis to investigate diet, many Native American communities in the United States are against destructive analyses, some including X-ray analysis in this category. The advantage of long-term commitment to community engagement is well illustrated in the work of Cybulski, as it has been for Walker in the United States. All the Canadian examples, however, specified reburial rather than curation for future generations of researchers. As in the United States, each approach reflects the cultural and geographic contexts in which the collaboration developed.

At the level of the national museum, even without a federal mandate, the NMC stopped accessioning aboriginal human remains in the early 1970s. Remains are repatriated according to a policy informed by the 1992 task force document promulgated by the Assembly of First Nations and the Canadian Museum Association. A case-by-case approach has been implemented, with validation requiring demonstration of ancestral–descendant relationship or historical connections. Prior to deaccessioning materials, the NMC reserved the right to inventory and study, for both scientific and heritage preservation purposes. The present director of the Archaeology and History Division of the NMC emphasizes that nearly all the collections are vulnerable to repatriation (Morrison, 2004).

The NMC and First Nations, such as the Inuit Heritage Trust (IHT), have, however, reached agreements that may ensure the long-term availability of existing collections. The NMC's agreement with the IHT, finalized in 1998, affects the Nunavut Territory human bones, approximately one-fourth of the NMC's human remains collection. Destructive analyses must receive explicit consent of the IHT; other studies require notification. By the end of 2004, two requests for destructive analyses had been received by the IHT and both were approved. Proposals submitted to the IHT must include not only clear statements concerning research design and analytical methods, but also must explain potential benefits to the community. As Cybulski, curator of physical anthropology at the NMC states, such practices "make our work more accessible and less mysterious to the public" (Cybulski, personal communication, Dec. 15, 2004).

In general, Canadian museums, physical anthropologists, and archaeologists appear to favor mediation over litigation in addressing repatriation issues (Ferris, 2003). Pressure from First Nations for additional repatriations continue, however, including activist rhetoric: "Kitigan Zibi, like all other First Nations across Canada, in our relationship with the federal government has been, and still is, impacted by a hierarchy of institutional racism" (Odjick, 2004; Whiteduck, 2004). It remains to be seen whether such pressure will lead to a federal mandate. As emphasized by Watkins (2003), collaborations in Canada appear to be proceeding without the necessity of federal legislation. Such laws of necessity gloss vast differences in cultural perspectives and risk creating a situation in which one size fits none.

IV. BIOARCHAEOLOGY IN THE 21ST CENTURY: CONSEQUENCES OF NAGPRA/NMAIA

Clearly, one of the intended consequences of federal repatriation legislation was increased communication between and active collaboration with Indian communities and those who excavate and study mortuary sites. For archaeologists, this has obviously occurred, especially in regions with large Native American resident populations. An unintended impact, however, appears to be the isolation of most skeletal biologists/bioarchaeologists from Indian people, from archaeologists, and from archaeological contexts.

Walker (2002:31), who is one of the few success stories in long-term bioarchaeology/Indian collaborative efforts, urges skeletal biologists to be proactive in communicating and lifting "the shroud of mystery" associated with what we actually do. Apparently, few have heeded Walker's sage advice.

One reason why physical anthropologists are relatively invisible in the debates surrounding NAGPRA and in published results of collaborative ventures is that they are frequently mistaken for archaeologists. One archaeologist who has been a key figure in NAGPRA/NMAIA-related activities suspects that in part this is due to a calculated masquerade by physical anthropologists.

> Although the situation has improved dramatically over the last 10 years, it is probably still accurate to say that 10% or fewer of the archaeologists in this country have ever sat down and talked with an Indian, particularly about the archaeologist's research. I have no doubt that the number of physical anthropologists making such contacts is even less, since several physical anthropologists have told me that they pretend to be archaeologists when placed in a situation where Indians might be present so that they can avoid having physical anthropology be directly attacked. (Goldstein, 1992:68)

In addition to conflict avoidance, Goldstein (Chapter 14) suggests several other reasons for the isolation of bioarchaeologists from archaeological contexts, archaeologists, archaeological problem solving, and Indian peoples. One of the reasons for a divergence of archaeological from bioarchaeological interests is inherent in NAGPRA. Since normally only endangered burials are excavated, mortuary features cannot enter a planned research design, as specified by Buikstra (1977) in her call for interdisciplinary integration in bioarchaeological study. Burials are avoided and thus students—whether archaeologists or bioarchaeologists—have very little experience in excavating and analyzing mortuary contexts as part of larger, problem-oriented research designs. Nuanced interpretations of ancient lifeways depend on an appreciation of the subtleties inherent in the archaeological record. While excavation projects continue, these are infrequently guided by the integrative research strategies that lead to a contextually sensitive bioarchaeology.

Second, since post-NAGPRA archaeologists seldom excavate and analyze mortuary sites, they are thus led to deemphasize data derived from the study of burials (Chapter 14). Meanwhile, bioarchaeologists and osteologists have become increasingly busy generating such data as another consequence of NAGPRA. Clearly, the law does not require research other than that necessary to create and inventory and to conduct the necessary archival review to determine cultural affinity. Neither does NAGPRA (or NMAIA) preclude research. As emphasized by Rose and colleagues (1996), only a small percentage of excavated remains had been studied prior to NAGPRA and data collection protocols varied widely. One of the profession's responses to NAGPRA has been to develop an extensive set of recommended standards for data collection, begun by the Paleopathology Association (Rose *et al.*, 1991) and later elaborated as *Standards for Data Collection from Human Skeletal Remains* (SOD), with National Science Foundation support (Buikstra and Ubelaker, 1994).

Federal agencies, e.g., the Army Corps of Engineers, CRM firms, and museums, are requiring detailed inventories following recommendations in SOD. While it is imperative that these data be collected, it is being gathered from remains far removed from their archaeological contexts and their living descendants. The quality of archaeological field notes varies, and these are infrequently consulted when the priority is data collection, using a time-consuming protocol. Thus, skeletal samples are commonly reported as aggregates, as if they were *de facto* representative samples of the populations from whom they were derived. Extensive and rich databases have been generated, for example, from regional surveys sponsored by the COE (Owsley and Rose, 1997; Rose, 1999). Importantly, these investigations include a number of non-Native American ethnic groups, such as Chinese and Euro-Americans. However, with few exceptions, the regional surveys link remains to geographic and environmental zones rather than nuanced considerations of archaeological and cultural contexts. To some degree, the latter is again a by-product of the way NAGPRA inventories have been implemented. Although funerary items must be inventoried and reported, there have been no extensive, standardized studies in the manner specified for skeletal remains.

A further factor relates to the manner in which different "bioarchaeologies" have developed. As Goldstein (Chapter 14) emphasizes, the original bioarchaeological emphasis was on the collaborative integration of skeletal biologists, other specialists (archaeobotanists, geomorphologists, faunal analysts, etc.), and archaeologists in the development and implementation of research designs. In contrast, Clark Larsen's influential approach to bioarchaeology, while also interdisciplinary, links skeletal study to the physical and natural sciences rather than to archaeology. Furthermore, recent (post-1980) theoretical approaches in archaeology are seldom referenced (Chapter 14).

In a summarizing statement on reburial, Reinhard (2000) dichotomizes the globe into reburial prone and nonprone regions. He contends that within the former, research will center on tribal issues and concerns and it is in such contexts that nontraditional, culturally sensitive approaches will develop. He argues that traditional scientific paradigms will shift to nonprone regions and it is here where training shall occur and where new methodologies will develop. Bioarchaeologists, as natural scientists, will move to nonprone regions or embrace forensic anthropology. Museum curators and archaeologists, as regionally based social scientists, are more likely to remain, to compromise, and to shift paradigms (Reinhard, 2000). It would be a shame if Reinhard's predictions come to pass, especially since his work with the Omaha combines scientific rigor and community sensitivity in an exemplary manner.

Reinhard (2000) indicates that the Omaha defined their study goals in collaboration with University of Nebraska officials. The research questions, of concern to living Omaha, required not only knowledge of oral traditions and contemporary medical problems, but also the manner in which analyses of bone chemistry, for example, could be used to implement the research design. Osteological analysis involved not simply standard descriptions of age, sex, and inherited features, but instead the definition of significant research problems drove the protocol. This is exactly the problem-oriented model that anchored "bioarchaeology" as originally conceived (Buikstra, 1977). This model advocated interdisciplinary study whereby all stakeholders collaborated in a spirit of mutual respect to develop anthropologically significant research designs. The conversation is now being extended to include Native Americans, who bring both special humanistic knowledge and defined contemporary concerns to the table. Their perspectives are no less complex and nuanced than the bioarchaeologists' assessment of variant chemical signatures in bone. The inclusion of living descendants in collaborative "bioarchaeological" initiatives thus poses new challenges, but the opportunities hold remarkable promise for advancing our knowledge of living peoples and their pasts.

Chapter 16

A View from Afar: Bioarchaeology in Britain

Charlotte A. Roberts

I. INTRODUCTION

> The United States has, to date, overshadowed the U.K. in the study of osteology and palaeopathology due to its deep grounding in anthropology . . . but we are catching up slowly. (Roberts, 2003:107)

The study of human remains from archaeological sites has a long history in Europe, as it has in North America. However, some parts of Europe have seen more rapid development than others. In Britain it is only fairly recently (since the mid-1980s) where we have seen a significant change in the quantity and quality of data produced from skeletal analysis, data that have been used from a bioarchaeological perspective. It is noticeable, but not surprising, that Europe as a whole (but particularly Britain) has lagged behind in the development of a biocultural/bioarchaeological approach to using skeletal data to contribute to our understanding of past human populations. First, this is probably because most work, until recently, had been undertaken by people working in other disciplines such as anatomy, dentistry, and medicine who had little background knowledge of archaeology to allow them to contextualize their biological data. Second, until the 1980s, there was no specific training for people wishing to work in the field of (palaeo) physical anthropology/bioarchaeology (i.e., archaeologically derived human remains *and not* early hominid remains), at both undergraduate and graduate levels, nor were there many people employed as (palaeo) physical anthropologists to teach in departments of archaeology, certainly in Britain. Key to this problem has been the emphasis on archaeology departments in universities rather than anthropology departments, as seen in North America.

Bioarchaeology: The Contextual Analysis of Human Remains, Buikstra and Beck (eds.)

The long-standing four-field approach taken in anthropology in North America (incorporating linguistics, cultural anthropology, physical/biological anthropology, and archaeology) has allowed the development of a truly integrated bioarchaeological approach to the study of past humans, and the production of graduates with a broad all-encompassing knowledge.

This chapter compares and contrasts the study of the bioarchaeology of past human populations (utilizing their skeletal remains) in North America and Britain, highlighting the major similarities and differences. Where relevant, examples from the rest of Europe are used to emphasize particular points. The chapter starts with a commentary on the use of terms in Britain to describe the study of human remains, follows with a brief history of development of study in Europe, focuses on the contribution of British people to bioarchaeology (recognizing that some non-British people have contributed and continue to do so), highlights recent impacts on the progress of the discipline, and makes predictions for the future.

II. DEFINITIONS

The study of skeletal remains from archaeological sites (the most common type of human remains most [palaeo] physical anthropologists deal with) in North America has been termed physical anthropology, biological anthropology, and more recently bioarchaeology. While all these terms have been used in Britain, at some time and in parallel with North America, there are many more that practitioners have decided are/were more appropriate at certain points. These include human skeletal biologist, osteologist, palaeopathologist, and osteoarchaeologist. Of interest here is the term "bioarchaeologist" used in Britain where it describes somebody who studies *any* biological materials (as opposed to North America where it relates *only* to human remains). In Britain, bioarchaeology could include the study of macroscopic/microscopic plant remains, animal bones, molluscs, or human remains. As the Preface to this book shows, however, the term bioarchaeology has seen a long history in Britain, which stretches back to Clark's work at Starr Carr (1972). At that time, the term was reserved for plant and animal remains, whereas "osteoarchaeology" has become (since the early 1980s) the word in Britain associated with the study of both human and animal remains (as seen in papers represented in the *International Journal of Osteoarchaeology*, founded in Britain in 1991, and in the names of a number of MSc courses). However, it should be noted that Vilhelm Møller-Christensen in Denmark had already referred to the term in the 1950s (1973; see Preface). Previous to these more recent developments, the term environmental archaeology encompassed the study of all archaeologically derived biological materials, and of interest here is the Association of Environmental Archaeology, based in Britain, whose members concentrate on any biological materials, but rarely human remains (Association for Environmental Archaeology home page). Another oddity is the lack of understanding of all these terms by the British archaeological community whereby, for example, a "palaeopathologist" commonly describes somebody who studies "human remains," although strictly speaking it refers only to the study of ancient disease. The term "bioarchaeology" is used where appropriate throughout this chapter, but it should be understood that it is not by any means a widely accepted term in Britain for the study of archaeologically derived human remains.

III. THE STUDY OF BIOARCHAEOLOGY IN BRITAIN TO THE 1950s

A. Europe

By the second half of the 19th century, physical anthropology as a discipline in its own right had been recognized in Europe (Shapiro, 1959). Work by scholars such as Paul Broca, a surgeon in France (1873), and Rudolph Virchow, a physician and anatomist in Germany (1872), pioneered research on human remains. However, in terms of the study of disease, the German naturalist Johann Friederich Esper had already identified a neoplastic lesion in a cave bear's femur (1774). The late 19th century also fixed the stereotype of the physical anthropologist with calipers in hand busily measuring heads (Shapiro, 1959), but by this time the wider implications of studying human remains had been realized in Europe. Interestingly, Shapiro (1959) believed that European physical anthropology then began to influence its development in North America; I shall argue later that the influence reversed in later years with reference to bioarchaeology. Of note, however, is the work of Marc Armund Ruffer (1859–1917), a French/German medical doctor, who appears to have taken the study of human remains beyond the curiosity stage to attempt to understand palaeoepidemiology (e.g., 1913). It was he that coined the term "palaeopathology" (Aufderheide and Rodríguez-Martín, 1998:6).

B. Britain

Early work in physical anthropology in Britain was done by many scholars. In the 19th century, biometric studies became prominent, with Karl Pearson (1857–1936) of University College, London, and Geoffrey Morant (1899–1964), a "disciple" of Pearson (Stepan, 1982:137), contributing in the area of the evolutionary significance of Neanderthals and the origin of modern humans

(Spencer, 1997d). However, skeletal remains of early humans in Britain have always been rare and overshadowed by finds elsewhere. Although late Upper Palaeolithic remains have been found in Britain, little can be said of this time period because of the scarcity of data (Roberts and Cox, 2003). A British interest in craniology was also established in the 19th century, as elsewhere in Europe, the impulse coming from phrenology (e.g., Davis and Thurnam, 1865). In 1800 all British scientists thought (like the rest of Europe) that there was a single human biological species united by a common humanity (Stepan, 1982). However, by the end of the 18th century there was doubt about a single created species. Polygenic thought specified that human "races" were separated from each other by mental, moral, and physical differences, in the manner of species. By the 1860s polygenism was a distinct but minority strand of British racial thought, specifying that the "races of humankind formed separate biological entities created independently of each other" (Stepan, 1982:3). By this time, too, whites were considered superior to nonwhites and "... culture and the social behaviour of man became epiphenomena of biology" (Stepan, 1982:4); science followed the public opinion of "race." During the 1860s, new sciences in Britain shaped the study of "race" and these included comparative anatomy, physiology, histology, and palaeontology. Information on human racial variation was gathered and, although there was less reliance on the Bible for ideas, there remained an insistence on the permanency of racial "types" (Stepan, 1982). By the close of the 19th century, "race" was firmly established in popular opinion and science in Britain. Between 1900 and 1925 the eugenics movement reinforced "race science," although in Europe this was never as extreme as in other parts of the world (Stepan, 1982:111). By the first years of the 20th century the eugenics movement became established and by the 1920s this was almost a worldwide movement. The eugenics movement aimed to explore the hereditary nature of traits in human populations that were desirable or undesirable and to establish variability in individuals, with the ultimate aim of classifying people.

Karl Pearson took a "statistical population approach" and explored the belief that evolution proceeded by small and continuous variations (Stepan, 1982:136). However, some physical anthropologists determined "racial" averages on just a few skulls, and Pearson was of the opinion that this was unacceptable; standards of recording the variables to determine "averages," which affected comparative studies, were criticized and environmental factors affecting these variables were also highlighted (Stepan, 1982:136). Pearson's interest in the people of Britain, and his obvious criticisms of other work to identify groups of people with the same biological features, led to his opinion that "there was no such thing as a physically and mentally 'typical' Englishman" (Stepan, 1982:137). His recognition of this fact early in the history of the study of physical anthropology underpins how the study of human remains from archaeological sites in Britain has developed, and his view that there was no relationship between physical and mental traits was inspirational for that time period. Nevertheless, his thoughts and opinions fell on deaf ears (Stepan, 1982). Pearson (1899) with Bell (Pearson and Bell, 1919) also wrote on the stature of prehistoric "races" and the character of the long bones of the English skeleton, both studies involving large numbers of skeletons/bones.

Following Pearson, the anatomist Arthur Keith (1924) published on a range of subjects, such as human remains from early cave deposits, and Parsons and Box (1905) wrote about the relationship of the cranial sutures to age. Cave studied a variety of subjects such as cervical ribs (1941) and trepanation (1940), and the anatomist Warren Dawson (1927) worked on mummified remains mainly from Egypt. Another British anatomist, Frederic Wood-Jones (Elliott-Smith and Wood-Jones, 1910), studied skeletal remains from Egypt; Elliott-Smith and Wood-Jones' work is undeniably an innovative piece of work because of the number of remains studied at that time. Later, Kenneth Oakley (1950) was involved with the relative dating of the Piltdown skull, but also considered subjects ranging from trepanning (Oakley *et al.*, 1959) to ancient brains. In the 1960s, Berry and Berry (1967) were instrumental in applying methods developed from the study of non-metrical variants in house mice to human skeletons, research that is used by many even today. Nineteenth- and early 20th-century work on archaeological human remains in Britain (and Europe) was very focused on specific variables that classified different groups of people and usually concentrated on normal rather than abnormal variation.

Early (palaeo) physical anthropology study in Britain had covered all aspects of the study of human remains: age and sex estimation and normal and abnormal variation. By the late 1800s Britain had established itself as one of the leading industrial nations of Europe and had created an unrivaled colonial empire (Spencer, 1997d). In the 1860s, the Anthropological Society of London (ASL) was established by James Hunt (1833–1869). Hunt also secured anthropology as a discipline in the British Association for the Advancement of Science; during the 19th and early 20th centuries much anthropological work was reported through this organization. By 1971 the ASL had merged with the Ethnological Society of London to form the Anthropological Institute of Great Britain and Ireland (Spencer, 1997d). The institute aimed to have a balance between its cultural and physical anthropological work, but the former always dominated.

At this point it is worth considering the many museums that were and are devoted to curating skeletal remains in Europe, skeletal remains that have been and will contribute to understanding our data. This of course includes the many museums that house skeletal remains from archaeological sites, but also those museums established in the 18th and 19th centuries as a result of collecting activities, principally by medical doctors and anatomists. Reviewed here are the main collections in Britain—examples of the tradition. The Royal College of Surgeons in Edinburgh Museum (The Royal College of Surgeons of Edinburgh home page) is one of the largest and most historic collections of surgical pathology in Europe, while the two Hunterian Museums in Glasgow and London are probably the most famous. The Hunterian Museum and Art Gallery was established in Glasgow in 1783; William Hunter (1718–1783) was born in Glasgow and studied medicine, becoming a well-respected anatomy and surgery teacher, and one of the first male midwives. He collected anatomical and pathological specimens plus other items, which were all bequeathed to Glasgow University (hunterian.gla.ac.uk/collections). The Hunterian Museum in the Royal College of Surgeons, London was established as a result of the collections of John Hunter (1728–1793), brother to William and a surgeon and anatomist. This private museum in the Royal College of Surgeons acquired his collection through the government on the day after his death in 1799 (Fforde, 1990). Following extensions to his original collection by scholars such as Arthur Keith, by the end of the 19th century 65,000 specimens existed. The collection includes human and animal pathology, physiological, and anatomical specimens. However, in the 1941 bombing of London, two-thirds of the collection were destroyed, but later the remaining collection was separated into "anatomical and pathological," "odontological," and "Hunterian" museums (The Royal College of Surgeons of England). Over the last few years the museum has been reorganized and refurbished, being opened by HRH Princess Anne in early 2005.

Perhaps of more relevance and direct interest to people working on human remains from archaeological sites is the Duckworth Osteological Collection at Cambridge (named after a former reader in anatomy, W. L. H. Duckworth). The collection has been assembled over the last 150 years by professors of anatomy (Foley, 1990; University of Cambridge, Duckworth Laboratory web site), and now there are over 17,000 human and nonhuman primate skeletal items (University of Cambridge, Duckworth Laboratory web site). The Natural History Museum in London also curates the famous collections of skeletons from Christchurch, Spitalfields. Three hundred and eighty-three skeletons are historically documented with age-at-death, sex, and date of death (Molleson and Cox, 1993). Undoubtedly these museum collections have been very valuable for educating people from different backgrounds about our ancestors. Important for bioarchaeology, these collections have facilitated the opportunity to observe the features useful for identifying age-at-death and sex, and dry bone pathology, in skeletons with known age, sex, and medical histories. However, in contrast to America [the "Robert Terry Collection," Smithsonian Institution, Washington, DC (Hunt and Albanese, 2005), and "Hamann Todd Collection," Cleveland, Ohio] and Portugal (the "Identified Skeletal Collection" in Coimbra and the "Luis Lopes" Collection in Lisbon), where there are large collections of complete skeletons that are documented, much of the skeletal collections in the museums in Britain described earlier that have "modern" collections mainly curate individual bones more often than complete skeletons.

IV. THE SECOND HALF OF THE 20TH CENTURY

A. EUROPE

It was not until the 1950s and 1960s that more extensive and innovative research on human remains was seen where a truly bioarchaeological approach was taken using large amounts of data collected from many skeletons. Turning back to the rest of Europe briefly, we should consider the work of Vilhelm Møller-Christensen (Bennike, 2002) and his seminal work on the bone changes of leprosy as seen in late medieval Danish skeletons from a number of sites (Møller-Christensen, 1961). Møller-Christensen (1903–1988) was a doctor in Denmark by trade but had a strong interest in archaeology (like Calvin Wells in Britain). His most famous work involved the excavation of the cemeteries of Æbelholt in the 1930s and the medieval leprosy hospital at Naestved between 1948 and 1968, including examination of the skeletal remains recovered. His excavation technique was termed "the osteoarchaeological technique" (Møller-Christensen, 1973), a term that has seen recent favor in the United Kingdom. He had criticized the methods archaeologists were using to excavate skeletons. His newly developed excavation technique involved excavating a "ditch" around the grave and then excavating the skeleton on top of the "pedestal" left (similar to the approach taken in an anatomy dissection room/operating theatre). Møller-Christensen's contribution to bioarchaeology during those early years was mainly in his detailed observations of leprous bone changes at Naestved and his corresponding confirmation of the changes in contemporaneous leprous patients in other parts of the world. The rest of Europe has also contributed to our understanding of the past from a bioarchaeological perspective, although the archaeological context has not always been of prime importance in the final interpretation. We now turn to focus in more detail on Britain.

B. BRITAIN

In Britain, two key people really advanced the study of human remains from archaeological sites from a bioarchaeological perspective: Calvin Wells (1908–1978) and Don Brothwell (born in 1933), the former a doctor from Norfolk. Both practitioners were unaware of the term "bioarchaeology" when they were working in the early years, and for Don Brothwell it will only have become a term used in more recent times. Nevertheless, their approach was bioarchaeological and, for Brothwell, remains so. The late Calvin Wells (Fig. 1) had a strong interest in archaeology and soon became *the* person in Norfolk who would produce reports on human remains for archaeologists (e.g., 1966). From this he consequently noticed interesting pathological lesions (e.g., 1965), not a

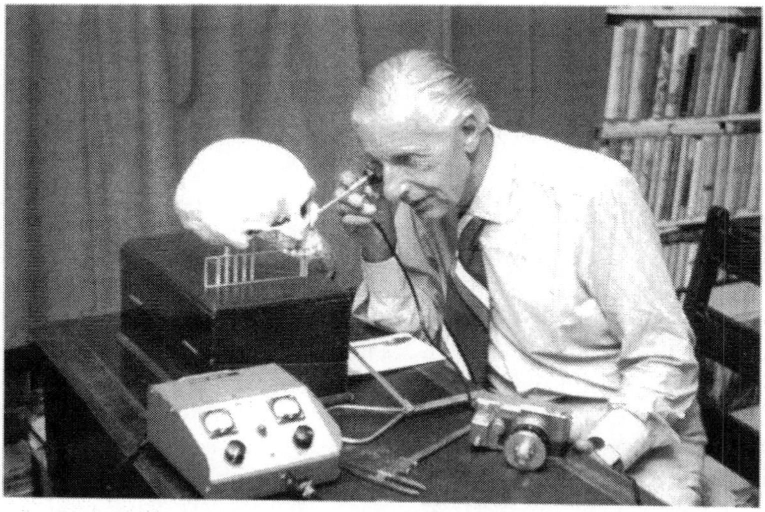

Figure 1 Calvin Wells (with permission of the Department of Archaeological Science, University of Bradford).

real new development in physical anthropology as "case studies" were already prominent and have been maintained throughout the 20th century and into the 21st in Britain and elsewhere. Of recent relevance here is Mays' (1997) review of the study of palaeopathology in Britain and the United States. Of seven journals considered between 1991 and 1995, 51/90 (57%) papers were on palaeopathology and 55% (28) were case studies compared to 15/53 or 28% in the United States. When population studies of palaeopathology were considered there were 14 (27%) from Britain compared to 23 (43%) from the United States. Although in recent years this has improved, people working in Britain still need to consider more hypothesis and question driven approaches to the study of past populations, including comparisons with other studies, and this includes any physical anthropological study.

Calvin Wells started publishing in the 1950s (Wells and Clarke, 1955) and quickly became known for his "stories" about people in the past that he inferred from what he observed from their skeletal remains. While many question his interpretations of the data (e.g., his paper on skeletal changes of "rape" in an Anglo-Saxon woman; Hawkes and Wells, 1975), he "brought alive" the people he was studying and considered not only biological data, but the relevant archaeological evidence.

> The pattern of disease and injury that affects any group of people is never a matter of chance. It is invariably the expressions of stresses and strains to which they were

> exposed, to everything in their environment and behaviour. It reflects their genetic inheritance (which is their internal environment), the climate in which they lived, the soil that gave them sustenance and the animals and plants that shared their homeland. It is influenced by their daily occupations, their habits of diet, their choice of dwellings and clothes, their social structure, even their folklore and mythology. (Wells, 1964a:17)

These statements should be followed by all bioarchaeologists attempting to reconstruct the health status of a past population. His work was phenomenal in breadth and included two books, most notably *Bones, Bodies and Disease* (1964a; Fig. 2), and eight chapters in books, e.g., the paper on radiography of human remains (1963) and that published on pseudopathology (1967a). He also published alone, and with others, over 100 papers in journals (fully listed in Hart, 1983). The papers are wide ranging and include population studies (almost 40 reports on inhumations and cremated burials from different periods and geographic locations in Britain). For example, his extremely detailed study of the burials from the Roman–British cemetery at Cirencester, Gloucester (1982) shows his attempt to place the burials in context to be able to interpret his findings. He also published on many pathological conditions affecting the skeleton. This includes fractures, Paget's disease, malignant disease, leprosy, and obstetrical problems, and he studied the link between pollution and health using maxillary sinus evidence (the first of its kind in the world; Wells, 1977), and the evidence for the treatment of long bones fractures (1974). He delved into diseases in other animals (e.g., 1964b), considered the history of disease from other perspectives such as sculpture (e.g., 1968) and surgical instruments (e.g., 1967b), and even made a set of videos as a teaching tool at the Castle Museum in Norwich with the University of East Anglia. How he managed to do all this and practice as a doctor is unknown! Wells should be considered, along with Brothwell, a bioarchaeologist in the American definition sense. Some criticize his background as being inappropriate in some respects to studying and interpreting human remains from archaeological sites (he was a doctor and not an archaeologist/anthropologist). However, he did consider his biological data from the point of view of context, even though this was usually only in a broad manner. In fact, he was of the very strong opinion that physical anthropologists should not be attempting to diagnose disease in skeletal remains if they were not medically trained. However, it is now widely accepted that any number of educational backgrounds can contribute in different ways to our knowledge of the human past.

Don Brothwell (Fig. 3) is the other key figure in bioarchaeological study in Britain [see a recent tribute to his work in Dobney and O'Connor (2002)]. He graduated in 1956 from Cambridge with a BSc in anthropology and archaeology (with geology and zoology). This perhaps explains his diverse interests in bioarchaeology. Commencing publishing also in the 1950s, he has made a huge impact on many a scholar's work from a global point of view, and still does. Unlike Calvin Wells, he has worked on a much wider range of

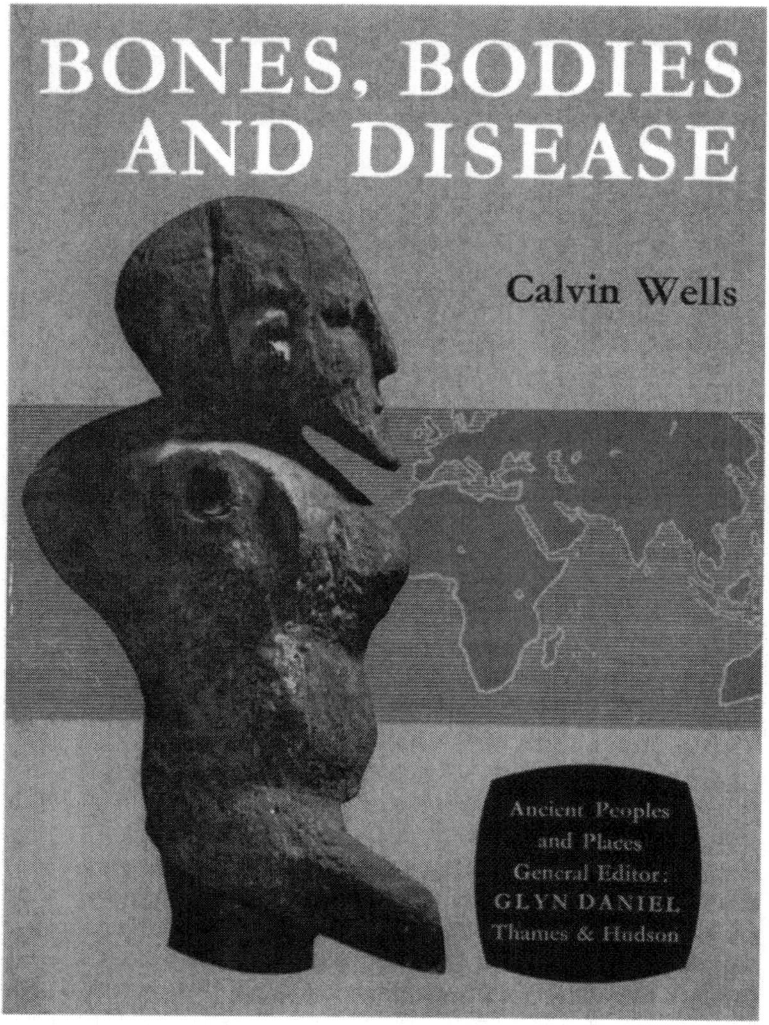

Figure 2 Cover of *Bones, Bodies and Disease*.

biological materials, including practically based considerations of biological evidence from human and nonhuman skeletal remains and mummified materials; he has also contributed to theoretical debates about the human and animal past. His 16 authored/edited books have proved influential throughout his lifetime, particularly *Diseases in Antiquity* (Brothwell and Sandison, 1967),

Figure 3 Don Brothwell (courtesy of the Department of Archaeology, University of York).

Digging Up Bones (1963c, 1981), *Dental Anthropology* (1963b), *Science in Archaeology* (Brothwell and Higgs, 1969, revised and enlarged edition), *Handbook of Archaeological Sciences*, (Brothwell and Pollard, 2001), and *Skeletal Biology of Earlier Human Populations* (1968). Brothwell and Sandison (1967) provided a survey of knowledge of disease and injury in the past from human remains at that time, a source book that many return to time and again today. His book in 1981 provides a general survey of the study of human remains from all aspects and puts data into archaeological context. "Digging up Bones" (Fig. 4) was a landmark book used by many, including field archaeologists, and is currently being revised for a new edition; it has been translated into other languages. Of course he has published many papers and book chapters on a wide variety of subjects (around 150). These include palaeodemography (1972–1973), dental wear as an indicator of age (1989), many on palaeopathology, e.g., neoplasms (1967), trauma (1999), the history of syphilis (1970), tuberculosis (Morse *et al.*, 1964), paralysis and possible diagnoses in skeletal remains (Brothwell, 2003a), and early humans (e.g., 1960), and the zoonoses (1991).

Figure 4 Cover of *Digging up Bones*.

He has also published on demography and health of skeletal samples from various sites of all periods in Britain, plus more focused studies on dental disease, leprosy, amputation, trepanation, trauma, congenital disease, cannibalism, epigenetics, metrical analysis, radiography, scanning electron microscopy, pollution in the past, and hair analysis. Unlike most people working in bioarchaeology he has, furthermore, contributed to the study of mummified material (e.g., 2003b).

Of all his contributions, the zoonoses and their impact on humans is an area that needs much more work in bioarchaeology and much more of an integrated approach between archaeozoologists and people studying human palaeopathology. Brothwell also published (with Baker) in 1980 the only review of animal diseases in the archaeological record, but this work has yet to stimulate much reaction in the archaeozoological world. Although many diseases can be transmitted to humans from animals (e.g., tuberculosis; Roberts and Buikstra, 2003), reflecting their close economic association and their clear relevance to appreciating impacts on human health, there has been little advance in this field of study. However, the problems of studying animal bones for evidence of disease have been outlined by O'Connor (2000), and it seems that the development of the methodology for recognizing and interpreting pathological changes in animal remains is badly needed. Britain is well placed to do this in the future.

Keith Manchester, another doctor, but this time from West Yorkshire, followed in the same vein as Calvin Wells and has published numerous papers from a bioarchaeological standpoint. Most notable is his work on the infectious diseases, especially the diagnostic features of leprosy (e.g., Andersen and Manchester, 1987), and his commentary on the cross-immunity hypothesis in relation to leprosy and tuberculosis (1984). His book (1983) was a landmark in the study of palaeopathology, where he considered the history of disease from a theoretical and practical standpoint, incorporating the history of medicine, where appropriate; this book has now evolved into its third edition (Roberts and Manchester, 2005). Cecil Hackett (1967) and Eric Hudson (1965) also contributed to our knowledge of the infectious diseases and focused on the history of syphilis. Both were physicians, like Wells and Manchester, who had a strong interest in the history of disease and attempted to explain the evidence using a socioeconomic context. Hackett used his experience of working with treponemal disease in other parts of the world to develop diagnostic criteria for the treponematoses (1976). Other people with medical training have, over the years, been part of the development of human bioarchaeological study, including the late Juliet Rogers, who contributed so much to our understanding of joint diseases and their diagnosis and interpretation in archaeological contexts (Rogers, 2000; Waldron and Rogers, 1991; Rogers and Waldron, 1995). Tony Waldron himself has made notable contributions to the literature on joint disease (e.g., 1992), made us think about palaeoepidemiology (1994), and emphasized that determining occupation from bone changes in the skeleton is by no means easy (Waldron and Cox, 1989).

Teeth have also been a focus for several people in Britain, most notably Dorothy Lunt, a dentist from Scotland who prepared many dental reports for skeletons excavated from archaeological sites (e.g., 1972), and Simon Hillson. Hillson's books have had a large impact on scholars around the world (1986, 1996) along with his papers focusing on diagnostic criteria (e.g., 2001; Hillson *et al.*, 1998). The 1980s, 1990s, and into the new millennium have also seen a number of scholars contributing to bioarchaeological studies, such as Chamberlain on palaeodemography (2000), the late Trevor Anderson on a variety of subjects (e.g., 1994), Lewis' work on medieval child health (2002), Brickley's work on metabolic disease (2000), Cox's particular contribution to post-Medieval skeletal analysis and interpretation (1996), Molleson's work on a wide range of subjects (e.g., 1989; Molleson and Cohen, 1990), and McKinley's monumental work on cremations and many other aspects of human remains study (e.g., 1994, 2000). Finally, Simon Mays of English Heritage has contributed tremendously to bioarchaeological studies of human populations in Britain, with his work ranging from bone reports to studies on osteoporosis (1996a), amputations (1996b), infanticide (Mays and Faerman, 2001), treponemal disease (Mays *et al.*, 2003), and biomolecular studies on tuberculosis (e.g., Mays *et al.*, 2001). His book (1998) and edited book with Cox (2000) have been particularly influential in a global sense and have contributed in part to putting the study of archaeological skeletal remains "on the map."

However, of particular note is Britain's contribution to biomolecular studies of archaeological human remains since the early 1990s. Analysis of ancient DNA (e.g., Taylor *et al.*, 2000; Bouwman and Brown, 2005) and mycolic acids (e.g., Gernaey *et al.*, 2001) to diagnose disease and stable isotopes to reconstruct palaeodiet (e.g., Richards *et al.*, 1998) and mobility of populations (e.g., Montgomery *et al.*, 2005) have all been used to answer specific questions about the archaeological past in Britain. Nevertheless, while some of this work in Britain may appear to some, in many respects, to be at the forefront in the world, the use of stable isotopes to address questions of palaeodiet has seen little attention until recently compared to the New World; this has mainly come through work by Mike Richards and his students (e.g., Richards *et al.*, 2000).

V. RECENT DEVELOPMENTS AFFECTING THE PROGRESS OF BIOARCHAEOLOGY AS A DISCIPLINE IN BRITAIN (LATE 20TH CENTURY TO DATE)

A. TRAINING

In the 1980s, a major development occurred in Britain that has become unique in the world: the establishment of 1-year masters courses, which involved taught courses and a research dissertation; surprisingly North America has not followed this lead. This was a key turning point from which the study of bioarchaeology was placed on a firm footing and encouraged "practitioners" to take a more holistic view of human remains study by considering the biological evidence for people within its archaeological context. Up until this time, the only people who had really taken a serious bioarchaelogical approach were Calvin Wells and Don Brothwell.

The Institute of Archaeology, University College, London, and the Department of Archaeology and Prehistory, University of Sheffield commenced masters courses in the study of human/animal remains and human remains, respectively, in the 1980s. Small numbers of students were given the chance to specialize in an area of archaeological study that had, until then, been the remit of doctors, dentists, and anatomists in their spare time. Until then too, archaeology departments had not, on the whole, provided any undergraduate training in human bioarchaeology, although anthropology departments naturally taught human evolution using skeletal casts. When looking back, it was therefore not surprising that work on human remains was usually devoid of any integration of biological data with archaeological context. By 1990, a joint Universities of Bradford and Sheffield course (MSc in Osteology, Palaeopathology, and Funerary Archaeology) had been initiated and ran very successfully with international recruitment for 10 years until the two universities decided to go it alone.

Since 1990 other MSc courses have been set up at the universities of Bournemouth, Durham, Edinburgh, and Southampton, with variations on the theme including MSc courses in forensic archaeology and anthropology and skeletal and dental bioarchaeology (Institute of Archaeology, University College, London). If the Bradford/Sheffield course was being set up now it would probably have been named an MSc in Bioarchaeology, especially to attract North American students! What's in a name? Universities in Britain would say "a lot of money potentially." Therefore, as of the early 21st century we now have a plethora of MSc courses running in Britain, training people from mainly archaeological and anthropological backgrounds; with this background they are able to approach the study of skeletal remains from a truly bioarchaeological standpoint, and some of us would have welcomed these courses when we were younger! It has also encouraged students to extend their studies into Ph.D. programs, where they are much better prepared to do their research. The MSc course in Britain is equivalent to the 2-year masters program in North America, which then leads on to a Ph.D. These masters courses have transformed the nature of how people in Britain study the skeletal past, have opened up the eyes of higher education to the potential of this area of study in archaeology, and have ultimately led to more posts being created for physical anthropologists in archaeology departments. However, creating more qualified people has produced a problem for employment of these graduates in jobs that use the skills they have acquired. Most people work in museums and with contract archaeological units as practicing field archaeologists with osteological expertise, but many go on to a Ph.D. program (with inevitable competition for limited funding). The units appear to now recognize the expertise that these graduates can bring with both their archaeological/anthropological and their osteological backgrounds.

Research students in bioarchaeology have also increased. However, it is pleasing to see that more wide-ranging themed projects, with hypotheses to test and questions to answer, are being tackled that use a bioarchaeological perspective in the interpretation of data (a clear move away from the "cottage industry"; Roberts, 1986). This can only be seen as a good thing and a positive development from the often narrowly focused projects undertaken back in the 1980s. As an extension to the masters courses, short courses in palaeopathology have developed, mainly because of Don Ortner's commitment to teaching. Don Ortner, along with Walter Putschar, taught a short course in palaeopathology from 1971 through 1974 at the Smithsonian Institution. After a break to write their book, this course ran again in 1985 at which the author was present (along with some other authors in this book). Through his research links through Keith Manchester and the author at the University of Bradford, short courses were held at Bradford in 1988, 1994, 1998, 2001, 2003, and his last in 2005. However, it is anticipated that these short courses will continue in the future.

B. The Media

Of interest is the parallel development in the 1990s of a very strong desire by the media in making television programs involving skeletal and mummified remains. The BBC's "Meet the Ancestors," Channel 4's programs in the series "To the Ends of the Earth" and "Secrets of the Dead" and programs in the "Timewatch" series have enthused the public and developed their interest in the study of human remains as a subdiscipline of archaeology. In Britain if a program has three to four million viewers, this is considered a success (which the aforementioned always have had).

C. Standards for Recording Data

There have been a number of developments in bioarchaeology in Britain and Europe, developments that are both positive and negative. Although, again, late in coming compared to North America, standards for data collection and reporting of skeletal material have been published for Britain (Brickley and McKinley, 2004; Fig. 5). Its stimulus was Roberts and Cox's experiences of collating palaeopathological data from published and unpublished sources for their book (2003). The lack of standards for data collection and reporting was clear, but acknowledged as a historical development that had not been addressed or discussed within the "bioarchaeological" community. Following a workshop between the authors of chapters in the publication (very like that which preceded Buikstra and Ubelaker, 1994), the final volume was produced.

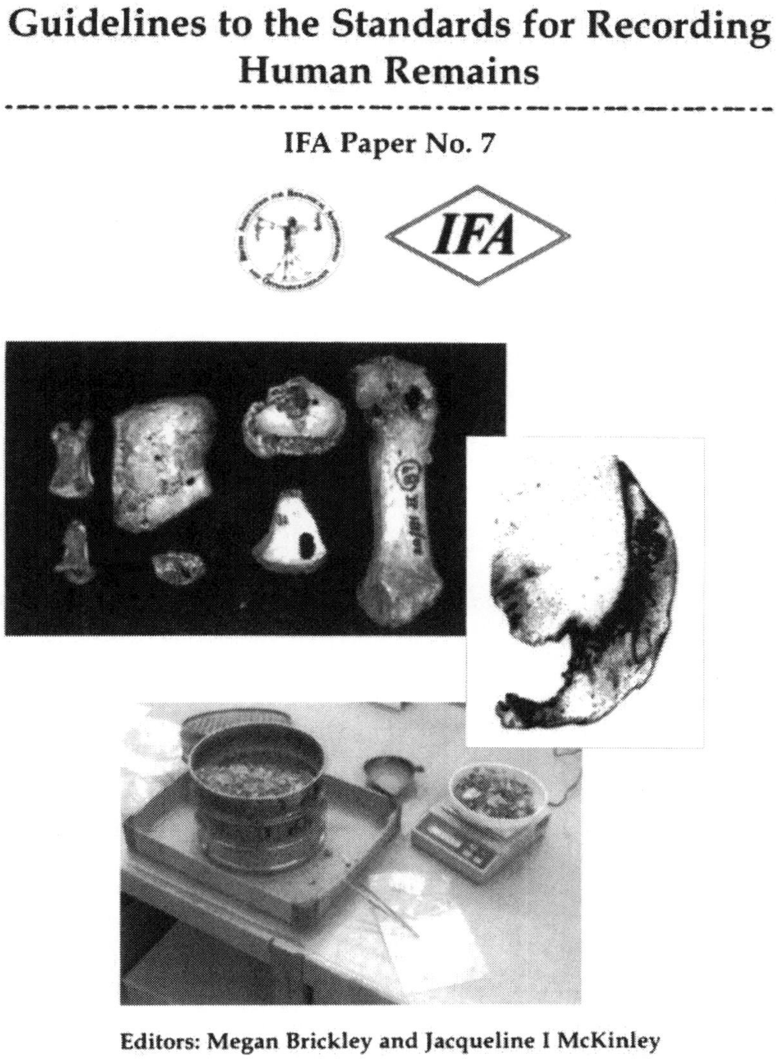

Figure 5 Cover of Brickley and McKinley (eds.) (with permission of the Institute of Field Archaeology and BABAO).

Along with the volume, and developments in training at masters level in Britain, it is anticipated that bioarchaeological data produced from Britain will improve markedly in the future. Of interest, however, are the different stimuli that led to both these publications; unlike for Britain, the repatriation and reburial act in the United States precipitated the need to record data in a standardized way before skeletons were repatriated and/or reburied (NAGPRA; Rose *et al.*, 1996). In Britain, we may be glad that we also have standards for recording now as we see more controls on our skeletal resource (see later). Another stimulus to the production of good quality data in Europe as a whole has been the establishment of the "Health in Europe Project" based at Ohio State University in Columbus and headed by Richard Steckel (Steckel *et al.*, 2002). In this project, an extension of the Western Hemisphere Health project (Steckel and Rose, 2002), a standard on-line recording form has been developed for participants to use in the coding of thousands of skeletons curated across Europe. This process should enable valid comparisons to be made of data both geographically and temporally; it will also incorporate contextual information that will allow a bioarchaeological approach to be taken once all data have been collected.

D. Databases and the Burgeoning "Grey Literature"

One of the major problems to advances in human bioarchaeological study in Britain and elsewhere in Europe is the lack of cohesion in knowledge of our resource base, although there have been some developments recently, such as the medieval cemeteries database (Gilchrist and Sloane, 2005). First, there remains no database of skeletal collections available for study in Britain, or where they are curated, which makes generating and carrying out research proposals difficult. Second, much of the work on skeletal remains that exists in Britain is published in the "grey literature" and may never come into the public domain. For example, of the 311 reports on human skeletal remains considered for Roberts and Cox (2003), 38% were unpublished, and access to their contents was enabled only through very cooperative colleagues willing to share data. Without knowledge of what has already been done, it is impossible to identify gaps in our knowledge about the past from a geographical and temporal point of view. Much of this problem has become more apparent since the introduction of Policy Planning Guidance 16 (1990), where archaeologists bid to gain archaeological work in advance of development ("contract archaeology"); the cheapest bids often get the work, including that for analyzing skeletal remains that are excavated, and the quality of the work can thus be poor. In addition, much of the work may never be published; for example, the author produced many reports in the 1980s but only a handful have been published.

E. Controls on Our Skeletal Resource: Repatriation/Reburial/Damage?

The second major problem that may prevent advances in bioarchaeological study in the future is the threat to survival above ground of our skeletal collections. In 2001 the British government (Department of Culture, Media, and Sport) established a working party to consider the future of skeletal remains in museums and other institutions in Britain. Initially this was aimed at remains from elsewhere in the world but it is clear that all remains could be at risk; in 2005, a report was produced (Department of Culture, Media, and Sport, 2005). This was supplemented by a report from the Cathedrals and Church Buildings Division of the Church of England and English Heritage's Human Remains Working Group (Church of England and English Heritage, 2005). This report looked at ethical, legal, and scientific issues to agree on guidelines covering the excavation and treatment of Christian burials in archaeological projects and their reburial. Ireland and Scotland already have guidance on the treatment of human remains (Historic Scotland, 1997; Heritage Council, 2002), in addition to guidance on the law and burial archaeology in England (Garratt-Frost, 1992). Clearly, the spotlight is on skeletal collections in Britain, and better guidelines for their treatment were badly needed, but it would be regrettable if eventually all skeletal remains were unavailable for study in the future in Britain. As Buikstra and Gordon (1981) have indicated, the retention of skeletal remains for future research is very beneficial, especially with the rapid advancement in analytical techniques (although destructive sampling for biomolecular analysis should be carefully controlled and restricted to proposals of real scientific value). What is happening in Britain appears to follow earlier developments in the United States (Rose *et al.*, 1996), as are so many aspects of our lives today!

Of relevance to maintaining the integrity of our resource is a recent study of damage that may occur through handling of skeletal remains and the need to limit that damage (Caffell *et al.*, 2001). In Britain, particularly, the proliferation of masters courses (and Ph.D. students) has led to pressure on the handling of skeletal remains curated in universities (and on skeletal samples in other institutions such as museums through masters and Ph.D. dissertations). This might be expected to be an inevitable outcome but up until recently little thought has been put into controlling access and limiting damage to such a valuable resource. As an extension of the work of Caffell and colleagues, Bowron (2003), a conservation masters student at Durham University, developed a box suitable for skeleton curation to prevent needless damage during use. This boxing system is slowly being recognized in museums as a solution to prevent damage, but the curatorial state of skeletal collections in many museums is far from ideal.

F. Research Funding

Funding for research in bioarchaeology in Britain has been reasonably generous over the years but the type of projects funded has changed. In the 1980s, basic research on palaeopathology was likely to get funded from a number of funding bodies (Natural Environmental Research Council, www.nerc.ac.uk; Wellcome Trust, www.wellcome.ac.uk). However, from the mid-1990s research funding is now more focused on biomolecular studies such as aDNA and isotope analysis, although the range of funding bodies supporting such research has increased. Of particular note is the bioarchaeology awards program from the Wellcome Trust, a program that ran for 10 years until 2005. Its aim was to fund research on the history of human disease, health, and human evolution and biocultural adaptation. It provided excellent career opportunities for many and granted thousands of pounds to bioarchaeological research, using cutting-edge biomedical techniques of analysis. This trend in concentrating funding on biomolecular work in some respects is unfortunate because there is much basic work in bioarchaeology to be done in Britain and many skeletal samples have not yet been studied using standard methods of data collection nor have data been interpreted in relation to context. Perhaps one exception to this funding anomaly has been the generous funding by the Wellcome Trust of the Centre for Human Bioarchaeology and the Spitalfields project at the Museum of London. This has enabled a number of people to be employed to undertake the basic analysis of several thousand skeletons from a late medieval site, and the creation of a database of human skeletal remains from London sites (Museum of London home page).

Despite restricted funding for bioarchaeology in Britain over the years, there have been many significant contributions to bioarchaeology in a global sense. While an integrated approach to studying human remains has been slow in arriving in Britain, and much work remains unpublished, some of the work over the years has been influential in the world of bioarchaeology. For example, Britain has set standards for recording specific diseases (e.g., Rogers and Waldron, 1995), generated hypotheses to test (e.g., Manchester, 1984), stimulated interest in animal palaeopathology (Baker and Brothwell, 1980), and undertaken hypothesis-driven population studies (e.g., Roberts *et al.*, 1998).

There have also been some recent bioarchaeological studies fully integrating mortuary analysis with biological data (e.g., Buckberry, 2004; Gowland, 2002, 2004; Loe, 2004; recent Ph.D. submissions by Groves and Redfern; and a critique by Tyrrell, 2000), although more work needs to be done in this area (a forthcoming book may address this gap; Gowland and Knüsel, for 2006). There have also been some recent stable isotopic studies that have utilized biological and funerary data such as grave goods (e.g., Montgomery *et al.*, 2005), although interpretations can be problematic (e.g., Privat *et al.*, 2002). Work up to now, when its exists, has been by "funerary archaeologists" who have tended to use both biological and funerary archaeological data but often with a poor understanding of the subtle nature of biological data (e.g., Harke, 1990; Stoodley, 1999), especially with respect to palaeopathology. This is seen as a problem of the science/theory "divide" in Britain.

Furthermore, even though much of it will never be seen by the rest of the world because it is published in inaccessible form (or not published at all), there are hundreds of "skeletal reports" on individual skeletons, small and large groups of individuals, from prehistory to the post-Medieval periods, in most areas of Britain; some of these reports are very much bioarchaeological studies (e.g., Molleson and Cox, 1993; Boylston *et al.*, 2001; Brickley *et al.*, 1999). However, and unfortunately, if studies such as these do not reach the international peer-reviewed literature they remain invisible to the global bioarchaeological community.

In Britain we are fortunate to have a wealth of skeletal collections, a huge range of archaeological data to make bioarchaeological interpretations, and contemporary historical records from the medieval period onward. Another strength, but one that is not used particularly well, is a very strong tradition in the study of the history of medicine, grounded at the Wellcome Trust in London and other centers around Britain. While both bioarchaeologists and medical historians in Britain recognize the strengths of their opposites, there is little attempt to develop complimentary research proposals or activities, although there have been recent attempts (Maehle and Roberts, 2002). However, there have been a series of successful conferences in Europe [France, on syphilis (Dutour *et al.*, 1994); Hungary, on tuberculosis (Pálfi *et al.*, 1999); Britain, on leprosy (Roberts *et al.*, 2002); and France, on the plague] where clinicians, medical historians, molecular biologists, and bioarchaeologists have come together to discuss the same disease from their own perspectives. Furthermore, the European Anthropological Association (European Anthropological Association home page), the European members of the Palaeopathology Association (Palaeopathology Association home page), and the British Association of Biological Anthropology and Osteoarchaeology (BABAO home page, Fig. 6) are very active in promoting bioarchaeological study. The BABAO has been a uniting force in Britain since its establishment in 1998, and with its excellent web site, annual conference, "Annual Review" publication, and an active email list, "bioarchaeological activity" has increased considerably.

VI. THE FUTURE FOR BIOARCHAEOLOGY IN BRITAIN

This chapter has perhaps painted a bleak picture of bioarchaeology in Europe, particularly Britain. However, the future looks particularly bright. We now

Figure 6 Logo of the British Association of Biological Anthropology and Osteoarchaeology (with permission of BABAO).

have many scholars with the "right" background to fully appreciate the need to integrate biological data from human remains with archaeological data to test hypotheses and answer questions about our past. We are moving toward developing a database of curated human remains in Britain that will help scholars locate the right skeletal samples for their research questions. There are now skeletal recording standards to ensure that collected data in the future conform to a set pattern and can therefore be used for comparison with other "population" studies. Furthermore, we have an increased awareness in the field archaeological community that the study of human remains can offer considerable insights into our past. As a country, we have moved from physical anthropologists who were interested in studying human remains for their own sake, with little attempt to contextualize their data, to a proliferation of people ("bioarchaeologists") who truly wish to do this and make their data matter to understanding our ancestors. However, there are real threats to our resource base of human remains, as discussed earlier. North American bioarchaeology grew up in a very different tradition when compared to Britain, a tradition (under the umbrella of "anthropology") that emphasized an integrated approach. In Britain we have faced, and are facing, the same problems that North America has faced, but we are now in a position to rapidly "catch up" with our neighbors across the Atlantic.

ACKNOWLEDGMENTS

I thank Jane Buikstra for inviting me to write this chapter and giving me a chance to reflect on how bioarchaeology has developed in Britain and the key factors in its "emergence." Both Jane and Elsevier are also thanked for their patience in waiting for the chapter due to the author's personal circumstances. Becky Gowland has also contributed to this chapter by enlightening the author to research linking funerary context with biological data from skeletal remains. Finally, Jeff Veitch at Durham University transformed Figs. 1, 2, and 4 into electronic versions, Martin Smith (Birmingham University), webmaster for BABAO, provided the BABAO logo, and Jackie McKinley (Wessex Archaeology) provided the cover image for Brickley and McKinley (2004).

Glossary of Acronyms

AA *American Antiquity*

AAFS American Academy of Forensic Sciences

AAM American Association of Museums

AAPA American Association of Physical Anthropologists

AID American Indians Against Desecration

AIM American Indian Movement

AJPA *American Journal of Physical Anthropology*

AMM Army Medical Museum (USA)

ARPA Archaeological Resources Protection Act

BABAO British Association for Biological Anthropology and Osteoarchaeology

CAFN The Champagne and Aishihik First Nations (Canada)

COE Army Corps of Engineers (USA)

DOI Department of the Interior (USA)

IHT Inuit Heritage Trust (Canada)

NAGPRA Native American Graves Protection and Repatriation Act

NARF Native American Rights Fund

NHPA National History Preservation Act of 1966

NMAI National Museum of the American Indian (USA)

NMAIA National Museum of the American Indian Act

NMC National Museum of Civilization (Canada)

NMM National Museum of Man (Canada)

NMNH National Museum of Natural History of the Smithsonian Institution (USA)

NPS National Park Service (USA)

NYU New York University

MAI Heye Foundation Museum of the American Indian (USA)

OSA Office of the State Archaeologist (USA; Iowa)

SAA Society for American Archaeology

SOD *Standards for Data Collection from Skeletal Remains*

SOPA Society

SI Smithsonian Institution (USA)

Bibliography

Abrams, George H. 1965. The Cornplanter Cemetery. *Pennsylvania Archaeologist* XXXV (2):59–73.

Ackerknecht, Erwin H. 1953. Paleopathology. In *Anthropology Today*, edited by A. L. Kroeber. Chicago: University of Chicago Press.

Ackerknecht, Erwin H. 1965. *History and Geography of the Most Important Diseases*. New York: Hafner.

Ackerknecht, Erwin H. 1971. Medicine & Ethnology: Selected Essays. edited by H. H. Walser, H. M. Koebling. Baltimore: Johns Hopkins Press.

Acsádi, G., and J. Nemeskéri. 1970. *History of Human Life Span and Mortality*. Budapest: Akadémiai Kiadó.

Adams, R. E. W. 1984. Annual Report of the President. *American Antiquity* 49:214–216.

Adams, W. Y., Dennis P. Van Gerven, and R. S. Levy. 1978. The Retreat from Migrationism. *Annual Review of Anthropology* 7:483–532.

Agarwal, Sabrina C., and Marc D. Grynpas. 1996. Bone Quantity and Quality in Past Populations. *Anatomical Record* 249:423–432.

Agarwal, Sabrina C, and Sam D. Stout, eds. 2003. *Bone Loss and Osteoporosis: An Anthropological Perspective*. New York: Kluwer Academic.

Aguiar, G. F. S., and Walter A. Neves. 1991. Postmarital Residence and within-Sex Genetic Diversity among the Urubu-Ka'apor Indians, Brazilian Amazon. *Human Biology* 63:467–488.

Akins, N. J. 1986. *A Biocultural Approach to Human Burials from Chaco Canyon, New Mexico*. Santa Fe, NM: National Park Service.

Alcock, Susan. 1993. *Graecia Capta: The Landscapes of Roman Greece*. Cambridge: Cambridge University Press.

Alexander, R. McNeill. 1977. Allometry of the Limbs of Antelopes (Bovidae). *Journal of Zoology, London* 183:125–146.

Alexander, Ralph W. 1971. The Hybrid Origin of the Arikara Indians of the Great Plains. Ph.D., Indiana University, Bloomington.

Allen, Harrison. 1867. The Jaw of Moulin Quignon. *Dental Cosmos* IX(4):169–180.

Allen, Harrison. 1870. Localization of Disease Action in the Osseous System. *American Journal of the Medical Sciences* XVI:401–405.

Allen, Harrison. 1893. Remarks on Congenital Defect of Face, with Exhibition of a Rare Form of Cleft Palate. *New York Medical Journal* LVIII:759–760.

Allen, Harrison. 1895a. Crania from the Mounds of the St. John's River, Florida: A Study Made in Connection with Crania from Other Parts of North America. *Journal of the Academy of Natural Sciences of Philadelphia* 10 (Series 2 Part 4):367–448.

Allen, Harrison. 1895b. Demonstration of Skulls Showing the Effects of Cretinism on the Shape of the Nasal Chambers. *New York Medical Journal* LXI:139–140.

Allen, Harrison. 1898. The Study of Skulls from the Hawaiian Islands, with an Introduction by D. G. Brinton, Wagner Institute of Science, Philadelphia. *Proceedings of the Academy of Natural Sciences of Philadelphia* V:1–55.

Allison, Marvin J., Daniel Mendoza, and Alejandro Pezzia. 1973. Documentation of a Case of Tuberculosis in Pre-Columbian America. *American Review of Respiratory Diseases* 107(6):985–991.

Allison, Marvin J., Alejandro Pezzia, Enrique Gerszten, and Daniel Mendoza. 1974. A Case of Carrion's Disease Associated with Human Sacrifice from the Huari Culture of Southern Peru. *American Journal of Physical Anthropology* 41(2):295–300.

Allison, Marvin J., Alejandro Pezzia, I. Hasegawa, and Enrique Gerszten. 1974. A Case of Hookworm Infestation in a Precolumbian American. *American Journal of Physical Anthropology* 41(1):103–106.

Allison, Marvin J., Daniel Mendoza, and Alejandro Pezzia. 1974. A Radiographic Approach to Childhood Illness in Precolumbian Inhabitants of Southern Peru. *American Journal of Physical Anthropology* 40(3):409–415.

Allison, Marvin J., Enrique Gerszten, H. Jean Shadomy, Juan Munizaga, and M. Gonzales. 1979. Paracoccidioidomycosis in a North Chilean Mummy. *Bulletin of the New York Academy of Medicine* 55:670–683.

Allison, P. J., D. Locker, and J. S. Feine. 1997. Quality of Life: A Dynamic Construct. *Social Science and Medicine* 45:221–230.

Alt, Kurt W., Friedrich W. Rösing, and Maria Teschler-Nicola, eds. 1998. *Dental Anthropology: Fundamentals, Limits, and Prospects*. Wein and New York: Springer.

American Association of Physical Anthropologists. 1982. On the Reburial of Human Skeletal Remains. *AAPA Proceedings* 1982:230.

Ames, Kenneth M. 2000. Chapter 2: Review of the Archaeological Data. In "Cultural Affiliation Report." Retrieved Dec. 31, 2005, from Department of the Interior, National Park Service, Archaeology Program Web Site: http://www.cr.nps.gov/archeology/kennewick/ames.htm.

Anderson, Duane, C., Michael Finnegan, John Hotopp, and Alton K. Fisher. 1978. The Lewis Central School Site (13PW5): A Resolution of Ideological Conflicts at an Archaic Ossuary in Western Iowa. *Plains Anthropologist* 23 (81):183–219.

Anderson, James E. 1960. The Development of the Tympanic Plate. *National Museum of Canada Bulletin* 180(Part 1):143–153.

Anderson, James E. 1961. The Montgomery and Boyle Osteology Collection. *R. O. M. Arts and Archaeology Division Annual Report*:67–72.

Anderson, James E. 1964. The People of Fairty: An Osteological Analysis of an Iroquois Ossuary. *Canadian National Museum Bulletin* 193:28–129.

Anderson, James E. 1965. The Human Skeletons of Tehuacan. *Science* 148(3669): 496–497.

Anderson, Johannes, G., and Keith Manchester. 1987. Grooving of the Proximal Phalanx in Leprosy: A Palaeopathological Study. *Journal of Archaeological Science* 14 (1):77–82.

Anderson, T. 1993. An Anglo-Saxon Case of Hyperostosis Frontalis Interna. *Journal of Paleopathology* 6:29–34.

Andrews, Robert R. 1893. Correspondence. *The Dental Practitioner and Advertiser* XXIV(4):206–208.

Angel, J. Lawrence. 1943. Ancient Cephallenians: The Population of a Mediterranean Island. *American Journal of Physical Anthropology* 1:229–260.

Angel, J. Lawrence. 1946b. Social Biology of Greek Culture Growth. *American Anthropolgist* 48:493–553.

Angel, J. Lawrence. 1946b. Skeletal Change in Ancient Greece. *American Journal of Physical Anthropology* 4:69–97.

Angel, J. Lawrence. 1947. The Length of Life in Ancient Greece. *Journal of Gerontology* 2:18–24.

Angel, J. Lawrence. 1952. The Human Remains from Hotu Cave, Iran. *Proceedings of the American Philosophical Society* 96:258–269.

Angel, J. Lawrence. 1954. Review: Archaeology and Race in the American Indian. *American Antiquity* 19 (2):172–173.

Angel, J. Lawrence. 1966a. *Early Skeletons from Tranquillity, California.* Vol. 2, *Smithsonian Contributions to Anthropology*. Washington: Smithsonian Press.

Angel, J. Lawrence. 1966b. Porotic Hyperostosis, Anemias, Malarias, and Marshes in the Prehistoric Eastern Mediterranean. *Science* 153 (3737):760–763.

Angel, J. Lawrence. 1969. The Bases of Paleodemography. *American Journal of Physical Anthropology* 30:427–438.

Angel, J. Lawrence. 1971. *The People of Lerna*. Washington, D.C.: Smithsonian Institution Press.

Angel, J. Lawrence. 1972. Late Bronze Age Cypriotes from Bamboula: The Skeletal Remains. In *Bamboula at Kourion: The Necropolis and the Finds Excavated by J. F. Daniel*, edited by J. L. Benson, pp. 148–158. Philadelphia: University of Pennsylvania Press.

Angel, J. Lawrence. 1973a. Biological Relations of Egyptians and Eastern Mediterranean Populations During Pre-Dynastic and Dynastic Times. In *Population Biology of the Ancient Egyptians*, edited by D. R. Brothwell and B. A. Chiarelli. 307–313. London and New York: Academic Press.

Angel, J. Lawrence. 1973b. Early Neolithic People of Nea Nikomedeia. In *Die Anfange des Neolithikums Vom Orient Bis Nordeuropa*, edited by I. Schwidetzky. 103–112. Cologne: Bohlau.

Angel, J. Lawrence. 1973c. Neolithic Human Remains. *Hesperia* 42:277–282.

Angel, J. Lawrence. 1974a. The Cultural Ecology of General versus Dental Health. In *Bevolkerungsbiologie*. 382–391. Stuttgart: Gustav Fischer.

Angel, J. Lawrence. 1974b. Bones Can Fool People. *FBI Law Enforcement Bulletin* 43 (1):17–20.

Angel, J. Lawrence. 1976a. Introduction to a Symposium in Honor of T. Dale Stewart. *American Journal of Physical Anthropology* 45 (3):521–530.

Angel, J. Lawrence. 1976b. Colonial to Modern Skeletal Change in the U.S.A. *American Journal of Physical Anthropology* 45 (3, Part 2):723–735.

Angel, J. Lawrence. 1978. Porotic Hyperostosis in the Eastern Mediterranean. *Medical College of Virginia Quarterly* 14 (1):10–16.

Angel, J. Lawrence. 1979. Osteoarthritis in Prehistoric Turkey and Medieval Byzantium. *Henry Ford Hospital Medical Journal* 27 (1):38–43.

Angel, J. Lawrence. 1980. Physical Anthropology: Determining Sex, Age, and Individual Features. In *Mummies, Disease, and Ancient Cultures*, edited by T. A. Cockburn and E. Cockburn. 241–257. Cambridge: Cambridge University Press.

Angel, J. Lawrence. 1981a. Physical Anthropological Analysis. In *Archaeological Data Recovery at Cacotin Furnace Cemetery, Frederic County, Maryland*, edited by S. A. Burston and R. A. Thomas. Appendix 2. Baltimore: Maryland Department of Transportation.

Angel, J. Lawrence. 1981b. History and Development of Paleopathology. *American Journal of Physical Anthropology* 56:509–515.

Angel, J. Lawrence. 1982. A New Measure of Growth Efficiency: Skull Base Height. *American Journal of Physical Anthropology* 58 (3):297–305.

Angel, J. Lawrence. 1984. Health as a Crucial Factor in the Changes from Hunting to Developed Farming in the Eastern Mediterranean. In *Paleopathology at the Origins of Agriculture*, edited by M. N. Cohen and G. J. Armelagos. 51–73. Orlando: Academic Press, Inc.

Angel, J. Lawrence, and Sara E. Bisel. 1985. Health and Stress in an Early Bronze Age Population. In *Ancient Anatolia, Aspects of Change and Cultural Development. Essays in Honor of Machteld J. Mellink*, edited by J. V. Canby, E. Porada, B. S. Ridgway and T. Stech. 12–30. Madison: University of Wisconsin Press.

Angel, J. Lawrence, and Jennifer Olsen Kelly. 1986. Description and Comparison of the Skeleton. In *The Wadi Kubbaniya Skeleton: A Late Paleolithic Burial From Southern Egypt*, edited by F. Wendorf and R. Schild. 53–70. Dallas: Southern Methodist University Press.

Angel, J. Lawrence, Jennifer Olsen Kelly, Michael Parrington, and Stephanie Pinter. 1987. Life Stresses of the Free Black Community as Represented by the First African Baptist Church, Philadelphia, 1823–1841. *American Journal of Physical Anthropology* 74(2):213–229.

Angel, J. Lawrence, Judy Myers Suchey, Mehmet Yaşar İşcan, and Michael R. Zimmerman. 1986. Age-at-Death from the Skeleton and Viscera. In *Dating and Age Determination of Biological Materials*, edited by M. R. Zimmerman and J. L. Angel. 179–220. London: Croom Helm.

Angel, J. Lawrence, and F. Wagner. 1945. Constitutional Type and Varicose Veins. Paper read at Annual Meeting of the American Association of Physical Anthropologists.

Anonymous. 1938. WPA Quarterly Report, October–December. W. S. Webb Museum of Anthropology, University of Kentucky, Lexington.

Anonymous. 1940a. In Memoriam: Arthur S. Hamilton. *Minnesota Medicine* 23 (7):488.

Anonymous. 1940b. WPA Quarterly Report, July–September. W. S. Webb Museum of Anthropology, University of Kentucky, Lexington.

Anonymous. 1941. WPA Quarterly Report, July–September. W. S. Webb Museum of Anthropology, University of Kentucky, Lexington.

Anonymous. 2001. AAHS Mourns the Loss of Katharine Bartlett. *Glyphs* 52 (2):12–13.

Anonymous. 2003. *American Men and Women of Science*. Detroit: Thompson Gale.

Anonymous. n.d.a. Manual for Labor, Kentucky WPA Archaeological Project. WPA Papers, 4844 Box 10, National Anthropological Archives, Smithsonian Institution, Washington, DC.

Anonymous. n.d.b. This Work Pays Your Community Week. W. S. Webb Museum of Anthropology, University of Kentucky, Lexington.

Anthony, D. W. 1990. Migration in Archaeology: The Baby and the Bathwater. *American Anthropolgist* 92:895–914.

Anyon, Roger. 1991. Protecting the Past, Protecting the Present: Cultural Resources and American Indians. In *Protecting the Past*. 215–222. Boca Raton, FL: CRC Press.

Anyon, Roger, and T. J. Ferguson. 1995. Cultural Resources Management at the Pueblo of Zuni. *Antiquity* 69:913–930.

Anyon, Roger, T. J. Ferguson, Loretta Jackson, and Lillie Lane. 2000. Native American Oral Traditions and Archaeology. In *Working Together: Native Americans & Archaeologists*, edited by K. E. Dongoske, M. Aldenderfer and K. Doehner. 61–66. Washington, DC: Society for American Archaeology.

Anyon, Roger, T. J. Ferguson, Loretta Jackson, Lillie Lane, and Philip Vicenti. 1997. Native American Oral Tradition and Archaeology: Issues of Structure, Relevance, and Respect. In *Native Americans and Archaeologists. Stepping Stones to Common Ground*, edited by N. Swidler, K. E. Dongoske, R. Anyon and A. S. Downer. 77–87. Walnut Creek, CA: AltaMira Press in Cooperation with the Society for American Archaeology.

Anyon, Roger, and Russell Thornton. 2002. Implementing Repatriation in the United States: Issues Raised and Lessons Learned. In *The Dead and Their Possessions: Repatriation in Principle, Policy and Practice*, edited by C. Fforde, J. Hubert and P. Turnbull. 190–198. New York: Routledge.

Anyon, Roger, and Jerome Zunie. 1989. Cooperation at the Pueblo of Zuni: Common Ground for Archaeology and Tribal Concerns. *Practicing Anthropology* 11 (3):13–15.

Araújo, Adauto, Ulisses Confalonieri, and Luis Fernando Ferreira. 1980. Oxyurid (Nematoda) Egg from Coprolites from Brazil. *Journal of Parasitology* 68(3): 511–512.

Araújo, Adauto, A. M. Jansen, F. Bouchet, Karl Reinhard, and Luis Fernando Ferreira. 2003. Parasitism, the Diversity of Life, and Paleoparasitology. *Memorias do Instituto Oswaldo Cruz* 98(Suppl 1):5–11.

Arensberg, B. 1995. On the Origin of Native Americans: de Acosta Versus la Peyrère. *Antropologia Biologica* 3:1–11.

Armelagos, George J. 1968. Paleopathology of Three Archaeological Populations from Sudanese Nubia. Ph.D. Dissertation, Department of Anthropology, University of Colorado, Boulder.

Armelagos, George J. 1969. Disease in Ancient Nubia. *Science* 163:255–259.

Armelagos, George J. 2003. Bioarchaeology as Anthropology. In *Archaeology Is Anthropology*, edited by S. D. Gillespie and D. Nichols. 27–41, *Archeological Papers of the American Anthropological Association Series, No. 13*.

Armelagos, George J., David S. Carlson, and Dennis P. Van Gerven. 1982. The Theoretical Foundations and Development of Skeletal Biology. In *A History of American Physical Anthropology*, edited by F. Spencer. 305–328. New York: Academic Press.

Armelagos, George J., and O. Donald Chrisman. 1988. Hyperostosis Frontalis Interna: A Nubian Case. *American Journal of Physical Anthropology* 76:25–28.

Armelagos, George J., James H. Mielke, and John Winters. 1971. *Bibliography of Human Paleopathology (Compilers), Research Reports Number 8*. Amherst: University of Massachusetts Department of Anthropology.

Armelagos, George J., and Dennis P. van Gerven. 2003. A Century of Skeletal Biology and Paleopathology: Contrasts, Contradictions, and Conflicts. *American Anthropologist* 105 (1):53–64.

Arnold, Bettina, and Nancy L. Wicker, eds. 2001. *Gender and the Archaeology of Death*. Walnut Creek, CA: AltaMira Press.

Asatru Folk Assembly. 1997. Kennewick Man. Retrieved Dec. 31, 2005, from Asatru Folk Assembly Web Site: http://www.runestone.org/kmrlg.html.

Asch, D. 1976. *The Middle Woodland Population of the Lower Illinois Valley: A Study in Paleodemographic Methods*. Evanston: Northwestern Archaeological Program.

Ashmore, Wendy, and Arthur Bernard Knapp, eds. 1999. *Archaeologies of Landscape: Contemporary Perspectives, Social Archaeology*. Malden, MA: Blackwell.

Aspillaga, Eugenio. 1996. *Juan Roverto Munizaga Villavicencio (1934–1996)*.

Association for Environmental Archaeology. n.d. Association for Environmental Archaeology home page. Retrieved Dec. 28, 2005, from Association for Environmental Archaeology web site: http://www.envarch.net/.

Atwater, Caleb. 1820. *Description of the Antiquities Discovered in the State of Ohio and Other Western States*. Vol. I, *American Antiquarian Society, Archaeologia Americana, Transactions and Collections*. Worcester: American Antiquarian Society.

Aufderheide, Arthur C. 2003. *The Scientific Study of Mummies*. Cambridge: Cambridge University Press.

Aufderheide, Arthur C., and Conrado Rodríguez-Martín. 1998. *The Cambridge Encyclopedia of Human Paleopathology*. Cambridge: Cambridge University Press.

Aufderheide, Arthur C., Wilmar Salo, Michael Madden, John Streitz, Jane E. Buikstra, Felipe Guhl, Bernardo Arriaza, Colleen Renier, Lorentz E. Wittmers Jr., Gino Fornaciari, and Marvin Allsion. 2004. A 9,000-Year Record of Chagas' Disease. *Proceedings of the National Academy of Science* 101(7): 2034–2039.

Aykroyd, R. G., D. Lucy, and A. M. Pollard. 1996. *Statistical Methods for the Estimation of Human Age-at-Death. Online Research Report No. STAT-96/08*: University of Leeds, Department of Statistics.

Aykroyd, R. G., D. Lucy, A. M. Pollard, and Charlotte A. Roberts. 1999. Nasty, Brutish, but Not Necessarily Short: A Reconsideration of the Statistical Methods Used to Calculate Age-at-Death from Adult Human Skeletal and Dental Age Indicators. *American Antiquity* 64:55–70.

Aykroyd, R. G., D. Lucy, A. M. Pollard, and T. Solheim. 1997. Regression Analysis in Adult Age Estimation. *American Journal of Physical Anthropology* 104:259–265.

Babbitt, Bruce. 2000. Letter from the Secretary of the Interior Bruce Babbitt to Secretary of the Army Louis Caldera Regarding Disposition of the Kennewick Human Remains, Retrieved Dec. 31, 2005, from U.S. Department of the Interior, National Park Service, Archaeology Program Web Site: http://www.cr.nps.gov/aad/kennewick/babb_letter.htm.

Babcock, Barbara A., and Nancy J. Parezo. 1988. *Daughters of the Desert: Women Anthropologists and the Native American Southwest, 1880–1980, an Illustrated Catalogue.* Albuquerque: University of New Mexico Press.

Baffi, Elvira Inez, and Maria F. Torres. 1997. Lehmann-Nitsche, (Paul Adolf) Robert (1872–1938). In *History of Physical Anthropology: An Encyclopedia*, edited by F. Spencer. 611. New York: Gartland.

Baker, Brenda J., and George J. Armelagos. 1988. Origin and Antiquity of Syphilis: Paleopathological Diagnosis and Interpretation. *Current Anthropology* 29 (5):703–737.

Baker, John R., and Don R. Brothwell. 1980. *Animal Diseases in Archaeology.* London and New York: Academic Press.

Baklanoff, Joy D., and Arthur F. Howington. 1989. *The Mounds Awaken: Mound State Monument and the Civilian Conservation Corps, Special Publications 3.* Tuscaloosa: State Museum of Natural History, University of Alabama.

Barbosa, M, V. C. Braz, L. G. L. Silva, R. Scheel-Ybert, M. Imazio, and L. M. C. Machado. 2003. Painel Simpósio: Projeto Saquarema: Resultados, Desdobramentose Perspectivas Futuras. In *Resumos do XII Congresso da Sociedade de Arqueologia Brasileira.* 74.

Barkan, Elazar. 1992. *The Rretreat of Scientific Racism: Changing Concepts of Race in Britain and the United States between the World Wars.* Cambridge and New York: Cambridge University Press.

Barnes, Ethne. 1994a. *Developmental Defects of the Axial Skeleton in Paleopathology*: University of Colorado Press.

Barnes, Ethne. 1994b. Polydactyly in the Southwest. *Kiva* 59:419–431.

Baron, Heike, Susanne Hummel, and Bernd Herrmann. 1996. *Mycobacterium Tuberculosis* Complex DNA in Ancient Human Bones. *Journal of Archaeological Science* 23 (5):667–671.

Barth, Fredrik, ed. 1969. *Ethnic Groups and Boundaries. The Social Organization of Culture Difference. Results of a Symposium Held at the University of Bergen, 23rd to 26th February 1967.* Bergen: Universitetsforlaget.

Bartlett, Katharine. 1931. Prehistoric Pueblo Foods. *Museum Notes of the Museum of Northern Arizona* 4:1–4.

Bartlett, Katharine. 1941. Skeletal Materials from Winona and the Ridge Ruin. In *Winona and the Ridge Ruin. Part I. Architecture and Material Culture*, edited by J. C. McGregor. 300–305, *Museum of Northern Arizona Bulletin 18.* Flagstaff: Northern Arizona Society of Science and Art.

Bartlett, Katharine. 1954. Crania from Montezuma Castle and Montezuma Well. In *Montezuma Castle Archaeology*, edited by E. Jackson. 52–63. Globe, Arizona: Southwestern Monuments Association.

Bartram, William. 1928. *The Travels of William Bartram.* New York: Macy-Masius.

Bass, William M. 1971. *Human Osteology: A Laboratory and Field Manual of the Human Skeleton*. Columbia: Missouri Archaeological Society.

Bass, William M. 1981. Skeletal Biology on the United States Great Plains: A History and Personal Narrative. *Plains Anthropologist* 26 (94):3–18.

Bass, William M. 1987. *Human Osteology: A Laboratory and Field Manual*. 3rd ed. Columbia: Missouri Archaeological Society.

Bass, William M., and David R. Evans. 1971. *The Leavenworth Site Cemetery: Archaeology and Physical Anthropology, University of Kansas Publications in Anthropology, No. 2*.

Bass, William M., John B. Gregg, and Pierre E. Provost. 1974. Ankylosing Spondylitis (Marie Strumpel Disease) in Historic and Prehistoric Northern Plains Indians. *Plains Anthropologist* 19(66, part 1):303–305.

Bass, William M., and W. L. Rhule II. 1976. Human Burials from Rodgers Shelter. In *Prehistoric Man and His Environments: A Case Study in the Ozark Highlands*, edited by W. R. Wood and R. B. McMillan. New York: Academic Press.

Bathurst, Rhonda R., and Jodi Lynn Barta. 2004. Molecular Evidence of Tuberculosis Induced Hypertrophic Osteopathy in a 16th-Century Iroquoian Dog. *Journal of Archaeological Science* 31:917–925.

Baxter, Sylvester. 1888. *The Old New World: An Account of the Explorations of the Hemenway Southwestern Archaeological Expedition in 1887–88, under the Direction of Frank Hamilton Cushing*. Salem, Mass.: Salem Press. Original edition, Reprinted from the Boston Herald, April 15, 1888.

Baxter, Sylvester. 1889. Archaeological Camping in Arizona. I–VI. *The American Architect and Building News* 25; 26 (680; 681; 682; 683; 714; 716):8–10; 15–16; 32–34; 43–44; 101–102; 120–122.

Beals, Kenneth L., Courtland L. Smith, and Stephen M. Dodd. 1984. Brain Size, Cranial Morphology, Climate, and Time Machines. *Current Anthropology* 25 (3):301–330.

Beattie, Owen, Brian Apland, Erik W. Blake, James A. Cosgrove, Sarah Gaunt, Sheila Greer, Alexander P. Mackie, Kjerstin E. Mackie, Dan Straathof, Valerie Thorp, and Peter M. Troffe. 2002. The Kwäday Dän Ts'inchi Discovery from a Glacier in British Columbia. *Journal Canadien d'Archéologie* 24:129–147.

Beck, Lane A., ed. 1995. *Regional Approaches to Mortuary Analysis*. New York: Plenum Press.

Beck, Lane A., and Lynn S. Teague. 2001. Reburial Law in Action: Case Studies from Arizona. In *Human Remains: Conservation, Retrieval and Analysis. Proceedings of a Conference Held in Williamsburg, VA, Nov. 7–11th 1999*, edited by E. Williams, *British Archaeological Reports. International Series 934*. Oxford: Archaeopress.

Beecher, Henry K., and Mark D. Altschule. 1977. *Medicine at Harvard: The First Three Hundred Years*. Hanover: The University Press of New England.

Begay, Richard M. 1997. The Role of Archaeology on Indian Lands: The Navajo Nation. In *Native Americans and Archaeologists. Stepping Stones to Common Ground*, edited by N. Swidler, K. E. Dongoske, R. Anyon and A. S. Downer. 161–166. Walnut Creek, CA: AltaMira Press in Cooperation with the Society for American Archaeology.

Bell, Catherine M. 1997. *Ritual: Perspectives and Dimensions*. New York: Oxford University Press.

Bement, Leland C. 1994. *Hunter-Gatherer Mortuary Practices during the Central Texas Archaic*. Austin: University of Texas Press.

Bendyshe, Thomas. 1865. *The Anthropological Treatises of Johann Friedrich Blumenbach*. London: Longman and Green.

Benedict, R. 1934. *Patterns of Culture*. Boston: Houghton Mifflin Company.

Benfer, R. A. 1968. An Analysis of a Prehistoric Skeletal Population, Casas Grandes, Chihuahua, Mexico. Ph.D., University of Texas at Austin, Austin.

Benfer, R. A. 1975. Morphometric Analysis of Cartesian Coordinates of the Human Skull. *American Journal of Physical Anthropology* 42:371–382.

Benfer, R. A., and D. S. Edwards. 1991. The Principal Axis Method for Measuring Rate and Amount of Dental Attrition: Estimating Juvenile or Adult Tooth Wear from Unaged Adult Teeth. In *Advances in Dental Anthropology*, edited by M. A. Kelley and C. S. Larsen. 325–340. New York: Wiley-Liss.

Bennett, K. A. 1973a. *The Indians of Point of Pines, Arizona: A Comparative Study of Their Physical Characteristics, Anthropological Papers of the University of Arizona, No. 23*. Tucson: University of Arizona Press.

Bennett, K. A. 1973b. On the Estimation of Some Demographic Characteristics Given Deaths by Age. *American Journal of Physical Anthropology* 39:223–232.

Bennett, M. E., and C. L. Phillips. 1999. Assessment of Health-Related Quality of Life for Patients with Severe Skeletal Disharmony. *International Journal of Adult Orthodontic and Orthognathic Surgery* 14:65–75.

Bennike, Pia. 2002. Vilhelm Møller-Christensen: His Work and Legacy. In *The Past and Present of Leprosy: Archaeological, Historical, Palaeopathological and Clinical Approaches*, edited by C. A. Roberts, M. E. Lewis and K. Manchester, *British Archaeological Reports, International Series 1054*. Oxford: Archaeopress.

Berg, Joseph M. 2003. Down Syndrome before Down: A Postscript. *American Journal of Medical Genetics* 116A:97–98.

Berry, A. C., and R. J. Berry. 1967. Epigenetic Variation in the Human Cranium. *Journal of Anatomy* 101:361–379.

Berryman, H. E. 1980. Mouse Creek, Dallas, and Middle Cumberland: A Multivariate Approach. In *The Skeletal Biology of Aboriginal Peoples in the Southeastern United States*, edited by P. Willey and F. H. Smith. 1–14. Chatanooga: Tennessee Anthropological Association.

Bertolazzo, Walter, and Marília Carvalho de Mello e Alvim. 1983/84. Os Seios Frontais em Grupos Indígenas Brasileiros — Construtores de Sambaquis e Botocudos. *Arqueivos do Museu de História Natural da Universidade Federal de Minas Gerais (Belo Horizonte)* 8/9:339–348.

Bertolazzo, Walter, and Marília Carvalho de Mello e Alvim. 1984. Os Seios Frontais em Grupos Indígenas Brasileiros — Construtores de Sambaquis e Botocudos. *Revista de Arqueologia (Belém)* 2 (2):41–45.

Bertolazzo, Walter, and Marília Carvalho de Mello e Alvim. 1985. Os Seios Frontais em Grupos Indígenas Brasileiros — O Homem de Lagoa Santa, Construtores de Sambaquise Botocudos. *Clio — Série Arqueológica (Recife)* 2 (7):113–130.

Bettinger, R. L. 1994. How, When and Why Numic Spread. In *Across the West: Human Population Movement and the Expansion of the Numa*, edited by D. B. Madsen and D. Rhode. 44–55. Salt Lake City: University of Utah Press.

Bice, Gillian. 2003. Reconstructing Behavior from Archaeological Skeletal Remains: A Critical Analysis of the Biomechanical Model. Ph.D. Dissertation, Department of Anthropology, Michigan State University, East Lansing.

Bieder, Robert E. 1986. *Science Encounters the Indian, 1820–1880: The Early Years of American Ethnology*. Norman, OK: University of Oklahoma Press.

Bieder, Robert E. 1990. *A Brief Historical Survey of the Expropriation of American Indian Remains*. Boulder, CO: Native American Rights Fund.

Billings, J. S., and Washington Matthews. 1885. On Composite Photography as Applied to Craniology; on Measuring the Cubic Capacity of Skulls. *Memoirs of the National Academy of Sciences* 8 (Memoir 13):105–116.

Billman, Brian R., Patricia M. Lambert, and Banks L. Leonard. 2000. Cannibalism, Warfare, and Drought in the Mesa Verde Region during the Twelfth Century A.D. *American Antiquity* 65 (1):145–178.

Binford, Lewis R. 1962. Archaeology as Anthropology. *American Antiquity* 28 (2): 217–225.

Binford, L. R. 1971. Mortuary Practices: Their Study and Their Potential. In *Approaches to the Social Dimensions of Mortuary Practices*, edited by J. A. Brown. 6–29. Washington, D.C.: Society for American Archaeology.

Binning, James. 1941. Letter Dated January 15 to Hooten, Hubbard, Barton, Duffey, Brooks, Loy, Hancock, Garrett and All Project Workers. Snow Letters, Laboratory for Human Osteology, Mary Harmon Bryant Scientific Collections Building, University of Alabama, Tuscaloosa.

Birkby, W. H. 1973. Discontinuous Morphological Traits of the Skull as Population Markers in the Prehistoric Southwest. Ph.D., University of Arizona.

Birkby, Walter H. 1982. Biosocial Interpretations from Cranial Nonmetric Traits of Grasshopper Pueblo Skeletal Remains. In *Multidisciplinary Research at Grasshopper Pueblo, Arizona*, edited by W. A. Longacre, S. J. Holbrook, M. W. Graves and L. D. Agenbroad. 36–41. Tucson, AZ: University of Arizona.

Black, T. K. III. 1978. Sexual Dimorphism in Tooth Crown Diameters of the Deciduous Teeth. *American Journal of Physical Anthropology* 48:77–82.

Black, T. K. III. 1979. *The Biological and Social Analyses of a Mississippian Cemetery from Southeast Missouri: The Turner Site, 23bu21a, Anthropological Papers No. 68*: University of Michigan, Museum of Anthropology.

Blackwelder, R. E., and Lucile E. Hoyme. 1955. Statistics, Experiment and the Future of Biology. *Scientific Monthly* 80:225–229.

Blakely, Robert L. 1971. Comparison of the Mortality Profiles of Archaic, Middle Woodland, and Middle Mississippi Skeletal Populations. *American Journal of Physical Anthropology* 34 (1):43–53.

Blakely, Robert L. 1973. Biological Variation among and between Two Prehistoric Populations at Dickson Mounds. Ph.D., Indiana University, Bloomington.

Blakely, Robert L, ed. 1977a. *Biocultural Adaptation in Prehistoric America, Southern Anthropological Society Proceedings, No. 11*. Athens: The University of Georgia Press.

Blakely, Robert L. 1977b. Introduction: Changing Strategies for the Biological Anthropologist. In *Biocultural Adaptation in Prehistoric America*, edited by R. L. Blakely. 1–9, *Southern Anthropological Society Proceedings, No. 11*. Athens: The University of Georgia Press.

Blakely, Robert L. 1977c. Sociocultural Implications of Demographic Data from Etowah, Georgia. In *Biocultural Adaptation in Prehistoric America*, edited by R. L. Blakely. 45–66, *Southern Anthropological Society Proceedings, No. 11*. Athens: The University of Georgia Press.

Blakely, Robert L. 1988a. The Life Cycle and Social Organization. In *The King Site: Continuity and Contact in Sixteenth-Century Georgia*, edited by R. L. Blakely. 17–34. Athens: University of Georgia Press.

Blakely, Robert L., ed. 1988b. *The King Site: Continuity and Contact in Sixteenth Century Georgia*. Athens: University of Georgia Press.

Blakely, Robert L., and Lane Beck. 1984. Tooth-Tool Use Versus Dental Mutilation: A Case Study from the Prehistoric Southeast. *Midcontinental Journal of Archaeology* 9 (2):269–284.

Blakey, Michael L. 1987. Skull Doctors: Intrinsic Social and Political Bias in the History of American Physical Anthropology. *Critique of Anthropology* 7:6–35.

Blakey, Michael L. 1998a. The New York African Burial Ground Project: An Examination of Enslaved Lives, a Construction of Ancestral Ties. *Transforming Anthropology* 7 (1):53–58.

Blakey, Michael L. 1998b. Beyond European Enlightenment: Toward a Critical and Humanistic Human Biology. In *Building a New Biocultural Synthesis: Political-Economic Perspectives on Human Biology*. 379–406. Ann Arbor: University of Michigan Press.

Blakey, Michael L. 2001. Bioarchaeology of the African Diaspora in the Americas: Its Origins and Scope. *Annual Review of Anthropology* 30:387–422.

Blom, D. E., Keng L. Hallgrímsson, C. M. C. Lozada, and Jane E. Buikstra. 1998. Tiwanaku "Colonization:" Bioarchaeological Implications for Migration in the Moquegua Valley, Peru. *World Archaeology* 30:238–261.

Blum, A. S. 1993. *Picturing Nature: American Nineteenth-Century Zoological Illustration*. Princeton: Princeton University Press.

Blumenbach, Johann Friedrich. 1775. De Generis Humani Varietate Nativa. M. D., Friedrich Schiller University, Jena.

Blumenbach, Johann Friedrich. 1790. *Decas Collectionis Sua Craniorum Diversarum Gentium Illustrata*. Göttingen: Dieterich.

Blumenbach, Johann Friedrich. 1793. *Decas Altera Collectionis Suae Craniorum Diversarum Gentium Illustratum*. Göttingen: Dieterich.

Blumenbach, Johann Friedrich. 1795. *De Generis Humani Varietate Nativa*. Third ed. Göttingen: Vandenhoek and Ruprecht.

Blumenbach, Johann Friedrich. 1795. *Decas Tertia Collectionis Suae Craniorum Diversarum Gentium Illustratum*. Göttingen: Dieterich.

Blumenbach, Johann Friedrich. 1800. *Decas Quarta Collectionis Suae Craniorum Diversarum Gentium Illustratum*. Göttingen: Dieterich.

Blumenbach, Johann Friedrich. 1806. *Decas Quinta Collectionis Suae Craniorum Diversarum Gentium Illustrata*. Göttingen: Dieterich.

Blumenbach, Johann Friedrich. 1816. *Decas Sexta Collectionis Suae Craniorum Diversarum Gentium Illustrata*. Göttingen: Dieterich.

Blumenbach, Johann Friedrich. 1828. *Nova Pentas Collectionis Suae Craniorum Diversarum Gentium Tanquam. Complementum Priorum Decadum*. Göttingen: Dieterich.

Boas, Franz. 1890. Deformation of Heads in British Columbia. *Science* 8:364–365.

Boas, Franz. 1972. *The Professional Correspondence of Franz Boas (Microform)*. Wilmington: Scholarly Resources.

Bocquet-Appel, J.-P. 1986. Once Upon a Time: Palaeodemography. *Mitteilungen der Berliner Gesellschaft fur Anthropolgie, Ethnologie und Urgeschichte* 7:127–133.

Bocquet-Appel, J.-P. 1989. L'anthropologie Physique en France et Ses Origines Institutionelles. *Gradhiva* 6:23–34.

Bocquet-Appel, J.-P. 1994. Estimating the Average for an Unknown Age Distribution in Anthropology. In *Statistical Tools in Human Biology*, edited by S. Borgognini-Tarli, M. Di Bacco and E. Pacciani. 197–202. Singapore: World Scientific.

Bocquet-Appel, J.-P., and J. N. Bacro. 1997. Brief Communication: Estimates of Some Demographic Parameters in a Neolithic Rock-Cut Chamber (Approximately 2000 BC) Using Iterative Techniques for Aging and Demographic Estimators. *American Journal of Physical Anthropology* 102:569–575.

Bocquet-Appel, J.-P., and C. Masset. 1977. Estimateurs en Paléodémographie. *L'Homme: Revue Française d'Anthropologie* 17 (4):65–90.

Bocquet-Appel, J.-P., and C. Masset. 1982. Farewell to Paleodemography. *Journal of Human Evolution* 11:321–333.

Bocquet-Appel, J.-P., and C. Masset. 1985. Paleodemography: Resurrection or Ghost? *Journal of Human Evolution* 14:107–111.

Bocquet-Appel, J.-P., and C. Masset. 1996. Paleodemography: Expectancy and False Hope. *American Journal of Physical Anthropology* 99:571–583.

Bodecker, C. F. 1930. Concerning Defects in the Enamel of Teeth of Ancient American Indians. *Journal of Dental Research* 10:313–322.

Boldsen, Jesper L. 1988. Two Methods for Reconstructing the Empirical Mortality Profile. *Human Evolution* 3:335–342.

Boldsen, Jesper L., George R. Milner, Lyle W. Konigsberg, and James W. Wood. 2002. Transition Analyses: A New Method for Estimating Age from Skeletons. In *Palaeodemography. I. Age Distributions from Skeletal Samples*, edited by R. D. Hoppa and J. W. Vaupel. 73–106. Cambridge: Cambridge University Press.

Bondioli, L., R. S. Corruccini, and R. Macchiarelli. 1986. Familial Segregation in the Iron Age Community of Alfedena, Abruzzo, Italy Based on Osteodental Traits. *American Journal of Physical Anthropology* 71:393–400.

Bourque, Bruce J. 2002. Maine Shell Midden Archaeology (1860–1910) and the Influence of Adolphe von Morlot. In *New Perspecitives on the Origins of Americanist Archaeology*, edited by D. L. Browman and S. Williams. 148–163. Tuscaloosa: University of Alabama Press.

Bouwman, A., and T. A. Brown. 2005. The Limits of Biomolecular Palaeopathology: Ancient DNA Cannot Be Used to Study Venereal Syphilis. *Journal of Archaeological Science* 32:703–713.

Bowles, Francis P. 1976. Measurement and Instrumentation in Physical Anthropology. *Yearbook of Physical Anthropology* 18:174–190.

Bowron, E. I. 2003. A New Approach to the Storage of Skeletal Remains. *The Conservator* 27:95–106.

Bowser, Brenda J. 2004. Prologue: Toward an Archaeology of Place. *Journal of Archaeological Method and Theory Special Issue: Recent Advances in the Archaeology of Place: Part I* 11 (1):1–3.

Boxberger, Daniel L., and Stacy Rasmus. 2000. Chapter 3: Review of Traditional Historical and Ethnographic Information. In "Cultural Affiliation Report." Retrieved Dec. 31, 2005, from Department of the Interior, National Park Service, Archaeology Program Web Site: http://www.cr.nps.gov/aad/kennewick/boxberger.htm.

Boyd, Robert. 1999. *The Coming of the Spirit of Pestilence: Introduced Infectious Diseases and Population Decline among the Northwest Indians, 1774–1874*. Seattle: University of Washington Press.

Boyle, David. 1894. *Notes on the Life of Dr. Joseph Workman*. Toronto: Arbuthnot Brothers.

Boyle, David. 1895. *Notes on Primitive Man in Ontario*. Toronto: Warwick Bros. & Rutter.

Boyle, Susanna P. 1892. *Cranial Measurements. Annual Archaeological Report and Canadian Institute, (Session 1891), Being an Appendix to the Report of the Minister of Education, Ontario*. Toronto: Warwick and Sons.

Boylston, A., M. Holst, and Charlotte A. Roberts. 2001. The Human Skeletal Remains from Hull Magistrates Court. Unpublished skeletal report.

Brace, C. Loring. 1962. Cultural Factors in the Evolution of the Human Dentition. In *Culture and the Evolution of Man*, edited by M. F. A. Montagu. 343–354. New York: Oxford University Press.

Brace, C. Loring. 1964. The Probable Mutation Effect. *American Naturalist* 98:453–455.

Brace, C. Loring. 1974. The "Ethnology" of Josiah Clark Nott. *Bulletin of the New York Academy of Medicine* 50:509–528.

Brace, C. Loring. 1981. Tales of the Phylogenetic Woods: The Evolution and Significance of Evolutionary Trees. *American Journal of Physical Anthropology* 56:411–430.

Brace, C. Loring. 1982. The Roots of the Race Concept in American Physical Anthropology. In *A History of American Physical Anthropology: 1930–1980*, edited by F. Spencer. 11–29. New York: Academic Press.

Brace, C. Loring, and Ashley Montagu. 1977. *Human Evolution: An Introduction to Biological Anthropology*. New York: Macmillan.

Brace, C. Loring, A. R. Nelson, N. Seguchi, H. Oe, Leslie Sering, P. Qifeng, L. Yongyi, and D. Tumen. 2001. Old World Sources of the First New World Human Inhabitants: A Comparative Craniofacial View. *Proceedings of the National Academy of Sciences* 98 (17):10017–10022.

Bradley, Richard. 1990. *Passages of Arms: An Archaeological Analysis of Prehistoric Hoards and Votive Deposits*. Cambridge: Cambridge University Press.

Bradley, Richard. 1993. *Altering the Earth: The Origins of Monuments in Britain and Continental Europe, Monograph Series 8*. Edinburgh: Society of Antiquities of Scotland.

Bradley, Richard. 1998a. Incised Motifs in the Passage-Graves at Quoyness and Cuween, Orkney. *Antiquity* 72:387–390.

Bradley, Richard. 1998b. Ruined Buildings, Ruined Stones: Enclosures, Tombs and Natural Places in the Neolithic of South-West England. *World Archaeology* 30 (1):13–22.

Braun, David P. 1981. A Critique of Some Recent North American Mortuary Studies. *American Antiquity* 46 (2):398–420.

Braun, Mark, Della Collins Cook, and Susan Pfeiffer. 1998. DNA from *Mycobacterium Tuberculosis* Complex Identified in North American Pre-Columbian Human Skeletal Remains. *Journal of Archaeological Science* 25:271–277.

Bray, Tamara L., and Thomas W. Killion, eds. 1994. *Reckoning with the Dead: The Larsen Bay Repatriation and the Smithsonian Institution*. Washington, DC: Smithsonian Institution Press.

Brennan, Matthew J., and William W. Howells. n.d. *A Craniometric and Linguistic Grouping of Selected North American and Asian Peoples. Manuscript Submitted for the Handbook of North American Indians*. Washington, DC: Smithsonian Press.

Brew, John Otis. 1946. *Archaeology of Alkali Ridge, Southeastern Utah, Papers of the Peabody Museum of American Archaeology and Ethnology*. Cambridge: Harvard University Press.

Brew, John Otis, ed. 1968. *One Hundred Years of Anthropology*. Cambridge: Harvard University Press.

Brickley, Megan. 2000. The Diagnosis of Metabolic Disease in Archaeological Bone. In *Human Osteology in Archaeology and Forensic Science*, edited by M. Cox and S. Mays. 183–198. London: Greenwich Medical Media.

Brickley, Megan, and Jaqueline I. McKinley, eds. 2004. *Guidelines to the Standards for Recording Human Remains. (Available on Line: Www.Babao.Org.Uk), Ifa Paper Number 7*. Reading: BABAO, Department of Archaeology, University of Southampton, and the Institute of Field Archaeologists.

Brickley, Megan, A. Miles, and H. Stainer. 1999. *The Cross Bones Burial Ground, Redcross Way, Southwark, London. Archaeological Excavations (1991–1998) for the London Underground Limited Jubilee Extension Project, Monograph 3*. London: Museum of London Archaeological Services.

Bridges, Patricia S. 1985. Changes in Long Bone Structure with the Transition to Agriculture: Implications for Prehistoric Activities. Ph.D., University of Michigan, Ann Arbor.

Bridges, Patricia S. 1989a. Changes in Activities with the Shift to Agriculture in the Southeastern United States. *Current Anthropology* 30 (385–394).

Bridges, Patricia S. 1989b. Spondylolysis and Its Relationship to Degenerative Joint Disease in the Prehistoric Southeastern United States. *American Journal of Physical Anthropology* 79:321–329.

Bridges, Patricia S. 1990. Osteological Correlates of Weapon Use. In *A Life in Science: Papers in Honor of J. Lawrence Angel*, edited by J. E. Buikstra. 87–98, *Scientific*

Papers Number 6. Kampsville, Illinois: American Anthropological Association/Center for American Archaeology.

Bridges, Patricia S. 1991a. Skeletal Evidence of Changes in Subsistence Activities between the Archaic and Mississippian Time Periods in Northwestern Alabama. In *What Mean These Bones? Studies in Southeastern Bioarchaeology*, edited by M. L. Powell, P. S. Bridges and A. M. W. Mires. 89–101. Tuscaloosa and London: The University of Alabama Press.

Bridges, Patricia S. 1991b. Degenerative Joint Disease in Hunter-Gatherers and Agriculturalists from the Southeastern United States. *American Journal of Physical Anthropology* 85:379–391.

Bridges, Patricia S. 1992. Prehistoric Arthritis in the Americas. *Annual Review of Anthropology* 21:67–91.

Bridges, Patricia S. 1994a. Warfare and Mortality at Koger's Island, Alabama. *International Journal of Osteoarchaeology* 6:66–75.

Bridges, Patricia S. 1994b. Vertebral Arthritis and Physical Activities in the Prehistoric Southeastern United States. *American Journal of Physical Anthropology* 93:83–93.

Bridges, Patricia S. 1997. The Relationship between Muscle Markings and Diaphyseal Strength in Prehistoric Remains from West-Central Illinois [Abstract]. *American Journal of Physical Anthropology* (Supplement 24):82.

Bridges, Patricia S., John H. Blitz, and Martin C. Solano. 2000. Changes in Long Bone Diaphyseal Strength with Horticultural Intensification in West-Central Illinois. *American Journal of Physical Anthropology* 112:217–238.

Bridges, Patricia S., Keith P. Jacobi, and Mary Lucas Powell. 2000. Warfare-Related Trauma in the Late Prehistory of Alabama. In *Bioarchaeological Studies of Life in the Age of Agriculture: A View from the Southeast*, edited by P. M. Lambert. 35–62. Tuscaloosa: University of Alabama Press.

British Association for Biological Anthropology and Osteoarchaeology. n.d. British Association for Biological Anthropology and Osteoarchaeology Home Page. Retrieved Dec. 28, 2005, from British Association for Biological Anthropology and Osteoarchaeology Web Site: http://www.babao.org.uk/.

British Association for Biological Anthropology and Osteoarchaeology. n.d. Conferences Page. Retrieved Dec. 28, 2005, from British Association for Biological Anthropology and Osteoarchaeology Web Site: http://www.babao.org.uk/.

Broca, Paul. 1873. The Troglodytes, or Cave Dwellers, of the Valley of the Vézère. *Annual Report of the Smithsonian Institution* 1872:310–347.

Broca, Paul. 1879. Instructions Relative À L'étude Anthropologique du Système Dentaire. *Bulletins de la Société d'Anthropologie de Paris* 2:128–152.

Brock, Sharon L., and Christopher B. Ruff. 1988. Diachronic Patterns of Change in Structural Properties of the Femur in the Prehistoric American Southwest. *American Journal of Physical Anthropology* 75:113–128.

Brooks, Richard H., and Sheilagh T. Brooks. 1985. Archaeological, Ethnological, and Historical Implications of the Game "Cuatro" in Northwestern Mexico. In *The Archaeology of West and Northwest Mexico*, edited by M. S. Foster and P. C. Wiegand. Boulder, CO: Westwood Press.

Brooks, Richard H., and Sheilagh T. Brooks. 1987. Recorded Pathologies of a Burial Cave Skeletal Series from a Burial Cave in North Central Nevada. In *Vth European Meeting, Paleopathology Association, 1984, Siena, Italy*, edited by V. Cappechi and R. Masso. 29–34. Italy: University of Siena, Italy.

Brooks, Richard H., and Sheilagh T. Brooks. 1988. Physical Anthropology. In *The Salina Sisters Burial Site. Archaeological Investigations Along the Mountain Fuel Supply Pipeline in Central and Southwestern Utah*, edited by L. B. Billat and S. C. Billat. *Brigham Young University, Museum of Peoples and Cultures, Technical Series No. 87–88.* Salt Lake City.

Brooks, Richard H., and Sheilagh T. Brooks. 1990. Who Were the Stillwater Marsh People? *Halcyon, A Journal of the Humanities* 12:63–74.

Brooks, Richard H., and Sheilagh T. Brooks. 1991a. Craniosynostoses [Abstract]. In *Papers on Paleopathology Presented at the Eighteenth Annual Meeting of the Paleopathology Association, Milwaukee, Wisconsin, 2–3 April 1991*, edited by E. Cockburn. 16–17.

Brooks, Richard H., and Sheilagh T. Brooks. 1991b. Pathological Traits Observed in a Historic Great Basin Native American Indian Skeleton. *International Journal of Anthropology* 6 (3):267–275.

Brooks, Richard H., Sheilagh T. Brooks, and J. P. King. 1981. *An Archaeological Inventory Report of the Owlshead/Amargosa-Mohave Basin Planning Units of the Southern California Desert Area, Cultural Resource Publications—Anthropology/History.* Riverside, California: U.S. Bureau of Land Management.

Brooks, Sheilagh T. 1947. A Comparative Study of Mammal Bone Recovered from Archaeological Sites in Marin County, California. M.A., University of California, Berkeley, Berkeley.

Brooks, Sheilagh T. 1951. A Comparison of the Criteria of Age Determination of Human Skeletons by Cranial and Pelvic Morphology. Ph.D., University of California, Berkeley, Berkeley.

Brooks, Sheilagh T. 1955 Skeletal Age-at-Death: The Reliability of Cranial and Pubic Age Indicators. *American Journal of Physical Anthropology* 13 (4):567–597.

Brooks, Sheilagh T. 1969. Skeletal Material from LAN-227, Appendix V. *Archaeological Survey Annual Report. Los Angeles: University of California Press* 10 (144–153).

Brooks, Sheilagh T. 1970. Obituary: Theodore Doney McCown. 1908-1969. A Biographical Sketch. *American Journal of Physical Anthropology* 32:165–168.

Brooks, Sheilagh T. 1975. Human or Not? A Problem in Skeletal Identification. *Journal of Forensic Sciences* 20 (1):149–153.

Brooks, Sheilagh T. 1976. A Tribute to Sherbourne Friend Cook. *The Journal of California Anthropology* 3:3–4.

Brooks, Sheilagh T. 1979. Obituary: Audrey J. Sublett. *American Journal of Physical Anthropology* 50:153–154.

Brooks, Sheilagh T. 1981. Teaching of Forensic Anthropology in the United States [Letter]. *Journal of Forensic Sciences* 26 (4):627–631.

Brooks, Sheilagh T. 1988. Obituary: Louise Marie Robbins. 1928–1987. *American Journal of Physical Anthropology* 75 (4):437–438.

Brooks, Sheilagh T. 1989. Description or Diagnosis? [Abstract]. In *Papers on Paleopathology Presented at the Sixteenth Annual Meeting of the Paleopathology Association, San Diego, California, 4–5 April 1989*, edited by E. Cockburn. 1.

Brooks, Sheilagh T., and Richard H. Brooks. 1969. Arm Position as Correlated with Sex Determination in the Niah Cave Extended Burial Series, Sarawak, Malaysia. *Sarawak Museum Journal* 16:67–87.

Brooks, Sheilagh T., and Richard H. Brooks. 1974. Excavation by Request: Physical Anthropological Identification of Pioneer Burials [Abstract]. *American Journal of Physical Anthropology* 40 (1):131.

Brooks, Sheilagh T., and Richard H. Brooks. 1978. Palaeoepidemiology as a Possible Interpretation of Multiple Child Burials near Zape Chico, Durango, Mexico. In *Across the Chichimec Sea: Papers in Honor of J. Charles Kelley*, edited by C. L. Riley and B. C. Hedrick. 96–101. Carbondale: Southern Illinois University Press.

Brooks, Sheilagh T., and Richard H. Brooks. 1980. Cranial Deformation: Possible Evidence of Pochteca Trading Movements. *Transactions of the Illinois State Academy of Science* 72 (4):1–12.

Brooks, Sheilagh T., and Richard H. Brooks. 1984. Problems of Burial Exhumation, Historical and Forensic Aspects. In *Human Identification: Case Studies in Forensic Anthropology*, edited by T. A. Rathbun and J. E. Buikstra. 64–86. Springfield, IL: Charles C Thomas.

Brooks, Sheilagh T., Richard H. Brooks, and Diane France. 1990. Alveolar Prognathism Contour, an Aspect of Racial Identification. In *Skeletal Attribution of Race, Methods for Forensic Anthropology*, edited by G. Gill and S. Rhine. 41–46. Vol. 4, *Maxwell Museum of Anthropology, Anthropological Papers*. Albuquerque: Maxwell Museum of Anthropology.

Brooks, Sheilagh T., Richard H. Brooks, and Dennis J. Jenkins. 1985. Pathology or the Borings of Invertebrates: A Case Study from Nevada. In *Health and Disease in the Prehistoric Southwest*, edited by C. F. Merbs and R. J. Miller. 9–17. Vol. 34, *Arizona State University Anthropological Research Papers*. Tempe: Arizona State University.

Brooks, Sheilagh T., Richard H. Brooks, Gail E. Kennedy, J. Austin, James R. Firby, Louis A. Payen, Peter J. Slota Jr., Christine A. Prior, and R. E. Taylor. 1990. The Haverty Human Skeletons: Morphological, Depositional, and Geochronological Characteristics. *Journal of California and Great Basin Anthropology* 12 (1):60–83.

Brooks, Sheilagh T., Richard H. Brooks, and W. R. K. Perizonius. 1994. Possible Syphilitic Lesions: Problem Cases from Holland and New Guinea. In *L'origine de la Syphilis en Europe: Avant Ou Aprés 1493?*, edited by O. Dutour, G. Pálfi, J. Berano and J.-P. Brun. 226–232. Paris: Editions Errance-Centre Archéologique du Var.

Brooks, Sheilagh T., K. Dixon, and B. L. Conrey. 1966. The Seal Beach Skull. *Papers of the Southern California Academy of Sciences* 64:229–241.

Brooks, Sheilagh T., M. Galliher, and Richard H. Brooks. 1977. A Proposed Model for Palaeodemography and Archaeology. In *Great Basin Models and Great Basin Prehistory*, edited by D. D. Fowler. 169–194, *Desert Research Institute Publications in the Social Sciences, No. 12*.

Brooks, Sheilagh T., Michele B. Haldeman, and Richard H. Brooks. 1988. *Osteological Analyses of the Stillwater Skeletal Series, Stillwater Marsh, Churchill County,*

Nevada: Final Report, United States Department of Interior, United States Fish and Wildlife Service, Cultural Resource Series. Number 2. Fallon, Nevada: United States Department of Interior.

Brooks, Sheilagh T., Michele B. Haldeman, and Richard H. Brooks. 1990. Unusual Eburnation Frequencies in a Skeletal Series from the Stillwater Marsh Area, Nevada. In *Wetland Adaptations in the Great Basin*, edited by J. C. Janetski and D. B. Madsen. 91–105, *Brigham Young University, Museum Peoples Cultures Occasional Paper, Number 1*. Salt Lake City: Brigham Young University Press.

Brooks, Sheilagh T., and Rodger Heglar. 1972. A Preliminary Report on the Palaeoserology of the Niah Cave Burials. *Asian Perspectives (Honolulu)* 15:87–88.

Brooks, Sheilagh T., Rodger Heglar, and Richard H. Brooks. 1979. Radiocarbon Dating and Palaeoserology of a Selected Burial Series from the Great Cave of Niah, Sarawak, Malaysia. *Asian Perspectives (Honolulu)* 20 (1):21–31.

Brooks, Sheilagh T., and William D. Hohenthal. 1963. Archaeological Defective Palate Crania from California. *American Journal of Physical Anthropology* 21:25–32.

Brooks, Sheilagh T., and F. Jerome Melbye. 1967. Skeletal Lesions Suggestive of Pre-Columbian Multiple Myeloma in a Burial from the Kane Mounds, near St. Louis, Missouri. In *Miscellaneous Papers in Paleopathology: 1*, edited by W. D. Wade. 23–29, *Technical Series Number 7. Museum of Northern Arizona*. Flagstaff, AZ: Northern Arizona Society of Science and Art.

Brooks, Sheilagh T., Caroline Stark, and Richard H. Brooks. 1984. John Reid's Redheaded "Giants" of Central Nevada: Fact or Fiction? *Nevada Historical Society Quarterly* 4:243–254.

Brooks, Sheilagh T., Caroline Stark, and Richard H. Brooks. 1985. Skeletal Analysis of an Historical Burial from Winnemucca, Nevada. *The Humboldt Historian* 8:4–9.

Brooks, Sheilagh T., D. Gentry Steele, and Richard H. Brooks. 1990. Formulae for Stature Estimation on Incomplete Long Bones: A Survey of Their Reliability. *Journal of Forensic Medicine (Adli Tip Dergisi)* 6 (1–2):167–170.

Brooks, Sheilagh T., and Judy Myers Suchey. 1990. Skeletal Age Determination Based on the Os Pubis: A Comparison of the Acsádi-Nemeskéri and Suchey-Brooks Methods. *Human Evolution* 5:227–238.

Brothwell, Don R. 1960. Upper Pleistocene Human Skull from Niah Caves. *Sarawak Museum Journal* 9:323–349.

Brothwell, Don R. 1963a. The Macroscopic Dental Pathology of Some Earlier Human Populations. In *Dental Anthropology*, edited by D. R. Brothwell. 271–288. Oxford: Pergamon Press.

Brothwell, Don R, ed. 1963b. *Dental Anthropology*. Oxford: Pergamon Press.

Brothwell, Don R. 1963c. *Digging up Bones*. 1st ed. London: British Museum (Natural History).

Brothwell, Don R. 1967. The Evidence for Neoplasms. In *Diseases in Antiquity*, edited by D. R. Brothwell and A. T. Sandison. 320–345. Springfield, IL: Charles C Thomas.

Brothwell, Don R, ed. 1968. *Skeletal Biology of Earlier Human Populations*. Vol. 8, *Symposium of the Society of Human Biology*. London: Pergamon Press.

Brothwell, Don R. 1970. The Real History of Syphilis. *Science* 6 (9):27–33.

Brothwell, Don R. 1972/73. Palaeodemography and Earlier British Populations. *World Archaeology* 4 (1):75–87.

Brothwell, Don R. 1981. *Digging up Bones*. 3rd ed. Ithaca; London: Cornell University Press; British Museum (Natural History).

Brothwell, Don R. 1989. The Relationship of Tooth Wear to Ageing. In *Age Markers in the Human Skeleton*, edited by M. Y. İşcan. Springfield, IL: Charles C Thomas.

Brothwell, Don R. 1991. On the Zoonoses and Their Relevance to Paleopathology. In *Human Paleopathology. Current Syntheses and Future Options*, edited by D. J. Ortner and A. C. Aufderheide. 18–22. Washington, DC: Smithsonian Institution Press.

Brothwell, Don R. 1999. Biosocial and Bioarchaeological Aspects of Conflict and Warfare. In *Ancient Warfare: Archaeological Perspectives*, edited by J. Carman and A. Harding. 25–39. Sutton: Stroud.

Brothwell, Don R. 2003a. Skeletal Atrophy and the Problem of Differential Diagnosis of Conditions Causing Paralysis. *Antropologia Portuguesa* 19:5–18.

Brothwell, Don R. 2003b. The Analysis of the Hair of the Neolithic Iceman. In *Mummies in a New Millennium: Proceedings of the 4th World Congress on Mummy Studies. Nuuk, Greenland, September 4th to 10th 2001*, edited by N. Lynnerup, C. Andreasen and J. Berglund. 66–68, *Danish Polar Center Publication, No. 11*. Copenhagen: Greenland National Museum and Archives, Danish Polar Center.

Brothwell, Don R., and E. S. Higgs. 1969. *Science in Archaeology: A Survey of Progress and Research*. Revised and enlarged ed. London: Thames & Hudson.

Brothwell, Don R., and A. M. Pollard. 2001. *Handbook of Archaeological Sciences*. Chichester: New York: J. Wiley.

Brothwell, Don R., and A. T. Sandison. 1967. *Diseases in Antiquity*. Springfield, IL: Charles C Thomas.

Browman, David L. 2002. Origins of Stratigraphic Excavation in North America: The Peabody Museum Method and the Chicago Method. In *New Perspectives on the Origins of Americanist Archaeology*, edited by D. L. Browman and S. Williams. 242–264. Tuscaloosa: University of Alabama Press.

Brown, James A. 1971. The Dimensions of Status in the Burials at Spiro. In *Approaches to the Social Dimensions of Mortuary Practices*, edited by J. A. Brown. 92–112, *Memoirs of the Society for American Archaeology 25*. Washington, DC: Society for American Archaeology.

Brown, J. A. 1981. The Search for Rank in Prehistoric Burials. In *The Archaeology of Death*, edited by R. Chapman, I. Kinnes and K. Randsborg. 25–37, *New Directions in Archaeology*. Cambridge: Cambridge University Press.

Brown, James A. 1995. On Mortuary Analysis — with Special Reference to the Saxe-Binford Research Program. In *Regional Approaches to Mortuary Analysis*, edited by L. Beck. 3–26. New York: Plenum Press.

Brown, James A. 1996. *The Spiro Ceremonial Center: The Archaeology of Arkansas Valley Caddoan Culture in Eastern Oklahoma, Memoirs of the Museum of Anthropology, No. 29*. Ann Arbor: University of Michigan.

Brown, James A. 1997. The Archaeology of Ancient Religion in the Eastern Woodlands. *Annual Review of Anthropology* 26:465–485.

Brown, P. J. 1993. *Measurement, Regression, and Calibration*. Oxford: Oxford University Press.

Brown, P. J., and R. Sundberg. 1987. Confidence and Conflict in Multivariate Calibration. *Journal of the Royal Statistical Society Series B* 49:46–57.

Brown, P. J., and R. Sundberg. 1989. Prediction Diagnostics and Updating in Multtivariate Calibration. *Biometrika* 76:349–361.

Brown, Thomas E. 2000. "Workman, Joseph." Retrieved Jan. 10, 2006, from Library and Archives Canada, Dictionary of Canadian Biography Online Web Site: http://www.biographi.ca/EN/ShowBio.asp?BioId=40632&query=workman.

Browne, S. H. 2000. Counter-Science: African American Historians and the Critique of Ethnology in Nineteenth-Century America. *Western Journal of Communication* 64:268–285.

Brownlee, Kevin, and E. Leigh Syms. 1999. *Kayasochi Kikawenow. Our Mother from Long Ago. An Early Cree Woman and Her Personal Belongings from Nagami Bay, Southern Indian Lake*. Winnipeg: The Manitoba Museum of Man and Nature.

Bruce, R. V. 1988. *The Launching of Modern American Science, 1846–1876*. New York: Cornell University Press.

Brues, Alice Mossie. 1940. Sibling Resemblances as Evidence for the Genetic Determination of Traits of the Eye, Skin and Hair in Man. Ph.D., Radcliffe College.

Brues, Alice Mossie. 1946a. A Genetic Analysis of Human Eye Color. *American Journal of Physical Anthropology* 4:1–36.

Brues, Alice Mossie. 1946b. Appendix A: Tables A.I–A.IV: Alkali Ridge Skeletal Material, Measurements, Indices, and Observations. In *Archaeology of Alkali Ridge, Southeastern Utah*, edited by J. O. Brew. 317–329. Vol. 21, *Papers of the Peabody Museum of American Archaeology and Ethnology*. Cambridge: Harvard University Press.

Brues, Alice Mossie. 1946c. Appendix B. Alkali Ridge Skeletons, Pathology and Anomaly. In *Archaeology of Alkali Ridge, Southeastern Utah*, edited by J. O. Brew. 327–329. Vol. 21, *Papers of the Peabody Museum of American Archaeology and Ethnology*. Cambridge: Harvard University Press.

Brues, Alice Mossie. 1946d. The San Simon Branch: Excavations at Cave Creek and in the San Simon Valley II. Skeletal Material. In *Medallion Papers No. XXXV*. 1–26. Globe, Arizona: Gila Pueblo.

Brues, Alice Mossie. 1950. Linkage of Body Build with Sex, Eye Color and Freckling. *American Journal of Human Genetics* 2:215–239.

Brues, Alice Mossie. 1957. Skeletal Material from the Nagle Site. *Bulletin of the Oklahoma Anthropological Society* 5:101–106.

Brues, Alice Mossie. 1958a. Identification of Skeletal Remains. *Journal of Criminal Law, Criminology and Police Science* 48 (5):551–563.

Brues, Alice Mossie. 1958b. Skeletal Material from the Horton Site. *Bulletin of the Oklahoma Anthropological Society* 6:27–32.

Brues, Alice Mossie. 1959a. Skeletal Material from the Morris Site (CK-39). *Bulletin of the Oklahoma Anthropological Society* 7:63–70.

Brues, Alice Mossie. 1959b. The Spearman and the Archer: An Essay on Selection in Body Build. *American Anthropologist* 61 (3):457–469.

Brues, Alice Mossie. 1960. Computer Analysis of Blood Group Polymorphism. In *Abstracts of 1960 Meeting of the American Society of Human Genetics*. 21.

Brues, Alice Mossie. 1962. Skeletal Material from the Mclemore Site. *Bulletin of the Oklahoma Anthropological Society* 10:69–78.

Brues, Alice Mossie. 1963a. Stochastic Tests of Selection in the Abo Blood Groups. *American Journal of Physical Anthropology* 21:287–299.

Brues, Alice Mossie. 1963b. Skeletal Material from the Fred Loomis Site. *Bulletin of the Oklahoma Anthropological Society* 11:128.

Brues, Alice Mossie. 1964. Genetic Drift in the Differentiation of the American Indigenes: Evidence from the Blood Groups. In *Actas y Memorias, Xxv, Congreso Internacional de Americanistas*. 119–120.

Brues, Alice Mossie. 1966. Discussion: Palaeoepidemiology. In *Human Palaeopathology*, edited by S. Jarcho. 107–112. New Haven and London: Yale University Press.

Brues, Alice Mossie. 1967. Comment on "On Pestilence, Diet, Natural Selection, and the Distribution of Microbial and Human Blood Group Antigens and Antibodies." *Current Anthropology* 8 (3):217.

Brues, Alice Mossie. 1975. Comment on "New Evidence for a Late Introduction of Malaria into the New World." *Current Anthropology* 16 (1):96.

Brues, Alice Mossie. 1977. *People and Races*. New York: Macmillan.

Brues, Alice Mossie. 1990a. The Once and Future Diagnosis of Race. In *Skeletal Attribution of Race*, edited by G. W. Gill and S. Rhine. 1–7, *Anthropological Papers, No. 4*. Albuquerque: Maxwell Museum of Anthropology.

Brues, Alice Mossie. 1990b. Sixty Years of Physical Anthropology, Including False Starts and Dead Ends. Paper Presented at the American Anthropological Association in a session on "Some Reflections by Senior Anthropologists on Direction and Change in Anthropology" sponsored by the Association of Senior Anthropologists.

Brues, Alice Mossie. 1992. Forensic Diagnosis of Race: General Race Vs Specific Populations. *Social Science and Medicine* 34 (2):125–128.

Brues, Alice Mossie. 1996. Skeletal Analysis. In *The Spiro Ceremonial Center: The Archaeology of Arkansas Valley Caddoan Culture in Eastern Oklahoma*, edited by J. A. Brown. 317–324. Vol. 2, *Memoirs of the Museum of Anthropology, University of Michigan, Number 29*. Ann Arbor: University of Michigan.

Brunson, Judy Lynn. 1989. The Social Organization of the Los Muertos Hohokam: A Reanalysis of Cushing's Hemenway Expedition Data. Ph.D., Arizona State University.

Buckberry, J. 2004. A Social and Anthropological Analysis of Conversion Period and Later Anglo-Saxon Cemeteries in Lincolnshire and Yorkshire. Ph.D. Dissertation, University of Sheffield.

Buikstra, Jane E. 1976. *Hopewell in the Lower Illinois Valley: A Regional Approach to the Study of Biological Variability and Mortuary Activity*. Vol. 2 (No. 2), *Scientific Papers*. Evanston: Northwestern University Archaeological Program.

Buikstra, Jane E. 1977. Biocultural Dimensions of Archeological Study: A Regional Perspective. In *Biocultural Adaptation in Prehistoric America*, edited by R. L. Blakely. 67–84, *Proceedings of the Southern Anthropological Society, No. 11*. Athens, GA: University of Georgia Press.

Buikstra, Jane E. 1979. Contributions of Physical Anthropologists to the Concept of Hopewell: A Historical Perspective. In *Hopewell Archaeology*, edited by D. S. Brose and N. Greber. 220–233. Kent, OH: Kent State University Press.

Buikstra, Jane E. 1980. Epigenetic Distance: A Study of Biological Variability in the Lower Illinois River Region. In *Early Native Americans*, edited by D. L. Browman. 271–299. Paris: Mouton.

Buikstra, Jane E. 1981a. *Bio-Archaeology at the Koster Site: The Multidisciplinary Approach of Modern Archaeology*. Vol. 2 (No. 1), *Koster*. Peoria, Illinois.

Buikstra, Jane E. 1981b. Mortuary Practices, Palaeodemography and Palaeopathology: A Case Study from the Koster Site (Illinois). In *The Archaeology of Death*, edited by R. Chapman, I. Kinnes and K. Randsborg. 123–132. London: Cambridge University Press.

Buikstra, Jane E., ed. 1981c. *Prehistoric Tuberculosis in the Americas*. Evanston: Northwestern University Archaeological Program.

Buikstra, Jane E. 1982. Reburial: How We All Lose. *California Archaeological Newsletter* No. 1:1–4.

Buikstra, Jane E, ed. 1990. *A Life in Science: Papers in Honor of J. Lawrence Angel, Scientific Papers 6*. Kampsville, IL: American Anthropological Association/Center for American Archaeology.

Buikstra, Jane E. 1991. Out of the Appendix and into the Dirt: Comments on Thirteen Years of Bioarchaeological Research. In *What Mean These Bones?* edited by M. L. Powell, P. S. Bridges and A. M. W. Mires. 172–188. Tuscaloosa: University of Alabama Press.

Buikstra, Jane E. 1992. Diet and Disease in Late Prehistory. In *Disease and Demography in the Americas: Changing Patterns before and after 1492*, edited by J. W. Verano and D. H. Ubelaker. 87–101. Washington, DC: Smithsonian Institution Press.

Buikstra, Jane E. 1995. Tombs for the Living . . . or . . . for the Dead: The Osmore Ancestors. In *Tombs for the Living: Andean Mortuary Practices*, edited by T. D. Dillehay. 229–280. Washington, DC: Dumbarton Oaks Research Library and Collection.

Buikstra, Jane E. 1997. Paleodemography: Context and Promise. In *Integrating Archaeological Demography: Multidisciplinary Approaches to Prehistoric Population*, edited by R. R. Paine. 367–380, *Southern Illinois University, Center for Archaeological Investigations No. 24*. Carbondale.

Buikstra, Jane E. 1999. Palaeoepidemiology of Tuberculosis in the Americas. In *Tuberculosis Past and Present*, edited by G. Pálfi, O. Dutour, J. Deák and I. Hutás. 479–494. Budapest: Golden Book Publisher Ltd./Szeged: Tuberculosis Foundation.

Buikstra, Jane E. 2000. Historical Bioarchaeology and the Beautification of Death. In *Never Anything So Solemn: An Archeological, Biological, and Historical Investigation of the Nineteenth-Century Grafton Cemetery*, edited by J. E. Buikstra, J. A. O'Gorman and C. Sutton. 15–20, *Kampsville Studies in Archaeology and History, No. 3*. Kampsville, IL: Center for American Archaeology.

Buikstra, Jane E., Douglas K. Charles, and Gordon F. M. Rakita. 1998. *Staging Ritual: Hopewell Ceremonialism at the Mound House Site, Greene County, Illinois, Kampsville Studies in Archaeology and History, No. 1*. Kampsville, IL: Center for American Archaeology.

Buikstra, Jane E., and Della Collins Cook. 1980. Palaeopathology: An American Account. *Annual Review of Anthropology* 9:433–470.

Buikstra, Jane E., and Diedre Fowler. 1975. An Osteological Study of the Human Skeletal Materials from the Bentsen-Clark Site. In *The Bentsen-Clark Site, Red River County, Texas: A Preliminary Report*, edited by L. D. Banks and J. Winters. 79–96. San Antonio: Texas Archeological Society.

Buikstra, Jane E., Susan R. Frankenberg, and Lyle W. Konigsberg. 1990. Skeletal Biological Distance Studies in American Physical Anthropology: Recent Trends. *American Journal of Physical Anthropology* 82:1–7.

Buikstra, Jane E., and Claire C. Gordon. 1981. The Study and Restudy of Human Skeletal Series: The Importance of Long-Term Curation. *Annals of the New York Academy of Sciences* 376 (January):449–465.

Buikstra, Jane E., Claire C. Gordon, and Lucile E. St. Hoyme. 1984. Case of the Severed Skull: Individuation in Forensic Anthropology. In *Human Identification*, edited by T. A. Rathbun and J. E. Buikstra. 121–135. Springfield, IL: Charles C Thomas.

Buikstra, Jane E., and Lisa M. Hoshower. 1990. Introduction. In *A Life in Science: Papers in Honor of J. Lawrence Angel*, edited by J. E. Buikstra. Kampsville, IL: American Anthropological Association/Center for American Archaeology.

Buikstra, Jane E., and Lisa M. Hoshower. 1994. Análisis de os Restos Humanos de la Necrópolis de Gatas. In *Proyecto Gatas: Sociedad y Economia en el Sudeste de España C.2500-900 Cal Ane*, edited by P. C. Martínez. 339–403. Seville: presentada a la consciena de Cultura de la Junta de Andalucía.

Buikstra, Jane E., and Lyle W. Konigsberg. 1985. Paleodemography: Critiques and Controversies. *American Anthropologist* 87:316–333.

Buikstra, Jane E., Lyle W. Konigsberg, and J. Bullington. 1986. Fertility and the Development of Agriculture in the Prehistoric Midwest. *American Antiquity* 51:528–546.

Buikstra, Jane E., and George R. Milner. 1989. *Fifty Years of Research at Dickson Mounds*: Illinois State Museum.

Buikstra, Jane E., Andrew Poznanski, Maria Lozada Cerna, Paul Goldstein, and Lisa M. Houshouer. 1990. A Case of Juvenile Rheumatoid Arthritis from Pre-Columbian Peru. In *A Life in Science: Papers in Honor of J. Lawrence Angel, Scientific Papers 6*. edited by J. E. Buikstra, pp. 99–137. Kampsville, ILCenter for American Archaeology: American Anthropological Association/Center for American Archaeology.

Buikstra, Jane E., T. Douglas Price, James H. Burton, and Lori E. Wright. 2003. Tombs from Copan's Acropolis: A Life History Approach. In *Understanding Early Classic Copan*, edited by E. E. Bell, M. A. Canuto and R. J. Sharer. Philadelphia: University of Pennsylvania Museum of Archaeology and Anthropology.

Buikstra, Jane E., and Douglas H. Ubelaker, eds. 1994. *Standards for Data Collection from Human Skeletal Remains: Proceedings of a Seminar at the Field Museum of Natural History, Arkansas Archeological Survey Research Series No. 44*. Fayetteville: Arkansas Archeological Survey.

Buikstra, Jane E., Michael Vadeboncouer, and Gary Behrend. 1971. The Lundy Site: Human Osteology. *Oklahoma Anthropological Society Bulletin* 22:160–170.

Bullen, Adelaide K. 1945. A Cross-Cultural Approach to the Problem of Stuttering. *Child Development* 16 (1–2):1–88.

Bullen, Adelaide K. 1947. Archaeological Theory and Anthropological Fact. *American Antiquity* 13:128–134.

Bullen, Adelaide K. 1948. Vocational Incidence of Seven Body Types in 567 U.S. Army Women. Prepared for Military Planning Division, Office of the Quartermaster General, under Contract W44-109-Qm-1078, Project No. 64-02-01, Work Phase 5, Department of Anthropology, Harvard University, Cambridge (Mimeographed).

Bullen, Adelaide K. 1952. Some Problems in the Practical Application of Somatotyping. *Florida Anthropologist* 5:17–20.

Bullen, Adelaide K. 1953a. Qualitative and Quantitative Theory as Applied to Body Build Research. *Quarterly Journal of the Florida Academy of Sciences* 16 (1):35–64.

Bullen, Adelaide K. 1953b. Introduction to Papers from "The Development of High Civilizations in Hot Climates Symposium." *Florida Anthropologist* 6 (4):101–102.

Bullen, Adelaide K. 1956. *New Answers to the Fatigue Problem*. Gainesville: University of Florida Press.

Bullen, Adelaide K. 1962a. Comments on the Aging Population of Florida. In *Aging in a Changing Society. Eleventh Annual Southern Conference on Gerontology, 1962. University of Florida*, edited by R. E. Albrecht. Gainesville: Published for the University of Florida Institute of Gerontology by the University of Florida Press.

Bullen, Adelaide K. 1962b. Report of Skeletal Findings. In *Ceramic Periods of St. Thomas and St. John Islands, Virgin Islands*, edited by R. P. Bullen. 25–27. *The William L. Bryant Foundation, American Studies. Report No. 4*. Maitland: Central Florida Museum.

Bullen, Adelaide K. 1963. Physical Anthropology of the Goodman Mound, McCormack Site, Duval County, Florida. In *Papers on the Jungerman and Goodman Sites, Florida*, edited by D. F. Jordan, E. S. Wing and A. K. Bullen. 61–70. Vol. 10, *Contributions of the Florida State Museum, Social Sciences*. Gainesville: University of Florida.

Bullen, Adelaide K. 1965. Florida Indians of Past and Present. In *Florida from Indian Trail to Space Age*, edited by C. W. Tebeau and R. L. Carson. 317–350. Vol. 1. Holly Hill, FL: Southern Publishing Company.

Bullen, Adelaide K. 1966. Report on Skeletal Findings. In *Burtine Island, Citrus County, Florida*, edited by R. P. Bullen. 11–12. Vol. 14, *Contributions of the Florida State Museum, Social Sciences*. Gainesville, FL: University of Florida.

Bullen, Adelaide K. 1967a. Human Variation in Industry. Hayka, Moscow. 1964 Conference. *VII Congres International des Sciences Anthropologiques et Ethnologiques Moscou* 2:66–74.

Bullen, Adelaide K. 1967b. Individual Variation in Nervous and Mental Fatigue on the Job. In *Cinquieme Congres Caraioe Pour la Sante Mentale, 1965*. 104–111. Fort-de-France, Martinique.

Bullen, Adelaide K. 1968. Field Comments on the Skull Excavated in 1967 at Caliviny Island, Grenada, W.I. In *Proceedings of the Second International Congress for the Study of Pre-Columbian Cultures of the Lesser Antilles, St. Ann's Garrison, Barbados, 1967*. 44–46. Barbados: The Barbados Museum and Historical Society.

Bullen, Adelaide K. 1970. Case Study of an Amerindian Burial with Grave Goods from Grand Anse, St. Lucia. In *Proceedings of the Third International Congress for the Study of Pre-Columbian Cultures of the Lesser Antilles, St. George's, Grenada, 1969*. 45–60. Grenada: Grenada National Museum.

Bullen, Adelaide K. 1972. Paleoepidemiology and Distribution of Prehistoric Treponemiasis (Syphilis) in Florida. *Florida Anthropologist* 25 (4):133–174.

Bullen, Adelaide K. 1973. Some Human Skeletal Remains from Amelia Island, Florida. In *Excavations on Amelia Island in Northeast Florida*, edited by E. T. Hemmings and K. A. Deagan. 72–87. Vol. 18, *Contributions of the Florida State Museum: Anthropology and History*. Gainesville: University of Florida.

Bullen, Adelaide K. 1975. *Jim Tall and Count Small*. Gainesville: Kendall Books.

Bullen, Adelaide K. 1978. Bibliography of Ripley P. Bullen. In *Proceedings of the Seventh International Congress for the Study of Pre-Columbian Cultures of the Lesser Antilles, Caracas, Venezuala, 1977*. 11–25. Montreal: Centre de Recherches Caraïbes.

Bullen, Adelaide K., and Ripley P. Bullen. 1942. A Pueblo Cave at Tres Piedras, New Mexico. *American Antiquity* 8:57–64.

Bullen, Adelaide K., and Ripley P. Bullen. 1945. Black Lucy's Garden. *Bulletin of the Massachusetts Archaeological Society* 6 (2):17–28.

Bullen, Adelaide K., and Ripley P. Bullen. 1946. Two Burials at Tiverton, Rhode Island. *Bulletin of the Massachusetts Archaeological Society* 8:5–6.

Bullen, Adelaide K., and Ripley P. Bullen. 1950. The Johns Island Site, Hernando County, Florida. *American Antiquity* 16:23–45.

Bullen, Adelaide K., and Ripley P. Bullen. 1953. The Battery Point Site, Bayport, Hernando County, Florida. *Florida Anthropologist* 6:85–92.

Bullen, Adelaide K., and Ripley P. Bullen. 1954. Further Notes on the Battery Point Site, Bayport, Hernando County, Florida. *Florida Anthropologist* 7:103–108.

Bullen, Adelaide K., and Ripley P. Bullen. 1961a. The Summer Haven Site, St. Johns County, Florida. *Florida Anthropologist* 14 (1–2):69–73.

Bullen, Adelaide K., and Ripley P. Bullen. 1961b. Wash Island in Crystal River. *Florida Anthropologist* 14:69–73.

Bullen, Adelaide K., and Ripley P. Bullen. 1963. The Wash Island Site, Crystal River, Florida. *Florida Anthropologist* 16:81–92.

Bullen, Adelaide K., and Ripley P. Bullen. 1966a. Barbados, a Carib Centre. In *The Bajan and South Caribbean*. 20–22. Vol. 155: Bridgetown, Barbados.

Bullen, Adelaide K., and Ripley P. Bullen. 1966b. Three Indian Sites of St. Martin. *Nieuwe West-Indische Gids* 45 (2–3):137–144.

Bullen, Adelaide K., and Ripley P. Bullen. 1966c. The Early Peoples of the Bahamas. (Mimeographed). In *Report of the First Bahamas Conference on Archaeology. San Salvador*. 19. New Haven: Sponsored by the Center for Latin American Studies, University of Florida.

Bullen, Adelaide K., and Ripley P. Bullen. 1970. The Lavoutte Site, St. Lucia: A Carib Ceremonial Center. In *Proceedings of the Third International Congress for the Study of Pre-Columbian Cultures of the Lesser Antilles, St. George's, Grenada, 1969*. 61–86. Grenada: Grenada National Museum.

Bullen, Adelaide K., Ripley P. Bullen, and Eric M. Branford. 1973a. The Giraudy Site, Beane Field, St. Lucia. In *Proceedings of the Fourth International Congress for the Study of Pre-Columbian Cultures of the Lesser Antilles, Castries, St. Lucia, 1971*. 197–198. Castries, St. Lucia: St. Lucia Archaeological and Historical Society.

Bullen, Adelaide K., and Harriet L. Hardy. 1946. Analysis of Body Build Photographs of 175 College Women. *American Journal of Physical Anthropology* 4:37–68.

Bullen, Adelaide K. contributing editor. 1969. Physical Anthropology. In *Handbook of Latin American Studies*. 179–198. Vol. 31. Gainesville: Library of Congress, University of Florida Press.

Bullen, Adelaide K. 1971. Physical Anthropology. In *Handbook of Latin American Studies*. 166–183. Vol. 33. Gainesville: Library of Congress, University of Florida Press.

Bullen, Ripley P., and Adelaide K. Bullen. 1956. *Excavations on Cape Haze Peninsula, Florida*. Vol. 1, *Contributions of the Florida State Museum, Social Sciences*. Gainesville: University of Florida.

Bullen, Ripley P., and Adelaide K. Bullen. 1963. The Lemon Bay School Mound. *Florida Anthropologist* 16:51–56.

Bullen, Ripley P., and Adelaide K. Bullen. 1967. Archaeology of the Lesser Antilles. *The Bajan and South Caribbean* 169:16–18.

Bullen, Ripley P., and Adelaide K. Bullen. 1968a. Barbados Archaeology: 1966. In *Proceedings of the Second International Congress for the Study of Pre-Columbian Cultures of the Lesser Antilles, St. Ann's Garrison, Barbados, 1967*. 134–144. Barbados: Barbados Museum and Historical Society.

Bullen, Ripley P., and Adelaide K. Bullen. 1968b. Salvage Archaeology at Caliviny Island, Grenada: A Problem in Typology. In *Proceedings of the Second International Congress for the Study of Pre-Columbian Cultures of the Lesser Antilles, St. Ann's Garrison, Barbados, 67*. 31–43. Barbados: Barbados Museum and Historical Society.

Bullen, Ripley P., and Adelaide K. Bullen. 1968c. Two Stratigraphic Tests at the Grand Anse Site, St. Lucia. In *The Amerindians in St. Lucia*, edited by C. Jesse. 21–41: St. Lucia Archaeological and Historical Society.

Bullen, Ripley P., and Adelaide K. Bullen. 1972. *Archaeological Investigations on St. Vincent and the Grenadines, West Indies, The William L. Bryant Foundation, American Studies. Report No. 8*. Orlando.

Bullen, Ripley P., and Adelaide K. Bullen. 1973a. Settlement Pattern and Environment in Pre-Columbian Eastern Dominican Republic. *Boletin del Museo del Hombre Dominicano* 3:315–324.

Bullen, Ripley P., and Adelaide K. Bullen. 1973b. Stratigraphic Tests at Two Sites on Guadeloupe. In *Proceedings of the Fourth International Congress for the Study of Pre-Columbian Cultures of the Lesser Antilles, Castries, St. Lucia, 1971*. 192–196. Castries, St. Lucia: St. Lucia Archaeological and Historical Society.

Bullen, Ripley P., and Adelaide K. Bullen. 1974a. Further Notes on the West Bay Site. *Florida Anthropologist* 27:119.

Bullen, Ripley P., and Adelaide K. Bullen. 1974b. Inferences from Cultural Diffusion to Tower Hill, Jamaica, and Cupernoy, St. Martin. In *Proceedings of the Fifth International Congress for the Study of Pre-Columbian Cultures of the Lesser Antilles, Antigua, 1973*. 48–60. Antigua: The Antigua Archaeological Society.

Bullen, Ripley P., and Adelaide K. Bullen. 1974c. Stone Mortars in Florida. *Florida Anthropologist* 27:169–170.

Bullen, Ripley P., and Adelaide K. Bullen. 1974d. Tests at Hacienda Grande, Puerto Rico. *Fundacion Arqueologica, Antropologicae y Historia de Puerto Rico, Boletin Informativo (Edicion Especial)* 1:1–6.

Bullen, Ripley P., and Adelaide K. Bullen. 1976a. Culture Areas and Climaxes in Antillean Prehistory. In *Proceedings of the Sixth International Congress for the Study of Pre-Columbian Cultures of the Lesser Antilles, Pointe Á Pitre, Guadeloupe, 1975*. 1–10. Pointe á Pitre, Guadeloupe: Société d'Histoire de la Guadeloupe.

Bullen, Ripley P., and Adelaide K. Bullen. 1976b. The Palmer Site. *Florida Anthropological Society Publication* 8.

Bullen, Ripley P., and Adelaide K. Bullen. 1976c. Three Stratigraphic Tests along the Eastern Shore of Trinidad. In *Proceedings of the Sixth International Congress for the Study of Pre-Columbian Cultures of the Lesser Antilles, Pointe Á Pitre, Guadeloupe, 1975*. 28–34. Pointe á Pitre, Guadeloupe: Société d'Histoire de la Guadeloupe.

Bullen, Ripley P., Adelaide K. Bullen, and W. J. Bryant. 1967. *Archaeological Investigations at the Ross Hammock Site, Florida, The William L. Bryant Foundation, American Studies. Report No. 7*. Orlando.

Bullen, Ripley P., Adelaide K. Bullen, and C. J. Clausen. 1968. The Cato Site near Sebastian Inlet, Florida. *Florida Anthropologist* 21:14–16.

Bullen, Ripley P., Adelaide K. Bullen, and I. A. E. Kirby. 1973b. Dating the Troumassee Decorated Cylinder: A Horizon Style. In *Proceedings of the Fourth International Congress for the Study of Pre-Columbian Cultures of the Lesser Antilles, Castries, St. Lucia, 1971*. 197–198. Castries, St. Lucia: St. Lucia Archaeological and Historical Society.

Bullington, J. 1991. Deciduous Dental Microwear of Prehistoric Juveniles from the Lower Illinois River Valley. *American Journal of Physical Anthropology* 84:59–73.

Bullington, J. 1991. Dental Microwear of Prehistoric Juveniles from the Lower Illinois River Valley. *American Journal of Physical Anthropology* 84:59–73.

Bullock, Peter Y. 1991. Reappraisal of Anasazi Cannibalism. *The Kiva* 57 (1):5–16.

Burke, J. 1998. Connections: Turkish Delight. *Scientific American* 278:105.

Burley, David V. 1994. A Never Ending Story: Historical Developments in Canadian Archaeology and the Quest for Federal Heritage Legislation. *Journal Canadien d'Archéologie* 18:77–98.

Burley, David V. 1995. Contexts of Meaning: Beer Bottles and Can in Contemporary Burial Practices in the Polynesian Kingdom of Tonga. *Historical Archaeology* 29 (1):75–83.

Butler, B. H. 1971. The People of Casas Grandes: Cranial and Dental Morphology through Time. Ph.D., Southern Methodist University.

Buxton, L. H. D. 1920. The Teeth and Jaws of Savage Man. *Transactions of the British Society for the Study of Orthodontics* 1916–1920:79–88.

Cadien, J. D., E. F. Harris, W. P. Jones, and L. J. Mandarino. 1974. Biological Lineages, Skeletal Populations, and Microevolution. *Yearbook of Physical Anthropology* 18:194–201.

Caffel, A. C., Charlotte A. Roberts, R. C. Janaway, and A. S. Wilson. 2001. Pressures on Osteological Collections: The Importance of Damage Limitation. In *Human Remains: Conservation, Retrieval and Analysis. Proceedings of a Conference Held in Williamsburg, Va, Nov. 7–11th 1999*, edited by E. Williams. 187–197, *British Archaeological Reports. International Series 934*. Oxford: Archaeopress.

Campbell, T. D. 1925. *Dentition and Palate of the Australian Aboriginal*. Adelaide: Hassell Press.

Cannon, Aubrey. 1989. The Historical Dimension in Mortuary Expressions of Status and Sentiment. *Current Anthropology* 30 (4):437–458.

Carbonell, V. 1966. The Paleodontal Pathology of Ancient Mesopotamians. *Journal of Dental Research* 45:413.

Cárdenas-Arroyo, F., and Conrado Rodríguez-Martín, eds. 2001. *Studies on Ancient Mummies and Burial Archaeology: Proceedings of the II World Congress on Mummy Studies, 1995, Cartagena de Indias, Columbia*. Cartagena, Colombia: Universidad De Los Andes, Centro De Estudios En Bioanthropologia.

Carli-Thiele, P. 1996. *Spuren von Mangelerkrankungen an Steinzeitlichen Kinderskeleten*. Göttingen: Verlag Erich Goltze.

Carr, H. G. 1960. Some Dental Characteristics of the Middle Minoans. *Man* 60:157.

Carr, Lucien. 1878. Observations on the Crania from the Stone Graves in Tennessee. *Reports of the Peabody Museum of American Archaeology and Ethnology* 2 (Eleventh Annual Report of the Trustees of the Peabody Museum of Armerican Archaeology and Ethnology):361–384.

Carr, Lucien. 1879. *Observations on the Crania of St. Barbara Island*. Washington, DC: Government Printing Office.

Carr, Lucien. 1880. Measurements of Crania from California. *Reports of the Peabody Museum of American Archaeology and Ethnology* 2 (Twelfth Annual Report of the Trustees of the Peabody Museum of Armerican Archaeology and Ethnology):497–505.

Cartmill, M., and K. Brown. 2003. Surveying the Race Concept: A Reply to Liberman, Kirk, and Littlefield. *American Anthropologist* 105 (1):114–115.

Carvalho, Eliana T., and L. M. C. Machado. 1975. Pesquisas Arqueológicas Na Região do Médio São Francisco Mineiro. *Boletim Do Instituto De Arqueologia Brasileira, Rio de Janeiro* 7:20–52.

Carvalho, Eliana T., L. M. C. Machado, and A. F. Costa. 1973. Notas Prévias Sobre Pesquisas Arqueológicas Na Região Norte Mineira. *Anais do I Congresso Brasileiro de Arqueologia*.

Case, P. 1988. Ruth Sawtell Wallis. In *Women Anthropologists: A Biographical Dictionary*, edited by U. Gacs, A. Khan, J. McIntyre and R. Weinburg. 361–366. Westport, CT: Greenwood Press.

Caselitz, Peter. 1998. Caries: Ancient Plague of Humankind. In *Dental Anthropology: Fundamentals, Limits, and Prospects*, edited by K. W. Alt, F. W. Rèosing and M. Teschler-Nicola. 203–226. Wein and New York: Springer.

Cassidy, Claire Monod. 1972. A Comparison of Nutrition and Health in Pre-Agricultural and Agricultural Amerindian Skeletal Populations. Ph.D. Dissertation, Department of Anthropology, University of Wisconsin.

Cavalli-Sforza, L. Luca. 1997. Genes, Peoples, and Languages. *Proceedings of the National Academy of Sciences* 94 (15):7719–7724.

Cave, A. J. E. 1940. Surgical Aspects of the Crichel Down Trepanation. *Proceedings of the Prehistoric Society* 6:131.

Cave, A. J. E. 1941. The Earliest English Example of Cervical Rib. *British Journal of Surgery* 29 (113):47–51.

Chamberlain, Andrew T. 2000. Problems and Prospects in Paleodemography. In *Human Osteology in Archaeology and Forensic Science*, edited by M. Cox and S. Mays. 101–115. London: Greenwich Medical Media.

Chapman, Robert, Ian Kinnes, and Klavs Randsborg. 1981. *The Archaeology of Death, New Directions in Archaeology*. Cambridge and New York: Cambridge University Press.

Charles, Douglas K. 1992. Woodland Demographic and Social Dynamics in the American Midwest: Analysis of a Burial Mound Survey. *World Archaeology* 24 (2):175–197.

Charles, Douglas K., and Jane E. Buikstra. 1983. Archaic Mortuary Sites in the Central Mississippi Drainage; Distribution, Structure, and Behavioral Implications. In *The Midwest Archaic: Paleoecology, Stratigraphy and Archaeology*, edited by J. L. Phillips and J. A. Brown, *New World Archaeological Record*. New York: Academic Press.

Chatters, James C. 1997. Encounter with an Ancestor. *Anthropology Newsletter* 38 (1):9, 10.

Chatters, James C. 2000. The Recovery and First Analysis of an Early Holocene Human Skeleton from Kennewick, Washington. *American Antiquity* 65 (2):291–316.

Chatters, James C. 2001. *Ancient Encounters: Kennewick Man and the First Americans*. New York: Simon & Schuster.

Chen, Xingbin, and Hao Chen. 1998. The Influence of Alveolar Structures on the Torsional Strain Field in a Gorilla Corporeal Cross-Section. *Journal of Human Evolution* 35:611–633.

Chesson, Meredith S. 1999. Libraries of the Dead: Early Bronze Age Charnel Houses and Social Identity at Urban Bab Edh-Dhra', Jordan. *Journal of Anthropological Archaeology* 18:137–164.

Chevalier, Alexandre. 1997. Bioarchéologie. Retrieved Dec. 28, 2005, from ILLAPA Web Site: http://biant.unige.ch/~chevalie/bioarcheo/bioarcheo.html.

Christensen, A. F. 1998. Colonization and Microevolution in Formative Oaxaca, Mexico. *World Archaeology* 30:262–285.

Christopherson, K. M., and P. O. Pedersen. 1939. Investigation into Dental Conditions in the Neolithic Period and in the Bronze Age in Denmark. *Dental Record* 59:575–585.

Chung, Sue Fawn, and Priscilla Wegers, eds. 2005. *Chinese American Death Rituals*. Lanham: AltaMira Press.

Church of England and English Heritage. 2005. *Guidance on Best Practice for Treatment of Human Remains Excavated from Christian Burials Grounds in England*. Swindon: English Heritage and Church of England.

Churchill, Steven E. 1996. Particulate Versus Integrated Evolution in the Upper Body in Late Pleistocene Humans: A Test of Two Models. *American Journal of Physical Anthropology* 100:559–583.

Churchill, Steven E., and Vincenzo Formicola. 1997. A Case of Marked Bilateral Asymmetry in the Upper Limbs of an Upper Palaeolithic Male from Barma Grande (Liguria), Italy. *International Journal of Osteoarchaeology* 7:18–38.

Churchill, Steven E., and Alan G. Morris. 1998. Muscle Marking Morphology and Labour Intensity in Prehistoric Khoisan Foragers. *International Journal of Osteoarchaeology* 8:390–411.

Churchill, Steven E., Osbjorn M. Pearson, Frederic E. Grine, Erik Trinkaus, and Trent W. Holliday. 1996. Morphological Affinities of the Proximal Ulna from Klasies River Mouth Main Site: Archaic or Modern? *Journal of Human Evolution* 31 (3):213–237.

Claassen, Cheryl, and Rosemary A. Joyce. 1997. *Women in Prehistory: North America and Mesoamerica, Regendering the Past.* Philadelphia: University of Pennsylvania Press.

Clark, George A. 1988. New Method for Assessing Changes in Growth and Sexual Dimorphism in Paleoepidemiology. *American Journal of Physical Anthropology* 77:105–116.

Clark, Geoffrey A. 1996. NAGPRA and the Demon-Haunted World. *SAA Bulletin* 14 (5):3.

Clark, George A., N. R. Hall, George J. Armelagos, G. A. Borkan, M. M. Panjabi, and F. T. Wetzel. 1986. Poor Growth Prior to Early Childhood: Decreased Health and Lifespan in the Adult. *American Journal of Physical Anthropology* 77:145–160.

Clark, J. G. D. 1972. *Starr Carr: A Case Study in Bioarchaeology, Modular Publications 10.* London: Addison-Wesley.

Clark, S. K. 1978. Markers of Metabolic Insult: The Association of Radiopaque Transverse Lines, Enamel Hypoplasias and Enamel Histopathologies in a Prehistoric Skeletal Sample. Ph.D. Dissertation, Department of Anthropology, University of Colorado.

Clark, W. G. 1981. Restricted Least-Squares Estimates of Age Composition from Length Composition. *Canadian Journal of Fisheries and Aquatic Science* 38:297–307.

Clement, A. J. 1963. Variations in the Microstructure and Biochemistry of Human Teeth. In *Dental Anthropology*, edited by D. R. Brothwell. 245–269. Vol. V. New York: The MacMillan Company.

Coale, A. J., and P. Demeny. 1966. *Regional Model Life Tables and Stable Populations.* Princeton: Princeton University Press.

Cockburn, Aidan, and Eve Cockburn, eds. 1980. *Mummies, Disease, and Ancient Cultures.* New York: Cambridge University Press.

Cockburn, Aidan, Eve Cockburn, and Theodore A. Reyman, eds. 1998. *Mummies, Disease, and Ancient Cultures.* 2nd ed. New York: Cambridge University Press.

Cockburn, Eve. 1994. The Paleopathology Association: Amortui Viventes Docent. *Eres (Arqueología)* 5 (1):135–147.

Cockburn, Eve. 1997. An Introduction to Paleopathology. In *Human Paleopathology and Related Subjects. An International Bibliography*, edited by D. V. Elerick, R. A. Tyson and (compilers). xv–xvi. San Diego: San Diego Museum of Man.

Cockburn, Eve, and Theodore A. Reyman. 1982. Paleopathology (Editorial). *Journal of the American Medical Association* 248 (4):472–473.

Cohen, Mark Nathan. 1989. *Health and the Rise of Civilization.* New Haven: Yale University Press.

Cohen, Mark Nathan. 1994. The Osteological Paradox Reconsidered. *Current Anthropology* 35 (5):629–637.

Cohen, Mark Nathan. 1998. The Emergence of Health and Social Inequalities in the Archaeological Record. In *Human Biology and Social Inequality*, edited by S. S. Strickland and P. S. Shetty. 249–271. Cambridge: Cambridge University Press.

Cohen, Mark Nathan, and George J. Armelagos, eds. 1984. *Paleopathology at the Origins of Agriculture.* New York: Academic Press.

Cole, Fay-Cooper, and Thorne Deuel. 1937. *Rediscovering Illinois: Archaeological Explorations in and around Fulton County*. Chicago: University of Chicago Press.

Cole, T. M. 1996. The Use of Matrix Permutation Tests for Evaluating Competing Hypotheses of Modern Human Origins. *Journal of Human Evolution* 31:477–484.

Collins, J. M. 1979. Ruth Sawtell Wallis 1895–1978. *American Anthropologist* 81:85–87.

Colquitt, W. T., and C. H. Webb. 1940. Dental Diseases in an Aboriginal Group. *Tri-State Medical Journal* 12 (4):2414–2417.

Comas, Juan. 1942a. Contribution à L'étude du Metopisme. *Archives Suisses d'Anthropologie Générale* 10:273–412.

Comas, Juan. 1942b. El Hueso Interparietal en Los en Los Cráneos Mexicanos. *Anales de la Escuela Nacional de Ciencias Biológicas (México)* 2(4):469–490.

Comas, Juan. 1943. El Metopismo: Sus Causas y Frecuencia en Los en Los Cráneos Mexicanos. *Anales del Instituto del Etnografía Americana, Universidad Nacional de Cuyo* 4:121.

Comas, Juan. 1960. *Manual of Physical Anthropology*. Springfield, IL: Charles C Thomas.

Comas, Juan. 1965. Crânes Mexicaíns Scaphocéphales. *L'Anthropologie* 69(3,4): 273–301.

Comas, Juan. 1968. *Dos Microcéfalos: Aztecas*: Instituto de Investigaciones Históricas, UNAM, Cuadernos.

Comas, Juan. 1974. *Antropologia de Los Pueblos Iberoamericanos*. Barcelona: Editorial Labor.

Comas, Juan. 1976. *El Metopismo: Sus Causas y Frecuencia en Los en Los Cráneos Mexicanos*.

Committee on Basic Needs in American Archaeology. 1939. Recommendations, June 24–25. National Research Council. Frank M. Seltzer Collection, Series 9, Box 30, National Anthropological Archives, Smithsonian Institution, Washington, DC.

Condon, K. W. 1981. Correspondence of Developmental Enamel Defects between the Mandibular Canine and the First Premolar. M.A. Thesis, Department of Anthropology, University of Arkansas.

Conkey, Margaret W., and Janet D. Spector. 1984. Archaeology and the Study of Gender. In *Advances in Archaeological Method and Theory*, edited by M. B. Schiffer. 1–38. Vol. 7. London and New York: Academic Press.

Conover, W. J., and R. L. Iman. 1980. Rank Transformations as a Bridge between Parametric and Nonparametric Statistics. *American Statistician* 35:124–129.

Cook, Della Collins. 1976. Pathologic States and Disease Process in Three Illinois Woodland Populations: An Epidemiologic Approach. Ph.D. Dissertation, Department of Anthropology, University of Chicago.

Cook, Della Collins. 1979. Subsistence Base and Health in Prehistoric Illinois Valley: Evidence from the Human Skeleton. *Medical Anthropology* 3 (1):109–124.

Cook, Della Collins. 1980a. Hereditary Enamel Hypoplasia in a Prehistoric Indian Child. *Journal of Dental Research* 59:1522.

Cook, Della Collins. 1980b. Paget's Disease and Treponematosis in Prehistoric Midwestern Indians: The Case for Misdiagnosis. *Ossa* 7:41–63.

Cook, Della Collins. 1981. Mortality, Age-Structure and Status in the Interpretation of Stress Indicators in Prehistoric Skeletons: A Dental Example from the Lower

Illinois Valley. In *The Archaeology of Death*, edited by R. Chapman, I. Kinnes and K. Randsborg. 133–144. London: Cambridge University Press.

Cook, Della Collins. 1984. Subsistence and Health in the Lower Illinois Valley: Osteological Evidence. In *Paleopathology at the Origins of Agriculture*, edited by M. N. Cohen and G. J. Armelagos. 237–269. Orlando, FL: Academic Press.

Cook, Della C., and Jane E. Buikstra. 1979. Health and Differential Survival in Pre-Historic Populations: Prenatal Dental Defects. *American Journal of Physical Anthropology* 51:649–664.

Cook, Della Collins, Jane E. Buikstra, Carol J. DeRousseau, and D. Carl Johanson. 1983. Vertebral Pathology in the Afar Australopithecines. *American Journal of Physical Anthropology* 60:83–101.

Cook, S. F., Sheilagh T. Brooks, and H. Ezra-Cohn. 1961. The Process of Fossilization. *Southwestern Journal of Anthropology* 17:335–364.

Cook, S. F., Sheilagh T. Brooks, and H. Ezra-Cohn. 1962. Histological Studies on Fossil Bone. *Journal of Paleontology* 36:483–495.

Coon, Carleton S. 1965. *The Living Races of Man*. New York: Knopf.

Corbett, M. E., and W. J. Moore. 1976. Distribution of Caries in Ancient British Populations. IV. The 19th Century. *Caries Research* 10:401–414.

Cordell, L. S., ed. 1980. *Tijeras Canyon: Analyses of the Past*. Albuquerque, NM: University of New Mexico Press.

Corruccini, R. S. 1972. The Biological Relationship of Some Prehistoric and Historic Pueblo Populations. *American Journal of Physical Anthropology* 37:373–388.

Corruccini, R. S. 1998. On Hawikku Cemetary Kin Groups. *American Antiquity* 63 (2):161–163.

Corruccini, R. S., E. M. Brandon, and J. S. Handler. 1989. Inferring Fertility from Relative Mortality in Historically Controlled Cemetery Remains from Barbados. *American Antiquity* 54:609–614.

Covert, H. H., and R. F. Kay. 1981. Dental Microwear and Diet: Implications for Determining the Feeding Behaviors of Extinct Primates. *American Journal of Physical Anthropology* 55:331–336.

Cox, M. 1996. *Life and Death at Spitalfields 1700–1850*. York: Council for British Archaeology.

Cox, M., and Simon Mays, eds. 2000. *Human Osteology in Archaeology and Forensic Science*. London: Greenwich Medical Media.

Crain, Jay B. 1971. *Human Palaeopathology: A Bibliographic List*. Sacramento Anthropological Society Paper No. 12.

Cran, J. A. 1959. The Relationship of Diet to Dental Caries. *Australian Dental Journal* 4:182–190.

Cran, J. A. 1960. The Histological Structure of the Teeth of Central Australian Aborigines. *Australian Dental Journal* 5:100–104.

Crawford, Michael H. 1998. *The Origins of Native Americans: Evidence from Anthropological Genetics*. New York: Cambridge University Press.

Crawford, Michael H., and P. L. Workman. 1973. *Methods and Theories of Anthropological Genetics*. Albuquerque, NM: University of New Mexico.

Crooks, Deborah L. 1995. American Children at Risk: Poverty and Its Consequences for Children's Health, Growth, and School Achievement. *Yearbook of Physical Anthropology* 38:57–86.

Crosby, Alfred W. 1997. *The Measure of Reality: Quantification and Western Society, 1250-1600*. Cambridge: Cambridge University Press.

Crow, J. F., and C. Denniston. 1974. *Genetic Distance*. New York: Plenum Press.

Crown, Patricia L., and Suzanne K. Fish. 1996. Gender and Status in the Hohokam Pre-Classic to Classic Transition. *American Anthropologist* 98 (4):803–817.

Crubézy, Eric, and Erik Trinkhaus. 1992. Shanidar 1: A Case of Hyperostotic Disease (Dish) in the Middle Paleolithic. *American Journal of Physical Anthropology* 89:411–420.

Cruikshank, Julie. 1981. Legend and Landscape: Convergence of Oral and Scientific Traditions in the Yukon Territory. *Arctic Anthropology* XVIII (2):67–93.

Cucina, Andrea, and Vera Tiesler. 2003. Dental Caries and Antemortem Tooth Loss in the Northern Peten Area, Mexico: A Biocultural Perspective on Social Status Differences among the Classic Maya. *American Journal of Physical Anthropology* 122:1–10.

Cummins, Harold, and Charles Midlo. 1943. *Finger Prints, Palms and Soles: An Introduction to Dermatoglyphics*. New York: Blakiston.

Cunha, Ernesto de Melo Salles, and Marília Carvalho de Mello e Alvim. 1969. Contribuição Para O Conhecimento da Morfologia Das Populações Indígenas da Guanabara: Notas Sobre a População do Sítio Arqueológico Cabeça do Índio. *Ciência e Cultura (São Paulo)* 21 (2):318–319.

Cunha, Ernesto de Melo Salles, and Marília Carvalho de Mello e Alvim. 1971. Contribuição Para O Conhecimento Das Populações Indígenas da Guanabara: Notas Sobre a População Dos Sítio Arqueológico Cabeça do Índio. In *O Homem Antigo Na América*. 21–24. São Paulo: Universidade de São Paulo.

Currey, John D. 1984. *The Mechanical Adaptations of Bones*. Princeton: Princeton University Press.

Currey, John D., and R. McNeill Alexander. 1985. The Thickness of the Walls of Tubular Bones. *Journal of Zoology, London* 206:453–468.

Cushing, Frank Hamilton. 1890. Preliminary Notes on the Origin, Working Hypothesis, and Primary Researches of the Hemenway Southwestern Archaeological Expedition. In *Congres International des Americanistes. Compte-Rendu de la Septieme Session, Berlin, 1888*. 151–194. Berlin.

Cuvier, G. 1836. *Le Règne Animal: Races Humaines*. Paris: Victor Masson.

Cybulski, Jerome S. 1976. Scientific Aspects of Archaeology in Canada: A Physical Anthropologist's View. In *Symposium on New Perspectives in Canadian Archaeology, 22–23 October 1976, Theatre Auditorium, Royal Ontario Museum, Toronto*, edited by A. G. McKay. 177–184. Ottawa: The Royal Society of Canada.

Cybulski, Jerome S. 1978. Modified Human Bones and Skulls from Prince Rupert Harbour, British Columbia. *Canadian Journal of Archaeology* 2:15–32.

Cybulski, Jerome S. 1988. Brachydactyly, a Possible Inherited Anomaly at Prehistoric Prince Rupert Harbour. *American Journal of Physical Anthropology* 76:363–375.

Cybulski, Jerome S. 1992. *A Greenville Burial Ground: Human Remains and Mortuary Elements in British Columbia Coast Prehistory, Archaeological Survey of Canada Mercury Series Paper 146*. Hull: Canadian Museum of Civilization.

Cybulski, Jerome S. 2001. Human Biological Relationships for the Northern Northwest Coast. In *Perspectives on Northern Northwest Coast Prehistory*, edited by

J. S. Cybulski. 107–144, *Mercury Series Paper 160*. Hull: Canadian Museum of Civilization, Archaeological Survey of Canada.

Cybulski, Jerome S., H. Harry, and A. McMillan. 2004. The China Lake and Big Bar Projects: Community Based Research in Physical Anthropology and Mortuary Archaeology. Paper read at 32nd Annual Meeting of the Canadian Association for Physical Anthropology, Symposium on Repatriation, October 27–31, at London, Ontario.

Cybulski, Jerome S., N. S. Ossenberg, and W. D. Wade. 1979. Committee Report: Statement on the Excavation, Treatment, Analysis and Disposition of Human Skeletal Remains from Archaeological Sites in Canada. *Canadian Review of Physical Anthropology* 1:32–36.

Daegling, David J. 1989. Biomechanics of Cross-Sectional Size and Shape in the Hominoid Mandibular Corpus. *American Journal of Physical Anthropology* 80:91–106.

Daegling, David J., and Frederic E. Grine. 1991. Compact Bone Distribution and Biomechanics of Early Hominid Mandibles. *American Journal of Physical Anthropology* 86:321–339.

Daegling, David J., and Jennifer L. Hotzman. 2003. Functional Significance of Cortical Bone Distribution in Anthropoid Mandibles: An in Vitro Assessment of Bone Strain under Combined Loads. *American Journal of Physical Anthropology* 122:38–50.

Daegling, David J., and William L. Hylander. 1998. Biomechanics of Torsion in the Human Mandible. *American Journal of Physical Anthropolology* 105:73–87.

Daegling, David J., and William L. Hylander. 2000. Experimental Observation, Theoretical Models, and Biomechanical Inference in the Study of Mandibular Form. *American Journal of Physical Anthropology* 112:541–551.

Dahlberg, Albert A. 1945. The Changing Dentition of Man. *Journal of the American Dental Association* 32:676–690.

Dahlberg, Albert A. 1951. The Dentition of the American Indian. In *Papers on the Physical Anthropology of the American Indian*, edited by W. S. Laughlin. 138–176. New York: Viking Fund.

Dahlberg, Albert A. 1963. Analysis of American Indian Dentition. In *Dental Anthropology*, edited by D. R. Brothwell. 149–177. New York: Pergamon Press.

Dahlberg, Albert A., and W. Kinsey. 1963. Étude Microscopique de L'abrasion et de L'attrition Sur la Surface des Dents. *Bull Group Int Rech Sci Stomatol* 5:242–251.

Dalhberg, Albert A. 1960. The Dentition of the First Agriculturalists (Jarmo, Iraq). *American Journal of Physical Anthropology* 18:243–256.

Danforth, M. E. 1999. Nutrition and Politics in Prehistory. *Annual Review of Anthropology* 28:1–25.

Danforth, M. E., D. C. Cook, and S. G. Knick. 1994. The Human Remains from Carter Ranch Pueblo, Arizona: Health in Isolation. *American Antiquity* 59 (1):88–101.

Darling, J. A. 1998. Mass Inhumation and the Execution of Witches in the American Southwest. *American Anthropologist* 100 (3):732–752.

Darling, J. A. 1999. From Hobbes to Rousseau and Back Again. *Science* 285 (5427):537.

Davis, J. B., and J. Thurnam. 1865. *Crania Britannica. Delineation and Descriptions of the Skulls of the Aboriginal and Early Inhabitants of the British Isles*. 2 vols. London: Printed Privately.

Dawson, W. R. 1927. On Two Egyptian Mummies Preserved in the Museums of Edinburgh. *Proceedings of the Society of Antiquaries of Scotland* 61:290.

Day, M. C. 1944. Nutritional Deficiencies and Dental Caries in Northern India. *British Dental Journal* 76:115–123, 143–147.

Deetz, J. 1968. The Inference of Residence and Descent Rules from Archaeological Data. In *New Prespectives in Archaeology*, edited by L. R. Binford and S. R. Binford. 41–48. Chicago: Aldine.

DeGusta, David. 2003. Aubesier 11 Is Not Evidence of Neanderthal Conspecific Care. *Journal of Human Evolution* 45:91–94.

Deignan, Stella L. 1942. Letter Dated January 28 to Strong. Webb Papers, Box 4, Library Archives, University of Kentucky, Lexington.

De la Cour, L., and Rose Sheinin. 1990. The Ontario Medical College for Women 1883–1906: Lessons from Gender Separatism in Education. In *Despite the Odds: Essays on Canadian Women and Science*, edited by M. G. Ainley. 112–120. Véhicule Press: Montreal.

de Laguna, Frederica. 1956. *Chugach Prehistory: The Archaeology of Prince William Sound, Alaska*. Vol. 13, *Publications in Anthropology*. Seattle: University of Washington.

Deloria, Vine Jr. 1969. *Custer Died for Your Sins: An Indian Manifesto*. New York: Macmillan.

Deloria, Vine Jr. 1989. A Simple Question of Humanity: The Moral Dimensions of the Reburial Issue. *Native American Rights Fund Legal Review* 4.

Deloria, Vine Jr. 1995. *Red Earth, White Lies: Native Americans and the Myth of Scientific Fact*. New York: Scribner.

De Waal Malefijt, Annemarie. 1974. *Images of Man: A History of Anthropological Thought*. New York: Knopf.

Demes, Brigitte, Jack T. Jr. Stern, Michael R. Hausman, Susan G. Larson, Kenneth J. McLeod, and Clinton T. Rubin. 1998. Patterns of Strain in the Macaque Tibia During Functional Activity. *American Journal of Physical Anthropology* 116:257–265.

Dempster, A. P., N. M. Laird, and D. B. Rubin. 1977. Maximum Likelihood from Incomplete Data Via the EM Algorithm. *Journal of the Royal Statistical Society Series B* 39:1–38.

Denninger, Henri Stearns. 1931a. The Pathology of the Prehistoric American Indians of the Illinois River Valley. *Transactions of the Illinois State Academy of Science* 24:371–376.

Denninger, Henri Stearns. 1931b. Cervical Ribs: A Prehistoric Example. *American Journal of Physical Anthropology* 16:211–214.

Denninger, Henri Stearns. 1931c. Osteitis Fibrosa in a Skeleton of a Prehistoric American Indian. *Archives of Pathology* 11:939–944.

Denninger, Henri Stearns. 1933. Paleopathological Evidence of Paget's Disease. *Annals of Medical History* 5(73–81).

Denninger, Henri Stearns. 1935. Prehistoric Syphilitic Lesions (an Example from North America). *Southwestern Medical Journal* 19:202–204.

Denninger, Henri Stearns. 1938. Syphilis of the Pueblo Skull before 1350. *Archives of Pathology* 26:724–726.

Department of Culture Media and Sport. 2005. *Guidance for the Care of Human Remains in Museums (Also see www.Culture.Gov.Uk, Select: Cultural Property)* London: DCMS.

Department of the Interior, Office of the Secretary. 1995. 43 Cfr Part 10. Native American Graves Protection and Repatriation Act Regulations; Final Rule. *Federal Register* 60 (232):62133–62169.

Department of the Interior, Office of the Secretary. 2000. Recommendations Regarding the Disposition of Culturally Unidentifiable Native American Human Remains. *Federal Register* 65 (111):36462–36464.

Dettwyler, Katherine A. 1991. Does Paleopathology Provide Evidence for "Compassion"? *American Journal of Physical Anthropology* 84:375–384.

Dias Jr., O., E. T. Carvalho, and L. M. C. Machado. 1975. Fase Belvedere: Uma Fase Tupiguarani no Sul de Minas Gerais. *Boletim do Instituto de Arqueologia Brasileira, Rio De Janeiro* 7:5–17.

Dias Jr., O., E. T. Carvalho, and L. M. C. Machado. 1976a. Pesquisas Arqueológicas no Vale do São Francisco em Minas Gerais, Brasil. *Anais Do XIII Congresso Internacional de Americanistas, Paris.*

Dias Jr., O., E. T. Carvalho, and L. M. C. Machado. 1976b. A Arte Rupestre do Vale do Rio São Francisco, Mg, Brasil. *Anais Do XIII Congresso Internacional de Americanistas, Paris.*

Di Bacco, M., V. Ardito, and E. Pacciani. 1999. Age-at-Death Diagnosis and Age-at-Death Distribution Estimate: Two Different Problems with Many Aspects in Common. *International Journal of Anthropology* 14:161–169.

Dickel, David N., and Glen H. Doran. 1989. Severe Neural Tube Defect Syndrome from the Early Archaic of Florida. *American Journal of Physical Anthropology* 80:325–334.

Dickson, James H., Michael P. Richards, Richard J. Hebda, Petra J. Mudie, Owen Beattie, Susan Ramsay, Nancy J. Turner, Bruce J. Leighton, John M. Webster, Niki R. Hobischak, Gail S. Anderson, Peter M. Troffe, and Rebecca J. Wigen. 2004. Kwäday Dän Ts'ínchí, the First Ancient Body of a Man from a North American Glacier: Reconstructing His Last Days by Intestinal and Biomolecular Analyses. *The Holocene* 14 (4):481–486.

Dillenius, Juliane A. 1909. Observaciones Arqueolgicas Sobre la Alfarera Funeraria de "La Poma." *Revista de la Universidad de Buenos Aires* XI.

Dillenius, Juliane A. 1910. *El Hueso Parietal Bajo la Influencia de la Deformacin Fronto-Occipital. Contribucin Al Estudio Somtico de Los Antiguos Calchaques, Publicaciones de la Seccion Antropologa de la Faculdad de Filosofia y Letras, No. 7.* Buenos Aires: Imprenta y Casa Editora Juan Alsina.

Dillenius, Juliane A. 1912a. *La Verdadera Forma Del Craneo Calchaqu Deformado, Separata de Actas Del Xvii Congreso Internacional de Americanistas.* Buenos Aires: Ediciones de la Imprenta Coni Hermanos.

Dillenius, Juliane A. 1912b. Das Scheitelbein Unter Dem Einfluss der Frontao-Occipitalen Schdeldeformation: Ein Beitrag Zur Somatischen Anthropologie der Calchaqustmme. *Archiv für Anthropologie* XI:113–139.

Dillenius, Juliane A. 1913. *Craniometra Comparativa de Los Antiguos Habitantes de la Isla y Del Pucar de Tilcara (Province de Jujuy), Publicaciones de la Seccion*

Antropologa de la Facultad de Filosofia y Letras, No. 12. Buenos Aires: Imprenta y Casa Editora Juan Alsina.

Ditch, Larry, and Jerome C. Rose. 1972. A Multivariate Dental Sexing Technique. *American Journal of Physical Anthropology* 37:61–64.

Dixon, Roland B. 1923. *The Racial History of Man* New York: Charles Scribner's Sons.

Dobney, K., and T. O'Connor, eds. 2002. *Bones and the Man. Studies in Honour of Don Brothwell*. Oxford: Oxbow Books.

Dobson, Seth D., and Erik Trinkaus. 2002. Cross-Sectional Geometry and Morphology of the Mandibular Symphysis in Middle and Late Pleistocene Homo. *Journal of Human Evolution* 43 (1):67–87.

Dongoske, Kurt E., Mark Aldenderfer, and Karen Doehner, eds. 2000. *Working Together: Native Americans & Archaeologists*. Washington, DC: Society for American Archaeology.

Dongoske, Kurt E., Debra L. Martin, and T. J. Ferguson. 2000. Critique of the Claim of Cannabalism at Cowboy Wash. *American Antiquity* 65 (1):179–190.

Droessler, Judith B. 1975. Late Woodland Maples Mills Focus Populations in the Illinois Valley: Cultural Affinities and Biological Distance. *Proceedings of the Indiana Academy of Science* 85:66–73.

Droessler, Judith B. 1981. *Craniometry and Biological Distance: Biocultural Continuity and Change at the Late-Woodland-Mississippian Interface*. Vol. 1, *Center for American Archaeology Research Series*. Evanston: Center for American Archaeology at Northwestern University.

Dublin, L. I., and A. J. Lotka. 1936. *Length of Life: A Study of the Life Table*. New York: Ronald Press.

Duday, Henri. 1978. Archaeologie Funeraire et Anthropologie. *Cahiers d'Anthropologie* 1:55–101.

Duday, Henri, and Claude Masset, eds. 1987. *Anthropologie Physique et Archeologie: Méthodes D'etude des Sépultures, Actes du Dolloque de Toulouse 4, 5, et 6 Novembre 1982*. Paris: Centre National de la Recherche Scientifique.

Duffield, L. F., and L. M. Robbins. 1973. Manganese Deficiency: A Prehistoric Disease? Paper presented at the 38th Annual Meeting of the Society for American Archaeology, at San Francisco, California.

Dufour, Darna L. 1988. Alice Mossie Brues. In *Women Anthropologists: A Biographical Dictionary*, edited by U. Gacs, A. Khan, J. McIntyre and R. Weinburg. 23–28. Westport, Conneticut: Greenwood Press.

Durkheim, Emile. 1965 [1912]. *The Elementary Forms of the Religious Life*. Translated by J. W. Swain. New York: Free Press.

Dutour, Olivier. 1986. Enthesopathy (Lesions of Muscle Insertions) as Indicators of Activities of Neolithic Saharan Populations. *American Journal of Physical Anthropology* 71:221–224.

Dutour, Olivier. 1992. Activités Physiques et Squelette Humain: Le Difficile Passage de L'actuel Au Fossile. *Bulletin et Mémoires de la Société d'Anthropologie de Paris* 4:233–241.

Dutour, Olivier, György Pálfi, Jacques Berano, and Jean-Pierre Brun, eds. 1994. *L'origine de la Syphilis en Europe: Avant Ou Aprés 1493?* Paris; Toulon: Editions Errance; Centre Archéologique du Var.

Dye, David H. 1991. New Deal Archaeology in the Middle Tennessee Valley: 1934–1942. *Louisiana Archaeology* 18:23–60.

Echo-Hawk, Roger. 1994. Kara Katit Pakutu: Exploring the Origins of Native America in Anthropology and Oral History. M.A., Department of History, University of Colorado, Boulder.

Echo-Hawk, Roger. 1997. Forging a New Ancient History for Native America. In *Native Americans and Archaeologists. Stepping Stones to Common Ground.*, edited by N. Swidler, K. E. Dongoske, R. Anyon and A. S. Downer. 88–102. Walnut Creek, CA: AltaMira Press in Cooperation with the Society for American Archaeology.

Echo-Hawk, Roger. 2000. Exploring Ancient Worlds. In *Working Together: Native Americans and Archaeologists*, edited by K. E. Dongoske, M. Aldenderfer and K. Doehner. 3–7. Washington, DC: Society for American Archaeology.

Eggan, Frederic Russell. 1967. *The American Indian: Perspectives for the Study of Social Change, Lewis Henry Morgan Lectures, 1964*. London: Weidenfeld & Nicolson.

Egnatz, Dennis G. 1983. Analysis of Human Skeletal Materials from Mound C, Appendix A. In *Excavations at the Lake George Site, Yazoo County, Mississippi, 1958–1960*, edited by S. Williams and J. P. Brain. 421–441. Cambridge: Harvard University Press.

Ehrenberg, Margaret. 1989. Interpretation of Regional Variability in British and Irish Bronze Age Metalwork. In *Bronze Age Studies: Transactions of the British-Scandinavian Colloquium in Stockholm, May 10–11, 1985*. 77–88. Vol. 6, *Studies, the Museum of National Antiquities, Stockholm*. Stockholm: Statens Historiska Museum.

Elerick, Daniel V., and Rose A. Tyson. 1997. *Human Paleopathology and Related Subjects. An International Bibliography*. San Diego: San Diego Museum of Man.

Elkins, Aaron. 1959. The Skeletal Material from Ck-44, the Smullins Site. *Oklahoma Anthropological Society Bulletin* VII:55–61.

Elliott-Smith, G., and F. Wood-Jones. 1910. *The Archaeological Survey of Nubia 1907–8. Volume 2. Report on the Human Remains*. Cairo: National Printing Department.

El-Najjar, M. Y. 1974. People of Canyon de Chelly: A Study of Their Biology and Culture. Ph.D., Arizona State University.

El-Najjar, M. Y. 1976. Maize, Malaria and the Anemias in the Pre-Columbian New World. *Yearbook of Physical Anthropology* 20:329–337.

El-Najjar, M. Y. 1978. Southwestern Physical Anthropology: Do the Cultural and Biological Parameters Correspond? *American Journal of Physical Anthropology* 48:151–158.

El-Najjar, M. Y., D. J. Ryan, C. G. Turner, and B. Lozoff. 1976. The Etiology of Porotic Hyperostosis among the Prehistoric and Historic Anasazi Indians of Southwestern United States. *American Journal of Physical Anthropology* 44:477–488.

Elzay, Richard P., Marvin J. Allison, and Alejandro Pezzia. 1977. A Comparative Study on the Dental Health Status of Five Precolumbian Peruvian Cultures. *American Journal of Physical Anthropology* 46(1):135–139.

Endo, Banri, and Tasuku Kimura. 1970. Postcranial Skeleton of the Amud Man. In *The Amud Man and His Cave Site*, edited by H. Suzuki and F. Takai. 231–406. Tokyo: Academic Press of Japan.

Epperson, B. K. 1993. Spatial and Space-Time Correlations in Systems of Subpopulations with Genetic Drift and Migration. *Genetics* 133:711–727.

Ericksen, Mary Francis. 1962a. Undeformed Pre-Columbian Crania from the North Sierra of Peru. *American Journal of Physical Anthropology* 20:209–222.

Ericksen, Mary Francis. 1962b. Dos Craneos Provenientes de Un Cementerio Indigena en Pica. *Boletin Museo Arqueologico de La Serena* 12:19–23.

Ericksen, Mary Francis. 1962c. Restos Oseos Encontrados en La Totorita. *Boletin Museo Arqueologico de La Serena* 12:47–53.

Ericksen, Mary Francis. 1973. Age-Related Bone Remodeling in Three Aboriginal American Populations. Ph.D., George Washington University, Washington, DC.

Ericksen, Mary Francis. 1976a. Cortical Bone Loss with Age in Three Native American Populations. *American Journal of Physical Anthropology* 45:443–452.

Ericksen, Mary Francis. 1976b. Some Aspects of Aging in the Lumbar Spine. *American Journal of Physical Anthropology* 45:575–580.

Ericksen, Mary Francis. 1977/78. Restos Oseos Provenientes de Punta de Piedra (Prov. de Elqui, IV Region). *Boletin Museo Arqueologico de La Serena* 16:189–210.

Ericksen, Mary Francis. 1978a. Aging in the Lumbar Spine. II. L1 and L2. *American Journal of Physical Anthropology* 48:241–246.

Ericksen, Mary Francis. 1978b. Aging in the Lumbar Spine. III. L5. *American Journal of Physical Anthropology* 48:247–250.

Ericksen, Mary Francis. 1979. Age Changes in the Medullary Cavity of the Proximal Femur in American Blacks and Whites. *American Journal of Physical Anthropology* 51:563–570.

Ericksen, Mary Francis. 1980. Patterns of Microscopic Bone Remodeling in Three Aboriginal American Populations. In *Early Native Americans: Prehistoric Demography, Economy and Technology*, edited by D. L. Browman. 239–270. The Hague: Mouton.

Ericksen, Mary Francis. 1982a. Aging Changes in the Thickness of the Proximal Femoral Cortex. *American Journal of Physical Anthropology* 59:121–130.

Ericksen, Mary Francis. 1982b. How "Representative" Is the Terry Collection? Evidence from the Proximal Femur. *American Journal of Physical Anthropology* 59:345–350.

Ericksen, Mary Francis. 1991. Histologic Estimation If Age-at-Death Using the Anterior Cortex of the Femur. *American Journal of Physical Anthropology* 84:171–179.

Ericksen, Mary Francis. 1995. Problems in Histological Aging of Archaeological Skeleton. *Rivista di Antropologia* 73:51–63.

Ericksen, Mary Francis. 1997. Comparison of Two Methods of Estimating Age-at-Death in a Chilean Preceramic Population. *International Journal of Osteoarchaeology* 7:65–70.

Ericksen, Mary Francis, and A. I. Stix. 1991. Histologic Examination of Age of the First African Baptist Church Adults. *American Journal of Physical Anthropology* 85:47–252.

Esper, J. F. 1774. *Ausfürliche Nachrichten von Neuentdeckten Zoolithen Unbekannter Vierfüssiger Thiere*. Nüremberg: Erben.

European Anthropological Association. n.d. European Anthropological Association Home Page. Retrieved Dec. 28, 2005, from European Anthropological Association Web Site: http://www.vub.ac.be/gst.

Evans, H. T. E. 1997. Memorial of George E. Ericksen 1920–1996. *American Mineralogist* 82:1046–1048.

Ezzo, J. A. 1993. *Human Adaptation at Grasshopper Pueblo, Arizona: Social and Ecological Perspectives, Archaeological Series 4*. Ann Arbor: International Monographs in Prehistory.

Faerman, M., R. Jankauskas, A. Gorski, H. Bercovier, and C. L. Greenblatt. 1999. Detecting *Mycobacterium Tuberculosis* DNA in Medieval Skeletal Remains from Lithuania. Tb Foundation. In *Tuberculosis: Past and Present*, edited by G. Pálfi, O. Dutour, J. Deák and I. Hutás. 371–376. Budapest: Golden Book Publisher Ltd./Szeged: Tuberculosis Foundation.

Falin, L. I. 1961. Histological and Histochemical Studies of Human Teeth of the Bronze and Stone Ages. *Archives of Oral Biology* 5:5–13.

Farber, Paul Lawrence. 1982. *Discovering Birds: The Emergence of Ornithology as a Scientific Discipline, 1760–1850*. Hardcover ed. Dordrecht and Boston: D. Reidel Publishing Company.

Fee, E. 1979. Nineteenth Century Craniology: The Study of the Female Skull. *Bulletin of the History of Medicine* 53:415–433.

Fenn, Elizabeth A. 2001. *Pox Americana: The Great Smallpox Epidemic of 1775–82*. New York: Hill and Wang.

Ferguson, T. J. 1996. Native Americans and the Practice of Archaeology. *Annual Review of Anthropology* 25:63–79.

Ferguson, T. J. 1997. The Changing Role of Native Americans in the Archaeological Study of the Past: Recent History and Future Prospects. Paper read at 62nd Annual Meeting of the Society for American Archaeology, April 2, at Nashville, Tennessee.

Ferguson, T. J., Roger Anyon, and Edmund Ladd. 1996. Repatriation at the Pueblo of Zuni: Diverse Solutions to Complex Problems. *American Indian Quarterly* 20 [2, Special Issue: Repatriation: An Interdisciplinary Dialogue (Spring 1996)]:251–273.

Ferguson, T. J., Kurt Dongoske, Leigh Jenkins, Eric Polingyouma, and Mike Yeatts. 1993. Working Together: The Roles of Archaeology and Ethnohistory in Hopi Cultural Preservation. *Cultural Resource Management* 16 (Special Issue):27–37.

Ferguson, T. J., Kurt E. Dongoske, Mike Yeatts, and Leigh J. Kuwanwisiwma. 2000. Hopi Oral History and Archaeology. In *Working Together: Native Americans & Archaeologists*, edited by K. E. Dongoske, M. Aldenderfer and K. Doehner. 45–60. Washington, DC: Society for American Archaeology.

Ferreira, Luis Fernando, Adauto Araújo, and Ulisses Confalonieri. 1980. The Finding of Eggs and Larvae of Parasitic Helminths in Archaeological Material from Unai, Minais Gerais, Brazil. *Transactions of the Royal Society of Tropical Medicine and Hygiene* 74(6):798–800.

Ferreira, Luis Fernando, Adauto Araújo, Ulisses. Confalonieri, and L. M. C. Machado. 1982. Estudo Prévio Das Práticas Funerárias e O Encontro de Parasitos Humanos Na Gruta do Gentio II, Unaí. *Arquivos Do Museu De História Natural, Ufmg. Belo Horizonte, Mg* 6–7:207–213.

Ferreira, Luis Fernando, Adauto Araújo, Ulisses Confalonieri, and Lautaro Nuñez. 1984. The Finding of Eggs of Diphyllobothrium in Human Coprolites (4,100–1,950 B.C.) from Northern Chile. *Memorias de Instituto Oswaldo Cruz* 79(2):175–180.

Ferreira, Luis Fernando, Adauto Araújo, Ulisses Confalonieri, and L. M. C. Machado. 1984. Estudo Prévio Das Práticas Funerárias e O Encontro de Parasitos Humanos Na Gruta do Gentio II, Unaí. *Boletim Série Ensaios, Instituto De Arqueologia Brasileira, Rio De Janeiro* 2:1–32.

Ferris, Neal. 2003. Between Colonial and Indigenous Archaeologies: Legal and Extra-Legal Ownership of the Archaeological Past in North America. *Canadian Journal of Archaeology* 27 (2):154–190.

Fforde, C. 1990. The Royal College of Surgeons of England: A Brief History of Its Collections and a Catalogue of Some Current Holdings. *World Archaeological Bulletin* 6:22–30.

Fiedel, S. J. 1999. Older Than We Thought: Implications of Corrected Dates for Paleoindians. *American Antiquity* 64:95–116.

Fienberg, S. E. 1977. *The Analysis of Cross-Classified Categorical Data*. Cambridge: MIT Press.

Fitzgerald, Charles M., and Jerome C. Rose. 2000. Reading between the Lines: Dental Development and Subadult Age Assessment Using the Microstructural Growth Markers of Teeth. In *Biological Anthropology of the Human Skeleton*, edited by M. A. Katzenberg and S. R. Saunders. 163–186. New York: Wiley-Liss.

Foley, R. 1990. The Duckworth Osteological Collection at the University of Cambridge. *World Archaeological Bulletin* 6:53–62.

Folmsbee, S. J., and Madeline Kneberg Lewis editors. 1965. Journals of the Juan Pardo Expeditions, 1566–1567, translated by Gerald W. Wade. *The East Tennessee Historical Society's Publications* 37:106–121.

Ford, James A. 1963. Hopewell Culture Burial Mounds near Helena, Arkansas. *Anthropological Papers of the American Museum of Natural History* 50 (Part 1).

Fowke, Gerard. 1902. *Archaeological History of Ohio: The Mound Builders and Later Indians*. Columbus, Ohio: Press of Fred. J. Heer.

Fowler, M. L. 1985. A Brief History of Illinois Archaeology. *Illinois Archaeology*. Illinois Archaeological Survey, Bulletin No. 1, Revised:3–11.

Frayer, D. 1981. Body Size, Weapon Use and Natural Selection in the European Upper Paleolithic and Mesolithic. *American Anthropologist* 83:57–73.

Frazer, James George. 1890. Some Popular Superstitions of the Ancients. *Folklore* 1:145–171.

Fresia, A. E., Christopher B. Ruff, and Clark Spencer Larsen. 1990. Temporal Decline in Bilateral Asymmetry of the Upper Limb on the Georgia Coast. In *The Archaeology of Santa Catalina de Gaule. 2. Biocultural Interpretations of a Population in Transition*. 121–131. Vol. 68, *Anthropological Papers of the American Museum of Natural History*. New York: American Museum of Natural History.

Frost, Harold M. 1986. *Organization of the Skeleton*. Boca Raton, FL: CRC Press.

Funkhouser, William D. 1938. A Study of the Physical Anthropology and Pathology of the Osteological Material from the Norris Basin. In *An Archaeological Survey of the Norris Basin in Eastern Tennessee*, edited by W. S. Webb. 225–251, *Bulletin 118*. Washington, DC: Bureau of American Ethnology, Smithsonian Institution.

Funkhouser, William D., and William S. Webb. 1928. *Ancient Life in Kentucky*. Vol. 34, *Geologic Reports, Series 6*. Frankfort, KY: Kentucky Geological Survey.

Gabel, N. E. 1950. Skeletal Remains of Ventana Cave. In *The Stratigraphy and Archaeology of Ventana Cave, Arizona*, edited by E. W. Haury. 473–520. Albuquerque, NM: University of New Mexico Press.

Gage, T. B. 1988. Mathematical Hazard Models of Mortality: An Alternative to Model Life Tables. *American Journal of Physical Anthropology* 76 (4):429–441.

Gage, T. B., and B. Dyke. 1986. Parameterizing Abridged Mortality Tables Tables: The Siler Three-Component Hazard Model. *Human Biology* 58:275–291.

Garcia, S. V., and I. Podgorny. 2000. El Sabiotiene Una Patria: la Gran Guerra y la Comunidad Cientifica Argentina. *Ciencia Hoy* 10 (55):32–41.

Garn, S. M., and E. Giles. 1995. Earnest Albert Hooton, 1887–1954. *Biographical Memoirs of the National Academy of Sciences* 68:1–15.

Garratt-Frost, S. 1992. *The Law and Burial Archaeology, Technical Paper 11.* Birmingham: Institute of Field Archaeologists.

Garson, J. G. 1881–1882. Pelvimetry. *Journal of Anatomy and Physiology* XVI:106–134.

Gebhard, Paul. n.d. Manuscript on file, Department of Anthropology, Indiana University.

Gejvall, Nils-Gustaf. 1960. *Westerhus: Medieval Population and Church in the Light of Skeletal Remains*. Lund: Håkon Ohlssons Boktrykeri.

Gernaey, Angela M., David E. Minnikin, Mark S. Copley, Ali M. S. Ahmed, Duncan J. Robertson, John Nolan, and Andrew T. Chamberlain. 1999. Correlation of the Occurrence of Mycolic Acids with Tuberculosis in an Archaeological Population. In *Tuberculosis, Past and Present*, edited by G. Pálfi, O. Dutour, J. Deák and I. Hutás. 275–282. Budapest: Golden Book Publisher Ltd./Szeged: Tuberculosis Foundation.

Gero, Joan M., and Margaret Wright Conkey, eds. 1991. *Engendering Archaeology: Women and Prehistory, Social Archaeology*. Oxford, UK: Blackwell.

Gilbert, Robert I., and James H. Mielke. 1985. *The Analysis of Prehistoric Diets*. Orlando, FL: Academic Press.

Gilchrist, R., and B. Sloane. 2005. *Requiem: The Medieval Monastic Cemetery in Britain, Monograph*. London: Museum of London Archaeological Service.

Giles, Eugene. 1997. Hooton, E(Arnest) a(Lbert) (1887–1954). In *History of Physical Anthropology: An Encyclopedia*, edited by F. Spencer. 499–501. Vol. 1. New York: Garland Publishing.

Giles, E., and H. K. Bleibtreu. 1961. Cranial Evidence in Archaeological Reconstructions: A Trial of Multivariate Techniques for the Southwest. *American Anthropologist* 63 (1):48–61.

Gillham, Nicholas W. 2001. *A Life of Sir Francis Galton: From African Exploration to the Birth of Eugenics*. New York: Oxford University Press.

Girouard, Laurent. 1976. Archaeology and the Native Peoples of Quebec. In *Symposium on New Perspectives in Canadian Archaeology, 22–23 October 1976, Theatre Auditorium, Royal Ontario Museum, Toronto*, edited by A. G. McKay. 159–161. Ottawa: The Royal Society of Canada.

Givens, Douglas R. 1992. *Alfred Vincent Kidder and the Development of Americanist Archaeology*. Albuquerque, NM: University of New Mexico Press.

Glenn, Elizabeth Jane. 1965. Historical and Physical Affinities of the Sauk Indians, 1790–1819. Master's, Indiana University, Bloomington.

Glenn, Elizabeth J. 1974. *Physical Affinities of the Oneota Peoples, Report No. 7.* Iowa City: Office of the State Archaeologist of Iowa.

Glower, Charlotte. 1923. A Contribution to the Morphology of the Apertura Piriformis. *American Journal of Physical Anthropology* 6 (1):27–36.

Goldstein, Lynne. 1992. The Potential for Future Relationships between Archaeologists and Native Americans. In *Quandries and Quests: Visions of Archaeology's Future*, edited by L. Wandsnider. 59–114. Carbondale, IL: Southern Illinois University.

Goldstein, Lynne, and Keith Kintigh. 1990. Ethics and the Reburial Controversy. *American Antiquity* 55 (3):585–591.

Goldstein, Lynne G. 1980. *Mississippian Mortuary Practices: A Case Study of Two Cemetaries in the Lower Illinois Valley, Scientific Papers, No. 4.* Evanston, IL: Northwestern University Archeological Program.

Goldstein, Lynne G. 1981. One-Dimensional Archaeology and Multi-Dimensional People: Spatial Organization and Mortuary Analysis. In *The Archaeology of Death*, edited by R. Chapman, I. Kinnes, and K. Randsborg. 53–69, Chapter 4: Cambridge University Press.

Goldstein, Lynne G. 1995. Landscapes and Mortuary Practices: A Case for Regional Perspectives. In *Regional Approaches to Mortuary Analysis*, edited by L. A. Beck. 101–121. New York: Plenum Press.

Goldstein, Lynne G. 2000. Mississippian Ritual as Viewed through the Practice of Secondary Disposal of the Dead. In *Mounds, Modoc, and Mesoamerica: Papers in Honor of Melvin L. Fowler*, edited by S. R. Ahler. 193–205. Vol. XXVIII, *Scientific Papers*. Springfield, IL: Illinois State Museum.

Goldstein, Lynne G. 2001. Ancient Southwest Mortuary Practices: Perspectives from Outside the Southwest. In *Ancient Burial Practices in the American Southwest: Archaeology, Physical Anthropology, and Native American Perspectives*, edited by D. R. Mitchell and J. L. Brunson-Hadley. 249–253. Albuquerque, NM: University of New Mexico Press.

Goldstein, Marcus S. 1932. Caries and Attrition in the Molar Teeth of the Eskimo Mandible. *American Journal of Physical Anthropology* 16:421–430.

Goldstein, Marcus S. 1948. Dentition of Indian Crania from Texas. *American Journal of Physical Anthropology* 6:63–84.

Goldstein, Marcus S. 1957. Skeletal Pathology of Early Indians in Texas. *American Journal of Physical Anthropology* 15:299–312.

Gonzáles-José, R., A. Gonzáles-Martín, M. Hernández, H. M. Pucciarelli, M. Sardi, A. Rosales, and S. Van der Molen. 2003. Craniometric Evidence for Palaeoamerican Survival in Baja California. *Nature* 425:62–65.

Goodman, Alan H. 1993. On the Interpretation of Health from Skeletal Remains. *Current Anthropology* 34 (3):281–288.

Goodman, Alan H. 1994. Cartesian Reductionism and Vulgar Adaptationism: Issues in the Interpretation of Nutritional Status in Prehistory. In *Paleonutrition: The Diet and Health of Prehistoric Americans*, edited by K. D. Sobolik. 163–177, *Occasional Paper No. 2*: Center for Archaeological Investigations, Southern Illinois University at Carbondale.

Goodman, Alan H., and George J. Armelagos. 1985. Disease and Death at Dr. Dickson's Mounds. *Natural History Magazine* 94 (9):12–18.

Goodman, Alan H., and George J. Armelagos. 1989. Infant and Childhood Morbidity and Morality Risks in Archaeological Populations. *World Archaeology* 21:225–243.

Goodman, Alan H., J. W. Lallo, George J. Armelagos, and George Clarke. 1984. Health Changes at Dickson Mounds (A.D. 950–1300). In *Paleopathology at the Origins of Agriculture*, edited by M. N. Cohen and G. J. Armelagos. 271–305. Orlando, FL: Academic Press.

Goodman, Alan H., and Thomas L. Leatherman, eds. 1998. *Building a New Biocultural Synthesis: Political-Economic Perspectives on Human Biology*. Ann Arbor: University of Michigan Press.

Goodman, Alan H., and Jerome C. Rose. 1990. Assessment of Systemic Physiological Perturbations from Dental Enamel Hypoplasias and Associated Histological Structures. *Yearbook of Physical Anthropology* 33:59–110.

Goodyear, George F. 1994. *Society and Museum: A History of the Buffalo Society of Natural Sciences, 1861–1993, and the Buffalo Museum of Science, 1928–1993.* Buffalo: The Society.

Goose, Denys H. 1963. Dental Measurement: An Assessment of Its Value in Anthropological Studies. In *Dental Anthropology*, edited by D. R. Brothwell. 125–148. Vol. V. New York: MacMillan.

Gopnik, Adam. 2000. What's Cooking? Two New Books Tell Us How the Restaurant Got Its Start. *The New Yorker* September 4:82–86.

Gordon, Claire C., and Jane E. Buikstra. 1992. Linear Models for the Prediction of Stature from Foot and Boot Dimensions. *Journal of Forensic Sciences* 37 (3):771–782.

Gordon, Kathleen D. 1982. A Study of Microwear on Chimpanzee Molars: Implications for Dental Microwear Analysis. *American Journal of Physical Anthropology* 59:195–215.

Gordon, Kathleen D. 1984. Hominoid Dental Microwear: Complications in the Use of Microwear Analysis to Detect Diet. *Journal of Dental Research* 63:1043–1046.

Gould, Circuit Judge Ronald M. 2004. Opinion. Appeal from the United States District Court for the District of Oregon John Jelderks, Magistrate Judge, Presiding. Bonnichsen V. United States. Retrieved Dec. 31, 2005, from Friends of America's Past Web Site: http://www.friendsofpast.org/kennewick-man/court/opinions/040204opinion.html. 1584–1608.

Gould, Stephen Jay. 1978a. Morton's Ranking of Races by Cranial Capacity. *Science* 200 (4341):503–509.

Gould, Stephen Jay. 1978b. Flaws in a Victorian Veil. *Natural History* 87 (6):16–26.

Gould, Stephen Jay. 1981. *The Mismeasure of Man*. New York: W. W. Norton.

Gould, Stephen Jay. 1994. The Geometer of Race. *Discover* 15 (11 Nov.):64–69.

Gould, Stephen Jay. 1996. *The Mismeasure of Man: Revised and Expanded*. New York: W. W. Norton.

Gowland, Rebecca L. 2002. Age as an Aspect of Social Identity in 4th–6th Century AD England: The Archaeological Funerary Evidence. Ph.D. Dissertation, University of Durham.

Gowland, Rebecca L. 2004. The Social Identity of Health in Late Roman Britain. In *Proceedings of the Thirteenth Annual Theoretical Roman Archaeology Conference,*

Leicester, 2003, edited by B. Croxford, H. Eckhardt, J. Meade, and J. Weekes. 135–146. Oxford: Oxbow.

Gowland, Rebecca L., and Andrew T. Chamberlain. 2002. A Bayesian Approach to Ageing Perinatal Skeletal Material from Archaeological Sites: Implications for the Evidence for Infanticide in Roman-Britain. *Journal of Archaeological Science* 29:677–685.

Gowland, Rebecca L., and C. Knüsel. In press. *The Social Archaeology of Human Remains*. Oxford: Oxbow.

Grant, William Robertson. 1852. *Sketch of the Life and Character of Samuel George Morton, M.D. Lecture Introductory to a Course on Anatomy and Physiology in the Medical Department of Pennsylvania College*. Philadelphia: J. Royer.

Grauer, A. L., and Patricia Stuart-Macadam, eds. 1998. *Sex and Gender in Paleopathological Perspective*. Cambridge: Cambridge University Press.

Green, David L. 1984. Fluctuating Dental Asymmetry and Measurement Error. *American Journal of Physical Anthropology* 65:283–289.

Greenblatt, Charles L., and Mark Spigelman. 2003. *Emerging Pathogens: The Archaeology, Ecology, and Evolution of Infectious Disease*. Oxford and New York: Oxford University Press.

Greene, David L. 1965. Dentition and the Taxonomy of Meroitic, X-Group and Christian Populations from Wadi Halfa, Republic of the Sudan. Ph.D. Dissertation, Department of Anthropology, University of Colorado, Boulder.

Greene, David L. 1966. Dentition and the Biological Relationship of Some Meroitic, X-Group, and Christian Populations from Wadi Halfa, Sudan. *Kush* 14:284–288.

Greene, David L. 1967. *Dentition of Meroitic, X-Group and Christian Populations from Wadi Halfa, Sudan, University of Utah Anthropological Papers, No. 85*. Salt Lake City, Utah.

Greene, David L., D. P. Van Gerven, and George J. Armelagos. 1986. Life and Death in Ancient Populations: Bones of Contention in Paleodemography. *Human Evolution* 1:193–207.

Gregg, John B., and Pauline S. Gregg. 1987. *Dry Bones: Dakota Territory Reflected*. Sioux Falls, SD: J. B. and P. S. Gregg.

Gregg, John B., James P. Steele, and William M. Bass. 1982. Unusual Osteolytic Defects in Ancient South Dakota Skulls. Roentgenographic Evaluation of Temporal Bones from South Dakota Indian Burials. *American Journal of Physical Anthropology* 58:243–254.

Gregg, John B., James P. Steele, and Ann Holzhueter. 1965. Roentgenographic Evaluation of Temporal Bones from South Dakota Indian Burials. *American Journal of Physical Anthropology* 23:51–61.

Gregory, W. K. 1922. *The Origin and Evolution of the Human Dentition*. Baltimore, MD: Williams and Wilkins.

Griffin, James B. 1972. Foreword. In *The Prehistoric People of the Fort Ancient Culture of the Central Ohio Valley*, edited by L. M. Robbins and G. K. Neumann. vii–ix, *Anthropological Papers, Museum of Anthropology, University of Michigan, No. 47*. Ann Arbor: Museum of Anthropology, University of Michigan.

Griffin, James B. 1978. A Brief Tribute to David L. Dejarnette. In *David L. Dejarnette. A Southeastern Archaeological Tribute*, edited by D. A. Peterson. 3–5, *Special*

Publication No. 5. Southeastern Archaeological Conference, Memphis State University.

Griffin, James B. 1996. Introduction. In *Spiro Ceremonial Center: The Archaeology of Arkansas Valley Caddoan Culture in Eastern Oklahoma*, edited by J. A. Brown, *Memoirs of the Museum of Anthropology, University of Michigan No. 29*. Ann Arbor: Museum of Anthropology, University of Michigan.

Griffin, James B., and Georg K. Neumann. 1940. Burial Terminology and Description No. 1. Committee on Archaeological Terminology, Society for American Archaeology. Bioarchaeology Laboratory, Pennsylvania State University, University Park.

Grine, Frederic E. 1977. Analysis of Early Hominid Deciduous Molar Wear by Scanning Electron Microscopy: A Preliminary Report. *Proceedings of the Electron Microscopy Society of South Africa* 7:157–158.

Grine, Frederic E., W. L. Jungers, P. V. Tobias, and Osbjorn M. Pearson. 1995. Fossil *Homo* Femur from Berg Aukas, Northern Nambia. *American Journal of Physical Anthropology* 97:151–185.

Gross, T. S., Kenneth J. McLeod, and Clinton T. Rubin. 1992. Characterizing Bone Strain Distributions in Vivo Using Three Triple Rosette Strain Gages. *Journal of Biomechanics* 25:1081–1087.

Groves, S. E. 2005. Spears or Ploughshares: Multiple Indicators of Activity Related Stress and Social Status in Four Early Medieval Populations from the North-East of England. Ph.D. Dissertation, University of Durham, UK.

Gruber, Jacob W. 1967. *The Philadelphia Anthropological Society: Papers Presented on Its Golden Anniversary*. New York: Columbia University Press.

Grupe, G. 1995. Reconstructing Migration in the Bell Beaker Period by ^{87}sr/^{86}sr Isotope Ratios in Teeth and Bones. In *Proceedings of the 10th International Symposium on Dental Morphology*, edited by R. J. Radlandski and H. Renz. 338–342. Berlin: "M" Marketing Services, C. & M. Brünne GbR.

Guagliardo, M. F. 1982. Tooth Crown Size Differences between Age Groups: A Possible New Indicator of Stress in Skeletal Samples. *American Journal of Physical Anthropology* 58:383–389.

Guerra, Francisco. 1993. The European-American Exchange. *History and Philosophy of the Life Sciences* 15:313–327.

Guhl, Felipe, C. Jaramillo, G. A. Vallejo, R. Yockteng, F. Cardenas-Arroyo, G. Forniciari, Bernardo Arriaza, and Arthur C. Aufderheide. 1999. Isolation of *Trypanosoma Cruzi* DNA in 4,000 Year-Old Mummified Remains from Northern Chile. *American Journal of Physical Anthropology* 108:401–408.

Guthe, Carl E. 1939. The Basic Needs of American Archaeology. *Science* 90:528–530.

Guyatt, G. H., D. H. Feeny, and D. L. Patrick. 1993. Measuring Health-Related Quality of Life: Basic Sciences Review. *Annals of Internal Medicine* 70:225–230.

Haag, William G. 1985. Federal Aid to Archaeology in the Southeast, 1933–1942. *American Antiquity* 50:272–280.

Haag, William G. 1986. Field Methods in Archaeology. In *American Archaeology Past and Future*, edited by D. J. Meltzer, D. D. Fowler, and J. A. Sabloff. 63–76. Washington, DC: Smithsonian University Press.

Haas, Christian J., Albert Zink, Erika Molnár, Ulrike Szeimies, Udo Reischl, Antonia Marcsik, Yann Ardagna, Olivier Dutour, György Pálfi, and Andreas G. Nerlich. 2000.

Molecular Evidence for Different Stages of Tuberculosis in Ancient Bone Samples from Hungary. *American Journal of Physical Anthropology* 113 (3):287–292.

Haas, Jonathan, and Winifred Creamer. 1993. *Stress and Warfare among the Kayenta Anasazi of the Thirteenth Century A.D.* Vol. 21, *Fieldiana, Anthropology New Series No. 21*. Chicago: Field Museum of Natural History.

Hackett, C. J. 1967. The Human Treponematoses. In *Diseases in Antiquity*, edited by D. R. Brothwell and A. T. Sandison. 152–170. Springfield, IL.

Hackett, C. J. 1976. *Diagnostic Criteria of Syphilis, Yaws and Treponarid (Treponematoses) and of Some Other Diseases in Dry Bones (for Use in Osteo-Archaeology), Sitzungsberichte der Heidelberger Akademie der Wissenschaften Mathematisch-Naturwissenschaftliche Klasse, Abhandlung 4*. Berlin: Springer-Verlag.

Hafner, Arthur W., ed. 1993. *Directory of Deceased American Physicians, 1804–1929*. Chicago: American Medical Association.

Haldeman, Michele B., and Sheilagh T. Brooks. 1987. Eburnation: Paleopathological Evidence of Severe Osteoarthritis from the Great Basin. *Nevada Archaeologist* 6 (1):31–33.

Haldeman, Michele B., Sheilagh T. Brooks, and Richard H. Brooks. 1987. The Palaeopathology of an Archaeologically Recovered Skeletal Series from the Stillwater Reservoir Area, Nevada. [Abstract]. In *Papers on Paleopathology Presented at the Fourteenth Annual Meeting of the Paleopathology Association, New York, New York, 1–2 April, 1987*, edited by E. Cockburn. 15–16. Detroit: Paleopathology Association.

Hallam, Elizabeth, and Jenny Hockey. 2001. Death, Memory & Material Culture. Oxford: Berg Publishers.

Hamilton, Mary E. 1982. Sexual Dimorphism in Skeletal Samples In *Sexual Dimorphism in Homo Sapiens: A Question of Size*, edited by R. L. Hall. 107–163. New York: Praeger.

Hanks, Christopher C. 1997. Ancient Knowledge of Ancient Sites: Tracing Gene Identity from the Late Pleistocene and Holocene. In *At a Crossroads: Archaeology and First Peoples of Canada*, edited by G. P. Nicholas and T. D. Andrews. 178–189. Burnaby: Archaeology Press.

Hann, J. H. 1988. Apalachee: The Land between the Rivers. Gainesville: University Press of Florida.

Hanna, B. L. 1962. The Biological Relationships among Indian Groups of the Southwest: Analysis of Morphological Traits. *American Journal of Physical Anthropology* 20 (4):499–508.

Hanna, B. L., A. A. Dahlberg, and H. H. Strandskov. 1953. A Preliminary Study of the Population History of the Pima Indians. *American Journal of Human Genetics* 5:377–388.

Hardwick, J. L. 1960. The Incidence and Distribution of Caries Throughout the Ages in Relation to the Englishmen's Diet. *British Dental Journal* 108:9–17.

Härke, H. 1990. Warrior Graves. The Background of the Anglo-Saxon Burial Rite. *Past and Present* 126:22–43.

Harmon, Anna M., and Jerome C. Rose. 1988. The Role of Dental Microwear Analysis in the Reconstruction of Prehistoric Diet. In *Diet and Subsistence: Current Archaeological*

Perspectives, edited by B. V. Kennedy and G. M. LeMoine. 267–272. Calgary: University of Calgary Arch. Assoc.

Harn, Alan D. 1971. *The Prehistory of Dickson Mounds: A Preliminary Report*. Springfield, IL: Illinois State Museum.

Harpending, Henry C., and R. Ward. 1982. Chemical Systematics and Human Evolution. In *Biochemical Aspects of Evolutionary Biology*, edited by M. Nitecki. 213–256. Chicago: University of Chicago.

Harper, A. B., and W. S. Laughlin. 1982. Inquiries into the Peopling of the New World: Development of Ideas and Recent Advance. In *A History of American Physical Anthropology 1930–1980*, edited by F. Spencer. 281–304. New York: Academic Press.

Harrison, William R., Charles F. Merbs, and C. R. Leathers. 1991. Evidence of Coccidioidomycosis in the Skeleton of an Ancient Arizona Indian. *Journal of Infectious Diseases* 164:4346–4347.

Hart, G. D., ed. 1983. *Disease in Ancient Man. An International Symposium*. Toronto: Clarke Irwin.

Hartl, D. L., and A. G. Clark. 1997. *Principles of Population Genetics*. 3rd ed. Sunderland, MA: Sinauer Associates.

Hartnady, Philip, and Jerome C. Rose. 1991. Abnormal Tooth-Loss Patterns among Archaic-Period Inhabitants of the Lower Pecos Region, Texas. In *Advances in Dental Anthropology*, edited by M. A. Kelley and C. S. Larsen. 267–278. New York: Wiley-Liss.

Haskell, J. Loring. 1987. *Southern Athapaskan Migration, A.D. 200–1750*. Tsaile: Navajo Community College Press.

Haury, Emil W. 1934. The Archaeology of the Salt River Valley, Arizona: A Study of the Interrelationships of Two Ethnic Groups. Ph.D., Harvard University.

Haury, Emil W. 1945. *The Excavation of Los Muertos and Neighboring Ruins in the Salt River Valley, Southern Arizona*. Vol. 24, *Papers of the Peabody Museum of American Archaeology and Ethnology*. Cambridge, MA: Peabody Museum of American Archaeology and Ethnology.

Hawkes, S. C., and Calvin Wells. 1975. Crime and Punishment in an Anglo-Saxon Cemetery. *Antiquity* 49:118–122.

Hawkey, Diane. 1988. Use of Upper Extremity Enthesopathies to Indicate Habitual Activity Patterns. M.A., Arizona State University, Tempe.

Hawkey, Diane, and Charles F. Merbs. 1995. Activity-Induced Musculoskeletal Stress Markers (MSM) and Subsistence Strategy Changes among Hudson Bay Eskimos. *International Journal of Osteoarchaeology* 5:324–338.

Heglar, R. 1974. The Prehistoric Population of Cochiti and Selected Inter-Population Biological Comparisons. Ph.D., University of Michigan.

Hendon, Julia A. 2000. Having and Holding: Storage, Memory, Knowledge, and Social Relations. *American Anthropologist* 102 (1):42–53.

Henry, Robert S. 1964. *The Armed Forces Institute of Pathology. Its First Century, 1862–1962*. Washington, DC: Office of the Surgeon General, Department of the Army.

Heritage Council. 2002. *Human Remains in Irish Archaeology. Legal, Scientific, Planning and Ethical Implications*. Dublin: The Heritage Council.

Herring, D. Ann, and Alan C. Swedlund. 2003. *Human Biologists in the Archives*. Cambridge: Cambridge University Press.

Hertz, Robert. 1960 [1907]. *Death and the Right Hand*. Translated by R. Needham and C. Needham. Glencoe, IL: The Free Press.

Hertzberg, H. T. E. 1940. Letter Dated November 26 to Snow. Snow Letters, Laboratory for Human Osteology, Mary Harmon Bryant Scientific Collections Building, University of Alabama, Tuscaloosa.

Hill, Erica. 1998. Death as a Rite of Passage: The Iconography of the Moche Burial Theme. *Antiquity* 72:528–538.

Hill, J. N. 1970. *Broken K Pueblo: Prehistoric Social Organization in the American Southwest, Anthropological Papers 18*. Tucson: University of Arizona.

Hillson, Simon. 1996. *Dental Anthropology*. Cambridge and New York: Cambridge University Press.

Hillson, Simon W. 1979. Diet and Dental Disease. *World Archaeology* 11 (2):147–162.

Hillson, Simon W. 1986. *Teeth*. Cambridge: Cambridge University Press.

Hillson, Simon W. 1996. *Dental Anthropology*. Cambridge: Cambridge University Press.

Hillson, Simon W. 2000. Editorial: Cannibalism and Violence. *International Journal of Osteoarchaeology* 10:1–3.

Hillson, Simon W. 2001. Recording of Dental Caries in Archaeological Human Remains. *International Journal of Osteoarchaeology* 11:249–289.

Hillson, Simon W., C. Grigson, and S. Bond. 1998. Dental Defects of Congenital Syphilis. *American Journal of Physical Anthropology* 107:25–40.

Hingley, Richard. 1996. Ancestors and Identity in the Later Prehistory of Atlantic Scotland: The Reuse and Reinvention of Neolithic Monuments and Material Culture. *World Archaeology* 28 (2):231–243.

Hinsley, Curtis M., and David R. Wilcox. 1996. *The Southwest in the American Imagination: The Writings of Sylvester Baxter, 1881–1889*. Vol. 1, *The Southwest Center Series. Frank Hamilton Cushing and the Hemenway Southwestern Archaeological Expedition: 1886–1889*. Tucson, AZ: University of Arizona Press.

Hinsley, Curtis M., and David R. Wilcox, eds. 2002. *The Lost Itinerary of Frank Hamilton Cushing*. Vol. 2, *The Southwest Center Series. Frank Hamilton Cushing and the Hemenway Southwestern Archaeological Expedition: 1886–1889*. Tucson, AZ: University of Arizona Press.

Hinton, R. J. 1981. Form and Patterning of Anterior Tooth Wear among Aboriginal Human Groups. *American Journal of Physical Anthropology* 54 (4):555–564.

Historic Scotland. 1997. *The Treatment of Human Remains in Archaeology, Operational Policy Paper 5*. Edinburgh: Historic Scotland.

Hockney, David. 2001. *Secret Knowledge: Rediscovering the Lost Techniques of the Old Masters*. New York: Viking Studio.

Hodder, Ian. 1980. Social Structure and Cemeteries: A Critical Appraisal. In *Anglo-Saxon Cemeteries*, edited by P. Rahtz, T. Dickinson and L. Watts. 161–169. *B. A. R. British Series 82*. Oxford: Archaeopress.

Hodder, Ian. 1982a. *Symbols in Action: Ethnoarchaeological Studies of Material Culture*. Cambridge: Cambridge University Press.

Hodder, Ian. 1982b. Theoretical Archaeology: A Reactionary View. In *Symbolic and Structural Archaeology*, edited by I. Hodder. 1–16. *New Directions in Archaeology*. Cambridge: Cambridge University Press.

Hodder, Ian. 1984. Burials, Houses, Women and Men in the European Neolithic. In *Ideology, Power and Prehistory*, edited by D. Miller and C. Tilley. 51–68. Cambridge: Cambridge University Press.

Hodder, Ian. 1985. Postprocessual Archaeology. *Advances in Archaeological Method and Theory* 8:1–26.

Hodder, Ian. 1986. *Reading the Past: Current Approaches to Interpretation in Archaeology*. Cambridge: Cambridge University Press.

Hodges, Denise C., Lee A. Harker, and Shirley J. Schermer. 1990. Atresia of the External Acoustic Meatus in Prehistoric Populations. *American Journal of Physical Anthropology* 83(1):77–81.

Hoenig, J. M., and D. M. Heisey. 1987. Use of a Log-Linear Model with the EM Algorithm to Correct Estimates of Stock Composition and Convert Length to Age. *Transactions of the American Fisheries Society* 116:232–243.

Hoffman, Douglas F. 1993. Arthritis and Exercise. *Primary Care* 20:895–910.

Hohenthal, William D., and Sheilagh T. Brooks. 1960. An Archaeological Scaphocephal from California. *American Journal of Physical Anthropology* 18 (1):59–67.

Hojo, T. 1989. Dietary Differences and Microwear on the Teeth of Late Stone Age and Early Modern People from Western Japan. *Scan Microscopy* 3:623–628.

Holland, T. D., and M. J. O'Brien. 1997. Parasites, Porotic Hyperostosis, and the Implications of Changing Perspectives. *American Antiquity* 62 (2):183–193.

Holliday, Trenton W. 1997a. Body Proportions in Late Pleistocene Europe and Modern Human Origins. *Journal of Human Evolution* 32:423–447.

Holliday, Trent W. 1997b. Postcranial Evidence of Cold Adaptation in European Neandertals. *American Journal of Physical Anthropology* 104:245–258.

Holloway, R. L. 2002. Head to Head with Boas: Did He Err on the Plasticity of Head Form? *Proceedings of the National Academy of Sciences* 99:14622–14623.

Holt, Brigitte M. 2003. Mobility in Upper Paleolithic and Mesolithic Europe: Evidence from the Lower Limb. *American Journal of Physical Anthropology* 122:200–215.

Holtorf, Cornelius J. 1998. The Life-Histories of Megaliths in Mecklenburg-Vorpommern (Germany). *World Archaeology* 30 (1):23–38.

Holzhueter, Ann M., John B. Gregg, and S. Clifford. 1965. A Search for Stapes Footplate Fixation in an Indian Population, Prehistoric and Historic. *American Journal of Physical Anthropology* 23:35–40.

Hooton, Earnest Albert. 1911. The Pre-Hellenistic Stage in the Evolution of the Literary Art at Rome. *Abstracts of Theses Submitted in Partial Fulfillment of the Requirements for the Degree of Doctor of Philosophy—University of Wisconsin, Madison* 1:37–57.

Hooton, Earnest Albert. 1918. On Certain Eskimo Characters in Icelandic Skulls. *American Journal of Physical Anthropology* 1:53–76.

Hooton, Earnest Albert. 1920. *Indian Village Site and Cemetery near Madisonville, Ohio*. Vol. 8, *Papers of the Peabody Museum of American Archaeology and Ethnology, Harvard University*. Cambridge, MA: Peabody Museum of American Archaeology and Ethnology.

Hooton, Earnest Albert. 1922. The Skeletal Remains. *Papers of the Peabody Museum of American Archaeology and Ethnology, Harvard University* 8 (3):99–132.

Hooton, Earnest Albert. 1925. *The Ancient Inhabitants of the Canary Islands*. Vol. VII, *Harvard African Studies*. Cambridge, MA: Peabody Museum of Harvard University.

Hooton, Earnest Albert. 1930. *The Indians of Pecos Pueblo: A Study of Their Skeletal Remains*. Vol. 4, *Papers of the Southwestern Expedition*. New Haven: Yale University Press.

Hooton, Earnest Albert. 1931. *Up from the Ape*. New York: Macmillan.

Hooton, Earnest Albert. 1935. Development and Correlation of Research in Physical Anthropology at Harvard University. *Proceedings of the American Philosophical Society* LXXV:499–516.

Hooton, Earnest Albert. 1940. Skeletons from the Cenote of Sacrifice at Chichen Itzá. In *The Maya and Their Neighbors: Essays on Middle American Anthropology and Archaeology*, edited by C. L. Hay, R. L. Linton, S. K. Lothrop, H. L.Shapiro, and G. C. Valliant. 272–280. New York: Appleton-Century.

Hooton, Earnest Albert. 1946. *Up from the Ape*. Rev. ed. New York: Macmillan.

Hooton, Earnest Albert. 1973. Racial Types in America and Their Relations to Old World Types. In *The American Aborigines: Their Origin and Antiquity. Fifth Pacific Science Congress, Canada, 1933*, edited by D. Jenness. New York: Cooper Square Publishers.

Hoppa, Robert D. 1996. Representativeness and Bias in Cemetery Samples: Implications for Paleodemographic Reconstruction of Past Populations. Ph.D. Dissertation, McMaster University, Hamilton, Ontario.

Hoppa, Robert D. 1999. Modeling the Effects of Selection-Bias on Paleodemographic Analyses. *Homo* 50:228–243.

Hoppa, Robert D. 2000. Population Variation in Osteological Aging Criteria: An Example from the Pubic Symphysis. *American Journal of Physical Anthropology* 111:185–191.

Hoppa, Robert D., and C. M. FitzGerald. 1999. From Head to Toe: Integrating Studies from Bones and Teeth in Biological Anthropology. In *Human Growth in the Past: Studies from Bones and Teeth*, edited by R. D. Hoppa and C. M. FitzGerald. 1–31. Cambridge: Cambridge University Press.

Hoppa, Robert D., and James W. Vaupel, eds. 2002. *Paleodemography: Age Distributions from Skeletal Samples*. Cambridge: Cambridge University Press.

Horowitz, S., and George J. Armelagos. 1988. On Generating Birth Rates from Skeletal Populations. *American Journal of Physical Anthropology* 76:189–196.

Horsman, Reginald. 1987. *Josiah Nott of Mobile: Southerner, Physician, and Racial Theorist, Southern Biography Series*. Baton Rouge: Louisiana State University Press.

Howell, N. 1976. Toward a Uniformitarian Theory of Human Paleodemography. In *The Demographic Evolution of Human Populations*, edited by R. H. Ward and K. M. Weiss. 25–40. New York: Academic Press.

Howell, N. 1982. Village Composition Implied by a Paleodemographic Life Table: The Libben Site. *American Journal of Physical Anthropology* 59:263–269.

Howell, T. L. 1994. Leadership at the Ancestral Zuni Village of Hawikku. Ph.D., Arizona State University.

Howell, T. L. 1995. Tracking Zuni Gender and Leadership Roles across the Contact Period. *Journal of Anthropological Research* 51:125–147.

Howell, T. L., and K. W. Kintigh. 1996. Archaeological Identification of Kin Groups Using Mortuary and Biological Data: An Example from the American Southwest. *American Antiquity* 61 (3):537–554.

Howell, T. L., and K. W. Kintigh. 1998. Determining Gender and Kinship at Hawikku: A Reply to Corruccini. *American Antiquity* 63 (1):164–167.

Howells, William White. 1957. The Cranial Vault: Factors of Size and Shape. *American Journal of Physical Anthropology* 15:19–48.

Howells, William White. 1959. Boas as Statistician. In *The Anthropology of Franz Boas: Essays on the Centennial of His Birth*, edited by W. R. Goldschmidt. 112–116. *American Anthropological Memoir No. 89*. Menasha: American Anthropological Association.

Howells, William W. 1960. Estimating Population Numbers through Archaeological and Skeletal Remains. In *The Application of Quantitative Methods in Archaeology*, edited by R. F. Heizer and S. F. Cook. 158–185. *Viking Fund Publications in Anthropology, No. 28*. Chicago: Quadrangle Books.

Howells, William White. 1989. *Skull Shapes and the Map: Craniometric Analyses in the Dispersion of Modern Homo*. Vol. 79, *Papers of the Peabody Museum of Archaeology and Ethnology, Harvard University*. Cambridge, MA: Peabody Museum of Archaeology and Ethnology.

Howells, William White. 1995. *Who's Who in Skulls: Ethnic Identification of Crania from Measurements*. Vol. 82, *Papers of the Peabody Museum of Archaeology and Ethnology, Harvard University*. Cambridge, MA: Peabody Museum of Archaeology and Ethnology Harvard University.

Howells, William White. n.d. Skeletal Remains and American Indian Population History: 1900–1975. In *Handbook of North American Indians*. Washington, DC: Smithsonian Institution.

Hoyme, Lucile E. 1950. The Role of Saliva in Inheritance by the Ability to Taste Phenyl-Thio-Carbamide. M. S., George Washington University, Washington, DC.

Hoyme, Lucile E. 1951. A Historical Survey of Anthropometric Instruments [Abstract]. *American Journal of Physical Anthropology* 9:246.

Hoyme, Lucile E. 1953. Physical Anthropology and Its Instruments: An Historical Study. *Southwestern Journal of Anthropology* 9:408–430.

Hoyme, Lucile E. 1954. Genetics, Physiology and PTC [Abstract]. *American Journal of Physical Anthropology* 12:289–290.

Hoyme, Lucile E. 1955. Genetics, Physiology and PTC. *Journal of Heredity* 46:167–175.

Hoyme, Lucile E. 1956a. New Landmark for Measuring Ilium, Ischium and Pubis Length [Abstract]. *American Journal of Physical Anthropology* 14:387.

Hoyme, Lucile E. 1956b. Review of "Francois Joseph Gall et Sa Collection" by E. H. Ackernecht and H. V. Vallois and "Phrenology: Fad and Science" by J. D. Davies. *American Journal of Physical Anthropology* 14:333–336.

Hoyme, Lucile E. 1957a. Sex Differences in the Circumauricular Area of the Human Ilium [Abstract]. *American Journal of Physical Anthropology* 15:434.

Hoyme, Lucile E. 1957b. The Earliest Use of Indices for Sexing Pelves. *American Journal of Physical Anthropology* 15:537–546.

Hoyme, Lucile E. 1958. Review of "Bones for the Archaeologist" by I. W. Cornwall. *Antiquity* 32:49–50.

Hoyme, Lucile E. 1959. Sex Differentiation in the Human Pelvis in Early Adult Life. In *Deutsche Gesellschaft Für Anthropologie*. 66–67. Vol. 6. Bericht: Göttingen.

Hoyme, Lucile E. 1963. The Relation of Age and Sex Ratios to Health, Longevity, and Culture in Aboriginal Skeletal Populations [Abstract]. *American Journal of Physical Anthropology* 21:402.

Hoyme, Lucile E. 1964a. Climate and Selection for Skull Shape [Abstract]. *American Journal of Physical Anthropology* 22:504.

Hoyme, Lucile E. 1964b. Science Fair Projects in Human Biology. *SI Leaflet* 408.

Hoyme, Lucile E. 1964c. Variation in Human Skeletal Characteristics. Ph.D., Anthropology, Oxford University, England.

Hoyme, Lucile E. 1965. The Nasal Index and Climate: A Spurious Case of Natural Selection in Man [Abstract]. *American Journal of Physical Anthropology* 23:336–337.

Hoyme, Lucile E. 1966. Are Some Wormian Bones the Cranial Analog of "Joint Mice"? [Abstract]. *American Journal of Physical Anthropology* 25:204.

Hoyme, Lucile E., and William M. Bass. 1962. Human Skeletal Remains from the Tolifero (Ha6) and Clarksville (Mc14) Sites, John H. Kerr Reservoir Basin, Virginia. *Bulletin of the Bureau of American Ethnology* 182:329–400.

Hrdlička, Aleš. 1894. A New Form of Abdominal Bandage for Use after Delivery. *New York Medical Journal* LIX:274–275.

Hrdlička, Aleš. 1895. Contribution to the General Pathology of the Insane (Physical Examinations and Measurements). In *24th Annual Report Middletown State Homeopathic Hospital*. 162–207. Albany: James B. Lyon.

Hrdlička, Aleš. 1896. Twenty Autopsies Held Upon the Cadavers of the Insane. In *25th Annual Report Middletown State Homeopathic Hospital*. 179–213.

Hrdlička, Aleš. 1897. A Few Words About Anthropometry. *American Journal of Insanity* LIII:521–533.

Hrdlička, Aleš. 1898. Study of the Normal Tibia. *American Anthropologist* 11:27–40.

Hrdlička, Aleš. 1899a. Description of an Ancient Anomalous Skeleton from the Valley of Mexico with Special Reference to Supernumerary and Bicipital Ribs in Man. *Bulletin of the American Museum of Natural History* XII: 81–107.

Hrdlička, Aleš. 1899b. The Needs of American Anthropologists. *American Naturalist* XXXIII (392):684–688.

Hrdlička, Aleš. 1900. Physical and Physiological Observations on the Navaho. *American Anthropologist* 2 (2):339–345.

Hrdlička, Aleš. 1901. An Eskimo Brain. *American Anthropologist* 3 (3):454–500.

Hrdlička, Aleš. 1902a. Anthropological Work in the Southwestern United States and Mexico. *Journal of the American Museum of Natural History, NY* II:68–72.

Hrdlička, Aleš. 1902b. The Aztecs of Yesterday and to-Day. *Harpers* CVI:37–42.

Hrdlička, Aleš. 1902c. The Crania of Trenton, New Jersey, and Their Bearing Upon the Antiquity of Man in That Region. *Bulletin of the American Museum of Natural History, NY* XVI:23–62.

Hrdlička, Aleš. 1903a. The Region of the Ancient "Chichimecs," with Notes on the Tepecanos and the Ruin of La Quemada, Mexico. *American Anthropologist* 5 (3):385–440.

Hrdlička, Aleš. 1903b. A Laguna Ceremonial Language. *American Anthropologist* 5 (4):730–732.

Hrdlička, Aleš. 1904a. Notes on the Indians of Sonora, Mexico. *American Anthropologist* 6 (1):51–89.

Hrdlička, Aleš. 1904b. Method of Preparing Tesvino among the White River Apaches. *American Anthropologist* 6 (1):190–191.

Hrdlička, Aleš. 1904c. Cora Dances. *American Anthropologist* 6 (5):744–745.

Hrdlička, Aleš. 1904d. Crow Burial in Montana [Abstract]. *American Anthropologist* 6:756–758.

Hrdlička, Aleš. 1904e. *Directions for Collecting Information and Specimens for Physical Anthropology, Bulletin of the United States National Museum, No. 39*. Washington, DC: Government Printing Office.

Hrdlička, Aleš. 1905a. Maricopa Weaving. *American Anthropologist* 7 (2):361.

Hrdlička, Aleš. 1905b. A Cora Cradle. *American Anthropologist* 7 (2):361–362.

Hrdlička, Aleš. 1905c. Notes on the San Carlos Apache. *American Anthropologist* 7 (3):480–495.

Hrdlička, Aleš. 1905d. Jay Feathers in a Cora Ceremony. *American Anthropologist* 7 (4):730.

Hrdlička, Aleš. 1906a. Notes on the Pima of Arizona. *American Anthropologist* 8 (1):39–46.

Hrdlička, Aleš. 1906b. Beauty among the American Indians. In *Boas Anniversary Volume. Anthropological Papers Written in Honor of Franz Boas*, edited by B. Laufer and H. A. Andrews. 38–42. New York: G. E. Stechert & Co.

Hrdlička, Aleš. 1907. The Skeletal Remains Suggesting or Attributed to Early Man in North America. In *Smithsonian Institution, Bureau of American Ethnology, Bulletin 33*. Washington DC: Government Printing Office.

Hrdlička, Aleš. 1908a. *Physiological and Medical Observations among the Indians of the Southwestern United States and Northern Mexico*. Vol. 34, *Bureau of American Ethnology Bulletin*. Washington, DC: Government Printing Office.

Hrdlička, Aleš. 1908b. Report on a Collection of Crania from Arkansas. *Academy of Natural Sciences of Philadelphia Journal* 13:558–563.

Hrdlička, Aleš. 1908c. *Physiological and Medical Observations among the Indians of the Southwestern United States and Northern Mexico, Bureau of American Ethnology Bulletin 34*. Washington, DC: Smithsonian Institution.

Hrdlička, Aleš. 1909a. *Tuberculosis among Certain Indian Tribes of the United States, Bureau of American Ethnology, Bulletin No. 42*. Washington: Government Print Office.

Hrdlička, Aleš. 1909b. The Civilization of Bohemia. *Science* XXX (781):880.

Hrdlička, Aleš. 1909c. Report on an Additional Collection of Skeletal Remains from Arkansas and Louisiana. *Academy of Natural Sciences of Philadelphia Journal* 14:171–249.

Hrdlička, Aleš. 1909d. *Tuberculosis among Certain Indian Tribes of the United States, Bureau of American Ethnology Bulletin 42*. Washington, DC: Smithsonian Institution.

Hrdlička, Aleš. 1910. Report of Skeletal Material from Missouri Mounds, Collected in 1906–1907 by Mr. Gerard Fowke. *Bureau of American Ethnology Bulletin* (37):103–112.

Hrdlička, Aleš. 1911. A Report on a Collection of Crania and Bones from Sorrel Bayou, Iberville Parish, Louisiana. *Journal of the Academy of Natural Sciences of Philadelphia* XVI:95–99.

Hrdlička, Aleš. 1912a. Early Man in America. *The American Journal of Science*, Fourth Series, 34:543–554.

Hrdlička, Aleš. 1912b. The Problems of the Unity or Plurality and the Probable Place of Origin of the American Aborigines: The Bearing of Physical Anthropology on the Problems under Consideration. *American Anthropologist* 14 (1):8–12.

Hrdlička, Aleš. 1912c. *The Natives of Kharga Oasis, Egypt*. Vol. 59, No.1, *Smithsonian Miscellaneous Collections*. Washington, DC: Smithsonian Institution.

Hrdlička, Aleš. 1912d. Report on Skeletal Remains from a Mound on Haley Place, Near Red River, Miller County, Arkansas. *Journal of the Academy of Natural Sciences of Philadelphia* 14:639–640.

Hrdlička, Aleš. 1913. A Report on a Collection of Crania and Bones from Sorrel Bayou, Iberville Parish, Louisiana. *Journal of the Academy of Natural Sciences of Philadelphia* 16:95–99.

Hrdlička, Aleš. 1914a. Physical Anthropology in America: An Historical Sketch. *American Anthropologist* 16 (4):508–554.

Hrdlička, Aleš. 1914b. Special Notes on Some Pathological Conditions Shown by the Skeletal Material of the Ancient Peruvians. *Smithsonian Miscellaneous Collections* 62:57–59.

Hrdlička, Aleš. 1916. Brief Notes on Recent Anthropological Explorations under the Auspices of the Smithsonian Institution and the U.S. National Museum. *Proceedings of the National Academy of Sciences of the United States of America* 2 (1):32–37.

Hrdlička, Aleš. 1917a. Preliminary Report on Finds of Supposedly Ancient Human Remains at Vero, Florida. *The Journal of Geology* 25:43–51.

Hrdlička, Aleš. 1917b. The Genesis of the American Indian. *Proceedings of the 19th International Congress of Americanists* 1915:559–568.

Hrdlička, Aleš. 1918. Physical Anthropology: Its Scope and Aims; Its History and Present Status in America. *American Journal of Physical Anthropology* 1 (1; 2; 3):3–23; 133–182; 267–304.

Hrdlička, Aleš. 1919. The Effects of the War on the American People. *The Scientific Monthly* VIII (6):542–545.

Hrdlička, Aleš. 1920a. *Anthropometry*. Philadelphia: The Wistar Institute of Anatomy and Biology.

Hrdlička, Aleš. 1920b. On the Relations of Anthropology and Psychology. *Science* LI (1313):199–201.

Hrdlička, Aleš. 1920c. Shovel-Shaped Teeth. *American Journal of Physical Anthropology* III:429–465.

Hrdlička, Aleš. 1921a. Art of Czechoslovakia. *Art and Archaeology* XI:178–220.

Hrdlička, Aleš. 1921b. Further Studies of Tooth Morphology. *American Journal of Physical Anthropology* IV:141–176.

Hrdlička, Aleš. 1924. *Catalogue of Human Crania in the United States National Museum Collections: The Eskimo, Alaska and Related Indians, North Eastern Asiatics*. Vol. 63(12), *Proceedings of the United States National Museum*. Washington, DC: U.S. Goverment Printing Office.

Hrdlička, Aleš. 1927. *Catalogue of Human Crania in the United States National Museum Collections: The Algonquin and Related Iroquois; Siouan, Caddoan, Salish and Sahaptim, Shoshonean, and Californian Indians*. Vol. 69(5), *Proceedings of the United States National Museum*. Washington, DC: U.S. Government Printing Office.

Hrdlička, Aleš. 1928. *Catalogue of Human Crania in the United States National Museum Collections: Australians, Tasmanians, South African Bushmen, Hottentots and Negro*. Vol. 71(24), *Proceedings of the United States National Museum*. Washington, DC: U.S. Government Printing Office.

Hrdlička, Aleš. 1928. The Origin and Antiquity of Man in America. *Bulletin of the New York Academy of Medicine* Ser. 2. 4 (7):802–816.

Hrdlička, Aleš. 1931. *Catalogue of Human Crania in the United States National Museum Collections: Pueblos, Southern Utah Basket-Makers and Navaho*. Vol. 78 (Art. 2), *Proceedings of the United States National Museum*. Washington, DC: U.S. Government Printing Office.

Hrdlička, Aleš. 1932. Disease, Medicine and Surgery among the American Aborigines. *Journal of the American Medical Association* 94:1661–1666.

Hrdlička, Aleš. 1933. Seven Prehistoric American Skulls with Complete Absence of the External Auditory Meatus. *American Journal of Physical Anthropology* 17:355–377.

Hrdlička, Aleš. 1934a. Contributions to the Study of the Femur: The Crista Aspera and the Pilaster. *American Journal of Physical Anthropology* 19:17–37.

Hrdlička, Aleš. 1934b. The Human Femur: Shape of the Shaft. *Anthropologie (Prague)* 12 (Supplement):129–163.

Hrdlička, Aleš. 1934c. The Hypotrochanteric Fossa of the Femur. *Smithsonian Miscellaneous Collections* 92:1–49.

Hrdlička, Aleš. 1934d. Shapes of the Femur: Additional Data. *Anthropologie (Prague)* 12:27–40.

Hrdlička, Aleš. 1935a. The Pueblos: With Comparative Data on the Builds of the Tribes of the Southwest and Northern Mexico. *American Journal of Physical Anthropology* 20 (3):235–460.

Hrdlička, Aleš. 1935b. *Ear Exostoses*. Vol. 93, *Smithsonian Miscellaneous Collections*. Washington, DC: Smithsonian Institution.

Hrdlička, Aleš. 1937a. The Minnesota "Man." *American Journal of Physical Anthropology* 22 (2):175–199.

Hrdlička, Aleš. 1937b. The Gluteal Ridge and Gluteal Tuberosities (3rd Trochanters). *American Journal of Physical Anthropology* 23:129–198.

Hrdlička, Aleš. 1939a. *Practical Anthropometry*. Philadelphia: Wistar Institute of Anatomy and Biology.

Hrdlička, Aleš. 1939b. Trepanation among Prehistoric People, Especially in America. *Ciba Symposia* I:170–177, 200.

Hrdlička, Aleš. 1940a. *Catalogue of Human Crania in the United States National Museum Collections: Indians of the Gulf States*. Vol. 87, *Proceedings of the United States National Museum*. Washington, DC: U.S. Government Printing Office.

Hrdlička, Aleš. 1940b. *Ritual Ablation of Front Teeth in Siberia and America*. Washington, DC.

Hrdlička, Aleš. 1941. The Races of Man. In *Scientific Aspects of the Race Problem*, edited by H. S. Jennings. Washington: Catholic University Press of America.

Hrdlička, Aleš. 1942. *Catalogue of Human Crania in the United States National Museum Collections: Eskimo in General*. Vol. 91, *Proceedings of the United States National Museum*. Washington, DC: U.S. Government Printing Office.

Hrdlička, Aleš. 1943a. Contribution to the History of Physical Anthropology in the United States of America, with special reference to Philadelphia. *Proceedings of the American Philosophical Society* 87 (1):61–64.

Hrdlička, Aleš. 1943b. Skull of a Midget from Peru. *American Journal of Physical Anthropology* 1:77–82.

Hrdlička, Aleš. 1944. *Catalog of Human Crania in the United States National Museum Collections: Non-Eskimo Peoples of the Northwest Coast, Alaska and Siberia, Proceedings of the United States National Museum*. Washington, DC: U.S. Government Printing Office.

Hudson, E. H. 1965. Treponematosis and Man's Social Evolution. *American Anthropologist* 67 (4):885–901.

Hulse, Fred S. 1941. The People Who Lived at Irene: Physical Anthropology. In *Irene Mound, Chatham County, Georgia*, edited by J. Caldwell and C. McCann. 57–68. Athens, GA: University of Georgia Press.

Hunn, Eugene S. 2000. Chapter 4: Review of Linguistic Information. In "Cultural Affiliation Report." Retrieved Dec. 31, 2005, from Department of the Interior, National Park Service, Archaeology Program Web Site: http://www.cr.nps.gov/aad/kennewick/hunn.htm.

Hunt, David R. 2002. Aleutian Remains at the Smithsonian Institution. In *To the Aleutians and Beyond: The Anthropology of William S. Laughlin*, edited by B. Froelich, A. B. Harper, R. Gilberg, pp. 137–167. Copenhagen: National Museum of Denmark.

Hunt, David R. 2004. Obituary: Lucile E. St. Hoyme (1924–2001). *American Journal of Physical Anthropology* 123 (2):103–105.

Hunt, David R., and J. Albanese. 2005. History and Demographic Composition of the Robert J. Terry Anatomical Collection. *American Journal of Physical Anthropology* 127:406–417.

Huntington, Richard, and Peter Metcalf. 1979. *Celebrations of Death: The Anthropology of Mortuary Ritual*. Cambridge: Cambridge University Press.

Hurlbut, S. A. 2000. The Taphonomy of Cannibalism: A Review of Anthropogenic Bone Modification in the American Southwest. *International Journal of Osteoarchaeology* 10:4–26.

Huss-Ashmore, Rebecca, Alan H. Goodman, and George J. Armelagos. 1982. Nutritional Inference from Paleopathology. In *Advances in Archaeological Method and Theory*, edited by M. B. Schiffer. 395–474. New York: Academic Press.

Hylander, William L. 1988. Implications of In Vivo Experiments for Interpreting the Functional Significance of "Robust" Australopithecine Jaws. In *Evolutionary History of the "Robust" Australopithecines*, edited by F. E. Grine. 55–81. New York: Aldine de Gruyter.

Hylander, William L., and Kirk R. Johnson. 1994. Jaw Muscle Function and Wishboning of the Mandible during Mastication in Macaques and Baboons. *American Journal of Physical Anthropology* 94:523–547.

Insoll, Timothy. 2004. *Archaeology, Ritual, Religion*. New York: Routledge.

Irish, J. D., and Christy G. Turner II. 1987. More Lingual Surface Attrition of the Maxillary Anterior Teeth in American Indians: Prehistoric Panamanians. *American Journal of Physical Anthropology* 73:209–213.

Isaac, Barbara. 2002. Implementation of NAGPRA: The Peabody Museum of Archaeology and Ethnology, Harvard. In *The Dead and Their Possessions. Repatriation in Principle, Policy and Practice*, edited by C. Fforde, J. Hubert and P. Turnbull. 160–170. New York: Routledge.

İşcan, Mehmet Yaşar, and Kenneth A. R. Kennedy, eds. 1989. *Reconstruction of Life from the Skeleton*. New York: Alan R. Liss.

Ishida, H., and Y. Dodo. 1997. Cranial Variation in Prehistoric Human Skeletal Remains from the Marianas. *American Journal of Physical Anthropology* 104:399–410.

Jablonski, K. 1981. Enamel Calcification Patterns of the Averbuch Site (40DV60): Nashville, Tennessee. M.A. Thesis, Department of Anthropology, University of Tennessee, Knoxville.

Jackes, Mary K. 1983. Osteological Evidence for Smallpox: A Possible Case from Seventeenth Century Ontario. *American Journal of Physical Anthropology* 60:75–81.

Jackes, M. K. 1985. Pubic Symphysis Age Distributions. *American Journal of Physical Anthropology* 68:281–299.

Jackes, M. K. 2000. Building the Bases for Paleodemographic Analysis: Adult Age Estimation. In *Biological Anthropology of the Human Skeleton*, edited by M. A. Katzenberg and S. R. Saunders. 417–466. New York: Wiley-Liss.

Jackes, M. K. 2003. Testing a Method: Paleodemography and Proportional Fitting. *American Journal of Physical Anthropology* 121:385–386.

Jackson, John Barnard Swett. 1847. *A Descriptive Catalogue of the Anatomical Museum of the Boston Society for Medical Improvement*. Boston: William D. Ticknor and Company.

Jackson, John Barnard Swett. 1870. *A Descriptive Catalogue of the Warren Anatomical Museum*. Boston: A. Williams and Company.

Jacobi, Keith P. 2002. A Time Capsule of Physical Anthropology: Charles E. Snow's WPA Letters, 1940–1941. *Southeastern Archaeology* 21:55–62.

Jacobi, Keith P., Della Collins Cook, R. Corruccini, and J. Handler. 1992. Congenital Syphilis in the Past: Slaves at Newton Plantation, Barbados. *American Journal of Physical Anthropology* 89:145–158.

Jacobi, Keith P., and M. Cassandra Hill. 2001. Prehistoric Treatment of the Dead: Bone Handling in the Southeastern United States [Abstract]. *American Journal of Physical Anthropology* (Supplement 32):85.

Jacobs, K. 1993. Human Postcranial Variation in the Ukranian Mesolithic-Neolithic. *Current Anthropology* 34:311–324.

Jacquard, A. 1974. *The Genetic Structure of Populations*. New York: Springer-Verlag.

Jahoda, G. 1999. *Images of Savages: Ancient Roots of Modern Prejudices in Western Culture*. London: Routledge.

Jannssens, Paul A. 1970. *Palaeopathology: Diseases and Injuries of Prehistoric Man*. London: John Baker.

Jantz, R. L. 1973. Microevolutionary Change in Arikara Crania: A Multivariate Analysis. *American Journal of Physical Anthropology* 38:15–26.

Jantz, Richard L., and Douglas W. Owsley. 1994. White Traders in the Upper Missouri: Evidence from the Swan Creek Site. In *Skeletal Biology in the Great Plains: Migration, Warfare, Health and Subsistence*. 189–201. Washington, DC: Smithsonian Institution Press.

Jantz, R. L., and P. Willey. 1983. Temporal and Geographic Patterning of Relative Head Height in the Central Plains and Middle Missouri Areas. *Plains Anthropologist* 28:59–67.

Jarcho, Saul. 1964. Lead in the Bones of Prehistoric Lead-Glaze Potters. *American Antiquity* 330(1):94–96.

Jarcho, Saul. 1965a. Anomaly of the Vertebral Column (Klippel-Feil Syndrome) in American Aborigines. *Journal of the American Medical Association* 193(10):843–844.

Jarcho, Saul. 1965b. Paleopathology. *Science* 147(3662):1160–1163.

Jarcho, Saul. 1966a. The Development and Present Condition of Human Palaeopathology in the United States. In *Human Palaeopathology*, edited by S. Jarcho. 3–30. New Haven: Yale University Press.

Jarcho, Saul, ed. 1966b. *Human Palaeopathology*. New Haven: Yale University Press.

Jarcho, Saul, Norman Simon, and Edgar M. Bick. 1963. A Fused Hip from Wupatki. *Plateau* 35(3):69–74.

Jarcho, Saul, Norman Simon, and Henry L. Jaffe. 1965. Symmetrical Osteoporosis (Spongy Hyperostosis) in a Prehistoric Skull from New Mexico. *El Palacio* 72(4):26–30.

Jefferson, Thomas. 1853. *Notes on the State of Virginia*. Richmond, VA: J. W. Randolph.

Jelderks, Magistrate Judge John. 2001. Court Opinion on Polynesian Motion to Intervene. Retrieved Dec. 31, 2005, from Friends of America's Past Web Site: http://www.friendsofpast.org/kennewick-man/court/opinions/opinion-polynesian.html.

Jelderks, Magistrate Judge John. 2002. In the United States District Court for the District of Oregon, Ronson Bonnichsen v. United States of America. Opinion and Order. Civil No. 96-1481-Je. Retrieved Dec. 31, 2005, from Friends of America's Past Web Site: http://www.friendsofpast.org/pdf/decision-020830.pdf, www.friendsofpast.org/kennewick-man/court/opinions/opinions.html.

Jerkic, Sonia M. 2001. The Influence of James E. Anderson on Canadian Physical Anthropology. In *Out of the Past: The History of Human Osteology at the University of Toronto*, edited by L. Sawchuk, S. Pfeiffer. Toronto: CITD Press.

Johnson, Elden. 1973. Professional Responsibilities and the American Indian. *American Antiquity* 38 (2):129–130.

Johnston, Basil. 1976. The Cultural and Ethical Aspects of Archaeology in Canada: An Indian Point of View. In *Symposium on New Perspectives in Canadian Archaeology, 22–23 October 1976, Theatre Auditorium, Royal Ontario Museum, Toronto*, edited by A. G. McKay. 173–175. Ottawa: The Royal Society of Canada.

Johnston, Francis E. 1962. Growth of the Long Bones of Infants and Young Children at Indian Knoll. *American Journal of Physical Anthropology* 20:249–254.

Johnston, Francis E., and Charles E. Snow. 1961. The Reassessment of the Age and Sex of the Indian Knoll Skeletal Population: Demographic and Methodological Aspects. *American Journal of Physical Anthropology* 19:237–244.

Jones, Joseph. 1876. *Explorations of the Aboriginal Remains of Tennessee*. Vol. 22(259), Article II, *Smithsonian Contributions to Knowledge*. Washington, DC: Smithsonian Institution.

Jones, Joseph. 1878. Explorations and Researches Concerning the Destruction of the Aboriginal Inhabitants of America by Various Diseases, as Syphilis, Matlazahuatl, Pestilence, Malarial Fever, and Small-Pox. *The New Orleans Medical and Surgical Journal* V (926–941).

Jones, Joseph. 1901. *Abstract of the Catalogue of the Archaeological Collection of Joseph Jones, MD.*: Searcy & Pfaff.

Jones, Joseph, and Susan Polk Jones. 1901–1909. Joseph Jones Papers, Louisiana and Lower Mississippi Valley Collections. Box 6. Baton Rouge, Louisiana: LSU Libraries.

Jones, Russell Moseley. 1978. *The Parisian Education of an American Surgeon: Letters of Jonathan Mason Warren, 1832–1835*. Philadelphia: American Philosophical Society.

Jones, V. H. 1976. James Bennett Griffin, Archaeologist. In *Cultural Change and Continuity: Essays in Honor of James Bennett Griffin*, edited by C. E. Cleland. xxxix–lxxvii. New York: Academic Press.

Joyce, B. A. 2001. *The Shaping of American Ethnography: The Wilkes Expedition, 1838–1842*. Lincoln: University of Nebraska Press.

Jurmain, Robert D. 1975. *Distribution of Degenerative Joint Disease in Skeletal Populations*. Ph.D. dissertation. Harvard University.

Jurmain, Robert D. 1977a. Stress and the Etiology of Osteoarthritis. *American Journal of Physical Anthropology* 46:353–366.

Jurmain, Robert D. 1977b. Paleoepidemiology of Degenerative Knee Disease. *Medical Anthropology* 1:1–14.

Jurmain, Robert D. 1980. The Pattern of Involvement of Appendicular Degenerative Joint Disease. *American Journal of Physical Anthropology* 53:143–150.

Jurmain, Robert D. 1990. Paleoepidemiology of a Central California Prehistoric Population from CA-Ala-329. II. Degenerative Disease. *American Journal of Physical Anthropology* 83:83–94.

Jurmain, Robert D. 1991. Paleoepidemiology of Trauma in a Central California Population. In *Paleopathology: Current Synthesis and Future Options*, edited by D. J. Ortner and A. C. Aufderheide. 241–248. Washington, DC: Smithsonian Institution Press.

Jurmain, Robert D. 1999. *Stories from the Skeleton: Behavioral Reconstruction in Human Osteology*. Vol. 1, *Interpreting the Remains of the Past*. Toronto and Amsterdam: Gordon and Breach.

Kaestle, Frederika. 2000. Chapter 2: Report on DNA Analysis of the Remains of "Kennewick Man" from Columbia Park, Washington. In "Report on the DNA Testing Results of the Kennewick Human Remains from Columbia Park, Kennewick, Washington." Retrieved Dec. 31, 2005, from Department of the Interior, National Park Service, Archaeology Program Web Site: http://www.cr.nps.gov/aad/kennewick/kaestle.htm.

Kaestle, Frederika A. 1995. Mitochondrial DNA Evidence for the Identity of the Descendants of the Prehistoric Stillwater Marsh Population. *Anthropological Papers of the American Museum of Natural History* 77:73–80.

Kaestle, Frederika A., and K. Ann Horsburgh. 2002. Ancient DNA in Anthropology: Methods, Applications, and Ethics. *Yearbook of Physical Anthropology* 45:92–130.

Kaestle, Frederika A., J. G. Lorenz, and David Glenn Smith. 1999. Molecular Genetics and the Numic Expansion: A Molecular Investigation of the Prehistoric Inhabitants of Stillwater Marsh. In *Prehistoric Lifeways in the Great Basin Wetlands: Bioarchaeological Reconstruction and Interpretation*, edited by B. M. Hemphill and C. S. Larsen. Salt Lake City: University of Utah Press.

Kannus, Pekka, Heidi Haapasalo, Marja Sankelo, Harri Sievänen, Matti Pasanen, Ari Heinonen, Pekka Oja, and Ilkka Vuori. 1995. Effect of Starting-Age of Physical Activity on Bone Mass in the Dominant Arm of Tennis and Squash Players. *Annals of Internal Medicine* 123:27–31.

Kantner, J. 1996. Political Competition among the Chaco Anasazi of the American Southwest. *Journal of Anthropological Archaeology* 15:41–105.

Katzenberg, Mary Anne. 2001. Destructive Analyses of Human Remains in the Age of NAGPRA and Related Legislation. Available from https://tspace.library.utoronto.ca/citd/Osteology/Katzenberg.html. In *Out of the Past: The History of Human Osteology at the University of Toronto. Proceedings of a Symposium Held at the University of Toronto, October 25, 2000. Available from https://tspace.library.utoronto.ca/citd/Osteology/index.html.*, edited by L. Sawchuk and S. Pfeiffer: CITD Press.

Katzenberg, Mary Anne, and Shelley Rae Saunders. 2000. *Biological Anthropology of the Human Skeleton*. New York: Wiley-Liss.

Kay, R. F. 1987. Analysis of Primate Dental Microwear Using Image Processing Techniques. *Scan Microscopy* 1:657–662.

Keeley, Lawrence H. 1996. *War before Civilization*. New York: Oxford University Press.

Keenleyside, Anne. 1998. Skeletal Evidence of Health and Disease in Pre-Contact Alaskan Eskimos and Aleuts. *American Journal of Physical Anthropology* 107(1):51–70.

Keenleyside, Anne. 2003. An Unreduced Dislocated Mandible in an Alaskan Eskimo: A Case of Altruism or Adaptation? *International Journal of Osteoarchaeology* 13:384–389.

Keenleyside, Anne, and Robert W. Mann. 1991. Unilateral Ectrodactyly, Metatarsal Synostosis, and Hypoplasia in an Eskimo. *Journal of the American Podiatric Medical Association* 81:18–21.

Kehoe, Alice Beck. 1998. *The Land of Prehistory: A Critical History of American Archaeology*. New York: Routledge.

Kehoe, Alice Beck. 2002. Europe's Prehistoric Dawn Reproduced: Daniel Wilson's Magisterial Archaeology. In *New Perspectives on the Origins of Americanist Archaeology*, edited by D. L. Browman and S. Williams. 133–147. Tuscaloosa: University of Alabama Press.

Keith, Arthur. 1924. Description of Three Crania from Aveline's Hole. *Proceedings of the Bristol Spelaeological Society* 2:16.

Keith, Arthur. 1931. *New Discoveries Relating to the Antiquity of Man*. London: W. W. Norton.

Keith, Arthur. 1940. Fifty Years Ago. *American Journal of Physical Anthropology* 26:251–267.

Keith, Kenneth D. 1973. The Paleostomatology at the Moore Site (Lf-31), Leflore County, Oklahoma. *Bulletin of the Oklahoma Anthropological Society* 22:149–157.

Kelley, Jennifer Olsen, and J. Lawrence Angel. 1987. Life Stresses of Slavery. *American Journal of Physical Anthropology* 74:199–211.

Kelley, Marc A., and Clark Spencer Larsen, eds. 1991. *Advances in Dental Anthropology*. New York: Wiley-Liss.

Kelly, Marjorie. 1943. Report on the Skeletal Material. In *The S.U. Site: Excavations at a Mogollon Village, Western New Mexico, Second Season, 1941*, edited by P. S. Martin. 250–251. Vol. 32 (2), *Anthropological Series*. Chicago: Field Museum of Natural History.

Kelly, Robert L. 2004. Kennewick Man Is Native American. *The SAA Archaeological Record* 4:32–37.

Kennedy, Brenda. 1981. *Marriage Patterns in an Archaic Population: A Study of Skeletal Remains from Port Au Choix, Newfoundland, Archaeological Survey of Canada, National Museum of Man, Mercury Series, Paper No. 104*. Ottawa: National Museums of Canada.

Kennedy, Kenneth A. R. 1983. Morphological Variations in Ulnar Supinator Crests and Fossae, as Identifying Markers of Occupational Stress. *Journal of Forensic Science* 28 (4):871–876.

Kennedy, Kenneth A. R. 1984. Biological Adaptations and Affinities of Mesolithic South Asians. In *The People of South Asia: The Biological Anthropology of India, Pakistan, and Nepal*, edited by J. R. Lukacs. 29–57. New York: Plenum.

Kennedy, Kenneth A. R. 1989. Skeletal Markers of Occupational Stress. In *Reconstruction of Life from the Skeleton*, edited by M. Y. İşcan and K. A. R. Kennedy. 129–160. New York: Alan R. Liss.

Kennedy, Kenneth A. R., and Sheilagh T. Brooks. 1984. Theodore D. McCown: A Perspective on a Physical Anthropologist. *Current Anthropology* 25:99–103.

Kent, Susan. 1986. The Influence of Sedentism and Aggregation on Porotic Hyperostosis and Anaemia: A Case Study. *Man* 21:605–636.

Kent, Susan, E. D. Weinberg, and Patricia Stuart-Macadam. 1990. Dietary and Prophylactic Iron Supplements: Helpful or Harmful? *Human Nature* 1:53–79.

Kent, Susan, E. D. 1994. The Etiology of the Anemia of Chronic Disease and Infection. *Journal of Clinical Epidemiology* 47:23–33.

Kenyon, Walter A. 1977. Some Bones of Contention. The Neutral Indian Burial Site at Grimsby. *Rotunda* 10:4–13.

Kenyon, Walter A. 1982. *The Grimsby Site, an Historic Neutral Cemetery*. Toronto: Royal Ontario Museum.

Kerley, Ellis R., and William M. Bass. 1967. Paleopathology: Meeting Ground for Many Disciplines. *Science* 157:638–644.

Key, P. J. 1983. *Craniometric Relationships among Plains Indians*. Knoxville: University of Tennessee.

Key, P. J., and R. L. Jantz. 1981. A Multivariate Analysis of Temporal Change in Arikara Craniometrics: A Methodological Approach. *American Journal of Physical Anthropology* 55:247–259.

Khan, Karim, Heather A. McKay, Heidi Haapasalo, Kim L. Bennell, Mark R. Forwood, Pekka Kannus, and John D. Wark. 2000. Does Childhood and Adolescence Provide a Unique Opportunity for Exercise to Strengthen the Skeleton? *Journal of Science and Medicine in Sport* 3:150–164.

Kidd, Kenneth E. 1954. A Note on the Paleopathology of Ontario. *American Journal of Physical Anthropology* 12:610–615.

Kidder, A. V. 1924. *An Introduction to the Study of Southwestern Archaeology with a Preliminary Account of the Excavations at Pecos*. New Haven: Pub. for the Department of Archaeology, Phillips Academy, Andover, MA. by the Yale University Press.

Kidder, A. V. 1962. *An Introduction to the Study of Southwestern Archaeology with a Preliminary Account of the Excavations at Pecos, Papers of the Southwestern Expedition*. New Haven: Yale University Press.

Kidder, M. A., and A. V. Kidder. 1917. Notes on the Pottery of Pecos. *American Anthropologist* 19 (3):325–360.

Killan, Gerald. 1983. *David Boyle: From Artisan to Archaeologist*. Toronto: Published in association with the Ontario Heritage Foundation by University of Toronto Press.

Killion, Thomas W., and Paula Molloy. 2000. Repatriation's Silver Lining. In *Working Together: Native Americans and Archaeologists*, edited by K. E. Dongoske, M. Aldenderfer, and K. Doehner. 111–117. Washington, DC: Society for American Archaeology.

Kimura, D. K. 1977. Statistical Assessment of the Age-Length Key. *Journal of the Fisheries Research Board of Canada* 34:317–324.

Kimura, D. K., and S. Chikuni. 1987. Mixtures of Empirical Distributions: An Iterative Application of the Age-Length Key. *Biometrics* 43:23–35.

Kimura, Tasuku. 1974. Mechanical Characteristics of Human Lower Leg Bones. *Journal of the Faculty of Science of the University of Tokyo, Section V. Anthropology* 4:319–393.

Kintigh, K. W. 1996. The Cibola Region in the Post-Chacoan Era. In *The Prehistoric Pueblo World, A.D. 1150–1350*, edited by M. A. Adler. 131–144. Tucson, AZ: University of Arizona Press.

Kiple, Kenneth F., ed. 1993. *The Cambridge World History of Human Disease*. Cambridge: Cambridge University Press.

Klatsky, M., and R. L. Fisher. 1953. *The Human Masticatory Apparatus: An Introduction to Dental Anthropology*. Brooklyn: Dental Items of Interest Publishing Co.

Klepinger, Linda. 1972. An Early Human Skeleton from the Snyder Site, 14BU9, Butler County Kansas. *Plains Anthropologist* 17:71–72.

Klesert, A. L. 1992. A View from Navajohand on the Reconciliation of Anthropologists and Native Americans. *Human Organization* 51:17–22.

Klesert, Anthony L., and Michael J. Andrews. 1988. Treatment of Human Remains on Navajo Lands. *American Antiquity* 53 (2):310–320.

Klesert, Anthony L., and Alan S. Downer, eds. 1990. *Preservation on the Reservation. Native Americans, Native American Lands and Archaeology, Navajo Nations Papers in Anthropology Number 26*: Navajo Nation Archaeology, Navajo Nation Historic Preservation Department.

Klesert, Anthony L., and Shirley Powell. 1993. A Perspective on Ethics and the Reburial Controversy. *American Antiquity* 58 (2):348–354.

Kluth, Rose, and Kathy Munnell. 1997. The Integration of Tradition and Scientific Knowledge on the Leech Lake Reservation. In *Native Americans and Archaeologists. Stepping Stones to Common Ground*, edited by N. Swidler, K. E. Dongoske, R. Anyon, and A. S. Downer. 112–119. Walnut Creek, CA: AltaMira Press in Cooperation with the Society for American Archaeology.

Kneberg, Madeline D. 1935. Improved Technique for Hair Examination. *American Journal of Physical Anthropology* 20:51–67.

Kneberg, Madeline D. 1936a. Differential Staining of Thick Sections of Tissues. *Science* 83 (2162):561–562.

Kneberg, Madeline D. 1936b. Hair Weight as a Racial Criterion. *American Journal of Physical Anthropology* 21:279–286.

Kneberg, Madeline D. 1945. The Persistent Potsherd. *Tennessee Archaeologist* 1 (4):4–5.

Kneberg, Madeline D. 1947. Improved Technique for Hair Examination. M.A., Department of Anthropology, University of Chicago.

Kneberg, Madeline D. 1951a. An Archaic Autobiography. *Tennessee Archaeologist* 7 (1):1–5.

Kneberg, Madeline D. 1951b. An Early Woodland Autobiography. *Tennessee Archaeologist* 7 (2):31–38.

Kneberg, Madeline D. 1952a. The Autobiography of a Memorial Mound Builder. *Tennessee Archaeologist* 8 (1):1–9.

Kneberg, Madeline D. 1952b. The Autobiography of a "Bone House Indian." *Tennessee Archaeologist* 8 (2):37–41.

Kneberg, Madeline D. 1952c. The Tennessee Area. In *Archaeology of Eastern United States*, edited by J. B. Griffin. 190–198. Chicago: University of Chicago Press.

Kneberg, Madeline D. 1954. The Duration of the Archaic Tradition in the Lower Tennessee Valley. *Southern Indian Studies* 5:40–44.

Kneberg, Madeline D. 1956. Some Important Projectile Point Types Found in the Tennessee Area. *Tennessee Archaeologist* 12 (1):17–28.

Kneberg, Madeline D. 1957. Chipped Stone Artifacts of the Tennessee Valley Area. *Tennessee Archaeologist* 13 (1):55–65.

Kneberg, Madeline D. 1959. Engraved Shell Gorgets and Their Associations. *Tennessee Archaeologist* 15 (1):1–39.

Kneberg, Madeline D. 1961. Four Southeastern Limestone-Tempered Pottery Complexes. *Southeastern Archaeological Conference Newsletter* 7 (2):3–15.

Kneberg, Madeline D. 1962. Woodland Fabric Marked Ceramic System. In *Proceedings of the Sixteenth Southeastern Archaeological Conference (Macon, Georgia 1959), Newsletter*, edited by S. Williams. 33–40. Vol. 8.

Kneip, L., L. Pallestrini, F. Crancio, and L. M. C. Machado. 1991. As Estruturas e Suas Inter-Relações em Sítios de Caçadores-Coletores do Litoral de Saquarema, RJ. *Resumos da VI Reunião Científica da Sociedade de Arqueologia Brasileira (Sab), Rio De Janeiro*:78.

Knight, Marian Vera. 1915. *The Craniometry of Southern New England Indians. With an Introduction by Harris Hawthorne Wiulder*. Vol. 4, *Memoirs of the Connecticut Academy of Arts and Sciences*. New Haven, CT: Yale University Press.

Kobori, L., Sheilagh T. Brooks, R. J. Miller, C. Stevens, M. Galliher, and D. H. Morris. 1980. Great Basin Occurrence of a Southwestern Dental Trait: The Uto-Aztecan Premolar. *Nevada Historical Quarterly* 23:236–245.

Kolman, C. J., and Noreen Tuross. 2000. Ancient DNA Analysis of Human Populations. *American Journal of Physical Anthropology* 111:5–23.

Konigsberg. 1988. Migration Models of Prehistoric Residence. *American Journal of Physical Anthropology* 77 (4):471–482.

Konigsberg, Lyle W. 1985. Demography and Mortuary Practice at Seip Mound One. *Midcontinental Journal of Archaeology* 10:123–148.

Konigsberg, Lyle W. 1987. Population Genetic Models for Interpreting Prehistoric Intra-Cemetery Biological Variation. Ph.D. Dissertation, Northwestern University.

Konigsberg, Lyle W. 1988. Migration Models of Prehistoric Residence. *American Journal of Physical Anthropology* 77 (4):471–482.

Konigsberg, Lyle W. 1990a. Analysis of Prehistoric Biological Variation under a Model of Isolation by Geographic and Temporal Distance. *Human Biology* 62:49–70.

Konigsberg, Lyle W. 1990b. Temporal Aspects of Biological Distance: Serial Correlation and Trend in a Prehistoric Skeletal Lineage. *American Journal of Physical Anthropology* 82:45–52.

Konigsberg, Lyle W. 1997. Comments on Matrix Permutation Tests in the Evaluation of Competing Models for Modern Human Origins. *Journal of Human Evolution* 32:479–488.

Konigsberg, Lyle W., and Jane E. Buikstra. 1995. Regional Approaches to the Investigation of Past Human Biocultural Structure. In *Regional Approaches to Mortuary Analysis*, edited by L. Beck. 191–219. New York: Plenum Press.

Konigsberg, Lyle W., Jane E. Buikstra, and J. Bullington. 1989. Paleodemographic Correlates of Fertility: A Reply to Corruccini, Brandon, and Handler and to Holland. *American Antiquity* 54:626–636.

Konigsberg, Lyle W., and Susan R. Frankenberg. 1992. Estimation of Age Structure in Anthropological Demography. *American Journal of Physical Anthropology* 89:235–256.

Konigsberg, Lyle W., and Susan R. Frankenberg. 1994. Paleodemography: "Not Quite Dead." *Evolutionary Anthropology* 3 (3):92–105.

Konigsberg, Lyle W., and Susan R. Frankenberg. 2002. Deconstructing Death in Paleodemography. *American Journal of Physical Anthropology* 117 (4):297–309.

Konigsberg, Lyle W., and S. M. Hens. 1998. The Use of Ordinal Categorical Variables in Skeletal Assessment of Sex from the Cranium. *American Journal of Physical Anthropology* 107:97–112.

Konigsberg, Lyle W., S. M. Hens, L. M. Jantz, and W. L. Jungers. 1998. Stature Estimation and Calibration: Bayesian and Maximum Likelihood Perspectives in Physical Anthropology. *Yearbook of Physical Anthropology* 41:65–92.

Konigsberg, Lyle W., and D. Holman. 1999. Estimation of Age-at-Death from Dental Emergence and Implications for Studies of Prehistoric Somatic Growth. In *Human*

Growth in the Past: Studies from Bones and Teeth, edited by R. D. Hoppa and C. M. FitzGerald. 264–289. New York: Cambridge University Press.

Konigsberg, Lyle W., L. A. P. Kohn, and J. M. Cheverud. 1993. Cranial Deformation and Nonmetric Trait Variation. *American Journal of Physical Anthropology* 90:35–48.

Konigsberg, Lyle W., A. Kramer, S. M. Donnelly, J. H. Relethford, and J. Blangero. 1994. Modern Human Origins. *Nature* 372:228–229.

Kontulainen, Saija, Pekka Kannus, Heidi Haapasalo, Harri Sievänen, Matti Pasanen, Ari Heinonen, Pekka Oja, and Ilkka Vuori. 2001. Good Maintenance of Exercise-Induced Bone Gain with Decreased Training of Female Tennis and Squash Players: A Prospective 5-Year Follow-up Study of Young and Old Starters and Controls. *Journal of Bone and Mineral Research* 16:195–201.

Korioth, Tom W. P., Douglas P. Romilly, and Alan G. Hannam. 1992. Three-Dimensional Finite Element Stress Analysis of the Dentate Human Mandible. *American Journal of Physical Anthropology* 88:69–96.

Koritzer, Richard T., A. H. Schwartz, R. R. Harris, and Lucile E. St. Hoyme. 1992. Previously Undescribed Disk-Capsule Innervation: Some Speculative Thoughts for Tmd Clinicians. *American Journal of Orthodontics and Dentofacial Orthopedics* 102:109–112.

Koritzer, Richard T., and Lucile E. St. Hoyme. 1970a. Age and Area in Dental Paleopathology [Abstract]. *American Journal of Physical Anthropology* 33:135.

Koritzer, Richard T., and Lucile E. St. Hoyme. 1970b. Unusual Dental Pathology in an Illinois Indian Ca. 500 A.D. *Dental Digest* 76:386–387.

Koritzer, Richard T., and Lucile E. St. Hoyme. 1971a. Primate Canine Relation to Jaw Function and Taxonomy [Abstract]. *American Journal of Physical Anthropology* 35:285.

Koritzer, Richard T., and Lucile E. St. Hoyme. 1971b. Canine Guidance of Primate Chewing. Georgetown University Teaching Film.

Koritzer, Richard T., and Lucile E. St. Hoyme. 1971c. New Evidence on Prehistoric Amerindian Dental Repair. *Biomedical News* (July).

Koritzer, Richard T., and Lucile E. St. Hoyme. 1971d. Zahnsplomben bei Praehistorischen Indianern. *Munch Med Wochenschr* 113 (4):4.

Koritzer, Richard T., and Lucile E. St. Hoyme. 1971e. Commentary on Prehistoric Dentistry. *Dental Radiography and Photography* 44 (3):65.

Koritzer, Richard T., and Lucile E. St. Hoyme. 1974. Sex and Age Comparisons between Two Sites: Accokeek and Potomac Creek [Abstract]. *American Journal of Physical Anthropology* 40:142.

Koritzer, Richard T., and Lucile E. St. Hoyme. 1979a. Rhodesian Man Elemental Analysis [Abstract]. *Journal of Dental Research* 58:232.

Koritzer, Richard T., and Lucile E. St. Hoyme. 1979b. Extensive Caries in Early Man Circa 110,000 Years before Present. *Journal of the American Dental Association* 99:642–643.

Koritzer, Richard T., and Lucile E. St. Hoyme. 1980. Caries and Elemental Composition of the Rhodesian Man Dentition. *Journal of the Washington Academy of Sciences* 70:74–79.

Koritzer, Richard T., and Lucile E. St. Hoyme. 1984. Cranial Base Variation: The Digastric Groove [Abstract]. *American Journal of Physical Anthropology* 63:179.

Koritzer, Richard T., and Lucile E. St. Hoyme. 1985. Dental Pathology of the Maroons at Accompong Town, Jamaica: A Discrete Village Study [Abstract]. *American Journal of Physical Anthropology* 66:191.

Koritzer, Richard T., and Lucile E. St. Hoyme. 1986. Primate Craniomandibular Relations [Abstract]. *American Journal of Physical Anthropology* 69:224.

Koritzer, Richard T., and Lucile E. St. Hoyme. 1988. Correlation between Facial and Incisor Dimensions in Gorilla and Man [Abstract]. *American Journal of Physical Anthropology* 75:233.

Koritzer, Richard T., and Lucile E. St. Hoyme. 1989. Time and Space Variation in Gorilla Face and Incisors [Abstract]. *American Journal of Physical Anthropology* 78:255.

Koritzer, Richard T., and Lucile E. St. Hoyme. 1992. Biophysical Model for Cranio-Mandibular Management. In *New Concepts in Craniomandibular and Chronic Pain Management*, edited by H. Gelb. 187–214. Tokyo: Shiyaqu Euro-American Inc.

Koritzer, Richard T., and Lucile E. St. Hoyme. 1993. The Function Association of TM Articular Disc and Cartilaginous Ear Tube in Humans [Abstract]. *American Journal of Physical Anthropology* Suppl. 16:198.

Koritzer, Richard T., and Lucile E. St. Hoyme. 1997. Current Dissections of the Pterygopalatine Fossa [Abstract]. *American Journal of Physical Anthropology* Suppl. 24:146.

Koritzer, Richard T., and Lucile E. St. Hoyme. 1999. Exploring the Relation between Muscle Insertion Dimensions and Tooth Use [Abstract]. *American Journal of Physical Anthropology* Suppl. 28:174.

Koritzer, Richard T., Lucile E. St. Hoyme, M. Martin, A. Martin, and J. Canary. 1982. Caries and Juvenile Onset Diabetes: A Metabolic Viewpoint. *Journal of the Washington Academy of Sciences* 72:98–104.

Koritzer, Richard T., Lucile E. St. Hoyme, A. Tylenda, and N. Papadopoulos. 1987. Lysozyme Concentration Variation with Age and Periodontal Disease [Abstract]. *American Journal of Physical Anthropology* 72:221.

Kraus, B. S. 1954. Comment on "Some Problems in the Physical Anthropology of the American Southwest" (by J. N. Spuhler). *American Anthropologist* 56 (4):620–623.

Kreshover, S. J. 1940. Histopathologic Studies of Abnormal Enamel Formation in Human Teeth. *American Journal of Orthodontia* 26:1083–1101.

Kreshover, S. J. 1960. Metabolic Disturbance in Tooth Formation. *Annual of the New York Academy of Sciences* 85:161–167.

Kreshover, S. J., and O. W. Clough. 1953. Prenatal Influences on Tooth Development: Artificially Induced Fever in Rats. *Journal of Dental Research* 32:565–577.

Kreutz, K. 1997. *Ätiologie und Epidemiologie von Erkrankungen des Kindesalters bei der Bajuwarischen Population von Straubing (Niederbayern).* Göttingen: Cuvillier Verlag.

Kristmanson, Helen. 1997. The Micmac and New Brunswick Archaeology: Historical Perspectives and Contemporary Experiences. In *At a Crossroads: Archaeology and First Peoples of Canada*, edited by G. P. Nicholas and T. D. Andrews. 19–32. Burnaby: Archaeology Press.

Kroeber, A. L. 1927. Disposal of the Dead. *American Anthropologist* 29:308–315.

Krogman, Wilton Marion. 1935. Life Histories Recorded in Skeletons. *American Anthropologist* 37 (1):92–103.

Krogman, Wilton Marion. 1938. The Role of Urbanization in the Dentitions of Various Population Groups. *Zeitschrift für Rassenkunde, Ferdinand Enke-Verlag* 7:41–72.

Krogman, Wilton Marion. 1939. Guide to the Identification of Human Skeletal Material. *FBI Law Enforcement Bulletin* 8:7–12.

Krogman, Wilton Marion. 1941. Aboriginal Physical Types in the Western Hemisphere. *Ciba Symposia* 3 (1):804–812.

Krogman, Wilton Marion. 1962. *The Human Skeleton in Forensic Medicine*. 3rd ed. Springfield, IL: Charles C Thomas.

Krogman, Wilton Marion. 1981. Fay-Cooper Cole, 1881–1961. *American Journal of Physical Anthropology* 56 (4):467–471.

Krupnik, Igor. 2003. The Surveyor of the "Bridge": Interview with William S. Laughlin. In *To the Aleutians and Beyond. The Anthropology of William S. Laughlin*, edited by B. Frohlich, A. B. Harper, and R. Gilberg. 207–226. Vol. 20, *Publications of the National Museum of Denmark, Ethnographical Series*. Copenhagen: Department of Ethnography, National Museum of Denmark.

Kuckelman, Kristin A., Ricky R. Lightfoot, and Debra L. Martin. 2000. Changing Patterns of Violence in the Northern San Juan Region. *The Kiva* 66 (1):147–165.

Kuckelman, Kristin A., Ricky R. Lightfoot, and Debra L. Martin. 2002. The Bioarchaeology and Taphonomy of Violence at Castle Rock and Sand Canyon Pueblos, Southwestern Colorado. *American Antiquity* 67 (3):486–513.

Kuijt, Ian. 1996. Negotiating Equality through Ritual: A Consideration of Late Natufian and Prepottery Neolithic Period Mortuary Practices. *Journal of Anthropological Archaeology* 15:313–336.

Kullen, D. 2000. Reminiscences of George Langford. *Illinois Archaeology* 12:110–124.

Kus, Susan, and Victor Raharijaona. 1998. Between Earth and Sky There Are Only a Few Large Boulders: Sovereignty and Monumentality in Central Madagascar. *Journal of Anthropological Archaeology* 17:53–79.

Lacerda Filho, João Batista, and Peixoto Rodrigues. 1876. Contribuições Para O Estudo Anthropologico Das Raças Indigenas do Brazil. *Archivos do Museu Nacional de Rio de Janeiro* I:47–75.

LaFantasie, Glenn W., ed. 1988. *The Correspondence of Roger Williams. Volume II. 1654–1682*. Providence, RI: Published for The Rhode Island Historical Society by Brown University Press/University Press of New England.

Lahr, M. M., and J. E. Bowman. 1992. Palaeopathology of the Kechipawan Site: Health and Disease in a South-Western Pueblo. *Journal of Archaeological Science* 19:639–654.

Lai, Ping, and Nancy C. Lovell. 1992. Skeletal Markers of Occupational Stress in the Fur Trade: A Case Study from a Hudson's Bay Company Fur Trade Post. *International Journal of Osteoarchaeology* 2:221–234.

Lallo, J. 1973. The Skeletal Biology of Three Prehistoric American Indian Societies from Dickson Mounds. Ph.D. Dissertation, Department of Anthropology, University of Massachusetts, Amherst.

Lallo, John W., Jerome C. Rose, and George J. Armelagos. 1980. An Ecological Interpretation of Variation in Mortality within Three Prehistoric American Indian Populations from Dickson Mounds. In *Early Native Americans: Prehistoric Demography, Economy, and Technology*, edited by D. L. Browman. 203–238, *World Anthropology*. New York: Mouton.

Lamb, D. S. 1915. The Army Medical Museum in American Anthropology. In *Proceedings of the Nineteenth International Congress of Americanists*. 625–632. Washington, DC: International Congress of Americanists.

Lamb, S. M. 1958. Linguistic Prehistory in the Great Basin. *International Journal of American Linguistics* 24:95–100.

Lambert, Patricia M. 1994. War and Peace on the Western Front: A Stufy of Violent Conflict and Its Correlates in Prehistoric Hunter-Gatherer Societies of Coastal Southern California. Ph.D., University of California, Santa Barbara, Santa Barbara, CA.

Lambert, Patricia M. 1997. Patterns of Violence in Prehistoric Hunter Gatherer Societies of Coastal Southern California. In *Troubled Times: Violence and Warfare in the Past*, edited by D. L. Martin and D. W. Frayer. Amsterdam: Gordon and Breach.

Lambert, Patricia M. 2002. The Archaeology of War: A North American Perspective. *Journal of Archaeological Research* 10:207–241.

Lande, R. 1976. Natural Selection and Random Genetic Drift in Phenotypic Evolution. *Evolution* 30:314–334.

Lande, R., and S. J. Arnold. 1983. The Measurement of Selection on Correlated Characters. *Evolution* 37:1210–1226.

Lane, Nancy E., Beat Michel, Ann Bjorkengren, John Oehlert, Hong Shi, Daniel A. Bloch, and James F. Fries. 1993. The Risk of Osteoarthritis with Running and Aging: A 5-Year Longitudinal Study. *Journal of Rheumatology* 20:462–468.

Lane, Rebecca A. 1977. The Allegany Seneca: A Test of the Genetic Reliability of Nonmetric Osteological Traits for Interpopulational Analysis. Ph.D. Dissertation, University of Texas, Austin.

Lane, Rebecca A., and Audrey J. Sublett. 1972. Osteology of Social Organization: Residence Pattern. *American Antiquity* 37 (2):186–201.

Langdon, F. W. 1881. The Madisonville Prehistoric Cemetery: Anthropological Notes. *Journal of the Cincinnati Society of Natural History* 4:237–257.

Langford, George. 1927. The Fisher Mound Group, Successive Aboriginal Occupations Near the Mouth of the Illinois River. *American Anthropologist* 29 (3):153–205, 352.

Lanphear, K. M. 1989. Testing the Value of Skeletal Samples in Demographic Research: A Comparison with Vital Registration Samples. *International Journal of Anthropology* 4:185–193.

Larsen, Clark Spencer. 1981. Functional Implications of Postcranial Size Reduction on the Prehistoric Georgia Coast, U.S.A. *Human Evolution* 10:489–502.

Larsen, Clark Spencer. 1982. The Anthropology of St. Catherine's Island. 3. Prehistoric Human Biological Adaptation. *Anthropological Papers, American Museum of Natural History* 57 (Part 3):155–270.

Larsen, Clark Spencer. 1985. Dental Modifications and Tool Use in the Western Great Basin. *American Journal of Physical Anthropology* 67 (4):393–402.

Larsen, Clark Spencer. 1987. Bioarchaeological Interpretation of Subsistance Economy and Behavior from Human Skeletal Remains. *Advances in Archaeological Method and Theory* 10:339–445.

Larsen, Clark Spencer. 1990. *The Archaeology of Mission Santa Catalina de Guale. 2. Biocultural Interpretations of a Population in Transition, Anthropological Papers of the American Museum of Natural History 68*. New York.

Larsen, Clark Spencer. 1995. Biological Changes in Human Populations with Agriculture. *Annual Review of Anthropology* 24:185–213.

Larsen, Clark Spencer. 1997. *Bioarchaeology: Interpreting Behavior from the Human Skeleton*. New York and Cambridge: Cambridge University Press.

Larsen, Clark Spencer. 2000. *Skeletons in Our Closet: Revealing Our Past through Bioarchaeology*. Princeton: Princeton University Press.

Larsen, Clark Spencer. 2001. *Bioarchaeology of Spanish Florida, the Impact of Colonialism*. Gainesville, FL: University Press of Florida.

Larsen, Clark Spencer. 2002. Bioarchaeology: The Lives and Lifestyles of Past People. *Journal of Archaeological Research* 10 (2):119–166.

Larsen, Clark Spencer, and Robert L. Kelly. 1995. *Bioarchaeology of the Stillwater Marsh: Prehistoric Human Adaptation in the Western Great Basin*. Vol. 77: 1–170, *Anthropological Papers of the American Museum of Natural History*. New York.

Larsen, Clark Spencer, and George R. Milner. 1994. *In the Wake of the Contact: Biological Responses to Conquest*. New York: Wiley-Liss.

Larsen, Clark Spencer, and Christopher B. Ruff. 1991. Biomechanical Adaptation and Behavior on the Prehistoric Georgia Coast. In *What Mean These Bones? Studies in Southeastern Bioarchaeology*, edited by M. L. Powell, P. S. Bridges and A. M. W. Mires. 102–113. Tuscaloosa: University of Alabama Press.

Larsen, Clark Spencer, Christopher B. Ruff, and M. C. Griffin. 1996. Implications of Changing Biomechanical and Nutritional Environments for Activity and Lifeway in the Eastern Spanish Borderlands. In *Bioarchaeology of Native American Adaptation in the Spanish Borderlands*, edited by B. J. Baker and L. L. Kealhofer. 95–125. Gainesville, FL: University Press of Florida.

Larsen, Clark Spencer, Christopher B. Ruff, and Robert L. Kelly. 1995. Structural Analysis of the Stillwater Postcranial Human Remains: Behavioral Implications of Articular Joint Pathology and Long Bone Diaphyseal Morphology. In *Bioarchaeology of the Stillwater Marsh: Prehistoric Human Adaptation in the Western Great Basin*, edited by C. S. Larsen and R. L. Kelly. 107–133, *Anthropological Papers of the American Museum of Natural History, No. 77*. New York: Wiley-Liss.

Larsen, Clark Spencer, M. J. Schoeninger, Christopher B. Ruff, and Dale L. Hutchinson. 1992. Population Decline and Extinction in La Florida. In *Disease and Demography in the Americas: Changing Patterns before and after 1492*, edited by J. W. Verano and D. H. Ubelaker. 25–39. Washington, DC: Smithsonian Institution Press.

Larsen, Clark Spencer, Mark F. Teaford, and Mary K. Sandford. 1998. Teeth as Tools at Tutu: Extramasticatory Behavior in Prehistoric St. Thomas, U.S. Virgin Islands. In *Human Dental Development, Morphology, and Pathology: A Tribute to Albert A. Dahlberg*, edited by J. A. Lukacs. 401–420, *University of Oregon Anthropological Papers No. 54*.

Larson, J. L. 1994. *Interpreting Nature: The Science of Living Form from Linneaus to Kant*. Baltimore: Johns Hopkins Press.

Laughlin, William S., ed. 1951. *Papers on the Physical Anthropology of the American Indian*. New York: Viking Fund.

Laughlin, William S., ed. 1963. Eskimos and Aleuts: Their Origins and Evolution. *Science* 142:633–645.

Lauriere, Christine. 2000. L'anthropologue et le Prognathe. *L'Homme: Revue Française d'Anthropologie* 163:195–204.

Lawson, John. 1709. *A New Voyage to Carolina, Containing the Exact Description and Natural History of That Country*. London. Original edition, Raleigh: Printed by Strother & Marcom, 1860.

Leatherman, Thomas. 1987. Illness, Work, and Social Relations in the Southern Peruvian Highlands. Ph.D. Dissertation, Department of Anthropology, University of Massachusetts, Amherst.

Leatherman, Thomas L., and Alan H. Goodman. 1997. Expanding the Biocultural Synthesis toward a Biology of Poverty. *American Journal of Physical Anthropology* 102(1):1–3.

LeBlanc, S. A. 1999. *Prehistoric Warfare in the American Southwest*. Salt Lake City: University of Utah Press.

Ledger, Michael, Lucy-May Holtzhausen, Deborah Constant, and Alan G. Morris. 2000. Biomechanical Beam Analysis of Long Bones from a Late 18th Century Slave Cemetery in Cape Town, South Africa. *American Journal of Physical Anthropology* 112:207–216.

Leek, F. Filce. 1972. Bite, Attrition and Associated Oral Conditions as Seen in Ancient Egyptian Skulls. *Journal of Human Evolution* 1:289–295.

Lehmann-Hartleben, Karl. 1943. Archaeological Notes: Thomas Jeffreson, Archaeologist. *American Journal of Archaeology* 47 (2):161–163.

Leigh, Rufus Wood. 1925a. Dental Pathology of the Eskimo. *Dental Cosmos* 67:884–898.

Leigh, Rufus Wood. 1925b. Dental Pathology of Indian Tribes of Varied Environmental and Food Conditions. *American Journal of Physical Anthropology* 8 (2):179.

Leigh, Rufus Wood. 1928. Dental Pathology of Aboriginal California. *University of California Publications in American Ethnology and Archaeology* 23:399–440.

Leigh, Rufus Wood. 1929. Dental Morphology and Pathology of Prehistoric Guam. *Memoirs of the Bernice P. Bishop Museum* 11:258–273.

Leigh, Rufus Wood. 1930. Dental Morphology and Pathology of Prehistoric Guam. *Memoirs of the Bernice P. Bishop Museum* 11(3):255–273.

Leigh, Rufus Wood. 1934. Notes on the Stomatology and Pathology of Ancient Egypt. *University of California Publications in American Ethnology and Archaeology* 34:1–54.

Leigh, Rufus Wood. 1937. Dental Morphology and Pathology of Pre-Spanish Peru. *American Journal of Physical Anthropology* 22:267–296.

Lekson, Stephen H. 2002. War in the Southwest, War in the World. *American Antiquity* 67 (4):607–624.

Leon, Luis A. 1977. *Paul Rivet, 1876–1976*. Quito: Editorial Casa de la Cultura Ecuatoriana.

Lerroi-Gourhan, Andre, G. Bailloud, and M. Brezillon. 1962. L'hypogée II des Mournouards (Mesnil-Sur-Oger, Marne). *Gallia Prehistoire* 5 (1):23–133.

Lewis, Barbara A. 1998. Prehistoric Juvenile Rheumatoid Arthritis in a Precontact Louisiana Native Population Reconsidered. *American Journal of Physical Anthropology* 106(2):229–248.

Lewis, Mary E. 2002. Impact of Industrialisation: Comparative Study of Child Health in Four Sites from Medieval and Post-Medieval England. *American Journal of Physical Anthropology* 119 (3):211–223.

Lewis, T. M. N., and Madeline D. Kneberg. 1941. *Prehistory of the Chickamauga Basin in Tennessee. University of Tennessee, Division of Anthropology, Anthropological Papers No. 1 (Mimeographed).* Knoxville: University of Tennessee.

Lewis, T. M. N., and Madeline D. Kneberg. 1946. *Hiwassee Island: An Archaeological Account of Four Tennessee Indian Peoples.* Knoxville: The University of Tennessee Press.

Lewis, T. M. N., and Madeline D. Kneberg. 1947. *The Archaic Horizon in Western Tennessee.* Vol. 23, No. 4, *Tennessee Anthropology Papers No. 2, the University of Tennessee Record, Extension Series.* Knoxville.

Lewis, T. M. N., and Madeline D. Kneberg. 1951. Early Projectile Point Forms and Examples from Tennessee. *Tennessee Archaeologist* 7 (1):6–9.

Lewis, T. M. N., and Madeline D. Kneberg. 1954a. *Ocunaluftee Indian Village: An Interpretation of a Cherokee Community of 1750.* Cherokee, North Carolina: Cherokee Historical Association.

Lewis, T. M. N., and Madeline D. Kneberg. 1954b. *Ten Years of the Tennessee Archaeologist, Selected Subjects.* Chattanooga: J. B. Graham.

Lewis, T. M. N., and Madeline D. Kneberg. 1955. *The First Tennesseans: An Interpretation of Tennessee Prehistory.* Knoxville: Department of Anthropology, The University of Tennessee.

Lewis, T. M. N., and Madeline D. Kneberg. 1956. The Paleo-Indian Complex on the Lecroy Site. *Tennessee Archaeologist* 12 (1):5–11.

Lewis, T. M. N., and Madeline D. Kneberg. 1957. The Camp Creek Site. *Tennessee Archaeologist* 13 (1):1–48.

Lewis, T. M. N., and Madeline D. Kneberg. 1958a. *Tribes That Slumber: Indians of the Tennessee Region.* Knoxville: The University of Tennessee Press.

Lewis, T. M. N., and Madeline D. Kneberg. 1958b. The Nuckolls Site. *Tennessee Archaeologist* 14 (2):60–79.

Lewis, T. M. N., and Madeline D. Kneberg. 1959. The Archaic Culture in the Middle South. *American Antiquity* 25 (2):161–183.

Lewis, T. M. N., and Madeline D. Kneberg. 1961. *Eva: An Archaic Site.* Knoxville: The University of Tennessee Press.

Lewis, T. M. N., and Madeline D. Kneberg. n.d. Manual of Field and Laboratory Techniques Employed by the Division of Anthropology, University of Tennessee, Knoxville, Tennessee, in Connection with the Investigation of Archaeological Sites within the TVA Dam Reservoirs. Ms. On File, Frank McClung Museum, University of Tennessee, Knoxville; W. S. Webb Museum of Anthropology, University of Kentucky, Lexington.

Lewis, T. M. N., Madeline D. Kneberg, and Lynne P. Sullivan, editor and compiler. 1995. *The Prehistory of the Chickamauga Basin in Tennessee*. 2 vols. Knoxville: The University of Tennessee Press.

Lewis, Barbara A. 1998. Prehistoric Juvenile Rheumatoid Arthritis in a Precontact Louisiana Native Population Reconsidered. *American Journal of Physical Anthropology* 106(2):229–248.

Lieberman, Daniel E., Osbjorn M. Pearson, John D. Polk, Brigitte Demes, and A. W. Crompton. 2003. Optimization of Bone Growth and Remodeling in Response to Loading in Tapered Mammalian Limbs. *Journal of Experimental Biology* 206:3125–3138.

Lieberman, Daniel E., John D. Polk, and Brigitte Demes. 2004. Predicting Long Bone Loading from Cross-Sectional Geometry 123 (2):156–171.

Liebersohn, H. 1998. *Aristocratic Encounters: European Travelers and North American Indian*. Cambridge: Cambridge University Press.

Lippert, Dorothy T. 1992. Science, Savages, and Sacrilege: A Native American Archeologist Examines the Reburial Issue. *Cultural Management Resources News and Views* 4 (2):18–21.

Lipschutz, A., Grete Mostny, H. Helfritz, F. Jeldes, and M. Lipschutz. 1947. Physical Characteristics of Fuegians. An Analysis Aided by Photography. *American Journal of Physical Anthropology* 5 (3):295–322.

Lipschutz, A., Grete Mostny, and L. Robin. 1946a. The Bearing of Ethnic and Genetic Conditions on the Blood Groups of Three Fuegian Tribes. *American Journal of Physical Anthropology* 4 (3):301–322.

Lipschutz, A., Grete Mostny, L. Robin, and A. Santiana. 1946b. Blood Groups in Tribes of Tierra Del Fuego and Their Bearing on Ethnic and Genetic Relationships. *Nature* 157:696.

Loe, L. 2004. Health and Socio-Economic Status in Early Medieval Wales. An Analysis of Health Indicators and Their Socio-Economic Implications in an Early Medieval Human Population from Llandough, Glamorgan. Ph.D. Dissertation, University of Bristol.

Logan, Michael H., Corey S. Sparks, and Richard L. Jantz. 2003. Cranial Modification among 19th Century Osages: Admixture and Loss of an Ethnic Marker. *Plains Anthropologist* 48(187):209–224.

Lofsvold, D. 1988. Quantitative Genetics of Morphological Differentiation in Peromyscus. II. Analysis of Selection and Drift. *Evolution* 42:54–67.

Long, Joseph K. 1966. A Test of Multiple-Discriminant Analysis as a Means of Determining Evolutionary Changes and Intergroup Relationships in Physical Anthropology. *American Anthropologist* 68 (No. 2, Part 1):444–464.

Long, Joseph K. 1973. Jamaican Medicine: Choices between Folk Healing and Modern Medicine. Ph.D. Dissertation, University of North Carolina, Chapel Hill.

Longacre, W. A. 1970. *Archaeology as Anthropology: A Case Study*. Vol. 17, *Anthropological Papers*. Tucson: University of Arizona. Original edition, *Science* 144:1454–1455.

Loring, S., and M. Prokopec. 1994. A Most Peculiar Man: The Life and Times of Aleš Hrdlička. In *Reckoning with the Dead*, edited by T. Bray and T. W. Killion. 26–42. Washington, DC: Smithsonian Institution Press.

Lotka, A. J. 1922. The Stability of the Normal Age Distribution. *Proceedings of the National Academy of Sciences* 8:339–345.

Lovejoy, C. Owen. 1985. Dental Wear in the Libben Population: Its Functional Pattern and Role in the Determination of Adult Skeletal Age-at-Death. *American Journal of Physical Anthropology* 68 (1):47–56.

Lovejoy, C. Owen, Albert H. Burnstein, and Kingsbury G. Heiple. 1976. The Biomechanical Analysis of Bone Strength: A Method and Its Application to Platycnemia. *American Journal of Physical Anthropology* 44:489–506.

Lovejoy, C. Owen, and Kingsbury G. Heiple. 1981. Analysis of Fractures in Skeletal Populations with an Example from the Libben Site, Ottawa County, Ohio. *American Journal of Physical Anthropology* 55:529–541.

Lovejoy, C. Owen, Richard S. Meindl, James C. Ohman, Kingsbury G. Heiple, and Tim D. White. 2002. The Maka Femur and Its Bearing on the Antiquity of Human Walking: Applying Contemporary Concepts of Morphogenesis to the Human Fossil Record. *American Journal of Physical Anthropology* 119:97–133.

Lovejoy, C. Owen, Richard S. Meindl, T. R. Pryzbeck, T. S. Barton, K. G. Heiple, and D. Kotting. 1977. Paleodemography of the Libben Site, Ottawa County, Ohio. *Science* 198:291–293.

Lovejoy, C. Owen, Robert P. Mensforth, and George J. Armelagos. 1982. Five Decades of Skeletal Biology as Reflected in the American Journal of Physical Anthropology. In *A History of American Physical Anthropology 1930–1980*, edited by F. Spencer. 329–336. New York: Academic Press.

Lovejoy, C. Owen, and Erik Trinkaus. 1980. Strength and Robusticity of the Neandertal Tibia. *American Journal of Physical Anthropology* 53:465–470.

Lovell, Nancy C., and Aaron A. Dublenko. 1999. Further Aspects of Fur Trade Life Depicted in the Skeleton. *International Journal of Osteoarchaeology* 9:248–256.

Lovis, William A., Keith W. Kintigh, Vincas P. Steponaitis, and Lynne G. Goldstein. 2004. Archaeological Perspectives on the NAGPRA: Underlying Principles, Legislative History, and Current Issues. In *Legal Perspectives on Cultural Resources*, edited by J. R. Richman and M. P. Forsythe. 165–184. Walnut Creek, CA: AltaMira Press.

Lucy, D. 1997. Human Age Estimation from Skeletal and Dental Evidence. Ph.D. Dissertation, University of Bradford.

Lucy, D., R. G. Aykroyd, A. M. Pollard, and T. Solheim. 1996. A Bayesian Approach to Adult Human Age Estimation from Dental Observations by Johanson's Age Changes. *Journal of Forensic Science* 41:189–194.

Lukacs, John R. 1983. Dental Anthropology and the Origins of Two Iron Age Populations from Northern Pakistan. *Homo* 34:1–15.

Lukacs, John R. 1987. Biological Relationships Derived from Morphology of Permanent Teeth: Recent Evidence from Prehistoric India. *Anthropologischer Anzeiger* 45 (2):97–116.

Lukacs, John R. 1988. Dental Morphology and Odontometrics of Early Agriculturalists from Neolithic Mehrgarh, Pakistan. In *Teeth Revisited, Mémoires du Muséum Nationale D'histoire Naturelle*, edited by D. E. Russell, J. P. Santoro, and D. Sigogneau-Russell. 245–258. Vol. 53, *Série C, Sciences de la Terre*. Paris.

Lukacs, John R. 1991. Localized Enamel Hypoplasia of Human Deciduous Canine Teeth: Prevalence and Pattern of Expression in Rural Pakistan. *Human Biology* 63(4):513–522.

Lumholtz, Carl. 1902. *Unknown Mexico: Explorations in the Sierra Madres and Other Regions, 1890–1898*. New York: Dover Publications.

Lumholtz, Carl, and Aleš Hrdlička. 1897. Trephining in Mexico. *American Anthropologist* X (12):389–396.

Lumholtz, Carl, and Aleš Hrdlička. 1898. Marked Human Bones from a Prehistoric Tarasco Indian Burial Place in the State of Michoacan, Mexico. *Bulletin of the American Museum of Natural History, NY* X:61–79.

Lumpkin, C. K. 1976. A Multivariate Craniometric Analysis of Selected Southwestern Archaeological Populations. Ph.D. Dissertation University of New Mexico.

Lundberg, Murray. 2001. Kwaday Dan Sinche, the Yukon Iceman. Retrieved Dec. 31, 2005, from Explore North Web Site: http://www.explorenorth.com/library/weekly/aa082599.htm.

Lundstrom, A., and L. Lysell. 1953. An Anthropological Examination of a Group of Medieval Danish Skulls, with Particular Regard to the Jaws and Occlusal Conditions. *Acta Odontologica Scandinavica* 11:111.

Lunt, D. A. 1978. Molar Attrition in Medieval Danes. In *Development Function and Evolution of Teeth*, edited by P. M. Butler and K. A. Joysey. 465–482. New York: Academic Press.

Lux, Konrad. 1935. Dental Pathology from Crania Found in Central Texas. *Central Texas Archeological Society Bulletin* (1):12–15.

Lux, Konrad. 1936. A Detailed Report of the Teeth and Supporting Structures as Found in Crania of Aycock Shelter. *Bulletin of the Central Texas Archeological Society* 2:39–42.

Lux, Konrad. 1937. A Detailed Report of the Teeth and Supporting Structures as Found in Crania Mass Burial Near Waco. *Central Texas Archeologist* (3):34–40.

Lynnerup, Niels, Claus Andreasen, and Joel Berglund, eds. 2003. *Mummies in a New Millennium: Proceedings of the 4th World Congress on Mummy Studies. Nuuk, Greenland, September 4th to 10th, 2001, Danish Polar Center Publication, No. 11*. Copenhagen: Greenland National Museum and Archives, Danish Polar Center.

Lyon, E. A. 1996. *New Deal Archaeology in the Southeast*. Tuscaloosa: University of Alabama Press.

MacCurdy, George Grant. 1905. Prehistoric Surgery—A Neolithic Survival. *American Anthropologist* 7(1):17–23.

MacCurdy, George Grant. 1918. Surgery among the Ancient Peruvians. *Art and Archaeology* 7(9):381–394.

MacCurdy, George Grant. 1923. Human Skeletal Remains from the Highlands of Peru. *American Journal of Physical Anthropology* 6(3):217–329.

Machado, Lilia M. C. 1975/1976. Arte Rupestre Das Cavernas e Abrigos da Região Mineira. Síntese Das Pesquisas. Anais do X Congresso Nacional de Espeleologia. *Revista Espeleológica*.

Machado, L. M. C. 1977. Colaboração No Trabalho "Os Antigos Habitantes Da Área Arqueológica De Lagoa Santa, Minas Gerais, Brasil. Estudo Morfológico." *Arquivos Do Museu De História Natural, Ufmg. Belo Horizonte, Mg* 2:119–162.

Machado, L. M. C. 1981/1982. Um Novo Padrão de Desgaste Dentário e Evidência de Alto Consumo de Carbohidratos Numa População Esqueletal Arcaica do Brasil. *Arquivos Do Museu De História Natural, Ufmg. Belo Horizonte, Mg* 6–7:201–206.

Machado, L. M. C. 1984a. *Análise de Remanescentes Ósseos Humanos do Sítio Arqueológico Corondó, RJ. Aspectos Biológicos e Culturais, Letras e Ciências Humanas da Universidade de São Paulo.* São Paulo: Tese de Doutorado, Faculdade de Filosofia.

Machado, L. M. C. 1984b. *Análise de Remanescentes Ósseos Humanos do Sítio Arqueológico Corondó, RJ. Aspectos Biológicos e Culturais, Boletim do Instituto de Arqueologia Brasileira, Série Monografias, N.1.* Rio de Janeiro.

Machado, L. M. C. 1985c. *A Paleodemografia do Sítio Corondó, RJ. Análise Preliminar, Anais Da Iii Reunião Científica Da Sociedade De Arqueologia Brasileira (Sab), Go.*

Machado, L. M. C. 1985d. *Paleodemografia, Considerações Gerais, Anais Da Iii Reunião Científica Da Sociedade De Arqueologia Brasileira (Sab), Go.*

Machado, L. M. C. 1986. *Antropologia Física: Sua Importância Para a Arqueologia. Pesquisa do Passado: Arqueologia no Brasil, Boletim do Instituto de Arqueologia Brasileira, Rio de Janeiro.*

Machado, L. M. C. 1988. Biologia de Grupos Humanos Pré-Históricos do Sudeste do Brasil. As Tradições Itaipu e Una. Anais do Simpósio "Americans before Columbus." In *Prehistoria Sudamericana, Nuevas Perspectivas*, edited by B. Meggers. 77–103. Washington: Taraxacum.

Machado, L. M. C. 1990a. Sobre as Práticas Funerárias de Cremação e Suas Variações em Grutas do Norte e Noroeste de Minas Gerais. *Anais da V Reunião Científica da Sociedade de Arqueologia Brasileira (Sab), Rs. Revista do Cepa, Santa Cruz do Sul* 20 (17):235–247.

Machado, L. M. C. 1990b. Paleopatologia: As Doenças Na América Pré-Colombiana. *Revista a Hora Veterinária* Ano 10 (58):39.

Machado, L. M. C. 1991a. As Estruturas e Suas Inter-Relações Em Sítios de Caçadores-Coletores do Litoral de Saquarema, RJ. *Boletim do Instituto de Arqueologia Brasileira, Série Ensaios, N.5*:1–42.

Machado, L. M. C. 1991b. Fratura Não Unida (Pseudo Artrose). Registro de Um Caso no Sambaqui Da Beirada, RJ. *Resumos da Vi Reunião Científica da Sociedade de Arqueologia Brasileira (Sab), Rio de Janeiro*:88.

Machado, L. M. C. 1992a. Paleodemografia e Saúde Em Perspectiva Populacional. In *Paleopatologia e Paleoepidemiologia. Estudos Preliminares*, edited by A. J. Araújo and L. F. Ferreira. 87–93: Fundação Oswaldo Cruz, Série Panorama.

Machado, L. M. C. 1992b. *Pescadores-Coletores Pré-Históricos de Saquarema, RJ: Estudo Dentário e Correlações Culturais. Resumos da I Reunião da Sociedade Brasileira de Paleopatologia.* Rio de Janeiro: Escola Nacional de Saúde Pública/Fundação Oswaldo Cruz.

Machado, L. M. C. 1995a. Inferências Sócio-Comportamentais a Partir de Dados Bioculturais de População Horticultora Pré-Histórica. *Coleção Arqueológica, Edipucrs, Porto Alegre* 1 (1):519–535.

Machado, L. M. C. 1995b. Estudo de Condições Patológicas Na População Esqueletal do Sambaqui da Beirada, RJ. *Resumos do III Congresso Latinoamericano de Antropologia Biológica, Rio de Janeiro*:28.

Machado, L. M. C. 1995c. Possível Associação Entre Idade, Patologia Esqueletal e Tratamento Funerário Diferenciado. Estudo de Caso Em População Horticultora Pré-Histórica, Sítio do Caju, Campos, RJ. *Resumos Do Iii Congresso Latinoamericano De Antropologia Biológica, Rio de Janeiro*:25.

Machado, L. M. C. 1995d. Contribuição do Estudo Dentário À Reconstituição da Dieta de Populações Humanas Pré-Históricas. A Importância da Abordagem Biocultural e Multidisciplinar. *Resumos Do II Simpósio De Arqueologia Da Região Sudeste (Mae/Usp), São Paulo*:26.

Machado, L. M. C. 1995f. *Inferências Sócio-Comportamentais a Partir de Dados Bioesqueletais de População Horticultora Pré-Histórica do Rio De Janeiro, Anais da VIII Reunião Científica da Sociedade de Arqueologia Brasileira, Porto Alegre.*

Machado, L. M. C. 1995g. Tendências À Continuidade e Mudança Em Ritos Funerários de Populações Humanas Pré-Históricas do Rio de Janeiro. In *Arqueologia do Estado do Rio de Janeiro*, edited by C. Beltrão. 111–118. Rio de Janeiro: Arquivo Público do Estado do Rio de Janeiro.

Machado, L. M. C. 1997. Tafonomia Humana: Alguns Problemas e Interpretações em Arqueologia Funerária. In *Arqueologia e Suas Interfaces Disciplinares. Resumos da IX Congresso da Sab.* 28. Rio de Janeiro.

Machado, L. M. C. 1999a. *Resultados de Pesquisas Bioculturais em Populações Humanas Pré-Históricas do Litoral do Rio de Janeiro, Anais da X Reunião Científica da Sociedade de Arqueologia Brasileira (Sab). Recife, Pe.*

Machado, L. M. C. 1999b. Resultados de Pesquisas Bioculturais em Populações Humanas Pré-Históricas do Litoral do Rio de Janeiro. In *Resumos da X Reunião Científica da Sociedade de Arqueologia Brasileira (Sab).* 182.

Machado, L. M. C. 2001a. Os Sepultamentos, Contextos Arqueológicos e Dados Bioesqueletais. *Documento de Trabalho, Rio de Janeiro, N.5*:71–75.

Machado, L. M. C. 2001b. *Os Sepultamentos e Correlações Bioculturais (Sambaqui Manitiba I—Saquarema, RJ), Resumos da XI Congresso da Sociedade de Arqueologia Brasileira, Rio de Janeiro.*

Machado, L. M. C., and M. B. Almeida. 2001. *Ossos no Fogo: Contribuições Aos Estudos de Arqueologia Experimental, Resumos da XI Congresso da Sociedade de Arqueologia Brasileira, Rio De Janeiro.*

Machado, L. M. C., and M. A. Alves. 1995. Estruturas Arqueológicas e Padrões e Sepultamento do Sítio Água Limpa, Monte Alto, Sp. *Anais da VIII Reunião Científica da Sociedade de Arqueologia Brasileira, Porto Alegre.*

Machado, L. M. C., and M. A. Alves. 1995/1996. Estruturas Arqueológicas e Padrões de Sepultamento do Sítio Água Limpa, Monte Alto, Sp. *Coleção Arqueológica, Edipucrs, Porto Alegre* 2 (1):295–319.

Machado, L. M. C., and F. Crancio. 2003. Tributo a Lina Kneip: Uma Vida a Arqueologia. Painel Simpósio: Estudando os Ritos Funerários. A Diversidade e Singularidade do Tratamentop Aos Mortos Nos Sambaquis de Saquarema. In *Resumos do XII Congresso da Sociedade de Arqueologia Brasileira.* 74.

Machado, L. M. C., and L. Kneip. 1991. Cremação e Outras Práticas Funerárias em Sítios de Pescadores-Coletores do Litoral de Saquarema, RJ. *Resumos da VI Reunião Científica da Sociedade de Arqueologia Brasileira (Sab), Rio de Janeiro*:77.

Machado, L. M. C., and L. Kneip. 1992. Cremação e Outras Práticas Funerárias em Sítios de Pescadores-Coletores do Litoral de Saquarema, RJ. *Anais da VI Reunião Científica da Sociedade de Arqueologia Brasileira (Sab), Rio De Janeiro*:457–465.

Machado, L. M. C., and L. Kneip. 1993a. Ritos Funerários Das Populações Pré-Históricas da Saquarema, RJ: Sambaquis da Beirada, Moa e Pontinha. *Documento De Trabalho, Museu Nacional, Ufrj, Departamento de Antropologia, N.1*:1–76.

Machado, L. M. C., and L. Kneip. 1993b. Padrões Dentários, Dieta e Subsisténcia Das Populações Dos Sambaquis de Saquarema, RJ. *João Pessoa Resumos da VII Reunião Científica da Sab*:30.

Machado, L. M. C., and L. Kneip. 1994. Padrões Dentários, Dieta e Subsisténcia Das Populações Dos Sambaquis de Saquarema, RJ. *Revista De Arqueologia. João Pessoa, Anais da VIIi Reunião Científica da Sab* 1 (8):45–57.

Machado, L. M. C., L. Kneip, and F. Crancio. 1995. Contribuição ao Estudo Das Práticas Funerárias: Os Ossos Humanos Trabalhados do Sambaqui de Saquarema, RJ. *Resumos Do III Congresso Latinoamericano De Antropologia Biológica, Rio De Janeiro*:24.

Machado, L. M. C., L. Kneip, and F. Crancio. 1996. Ossos Humanos Trabalhados e Biologia Esqueletal do Sambaqui de Saquarema, RJ. *Documento De Trabalho, Museu Nacional, Ufrj, Departamento De Antropologia, N.3*:3–39.

Machado, L. M. C., and Cleber Bidegain Pereira. 1985. *Análise da Morfologia Dentária. Método Turner. Nota Prévia, Anais da III Reunião Científica da Sociedade de Arqueologia Brasileira (Sab), Go.*

Machado, L. M. C., E. Pons, and L. R. Silva. 1987a. Os Sítios Massambaba (Rj-Jc-56) e Boqueirão (Rj-Jc-57), Arraial do Cabo: Os Padrões de Sepultamento. *Resumos da IV Reunião Científica da Sociedade de Arqueologia Brasileira (Sab), Santos, Sp*:83.

Machado, L. M. C., E. Pons, and L. R. Silva. 1987b. Adaptação Biocultural do Litoral Fluminense. Os Restos Ósseos Humanos de Dois Sítios Arqueológicos de Arraial do Cabo, RJ. *Resumos da IV Reunião Científica da Sociedade de Arqueologia Brasileira (Sab), Santos, Sp*:84.

Machado, L. M. C., E. Pons, and L. R. Silva. 1989a. Os Sítios Massambaba (Rj-Jc-56) e Boqueirão (Rj-Jc-57), Arraial do Cabo: Os Padrões de Sepultamento. *Anais da IV Reunião Científica da Sociedade de Arqueologia Brasileira (Sab), Santos, Sp. Dédalo, Museu de Arqueologia e Etnologia, São Paulo (Publ. Av. 1)*:447–454.

Machado, L. M. C., E. Pons, and L. R. Silva. 1989b. Adaptação Biocultural do Litoral Fluminense. Os Restos Ósseos Humanos de Dois Sítios Arqueológicos de Arraial do Cabo, RJ. *Anais da IV Reunião Científica da Sociedade de Arqueologia Brasileira (Sab), Santos, Sp. Dédalo, Museu de Arqueologia e Etnologia, São Paulo (Publ. Av. 1)*:429–446.

Machado, L. M. C., and G. M. Sene. 1999. Um Exemplo de Preservação Arqueológica no Litoral Fluminense: da Análise Biológica À Identidade Cultural. *Resumos Da X Reunião Científica Da Sociedade De Arqueologia Brasileira (Sab)*:183.

Machado, L. M. C., and G. M. Sene. 1999/2000. Um Exemplo de Preservação Arqueológica no Litoral Fluminense: da Análise Biológica À Identidade Cultural? *Revista de Arqueologia* 12–13:81–92.

Machado, L. M. C., and G. M. Sene. 2001. *Simbolismo, Identidade Cultural e Análise Biológica: Os Remanescentes Esqueletais Humanos do Sambaqui da Tarioba. A Pré-História De Rio Das Ostras — O Sítio Arqueológico Sambaqui Da Tarioba.*

Machado, L. M. C., G. M. Sene, and L. R. Silva. 1992. Padrões Mortuários e Adaptação Ecológica. Tradição Una, RJ. *Anais da VI Reunião Científica da Sociedade de Arqueologia Brasileira (Sab), Rio De Janeiro*:476–483.

Machado, L. M. C., G. M. Sene, and L. R. Silva. 1994. Estudo Preliminar Dos Ritos Funerários do Sítio do Caju, RJ. *Revista De Arqueologia. João Pessoa, Anais Da VII Reunião Científica Da Sab* 8 (1):75–90.

Machado, L. M. C., and P. J. Tsaknis. 1985. Oral Pathology of Precolumbian Indians of Brazil. *Annals of the 37th Annual Meeting of the American Academy of Forensic Sciences, Nevada.*

Mackey, J. 1977. A Multivariate, Osteological Approach to Towa Culture History. *American Journal of Physical Anthropology* 46:477–482.

Maehle, H., and Charlotte A. Roberts. 2002. History of Medicine Meets Palaeopathology: A Seminar Series at Durham. *Wellcome History* 20:20–21.

Malvido, E., G. Pereira, and Vera Tiesler, eds. 1997. *El Cuerpo Humano y su Tratamiento Mortuorio*. México City: INAH (Colección Científica).

Manchester, Keith. 1983. *The Archaeology of Disease*. Bradford: University Press.

Manchester, Keith. 1984. Tuberculosis and Leprosy in Antiquity. *Medical History* 28: 162–173.

Mann, Robert W., Albert A. Dahlberg, and T. Dale Stewart. 1990. Anomalous Morphologic Formation of the Deciduous and Permanent Teeth in a 5-Year-Old 15th Century Child: A Variant of the Ekman-Westborg-Julin Syndrome. *Oral Surgery, Oral Medicine, Oral Pathology* 70:90–94.

Mann, Robert W., and Sean P. Murphy. 1990. *Regional Atlas of Bone Disease*. Springfield, IL: Charles C Thomas.

Manouvrier, Leonce Pierre. 1888. Memoire Sur la Platycnémie Chez L'homme et Chez les Anthropoides. *Mémoires de la Société d'Anthropologie de Paris, Série 2* 3 (3–4): 469–548.

Maples, William R. 1962. A Morphological Comparison of Skeletal Material from Sanders Focus and from Fulton Aspect. M.A. Thesis, Department of Anthropology, University of Texas, Austin.

Mark, Joan T. 1980. *Four Anthropologists: An American Science in Its Early Years*. New York: Science History Publications/USA.

Marks, Jonathan. 1995. *Human Biodiversity: Genes, Race, and History, Foundations of Human Behavior*. New York: Aldine de Gruyter.

Marlar, Richard A., Banks L. Leonard, Brian R. Billman, Patricia M. Lambert, and Jennifer E. Marlar. 2000. Biochemical Evidence of Cannibalism at a Prehistoric Puebloan Site in Southwestern Colorado. *Nature* 407:74–78.

Marrinan, R. A. 1999. Best Supporting Actress? The Contributions of Adelaide K. Bullen. In *Grit-Tempered: Early Women Archaeologists in the Southeastern United States*, edited by N. M. White, L. P. Sullivan, and R. A. Marrinan. 148–162. Gainesville, FL: University Press of Florida.

Martin, Debra L. 1994. Patterns of Diet and Disease: Health Profiles for the Prehistoric Southwest. In *Themes in Southwest Prehistory*, edited by G. J. Gumerman. 87–108. Santa Fe: School of American Research Press.

Martin, Debra L. 2000. Review. Man Corn: Cannibalism and Violence in the Prehistoric American Southwest. *American Antiquity* 65:199–201.

Martin, Debra L., George J. Armelagos, Alan H. Goodman, and Dennis P. Van Gerven. 1984. The Effects of Socioeconomic Change in Prehistoric Africa. In *Paleopathology at the Origins of Agriculture*, edited by M. N. Cohen and G. J. Armelagos. 193–214. New York: Academic Press.

Martin, Debra L., and David W. Frayer. 1997. *Troubled Times: Violence and Warfare in the Past*. Amsterdam: Gordon and Breach.

Martin, Debra L., A. H. Goodman, George J. Armelagos, and A. L. Magennis. 1991. *Black Mesa Anasazi Health: Reconstructing Life from Patterns of Death and Disease*. Vol.14, *Occasional Paper No. 14*. Carbondale: Southern Illinois University at Carbondale, Center for Archaeological Investigations.

Martin, Paul S., George I. Quimby, and Donald Collier. 1947. *Indians before Columbus. Twenty Thousand Years of North American History Revealed by Archaeology*. Chicago: University of Chicago Press.

Martin, Rudolf. 1914. *Lehrbuch der Anthropologie in Systematischer Darstellung mit Besonderer Berücksichtigung der Anthropologischen Methoden*. Jena: G. Fischer.

Martin, Rudolf, and Karl Saller. 1957. *Lehrbuch der Anthropologie in Systematischer Darstellung mit Besonderer Berücksichtigung der Anthropologischen Methoden*. 3rd ed. Stuttgart: G. Fischer.

Martin, R. Bruce, and David B. Burr. 1989. *Structure, Function, and Adaptation of Compact Bone*. New York: Raven Press.

Martin, R. Bruce, David B. Burr, and Neil A. Sharkey. 1998. *Skeletal Tissue Mechanics*. New York: Springer-Verlag.

Mason, Ronald J. 1997. Letter to the Editor. *SAA Bulletin* 15 (1):3.

Mason, Ronald J. 2000. Archaeology and Native North American Oral Traditions. *American Antiquity* 65 (2):239–266.

Masset, Claude. 1972. The Megalithic Tomb of La Chassée Tirancourt. *Antiquity* 46: 297–300.

Masset, Claude. 1989. Age Estimation on the Basis of Cranial Sutures. In *Age Markers in the Human Skeleton*, edited by M. Y. İşcan. 71–103. Springfield: Charles C Thomas.

Masset, Claude. 1993. Encore l'age des Adultes. *Bull. et Mem. de la Societe d'Anthropologie de Paris* 5:217–224.

Masset, Claude, and B. Parzysz. 1985. Démographie des Cimetières: Incertitude Statistique des Estimateurs en Paléodémographie. *L'Homme: Revue Française d'Anthropologie* 25 (2):147–154.

Massin, Benoit. 1996. From Virchow to Fischer: Physical Anthropology and "Modern Race Theories." In Wilhelmine Germany. In *Volksgeist as Method and Ethic: Essays on Boasian Ethnography and the German Anthropological Tradition*, edited by G. W. Stocking. 79–154. Vol. 8, *History of Anthropology*. Madison: University of Wisconsin Press.

Massler, M., I. Schour, and H. Poncher. 1941. Developmental Pattern of the Child as Reflected in the Calcification Pattern of the Teeth. *American Journal of Disease in Children* 62:33–67.

Matson, R. G., and B. Chisholm. 1991. Basketmaker II Subsistence: Carbon Isotopes and Other Dietary Indicators from Cedar Mesa, Utah. *American Antiquity* 56:444–459.

Matthews, Washington. 1886. Consumption among the Indians. *Transactions of the American Climatological Association*:234–241.

Matthews, Washington. 1887. The Study of Consumption among the Indians. *The New York Medical Journal* 46 (July 30):127–128.

Matthews, Washington. 1888. A Further Contribution to the Study of Consumption among the Indians. *Transactions of the American Climatological Association* 5:136–155.

Matthews, Washington. 1889. The Inca Bone and Kindred Formations among the Ancient Arizonians. *American Anthropologist* 2 (4):337–346.

Matthews, Washington. 1898. Use of Rubber Bands in Gauging Cranial Capacity. *American Anthropologist* 11 (6):171–176.

Matthews, Washington, J. L. Wortman, and J. S. Billings. 1893. Human Bones of the Hemenway Collection in the United States Army Medical Museum. *Memoirs of the National Academy of Sciences* 6 (7th Memoir):139–286.

Mays, Simon. 1996a. Age-Dependent Bone Loss in a Medieval Population. *International Journal of Osteoarchaeology* 6 (2):144–154.

Mays, Simon. 1996b. Healed Limb Amputations in Human Osteoarchaeology and Their Causes. *International Journal of Osteoarchaeology* 6 (3):101–113.

Mays, Simon. 1997. A Perspective on Human Osteoarchaeology in Britain. *International Journal of Osteoarchaeology* 7:600–604.

Mays, Simon. 1998. *The Archaeology of Human Bones*. London: Routledge.

Mays, Simon, G. Crane-Kramer, and A. Bayliss. 2003. Two Probable Cases of Treponemal Disease of Medieval Date from England. *American Journal of Physical Anthropology* 120:133–143.

Mays, Simon, and M. Faerman. 2001. Sex Identification in Some Putative Infanticide Remains from Roman Britain Using Ancient DNA. *Journal of Archaeological Science* 28:555–559.

Mays, Simon, G. Michael Taylor, A. J. Legge, D. B. Young, and G. Turner-Walker. 2001. Paleopathological and Biomolecular Study of Tuberculosis in a Medieval Skeletal Collection from England. *American Journal of Physical Anthropology* 114:298–311.

McCutchen, D. 1993. *The Red Record: The Wallam Olum. The Oldest Native North American History*. Garden City Park: Avery Publishing Group.

McGregor, John C. 1943. Burial of an Early American Magician. *Proceedings of the American Philosophical Society* 86 (2):270–298.

McGuigan, Dorothy Gies. 1970. *A Dangerous Experiment: 100 Years of Women at the University of Michigan*. Ann Arbor: Center for Continuing Education of Women.

McGuire, Randall H. 1988. Dialogues with the Dead: Ideology and the Cemetery. In *The Recovery of Meaning: Historical Archaeology in the United States*, edited by M. P. Leone and P. B. Potter. 435–480. Washington, DC: Smithsonian Institution Press.

McGuire, Randall H. 1992. Archaeology and the First Americans. *American Anthropologist* 94 (4):816–836.

McGuire, Randall H., and D. J. Saitta. 1996. Althought They Have Petty Captains, They Obey Them Badly: The Dialectics of Prehispanic Western Pueblo Social Organization. *American Antiquity* 61 (2):197–216.

McKee, J. K. 1984. A Genetic Model of Dental Reduction through the Probable Mutation Effect. *American Journal of Physical Anthropology* 65:231–241.

McKeown, A. H. 2000. Investigating Variation among Arikara Crania Using Geometric Morphometry. Ph.D. Dissertation, University of Tennessee, Knoxville.

McKeown, C. Timothy. 2002. Implementing a "True Compromise": The Native American Graves Protection and Repatriation Act after Ten Years. In *The Dead and Their Possessions. Repatriation in Principle, Policy and Practice*, edited by C. Fforde, J. Hubert and P. Turnbull. 108–132. New York: Routledge.

McKern, Thomas W., and T. Dale Stewart. 1957. *Skeletal Changes in Young American Males. Technical Report No. EP-45*. Natick: U.S. Quartermaster Research and Development Center, Environmental Protection Research Division.

McKern, William C. 1940. Letter Dated January 8 to Snow. Snow Letters, Laboratory for Human Osteology, Mary Harmon Bryant Scientific Collections Building, University of Alabama, Tuscaloosa.

McKinley, Jaqueline I. 1994. *Spong Hill. Part VIII. The Cremations, East Anglian Archaeology Report 69*. Norwich, Norfolk: Field Archaeology Division, Norfolk Museums Servic.

McKinley, Jaqueline I. 2000. The Analysis of Cremated Bone. In *Human Osteology in Archaeology and Forensic Science*, edited by M. Cox and S. Mays. 403–421. London: Greenwich Medical Media.

McLaughlin, Robert H. 2004. NAGPRA, Dialogue, and the Politics of Historical Authority. In *Legal Perspectives on Cultural Resources*, edited by J. R. Richman and M. P. Forsythe. 185–201. Walnut Creek, CA: AltaMira Press.

McMahon, Thomas A., and John Tyler Bonner. 1983. *On Size and Life*. New York: Scientific American Books.

McManamon, Francis P. 2000. Enclosure 1: Determination That the Kennewick Human Skeleton Remains Are "Native American" for the Purposes of the Native American Graves and Repatriation Act (NAGPRA). Memorandum from the Departmental Consulting Archeologist to the Assistant Secretary, Fish and Wildlife and Parks. Retrieved Dec. 31, 2005, from U.S. Department of the Interior, National Park Service, Archaeology Program Web Site: http://www.cr.nps.gov/archeology/kennewick/c14memo.htm.

McManamon, Francis P. 2004. Kennewick Man. Retrieved Dec. 31, 2005, from U.S. Department of the Interior, National Park Service, Archaeology Program Web Site: http://www.cr.nps.gov/archeology/kennewick/.

McWilliams, Dick. 1968. Appendix: Skeletal Material from Site Mc-84, McCurtain County, Oklahoma. *Oklahoma Anthropological Society Bulletin* 16:122–123.

McWilliams, K. R. 1974. Gran Quivira Pueblo and Biological Distance in the U.S. Southwest. Ph.D., Arizona State University.

McWilliams, R. 1965. Periodontal Pathologies of a Sample of the Skeletal Material from the Sully Site, 39SL4, Sully County, South Dakota. Report on File, Department of Anthropology, University of Kansas, Lawrence.

MedicineNet.com. n.d. Definition of Bioarchaeology. Retrieved Dec. 28, 2005, from Medterms Medical Dictionary Web Site: http://www.medterms.com/script/main/art.asp?articlekey=33158.

Mehta, Jim D., and E. C. Sensenig. 1966. Some Observations on Arkansas Indian Skulls. *Australian Dental Journal* 11:410–414.

Meighan, Clement W. 1984. Archaeology: Science or Sacrilege. In *Ethics and Values in Archaeology*, edited by E. L. Green. 208–223. New York: Free Press.

Meighan, Clement W. 1992. Some Scholars' View on Reburial. *American Antiquity* 57 (4):704–710.

Meigs, Charles Delucena. 1851. *A Memoir of Samuel George Morton, M.D., Late President of the Academy of Natural Sciences of Philadelphia*. Philadelphia: T. K. and P. G. Collins Printers.

Meigs, James Aitken. 1857. The Cranial Characteristics of the Races of Men. In *Indigenous Races of the Earth*, edited by J. C. Nott and G. R. Gliddon. 201–352. Philadelphia: J. B. Lippincott & Co.

Meigs, James Aitken. 1866. Observations Upon the Cranial Forms of the American Aborigines, Based Upon Specimens Contained in the Collection of the Academy of Natural Sciences of Philadelphia. *Proceedings of the Academy of Natural Sciences of Philadelphia* 1866:197–235.

Meigs, James Aitken. 1867. Description of a Human Skull in the Collection of the Smithsonian Institution. *Annual Report of the Smithsonian Institution* 1867: 412–415.

Meiklejohn, Christopher, J. M. Wyman, and C. T. Schentag. 1992. Caries and Attrition: Dependent or Independent Variables? *International Journal of Anthropology* 7(1):17–22.

Meindl, Richard S., and C. Owen Lovejoy. 1985. Ectocranial Suture Closure: A Revised Method of the Determination of Skeletal Age-at-Death Based on the Lateral-Anterior Surface. *American Journal of Physical Anthropology* 68:57–66.

Meindl, Richard S., and C. Owen Lovejoy. 1989. Age Changes in the Pelvis: Implications for Paleodemography. In *Age Markers in the Human Skeleton*, edited by M. Y. İşcan. 137–168. Springfield, IL: Charles C Thomas.

Meindl, Richard S., C. Owen Lovejoy, Robert P. Mensforth, and Robert A. Walker. 1985. A Revised Method of Age Determination Using the Os Pubis, with a Review and Tests of Accuracy of Other Current Methods of Pubic Symphyseal Aging. *American Journal of Physical Anthropology* 68:29–45.

Meindl, Richard S., and K. F. Russell. 1998. Recent Advances in Method and Theory in Paleodemography. *Annual Review of Anthropology* 27:375–399.

Melbye, Jerry, and Scott I. Fairgrieve. 1994. A Massacre and Possible Cannibalism in the Canadian Arctic: New Evidence from the Saunaktuk Site (NgTn-1). *Arctic Anthropology* 31(2):57–77.

Mellanby, M. 1929. *Diet and the Teeth: An Experimental Study. Part I. Dental Structure in Dogs, Medical Research Council, Special Report Series, No. 140*. London: His Majesty's Stationery Office.

Mello e Alvim, Marília Carvalho de. 1962. O Negro. In *Enciclopédia Delta Larousse*. 2297–2301. Vol. 5. Rio de Janeiro: Larousse.

Mello e Alvim, Marília Carvalho de. 1963a. Paleontologia Humana — Sinopse. In *Enciclopédia Delta Larousse*. 6429–6633. Vol. 12. Rio de Janeiro: Larousse.

Mello e Alvim, Marília Carvalho de. 1963b. Diversidade Morfológica Entre os Índios "Botocudos" do Leste Brasileiro (Séc. XIX) e O Homem de Lagoa Santa. *Boletim do Museu Nacional — Antropologai-Nova Série (Rio de Janeiro)* 23:1–70.

Mello e Alvim, Marília Carvalho de. 1963c. Estudo do Homem de Lagoa Santa, Minas Gerais, Brasil-Morfologia do Astrágalo. *Boletim do Museu Nacional—Antropologia Nova Série (Rio de Janeiro)* 21:1–42.

Mello e Alvim, Marília Carvalho de. 1966. Aspectos Das Diretivas Atuais do Estudo da Evolução e Das Diferenciações Raciais. *Delfos (Rio de Janeiro)* 6:68–72.

Mello e Alvim, Marília Carvalho de. 1971a. *Charles Darwin: The Descent of Man (1871), Reavaliada, Publicações Avulsas do Museu Nacional 58.*

Mello e Alvim, Marília Carvalho de. 1971b. As Características e as Variedades Dos Grupos Indígenas. *História da Cultura Brasileira (Rio de Janeiro)* 1:31–46.

Mello e Alvim, Marília Carvalho de. 1972a. *Populações e Culturas Pré-Históricas do Brasil.* Brasilia: Alvorada.

Mello e Alvim, Marília Carvalho de. 1972b. Contribuição ao Estudo Biológico e Sóciocultural Das Populações Indígenas do Estado do Rio de Janeiro: O Achado do Sítio Arqueológico Duas Barras. *Ciência e Cultura—Suplemento (São Paulo)* 24 (6):481.

Mello e Alvim, Marília Carvalho de. 1977a. Populações e Culturas Pré-Históricas do Brasil (Parte I). *Atualidade Indígena (Brasília)* 1 (5):14–22.

Mello e Alvim, Marília Carvalho de. 1977b. Populações e Culturas Pré-Histõricas do Brasil (Parte II) Sambaquis. *Atualidade Indígena (Brasília)* 1 (6):37–40.

Mello e Alvim, Marília Carvalho de. 1977c. Populações e Culturas Pré-Histõricas do Brasil (Parte III) a Busca Das Origens Na Ceramica. *Atualidade Indígena (Brasília)* 2 (7):25–31.

Mello e Alvim, Marília Carvalho de. 1977d. O Homem de Lagoa Santa: Estudo Morfológico. *Arquivos de Anatomia e Antropologia da Universidade Souza Marques (Rio de Janeiro)* 3:333–339.

Mello e Alvim, Marília Carvalho de. 1977e. Os Antigos Habitantes da Área Arqueológica de Lagoa Santa, MG-Brasil: Estudo Morfológico. *Arquivos do Museu de História Natural da Universidade Federal de Minas Gerais (Belo Horizonte)* 2:119–174.

Mello e Alvim, Marília Carvalho de. 1978. Caracterização da Morfologia Craniana Das Populações Pré-Históricas do Litoral Meridional do Brasil: Parana e Santa Catarina. *Arquivos de Anatomia e Antropologia da Universidade Souza Marques (Rio de Janeiro)* 3 (3):293–322.

Mello e Alvim, Marília Carvalho de. 1986. Antropometria (Verbete). In *Dicionário de Ciências Sociais.* 69–71. Rio de Janeiro: Fundação Getúlio Vargas.

Mello e Alvim, Marília Carvalho de. 1991. O Grupo Pré-Histórico da Furna do Estrago—PE e Suas Relações Biológicas Com Outras Populações Pré-Histõricas do Nordeste Brasileiro. *Clio Série Arqueológica—Número Extraordinário (Recife)* 4:81–83.

Mello e Alvim, Marília Carvalho de. 1993. As Populações Indígenas do Brasil no Século do Descobrimento: Aspectos e Problemas. *Revista de Arqueologia* 7:11–31.

Mello e Alvim, Marília Carvalho de. 1996. Povoamento da América Indígena: Questões Controversas. *Revista Fundhamentos (Recife)* 1 (1):139–146.

Mello e Alvim, Marília Carvalho de, and Pessoa de Barros. 1971/1972. População Brasileira—O Negro e Seus Descendentes. *Delfos (Rio de Janeiro)* 11/12:60–68.

Mello e Alvim, Marília Carvalho de, and Ondemar Ferreira Dias Júnior. 1972. Relações Biológicas e Culturais da Fase Mucuri, Com Grupos do Tronco Macro-Jê. *Anais da Academia Brasileira de Odontologia (Rio de Janeiro)*:70–73.

Mello e Alvim, Marília Carvalho de, Ondemar Ferreira Dias Júnior, and Maria Tereza Toríbio. 1973/1974a. Relações Culturais e Biológicas de Grupos Indígenas da Fase Mucuri: os Sítios Arqueológicos do Município de Santa Maria Madalena, RJ. *Cinêcia e Cultura (São Paulo)* 25 (6):386.

Mello e Alvim, Marília Carvalho de, Ondemar Ferreira Dias Júnior, and Maria Tereza Toríbio. 1973/1974b. Relações Culturais e Biológicas de Grupos Indígenas da Fase Mucuri: os Sítios Arqueológicos do Município de Santa Maria Madalena, RJ. *Delfos (Rio de Janeiro)* 13/14:55–62.

Mello e Alvim, Marília Carvalho de, and F. J. L. da C. F. Ferreira. 1984. Os Esqueletos do Abrigo Toca do Paraguaio, Município de São Raimundo Nonato, Piauí: Estudo Antropofísico. *Caderno Pesquisas–Série Antropológica (Terezina)* 4 (3):239–261.

Mello e Alvim, Marília Carvalho de, and João Carlos de Oliveira Gomes. 1989. Análise e Interpretação de Condições Patológicas: Orbita Crivosa, Osteoporose Puntiforme e Hiperostose Esponjosa — em Crânios Humanos Provenientes de Sítio Arqueológico — Sambaqui de Cabeçuda, Laguna, Santa Catarina, Brasil. *Revista de Pré-história (São Paulo)* 7:127–145.

Mello e Alvim, Marília Carvalho de, João Carlos de Oliveira Gomes, and Helio Vianna. 1987. Exame Morfológico. In Mumificações Naturais Na Pré-História Brasileira: Um Estudo de Caso. *Revista de Arqueologia* 3 (1):30–31.

Mello e Alvim, Marília Carvalho de, and L. M. C. Machado. 1987. Antropologia Física e Sua Importância Para a Arqueologia. In *Catálogo Comemorativo Dos 25 Anos do Instituto de Arqueologia Brasileira. Série Catálogos* 36–39. Vol. 3.

Mello e Alvim, Marília Carvalho de, and Denizart Pereira Mello Filho. 1965. Morfologia Craniana da População do Sambaqui da Cabeçuda (Laguna-Santa Catarina) e Sua Relação Com Outras Populações de Paleoameríndios do Brasil. In *Homenage a Juan Comas en su 65 Aniversário*. Vol. II. México.

Mello e Alvim, Marília Carvalho de, and Denizart Pereira Mello Filho. 1967/1968. Morfologia da População do Sambaqui do Forte Marechal Luz (Santa Catarina). *Revista de Antropologia da Universidade de São Paulo (São Paulo)* 15/16:5–12.

Mello e Alvim, Marília Carvalho de, and Sheila Maria Ferraz Mendonça de Souza. 1984. Os Esqueletos Humanos da Furna do Estrago — Pernambuco, Brasil (Nota Prévia). *Clio (Recife)* 6:95–98.

Mello e Alvim, Marília Carvalho de, and Sheila Maria Ferraz Mendonça de Souza. 1986. Relações Biológicas Entre Populações Indígenas Atuais e Pré-Históricas do Brasil. *Clio — Série Arqueológica (Recife)* 1 (6):69–80.

Mello e Alvim, Marília Carvalho de, and Sheila Maria Ferraz Mendonça de Souza. 1990. Relações Biológicas Entre Populações Indígenas Atuais e Pré-Históricas do Brasil. *Clio (Recife)* 1 (6):69–79.

Mello e Alvim, Marília Carvalho de, and Giralda Seiferth. 1967. Estudos Morfológicos da População do Sambaqui do Forte Marechal Luz (Santa Catarina). *Anais da Academia Brasileira de Cinêcias (Rio de Janeiro)* 39 (3/4):549.

Mello e Alvim, Marília Carvalho de, and Giralda Seiferth. 1969a. Estudo Morfológico do Úmero da População do Sambaqui de Cabeçuda (Laguna — Santa Catarina). *Ciência e Cultura (São Paulo)* 21 (2):209–210.

Mello e Alvim, Marília Carvalho de, and Giralda Seiferth. 1969b. Estudo Morfológico do Úmero da População do Sambaqui de Cabeçuda (Laguna — Santa Catarina). *Revista do Museu Paulista — Nova Série (São Paulo)* 18:119–126.

Mello e Alvim, Marília Carvalho de, and Giralda Seiferth. 1971a. O Femur Na População do Sambaqui de Cabeçuda (Laguna), Estado de Santa Catarina, Brasil. *Boletim do Museu Nacional — Antropologia Nova Série (Rio de Janeiro)* 24:1–14.

Mello e Alvim, Marília Carvalho de, and Giralda Seiferth. 1971b. Estudo Morfológico do Úmero da População do Sambaqui de Cabeçuda (Laguna, Santa Catarina). In *O Homem Antigo Na América*. 25–28. São Paulo: Universidade de São Paulo.

Mello e Alvim, Marília Carvalho de, and Margareth Carvalho Soares. 1981. Estudo Comparativo de Traços Não-Métricos em Populações Pré-Históricas do Brasil. *Boletim do Museu Nacional — Antropologia Nova Série (Rio de Janeiro)* 38:14.

Mello e Alvim, Marília Carvalho de, and Margareth Carvalho Soares. 1981/1982. Incidência de Traços Não-Métricos em Material de Sambaqui do Acervo do Museu Nacional da Universidade Federal do Rio de Janeiro. *Arquivos de Anatomia e Antropologia da Universidade Souza Marques (Rio de Janeiro)* 6/7:35–333.

Mello e Alvim, Marília Carvalho de, and Margareth Carvalho Soares. 1984. Incidência de Traços Não-Métricos em Material de Sambaqui do Acervo do Museu Nacional da Universidade Federal do Rio de Janeiro. *Revista de Arqueologia (Belém)* 2 (1):3–12.

Mello e Alvim, Marília Carvalho de, Margareth Carvalho Soares, and Paulo S. Pringshein Cunha. 1983/1984. Traços Não Métricos Cranianos e Distâcia Biológica em Grupas Indígenas Interioranos e do Litoral do Brasil: 'Homem de Lagoa Santa,' Índios Botocudos e Construtores de Sambaquis. *Arquivos do Museu de História Natural da Universidade Federal de Minas Gerais (Belo Horizonte)* 8/9:323–338.

Mello e Alvim, Marília Carvalho de, Margareth Carvalho Soares and Paulo S. Pringshein Cunha. 1984. Traços Não-Métricos Cranianos e Distâncias Biológicas em Grupos Indígenas do Brasil: Botocudos e Construtores de Sambaquis. *Revista de Pré-historia (São Paulo)* 6:107–117.

Mello e Alvim, Marília Carvalho de, and Dorath Pinto Uchoa. 1972. Contribuição ao Estudo da Morfologia Das Populações do Sambaqui de Piaçaguera, SP. *Ciência e Cultura (São Paulo)* 25 (6):385.

Mello e Alvim, Marília Carvalho de, and Dorath Pinto Uchoa. 1973. Subsídios Para O Estudo Das Populações e Culturas do Litoral Paulista: O Sítio Arqueológico de Tenório. *Ciência e Cultura — Suplemento (São Paulo)* 26 (7):690.

Mello e Alvim, Marília Carvalho de, and Dorath Pinto Uchoa. 1975. Aditamentos ao Estudo Das Populações e Culturas Pré-Históricas do Litoral Paulista: os Ambaqui do Buracão. *Ciência e Cultura — Suplemento (São Paulo)* 27 (7):690.

Mello e Alvim, Marília Carvalho de, and Dorath Pinto Uchoa. 1976. *Contribução ao Estudo Das Populações de Sambaquis: Os Construtores de Sambaqui de Piaçaguera, São Paulo. Instituto de Pré-História, Pesquisas — Série Sambaqui de Piaçaguera.* São Paulo: Universidade de São Paulo.

Mello e Alvim, Marília Carvalho de, and Dorath Pinto Uchoa. 1978. O Sambaqui de Buracão: Uma Contribuição ao Estudo da Pré-História do Litoral Paulista. *Ciência e Cultura (São Paulo)* 31:97–98.

Mello e Alvim, Marília Carvalho de, and Dorath Pinto Uchoa. 1979. O Sambaqui de Buracão: Uma Contribuição ao Estudo da Pré-História do Litoral Paulista. *Arquivos de Anatomia e Antropologia da Universidade Souza Marques (Rio de Janeiro)* 6/7:315–333.

Mello e Alvim, Marília Carvalho de, and Dorath Pinto Uchoa. 1980. O Sambaqui de Buracão: Uma Contribuição ao Estudo da Pré-História do Litoral Paulista. *Arquivos de Anatomia e Antropologia da Universidade Souza Marques (Rio de Janeiro)* 4/5:337–393.

Mello e Alvim, Marília Carvalho de, and Dorath Pinto Uchoa. 1990. Proposta Para a Área de Antropologia Física. *Clio Série Arqueológica (Recife)* 1 (6):132.

Mello e Alvim, Marília Carvalho de, and Dorath Pinto Uchoa. 1993. Efeitos do Hábito de Cócoras no Tálus e Na Tíbia de Indígenas Pré-Históricos e de Um Grupo Atual do Brasil. *Revista do Museu de Arqueologia e Etnologia (São Paulo)* 3:35–53.

Mello e Alvim, Marília Carvalho de, and Dorath Pinto Uchoa. 1995. As Populações Indígenas do Brasil no Século do Descobrimento. *SBPC Documento 1 (São Paulo)*:3–23.

Mello e Alvim, Marília Carvalho de, and Dorath Pinto Uchoa. 1995/1996a. Osteobiograia da População Pré-Historica do Abrigo Pedra do Alexandre, Carnaúba Dos Dantas, Rn. *Clio Série Arqueológica (Recife)* 1 (11):17–42.

Mello e Alvim, Marília Carvalho de, and Dorath Pinto Uchoa. 1995/1996b. Povoamento da América Indígena: Questões Controversas. *Clio Série Arqueológica (Recife)* 1 (11):9–15.

Mello e Alvim, Marília Carvalho de, and Dorath Pinto Uchoa. 1996. Hiperostose Porosa em Crânios de Índios e Mulatos do Sudeste Brasileiro: Correlações Entre as Lesões Na Calvária e Na Órbita. *Revista do Museu de Arqueologia e Etnologia (São Paulo)* 6:115–168.

Mello e Alvim, Marília Carvalho de, Dorath Pinto Uchoa, and João Carlos de Oliveira Gomes. 1991. Cribra Orbitalia e Lesões Cranianas Congêneres em Populações Pré-Históricas do Brasil. *Revista do Museu de Arqueologia e Etnologia (São Paulo)* 1:21–53.

Mello e Alvim, Marília Carvalho de, Marcus Infante Vieira, and Lilia Maria Cheuiche. 1975. Os Construtores do Sambaqui de Cabeçuda, SC e de Piaçaguera, SP: Estudo Morfológico Comparativo. *Arquivos de Anatomia e Antropologia da Universidade Souza Marques (Rio de Janeiro)* 1:393–406.

Mello e Alvim, Marília Carvalho de, M. I. Vieira, and L. M. C. Machado. 1974. Os Construtores Dos Sambaquis de Cabeçuda, SC e Piaçagüera, SP. Estudo Morfométrico Comparativo. *Arquivos de Anatomia e Antropologia, Instituto de Antropologia Prof. Souza Marques, Rio De Janeiro* 1:393–405.

Mendonça de Souza, Sheila Maria Ferraz, and Marília Carvalho de Mello e Alvim. 1992. A População Pré-Histórica da Furna do Estrago: Adaptação Humana ao Agreste Pernambucano. *Symposium (Recife)* 34 (2):123–145.

Mendonça de Souza, Sheila Maria Ferraz, and Marília Carvalho de Mello e Alvim. 1992/1993. Os Antigos Habitantes da Serra do Cipó (MG-Brasil): Estudo Morfológico

Preliminar. *Arquivos do Museu Nacional de História Natural da Universidade Federal de Minas Gerais, Belo Horizonte* 13/14:107–128.

Mensforth, Robert P. 1990. Paleodemography of the Carlston Annis (Bt-5) Late Archaic Skeletal Population. *American Journal of Physical Anthropology* 82:81–99.

Mensforth, Robert P. 2001. Warfare and Trophy Taking in the Archaic Period. In *Archaic Transitions in Ohio and Kentucky Prehistory*, edited by O. H. Prufer, S. E. Pedde, and R. S. Meindl. 110–138. Kent, OH: Kent State University Press.

Merbs, Charles F. 1969. Patterns of Activity-Induced Pathology in a Canadian Eskimo Isolate. Ph.D. Dissertation, The University of Wisconsin, Madison.

Merbs, Charles F. 1980. The Pathology of a La Jollan Skeleton from Punta Minitas, Bajas California. *Pacific Coast Archaeological Society Quarterly* 16 (4):37–43.

Merbs, Charles F. 1983. *Patterns of Activity-Induced Pathology in a Canadian Inuit Population, Archaeological Survey of Canada, National Museum of Man, Mercury Series, No. 119*. Ottawa: National Museums of Canada.

Merbs, Charles F. 1992. A New World of Infectious Diseases. *Yearbook of Physical Anthropology* 35:3–42.

Merbs, Charles F. 1995. Incomplete Spondylolysis and Healing: A Study of Ancient Canadian Eskimo Skeletons. *Spine* 20:2328–2334.

Merbs, Charles F. 1996a. Spondylolysis of the Sacrum in Alaskan and Canadian Inuit Skeletons. *American Journal of Physical Anthropology* 101:357–367.

Merbs, Charles F. 1996b. Spondylolysis and Spondylolisthesis: A Cost of Being an Erect Biped or a Clever Adaptation? *Yearbook of Physical Anthropology* 39:201–228.

Merbs, Charles F. 2002a. Asymmetrical Spondylolysis. *American Journal of Physical Anthropology* 119(2):156–174.

Merbs, Charles F. 2002b. Washington Matthews and the Hemenway Expedition of 1887–88. *Journal of the Southwest* 44:303–335.

Merbs, Charles F., and Robert C. Euler. 1985. Atlanto-Occipital Fusion and Spondylolisthesis in an Anazasi Skeleton from Bright Angel Ruin National Park, Arizona. *American Journal of Physical Anthropology* 67:381–392.

Merbs, Charles F., and R. J. Miller, eds. 1985. *Health and Disease in the Prehistoric Southwest*. Tucson, AZ: Arizona State University.

Merriwether, D. Andrew, Graciela S. Cabana, and David M. Reed. 2000. Chapter 3: Kennewick Man Ancient DNA Analysis: Final Report Submitted to the Department of the Interior, National Park Service. In "Report on the DNA Testing Results of the Kennewick Human Remains from Columbia Park, Kennewick, Washington." Retrieved Dec. 31, 2005, from Department of the Interior, National Park Service, Archaeology Program Web Site: http://www.cr.nps.gov/aad/kennewick/Merriwether_Cabana.htm.

Meskell, Lynn. 2000. Cycles of Life and Death: Narrative Homology and Archaeological Realities. *World Archaeology* 31 (3):423–441.

Meskell, Lynn. 2000. Cycles of Life and Death: Narrative Homology and Archaeological Realities. *World Archaeology* 31 (3):423–441.

Messias, T. T., and Marília Carvalho de Mello e Alvim. 1962. Contribução ao Estudo do Homem de Lagoa Santa. *Boletim do Museu Nacional (Antropologia Nova Série)* 20:1–55.

Metcalf, Peter, and Richard Huntington. 1991. *Celebrations of Death: The Anthropology of Mortuary Ritual.* 2nd ed. Cambridge: Cambridge University Press.

Metress, James F. 1971. An Osteobiography of the Buffalo Site, Fulton County, West Virginia. Ph.D., Indiana University, Bloomington.

Michael, J. A. 1988. A New Look at Morton's Craniological Research. *Current Anthropology* 29 (2):349–354.

Micheli, L. J., and J. D. Klein. 1991. Sports Injuries in Children and Adolescents. *British Journal of Sports Medicine* 25:6–9.

Miles, A. E. W. 1958. The Assessment of Age from the Dentition. *Proceedings of the Royal Society of Medicine* 51:1057–1060.

Miles, A. E. W. 1963. Dentition in the Assessment of Individual Age in Skeletal Material. In *Dental Anthropology*, edited by D. R. Brothwell. 191–209. Oxford: Pergamon Press.

Miles, J. S. 1966. Diseases Encountered at Mesa Verde, Colorado II. Evidences of Diseases. In *Human Palaeopathology*, edited by S. Jarcho. 91–97. New Haven: Yale University Press.

Miller, Daniel, and Christopher Y. Tilley, eds. 1984. *Ideology, Power, and Prehistory, New Directions in Archaeology*. Cambridge: Cambridge University Press.

Miller, W. 1883a. Der Einfluss der Mikroorganismen Auf Die Caries der Menschlichen Zähne. *Arch. Exp. Pathol. Pharmakol.* 16:291–304.

Miller, W. 1883b. Agency of Micro-Organisms in Decay of Human Teeth. *Dental Cosmos* 25:1–12.

Milner, George R. 1995. Osteological Evidence for Prehistoric Warfare. In *Regional Approaches to Mortuary Analysis*, edited by L. A. Beck. 221–244. New York: Plenum Press.

Milner, George R. 1999. Warfare in Prehistoric and Early Historic Eastern North America. *Journal of Archaeological Research* 7:105–151.

Milner, George R., Eve Anderson, and Virginia G. Smith. 1991. Warfare in Late Prehistoric West-Central Illinois. *American Antiquity* 56 (4):581–603.

Milner, George R., and Richard W. Jeffries. 1987. A Reevaluation of the WPA Excavation of the Robbins Mound in Boone County, Kentucky. In *Current Archaeological Research in Kentucky: Volume One*, edited by D. Pollack. 33–42. Frankfort: Kentucky Heritage Council.

Milner, George R., and Clark Spencer Larsen. 1991. Teeth as Artifacts of Human Behavior: Intentional Mutilation and Accidental Modification. In *Advances in Dental Anthropology*, edited by M. A. Kelley and C. S. Larsen. 357–378. New York: Wiley-Liss.

Milner, George R., and Virginia G. Smith. 1986. *New Deal Archaeology in Kentucky: Excavations, Collections, and Research, Occasional Papers in Anthropology 5.* Lexington, KY: Program for Cultural Resource Assessment, University of Kentucky.

Milner, George R., James W. Wood, and Jesper L. Boldsen. 2000. Paleodemography. In *Biological Anthropology of the Human Skeleton*, edited by M. A. Katzenberg and S. R. Saunders. 467–497. New York: Wiley-Liss.

Mitchell, D. R. 1992. Burial Practices and Paleodemographic Reconstructions at Pueblo Grande. *Kiva* 58 (1):89–104.

Mitchell, D. R. 1994. The Pueblo Grande Burial Artifact Analysis: A Search for Wealth, Ranking, and Prestige. In *The Pueblo Grande Project*, edited by D. R. Mitchell. 129–180. Vol. 7, *Soil Systems Publications in Archaeology*. Phoenix: Soil Systems.

Mitchell, D. R., and J. L. Brunson-Hadley, eds. 2001. *Ancient Burial Practices in the American Southwest: Archaeology, Physical Anthropology, and Native American Perspectives*. Albuquerque: University of New Mexico Press.

Mobley, C. 1980. Demographic Structure of Pecos Indians: A Model Based on Life Tables. *American Antiquity* 45:518–530.

Møller-Christensen, Vilhelm. 1961. *Bone Changes in Leprosy*. Copenhagen: Munksgaard.

Møller-Christensen, Vilhelm. 1973. Osteo-Archaeology as a Medico-Historical Auxiliary Science. *Medical History* 17:411–418.

Møller-Christensen, Vilhelm. 1978. *Leprosy Changes of the Skull*. Odense: Odense University Press.

Molleson, T. I. 1989. Seed Preparation in the Neolithic: The Osteological Evidence. *Antiquity* 63:356–362.

Molleson, T. I., and P. Cohen. 1990. The Progression of Dental Attrition Stages Used for Age Assessment. *Journal of Archaeological Science* 17 (4):363–371.

Molleson, T. I., and M. Cox. 1993. *The Spitalfields Project: The Anthropology. The Middling Sort, Research Report 86*: York Council for British Archaeology.

Molleson, T. I., K. Jones, and S. Jones. 1993. Dietary Change and the Effects of Food Preparation on Microwear Patterns in the Late Neolithic of Abu Hureyra, Northern Syria. *Journal of Human Evolution* 24 (6):455–468.

Molnar, P. 2003. Tracing Prehistoric Activities [Abstract]. *American Journal of Physical Anthropology* (Suppl. 36):153.

Molnar, Stephen. 1971. Human Tooth Wear, Tooth Function and Cultural Variability. *American Journal of Physical Anthropology* 34:175–189.

Molnar, Stephen. 1972. Tooth Wear and Culture: A Survey of Tooth Functions among Some Prehistoric Populations. *Current Anthropology* 13 (5):511–526.

Molnar, S., and S. C. Ward. 1975. Mineral Metabolism and Microstructural Defects in Primate Teeth. *American Journal of Physical Anthropology* 43:3–17.

Molto, Joseph Eldon. 1983. *Biological Relationships of Southern Ontario Woodland Peoples: The Evidence of Discontinuous Cranial Morphology, Archaeological Survey of Canada, Paper No. 117*. Ottawa: National Museums of Canada.

Montagu, Ashley. 1960. *An Introduction to Physical Anthropology*. Springfield, IL: Charles C Thomas.

Montgomery, J., J. Evans, D. Powesland, and Charlotte A. Roberts. 2005. British Continuity or Immigrant Replacement at West Heslerton Anglian Settlement: Lead and Strontium Isotope Evidence for Mobility, Subsistence Practice and Status. *American Journal of Physical Anthropology* 126 (2):123–138.

Moodie, Roy L. 1917. Studies in Paleopathology. I. General Consideration of the Evidences of Pathological Conditions Found among Fossil Animals. *Anals of Medical History* 1 (4):374–393.

Moodie, Roy L. 1923a. *The Antiquity of Disease*. Chicago: University of Chicago Press.

Moodie, Roy L. 1923b. Paleopathology: An Introduction to the Study of Ancient Evidences of Disease. Urbana, IL: University of Illinois Press.

Moodie, Roy L. 1928. Studies in Paleo-Odontology, IX. Definite Association of Rheumatic Lesions with Disease of the Teeth in an Ancient Peruvian. *Pacific Dental Gazette* 36:757–759.

Moodie, Roy L. 1929. Studies in Paleodontology. XI. Dental Attrition and Its Results among Certain Ancient Indians from Southern California. *Pacific Dental Gazette* 37:217–227.

Moodie, Roy L. 1931. *Roentgenologic Studies of Egyptian and Peruvian Mummies*. Chicago.

Moore, J. A., A. C. Swedlund, and George J. Armelagos. 1975. The Use of Life Tables in Paleodemography. In *Population Studies in Archaeology and Biological Anthropology: A Symposium*, edited by A. C. Swedlund. 57–70. Washington, DC: Society for American Archaeology.

Moore, W. J., and M. E. Corbett. 1971. The Distribution of Dental Caries in Ancient British Populations. I. Anglo-Saxon Period. *Caries Research* 5:151–168.

Moore, W. J., and M. E. Corbett. 1975. Distribution of Dental Caries in Ancient British Populations. III. The 17th Century. *Caries Research* 9:163–175.

Moorrees, Coenraad F. A. 1957. *The Aleut Dentition: A Correlative Study of Dental Characteristics in an Eskimoid People*. Cambridge: Harvard University Press.

Morris, Ian. 1991. The Archaeology of Ancestors: The Saxe/Goldstein Hypothesis Revisited. *Cambridge Archaeological Journal* 1:147–169.

Morris, Ian. 1992. *Death-Ritual and Social Structure in Classical Antiquity*. Cambridge: Cambridge University Press.

Morrison, David. 2004. Human Remains Repatriation at the Canadian Museum of Civilization. Paper read at CAPA Conference, October 28, 2004, at London, Ontario.

Morse, D. 1961. Prehistoric Tuberculosis in America. *American Review of Respiratory Diseases* 83:489–504.

Morse, D., Don R. Brothwell, and P. J. Ucko. 1964. Tuberculosis in Ancient Egypt. *American Review Respiratory Disease* 90 (4):526–541.

Morton, N. E. 1973. *Genetic Structure of Human Populations*. Honolulu: University of Hawaii.

Morton, N. E. 1975. Kinship, Information, and Biological Distance. *Theoretical Population Biology* 7:246–255.

Morton, Samuel George. 1823. Tentamen Inaugurale de Corporis Dolore, Etc. M.D., University of Edinburgh, Edinburgh.

Morton, Samuel George. 1839. *Crania Americana; or, a Comparative View of the Skulls of Various Aboriginal Nations of North and South America, to Which Is Prefixed an Essays on the Varieties of the Human Species*. Philadelphia: J. Dobson. Original edition, London: Simpkin, Marshall and Co.

Morton, Samuel George. 1841. Remarks on the So Called Pigmy Race of the Valley of the Mississippi. *Journal of the Academy of Natural Sciences of Philadelphia* 8:205–207.

Morton, Samuel George. 1844a. *Crania Aegyptica*. Philadelphia: John Pennington.

Morton, Samuel George. 1844b. *An Inquiry into the Distinctive Characteristics of the Aboriginal Race of America*. 2nd ed. Philadelphia: John Pennington.

Morton, Samuel George. 1847. Remarks on an Indian Cranium near Richmond, and a Chinook Infant Mummy. *Proceedings of the Academy of Natural Sciences of Philadelphia* III:330.

Morton, Samuel George. 1849. Observations on the Size of the Brain in Various Races and Families of Man. *Proceedings of the Academy of Natural Sciences of Philadelphia* 1849:221–224.

Morton, Samuel George. 1852. Physical Type of the American Indians. In *Information Respecting the History, Condition and Prospects of the Indian Tribes of the United States*, edited by H. R. Schoolcraft. 316–330. Philadelphia: Lippincott Grambo.

Moss, Madonna L., and George B. Wasson. Jr. 1998. Intimate Relations with the Past: The Story of an Athapaskan Village on the Southern Northwest Coast of North America. *World Archaeology* 29 (3):317–332.

Mostny, Grete. 1940. Las Momias Egipcias Conservadas en el Museo. *Boletín de Museo Nacional de Historia Natural, Santiago, Chile* 18:87–102.

Mostny, Grete. 1947. Un Cementerio Incásico en Chile Central. *Boletín de Museo Nacional de Historia Natural, Santiago* 23:17–39.

Mostny, Grete. 1952. Una Tumba en Chiu-Chiu. *Boletín de Museo Nacional de Historia Natural, Santiago* 26 (1):55.

Mostny, Grete. 1954. *Culturas Precolombinas de Chile*. Santiago: Editorial del Pacífico.

Mostny, Grete. 1955. El Niño Del Cerro El Plomo. *31° Congreso Internacional de Americanistas, Sao Paulo, Brazil* 2:847–862.

Mostny, Grete. 1956. La Momia Del Cerro El Plomo. *Museo Nacional de Historia Natural Mensual* 1 (1):5.

Mostny, Grete, ed. 1957a. La Momia Del Cerro El Plomo. *Boletín de Museo Nacional de Historia Natural, Santiago* 27 (1):1–119.

Mostny, Grete. 1957b. Estado de Conservación Del Conjunto Arqueológico. In *La Momia Del Cerro El Plomo*, edited by G. Mostny. 11–13. Vol. 27(1), *Boletín de Museo Nacional de Historia Natural, Santiago*.

Mostny, Grete. 1957c. Protocolo de Arqueología. Estudio de Los Objetos. In *La Momia Del Cerro El Plomo*, edited by G. Mostny. 31–63. Vol. 27(1), *Boletín de Museo Nacional de Historia Natural, Santiago*.

Mostny, Grete. 1957d. Los Incas en Chile. In *La Momia Del Cerro El Plomo*, edited by G. Mostny. 109–112. Vol. 27(1), *Boletín de Museo Nacional de Historia Natural, Santiago*.

Mostny, Grete. 1957e. Conclusiones Finales. In *La Momia Del Cerro El Plomo*, edited by G. Mostny. 113–115. Vol. 27(1), *Boletín de Museo Nacional de Historia Natural, Santiago*.

Mostny, Grete. 1964. Tsantas. Museo Nacional de Historia Natural, Noticas. *Mensual* 8 (94):1–4.

Mostny, Grete. 1967a. The Little Prince from El Plomo. In *Conquistadores without Swords: Archaeologists in the Americas*, edited by L. Deuel. 113–120. New York: St. Martin's Press.

Mostny, Grete. 1967b. Boy from the Lead Mountain. In *History Was Buried*, edited by M. Wheeler. 381–388. New York: Hart Publishing Company, Inc.

Müller, H.-G., B. Love, and Robert D. Hoppa. 2002. Semiparametric Method for Estimating Paleodemographic Profiles from Age Indicator Data. *American Journal of Physical Anthropology* 117:1–14.

Mullis, Kary B., and F. A. Faloona. 1987. Specific Synthesis of DNA in Vitro Via a Polymerase-Catalyzed Chain Reaction. *Methods in Enzymology* 155:335–350.

Mummery, John R. 1870. On the Relations Which Dental Caries—as Discovered amongst the Ancient Inhabitants of Britain and amongst Existing Aboriginal Races—May Be Supposed to Hold to Their Food and Social Condition. *Transactions of the Odontological Society of Great Britain (New Series)* 2:27–102.

Munizaga, Juan R. 1965. Espongio Hyperostosis (Hamperl y Weiss) U Osteoporosis Simétrica (Hrdlička). *Rivista de Antropología* 3:31–63.

Munizaga, Juan R. 1991. Human Skeletal Pathology in Pre-Columbian Populations of Northern Chile. In *Human Paleopathology: Current Syntheses and Future Options*, edited by D. J. Ortner, A. Aufderheide, pp. 145–150. Washington, D.C.: Smithsonian Institution Press.

Munizaga, Juan R. 1993. More on Piltdown: The Keith–Shattock Discrepancy Reconsidered. *Current Anthropology* 34(3):279–281.

Munizaga, Juan R., Marvin J. Allison, and Eugenio Aspillaga. 1978. Diaphragmatic Hernia Associated with Strangulation of the Small Bowel in an Atacamena Mummy. *American Journal of Physical Anthropology* 48(1):17–19.

Munizaga, Juan R., Marvin J. Allison, and Claudio Paredes. 1978. Cholelithiasis and Cholecystitis on Pre-Columbian Chileans. *American Journal of Physical Anthropology* 48(2):209–212.

Munson Chapman, Nancy E. 1997. Evidence for Spanish Influence on Activity Induced Musculoskeletal Stress Markers at Pecos Pueblo. *International Journal of Osteoarchaeology* 7:497–506.

Murphy, T. 1959a. The Changing Pattern of Dentine Exposure in Human Tooth Attrition. *American Journal of Physical Anthropology* 17 (3):167–178.

Murphy, T. 1959b. Gradients of Dentine Exposure in Human Molar Teeth Attrition. *American Journal of Physical Anthropology* 17 (3):179–186.

Murphy, Kimmarie A. 1999. A Prehistoric Example of Polydactyly from the Iron Age Site of Simbusenga, Zambia. *American Journal of Physical Anthropology* 108:311–319.

Museum of London. n.d. Museum of London Home Page. Retrieved Dec. 28, 2005, from Museum of London Web Site: http://www.museumoflondon.org.uk/.

NAGPRA Review Committee. 2004. Minutes. Native American Graves Protection and Repatriation Review Committee. Twenty-Sixth Meeting. Teleconference. July 19, 2004.

Nagura, Shigeo. 1960. The So-Called Osteochondritis Dissecans of König. *Clinical Orthopaedics* 18:100–122.

Nagurka, Mark L., and Wilson C. Hayes. 1980. An Interactive Graphics Package for Calculating Cross-Sectional Properties of Complex Shapes. *Journal of Biomechanics* 13:59–64.

Nathan, Hilel. 1962. Osteophytes of the Vertebral Column: Anatomical Study of Their Development According to Age, Race, and Sex with Considerations of Their Etiology and Significance. *Journal of Bone and Joint Surgery* 44A (2):243–268.

National Environment Research Council. n.d. NERC Home Page. Retrieved Dec. 28, 2005, from National Environment Research Council Web Site: http://www.nerc.ac.uk/.

National Museum of Health & Medicine. n.d. A Brief History of the Collecting of Anatomical Specimens. Retrieved from http://www.nmhm.washingtondc.museum/collections/anatomical/articles/brief_history.html.

National Park Service, U.S. Department of the Interior. n.d. National NAGPRA Retrieved Dec. 31, 2005, from National NAGPRA Home Page Web Site: http://www.cr.nps.gov/nagpra/.

Neiburger, Ellis J., ed. 1988. *The Preservation Book.* Grays Lake: Ethnic Minority Council of America.

Nelson, B. A., D. L. Martin, A. C. Swedlund, P. R. Fish, and George J. Armelagos. 1994. Studies in Disruption: Demography and Health in the Prehistoric American Southwest. In *Understanding Complexity in the Prehistoric Southwest*, edited by G. J. Gumerman and M. Gell-Mann. 59–112. Reading, MA: Addison-Wesley.

Nelson, C. T. 1938. The Teeth of *the Indians of Pecos Pueblo. American Journal of Physical Anthropology* 23:261.

Neumann, Georg K. 1937. Preliminary Notes on the Crania from Fulton County, Illinois. In *Rediscovering Illinois. Archaeological Explorations in and around Fulton County*, edited by F.-C. Cole and T. Deuel. 227–264. Chicago: University of Chicago Press.

Neumann, Georg K. 1940. Letter Dated January 8 to Snow. Snow Letters, Laboratory for Human Osteology, Mary Harmon Bryant Scientific Collections Building, University of Alabama, Tuscaloosa.

Neumann, Georg K. 1941. The Relationship between Archaeological Cultures and Physical Types in the Eastern United States. *American Journal of Physical Anthropology* 28:9–10.

Neumann, Georg K. 1942. Types of Artificial Cranial Deformation in the Eastern United States. *American Antiquity* 7 (3):306–310.

Neumann, Georg K. 1950. Racial Differentiation in the American Indian. Ph.D., Department of Anthropology, University of Chicago.

Neumann, Georg K. 1952. Archaeology and Race in the American Indian. In *Archaeology of Eastern United States*, edited by J. B. Griffin. 13–34. Chicago: University of Chicago Press.

Neumann, Georg K. 1954a. Archaeology and Race in the American Indian. *Yearbook of Physical Anthropology* 8:213–242.

Neumann, Georg K. 1954b. Measurements and Indices of American Indian Varieties. *Yearbook of Physical Anthropology* 8:243–255.

Neumann, Georg K. 1960. Origins of the Indians of the Middle Mississippian Area. *Proceedings of the Indiana Academy of Science* 69:66–68.

Neumann, Georg K. 1966. Racial History of the Indians. In *Natural Features of Indiana; a Symposium Held April 22–23, 1966, at Wabash College, Crawfordsville, Indiana*, edited by A. A. Lindsey. 505–518. Indianapolis: Indiana Academy of Science.

Neumann, Holm Wolfram. 1960a. Diagnostic Morphological Traits for the Walcolid Variety of American Indians. *Proceedings of the Indiana Academy of Science* 69:69–72.

Neumann, Holm Wolfram. 1960b. Delimitation of the Walcolid Variety of Man. Master's, Indiana University, Bloomington.

Neumann, Holm Wolfram. 1967. *The Paleopathology of the Archaic Modoc Rock Shelter Inhabitants*. Springfield, IL: Illinois State Museum.

Neville, Gwen Kennedy. 1977. Preface. In *Biocultural Adaptation in Prehistoric America*, edited by R. L. Blakely. vii. Athens: University of Georgia Press.

Newell, C. 1988. *Methods and Models in Demography*. New York: The Guilford Press.

Newman, Marshall Thornton. 1951. Report on Skeletal Material. In *Archaeology of the Bynum Mounds, Mississippi*, edited by J. L. Cotter and J. M. Corrett. 43–47, *United States National Park Service Archaeology Research Series, No. 1*. Washington.

Newman, Marshall Thornton. 1953. The Application of Ecological Rules to the Racial Anthropology of the Aboriginal New World. *American Anthropologist* 55 (3): 311–327.

Newman, Marshall Thornton. 1962. Evolutionary Changes in Body Size and Head Form in American Indians. *American Anthropologist* 64 (2):237–257.

Newman, Marshall Thornton, and Charles E. Snow. 1942. Preliminary Report on the Skeletal Material from Pickwick Basin, Alabama. In *An Archaeological Survey of Pickwick Basin in the Adjacent Portions of the States of Alabama, Mississippi, and Tennessee*, edited by W. S. Webb and D. L. DeJarnette. 393–507, *Bureau of American Ethnology Bulletin 129*. Washington, DC: Smithsonian Istitution.

Nicholas, George P., and Thomas D. Andrews. 1997. Indigenous Archaeology in the Post-Modern World. In *At a Crossroads: Archaeology and First Peoples of Canada*, edited by G. P. Nicholas and T. D. Andrews. 1–18. Burnaby: Archaeology Press.

Nicholls, B. 1914. A Contribution to the Study of the Teeth of the Australian and Tasmanian Aboriginals. In *Transactions of the 6th International Dental Congress*. 67–102: London.

Nott, Josiah Clark, and George Robins Gliddon. 1857. *Indigenous Races of the Earth*. Philadelphia: Lippincott.

Oakley, K. P. 1950. Relative Dating of the Piltdown Skull. *Advancement of Science* 6:343–344.

Oakley, K. P. 1960. Ancient Preserved Brains. *Man* 60:90–91.

Oakley, K. P., M. A. Winnifred, A. Brooke, R. Akester, and Don R. Brothwell. 1959. Contributions on Trepanning or Trephination in Ancient and Modern Times. *Man* 59 (June):93–96.

O'Connor, Kathleen A. 1995. The Age Pattern of Mortality: A Micro-Analysis of Tipu and a Meta-Analysis of Twenty-Nine Paleodemographic Samples. Ph.D. Dissertation, University of Albany, SUNY, Albany.

O'Connor, T. P. 2000. *The Archaeology of Animal Bones*. Stroud, Gloucester: Sutton Publishing.

Odjick, Brenda. 2004. Repatriation — Resisting Assimilation: Kitigan Zibi's Journey from Honouring Our Past to Protecting Our Future — Part 2 [Abstract]. *CAPA Selected Abstracts* — 32nd Annual Meeting of the Canadian Association for Physical Anthropology, London, Ontario:28.

Oestreicher, David M. 1996. Unraveling the *Walam Olum*. *Natural History* 105 (10): 14–20.

Oestreicher, David M. 2002. Roots of the Walam Olum: Constantine Samuel Rafinesque and the Intellectual Heritage of the Early Nineteenth Century. In *New Perspectives on the Origins of Americanist Archaeology*, edited by D. L. Browman and S. Williams. 60–86. Tuscaloosa: University of Alabama Press.

Oetteking, Bruno. 1925. *Skeletal Remains from Santa Barbara, California. I. Craniology, Indian Notes and Monographs No. 39*. New York: Museum of the American Indian, Heye Foundation.

Oetteking, Bruno. 1928. Discussion of Hrdlička, Aleš: The Origin and Antiquity of Man in America. *Bulletin of the New York Academy of Medicine* Ser. 2. 4 (7):816–820.

Oetteking, Bruno. 1930a. Craniology of the North Pacific Coast. In *The Jesup North Pacific Expedition*, edited by F. Boas, *Memoirs of the American Museum of Natural History, Vol. 15, Pt. 1; Publications of the Jesup North Pacific Expedition, Vol. 11, Pt. 1*. New York: E. J. Brill, Leiden, and G. E. Stechert.

Oetteking, Bruno. 1930b. Skeletal Remains from Texas. *Indian Notes* VII (3):336–346.

Oetteking, Bruno. 1931. A Contribution to the Physical Anthropology of Baffin Island Based on Somatometrical Data and Skeletal Material Collected by the Putnam Baffin Island Expedition of 1927. *American Journal of Physical Anthropology* 15:421–468.

Oetteking, Bruno. 1932. Morphologie und Menschliches Alterum in Amerika. *Anthropos* 27:899–903.

Oetteking, Bruno. 1934. Skeletal Remains from Cook Inlet and Prince William Sound. In *The Archaeology of Cook Inlet, Alaska*, edited by F. de Laguna. Philadelphia: University of Pennsylvania Press for the University Museum.

Oetteking, Bruno. 1945. Skeletal Remains from Prince William Sound, Alaska. *American Journal of Physical Anthropology* 3:57–96, 177–204, 277–311.

Oetteking, Bruno. 1975. *Craniology of the North Pacific Coast*. New York: AMS Press. Original edition. Reprint of the 1930 ed. published by E. J. Brill, Leiden, and G. E. Stechert, New York, which was issued as v. 15, pt. 1 of the Memoirs of the American Museum of Natural History, and as v. 11, pt. 1 of the Publications of the Jesup North Pacific Expedition.

Ogilvie, Marsha D., and Charles E. Hilton. 1993. Analysis of Selected Human Skeletal Material from Sites 423-124 and -131. In *Across the Colorado Plateau: Anthropological Studies for the Transwestern Pipeline Expansion Project, Volume XVIII, Human Remains and Burial Goods*, edited by N. P. Herrmann, M. D. Ogilvie, C. E. Hilton, and K. L. Brown. 97–128. Albuquerque, NM: Office of Contract Archaeology and Maxwell Museum of Anthropology, University of New Mexico.

Ogilvie, M. D., and C. E. Hilton. 2000. Ritualized Violence in the Prehistoric American Southwest. *International Journal of Osteoarchaeology* 10:27–48.

Omoto, Constance E. 1960. The Identification of a Prehistoric Skeletal Series from Will County, Illinois. Master's, Indiana University, Bloomington.

Oranje, P., J. N. Noriskin, and T. W. B. Osborn. 1935. The Effect of Diet Upon Dental Caries in the South African Bantu. *South African Journal of Medical Science* 1:57–62.

O'Rourke, Dennis H., S. W. Carlyle, and R. L. Parr. 1996. Ancient DNA: A Review of Methods, Progress, and Perspectives. *American Journal of Human Biology* 8:557–571.

O'Rourke, Dennis H., M. G. Hayes, and S. W. Carlyle. 2000. Ancient DNA Studies in Physical Anthropology. *Annual Review of Anthropology* 29:217–242.

O'Rourke, Dennis H., R. L. Parr, and S. W. Carlyle. 1999. Molecular Genetic Variation in Prehistoric Inhabitants of the Eastern Great Basin. In *Prehistoric Lifeways in the Great Basin Wetlands: Bioarchaeological Reconstruction and Interpretation*, edited by B. M. Hemphill and C. S. Larsen. 84–102. Salt Lake City: University of Utah Press.

Orr, J. B., and K. L. Gilkes. 1931. *The Physique and Health of Two African Tribes, Studies of Nutrition, Medical Research Council Special Reports, Series 155*. London.

Ortner, Donald J. 1968. Description and Classification of Degenerative Bone Changes in the Distal Joint Surfaces of Humerus. *American Journal of Physical Anthropology* 28 (2):139–155.

Ortner, Donald J. 2003. *Identification of Pathological Conditions in Human Skeletal Remains*. 2nd ed. New York: Academic Press.

Ortner, Donald J., and Arthur C. Aufderheide, eds. 1991. *Human Paleopathology: Current Syntheses and Future Options*. Washington, DC: Smithsonian Institution Press.

Ortner, Donald J., Whitney Butler, Jessica Cafarella, and Lauren Milligan. 2001. Evidence of Probable Scurvy in Subadults from Archeological Sites in North America. *American Journal of Physical Anthropology* 114:343–351.

Ortner, Donald J., and M. F. Ericksen. 1997. Bone Changes in the Human Skull Probably Resulting from Scurvy in Infancy and Childhood. *International Journal of Osteoarchaeology* 7:212–220.

Ortner, Donald J., and Jennifer Olsen Kelley. 1990. Contemporary Trends and Future Directions in Human Osteology and Paleopathology. In *A Life in Science: Papers in Honor of J. Lawrence Angel*, edited by J. E. Buikstra. 17–24, *Scientific Papers 6*. Kampsville, IL: American Anthropological Association/Center for American Archaeology.

Ortner, Donald J., and Mary Lucas Powell. 2006. Paleopathology. In *Environment, Origins, and Population*, edited by D. Ubelaker. Vol. 3, *Handbook of North American Indians*, edited by W. C. Sturtevant. Washington, DC: Smithsonian Institution.

Ortner, Donald J., and Walter G. J. Putschar. 1981. *Identification of Pathological Conditions in Human Skeletal Remains, Smithsonian Contributions to Anthropology, No. 28*. Washington, DC: Smithsonian Institution Press.

Ortner, Donald J., and Walter G. J. Putschar. 1985. *Identification of Pathological Conditions in Human Skeletal Remains, Reprint Edition of Smithsonian Contributions to Anthropology, No. 28*. Washington, DC: Smithsonian Institution Press.

Ortner, Donald J., Noreen Tuross, and Agnes I. Stix. 1992. New Approaches to the Study of Disease in Archaeological Populations. *Human Biology* 64:337–360.

Ortner, Donald J., and Charles J. Utermohle. 1981. Polyarticular Inflammatory Arthritis in a Pre-Columbian Skeleton from Kodiak Island. *American Journal of Physical Anthropology* 56:23–31.

Ousley, Stephen D. 1995. Relationships between Eskimos, Amerindians, and Aleuts: Old Data, New Perspectives. *Human Biology* 67 (3):427–458.

Ousley, Stephen D., Douglas W. Owsley, and Dawn Mulhern. 2000. Lost and Found in the Museum: Repatriation, Ancestry, Ethnicity, and History [Abstract]. *American Journal of Physical Anthropology* 111 (Suppl. 30):243.

Owsley, Douglas W. 1984. Human Bones from Archaeological Context: An Important Source of Information. *Council for Museum Anthropology Newsletter* 8:2–8.

Owsley, Douglas W. 1999. From Jamestown to Kennewick: An Analogy Based on Early Americans. In *Who Were the First Americans*, edited by R. Bonnichsen. 127–140. Corvallis: Center for the Study of the First Americans, Oregon State University.

Owsley, Douglas W., and William M. Bass. 1979. A Demographic Analysis of Skeletons from the Larson Site (39WW2). *American Journal of Physical Anthropology* 51: 145–154.

Owsley, Douglas W., S. M. Bennett, and Richard L. Jantz. 1982. Intercemetery Morphological Variation in Arikara Crania from the Mobridge Site (39WW1). *American Journal of Physical Anthropology* 58:179–185.

Owsley, Douglas W., H. E. Berryman, and William M. Bass. 1977. Demographic and Osteological Evidence for Warfare at the Larson Site, South Dakota. *Plains Anthropologist* 13:119–131.

Owsley, Douglas W., and Richard L. Jantz. 1978. Intracemetery Morphological Variation in Arikara Crania from the Sully Site (39SL4), Sully County, South Dakota. *Plains Anthropologist* 23:139–147.

Owsley, Douglas W., and Richard L. Jantz, eds. 1994. *Skeletal Biology in the Great Plains: Migration, Warfare, Health, and Subsistence*. Washington, D.C.: Smithsonian Institution Press.

Owsley, Douglas W., and Richard L. Jantz. 2001. Archaeological Politics and Public Interest in Paleoamerican Studies: Lessons from Gordon Creek Woman and Kennewick Man. *American Antiquity* 66 (4):565–575.

Owsley, Douglas W., and Helen L. O'Brien. 1982. Stature of Adult Cherokee Indians during the Eighteenth Century. *Journal of Cherokee Studies* 1982 (Fall):74–78.

Owsley, Douglas W., and Jerome Carl Rose. 1997. *Bioarchaeology of the North Central United States, Arkansas Archeological Survey Research Series no. 49*. Fayetteville, AR: Arkansas Archeological Survey.

Pader, Ellen. 1982. *Symbolism, Social Relations and the Interpretation of Mortuary Remains, British Archaeological Reports. International Series 130*. Oxford: Archaeopress.

Paine, R. R. 1989. Model Life Table Fitting by Maximum Likelihood Estimation: A Procedure to Reconstruct Paleodemographic Characteristics from Skeletal Age Distributions. *American Journal of Physical Anthropology* 79:51–56.

Paine, R. R. 2000. If a Population Crashes in Prehistory, and There Is No Paleodemographer to Hear It, Does It Make a Sound? *American Journal of Physical Anthropology* 112:181–190.

Paine, R. R., and Henry C. Harpending. 1998. Effect of Sample Bias on Paleodemographic Fertility Estimates. *American Journal of Physical Anthropology* 105:231–240.

Paleopathology Association. 2000. Paleopathology Association Home Page. Retrieved Dec. 28, 2005, from Paleopathology Association Web Site: http://www.paleopathology.org/.

Pálfi, György, Olivier Dutour, Judith Deák, and Imre Hutás, eds. 1999. *Tuberculosis: Past and Present*. Budapest; Szeged: Golden Book Publisher Ltd./Tuberculosis Foundation.

Palkovich, A. M. 1980. *Pueblo Population and Society: The Arroyo Hondo Skeletal and Mortuary Remains*. Vol. 3, *Arroyo Hondo Archaeological Series*. Santa Fe, NM: School of American Research Press.

Palkovich, A. M. 1983. A Comment on Mobley's "Demographic Structure of Pecos Indians." *American Antiquity* 48:142–147.

Panush, Richard S., and Nancy E. Lane. 1994. Exercise and the Musculoskeletal System. *Baillierés Clinical Rheumatology* 8:79–102.

Pany, D., M. Teschler-Nicola, and H. Wilfing. 2003. Miners or Mine Owners: Do the Hallstatt Skeletons Reflect Occupation and Social Structure? [Abstract]. *American Journal of Physical Anthropology* (Suppl. 36):164.

Parker, Charles A. 1904. Evidences of Rheumatoid Arthritis in the Lansing Man. *American Geologist* 33(42):39–42.

Parker Pearson, Mike. 1982. Mortuary Practices, Society and Ideology: An Ethnoarchaeological Study. In *Symbolic and Structural Archaeology*, edited by I. Hodder. 99–113. Cambridge: Cambridge University Press.

Parker Pearson, Mike. 1999. *The Archaeology of Death and Burial*. Great Britain: Sutton Publishing.

Parker Pearson, Michael. 2001. *The Archaeology of Death and Burial*. U.S. ed., *Texas A&M University Anthropology Series, No. 3*. College Station, TX: Texas A&M University Press.

Parr, R. L., S. W. Carlyle, and Dennis H. O'Rourke. 1996. Ancient DNA Analysis of Fremont Amerindians of the Great Salt Lake Wetlands. *American Journal of Physical Anthropology* 99:507–518.

Parsons, F. G., and C. R. Box. 1905. The Relationship of the Cranial Sutures to Age. *Journal of the Royal Archaeological Institute* 35:30–38.

Pastor, R. F. 1992. Dietary Adaptations and Dental Microwear in Mesolithic and Chalcolithic South Asia. In *Culture, Ecology and Dental Anthropology*, edited by J. R. Lukacs. 215–228, *Journal of Human Ecology, Special Issue 2*.

Pastor, R. F. 1994. Dental Microwear among Prehistoric Inhabitants of the Indian Subcontinent: A Quantitative and Comparative Analysis. Ph.D. Dissertation, University of Oregon.

Patrick, John. J. R. 1895. Examination of Pre-Historic Crania. *Dental Cosmos* XXXVII(1):1–24.

Patterson, Henry S. 1854. Memoir of the Life and Scientific Labors of Samuel George Morton. In *Types of Mankind or, Ethnological Researches: Based upon the Ancient Monuments, Paintings, Sculptures, and Crania of Races, and upon Their Natural, Geographical, Philological and Biblical History, Illustrated by Selections from the Inedited Papers of Samuel George Morton and by Additional Contributions from L. Agassiz; W. Usher; and H. S. Patterson*, edited by J. C. Nott and G. R. Gliddon. xvii–lvii. Philadelphia: J.B. Lippincott, Grambo & Co.

Pauwels, Friedrich. 1980. *Biomechanics of the Locomotor Apparatus*. Berlin: Springer Verlag.

Peabody Museum. 2001. Catalogue and Accession Records of the Peabody Museum. In *Hemenway Expedition Records 1886–1914: A Finding Aid*. Peabody Museum Archives. Harvard University: The President and Fellows of Harvard College.

Pearson, Karl. 1899. On the Reconstruction of Prehistoric Races. *Philosophical Transactions of the Royal Society of London Series A* 192:169–244.

Pearson, Karl. 1926. On the Coefficient of Racial Likeness. *Biometrika* 18 (1/2):105–117.

Pearson, Karl, and J. A. Bell. 1919. *A Study of the Long Bones of the English Skeleton*. Cambridge: Cambridge University Press.

Pearson, Maria D. 1990. The Indian Point of View: Exhumation and Reburial. In *Preservation on the Reservation. Native Americans, Native American Lands and Archaeology*, edited by A. L. Klesert and A. S. Downer. 397–408: Navajo Nation Archaeology, Navajo Nation Historic Preservation Department.

Pearson, Osbjorn M. 2000. Activity, Climate, and Postcranial Robusticity: Implications for Modern Human Origins and Scenarios of Adaptive Change. *Current Anthropology* 41:569–607.

Pearson, Osbjorn M., and Frederic E. Grine. 1996. Morphology of the Border Cave Hominid Ulna and Humerus. *South African Journal of Science* 92 (5):231–236.

Pearson, Osbjorn M., and Frederic E. Grine. 1997. Re-Analysis of the Hominid Radii from Cave of Hearths and Klasies River Mouth, South Africa. *Journal of Human Evolution* 32:577–592.

Pearson, Osbjorn M., and Daniel E. Lieberman. 2004. The Aging of Wolff's "Law:" Ontogeny and Responses to Mechanical Loading in Cortical Bone. *Yearbook of Physical Anthropology* 47:63–99.

Pedde, Sara E., and Olaf H. Prufer. 2001. The Kentucky Green River Archaic as Seen from the Ward Site. In *Archaic Transitions in Ohio and Kentucky Prehistory*, edited by O. H. Prufer, S. E. Pedde, and R. S. Meindl. 59–86. Kent, OH: Kent State University Press.

Pedersen, Scott C., and Susan C. Anton. 1998. Bicoronal Synostosis in a Child from Historic Omaha Cemetery 25DK10. *American Journal of Physical Anthropology* 105:369–376.

Pendergast, David M., and Clement W. Meighan. 1959. Folk Traditions as Historical Fact: A Paiute Example. *Journal of American Folklore* 72:128–133.

Pennefather-O'Brien, E. E., and S. Strezewski. 2002. An Initial Description of the Archaeology and Morphology of the Clark's Fork Skeletal Material, Bonner County, Idaho. *North American Archaeologist* 23:101–115.

Pereira, Cleber Bidegain, and Marília Carvalho de Mello e Alvim. 1979. *Manual Para Estudos Craniométricos e Cranioscópicos*. Santa Maria: Universidade de Santa Maria.

Pereira, Cleber Bidegain, and Marília Carvalho de Mello e Alvim. 1985. Craniometria Radiográfica em População Pré-Histõrica Brasileira. *Pesquisas—Antropologia (São Leopoldo)* 40:34–44.

Pereira, Grégory. 1999. *Potrero de Guadalupe. Anthropologie Funéraire D'une Communauté Pré-Tarasque du Nord du Michoacán, Mexique, British Archaeological Reports. International Series 816*. Oxford: Archaeopress.

Perizonius, W. R. K., Sheilagh T. Brooks, and Richard H. Brooks. 1991. A Suture Connecting Parietal Foramina. *Paleopathology Newsletter No. 76*:11–12.

Perzigian, A. J. 1975. Natural Selection on the Dentition of an Arikara Population. *American Journal of Physical Anthropology* 42:63–70.

Peters, Charles R. 1982. Electron-Optical Microscopic Study of Incipient Dental Microdamage from Experimental Seed and Bone Crushing. *American Journal of Physical Anthropology* 57:283–301.

Petersen, W. 1975. A Demographer's View of Prehistoric Demography. *Current Anthropology* 16:227–245.

Peterson, Jane. 1998. The Natufian Hunting Conundrum: Spears, Atlatls, or Bows? Musculoskeletal and Armature Evidence. *International Journal of Osteoarchaeology* 8:378–389.

Phenice, Terrell W. 1969a. *An Analysis of the Human Skeletal Material from the Burial Mounds in North Central Kansas, University of Kansas Publications in Anthropology 1*. Lawrence.

Phenice, T. W. 1969b. A Newly Developed Visual Method of Sexing the Os Pubis. *American Journal of Physical Anthropology* 30:297–302.

Phillips, Philip, and James A. Brown. 1983. *Pre-Columbian Shell Engravings from the Craig Mound at Spiro, Oklahoma*. Cambridge: Harvard University Press.

Pickering, Robert Bruce. 1984. Patterns of Degenerative Joint Disease in Middle Woodland, Late Woodland, and Mississippian Skeletal Series from the Lower Illinois Valley. Ph.D. Dissertation, Northwestern University.

Pietrusewsky, M. 2000. Metric Analysis of Skeletal Remains: Methods and Applications. In *Biological Anthropology of the Human Skeleton*, edited by M. A. Katzenberg and S. R. Saunders. 375–415. New York: Wiley-Liss.

Piontek, J., and A. Weber. 1990. Controversy on Paleodemography. *International Journal of Anthropology* 5:71–83.

Plog, S. 2003. Book Review Essay: Social Conflict, Social Structure and Processes of Culture Change. *American Antiquity* 68 (1):182–186.

Podgorny, I. 2001. *El Espaol de la Arqueologia: "no Se Ve en Ella Un Trabajo Literario." II Congreso Internacional de la Lengua Espaola: El Espaol en la Sociedad de la Informacin: Nuevas Fronteras Del Espaol*. Valladolid: Centro Virtual Cervantes.

Policy Planning Guidance. 1990. *Policy Planning Guidance 16: Archaeology and Planning*. London: Her Majesty's Stationary Office.

Pollex, Axel. 1999. Comments on the Interpretation of the So-Called Cattle Burials of Neolithic Central Europe. *Antiquity* 73 (281):542–550.

Porter, C. M. 1986. *The Eagle's Nest: Natural History and American Ideas, 1812–1842*. Alabama: University of Alabama Press.

Pourchet, Maria Júlia, and Marília Carvalho de Mello e Alvim. 1975. Um Estudo de Adaptabilidade Humana em Microregião Pesqueira (Nota Prévia). *Arqueivos de Anatomia e Antropologia da Universidade Souza Marques (Rio de Janeiro)* 1 (1):389–391.

Powell, Joseph F., and Walter A. Neves. 1999. Craniofacial Morphology of the First Americans: Pattern and Process in the Peopling of the New World. *Yearbook of Physical Anthropology* 42:153–188.

Powell, Mary Lucas. 1985. The Analysis of Dental Wear and Caries for Dietary Reconstruction. In *The Analysis of Prehistoric Diets*, edited by R. I. Gilbert and J. H. Mielke. 307–338. Orlando, FL: Academic Press.

Powell, Mary Lucas. 1988. *Status and Health in Prehistory: A Case Study of the Moundville Chiefdom, Smithsonian Series in Archaeological Inquiry*. Washington, DC: Smithsonian Institution Press.

Powell, Mary Lucas. 1992. Endemic Treponematosis and Tuberculosis in the Prehistoric Southeastern United States: The Biological Costs of Chronic Endemic Disease.

In *Human Paleopathology: Current Syntheses and Future Options*, edited by D. J. Ortner and A. C. Aufderheide. 173–180. Washington, DC: Smithsonian Institution Press.

Powell, Mary Lucas. 1994. Treponematosis before 1492 in the Southeastern United States: Why Call It Syphilis? In *L'origine de la Syphilis en Europe: Avant Ou Aprés 1493?* edited by O. Dutour, G. Pálfi, J. Berano, and J.-P. Brun. 158–163. Paris: Editions Errance-Centre Archaelogique du Var.

Powell, Mary Lucas. 2000. Ancient Diseases, Modern Perspectives: Tuberculosis and Treponematosis in the Age of Agriculture. In *Bioarchaeological Studies of Life in the Age of Agriculture, a View from the Southeast*, edited by P. M. Lambert. 6–34. Tuscaloosa: The University of Alabama Press.

Powell, Mary Lucas, Patricia S. Bridges, and Ann Marie Wagner Mires. 1991. *What Mean These Bones? Studies in Southeastern Archaeology*. Tuscaloosa: University of Alabama Press.

Powell, Mary Lucas, and Della Collins Cook. 2005. *The Myth of Syphilis: The Natural History of Treponematosis in North America*. Gainesville, FL: University Press of Florida.

Praeger, W. 1925. Das Gebiß des Menschen in der Altsteinzeit und Die Anfänger der Zahnkaries. *Deutsche Zahnärztliche Wochenschrift* 28:88-99–112-122.

Preston, D. 1998. Cannibals of the Canyon. *The New Yorker* (November 30):76–89.

Preuschoft, Holger. 1971. Body Posture and Mode of Locomotion in Early Pleistocene Hominids. *Folia Primatologica* 14:209–240.

Price, T. Douglas. 1989. *The Chemistry of Prehistoric Human Bone*. New York: Cambridge University Press.

Price, T. Douglas, Gisela Grupe, and Peter Schröter. 1994a. Reconstruction of Migration Patterns in the Bell Beaker Period by Stable Strontium Isotope Analysis. *Applied Geochemistry* 9:413–417.

Price, T. Douglas, Gisela Grupe, and Peter Schröter. 1998. Migration in the Bell Beaker Period of Central Europe. *Antiquity* 72:405–411.

Price, T. Douglas, C. M. Johnson, J. A. Ezzo, J. Ericson, and James H. Burton. 1994b. Residential Mobility in the Prehistoric Southwest United States: A Preliminary Study Using Strontium Isotope Analysis. *Journal of Archaeological Science* 21:315–330.

Price, W. A. 1933. Why Dental Caries with Modern Civilization? IV. Field Studies in the Primitive and in Modern Outer Hebrides, Scotland. *Dental Digest* 39:225–229.

Privat, K. L., T. O'Connell, and M. P. Richards. 2002. Stable Isotope Analysis of Human and Faunal Remains from the Anglo-Saxon Cemetery at Berinsfield, Oxfordshire: Dietary and Social Implications. *Journal of Archaeological Science* 29:779–790.

Puech, P. F., C. Serratrice, and F. F. Leek. 1983. Tooth Wear as Observed in Ancient Egyptian Skulls. *Journal of Human Evolution* 12:617–629.

Pullar, Gordon L. 2001. Foreword. In *Looking Both Ways: Heritage and Identity of the Alutiiq People*, edited by A. L. Crowell, A. F. Steffian, and G. L. Pullar. Fairbanks: University of Alaska Press.

Putnam, Frederic Ward. 1876. Report of the Curator. *Reports of the Peabody Museum of American Archaeology and Ethnology* 1 (Ninth Annual Report of the Trustees):7–15.

Putnam, Frederic Ward. 1878a. Report of the Curator. *Reports of the Peabody Museum of American Archaeology and Ethnology* 2 (Eleventh Annual Report of the Trustees of the Peabody Museum of Armerican Archaeology and Ethnology):191–224.

Putnam, Frederic Ward. 1878b. Archaeological Explorations in Tennessee. *Reports of the Peabody Museum of American Archaeology and Ethnology* 2 (Eleventh Annual Report of the Trustees of the Peabody Museum of Armerican Archaeology and Ethnology):305–360.

Putnam, Frederic Ward. 1884. Abnormal Human Skull from Stone Graves in Tennessee. *Proceedings of the American Association for the Advancement of Science* 32:390–392.

Putnam, Frederic Ward. 1886. Report of the Curator. *Reports of the Peabody Museum of American Archaeology and Ethnology* 3 (Eighteenth and Nineteenth Annual Reports of the Peabody Museum of American Archaeology and Ethnology):401–418.

Putnam, Frederic Ward. 1888. [Memorial Notice for Cordelia Adelaide Studley]. *Proceedings of the Boston Society of Natural History* 23:419–420.

Putnam, Frederic Ward. 1889–1893. Diseases Affecting the Teeth and Bone of the Mouth of Pre-Columbian Indians. *Transactions of the American Academy of Dental Science* 1889–1893:19–24.

Putnam, Frederic Ward. 1891. Report of the Curator. *Reports of the Peabody Museum of American Archaeology and Ethnology* 4 (Twenty-Third and Twenty-Fourth Annual Reports of the Trustees of the Peabody Museum of Armerican Archaeology and Ethnology):87–107.

Putnam, Frederic Ward. 1899. A Problem in American Anthropology. *Proceedings of the American Academy for the Advancement of Science* 48:1–17.

Qetel, Claude. 1990. *History of Syphilis*. Baltimore, MD: Johns Hopkins University Press.

Quatrefages, Armand de, and Ernest T. Hamy. 1882. *Crania Ethnica*. Paris: J.B. Baillière et Fils.

Quigley, Christine. 1998. *Modern Mummies: The Preservation of the Human Body in the Twentieth Century*. Jefferson, NC: McFarland and Company, Inc.

Quigley, Christine. 2001. *Skulls and Skeletons: Human Bone Collections and Accumulations*. Jefferson, NC: McFarland and Company, Inc.

Rabkin, Samuel. 1942. Dental Conditions among Prehistoric Indians of Northern Alabama. *Journal of Dental Research* 21:211–222.

Rabkin, Samuel. 1943. Dental Conditions among Prehistoric Indians of Kentucky. *Journal of Dental Research* 22:355–366.

Rafi, A., Mark Spigelman, J. Stanford, Eshetu Lemma, H. D. Donoghue, and J. Zias. 1994. DNA of *Mycobacterium Leprae* Detected by PCR in Ancient Bone. *International Journal of Osteoarchaeology* 4:287–290.

Rakita, Gordon F. M., and Jane E. Buikstra. 2005. Corrupting Flesh: Reexamining Hertz's Perspective on Mummification and Cremation. In *Interacting with the Dead: Perspectives on Mortuary Archaeology for the New Millennium.*, edited by G. F. M. Rakita, J. E. Buikstra, L. Beck, and S. R. Williams.

Rakita, Gordon F. M., Jane E. Buikstra, Lane A. Beck, and Sloan R. Williams, eds. 2005. *Interacting with the Dead: Perspecitves on Mortuary Archaeology for the New Millennium*. Gainesville: University Press of Florida.

Ramos, L. Scott. 1994. Characterisation of Mycobacterial Species by High Performance Liquid Chromatography and Pattern Recognition. *Journal of Chromatographic Science* 32:219–228.

Ratcliffe, Susan. 2000. *The Oxford Dictionary of Thematic Quotations*. New York: Oxford University Press.

Raven, C. 1994. Invisible from the West: Numic Expansion from the Perspective of the Carson Desert. In *Across the West: Human Population Movement and the Expansion of the Numa*, edited by D. B. Madsen and D. Rhode. 152–156. Salt Lake City: University of Utah Press.

Ravesloot, J. C. 1988. *Mortuary Practices and Social Differentiation at Casas Grandes, Chihuahua, Mexico, Anthropological Papers of the University of Arizona, No. 49*. Tucson, AZ: The University of Arizona Press.

Rawson, R., and Sheilagh T. Brooks. 1984. Classification of Human Breast Morphology Important to Bite Mark Investigation. *American Journal of Forensic Medicine and Pathology* 5:19–24.

Redfern, R. 2005. A Gendered Analysis of Health from the Iron Age to the End of the Romano-British Period in Dorset, England (Mid to Late 8th Century BC to the End of the 4th Century AD). Ph.D. Dissertation, University of Birmingham.

Reed, E. K. 1953. Human Skeletal Material from Te'ewi. In *Salvage Archaeology in the Chama Valley, New Mexico*, edited by F. Wendorf. 104–118. Santa Fe: Museum of New Mexico.

Reed, E. K. 1966. The 14th Century Population of the Galisteo Bason: Skeletal Remains and Burial Methods. Unpublished manuscript on file at the Laboratory of Anthropology, Santa Fe, New Mexico.

Reed, E. K. 1967. Human Skeletal Material from the Gran Quivira District. Unpublished manuscript on file at the Laboratory of Anthropology, Santa Fe, New Mexico.

Reichs, Kathy J. 1984. Pearls or People: A Biometric Analysis of Interregional Exchange During Hopewell Times. *Central Issues Anthropol* 5:47–65.

Reinhard, Karl J. 2000. Reburial, International Perspectives. In *Archaeological Method and Theory: An Encyclopedia*, edited by L. Ellis. 512–518. New York & London: Garland Publishing, Inc.

Reinhard, Karl J., Sheilagh T. Brooks, and Richard H. Brooks. 1989. Diet and Environment Determined from Analysis of Prehistoric Coprolites from an Archaeological Site near Zape Chico, Durango, Mexico. In *Advances in Paleopathology*, edited by L. Capasso. 151–155. Chieti: Journal of Paleopathology Monographs.

Reinhard, Karl J., and Jane E. Buikstra. 2003. Louse Infestation of the Chiribaya Culture, Southern Peru: Variation in Prevalence by Age and Sex. *Memorias de Instituto Oswaldo Cruz* 98(Suppl. 1):173–179.

Reinhard, Karl J., T. Michael Fink, and Jack Skiles. 2003. A Case of Megacolon in Rio Grande Valley as a Possible Case of Chagas Disease. *Memorias do Instituto Oswaldo Cruz* 98(Suppl. 1):165–172.

Reinhard, Karl J., Larry L. Tieszen, Karin L. Sandness, Lynae M. Beiningen, Elizabeth Miller, A. Mohammad Ghazi, Christiana E. Miewald, and Sandra V. Barnum. 1994. Trade, Contact, and Female Health in Northeast Nebraska. In *In the Wake of*

Contact: Biological Responses to Conquest, edited by C. S. Larsen and G. R. Milner. 63–74. New York: Wiley-Liss.

Relethford, John H. 1999. Models, Predictions, and the Fossil Record of Modern Human Origins. *Evolutionary Anthropology* 8:7–10.

Relethford, John H. 2001. *Genetics and the Search for Modern Human Origins*. New York: Wiley-Liss.

Relethford, John H., and John Blangero. 1990. Detection of Differential Gene Flow from Patterns of Quantitative Variation. *Human Biology* 62:5–25.

Relethford, John H., and Henry C. Harpending. 1994. Craniometric Variation, Genetic Theory, and Modern Human Origins. *American Journal of Physical Anthropology* 95:249–270.

Relethford, John H., and Francis C. Lees. 1982. The Use of Quantitative Traits in the Study of Human Population Structure. *Yearbook of Physical Anthropology* 25:113–132.

Retzius, Anders. 1859. Present State of Ethnology in Relation to the Form of the Human Skull. *Annual Report of the Smithsonian Institution* 1859:251–270.

Reycraft, Richard. 2004. *Us and Them: Archaeology and Ethnicity in the Andes*. Los Angeles: Cotsen Institute of Archaeology University of California.

Reynaud-Paligaut, C. 2001. Paul Rivet: Contradictions et Ambiguities D'un Intellectuel Antifasciste Dans les Annees Trente. In *Transmettre les Passes*, edited by M.-C. Hoock-Demarle and C. Liauzu. 161–174. Paris: Editions Syllepse.

Rhodd, Ben K. 1990. Reburial in South Dakota: Two Perspectives. In *Preservation on the Reservation. Native Americans, Native American Lands and Archaeology*, edited by A. L. Klesert and A. S. Downer. 369–377: Navajo Nation Archaeology, Navajo Nation Historic Preservation Department.

Rhode Island. Court of Trials. 1920. *Rhode Island Court Records. Volume I. Records of the Court of Trials of the Colony of Providence Plantations 1647–1662*. Providence, RI: Rhode Island Historical Society.

Richards, M. P., R. E. M. Hedges, R. Jacobi, A. Currant, and C. Stringer. 2000. FOCUS: Gough's Cave and Sun Hole Cave Human Stable Isotope Values Indicate a High Animal Protein Diet in the British Upper Palaeolithic. *Journal of Archaeological Science* 27:1–3.

Richards, M. P., R. E. M. Hedges, T. I. Molleson, and J. C. Vogel. 1998. Stable Isotope Analysis Reveals Variations in Human Diet at Poundbury Camp Cemetery Site. *Journal of Archaeological Science* 25:1247–1252.

Richmond, Brian G., Barth W. Wright, Ian Grosse, Paul C. Dechow, Callum F. Ross, Mark A. Spencer, and David S. Strait. 2005. Finite Element Analysis in Functional Morphology. *The Anatomical Record Part A* 238A:259–274.

Riding In, James. 1990. Report Verifying the Identity of Six Pawnee Scout Crania at the Smithsonian Institute and the National Museum of Health and Medicine. In *Hearings on S. 1021 and S. 1980 before the Senate Select Committee on Indian Affairs, 101st Congress, 2d Session (May 1990)*. 211–229.

Riding In, James. 1992. Without Ethics and Morality: A Historical Overview of Imperial Archaeology and American Indians. *Arizona State Law Journal* 24 (1):11–34.

Rightmire, G. P. 1970. Bushman, Hottentot and South African Negro Crania Studied by Distances and Discrimination. *American Journal of Physical Anthropology* 33:169–196.

Ritchie, William A. 1952. Paleopathological Evidence Suggesting Pre-Columbian Tuberculosis in New York State. *American Journal of Physical Anthropology* 10:305–317.

Ritchie, William A., and S. L. Warren. 1932. The Occurrence of Multiple Bony Lesions Suggesting Myeloma in the Skeleton of a Pre-Columbian Indian. *American Journal of Roentgenology and Radium Therapy* 28:622–628.

Rivet, Paul. 1908. La Race de Lagoa-Santa Chez les Populations Precolumbiennes de L'equateur. *Bulletins et Memoires de la Société d'Anthropologie de Paris* 9:209–223.

Rivet, Paul. 1909a. Recherches Anthropologiques Sur la Basse Californie. *Journal de la Societe de Americanistes (n.s.)* VI:147–253.

Rivet, Paul. 1909b. Recherches Sur le Prognathisme. *L'Anthropologie* XX:35–49.

Rivet, Paul. 1909c. Recherches Sur le Prognathisme. *L'Anthropologie* XX:175–187.

Rivet, Paul. 1910a. Recherches Sur le Prognathisme. *L'Anthropologie* XXI:505–518.

Rivet, Paul. 1910b. Recherches Sur le Prognathisme. *L'Anthropologie* XXI:637–666.

Rivet, Paul. 1943. *Les Origines de L'homme Americain*. Montreal: L'Arbre.

Robb, John. 1994. Skeletal Signs of Activity in the Italian Metal Ages: Methodology and Interpretive Notes. *Human Evolution* 9:215–229.

Robb, John. 1998. The Interpretation of Skeletal Muscle Sites: A Statistical Approach. *International Journal of Osteoarchaeology* 8:363–377.

Robbins, D. M., K. R. Rosenberg, and Christopher B. Ruff. 1989. Activity Patterns in the Late Woodland, Delaware [Abstract]. *American Journal of Physical Anthropology* 78:290–291.

Robbins, Louise Marie. 1962. An Investigation into the Physical and Cultural Basis of Personality in College Women. *Proceedings of the Indiana Academy of Science* 71:69–70.

Robbins, Louise Marie. 1963. Personality, Race, and Physique in College Women. *Proceedings of the Indiana Academy of Science* 72:63–66.

Robbins, Louise Marie. 1964. Physical and Cultural Relationships of the Late Archaic Red Ochre People of the Illinois Valley. M.A., Indiana University, Bloomington.

Robbins, Louise M. 1965. Somatology. *Report from the University to Kentucky Schools* 9 (6).

Robbins, Louise Marie. 1968. The Identification of the Prehistoric Shawnee Indians: The Description of the Population of the Fort Ancient Aspect. Ph.D., Indiana University, Bloomington.

Robbins, Louise Marie. 1969a. The Origin of the Shawnee Indians. *Proceedings of the Indiana Academy of Science* 78.

Robbins, Louise Marie. 1969b. The Identification of the Prehistoric Shawnee Indians: The Description of the Population of the Fort Ancient Aspect. *Tennessee Archaeological Society. Newsletter* 14 (6):11–13.

Robbins, Louise Marie, ed. 1970. The History of Anthropology Departments in Kentucky Universities. *Transactions of the Kentucky Academy of Science* 31 (1–2).

Robbins, Louise Marie. 1971a. A Woodland "Mummy" from Salts Cave, Kentucky. *American Antiquity* 36 (2):200–206.

Robbins, Louise Marie. 1971b. The High Incidence of Bone Pathologies in Fort Ancient Peoples in Kentucky. Paper presented at the 40th Annual Meeting of the American Association of Physical Anthropologists, Boston.

Robbins, Louise Marie. 1972. Obituary: Georg Karl Neumann. June 6, 1907–April 13, 1971. *American Journal of Physical Anthropology* 36 (1):1–8.

Robbins, Louise Marie. 1974a. Behavioral Modifications Resulting from Pathological Deformity in a Prehistoric Skeleton. [Abstract]. *American Journal of Physical Anthropology* 40 (1):149.

Robbins, Louise Marie. 1974b. Prehistoric People of the Mammoth Cave Area. In *Archaeology of the Mammoth Cave Area*, edited by P. J. Watson. 137–162. New York: Academic Press.

Robbins, Louise Marie. 1975. The Investigation of Infanticide in an Ohio Fort Ancient Site. Paper presented at the 40th Annual Meeting of the Society for American Archaeology, Dallas.

Robbins, Louise Marie. 1976. Analysis of Human Skeletal Material from Morton Shell Mound (16IB3), Iberia Parish, Louisiana. Manuscript on file in the Department of Anthropology, Louisiana State University, Baton, Rouge.

Robbins, Louise Marie. 1977a. The Story of Life Revealed by the Dead. In *Biocultural Adaptation in Prehistoric America*, edited by R. L. Blakely. 10–26, *Proceedings of the Southern Anthropological Society*. Athens: University of Georgia Press.

Robbins, Louise Marie. 1977b. Yaws-like Disease Processes in a Louisiana Shell Mound Population [Abstract]. *American Journal of Physical Anthropology* 47 (1):156.

Robbins, Louise Marie. 1978a. The Antiquity of "Tuberculosis" in Prehistoric Peoples of Kentucky. [Abstract]. *American Journal of Physical Anthropology* 48 (3):429.

Robbins, Louise Marie. 1978b. Yawslike Disease Processes in a Louisiana Shell Mound Population. *Medical College of Virginia Quarterly* 14 (1):24–31.

Robbins, Louise Marie. 1978c. The Individuality of Human Footprints. *Journal of Forensic Sciences* 23 (4):778–785.

Robbins, Louise Marie. 1982. Step-by-Step: Foot Impressions and Suspects. Workshop Session, 1st Inter-American Congress of Forensic Sciences, Sacramento, California.

Robbins, Louise Marie. 1984. Making Tracks. *Law Enforcement Communications* 12 (14–15).

Robbins, Louise Marie. 1985. *Footprints, Collections, Analysis and Interpretation*. Springfield, IL: Charles C Thomas.

Robbins, Louise Marie. 1986. Estimating Height and Weight from Size of Footprints. *Journal of Forensic Sciences* 31 (1):143–152.

Robbins, Louise Marie. 1987. Footprints. In *Laetoli*, edited by M. Leakey. Oxford: Oxford University Press.

Robbins, Louise Marie, and Georg K. Neumann. 1969. The Origin of the Shawnee Indians. *Proceedings of the Indiana Academy of Science* 78:93–96.

Robbins, Louise M., and Georg Karl Neumann. 1972. *The Prehistoric People of the Fort Ancient Culture of the Central Ohio Valley, Anthropological Papers, No. 47*. Ann Arbor: Museum of Anthropology, University of Michigan.

Roberts, Charlotte A. 1986. Cottage Industry or Interacting Discipline. In *Archaeology at the Interface. Studies in Archaeology's Relationship with History, Geography, Biology and Physical Science*, edited by J. L. Bintliff and C. F. Gaffney. 110–128, *British Archaeological Reports. International Series 300*. Oxford: Archaeopress.

Roberts, Charlotte A. 2003. Bees in My Bonnet. Reflections on Biological Anthropology in the UK. *Archaeological Reviews from Cambridge* 16:96–116.

Roberts, Charlotte A., and Jane E. Buikstra. 2003. *The Bioarchaeology of Tuberculosis. A Global View on a Reemerging Disease*. Gainesville, FL: University Press of Florida.

Roberts, Charlotte A., and M. Cox. 2003. *Health and Disease in Britain: Prehistory to the Present Day*. Gloucester: Sutton Publishing.

Roberts, Charlotte A., and Roland B. Dixon. 1994. The Detection of *Mycobacterium Tuberculosis* by PCR from Ancient Human Bone. Poster Presented at the Conference, "Ancient Biomolecule Initiative — First Workshop," Cambridge University, Cambridge.

Roberts, Charlotte A., Mary E. Lewis, P. Boocock, and Keith Manchester. 1998. Infectious Disease, Sex and Gender: The Complexity of It All. In *Sex and Gender in Paleopathological Persepective*, edited by A. L. Grauer and P. Stuart-Macadam. 93–113. Cambridge: Cambridge University Press.

Roberts, Charlotte A., Mary E. Lewis, and Keith Manchester, eds. 2002. *The Past and Present of Leprosy: Archaeological, Historical, Palaeopathological and Clinical Approaches, British Archaeological Reports. International Series 1054*. Oxford: Archaeopress.

Roberts, Charlotte A., and Keith Manchester. 2005. *The Archaeology of Disease*. 3rd ed. Stroud, Gloucester; Ithaca, NY: Sutton Publishing; Cornell University Press.

Roberts, Frank H. H. 1939. *Archaeological Remains in the Whitewater District Eastern Arizona, Part I: House Types, Bureau of American Ethnology Bulletin, No. 121*. Washington, DC: Government Printing Office.

Roberts, Frank H. H. 1940. *Archaeological Remains in the Whitewater District Eastern Arizona, II. Artifacts and Burials, Bureau of American Ethnology Bulletin, No. 121*. Washington, DC: Government Printing Office.

Robling, A. G., and S. D. Stout. 2000. Methods of Determining Age-at-Death Using Bone Microstructure. In *Biological Anthropology of the Human Skeleton*, edited by M. A. Katzenberg and S. R. Saunders. 187–205. New York: Wiley-Liss.

Rochon Ford, Anne. 1985. *A Path Not Strewn with Roses: One Hundred Years of Women at the University of Toronto 1884–1984*. Toronto: University of Toronto Press.

Rogers, A. R. 1995. How Much Can Fossils Tell Us About Regional Continuity? *Current Anthropology* 36:674–676.

Rogers, Juliet. 2000. The Palaeopathology of Joint Disease. In *Human Osteology in Archaeology and Forensic Science*, edited by M. Cox and S. Mays. 163–182. London: Greenwich Medical Media.

Rogers, Juliet, and Tony Waldron. 1995. *A Field Guide to Joint Disease in Archaeology*. New York: Wiley.

Rohlf, F. J., and L. F. Marcus. 1993. A Revolution in Morphometrics. *Tree* 8:129–132.

Roksandic, Mirjana. 2002. Position of Skeletal Remains as a Key to Understanding Mortuary Behavior. In *Advances in Forensic Taphonomy: Method, Theory, and Archaeological Perspectives*, edited by W. D. Haglund and M. H. Sorg. 95–113. Boca Raton, FL: CRC Press.

Roosevelt, A. C. 1991. Moundbuilders of the Amazon: Geophysical Archaeology on Marajo Island, Brazil. New York: Academic Press.

Rose, Jerome C. 1973. Analysis of Dental Micro-Defects of Prehistoric Populations from Illinois. Ph.D. Dissertation, University of Massachusetts, Amherst.

Rose, Jerome C. 1977. Defective Enamel Histology of Prehistoric Teeth from Illinois. *American Journal of Physical Anthropology* 46:439–446.

Rose, Jerome C. 1979. Morphological Variation of Enamel Prisms within Abnormal Striae of Retzius. *Human Biology* 51:139–151.

Rose, Jerome Carl. 1999. *Bioarchaeology of the South Central United States*. Fayetteville: Arkansas Archeological Survey.

Rose, Jerome C., Susan C. Anton, Arthur C. Aufderheide, Leslie Eisenberg, John B. Gregg, Ellis J. Neiburger, and Bruce Rothschild. 1991. *Skeletal Database Committee Recommendations*. Detroit: Paleopathology Association.

Rose, Jerome C., George J. Armelagos, and John W. Lallo. 1978. Histological Enamel Indicator of Childhood Stress in Prehistoric Skeletal Samples. *American Journal of Physical Anthropology* 49 (4):511–516.

Rose, Jerome C., Thomas J. Green, and Victoria D. Green. 1996. NAGPRA Is Forever: Osteology and the Repatriation of Skeletons. *Annual Review of Anthropology* 25:81–103.

Rose, Jerome C., and M. K. Marks. 1985. Bioarchaeology of the Alexander Site (Chapter 7). In *The Alexander Site*, edited by E. T. Hemmings and J. H. House. 76–98. Vol. 24, *Research Series*. Fayetteville, AR: Arkansas Archeological Survey.

Rose, Jerome C., and Ted A. Rathbun. 1987. Preface: Afro-American Biohistory Symposium. *American Journal of Physical Anthropology* 74:177.

Rose, Jerome C., and Peter S. Ungar. 1998. Gross Dental Wear and Dental Microwear in Historical Perspective. In *Dental Anthropology*, edited by K. W. Alt, F. W. Rosing, and M. Teschler-Nicola. 350–386. New York: Springer-Verlag Wien.

Ross, Callum F., Biren A. Patel, Dennis E. Slice, David S. Strait, Paul C. Dechow, Brian G. Richmond, and Mark A. Spencer. 2005. Modeling Masticatory Muscle Force in Finite Element Analysis: Sensitivity Analysis Using Principal Coordinates Analysis. *The Anatomical Record Part A* 283A (2):288–299.

Rothschild, Bruce M. 1995. Paleopathology, Its Character and Contribution to Understanding and Distinguishing among Rheumatic Diseases: Perspectives on Rheumatoid Arthritis and Spondyloarthropathy. *Clinical and Experimental Rheumatology* 13:657–662.

Rothschild, Bruce M., Larry D. Martin, Galit Lev, Helen Bercovier, Gila Kahila Bar-Gal, Charles Greenblatt, Helen Donoghue, Mark Spigelman, and David Brittain. 2001. Mycobacterium Tuberculosis Complex DNA from an Extinct Bison Dated 17,000 Years before the Present. *Clinical Infectious Diseases* 33(3):305–311.

Rothschild, Nan A. 1979. Mortuary Behavior and Social Organization at Indian Knoll and Dickson Mounds. *American Antiquity* 44:658–675.

Rudney, Joel D. 1981. The Paleoepidemiology of Early Childhood Stress in Two Ancient Nubian Populations. Ph.D. Dissertation, University of Colorado, Boulder.

Ruff, Christopher B. 1981. A Reassessment of Demographic Estimates for Pecos Pueblo. *American Journal of Physical Anthropology* 54:147–151.

Ruff, Christopher B. 1991. *Ageing and Osteoporosis in Native Americans from Pecos Pueblo, New Mexico: Behavioral and Biomechanical Effects*. New York: Garland.

Ruff, Christopher B. 1992. Biomechanical Analyses of Archaeological Human Skeletal Samples. In *Skeletal Biology of Past Peoples: Research Methods*, edited by S. R. Saunders and M. A. Katzenberg. 37–58. New York: Wiley-Liss.

Ruff, Christopher B. 1994. Biomechanical Analysis of Northern and Southern Plains Femora: Behavioral Implications. In *Skeletal Biology in the Great Plains: Migration, Warfare, Health, and Subsistence*, edited by D. W. Owsley and R. L. Jantz. 235–245. Washington, DC: Smithsonian Institution Press.

Ruff, Christopher B. 1999. Skeletal Structure and Behavioral Patterns of Prehistoric Great Basin Populations. In *Understanding Prehistoric Lifeways in the Great Basin Wetlands: Bioarchaeological Reconstruction and Interpretation*, edited by B. E. Hemphill and C. S. Larsen. 290–320. Salt Lake City: University of Utah Press.

Ruff, Christopher B. 2000. Biomechanical Analyses of Archaeological Human Skeletons. In *Biological Anthropology of the Human Skeleton*, edited by M. A. Katzenberg and S. R. Saunders. 71–102. Vol. 38. New York: Wiley-Liss.

Ruff, Christopher B., and W. C. Hayes. 1982. Subperiosteal Expansion and Cortical Remodeling of the Human Femur and Tibia with Aging. *Science* 217:945–948.

Ruff, Christopher B., and W. C. Hayes. 1983a. Cross-Sectional Geometry of Pecos Pueblo Femora and Tibiae: A Biomechanical Investigation. I. Method and General Patterns of Investigation. *American Journal of Physical Anthropology* 60:359–381.

Ruff, Christopher B., and Wilson C. Hayes. 1983b. Cross-Sectional Geometry of Pecos Pueblo Femora and Tibiae: A Biomechanical Investigation. II. Sex, Age, Side Differences. *American Journal of Physical Anthropology* 60 (3):383–400.

Ruff, Christopher B., and Clark Spencer Larsen. 2001. Reconstructing Behavior in Spanish Florida: The Biomechanical Evidence. In *Bioarchaeology of La Florida: Human Biology of Frontier Northern New Spain*, edited by C. S. Larsen. 113–145. Gainesville, FL: University Press of Florida.

Ruff, Christopher B., Clark Spencer Larsen, and Wilson C. Hayes. 1984. Structural Changes in the Femur with the Transition to Agriculture on the Georgia Coast. *American Journal of Physical Anthropology* 64 (2):125–136.

Ruff, Christopher B., and Frank P. Leo. 1986. Use of Computed Tomography in Skeletal Structure Research. *Yearbook of Physical Anthropology* 29:181–196.

Ruff, Christopher B., Henry M. McHenry, and J. Francis Thackeray. 1999. Cross-Sectional Geometry of the SK 82 and 97 Proximal Femora. *American Journal of Physical Anthropology* 109:509–521.

Ruff, Christopher B., Erik Trinkaus, Alan Walker, and Clark Spencer Larsen. 1993. Postcranial Robusticity in Homo. I. Temporal Trends and Mechanical Interpretation. *American Journal of Physical Anthropology* 91:21–53.

Ruff, Christopher B., Alan Walker, and Erik Trinkaus. 1994. Postcranial Robusticity in Homo. III. Ontogeny. *American Journal of Physical Anthropology* 93:35–54.

Ruffer, Marc Armand. 1913. On Pathological Lesions Found in Coptic Bodies. *Journal of Pathology and Bacteriology* 18:149–162.

Ruffer, Marc Armand. 1920. Study of Abnormalities and Pathology of Ancient Egyptian Teeth. *American Journal of Physical Anthropology* 3:335–382.

Runestad, Jacqueline A., Christopher B. Ruff, James C. Nieh, Richard W. Thorington Jr., and Mark F. Teaford. 1993. Radiographic Estimation of Long Bone Cross-Sectional Geometric Properties. *American Journal of Physical Anthropology* 90:207–213.

Russell, Frank. 1900. Studies in Cranial Variation. *American Naturalist* 34 (405):737–745.

Russell, Frank, and H. M. Huxley. 1899. A Comparative Study of the Physical Structure of the Labrador Eskimos and the New England Indians. *Proceedings of the American Association for the Advancement of Science* 48:365–379.

Ryan, A. S. 1980. Anterior Dental Microwear in Hominoid Evolution: Comparisons with Human and Nonhuman Primates. Ph.D. Dissertation, University of Michigan.

Salo, Wilmar L., Arthur C. Aufderheide, Jane Buikstra, and Todd A. Holcomb. 1994. Identification of *Mycobacterium tuberculosis* DNA in a Pre-Columbian Peruvian Mummy. *Proceedings of the National Academy of Sciences* 91:2091–2094.

San Vincenzo al Volturno. 2001. Ricerca Sui Materiali: Bioarcheologia. Retrieved Dec. 28, 2005, from San Vincenzo Al Volturno Web Site: http://www.sanvincenzoalvolturno.it/ricerca/bioarcheologia/bioarcheologia.htm.

Sandford, Mary K., ed. 1993. *Investigations of Ancient Human Tissue: Chemical Analyses in Anthropology*. Vol. 10, *Food and Nutrition in History and Anthropology*. Philadelphia: Gordon and Breach.

Santoro, Calogero, ed. 2001. *Proceedings of the III World Congress on Mummy Studies. Chungará*. Vols. 32(1,2) & 33(1,2). Arica, Chile: Universidad de Tarapacá.

Sarnat, B. G., and I. Schour. 1941. Enamel Hypoplasias (Chronic Enamel Aplasia) in Relationship to Systemic Diseases: A Chronological, Morphological and Etiological Classification. *Journal of Americal Dental Association* 28:1989–2000.

Sattenspiel, L, and Henry C. Harpending. 1983. Stable Populations and Skeletal Age. *American Antiquity* 48:489–498.

Saul, Frank P. 1972. *The Human Skeletal Remains of Altar de Sacrificios: An Osteobiographic Analysis*. Vol. 63(2), *Papers of the Peabody Museum of Archaeology and Ethnology*. Cambridge, MA: Peabody Museum of Archaeology and Ethnology, Harvard University.

Saul, Frank P., and N. Hammond. 1974. A Classic Maya Tooth Cache from Lubaantun, Belize. *Man* 9:123–127.

Saul, Frank P., and Julie M. Saul. 1989. Osteobiography: A Maya Example. In *Reconstruction of Life from the Skeleton*, edited by M. Y. İşcan and K. A. R. Kennedy. 287–302. New York: Alan R. Liss.

Saul, Frank P., and Julie M. Saul. 1991. The Preclassic Population of Cuello. In *Cuello: An Early Maya Population in Belize*, edited by N. Hammond, pp. 134–158. Cambridge: Cambridge University Press.

Saunders, Shelley R., Carol DeVito, and M. Anne Katzenberg. 1997. Dental Caries in Nineteenth Century Upper Canada. *American Journal of Physical Anthropology* 104:71–87.

Saunders, Shelley R., and D. Ann Herring, eds. 1995. *Grave Reflections: Portraying the Past through Cemetery Studies*. Toronto: Canadian Scholars' Press.

Saunders, Shelley R., D. Ann Herring, and Gerald Boyce. 1995. Can Skeletal Samples Accurately Represent the Living Populations They Come From? The St. Thomas'

Cemetery Site, Belleville, Ontario. In *Bodies of Evidence: Reconstructing History through Skeletal Analysis*. 69–89. New York: Wiley-Liss.

Sawtell, Ruth Otis. 1931. Azilian Skeletal Remains from Montardit (Ariege) France. *Papers of the Peabody Museum of American Archaeology and Ethnology, Harvard University* 11 (4):213–253.

Saxe, A. A. 1970. Social Dimensions of Mortuary Practices. Ph.D. Dissertation, University of Michigan.

Saxe, Arthur A. 1971. Social Dimensions of Mortuary Practices in a Mesolithic Population from Wadi Halfa, Sudan. *Society for American Archaeology Memoirs* 25:39–57.

Scarre, Chris. 2004. Review of Les Pratiques Funéraires Néolithiques Avant 3500 av. J.-C. en France et Dans les Régions Limitrophes. Table Ronde Spf, Saint-Germain-en-Laye 15–17 Juin 2001, edited by Philippe Chambon and Jean Leclerc Mémoire Xxxiii de la Société Préhistorique Française Société. Préhistorique Française, Paris, 2003. Published Online, the Prehistoric Society Web Page: http://www.ucl.ac.uk/prehistoric/.

Scheuer, Louise, and Sue Black. 2000. *Developmental Juvenile Osteology*. San Diego: Academic Press.

Schevill, M. E. 1948/1949. Dr. Washington Matthews, 1843–1905. *Kiva* 14 (1–4):2–6.

Schiebinger, L. 1993. *Nature's Body: Gender in the Making of Modern Science*. Boston, MA: Beacon Press.

Schillaci, M. A., and C. M. Stojanowski. 2000. Postmarital Residence and Population Structure at Pueblo Bonito [Abstract]. *American Journal of Physical Anthropology* (Suppl. 30):271.

Schillaci, M. A., and C. M. Stojanowski. 2002. A Reassessment of Matrilocality in Chacoan Culture. *American Antiquity* 67 (2):343–356.

Schlossberg, David, ed. 1994. *Tuberculosis*. New York: Springer-Verlag.

Schmidt-Nielsen, Knut. 1984. *Scaling: Why Is Animal Size So Important?* Cambridge: Cambridge University Press.

Schneider, Alan L. 2002. Kennewick Man: The Three-Million Dollar Man. In *Legal Perspectives on Cultural Resources*, edited by J. R. Richman and M. P. Forsyth. 202–215. Walnut Creek, CA: AltaMira Press.

Schnell, Frank T., ed. 1999. *Antiquities of the Southern Indians, Particularly the Georgia Tribes, by Charles C. Jones*. Tuscaloosa: University of Alabama Press.

Schoeninger, M. J. 1995. Stable Isotope Studies in Human Evolution. *Evolutionary Anthropology* 4:83–98.

Schour, I., and M. Massler. 1941. The Development of the Human Dentition. *Journal of the American Dental Association* 28:1153–1160.

Schour, I., and H. Van Dyke. 1932. Changes in the Teeth Following Hypophysectomy. *American Journal of Anatomy* 50:397.

Schultz, Adolph H. 1945. Biographical Memoir of Aleš Hrdlička 1869–1943. *National Academy of Sciences of the United States of America Biographical Memoirs* 23:305–338.

Schultz, M. 1993. Initial Stages of Systematic Bone Disease. In *Histology of Ancient Human Bone: Methods and Diagnosis*, edited by G. Grupe and A. N. Garland. 185–203. Berlin: Springer-Verlag.

Schultz, M., Clark Spencer Larsen, and K. Kreutz. 2001. Disease in Spanish Florida: Microscopy of Porotic Hyperostosis and Inferences about Health. In *Bioarchaeology of La Florida: Human Biology in Frontier Northern New Spain*, edited by C. S. Larsen. 207–225. Gainesville, FL: University Press of Florida.

Schulz, Peter D. 1977. Task Activity and Anterior Tooth Grooving in Prehistoric California Indians. *American Journal of Physical Anthropology* 46:87–91.

Schulz, P. D., and H. McHenry. 1975. Age Distribution of Enamel Hypoplasias in Prehistoric California Indians. *Journal of Dental Research* 54:913.

Schurr, Mark R. 1997. Weaning Behavior in Past Populations. *World Archaeology* 30 (2):327–342.

Schurr, Mark R., and Margaret J. Schoeninger. 1995. Associations between Agricultural Intensification and Social Complexity: An Example from the Prehistoric Ohio Valley. *Journal of Anthropological Archaeology* 14:315–339.

Schwartz, Douglas W. 1967. *Conceptions of Kentucky Prehistory*. Lexington: University of Kentucky Press.

Schwartz, Douglas Wright. 2000. Kidder and the Synthesis of Southwestern Archaeology. In *An Introduction to the Study of Southwestern Archaeology*, edited by A. V. Kidder. 1–55. New Haven: Yale University Press. Original edition, 1924 (1962 second edition).

Schwartz, J. 1946. The Teeth of the Masai. *Journal of Dental Research* 25:17–20.

Sciulli, Paul W. 1978. Developmental Abnormalities of the Permanent Dentition in Prehistoric Ohio Valley Amerindians. *American Journal of Physical Anthropology* 48:193–198.

Sciulli, Paul W., and Michael C. Mahaney. 1986. Evidence of Local Biological Continuity for an Ohio Hopewell Complex Population. *Midcontinental Journal of Archaeology* 11 (2):181–199.

Sciulli, Paul W., and Michael C. Mahaney. 1991. Phenotypic Evolution in Prehistoric Ohio Valley Amerindians: Natural Selection Versus Random Genetic Drift in Tooth Size Reduction. *Human Biology* 63:499–511.

Sciulli, Paul W., and Kim N. Schneider. 1986. Dental Caries and Horticulture in Prehistoric Ohio. *Pennsylvania Archaeologist* 56(1,2):21–28.

Scott, Douglas D., and Terje G. Birkedal. 1972. The Archaeology and Physical Anthropology of the Gahagan-Lipe Site with Comments on Colorado Woodland Mortuary Practices. *Southwestern Lore* 38 (3):1–18.

Scott, Douglas D., P. Willey, and Melissa Connor, eds. 2002. *They Died with Custer: Soldiers' Bones from the Battle of Little Bighorn*. Norman: University of Oklahoma Press.

Scott, E. C. 1979a. Dental Wear Scoring Techniques. *American Journal of Physical Anthropology* 51:213–218.

Scott, E. C. 1979b. Principal Axis Analysis of Dental Attrition Data. *American Journal of Physical Anthropology* 51:203–212.

Scott, G. Richard. 1992. Affinities of Prehistoric and Modern Kodiak Islanders and the Question of Kachemak-Koniag Biological Continuity. *Arctic Anthropology* 29 (2):150–166.

Scott, G. Richard, and Christy G. Turner. 1997. *The Anthropology of Modern Human Teeth*. Cambridge: Cambridge University Press.

Scott, G. Richard, and Christy G. Turner II. 1988. Dental Anthropology. In *Annual Review of Anthropology*, edited by B. J. Siegel, A. R. Beals, and S. A. Tyler. 99–126. Vol. 17.

Sealy, Judith C., Nikolaas J. van der Merwe, Andrew Sillen, F. J. Kruger, and Harold Kruger. 1991. ^{87}sr/^{86}sr as a Dietary Indicator in Modern and Archaeological Bone. *Journal of Archaeological Science* 18 (3):399–416.

Seckler, David. 1980. "Malnutrition": An Intellectual Odyssey. *Western Journal of Agricultural Economics* 5:219–227.

Seckler, D. 1982. Small but Healthy? Some Basic Problems in the Concept of Malnutrition. In *Newer Concepts in Nutrition and Their Implications for Policy*, edited by P. V. Sukhatme. 139–148. Pune, India: Maharashtra Association for the Cultivation of Science.

Seltzer, C. C. 1936. New Light on the Racial History of the Southwest Area [Abstract]. *American Journal of Physical Anthropology* 21 (2, Supp.):17.

Seltzer, C. C. 1944. *Racial Prehistory in the Southwest and the Hawikuh Zunis*. Vol. 23(1), *Papers of the Peabody Museum of American Archaeology and Ethnology, Harvard University*. Cambridge, MA: Peabody Museum of American Archaeology and Ethnology.

Seltzer, Frank M. 1942. Archeological Accomplishments during the Past Decade in the United States. *Journal of the Washington Academy of Sciences* 32:253–259.

Seltzer, Frank M. 1943. Archaeological Explorations in the United States, 1930–1942. *Acta Americana* 1:206–220.

Seltzer, Frank M., and W. D. Strong. 1936. Archaeology and Relief. *American Antiquity* 4:301–309.

Semple, Sarah. 1998. A Fear of the Past: The Place of the Prehistoric Burial Mound in the Ideology of the Middle and Later Anglo-Saxon England. *World Archaeology* 30 (1):109–126.

Senut, Brigitte. 1985. Computerized Tomography of a Neandertal Humerus from Le Régourdou (Dordogne, France): Comparisons with Modern Man. *Journal of Human Evolution* 14:717–723.

Settipane, Guy A., ed. 1995. *Columbus and the New World: Medical Implications*. Providence: OceanSide Press.

Shanks, Michael, and Christopher Y. Tilley. 1982. Ideology, Symbolic Power and Ritual Communication: A Reinterpretation of Neolithic Mortuary Practices. In *Symbolic and Structural Archaeology*, edited by I. Hodder. 129–154. Cambridge: Cambridge University Press.

Shanks, Michael, and Christopher Y. Tilley. 1987. *Social Theory and Archaeology*. Cambridge: Polity.

Shanks, Michael, and Christopher Y. Tilley. 1989. Archaeology into the 1990s. *Norwegian Archaeological Review* 22 (1):1–12.

Shapiro, Harry L. 1959. The History and Development of Physical Anthropology. *American Anthropologist* 61:371–379.

Shapiro, Harry L. 1981. Earnest A. Hooton, 1887–1954, *in Memoriam Cum Amore*. *American Journal of Physical Anthropology* 56 (4):431–434.

Sharpe, F. R., and A. J. Lotka. 1911. A Problem in Age Distribution. *Philosophical Magazine* 21:435–438.

Sheinin, Rose, and Alan Bakes. 1987. *Women and Medicine in Toronto since 1883: A Who's Who*. Toronto: University of Toronto.

Sheldon, Marvin, Basil G. Bibby, and Michael S. Bales. 1945. The Relationship between Microscopic Enamel Defects and Infantile Debilities. *Journal of Dental Research* 24:109–116.

Shryock, Andrew J. 1987. The Wright Mound Reexamined: Generative Structures and the Political Economy of a Simple Chiefdom. *Midcontinental Journal of Archaeology* 12:243–268.

Shufeldt, R. W. 1892. Notes on Palaeopathology. *Popular Science Monthly* 42:679–684.

Siegel, Peter E. 1996. Ideology and Culture Change in Prehistoric Puerto Rico: A View from the Community. *Journal of Field Archaeology* 23:313–333.

Silverberg, Robert. 1968. *Mound Builders of Ancient America: The Archaeology of a Myth*. Greenwich, CT: New York Graphic Society.

Simpson, Scott W. 1999. Reconstructing Patterns of Growth Disruption from Enamel Microstructure. In *Human Growth in the Past: Studies from Bones and Teeth*, edited by R. D. Hoppa and C. M. FitzGerald. 241–263. Cambridge: Cambridge University Press.

Simpson, Scott W. 2001. Patterns of Growth Perturbation in La Florida: Evidence from Enamel Microstructure. In *Bioarchaeology of La Florida: Human Biology in Frontier Northern New Spain*, edited by C. S. Larsen. Gainesville, FL: University Press of Florida.

Simpson, Scott W., Dale L. Hutchinson, and Clark Spencer Larsen. 1990. Coping with Stress: Tooth Size, Dental Defects, and Age-at-Death. In *The Archaeology of Mission Santa Catalina de Guale. 2. Biocultural Interpretations of a Population in Transition*, edited by C. S. Larsen. 66–77, *Anthropological Papers of the American Museum of Natural History 68*. New York.

Siven, C. H. 1991. On Estimating Mortalities from Osteological Age Data. *International Journal of Anthropology* 6:97–110.

Skarland, Ivar. 1939. The Skeletal Material. In *The Chiggerville Site, Site 1, Ohio County, Kentucky*, edited by W. S. Webb and W. G. Haag. 28–49, *University of Kentucky Reports in Anthropology 4*. Lexington: University of Kentucky.

Skinner, Mark F., and J. T. W. Hung. 1989. Social and Biological Correlates of Localized Enamel Hypoplasia of the Human Deciduous Canine Tooth. *American Journal of Physical Anthropology* 79:159–175.

Skomp, David. 1965. Racial Continuity in the American Southwest: An Examination of Carter Ranch Pueblo Physical Relationships. Master's, Indiana University, Bloomington.

Skytthe, A., and Jesper L. Boldsen. 1993. A Method for Construction of Standards for Determination of Skeletal Age-at-Death [Abstract]. *American Journal of Physical Anthropology* (Suppl. 16):182.

Sledzik, Paul S., and Lenore Barbian. 2001. From Privates to Presidents: Past and Present Memoirs from the Anatomical Collections of the National Museum of Health and Medicine. In *Human Remains: Conservation, Retrieval and Analysis. Proceedings of a Conference Held in Williamsburg, Va, Nov. 7–11th 1999*, edited by E. Williams. 227–236, *British Archaeological Reports. International Series 934*. Oxford: Archaeopress.

Sledzik, Paul S., and Peer H. Moore-Jansen. 1991. Dental Disease in Nineteenth Century Military Skeletal Samples. In *Advances in Dental Anthropology*, edited by M. A. Kelley, C. S. Larsen, pp. 215–224. New York: Wiley-Liss.

Sleight, Frederic W. 1962. *Archaeological Reconnaissance of the Island of St. John, United States Virgin Islands, The William L. Bryant Foundation, American Studies. Report No. 3*. Maitland: Central Florida Museum.

Smail, J. Kenneth. 1964. The Use of Female Crania in Demonstrating Racial Relationships as Exemplified in Two Upper Mississippi Amerind Groups. *Proceedings of the Indiana Academy of Science* 74:72–80.

Smith, B. D. 1989. Origins of Agriculture in Eastern North America. *Science* 246: 1566–1571.

Smith, B. Holly. 1984. Patterns of Molar Wear in Hunter-Gatherers and Agriculturalists. *American Journal of Physical Anthropology* 63:39–56.

Smith, B. Holly. 1991. Standards of Human Tooth Formation and Dental Age Assessment. In *Advances in Dental Anthropology*, edited by M. A. Kelley and C. S. Larsen. 143–168. New York: Wiley-Liss.

Smith, Courtland L., and Kenneth L. Beals. 1990. Cultural Correlates with Cranial Capacity. *American Anthropologist* 92 (1):193–200.

Smith, D., and N. Keyfitz. 1977. *Mathematical Demography: Selected Papers*. Berlin: Springer-Verlag.

Smith, David Glenn. 2000. Chapter 1. Attachment A. A Review of Documents and Evidence Pertaining to the Suitability of the Skeletal Remains from Kennewick, Washington, for DNA Studies. In "Report on the DNA Testing Results of the Kennewick Human Remains from Columbia Park, Kennewick, Washington." Retrieved Dec. 31, 2005, from Department of the Interior, National Park Service, Archaeology Program Web Site: http://www.cr.nps.gov/archeology/kennewick/GlennSmith.htm.

Smith, David Glenn, R. L. Bettinger, and B. K. Rolfs. 1995. Serum Albumin Phenotypes at Stillwater: Implications for Population History in the Great Basin. In *Bioarchaeology of the Stillwater Marsh: Prehistoric Human Adaptation in the Western Great Basin*, edited by C. S. Larsen and R. L. Kelly. 68–72, *Anthropological Papers of the American Museum of Natural History No. 77*.

Smith, David Glenn, R. S. Malhi, J. Eshleman, J. G. Lorenz, and Frederika A. Kaestle. 1999. Distribution of mtDNA Haplogroup X among Native North Americans. *American Journal of Physical Anthropology* 110:271–284.

Smith, Maria Ostendorf. 1991. *The Earliest Scalping Victim in America, McClung Museum Research Notes 9*. Knoxville: Frank H. McClung Museum.

Smith, Maria Ostendorf. 1995. Scalping in the Archaic Period: Evidence from the Western Tennessee Valley. *Southeastern Archaeology* 14:60–68.

Smith, Maria Ostendorf. 1997. Osteological Indications of Warfare in the Archaic Period of the Western Tennessee Valley. In *Troubled Times: Violence and Warfare in the Past*, edited by D. L. Martin and D. W. Frayer. 241–265. Amsterdam: Gordon and Breach.

Smith, Maria Ostendorf. 2003. Beyond Palisades: The Nature and Frequency of Late Prehistoric Deliberate Violent Trauma in the Chickamauga Reservoir of East Tennessee. *American Journal of Physical Anthropology* 121:303–318.

Smith, Patricia. 1972. Diet and Attrition in the Natufians. *American Journal of Physical Anthropology* 37:233–238.

Smith, W., R. B. Woodbury, and N. F. S. Woodbury. 1966. *The Excavation of Hawikuh by Frederic Webb Hodge: Report of the Hendricks-Hodge Expedition, 1917–1923*. New York: Museum of the American Indian, Heye Foundation.

Smithsonian Institution. n.d. Department of Anthropology: A History of the Department, 1897–1997. Retrieved Dec. 28, 2005, from National Museum of Natural History Web Site: http://www.nmnh.si.edu/anthro/outreach/depthist.html#Beginning.

Snow, Charles E. 1940a. Letter Dated April 10 to Hertzberg. Snow Letters, Laboratory for Human Osteology, Mary Harmon Bryant Scientific Collections Building, University of Alabama, Tuscaloosa.

Snow, Charles E. 1940b. Letter Dated June 10 to Webb. Snow Letters, Laboratory for Human Osteology, Mary Harmon Bryant Scientific Collections Building, University of Alabama, Tuscaloosa.

Snow, Charles E. 1940c. Letter Dated August 8 to Carlson. Snow Letters, Laboratory for Human Osteology, Mary Harmon Bryant Scientific Collections Building, University of Alabama, Tuscaloosa.

Snow, Charles E. 1941a. Indian Skeletons from the Museum Burials and Moundville. In *Anthropological Studies at Moundville*. 5–52, *Alabama Museum of Natural History, Paper No. 15, Part 1*. Birmingham: Geological Survey of Alabama.

Snow, Charles E. 1941b. Possible Evidence of Scalping at Moundville. In *Anthropological Studies at Moundville*. 55–59, *Alabama Museum of Natural History, Paper No. 15, Part 2*. Birmingham: Geological Survey of Alabama.

Snow, Charles E. 1941c. Letter Dated January 20 to Webb. Snow Letters, Laboratory for Human Osteology, Mary Harmon Bryant Scientific Collections Building, University of Alabama, Tuscaloosa.

Snow, Charles E. 1941d. Letter Dated February 5 to German. Snow Letters, Laboratory for Human Osteology, Mary Harmon Bryant Scientific Collections Building, University of Alabama, Tuscaloosa.

Snow, Charles E. 1941e. Letter Dated February 11 to Newman. Snow Letters, Laboratory for Human Osteology, Mary Harmon Bryant Scientific Collections Building, University of Alabama, Tuscaloosa.

Snow, Charles E. 1941f. Letter Dated February 17 to Neumann. Snow Letters, Laboratory for Human Osteology, Mary Harmon Bryant Scientific Collections Building, University of Alabama, Tuscaloosa.

Snow, Charles E. 1941g. Letter Dated February 19 to the Shawinigan Products Corporation. Snow Letters, Laboratory for Human Osteology, Mary Harmon Bryant Scientific Collections Building, University of Alabama, Tuscaloosa.

Snow, Charles E. 1941h. Letter Dated April 24 to Goldstein. Snow Letters, Laboratory for Human Osteology, Mary Harmon Bryant Scientific Collections Building, University of Alabama, Tuscaloosa.

Snow, Charles E. 1942. Additional Evidence of Scalping. *American Antiquity* 7 (4): 398–400.

Snow, Charles E. 1943. *Two Prehistoric Indian Dwarf Skeletons from Moundville, Papers of the Alabama Museum of Natural History, No. 21*. Birmingham: Geological Survey of Alabama.

Snow, Charles E. 1945a. Tchefuncte Skeletal Remains. In *The Tchefuncte Culture: An Early Occupation of the Lower Mississippi Valley (Memoir 2)*, edited by J. A. Ford and J. G. I. Quimby. 99–109. Menasha, WI: Society for American Archaeology Memoirs.

Snow, Charles E. 1945b. Physical Anthropology of Adena. *University of Kentucky Reports in Anthropology and Archaeology* 6:247–309.

Snow, Charles E. 1948. Indian Knoll Skeletons of Site Oh2, Ohio County, Kentucky. *University of Kentucky Reports in Anthropology* 4 (3, Part 2):371–555.

Snow, Charles E. 1951. Adena Skull Deformation Redefined [Abstract]. *American Journal of Physical Anthropology* 9 (2):238.

Snow, Charles E. 1962. *Indian Burials from St. Petersburg, Florida, Contributions of the Florida State Museum, Social Sciences, No. 8*. Gainesville, FL: University of Florida.

Snow, Charles E. 1974. *Early Hawiians: An Initial Study of Skeletal Remains from Mokapu, Oahu*. Lexington: University of Kentucky Press.

Snyder, R. G. 1959. The Dental Morphology of the Point of Pines Indians. Ph.D. Dissertation, University of Arizona.

Sobilik, Kristin D., ed. 1994. *Paleonutrition: The Diet and Health of Prehistoric Americans, Occasional Paper No. 22*. Carbondale: Southern Illinois University at Carbondale, Center for Archaeological Investigations.

Society for American Archaeology. n.d. Kennewick Legal Briefs. Retrieved Dec. 31, 2005, from Society for American Archaeology Web Site: http://www.saa.org/repatriation/kennewickbriefs.html.

Sofaer Derevenski, Joanna. 1997. Age and Gender at the Site of Tiszapolgar-Basatanya, Hungary. *Antiquity* 71:875–889.

Sofaer Derevenski, Joanna. 2000. Rings of Life: The Role of Early Metalwork in Mediating the Gendered Life Course. *World Archaeology* 31 (3):389–406.

Sognnaes, Reidar F. 1955. Post-Mortem Microscopic Defects in the Teeth of Ancient Man. *Archives of Pathology* 59:559–570.

Sognnaes, Reidar F. 1956. Histological Evidence of Developmental Lesions in Teeth Originating from Paleolithic, Prehistoric, and Ancient Man. *American Journal of Pathology* 32:547–577.

Sokal, R. R., N. L. Oden, J. Walker, and D. M. Waddle. 1997. Using Distance Matrices to Choose between Competing Theories and an Application to the Origin of Modern Humans. *Journal of Human Evolution* 32:501–522.

Sozinskey, Thomas S. 1878. The Teeth of the Mound-Builders. *Dental Cosmos* 20:496–499.

Speaker, Stuart. 1994. The Red, Black, and Blue: Context and Mortuary Data at Larsen Bay, Kodiak Island (Chapter 6). In *Reckoning with the Dead: The Larson Bay Repatriation and the Smithsonian Institution*, edited by T. L. Bray and T. W. Killion. 54–66. Washington, DC: Smithsonian Institution Press.

Spence, M. W. 1974a. Residential Practices and the Distribution of Skeletal Traits in Teotihuacan, Mexico. *Man* 9 (262–273).

Spence, M. W. 1974b. The Study of Residential Practices among Prehistoric Hunters and Gatherers. *World Archaeology* 5:346–357.

Spencer, Frank. 1979. Aleš Hrdlička, M.D., 1869–1943: A Chronicle of the Life and Work of an American Physical Anthropologist. Ph.D., University of Michigan.

Spencer, Frank. 1982a. Introduction. In *A History of American Physical Anthropology: 1930–1980*, edited by F. Spencer. 1–10. New York: Academic Press.

Spencer, Frank, ed. 1982b. *A History of American Physical Anthropology 1930–1980*. New York: Academic Press.

Spencer, Frank. 1983. Samuel George Morton's Doctoral Thesis on Bodily Pain: The Probable Source of Morton's Polygenism. *Transactions and Studies of the College of Physicians of Philadelphia Series* 5 (4):497–527.

Spencer, Frank. 1986. *Ecce Homo: An Annotated Bibliographic History of Physical Anthropology*. New York: Greenwood Press.

Spencer, Frank. 1997a. Blumenbach, Johann Friedrich (1752–1840). In *History of Physical Anthropology: An Encyclopedia*, edited by F. Spencer. 183–186. New York: Garland.

Spencer, Frank. 1997b. Broca, Paul (1824–1880). In *History of Physical Anthropology: An Encyclopedia*, edited by F. Spencer. 221–222. New York: Garland Publishing.

Spencer, Frank, ed. 1997c. *History of Physical Anthropology: An Encyclopedia*. New York: Garland.

Spencer, Frank. 1997d. United Kingdom. In *Encyclopedia of Physical Anthropology*, edited by F. Spencer. 1054–1099. New York and London: Garland.

Spielmann, K. A., M. J. Schoeninger, and W. J. Moore. 1990. Plains-Pueblo Interdependence and Human Diet at Pecos Pueblo, New Mexico. *American Antiquity* 55 (4):745–765.

Spigelman, Mark, and H. D. Donoghue. 1999. *Mycobacterium tuberculosis* DNA in Archaeological Specimens. In *Tuberculosis: Past and Present*, edited by G. Pálfi, O. Dutour, J. Deák, and I. Hutás. 353–360. Budapest: Golden Book Publisher Ltd./Szeged: Tuberculosis Foundation.

Spigelman, Mark, and Eshetu Lemma. 1993. The Use of the Polymerase Chain Reaction (PCR) to Detect *Mycobacterium tuberculosis* in Ancient Skeletons. *International Journal of Osteoarchaeology* 3 (2):137–143.

Sprague. 2005. *Burial Terminology: A Guide for Researchers*. Lanham, MD: AltaMira Press.

Spuhler, J. N. 1954. Some Problems in the Physical Anthropology of the American Southwest. *American Anthropologist* 56 (4):604–619.

Squier, Ephraim G., and Edwin Hamilton Davis. 1848. *Ancient Monuments of the Mississippi Valley*. Vol. 1, *Smithsonian Contributions to Knowledge*. Washington, DC: Smithsonian Institution.

St. Hoyme, Lucile E. 1968a. Review of Homenaje a Juan Comas en su 65 Aniversario. Vol. II Antropología Fisíca. Santiago Genoves (Ed.). *Human Biology* 40:102–103.

St. Hoyme, Lucile E. 1968b. Review of "Les Amerindiens de la Haute-Guyane Française: Anthropologie, Pathologie, Biologie" by Etienne P. Bois. *Man, New Series* 3: 664–665.

St. Hoyme, Lucile E. 1968c. Review of "The Human Skull: A Cultural History" by Kenneth Oakley (ed); and "Der Menschliche Schädel in der Kulturgeschichte" by F. Henschel. *American Anthropologist* 70:168–169.

St. Hoyme, Lucile E. 1969a. On the Origins of New World Paleopathology. *American Journal of Physical Anthropology* 31:295–302.

St. Hoyme, Lucile E. 1969b. Review of "Nomogenesis: Evolution Determined by Law," by Leo S. Berg and J. N. Rostovstov. *Man, New Series* 4:652.

St. Hoyme, Lucile E. 1969c. Review of "Evolution of Man" by H. Nelson. *American Anthropologist* 71:1008–1009.

St. Hoyme, Lucile E. 1972. Dental Pathology in Crania of Mongolians, Egyptians and American Indians [Abstract]. *American Journal of Physical Anthropology* 37:451.

St. Hoyme, Lucile E. 1976. Ecology of Dental Disease [Abstract]. *American Journal of Physical Anthropology* 44:208.

St. Hoyme, Lucile E. 1980. A Simple Statistical Method for Estimating Sex Distribution and Dimensions in Dissociated Long Bones. *International Journal of Skeletal Research* 7:119–128.

St. Hoyme, Lucile E. 1982. Causes of Tooth Loss in Prehistoric Populations. In *Second Anthropological Congress of Aleš Hrdlička. Universita Carolina Pragensis*. 463–465.

St. Hoyme, Lucile E. 1984a. Sex Differences in the Posterior Pelvis [Abstract]. *American Journal of Physical Anthropology* 57:222–223.

St. Hoyme, Lucile E. 1984b. Sex Differentiation in the Posterior Pelvis. *Collegium Antropologicum (Zagreb)* 8:139–153.

St. Hoyme, Lucile E. 1988. Obituary: J. Lawrence Angel, 1915–1986. *American Journal of Physical Anthropology* 75:291–301.

St. Hoyme, Lucile E. 1993. Thomas Dale Stewart (Winner of the AAPA's Lifetime Achievement Award). *American Journal of Physical Anthropology* 92 (4):556–557.

St. Hoyme, Lucile E., and Mehmet Yaşar İşcan. 1989. Determination of Sex and Race: Accuracy and Assumption. In *Reconstruction of Life from the Skeleton*, edited by M. Y. İşcan and K. A. R. Kennedy. 53–93. New York: Alan R. Liss.

St. Hoyme, Lucile E., and Richard T. Koritzer. 1971. Significance of Canine Wear in Pongid Evolution. *American Journal of Physical Anthropology* 35:145–148.

St. Hoyme, Lucile E., and Richard T. Koritzer. 1976. Ecology of Dental Disease. *American Journal of Physical Anthropology* 45:673–685.

Standen, Vivien G., and Bernardo T. Arriaza. 2000. Trauma in the Preceramic Coastal Populations of Northern Chile: Violence or Occupational Hazards. *American Journal of Physical Anthropology* 112:239–249.

Stanton, William Ragan. 1960. *The Leopard's Spots: Scientific Attitudes toward Race in America 1815–59*. Chicago: University of Chicago Press.

Stark, Caroline, and Sheilagh T. Brooks. 1984. Regional Paleopathology of Prehistoric Peoples in the Great Basin, Nevada. In *Proceedings of the Paleopathology*

Association's Fourth European Members Meeting, Middelburg-Antwerpen, edited by G. T. Haneveld and W. R. K. Perizonius. 17–18. Utrecht (Netherlands): Elinwijk B.V.

Stark, Caroline, and Sheilagh T. Brooks. 1985. A Survey of Prehistoric Paleopathology in the Nevada Great Basin. In *Health and Disease in the Prehistoric Southwest*, edited by C. F. Merbs and R. J. Miller. 65–78. Vol. 34, *Arizona State University Anthropological Research Papers (Tempe)*.

Staz, J. 1938. Dental Caries in South Africa. *South African Journal of Medical Sciences* 3:1–63.

Steadman, Dawnie Wolfe. 1998. The Population Shuffle in the Central Illinois Valley: A Diachronic Model of Mississippian Biocultural Interactions. *World Archaeology* 30 (2):306–326.

Steadman, Dawnie Wolfe. 2001. Mississippians in Motion? A Population Genetic Analysis of Interregional Gene Flow in West-Central Illinois. *American Journal of Physical Anthropology* 114 (1):61–73.

Steckle, Richard H., Clark Spencer Larsen, Paul W. Sciulli, and Phillip L. Walker. 2002. A History of Health in Europe over the Past 10,000 Years: Summary of a Research Proposal. Retrieved Dec. 28, 2005, from the Ohio State University Global History of Health Project Web Site: http://global.sbs.ohio-state.edu/project_overview.htm.

Steckle, Richard H., and Jerome C. Rose, eds. 2002. *The Backbone of History: Health and Nutrition in the Western Hemisphere*. New York and Cambridge: Cambridge University Press.

Steele, D. Gentry, and Claud A. Bramblett. 1988. *The Anatomy and Biology of the Human Skeleton*. 1st ed. College Station, TX: Texas A&M University Press.

Steen, Susan L., and Robert W. Lane. 1998. Evaluation of Habitual Activities among Two Alaskan Eskimo Populations Based on Musculoskeletal Stress Markers. *International Journal of Osteoarchaeology* 8:341–353.

Steinbock, R. Ted. 1976. *Paleopathological Diagnosis and Interpretation: Bones Diseases in Ancient Human Populations*. Springfield, IL: Charles C Thomas.

Stepan, Nancy L. 1982. *The Idea of Race in Science. Great Britain 1800–1960*. Hammden, CT: Archon Books.

Stepan, Nancy L. 1985. Biological Degeneration: Races and Proper Places. In *Degeneration: The Dark Side of Progress*, edited by J. E. Chamberlin and S. L. Gilman. 97–120. New York: Columbia University Press.

Stepan, Nancy L. 1991. *The Hour of Eugenics: Race, Gender, and Nation in Latin America*. Ithaca, NY: Cornell University Press.

Stewart, T. Dale. 1930. Anthropology and Dental Caries [Abstract]. *American Journal of Physical Anthropology* XIV (1):89.

Stewart, T. Dale. 1931a. Dental Caries in Peruvian Skulls. *American Journal of Physical Anthropology* 15 (2):315–326.

Stewart, T. Dale. 1931b. Incidence of Separate Neural Arch in the Lumbar Vertebrae of Eskimos. *American Journal of Physical Anthropology* XVI (1):51–62.

Stewart, T. Dale. 1935. Spondylolisthesis without Separate Neural Arch (Pseudo Spondylolisthesis of Junghanns). *Journal of Bone and Joint Surgery* 17 (3):640–648.

Stewart, T. Dale. 1937a. Different Types of Cranial Deformation in the Pueblo Area. *American Anthropologist* 39 (1): 169–171.

Stewart, T. Dale. 1937b. Did the American Indians Use the Cautery in Bone Surgery? *American Journal of Physical Anthropology* 23 (1):83–89.

Stewart, T. Dale. 1939a. A New Type of Artificical Cranial Deformation from Florida. *Journal of the Washington Academy of Sciences* 29 (10):460–465.

Stewart, T. Dale. 1939b. Excavating the Indian Village of Patawomeke (Potomac). In *Explorations and Field Work of the Smithsonian Institution in 1938*. 87–90. Washington, DC: Smithsonian Institution.

Stewart, T. D. 1940a. The Life and Writings of Dr. Aleš Hrdlička (1869–1939). *American Journal of Physical Anthropology* 26 (1):3–40.

Stewart, T. Dale. 1940b. A Report on the Skeletal Remains. In *An Ossuary Near Piscataway Creek. American Antiquity* 6 (1):13–18.

Stewart, T. Dale. 1940c. Skeletal Remains from the Whitewater District, Eastern Arizona. Appendix B. In *Archeological Remains in the Whitewater District, Eastern Arizona. II. Artifacts and Burials*, edited by F. H. H. Roberts. 153–166, *Bureau of American Ethnology Bulletin, No. 126*. Washington, DC: Smithsonian Institution.

Stewart, T. Dale. 1940d. The Finding of an Indian Ossuary on the York River in Virginia. *Journal of the Washington Academy of Sciences* 30 (8):356–364.

Stewart, T. Dale. 1940e. Further Excavations at the Indian Village Site of Patawomeke (Potomac). In *Explorations and Field Work of the Smithsonian Institution in 1939*. 79–82. Washington, DC: Smithsonian Institution.

Stewart, T. Dale. 1940f. Some Historical Implications of Physical Anthropology in North America. *Smithsonian Miscellaneous Collections* 100:15–50.

Stewart, T. Dale. 1941a. The Circular Type of Cranial Deformity in the United States. *American Journal of Physical Anthropology* XXVIII (3):343–351.

Stewart, T. Dale. 1941b. New Examples of Tooth Mutilation from Middle America. *American Journal of Physical Anthropology* XXVIII (1):117–124.

Stewart, T. Dale. 1941c. Skeletal Remains from the Buena Vista Sites, California. Appendix A. In *Archeological Investigations at Buena Vista Lake, Kern County, California*, edited by W. R. Wedel. 172–188, *Bureau of American Ethnology Bulletin 130*. Washington, DC: Smithsonian Institution.

Stewart, T. Dale. 1941d. Skeletal Remains from the Peachtree Site, North Carolina. Appendix D. In *Peachtree Mound and Village Site, Cherokee County, North Carolina*, edited by F. M. Setzler and J. D. Jennings. 81–99. Washington, DC: Smithsonian Institution.

Stewart, T. Dale. 1941e. An Ossuary at the Indian Village Site of Patawomeke (Potomac). In *Explorations and Field Work of the Smithsonian Institution in 1940*. 67–70. Washington, DC: Smithsonian Institution.

Stewart, T. Dale. 1942. Persistence of the African Type of Tooth Pointing in Panama. *American Anthropologist* 44 (2):328–330.

Stewart, T. Dale. 1943a. Skeletal Remains from Platte and Clay Counties, Missouri. Appendix. In *Archaeological Investigations in Platte and Clay Counties, Missouri*, edited by W. R. Wedel. 245–273, *United States National Museum Bulletin 183*. Washington, DC: Smithsonian Institution.

Stewart, T. Dale. 1943b. Skeletal Remains from Tajumulco, Guatemala. Appendix 1. In *Excavations at Tajumulco, Guatemala*, edited by B. P. Dutton and H. R. Hobbs.

111–114, *Archaeological Institute of America. School of American Research, Santa Fe Monographs 9*. Santa Fe, NM: University of New Mexico Press.

Stewart, T. Dale. 1943c. Skeletal Remains from Paracas, Peru (Based on Material Provided by Dr. Julio C. Tello, Lima, Peru). *American Journal of Physical Anthropology* 1:47–61.

Stewart, T. Dale. 1947. Racial Patterns in Vertebral Osteoarthritis. *American Journal of Physical Anthropology* 5:230–231.

Stewart, T. Dale. 1948a. Medico-Legal Aspects of the Skeleton. I. Sex, Age, Race and Stature. *American Journal of Physical Anthropology* 6 (3):315–321.

Stewart, T. Dale. 1948b. The True Form of the Cranial Deformity Originally Described under the Name "Tête Trilobée." *Journal of the Washington Academy of Sciences* 38 (2):66–72.

Stewart, T. Dale. 1950a. Pathological Changes in South American Indian Skeletal Remains. In *Handbook of South American Indians*, edited by J. H. Steward. 49–52. Vol. 6, *Bureau of American Ethnology Bulletin 143*. Washington, DC: Smithsonian Institution.

Stewart, T. Dale. 1950b. Report on Skeleton from the Butte Creek Cave, John Day River Region, Oregon. In *Archaeological Research in the John Day Region of North Central Oregon*, edited by L. S. Cressman. 385–387. Vol. 94(4), *Proceedings of the American Philosophical Society*.

Stewart, T. Dale. 1951a. Notes on Skeletal Material. Appendix B. In *Excavations at Nebaj, Guatemala*, edited by A. L. Smith and A. V. Kidder. 86–87, *Carnegie Institution of Washington Publication 594*. Washington, DC.

Stewart, T. Dale. 1951b. What Bones Tell Us. *FBI Law Enforcement Bulletin* 20 (2):2–5.

Stewart, T. Dale. 1953. The Location and Terminal Date of Historic Patawomeke. *Eastern States Archeological Federation Bulletin* 12:8.

Stewart, T. Dale. 1954a. A Method for Analyzing and Reproducing Pipe Decorations. *Quarterly Bulletin Archaeological Society of Virginia* 9 (1).

Stewart, T. Dale. 1954b. Evaluation of Evidence from the Skeleton. In *Legal Medicine*, edited by R. B. H. Gradwohl. 407–450. St. Louis, MD: C.V. Mosby.

Stewart, T. Dale. 1954c. Sex Determination of the Skeleton by Guess and by Measurement. *American Journal of Physical Anthropology* 12 (3):385–392.

Stewart, T. Dale. 1954d. Comment on "Some Problems in the Physical Anthropology of the American Southwest" (by J. N. Spuhler). *American Anthropologist* 56 (4):619–620.

Stewart, T. Dale. 1956a. Examination of the Possibility That Certain Skeletal Characters Predispose to Defects in the Lumbar Neural Arches. *Clinical Orthopaedics* 8:44–60.

Stewart, T. Dale. 1956b. Significance of Osteitis in Ancient Peruvian Trephining. *Bulletin of the History of Medicine* 30 (4):293–320.

Stewart, T. Dale. 1958a. The Rate of Development of Vertebral Osteoarthritis in American Whites and Its Significance in Skeletal Age Identification. *The Leech* 28:144–151.

Stewart, T. Dale. 1958b. Stone Age Skull Surgery: A General Review, with Emphasis on the New World. In *Annual Report to the Board of Regents of the Smithsonian Institution, Publication 4314*. 469–491. Washington, DC: Government Printing Office.

Stewart, T. Dale. 1959a. Restoration and Study of the Shanidar I Neanderthal Skeleton in Baghdad, Iraq. In *American Philosophical Society Year Book 1958*. 274–278. Philadelphia: American Philosophical Society.

Stewart, T. Dale. 1959b. The Restored Shanidar I Skull. In *Annual Report of the Smithsonian Institution for 1958*. 473–480. Washington, DC: Smithsonian Institution.

Stewart, T. Dale. 1959c. Bear Paw Remains Closely Resemble Human Bones. *FBI Law Enforcement Bulletin* 28 (11):18–21.

Stewart, T. Dale. 1959d. Description of the Skeletal Remains from Doniphan and Scott Counties, Kansas. In *An Introduction to Kansas Archaeology*, edited by W. R. Wedel. 669–683, *Bureau of American Ethnology Bulletin 174*. Washington, DC: Smithsonian Institution.

Stewart, T. Dale. 1959e. Skeletal Remains from the Vicinity of Point Barrow, Alaska. Appendix. In *Eskimo Prehistory in the Vicinity of Point Barrow, Alaska*, edited by J. A. Ford. 245–255, *Anthropological Papers of the American Museum of Natural History 47(Pt 1)*. New York.

Stewart, T. Dale. 1959f. Skeletal Remains from Venado Beach, Panama: Cranial Deformity. In *Actas Del Xxxiii Congreso Internacional de Americanistas, San José, 20–27 Julio 1958*. 45–54. Vol. 2. San José, Costa Rica: Lehmann.

Stewart, T. Dale. 1960a. A Physical Anthropologist's View of the Peopling of the New World. *Southwestern Journal of Anthropology* 16 (3):259–273.

Stewart, T. Dale. 1960b. Form of the Pubic Bone in Neanderthal Man. *Science* 131 (3411):1437–1438.

Stewart, T. Dale. 1961a. A Neglected Primitive Feature of the Swanscombe Skull. In *Homenaje a Pablo Martínez Del Río en el Vigésimoquinto Aniversario de la Primera Edición de Los Orígenes Americanos*. 207–217. Mexico: Instituto Nacional de Antropologia e Historica.

Stewart, T. Dale. 1961b. The Skull of Shanidar II. *Sumer* 17 (1, 2):97–106.

Stewart, T. Dale. 1962a. Neanderthal Cervical Vertebrae, with Special Attention to the Shanidar Neanderthals from Iraq. *Bibl. Primatol.* 1:130–154.

Stewart, T. Dale. 1962b. Neanderthal Scapulae, with Special Attention to the Shanidar Neanderthals from Iraq. *Anthropos* 57:779–800.

Stewart, T. Dale. 1962c. The Skull of Shanidar II. In *Annual Report of the Smithsonian Institution for 1961*. 521–533. Washington, DC: Smithsonian Institution.

Stewart, T. Dale. 1962d. Anterior Femoral Curvature: Its Utility for Race Identification. *Human Biology* 34 (1):49–62.

Stewart, T. Dale. 1962e. Comments on the Reassessment of the Indian Knoll Skeletons. *American Journal of Physical Anthropology* 20:142–148.

Stewart, T. Dale. 1963a. Shanidar Skeletons IV and VI. *Sumer* 19 (1, 2):8–26.

Stewart, T. Dale. 1963b. Deformity, Trephining, and Mutilation in South American Indian Skeletal Remains. In *Handbook of South American Indians*, edited by J. H. Steward, pp. 43–48. Washington, D.C.: Smithsonian Institution.

Stewart, T. Dale. 1963c. Pathological Changes in South American Indian Skeletal Remains. In *Handbook of South American Indians*, edited by J. H. Steward, pp. 49–52. Washington, D.C.: Smithsonian Institution.

Stewart, T. Dale. 1964. The Scapula of the First Recognized Neanderthal Skeleton. *Bonner Jahrbuch* 164:1–14.

Stewart, T. Dale. 1966. Some Problems in Human Paleopathology. In *Human Palaeopathology*, edited by S. Jarcho. 43–55. New Haven: Yale University Press.

Stewart, T. Dale. 1968. Identification by the Skeletal Structures. In *Gradwohl's Legal Medicine*, edited by F. E. Camps. 123–154. Bristol: J. Wright.

Stewart, T. Dale. 1969. The Effects of Pathology on Skeletal Populations. *American Journal of Physical Anthropology* 30 (3):443–449.

Stewart, T. Dale. 1970a. *Personal Identification in Mass Disasters*. Washington, DC: Smithsonian Institution.

Stewart, T. Dale, ed. 1970b. *Handbook of Middle American Indians: Physical Anthropology*. Edited by R. Wauchope. 16 vols. Vol. 9, *Handbook of Middle American Indians*. Austin, TX: University of Texas Press.

Stewart, T. Dale. 1972. What the Bones Tell — Today. *FBI Law Enforcement Bulletin* 41 (2):16–20, 30–31.

Stewart, T. Dale. 1973a. Recent Improvements in Estimating Stature, Sex, Age and Race from Skeletal Remains. In *Modern Trends in Forensic Medicine*, edited by A. K. Mant. 193–211. London: Butterworths.

Stewart, T. Dale. 1973b. *The People of America*. New York: Charles Scribner's Sons.

Stewart, T. Dale. 1974. Nonunion of Fractures in Antiquity, with Descriptions of Five Cases from the New World Involving the Forearm. *Bulletin of the New York Academy of Medicine* 50 (8):875–891.

Stewart, T. Dale. 1975. Cranial Dysraphism Mistaken for Trephination. *American Journal of Physical Anthropology* 42:435–438.

Stewart, T. Dale. 1976. Are Supra-Inion Depressions Evidence of Prophylactic Trephination? *Bulletin of the History of Medicine* 50:414–434.

Stewart, T. Dale. 1978. George A. Dorsey's Role in the Luetgert Case: A Significant Episode in the History of Forensic Anthropology. *Journal of Forensic Sciences* 23 (4):786–791.

Stewart, T. Dale. 1979a. *Essentials of Forensic Anthropology, Especially as Developed in the United States*. Springfield, IL: Charles C Thomas.

Stewart, T. Dale. 1979b. Patterning of Skeletal Pathologies and Epidemiology. In *The First Americans: Origins, Affinities, and Adaptations*, edited by W. S. Laughlin and A. B. Harper. 257–274. New York: Gustav Fischer.

Stewart, T. Dale. 1979c. Forensic Anthropology. In *The Uses of Anthropology*, edited by W. Goldschmidt. 169–183, *Special Publication of the Aaa, No. 11*. Washington, DC: American Anthropological Association.

Stewart, T. Dale. 1979d. A Tribute to French Forensic Anthropologist, Georges Fully (1926–1973). *Journal of Forensic Sciences* 24 (4):916–924.

Stewart, T. Dale. 1981. The Evolutionary Status of the First Americans. *American Journal of Physical Anthropology* 56 (4):461–466.

Stewart, T. Dale. 1982. Background of American Forensic Anthropology. *Criminal Justice Review* 7:4–7.

Stewart, T. Dale. 1983. The Points of Attachment of the Palpebral Ligaments: Their Use in Facial Reconstructions of the Skull. *Journal of Forensic Sciences* 28 (4): 858–863.

Stewart, T. Dale. 1984a. Pathologic Changes in Aging Sacroiliac Joints. A Study of Dissecting Room Skeletons. *Clinical Orthopedics* 183:188–196.

Stewart, T. Dale. 1984b. Perspective on the Reporting of Forensic Cases. In *Human Identification: Case Studies in Forensic Anthropology*, edited by T. A. Rathbun and J. E. Buikstra. 15–18. Springfield, IL: Charles C Thomas.

Stewart, T. Dale. 1992. *Archeological Exploration of Patawomeke: The Indian Town Site (44St2) Ancestral to the One (44St1) Visited in 1608 by Captain John Smith*. Vol. 36, *Smithsonian Contributions to Anthropology*. Washington, DC: Smithsonian Institution Press.

Stewart, T. Dale. n.d. *A Bibliography of Physical Anthropology in Latin America: 1937–1948*. New York: The Wenner-Gren Foundation for Anthropological Research, Inc.

Stewart, T. Dale, and John R. Groome. 1968. The African Custom of Tooth Mutilation in America. *American Journal of Physical Anthropology* 28 (1):31–42.

Stewart, T. Dale, and Marshall Thornton Newman. 1951. A Historical Resume of the Concept of Differences in Indian Types. *American Anthropologist* 53:19–36.

Stewart, T. Dale, and Marshall Thornton Newman. 1954. Review of "Archaeology and Race in the American Indian." *American Journal of Physical Anthropology* 12: 137–141.

Stewart, T. Dale, and Lawrence G. Quade. 1969. Lesions of the Frontal Bone in American Indians. *American Journal of Physical Anthropology* 30 (1):89–109.

Stewart, T. Dale, and Alexander Spoehr. 1952. Evidence on the Paleopathology of Yaws. *Bulletin of the History of Medicine* 29 (6):538–553.

Stewart, T. Dale, and P. F. Titterington. 1944. Filed Indian Teeth from the United States. *Journal of the Washington Academy of Sciences* 34 (10):317–321.

Stewart, T. Dale, and P. F. Titterington. 1946. More Filed Teeth from the United States. *Journal of the Washington Academy of Sciences* 36 (8):259–261.

Stewart, T. Dale, and Mildred Trotter. 1955. Role of Physical Anthropology in the Field of Human Identification. *Science* 122 (3175):883–884.

Stewart, T. Dale, and W. R. Wedel. 1937. The Finding of Two Ossuaries on the Site of the Indian Village of Nacotchtanke (Anacostia). *Journal of the Washington Academy of Sciences* 27 (5):213–219.

Stiner, Mary C., N. D. Munro, and T. A. Surovell. 2000. The Tortise and the Hare: Small-Game Use, the Broad Spectrum Revolution, and Paleolithic Demography. *Current Anthropology* 41:39–73.

Stiner, Mary C., N. D. Munro, T. A. Surovell, E. Tchernov, and O. Bar-Yosef. 1999. Paleolithic Population Growth Pulses Evidenced by Small Animal Exploitation. *Science* 283:190–194.

Stinson, Sara. 1992. Nutritional Adaptation. *Annual Review of Anthropology* 21:143–170.

Stock, Jay, and Susan Pfeiffer. 2001. Linking Structural Variability in Long Bone Diaphyses to Habitual Behaviors: Foragers from the Southern African Later Stone Age and the Andaman Islands. *American Journal of Physical Anthropology* 115:337–348.

Stodder, A. L. W. 1987. The Physical Anthropology and Mortuary Practices of the Dolores Anasazi: An Early Pueblo Population in Local and Regional Perspective. In *Dolores Archaeological Program Supporting Studies: Settlement and Environment*, edited by K. L. Peterson and J. D. Orcutt. 339–504. Denver: U.S. Bureau of Reclamation Engineering and Research Center.

Stodder, A. L. W. 1990. Paleoepidemiology of Eastern and Western Pueblo Communities in Protohistoric New Mexico. Ph.D. Dissertation, University of Colorado.

Stojanowski, Christopher M., and Jane E. Buikstra. 2004. Biodistance Analysis, a Biocultural Enterprise: A Rejoinder to Armelagos and Van Gerven (2003). *American Anthropologist* 106 (2):430–431.

Stone, Anne C. 2000. Ancient DNA from Skeletal Remains. In *Biological Anthropology of the Human Skeleton*, edited by M. A. Katzenberg and S. R. Saunders. 351–371. New York: Wiley-Liss.

Stoodley, N. 1999. *The Spindle and the Spear: A Critical Enquiry into the Construction and Meaning of Gender in the Early Anglo-Saxon Burial Rite, British Archaeological Reports British Series 288*. Oxford: Archaeopress.

Storey, Rebecca. 1992. *Life and Death in the Ancient City of Teotihuacan: A Modern Paleodemographic Synthesis*. Tuscaloosa: University of Alabama.

Storey, Rebecca. 1997. Individual Frailty, Children of Privilege, and Stress in Late Classic Copán. In *Bones of the Maya: Studies of Ancient Skeletons*, edited by S. Whittington and D. Reed. 116. Washington, DC: Smithsonian Institution Press.

Strait, David S., Qian Wang, Paul C. Dechow, Callum F. Ross, Brian G. Richmond, Mark A. Spencer, and Biren A. Patel. 2005. Modeling Elastic Properties in Finite-Element Analysis: How Much Precision Is Needed to Produce an Accurate Model? *The Anatomical Record Part A* 283A (2):275–287.

Strickland, S. S., and P. S. Shetty, eds. 1998. *Human Biology and Social Inequality*. Cambridge: Cambridge University Press.

Straus, William L., and A. J. E. Cave. 1957. Pathology and Posture of Neanderthal Man. *Quarterly Review of Biology* 32:348–363.

Stuart-Macadam, Patricia. 1992a. Anemia in Past Human Populations. In *Diet, Demography, and Disease: Changing Perspectives on Anemia, edited by P. Stuart-Macadam* and S. Kent. 151–170. New York: Adline de Gruyter.

Stuart-Macadam, Patricia. 1992b. Porotic Hyperostosis: A New Perspective. *American Journal of Physical Anthropology* 87:39–47.

Stuart-Macadam, Patricia, and Susan Kent. 1992. *Diet, Demography, and Disease: Changing Perspectives on Anemia*. New York: Aldine De Gruyter.

Studley, Cordelia A. 1884. Notes Upon Human Remains from the Caves of Coahuila, Mexico. *Reports of the Peabody Museum of American Archaeology and Ethnology* 3 (Sixteenth and Seventeenth Annual Reports of the Peabody Museum of American Archaeology and Ethnology):233–259.

Studley, Cordelia A. 1885. Description of the Human Remains Found in The "Intrustive Pit" in the Large Mound of the Turner Group, Little Miami Valley, Ohio, During the Explorations of Messrs. Putman and Metz. *Proceedings of the American Association for the Advancement of Science* 33:618.

Studley, Cordelia A. n.d. Shellheaps; Extracts from Various Writers. Manuscript in the F. W. Putnam Papers, Tozzer Library, Harvard University.

Sublett, Audrey J. 1965. The Cornplanter Cemetery: Skeletal Analyses. *Pennsylvania Archaeologist* XXXV (2):74–91.

Sublett, Audrey J. 1966. Seneca Physical Type and Change through Time. Ph.D., State University of New York, Buffalo, University Microfilms, Ann Arbor, Michigan.

Sublett, Audrey J. n.d. The Allegany Project: An Osteological Analysis of a Recent Population.

Sublett, Audrey J., Louise Marie Robbins, and Rebecca A. Lane. n.d. Skeletal Populations; the East. Manuscript on File (Handbook of North American Indians, Smithsonian Institution). Smithsonian Institution, Washington, DC.

Sublett, Audrey J., and C. F. Wray. 1970. Some Examples of Accidental and Deliberate Human Skeletal Modification in the Northeast. *The Bulletin. Publication of the New York State Archaeological Association No. 50*:14–26.

Suchey, Judy M. 1975. Biological Distance of Prehistoric Central California Populations Derived from Non-Metric Traits of the Cranium. Ph.D. Dissertation, University of California, Riverside.

Suchey, Judy Myers, Sheilagh T. Brooks, and Darryl Katz. 1988. Instructions for Use of the Suchey-Brooks System for Age Determination of the Female Os Pubis. Instructional Material Accompanying the Female Pubic Symphyseal Models of the Suchey-Brooks System. Distributed by Diane France Casting, Bellevue, Colorado.

Sullivan, Lynne P. 1994. Madeline Kneberg Lewis: An Original Southeastern Archaeologist. In *Women in Archaeology*, edited by C. Claassen. 110–119. Philadelphia: University of Pennsylvania Press.

Sullivan, Lynne P. 1999. Madeline D. Kneberg Lewis: Leading Lady of Tennessee Archaeology. In *Grit-Tempered: Early Women Archaeologists in the Southeastern United States*, edited by N. M. White, L. P. Sullivan, and R. A. Marrinan. 57–91. Gainesville, FL: University Press of Florida.

Swärdstedt, T. 1966. *Odontological Aspects of a Medieval Population from the Province of Jamtland/Mid-Sweden*. Stockholm: Tiden Barnangen, AB.

Swedlund, Alan, and Duane Anderson. 1999. Gordon Creek Woman Meets Kennewick Man: New Interpretations and Protocols Regarding the Peopling of the Americas. *American Antiquity* 64 (4):569–576.

Swedlund, A. C., and George J. Armelagos. 1976. *Demographic Anthropology*. Dubuque: Wm. C. Brown Company.

Swinton, George. 1976. Archaeology as a Concern of the Inuit Community. In *Symposium on New Perspectives in Canadian Archaeology, 22–23 October 1976, Theatre Auditorium, Royal Ontario Museum, Toronto*, edited by A. G. McKay. 163–171. Ottawa: The Royal Society of Canada.

Syms, E. Leigh. 1997. Increasing Awareness and Involvement of Aboriginal People in Heritage Preservation: Recent Developments at the Manitoba Museum of Man and Culture. In *At a Crossroads: Archaeology and First Peoples of Canada*, edited by G. P. Nicholas and T. D. Andrews. 53–68. Burnaby: Archaeology Press.

Szathmary, E. J. E., and N. S. Ossenberg. 1978. Are the Biological Differences between North American Indians and Eskimos Truly Profound? *Current Anthropology* 19: 673–701.

Tainter, Joseph A. 1978. Mortuary Practices and the Study of Prehistoric Social Systems. *Advances in Archaeological Method and Theory* 1:106–141.

TallBear, Kimberly. 2000. Genetics, Culture and Identity in Indian Country. Paper read at Seventh International Congress of Ethnobiology, October 23–27, at Athens, Georgia.

Tanner, J. M. 1959. Boas' Contribution to Knowledge of Human Growth and Form. In *The Anthropology of Franz Boas: Essays on the Centennial of His Birth*, edited by

W. R. Goldschmidt. 76–111, *American Anthropological Memoir No. 89*. Menasha: American Anthropological Association.

Tarlow, Susan. 1999. *Bereavement and Commemoration: An Archaeology of Mortality*. Oxford: Blackwell Publishers.

Tarpy, C. 2000. Pueblo Ancestors Return Home. *National Geographic* November: 118–125.

Taylor, G. Michael, Mary Crossey, John Saldanha, and Tony Waldron. 1996. DNA from *Mycobacterium tuberculosis* Identified in Mediaeval Human Skeletal Remains Using Polymerase Chain Reaction. *Journal of Archaeological Science* 23:789–798.

Taylor, G. Michael, S. Widdison, I. N. Brown, and D. Young. 2000. A Mediaeval Case of Lepromatous Leprosy from 13th–15th Century Orkney. *Journal of Archaeological Science* 27:1133–1138.

Taylor, R. 1963. Cause and Effect of Wear on Teeth. Further Non-Metrical Studies on the Teeth and Palate in Moriori and Maori Skulls. *Acta Anatomica* 53:97–157.

Taylor, William E. Jr. 1976. Introduction (Fourth Session: Archaeology: New Motivations and Attitudes). In *Symposium on New Perspectives in Canadian Archaeology, 22–23 October 1976, Theatre Auditorium, Royal Ontario Museum, Toronto*, edited by A. G. McKay. 151–158. Ottawa: The Royal Society of Canada.

Taylor, Waltor W., Jr. 1948. *A Study of Archaeology*. Vol. 69, *Memoir Series of the American Anthropologist Association*. Menasha, Wiscoxsin.

Teaford, Mark F. 1985. Molar Microwear and Diet in the Genus Cebus. *American Journal of Physical Anthropology* 66:363–370.

Teaford, Mark F. 1986. Dental Microwear and Diet in Two Species of Colobus. In *Proceedings of the 10th Annual International Primatology Conference, Volume II: Primate Ecology and Conservation*, edited by J. Else and P. Lee. 63–66. Cambridge: Cambridge University Press.

Teaford, Mark F. 1991. Dental Microwear: What Can It Tell Us About Diet and Dental Function. In *Advances in Dental Anthropology*, edited by M. A. Kelley and C. S. Larsen. 341–356. New York: Wiley-Liss.

Teaford, Mark F., Clark Spencer Larsen, R. F. Pastor, and V. E. Noble. 2001. Pits and Scratches: Microscopic Evidence of Tooth Use and Masticatory Behavior in La Florida. In *Bioarchaeology of La Florida: Human Biology in Northern Frontier New Spain*, edited by C. S. Larsen. 82–112. Gainesville, FL: University Press of Florida.

Teaford, Mark F., and J. D. Lytle. 1996. Diet-Induced Changes in Rates of Human Tooth Microwear: A Case Study Involving Stone-Ground Maize. *American Journal of Physical Anthropology* 100:143–147.

Teaford, Mark F., and A. Walker. 1984. Quantitative Differences in Dental Microwear between Primate Species with Different Diets and a Comment on the Presumed Diet of Sivapithecus. *American Journal of Physical Anthropology* 64:191–200.

Teague, Lynn S. 1993. Prehistory and the Traditions of the O'odham and Hopi. *Kiva* 58 (4):435–454.

Tello, Julio Caesar. 1909. *La Antigüedad de La Sifilis en el Peru*. Lima: Sanmartí.

Tello, Julio Caesar, and Herbert U. Williams. 1930. Ancient Syphilitic Skull from Paracas in Peru. *Annals of Medical History* 2:515–529.

ten Kate, Herman F. C. 1892. Somatological Observations on Indians of the Southwest. *Journal of American Archaeology and Ethnology* 3:117–144.

ten Kate, Herman F. C., and J. L. Wortman. 1888. On an Anatomical Characteristic of the Hyoid Bone of Precolumbian Pueblo Indians. In *Congressa International D'americanistes*. Vol. VII. Berlin.

The Royal College of Surgeons of Edinburgh. n.d. The Royal College of Surgeons of Edinburgh Home Page. Retrieved Dec. 28, 2005, from the Royal College of Surgeons of Edinburgh Web Site: http://www.rcsed.ac.uk/.

The Royal College of Surgeons of England. n.d. Museums. Retrieved Dec. 28, 2005, from the Royal College of Surgeons of England Web Site: http://www.rcseng.ac.uk/museums/.

Thomas, Cyrus. 1894. Report on the Mound Explorations of the Bureau of Ethnology. In *Twelfth Annual Report of the Bureau of Ethnology, 1890–1891*. Washington, DC: Government Printing Office.

Thomas, David Hurst. 1989. *Columbian Consequences: Volume 1. Archaeological and Historical Perspectives on the Spanish Borderlands West*. Washington, D.C.: Smithsonian Institution Press.

Thomas, David Hurst. 1990. *Columbian Consequences: Volume 2. Archaeological and Historical Perspectives on the Spanish Borderlands East*. Washington, D.C.: Smithsonian Institution Press.

Thomas, David Hurst. 1991. *Columbian Consequences: Volume 3. The Spanish Borderlands in Pan-American Perspective*. Washington, D.C.: Smithsonian Institution Press.

Thomas, David Hurst. 2000. *Skull Wars: Kennewick Man, Archaeology, and the Battle for Native American Identity*. New York: Basic Books.

Thomas, David Hurst. 2001. Postscript. In *Societies in Eclipse: Archaeology of the Eastern Woodland Indians, A.D. 1400–1700*, edited by D. S. Brose, C. W. Cowan, and R. C. J. Mainfort. 229–232. Washington and London: Smithsonian Institution Press.

Thomas, David Hurst. 2004. Finders Keepers and Deep American History: Some Lessons in Dispute Resolution (Revision). Paper read at "Imperialism, Art & Restitution," 26–27 March 2004, at Whitney R. Harris Institute for Global Legal Studies. Washington University, School of Law, Anheuser-Busch Hall.

Thompson, D. D. 1982. Forensic Anthropology. In *A History of American Physical Anthropology, 1930–1980*, edited by F. Spencer. 357–369. New York: Academic Press.

Thorpe, I. J. 2002. *Archaeology of Death*. London: Routledge.

Tiesler Blos, Vera. 2004. Vida y Muerte de Janaab' Pakal de Palenque: Hallazogos Bioarqueológicos Reciente. In *Janaab' Pakal de Palenque. Vida y Muerte de Un Gobernante Maya. Chapter 2*, edited by V. Tiesler and A. Cucina. 37–67. Mexico City: UNAM.

Tiffany, Joseph A. 1990. Reburial and the Law: A Review of Iowa's Burial Law Protecting Ancient Human Remains. In *Preservation on the Reservation. Native Americans, Native American Lands and Archaeology*, edited by A. L. Klesert and A. S. Downer. 353–367. Navajo Nation Archaeology, Navajo Nation Historic Preservation Department.

Tilley, Christopher Y. 1984. Ideology and the Legitimization of Power in the Middle Neolithic of Southern Sweden. In *Ideology, Power, and Prehistory*, edited by D. Miller and C. Y. Tilley. 111–146. Cambridge: Cambridge University Press.

Todd, Thomas Wingate. 1920. Age Changes in the Pubic Bone: I. The Male White Pubis. *American Journal of Physical Anthropology* 3 (3):285–334.

Todd, Thomas Wingate. 1927. Skeletal Records of Mortality. *The Scientific Monthly* 24 (6):481–496.

Tomes, C. S. 1892. Casual Communication. *Transactions of the Odontological Society* 24:90.

Topinard, Paul. 1876. *L'anthropologie*. Paris: Reinwald.

Toyne, J. M. 2003. Musculoskeletal Stress Markers (MSM) and Weaving Activities at a Prehistoric Coastal Site in Peru [Abstract]. *American Journal of Physical Anthropology* (Supplement 36):211.

Tozzer, A. M., and A. L. Kroeber. 1945. Roland Burrage Dixon. *American Anthropologist* 47:104–118.

Tri-City Herald. 2005. Kennewick Man Virtual Interpretive Center: Legal Documents. Retrieved Dec. 31, 2005, from Kennewick Man Virtual Interpretive Center Web Site: http://www.kennewick-man.com/kman/legal/index.html.

Trigger, Bruce Graham. 1966. Sir Daniel Wilson: Canada's First Anthropologist. *Anthropologica* 8 (1):3–28.

Trigger, Bruce G. 1980. Archaeology and the Image of the American Indian. *American Antiquity* 45 (4):662–676.

Trigger, Bruce G. 1989. *A History of Archaeological Thought*. Cambridge: Cambridge University Press.

Trigger, Bruce G. 1997. Foreword. In *At a Crossroads: Archaeology and First Peoples in Canada*, edited by G. P. Nicholas and T. D. Andrews. vii–xiii. Burnaby, B.C.: Archaeology Press.

Trinkaus, Erik, Steven E. Churchill, Isabelle Villemeur, K. G. Riley, Julie A. Heller, and Christopher B. Ruff. 1991. Robusticity Versus Shape: The Functional Interpretation of Neandertal Appendicular Morphology. *Journal of the Anthropological Society of Nippon* 99:257–278.

Trinkaus, Erik, and Christopher B. Ruff. 1999a. Diaphyseal Cross-Sectional Geometry of Near Eastern Middle Paleolithic Humans: The Femur. *Journal of Archaeological Science* 26:409–424.

Trinkaus, Erik, and Christopher B. Ruff. 1999b. Diaphyseal Cross-Sectional Geometry of Near Eastern Middle Palaeolithic Humans: The Tibia. *Journal of Archaeological Science* 26:1289–1300.

Trope, Jack F., and Walter R. Echo-Hawk. 1992. The Native American Graves Protection and Repatriation Act: Background and Legislative History. *Arizona State Law Journal* 24 (35):35–77.

Trotter, Mildred. 1926. The Movable Segments of the Vertebral Column in Old Egyptians. *American Journal of Physical Anthropology* IX (2):457–466.

Trotter, Mildred. 1943. Hair from Paracas Indian Mummies. *American Journal of Physical Anthropology* 1:69–75.

Trotter, Mildred, and Goldine C. Gleser. 1952. Estimation of Stature from Long Bones of American Whites and Negros. *American Journal of Physical Anthropology* 10: 463–514.

Turner, Christy G. II. 1979. Dental Anthropological Indications of Agriculture among the Jomon People of Central Japan. *American Journal of Physical Anthropology* 51: 619–636.

Turner, Christy G. II. 1983. Taphonomic Reconstructions of Human Violence and Cannibalism Based on Mass Burials in the American Southwest. In *Carnivores, Human Scavengers and Predators: A Question of Bone Technology: Proceedings of the Fifteenth Annual Conference, the Archaeological Association of the University of Calgary*, edited by G. M. LeMoine and A. S. MacEachern. 219–240. Calgary: University of Calgary Archaeological Association.

Turner, Christy G. II. 1983a. Dental Evidence for the Peopling of the Americas. In *Early Man in the New World*. 147–157. Beverly Hills: Sage Publications.

Turner, Christy G. II. 1983b. Sinodonty and Sundadonty: A Dental Anthropological View of Mongoloid Microevolution, Origin, and Dispersal into the Pacific Basin, Siberia, and the Americas. In *Late Pleistocene and Early Holocene Cultural Connections of Asia and America*, edited by R. S. Vasilievsky. 72–76. Novosibirsk: USSR Academy of Sciences, Siberian Branch.

Turner, Christy G. II. 1984. Advances in the Dental Search for Native American Origins. *Acta Anthropogenetica* 8 (1):23–78.

Turner, Christy G. II. 1985. Dental Evidence for Peopling of the Americas. *National Geographic Society Research Reports* 19:573–596.

Turner, Christy G. II. 1986. The First Americans: The Dental Evidence. *National Geographic Research* 2:37–46.

Turner, Christy G. II. 2002. So, Where Are All the Missing Bones? *The Review of Archaeology* 23 (3):51–55.

Turner, Christy G. II, J. D. Irish, and L. M. C. Machado. 1992. Reply to Robby, Cruwys and Smith, with Additional Remarks on LMSAT. *American Journal of Physical Anthropology* 85:348–351.

Turner, Christy G. II, and L. M. C. Machado. 1983. New Dental Wear Pattern and Evidence for High Carbohydrate Consumption in a Brazilian Archaic Skeletal Population. *American Journal of Physical Anthropology* 61 (1):125–130.

Turner, Christy G. II, C. R. Nichol, and G. R. Scott. 1991. Scoring Procedures for Key Morphological Traits of the Permanent Dentition: The Arizona State University Dental Anthropology System. In *Advances in Dental Anthropology*, edited by M. A. Kelley and C. S. Larson. 13–31. New York: Wiley-Liss.

Turner, Christy G. II, and J. A. Turner. 1992. The First Claim for Cannibalism in the Southwest: Walter Hough's 1901 Discovery at Canyon Butte Ruin 3, Northeastern Arizona. *American Antiquity* 57 (4):661–682.

Turner, Christy G. II, and Jacqueline A. Turner. 1999. *Man Corn: Cannibalism and Violence in the Prehistoric American Southwest*. Salt Lake City: University of Utah Press.

Turner, J. G., and F. J. Bennett. 1913. Teeth in Ancient Egypt. *The Dental Surgeon* 9:475.

Turner, K. R. 1980. Affinities of the Copena Skeletal Series from Site 1Sc42. In *The Skeletal Biology of Aboriginal Populations in the Southeastern United States*, edited by P. Willey and F. H. Smith. 15–27. Chattanooga: Tennessee Anthropological Association.

Tylor, Edward Burnett. 1871. *Primitive Culture: Researches into the Development of Mythology, Philosophy, Religion, Art, and Custom*. London: J. Murray.

Tyrrell, A. J. 2000. Skeletal Non-Metric Traits and the Assessment of Inter- and Intra-Population Diversity: Past Problems and Future Potential. In *Human Osteology in Archaeology and Forensic Science*, edited by M. Cox and S. Mays. 289–306. London: Greenwich Medical Media.

Tyrrell, A. J. 2000. Corpus Saxonum: Early Medieval Bodies and Corporeal Identity. In *Social Identity in Early Medieval Europe*, edited by W. O. Frazer and A. J. Tyrrell. 137–156. London and New York: Leicester University Press.

Tyrrell, A. J., and Andrew T. Chamberlain. 1998. Non-Metric Trait Evidence for Modern Human Affinities and the Distinctiveness of Neanderthals. *Journal of Human Evolution* 34:549–554.

Tyson, Rose A., and Elizabeth S. Dyer Alcauskas, eds. 1980. *Catalogue of the Hrdlička Paleopathology Collection*. San Diego: San Diego Museum of Man.

Ubelaker, Douglas, and Lauryn Guttenplan Grant. 1989. Human Skeletal Remains: Preservation or Reburial? *Yearbook of Physical Anthropology* 32:249–287.

Ubelaker, Douglas H. 1974. *Reconstruction of Demographic Profiles from Ossuary Skeletal Samples, Smithsonian Contributions to Anthropology, No. 18*. Washington, DC: Smithsonian Institution Press.

Ubelaker, Douglas H. 1982. The Development of American Paleopathology. In *A History of American Physical Anthropology 1930–1980*, edited by F. Spencer. 337–356. New York: Academic Press.

Ubelaker, Douglas H. 1989. J. Lawrence Angel 1915–1986. *American Antiquity* 54 (1):5–8.

Ubelaker, Douglas H. 1994. The Biological Impact of European Contact in Ecuador. In *In the Wake of Contact: Biological Responses to Conquest*, edited by C. S. Larsen and G. R. Milner. 147–160. New York: Wiley-Liss.

Ubelaker, Douglas H. 1999. Aleš Hrdlička's Role in the History of Physical Anthropology. *Journal of Forensic Sciences* 44 (4):708–714.

Ubelaker, Douglas H., and Lauryn Guttenplan Grant. 1989. Human Skeletal Remains: Preservation or Reburial? *Yearbook of Physical Anthropology* 32:249–287.

Uchoa, Dorath Pinto, and Marília Carvalho de Mello e Alvim. 1984. Morfologia e Incidência do Toro Mandibular Nos Construtores de Sambaquis da Costa Meridional do Brasil. *Revista de Pré-história (São Paulo)* 6:434–454.

Uchoa, Dorath Pinto, and Marília Carvalho de Mello e Alvim. 1989. Demografia Esqueletal Dos Construtores do Sambaqui de Piaçaguera, São Paulo-Brasil. *Dédalo—Publicação Avulsa (São Paulo)* 1:455–472.

Ucko, Peter J. 1969. Ethnography and Archaeological Interpretation of Funerary Remains. *World Archaeology* 1 (2):262–280.

Ungar, Peter S. 1995. A Semiautomated Image Analysis Procedure for the Quantification of Dental Microwear II. *Scanning* 17:57–59.

University College London. n.d. Graduate Prospectus 2006 Entry: MSc in Skeletal and Dental Bioarchaeology. Retrieved Dec. 28, 2005, from UCL: London's Global University Web Site: http://www.ucl.ac.uk/prospective-students/graduate-degrees/shs/archaeology/taught/msc-skeletal-dental-bioarchaeology/index.shtml.

University of Cambridge. n.d. Duckworth Laboratory. Retrieved Dec. 28, 2005, from Leverhulme Centre for Human Evolutionary Studies Web Site: http://www.human-evol.cam.ac.uk/Duckworth/duckworth.htm.

University of Cambridge. n.d. Research: Bioarchaeology. Retrieved Dec. 28, 2005, from Department of Archaeology Web Site: http://www.arch.cam.ac.uk/themes/bioarch.html.

University of Texas at El Paso. n.d. Bioarchaeological References. Retrieved Dec. 28, 2005, from Laboratory for Environmental Biology Web Site: http://www.utep.edu/leb/baref/biblio.htm.

Vallois, H. V. 1936. Les Ossements Natoufiens D'erq el-Ahmar (Palestine). *L'Anthropologie* 46:526–539.

Vallois, H. V. 1960. Vital Statistics in Prehistoric Population as Determined from Archaeological Data. In *The Application of Quantitative Methods in Archaeology*, edited by R. F. Heizer and S. F. Cook. 186–222, *Viking Fund Publications in Anthropology, No. 28*. Chicago: Quadrangle Books.

van der Merwe, Nikolaas J., and J. C. Vogel. 1978. ^{13}C Content of Human Collagen as a Measure of Prehistoric Diet in Woodland North America. *Nature* 276:815–816.

Van Gennep, Arnold. 1960. *Rites of Passage*. Translated by M. B. Vizedom and G. L. Caffee. Chicago: University of Chicago Press.

Van Gerven, D. P., and George J. Armelagos. 1983. "Farewell to Paleodemography?" Rumors of Its Death Have Been Greatly Exaggerated. *Journal of Human Evolution* 12:353–360.

Van Gerven, Dennis P., David S. Carlson, and George J. Armelagos. 1973. Racial History and Bio-Cultural Adaptation of Nubian Archaeological Populations. *Journal of African History* 14 (4):555–564.

Vansina, Jan. 1985. *Oral Tradition as History*. Madison, WI: University of Wisconsin Press.

van Valkenburgh, Blaire, and Fritz Hertel. 1993. Tough Times at La Brea: Tooth Breakage in Large Carnivores of the Late Pleistocene. *Science* 261:456–459.

Verano, John W., Laurel S. Anderson, and Régulo Franco. 2000. Foot Amputation by the Moche of Ancient Peru: Osteological Evidence and Archaeological Context. *International Journal of Osteoarchaeology* 10:177–188.

Verano, John W., and Douglas H. Ubelaker. 1992. *Disease and Demography in the Americas*. Washington, DC.

Verneau, Renee. 1875. *Le Bassin Suivant les Sexes et les Races*. Paris: J.B. Baillière.

Virchow, Rudolph. 1872. Untersuchung des Neanderthal-Schädels. *Zeitschrift fuer Ethnologie (Berlin)* 4:157–165.

Voegelin, C. F., and C. S. Rafinesque. 1954. *Walam Olum or Red Score: The Migration Legend of the Lenni Lenape or Delaware Indians*. Indianapolis: Indianapolis Historical Society.

Vogel, John C., and Nikolaas J. van der Merwe. 1977. Isotopic Evidence for Early Maize Cultivation in New York State. *American Antiquity* 42 (2):238–242.

von Bonin, Gerhardt. 1936. *Skeletal Material from the Lowry Area*. Vol. 23, *Anthropology Series, No. 1*. Chicago: Field Museum of Natural History.

von Bonin, Gerhardt. 1937. Cranial Deformity in the Pueblo Area. *American Anthropologist* 39 (4):720–721.

von Bonin, Gerhardt, and G. M. Mourant. 1938. Indian Races in the United States. A Survey of Previously Published Cranial Measurements. *Biometrika* XXX (I and II): 94–129.

von Eickstedt, Egon. 1937. *Rassenkunde und Rassengeschichte der Menschheit*. 2 ed. Stuttgart: F. Enke.

von Eickstedt, Egon. 1940. *Die Forschung Am Menschen. Teil 1*. Stuttgart: Ferdinand Enke Verlag.

Waddle, D. M. 1994a. Matrix Correlation Tests Support a Single Origin for Modern Humans. *Nature* 368:452–454.

Waddle, D. M. 1994b. Reply to Konigsberg et al. *Nature* 372:229.

Wagner, Konrad. 1927. Mittelalter-Knochen aus Oslo. Eine Untersuchung 3534 Langer Extremitätknochen Nebst 73 Ganzer Skelette. Skrifter Utgitt av Det Norske Videnskaps-Akadami I Oslo. I. Mat.-Naturv. *Klasse 1926* 7:1–137.

Wainwright, S. A., B. A. Biggs, John D. Currey, and J. M. Gosline. 1976. *Mechanical Design in Organisms*. Princeton: Princeton University Press.

Waldron, Tony. 1992. Osteoarthritis in a Black Death Cemetery in London. *International Journal of Osteoarchaeology* 2:235–240.

Waldron, Tony. 1994. *Counting the Dead: The Epidemiology of Skeletal Populations*. Chichester and New York: Wiley.

Waldron, Tony, and M. Cox. 1989. Occupational Arthropathy: Evidence from the Past. *Journal of Industrial Medicine* 46:420–422.

Waldron, Tony, and Juliet Rogers. 1991. Inter-Observer Variation in Coding for Osteoathritis in Human Skeletal Remains. *International Journal of Osteoarchaeology* 1:49–56.

Walker, A. C. 1981. Diet and Teeth: Dietary Hypothesis and Human Evolution. *Phil. Trans. Roy. Soc. Lond.* 292:57–64.

Walker, A. C., H. N. Hoeck, and L. Perez. 1978. Microwear of Mammalian Teeth as an Indicator of Diet. *Science* 201:808–810.

Walker, Phillip L. 1978. A Quantitative Analysis of Dental Attrition Rates in the Santa Barbara Channel Area. *American Journal of Physical Anthropology* 48 (1):101–106.

Walker, Phillip L. 1989. Cranial Injuries as Evidence of Violence in Prehistoric Southern California. *American Journal of Physical Anthropology* 80(3):313–323.

Walker, Phillip L. 1992. Update on the Implementation of the Native American Graves Protection Act. *Society for American Archaeology Bulletin* 10 (1):6.

Walker, Phillip L. 1995. Problems of Preservation and Sexism in Sexing: Some Lessons from Historical Collections for Paleodemographers. In *Grave Reflections: Portraying the Past through Skeletal Studies*, edited by A. Herring and S. R. Saunders. 31–47. Toronto: Canadian Scholars' Press.

Walker, Phillip L. 1996. Integrative Approaches to the Study of Ancient Health: An Example from the Santa Barbara Channel Area of Southern California. In *Notes on Populational Significance of Paleopathological Conditions: Health, Illness and Death in the Past*, edited by A. Pérez-Pérez. 97–105. Barcelona: Fundació Uriach.

Walker, Phillip L. 2001. Bioarchaeological Perspective on the History of Violence. *Annual Review of Anthropology* 30:573–596.

Walker, Phillip L. 2002. Bioarchaeological Ethics: A Historical Perspective on the Value of Human Remains. In *Biological Anthropology of the Human Skeleton*, edited by M. A. Katzenberg and S. R. Saunders. 3–39. New York: Wiley Liss.

Walker, Phillip L. 2004. Caring for the Dead: Finding a Common Ground in Disputes over Museum Collections of Human Remains. In *Conservation Policy and Current Research*, edited by G. Grupe and J. Peters. 13–27. Rahden: Verlag Marie Leidorf.

Walker, Phillip L., and Della Collins Cook. 1998. Gender and Sex: Vive la Difference. *American Journal of Physical Anthropology* 106 (2):255–259.

Walker, Phillip L., and Barry S. Hewlett. 1990. Dental Health, Diet, and Social Status among Central African Foragers and Farmers. *American Anthropologist* 92:383–398.

Walker, Phillip L., J. R. Johnson, and Patricia M. Lambert. 1988. Age and Sex Biases in the Preservation of Human Skeletal Remains. *American Journal of Physical Anthropology* 76:183–188.

Walker, W. H. 1998. Where Are the Witches of Prehistory? *Journal of Archaeological Method and Theory* 5 (3):245–308.

Wallis, R. S. 1934. Harmonic Types among Western European Crania. In *A Decade of Progress in Eugenics: Scientific Papers of the Third International Congress of Eugenics*. 99–104. Baltimore: Williams & Wilkins.

Ware, J. E. 1987. Standards for Validating Health Measures: Definition and Content. *Journal of Chronic Disease* 40:473–480.

Warren, John Collins. 1822. *A Comparative View of the Sensorial and Nervous Systems in Men and Animals; with Appendix: Account of the Crania of Some of the Aborigines of the United States*. Boston: Joseph W. Ingram.

Warren, John Collins. 1838. North American Antiquities. *American Journal of Science and Arts* 34:47–49.

Warren, Leonard. 1998. *Joseph Leidy: The Last Man Who Knew Everything*. New Haven: Yale University Press.

Warren, Lyman O. 1934. Cranial Relationships and Correlation. *Human Biology* 6: 308–323.

Warren, Lyman O. 1972. Treponematosis. *Florida Anthropologist* 25 (4):175–188.

Washburn, Sherwood L. 1951. The New Physical Anthropology. *Transactions of the New York Academy of Sciences, Series II* 13 (7):298–304.

Washburn, Sherwood L. 1953. The New Physical Anthropology. *Yearbook of Physical Anthropology* 7:124–130.

Watkins, Joe E. 2003. Beyond the Margin: American Indians, First Nations, and Archaeology in North America. *American Antiquity* 68 (2):273–285.

Watkins, Joe E., Lynne Goldstein, Karen D. Vitelli, and Leigh Jenkins. 1995. Accountability: Responsibilities of Archaeologists to Other Interest Groups. In *Ethics in American*

Archaeology: Challenges for the 1990s, edited by M. J. Lynott and A. Wylie. 33–37. Washington, DC: Society for American Archaeology Special Report.

Watson, P. J., S. A. LeBlanc, and C. L. Redman. 1980. Aspects of Zuni Prehistory: Preliminary Report on Excavations and Survey in the El Morro Valley of New Mexico. *Journal of Field Archaeology* 7:201–218.

Waugh, L. M. 1928. Nutrition and Health of the Labrador Eskimo with Special Reference to the Mouth and Teeth. *Journal of Dental Research* 8:428–429.

Waugh, L. M. 1931. A Study of the Nutrition and Teeth of the Eskimos of North Bering Sea and Arctic Alaska. *Journal of Dental Research* 11:450–457.

Waugh, L. M. 1933. Influence of Diet Upon the Jaws and Face of the American Eskimo. *Journal of Dental Research* 13:149–151.

Webb, Clarence H. 1944. Dental Abnormalities as Found in the American Indian. *American Journal of Orthodontics and Oral Surgery* 30 (9):386–394.

Webb, Clarence H. 1959. *The Belcher Mound, Society for American Archaeology (Memoir 16)*. Menasha, WI: Society for American Archaeology Memoirs.

Webb, William S. 1934. Letter Dated June 1 to Funkhouser. Webb Papers, Box 1, Library Archives, University of Kentucky, Lexington.

Webb, William S. 1938. *An Archaeological Survey of the Norris Basin in Eastern Tennessee, Bureau of American Ethnology, Bulletin 118*. Washington, DC: Smithsonian Institution.

Webb, William S. 1939. *An Archaeological Survey of Wheeler Basin on the Tennessee River in Northern Alabama, Bureau of American Ethnology, Bulletin 122*. Washington, DC: Smithsonian Institution.

Webb, William S. 1940. Letter Dated July 25 to Snow. Snow Letters, Laboratory for Human Osteology, Mary Harmon Bryant Scientific Collections Building, University of Alabama, Tuscaloosa.

Webb, William S. 1942. Letter Dated February 2 to Storey. Johnson Papers, National Anthropological Archives, Smithsonian Institution, Washington, DC

Webb, William S. 1945. Letter Dated March 30 to Willey. Webb Papers, Box 4, Library Archives, University of Kentucky, Lexington.

Webb, William S. 1946. Indian Knoll, Site Oh 2, Ohio County, Kentucky. *University of Kentucky Reports in Anthropology* 4 (3, Part 1).

Webb, William S., and William D. Funkhouser. 1932. *Archaeological Survey of Kentucky*. Vol. 2, *Reports in Archaeology and Anthropology*. Lexington: Department of Anthropology, University of Kentucky.

Webb, William S., and Charles E. Snow. 1945. *The Adena People*. Vol. 6, *Reports in Anthropology and Archaeology*. Lexington: Department of Anthropology, University of Kentucky.

Weiant. 1960. Bruno Oetteking 1871–1960. *American Anthropologist* 62:675–680.

Weidenreich, Franz. 1939. The Duration of Life in Fossil Man in China and the Pathological Lesions Found in His Skeleton. *Chinese Medical Journal* 55:34–44.

Weinberg, Eugene D. 1974. Iron and Susceptibility to Infectious Disease. *Science* 184 (4140):952–956.

Weinberg, Eugene D. 1992. Iron Withholding in Prevention of Disease. In *Diet, Demography, and Disease: Changing Perspectives on Anemia*, edited by P. Stuart-Macadam and S. Kent. 105–150. New York: Adline de Gruyter.

Weisensee, K. E. 2001. Pecos Revisited: A Modern Analysis of Earnest Hooton's *The Indians of Pecos Pueblo*. M.A., Department of Anthropology, University of Tennessee, Knoxville.

Weiss, Elizabeth. 2003a. Effects of Rowing on Humeral Strength. *American Journal of Physical Anthropology* 121:293–302.

Weiss, Elizabeth. 2003b. Understanding Muscle Markers: Aggregation and Construct Validity. *American Journal of Physical Anthropology* 121:230–240.

Weiss, Elizabeth. 2004. Understanding Muscle Markers: Lower Limbs. *American Journal of Physical Anthropology* 125:232–238.

Weiss, Kenneth M. 1972. On the Systematic Bias in Skeletal Sexing. *American Journal of Physical Anthropology* 37 (2):239–250.

Weiss, Kenneth M. 1973. *Demographic Models for Anthropology. Memoirs 27.* Washington, DC: Society for American Archaeology.

Weiss, Kenneth M., and R. Chakraborty. 1982. Genes, Populations, and Disease, 1930–1980: A Problem-Oriented Review. In *A History of American Physical Anthropology 1930–1980*, edited by F. Spencer. 371–404. New York: Academic Press.

Wellcome Trust. n.d. Wellcome Trust Home Page. Retrieved Dec. 28, 2005, from Wellcome Trust Web Site: http://www.wellcome.ac.uk/.

Wells, Calvin. 1963. The Radiological Examination of Human Remains. In *Science in Archaeology: A Comprehensive Survey of Progress and Research*, edited by D. R. Brothwell and E. S. Higgs. 401–411. London: Thames and Hudson.

Wells, Calvin. 1964a. *Bones, Bodies, and Disease: Evidence of Disease and Abnormality in Ancient Man*. London: Thames and Hudson.

Wells, Calvin. 1964b. Pathological Epipodials and Tarsus in Streptosuarus Macromerus from Kimmeridge Clay, Stretham, Cambridge. *The Quarterly Journal of the Geological Society, London* 120:299–304.

Wells, Calvin. 1965. A Pathological Anglo-Saxon Femur. *British Journal of Radiology* 38:393–394.

Wells, Calvin. 1966. Report on the Human Remains from Nethercourt Farm, Ramsgate. In *Neolithic Occupation Sites in East Kent*, edited by C. Dunning. 24–25. Vol. 66, *Antiquaries Journal*.

Wells, Calvin. 1967a. Pseudopathology. In *Diseases in Antiquity*, edited by D. R. Brothwell and A. T. Sandison. 5–19. Springfield, IL: Charles C Thomas.

Wells, Calvin. 1967b. A Roman Surgical Instrument from Norfolk. *Antiquity* 41:139–141.

Wells, Calvin. 1968. Goitre in Two African Sculptures. *Medical History* 12:89–91.

Wells, Calvin. 1974. The Results of Bone Setting in Anglo-Saxon Times. *Medical and Biological Illustration* 24:215–220.

Wells, Calvin. 1977. Disease of the Maxillary Sinus in Antiquity. *Medical and Biological Illustration* 27:173–178.

Wells, Calvin. 1982. The Human Remains. In *Romano-British Cemeteries at Cirencester*, edited by A. McWhirr, L. Viner, and C. Wells. 135–202. Cirencester: Excavations Committee.

Wells, Calvin, and R. R. Clarke. 1955. Early Bronze Age Burials at Barton Bendish. *Norfolk Archaeology* 31:224–230.

Wentworth, Bert, and Harris Hawthorne Wilder. 1918. *Personal Identification; Methods for the Identification of Individuals, Living or Dead.* Boston: R. G. Badger.

Wentworth, Bert, and Harris Hawthorne Wilder. 1932. *Personal Identification; Methods for the Identification of Individuals, Living or Dead. New and Rev. Ed., with Additional Text and Illustrations.* Chicago: T. G. Cooke, The Finger Print Publishing Association.

Westrheim, S. J., and W. E. Ricker. 1978. Bias in Using an Age-Length Key to Estimate Age-Frequency Distributions. *Journal of the Fisheries Research Board of Canada* 35:184–189.

Wheeler, Robert Eric Mortimer. 1954. *Archaeology from the Earth.* Oxford: Clarendon Press.

Wheeler, R. Linda. 1985. Pathology in Late Thirteenth Century Zuni from the El Morro Valley, New Mexico. In *Health and Disease in the Prehistoric Southwest*, edited by C. F. Merbs and R. J. Miller. 79–83. Tucso, AZ: Arizona State University.

Whipple, Inez L. 1904. The Ventral Surface of the Mammalian Chiridium with Special Reference to the Conditions Found in Man. *Zeitschrift für Morphologie Und Anthropologie* 7:261–368.

White, Christine D. 1999. *Reconstructing Ancient Maya Diet.* Salt Lake City: University of Utah Press.

White, M. E. 1967. *An Early Historic Niagara Frontier Iroquois Cemetery in Erie County, New York. Appendix by W. S. Cornwell and A. J. Sublett.* Vol. XVI(1), *Researches and Transactions of the New York State Archaeological Association.*

White, Tim D. 1992. *Prehistoric Cannibalism at Mancos 5MTUMR-2346.* Princeton: Princeton University Press.

White, Tim D., and Pieter A. Folkens. 1991. *Human Osteology.* San Diego: Academic Press.

Whiteduck, Gilbert. 2004. Resisting Assimilation: Kitigan Zibi's Journey from Honouring Our Past to Protecting Our Future. Part One. *CAPA Selected Abstracts* 32nd Annual Meeting of the Canadian Association for Physical Anthropology, London, Ontario:34.

Whiteley, Peter M. 2002. Archaeology and Oral Traditions: The Scientific Importance of Dialogue. *American Antiquity* 67 (3):405–415.

Whitney, William F. 1883. On the Existence of Syphilis in America before the Discovery by Columbus. *The Boston Medical and Surgical Journal* 108:365–366.

Whitney, William F. 1886. Notes on the Anomalies, Injuries and Diseases of the Bones of the Native Races of North America. *Reports of the Peabody Museum of American Archaeology and Ethnology* 3 (Eighteenth and Nineteenth Annual Reports of the Curator): 433–448.

Whittington, Stephen L., and David M. Reed, eds. 1997. *Bones of the Maya: Studies of Ancient Skeletons.* Smithsonian Institution Press. Washington, DC.

Whittlesey, Stephanie M., and J. Jefferson Reid. 2001. Mortuary Ritual and Organizational Inferences at Grasshopper Pueblo, Arizona. In *Ancient Burial Practices in the American Southwest: Archaeology, Physical Anthropology, and Native American Perspectives*, edited by D. R. Mitchell and J. L. Brunson-Hadley. 68–96. Albuquerque, NM: University of New Mexico Press.

Widdis, R. W. 1988. Scale and Context: Approaches to the Study of Canadian Migration Patterns in the Nineteenth Century. *Social Science History* 12:269–303.

Wiget, Andrew O. 1982. Truth and the Hopi: An Historiographic Study of Documented Oral Tradition Concerning the Coming of the Spanish. *Ethnohistory* 29 (1–4):181–199.

Wilbur, Alicia K. 2000. A Possible Case of Rubenstein-Taybi Syndrome in a Prehistoric Skeleton from West-Central Illinois. *American Journal of Medical Genetics* 91:56–61.

Wilcox, D. R., and Christian J. Haas. 1994. The Scream of the Butterfly: Competition and Conflict in the Prehistoric Southwest. In *Themes in Southwest Prehistory*, edited by G. J. Gumerman. 211–238. Santa Fe, New Mexico: School of American Research Press.

Wilczak, Cynthia A. 1998. Consideration of Sexual Dimorphism, Age, and Asymmetry in Quantitative Measurements of Muscle Insertion Sites. *International Journal of Osteoarchaeology* 8:311–325.

Wilczak, Cynthia A., and Kenneth A. R. Kennedy. 1998. Mostly Mos: Technical Aspects of Identification of Skeletal Markers of Occupational Stress. In *Forensic Osteology: Advances in the Identification of Human Remains*, edited by K. J. Reichs. 461–489. Springfield, IL: Charles C Thomas.

Wilder, Harris Hawthorne. 1902. *Palms and Soles*. Northampton: Zoological Laboratory of Smith College.

Wilder, Harris Hawthorne. 1908. *The Morphology of Cosmobia; Speculations Concerning the Significance of Certain Types of Monsters. Reprinted from American Journal of Anatomy Vol. VIII. No. 4. Dec. 355–440.* Baltimore.

Wilder, Harris Hawthorne. 1912. The Physiognomy of the Indians of Southern New England. *American Anthropologist* 14:415–436.

Wilder, Harris Hawthorne. 1920. *A Laboratory Manual of Anthropometry*. Philadelphia: P. Blakiston's Son.

Wilder, Harris Hawthorne. 1923a. Notes on the Indians of Southern Massachusetts. *American Anthropologist* 25 (2):197–218.

Wilder, Harris Hawthorne. 1923b. *The History of the Human Body*. New York: H. Holt and Company.

Wilder, Harris Hawthorne. 1923c. *Man's Prehistoric Past*. New York: Macmillan.

Wilder, Harris Hawthorne. 1926. *The Pedigree of the Human Race*. New York: H. Holt and Company.

Wilder, Harris Hawthorne, and Margaret Washington Pfeiffer. 1924. *The Bodily Proportions of Women in the United States; Based Upon Measurements Taken from One Hundred Smith College Students. Reprinted from the Proceedings of the American Academy of Arts and Sciences, V. Lix, No. 16, pp. 441–603, Dec. 1924 pp. 599–603.* Northampton.

Willey, Gordon R., and Jeremy A. Sabloff. 1974. *A History of American Archaeology*. San Francisco: W. H. Freeman.

Willey, Gordon R., and Jeremy A. Sabloff. 1980. *A History of American Archaeology*. 2nd ed. San Francisco: W. H. Freeman.

Willey, Gordon R., and Jeremy A. Sabloff. 1993. *A History of American Archaeology*. 3rd ed. New York: W. H. Freeman.

Willey, P. 1990. *Prehistoric Warfare on the Great Plains*. New York: Garland.

Willey, P., and Thomas E. Emerson. 1993. The Osteology and Archaeology of the Crow Creek Massacre. *Plains Anthropologist* 38 (145):227–269.

Williams, Howard. 2003. *Archaeologies of Remembrance: Death and Memory in Past Societies*. New York: Kluwer Academic/Plenum.

Williams, Herbert Upham. 1929. *Human Paleopathology, with Some Original Observations on Symmetrical Osteoporosis of the Skull*. Chicago: American Medical Association.

Williams, Herbert Upham. 1932. The Origin and Antiquity of Syphilis. The Evidence from Diseased Bones. *Archives of Pathology* 13:779–814, 931–983.

Williams, Herbert Upham. 1936. The Origin of Syphilis: Evidence from Diseased Bones: A Supplementary Report. *Archives of Dermatology and Syphilology* 33:783–787.

Williams, S., ed. 1962. Proceedings of the Sixteenth Southeastern Archaeological Conference, Macon, Georgia, 1959. *Southeastern Archaeological Conference Newsletter* 8:33–40.

Williams-Blangero, S., and J. Blangero. 1989. Anthropometric Variation and the Genetic Structure of the Jirels of Nepal. *Human Biology* 61:1–12.

Williams-Blangero, S., and J. Blangero. 1990. Effects of Population Structure on within-Group Variation in the Jirels of Nepal. *Human Biology* 62:131–146.

Williams-Blangero, S., J. Blangero, and B. Towne. 1990. Quantitative Traits and Population Structure: Introduction. *Human Biology* 62:1–4.

Williamson, Matthew A., Cheryl A. Johnston, Steven A. Symes, and John J. Schultz. 2003. Interpersonal Violence between 18th Century Native Americans and Europeans in Ohio. *American Journal of Physical Anthropology* 122:113–122.

Williamson, Ronald F., and Susan Pfeiffer, eds. 2003. *Bones of the Ancestors: The Archaeology and Osteobiography of the Moatfield Ossuary, Mercury Series Archaeology Paper 163*. Gatineau, Quebec: Canadian Museum of Civilization.

Wills, Wirt H. 1994. Evolutionary and Ecological Modeling in Southwestern Archaeology. In *Understanding Complexity in the Prehistoric Southwest*, edited by G. Gummerman and M. Gell-Mann. Vol. XVI, *Proceedings*. Reading, MA: Santa Fe Institute, Addison-Wesley.

Wilson, Daniel. 1863. Physical Ethnology. *Smithsonian Report* 1862:240–302.

Wilson, Daniel. 1876. *Prehistoric Man: Researches into the Origin of Civilisation in the Old and the New World*. 3d ed. London: Macmillan.

Wing, Elizabeth S., and A. B. Brown. 1979. *Paleonutrition*. New York: Academic Press.

Winski, John B. 1992. There Are Skeletons in the Closet: The Repatriation of Native American Human Remains and Burial Objects. *Arizona Law Review* 34:187–214.

Winter, Joseph C. 1980. Indian Heritage Preservation and Archaeologists. *American Antiquity* 45 (1):121–131.

Winters, Howard D. 1968. Value Systems and Trade Cycles of the Late Archaic in the Midwest. In *New Perspectives in Archaeology*, edited by S. R. Binford and L. R. Binford. 175–221. Chicago: Aldine.

Wolf, D. J. 1977. Middle Mississippian: A Prehistoric Cultural System Viewed from a Biological Perspective. In *Biocultural Adaptation in Prehistoric America*, edited by R. L. Blakely. 27–44. Athens: University of Georgia Press.

Wolpoff, Milford H., and Rachel Caspari. 1997. *Race and Human Evolution*. New York: Simon & Schuster.

Wood, George B. 1853. *A Biographical Memoir of Samuel George Morton, M.D.* Philadelphia: T.K. & P.G. Collins.

Wood, James W. 1986. Convergence of Genetic Distances in a Migration Matrix Model. *American Journal of Physical Anthropology* 71:209–219.

Wood, James W., Darryl J. Holman, Kathleen A. O'Connor, and Rebecca J. Ferrell. 2002. Mortality Models for Paleodemography. In *Paleodemography: Age Distributions from Skeletal Samples*, edited by R. D. Hoppa and J. W. Vaupel. 129–168. Cambridge: Cambridge University Press.

Wood, James W., Darryl J. Holman, Kenneth M. Weiss, Anne V. Buchanan, and Belinda LeFor. 1992. Hazards Models for Human Population Biology. *Yearbook of Physical Anthropology* 35:43–87.

Wood, James W., George R. Milner, Henry C. Harpending, and Kenneth M. Weiss. 1992. The Osteological Paradox: Problems of Inferring Prehistoric Health from Skeletal Samples. *Current Anthropology* 33 (4):343–370.

Worden, G. 2002. *Mütter Museum of the College of Physicians of Philadelphia*. New York: Blast.

World Congress on Mummy Studies. 1995. *Proceedings of the World I Congress on Mummy Studies, 1992, Tenerife, Canary Islands*. Tenerife: Museo Arquelâogico y Etnogrâafico de Tenerife; Organismo Autâonomo de Museos y Centros.

Wright, Lori E. 1987. Stresses of Conquest: A Scanning Electron Microscope Study of Wilson Bands and Enamel Hypoplasias in the Maya of Lauranai, Belize. B.A., Departments of Anthropology and Biology, Trent University, Peterborough, Ontario, Canada.

Wright, L. E. 1990. Stresses of Conquest: A Study of Wilson Bands and Enamel Hypoplasias in the Maya of Lamanai, Belize. *American Journal of Human Biology* 2:25–35.

Wright, Lori E., and Francisco Chew. 1998. Porotic Hyperostosis and Paleoepidemiology: A Forensic Perspective on Anemia among the Ancient Maya. *American Anthropologist* 100 (4):924–939.

Wright, Lori E., and Cassady J. Yoder. 2003. Recent Progress in Bioarchaeology: Approaches to the Osteological Paradox. *Journal of Archaeological Research* 11 (1):43–70.

Wright, S. 1969. *Evolution and the Genetics of Populations: The Theory of Gene Frequencies*. Chicago: University of Chicago.

Wylie, A. 1991. Gender Theory and the Archaeological Record: Why Is There no Archaeology of Gender. In *Engendering Archaeology: Women and Prehistory*, edited by J. M. Gero and M. W. Conkey. 31–54. Oxford: Basil Blackwell.

Wyman, Jeffries. 1866. On the Distorted Skull of a Child from the Hawaiian Islands. *Proceedings of the Boston Society of Natural History* 11:70–71.

Wyman, Jeffries. 1868a. Report of the Curator. *Reports of the Peabody Museum of American Archaeology and Ethnology* 1 (First Annual Report of the Trustees):5–18.

Wyman, Jeffries. 1868b. *Observations on Crania*. Boston: A.A. Kingman.

Wyman, Jeffries. 1869. Report of the Curator. *Reports of the Peabody Museum of American Archaeology and Ethnology* 1 (Second Annual Report of the Trustees):5–20.

Wyman, Jeffries. 1871. Report of the Curator. *Reports of the Peabody Museum of American Archaeology and Ethnology* 1 (Fourth Annual Report of the Trustees):5–24.

Wyman, Jeffries. 1873. Report of the Curator. *Reports of the Peabody Museum of American Archaeology and Ethnology* 1 (Sixth Annual Report of the Trustees):5–23.

Wyman, Jeffries. 1874. Report of the Curator. *Reports of the Peabody Museum of American Archaeology and Ethnology* 1 (Seventh Annual Report of the Trustees):6–37.

Wyman, Jeffries. 1875. *Fresh-Water Shell Mounds of the St. John's River, Florida*. Vol. 1, Fourth Memoir, *Memoirs of the Peabody Academy of Science*. Salem: Peabody Academy of Science.

Young, Robert M. 1990. *Mind, Brain and Adaptation in the Nineteenth Century: Cerebral Localization and Its Biological Context from Gall to Ferrier, History of Neuroscience, No. 3*. New York: Oxford University Press.

Zaino, E. C. 1968. Elemental Bone Iron in the Anasazi Indians. *American Journal of Physical Anthropology* 29:433–436.

Zeder, Melinda A. 1994. Repatriation at the National Museum of Natural History. In *Reckoning with the Dead: The Larsen Bay Repatriation and the Smithsonian Institution*, edited by T. L. Bray and T. W. Killion. 166–171. Washington, DC: Smithsonian Institution Press.

Zimmerman, Larry. 1989. Made Radical by My Own: An Archaeologist Learns to Accept Reburial. In *Conflict in the Archaeology of Living Traditions*, edited by R. Layton, *One World Archaeology 8*. London: Unwin Hyman.

Zimmerman, Larry. 1990. "This Has Bothered Me for 500 Years": The Impact of Concepts of Time and the Past on the Reburial Issue. In *Preservation on the Reservation. Native Americans, Native American Lands and Archaeology*, edited by A. L. Klesert and A. S. Downer. 409–424: Navajo Nation Archaeology, Navajo Nation Historic Preservation Department.

Zimmerman, Larry. 2002. A Decade after the Vermillion Accord: What Has Changed and What Has Not? In *The Dead and Their Possessions. Repatriation in Principle, Policy and Practice*, edited by C. Fforde, J. Hubert, and P. Turnbull. 91–107. New York: Routledge.

Zimmerman, Larry J. 1997. Remythologizing the Relationship between Indians and Archaeologists. In *Native Americans and Archaeologists: Stepping Stones to a Common Ground*, edited by N. Swindler, K. E. Dongoske, R. Anyon, and A. S. Downer. 44–56. Walnut Creek: Altamira Press.

Zimmerman, Larry J., T. Emerson, P. Willey, M. Swegle, J. Gregg, P. Gregg, E. White, C. Smith, T. Haberman, and M. Bumsted. 1981. *The Crow Creek Site (39BF11) Massacre: A Preliminary Report*. Omaha: Omaha District U.S. Army Corps of Engineers.

Zimmerman, Michael R. 1980. Annotated Bibliography of Paleopathology: 1970–79. *Transactions and Studies of the College of Physicians of Philadelphia* 2(1,2,3,4):41–66, 12–143, 187–222, 289–314.

Zimmerman, Michael R., and Marc A. Kelley. 1982. *Atlas of Human Paleopathology*. New York: Praeger.

Zimmerman, Michael R., Erik Trinkaus, Marjorie LeMay, Arthur C. Aufderheide, Theodore A. Reyman, Guy R. Marrocco, R. Wade Ortel, Jaime T. Benitez, William S. Laughlin, Patrick D. Horne, Richard E. Schultes, and Elizabeth A. Coughlin. 1981.

The Paleopathology of an Aleutian Mummy. *Archives of Pathology and Laboratory Medicine* 105:638–641.

Zink, Albert, and Andreas G. Nerlich. 2003. Brief Communication: Molecular Analysis of the "Pharaos": Feasibility of Molecular Studies in Ancient Egyptian Material. *American Journal of Physical Anthropology* 121(2):109–111.

Zuhrt, R. 1955. Stomatologische Utersuchungen an Spätmittelalterlichen Funden von Reckkahn (12.–14.Jh.) I. Die Zahnkaries und Ihre Folgen. *Dtsche Zahn Mund Kieferheilk* 25:1–15.

Zumwalt, A. C. 2004. A New Method to Quantify the 3D Morphology of Bone Surfaces, with Application to Muscle Enthesis Rugosity [Abstract]. *American Journal of Physical Anthropology* (Suppl. 38):216.

Zumwalt, A. C., Christopher B. Ruff, and Cynthia A. Wilczak. 2000. Primate Muscle Insertions: What Does Size Tell You? [Abstract]. *American Journal of Physical Anthropology* (Suppl. 30):331.

Index

M

O

P